BURGER'S MEDICINAL CHEMISTRY, DRUG DISCOVERY AND DEVELOPMENT

BURGER'S MEDICINAL CHEMISTRY, DRUG DISCOVERY AND DEVELOPMENT

Editors-in-Chief
Donald J. Abraham
Virginia Commonwealth University

David P. Rotella
Wyeth Research

Consulting Editor
Al Leo
BioByte Corp

Editorial Board
John H. Block
Oregon State University

Robert H. Bradbury
AstraZeneca

Robert W. Brueggemeier
Ohio State University

John W. Ellingboe
Wyeth Research

William R. Ewing
Bristol-Myers Squibb Pharmaceutical Research Institute

Richard A. Gibbs
Purdue University

Richard A. Glennon
Virginia Commonwealth University

Barry Gold
University of Pittsburgh

William K. Hagmann
Merck Research Laboratories

Glen E. Kellogg
Virginia Commonwealth University

Christopher A. Lipinski
Melior Discovery

John A. Lowe III
JL3Pharma LLC

Jonathan S. Mason
Lundbeck Research

Andrea Mozzarelli
University of Parma

Bryan H. Norman
Eli Lilly and Company

John L. Primeau
AstraZeneca

Paul J. Reider
Princeton University

Albert J. Robichaud
Lundbeck Research

Alexander Tropsha
University of North Carolina

Patrick M. Woster
Wayne State University

Jeff Zablocki
CV Therapeutics

Editorial Staff
VP & Director, STMS Book Publishing:
Janet Bailey
Editor: **Jonathan Rose**
Production Manager: **Shirley Thomas**
Production Editor: **Kris Parrish**
Illustration Manager: **Dean Gonzalez**
Editorial Program Coordinator: **Surlan Alexander**

BURGER'S MEDICINAL CHEMISTRY, DRUG DISCOVERY AND DEVELOPMENT

Seventh Edition

Volume 6: Cancer

Edited by

Donald J. Abraham
Virginia Commonwealth University

David P. Rotella
Wyeth Research

Burger's Medicinal Chemistry, Drug Discovery and Development
is available Online in full color at
http://mrw.interscience.wiley.com/emrw/9780471266945/home/

A JOHN WILEY & SONS, INC., PUBLICATION

Copyright © 2010 by John Wiley & Sons, Inc. All rights reserved

Published by John Wiley & Sons, Inc., Hoboken, New Jersey
Published simultaneously in Canada

No part of this publication may be reproduced, stored in a retrieval system, or transmitted in any form or by any means, electronic, mechanical, photocopying, recording, scanning, or otherwise, except as permitted under Section 107 or 108 of the 1976 United States Copyright Act, without either the prior written permission of the Publisher, or authorization through payment of the appropriate per-copy fee to the Copyright Clearance Center, Inc., 222 Rosewood Drive, Danvers, MA 01923, (978) 750-8400, fax (978) 750-4470, or on the web at www.copyright.com. Requests to the Publisher for permission should be addressed to the Permissions Department, John Wiley & Sons, Inc., 111 River Street, Hoboken, NJ 07030, (201) 748-6011, fax (201) 748-6008, or online at http://www.wiley.com/go/permission.

Limit of Liability/Disclaimer of Warranty; While the publisher and author have used their best efforts in preparing this book, they make no representations or warranties with respect to the accuracy or completeness of the contents of this book and specifically disclaim any implied warranties of merchantability or fitness for a particular purpose. No warranty may be created or extended by sales representatives or written sales materials. The advice and strategies contained herein may not be suitable for your situation. You should consult with a professional where appropriate. Neither the publisher nor author shall be liable for any loss of profit or any other commercial damages, including but not limited to special, incidental, consequential, or other damages.

For general information on our other products and services or for technical support, please contact our Customer Care Department within the United States at (800) 762-2974, outside the United States at (317) 572-3993 or fax (317) 572-4002.

Wiley also publishes its books in a variety of electronic formats. Some content that appears in print may not be available in electronic formats. For more information about Wiley products, visit our web site at www.wiley.com.

Library of Congress Cataloging-in-Publication Data:

Abraham, Donald J., 1936-
  Burger's medicinal chemistry, drug discovery, and development/Donald
J. Abraham, David P. Rotella. – 7th ed.
      p. ; cm.
  Other title: Medicinal chemistry, drug discovery, and development
  Rev. ed. of: Burger's medicinal chemistry and drug discovery. 6th ed. /
edited by Donald J. Abraham. c2003.
    Includes bibliographical references and index.
    ISBN 978-0-470-27815-4 (cloth)
1. Pharmaceutical chemistry. 2. Drug development. I. Rotella, David P.
II. Burger, Alfred, 1905-2000. III. Burger's medicinal chemistry and drug
discovery. IV. Title. V. Title: Medicinal chemistry, drug discovery, and
development.
  [DNLM: 1. Chemistry, Pharmaceutical–methods. 2. Biopharmaceutics–
methods. 3. Drug Compounding–methods. QV 744 A105b 2010]
  RS403.B8 2010
  615'.19–dc22   2010010779

Printed in Singapore

10 9 8 7 6 5 4 3 2 1

CONTENTS

PREFACE		vii
CONTRIBUTORS		ix
1	Natural Products as Cytotoxic Agents	1
2	Histone Deacetylase Inhibitors: A Brief Overview of Their Role and Medicinal Chemistry	49
3	Synthetic DNA-Targeted Chemotherapeutic Agents and Related Tumor-Activated Prodrugs	83
4	PARP Inhibitors as Anticancer Agents	151
5	Proteasome Inhibitors	175
6	Inhibitors of Kinesin Spindle Protein for the Treatment of Cancer	191
7	CNS Cancers	223
8	Kinase Inhibitors: Approved Drugs and Clinical Candidates	295
9	Structure-Based Design of Kinase Inhibitors: Molecular Recognition of Protein Multiple Conformations	345
10	Cancer Drug Resistance: Targets and Therapies	361
11	Hsp90 Inhibitors	383
12	Gene Therapy with Plasmid DNA	457
INDEX		501

PREFACE

The seventh edition of Burger's Medicinal Chemistry resulted from a collaboration established between John Wiley & Sons, the editorial board, authors, and coeditors over the last 3 years. The editorial board for the seventh edition provided important advice to the editors on topics and contributors. Wiley staff effectively handled the complex tasks of manuscript production and editing and effectively tracked the process from beginning to end. Authors provided well-written, comprehensive summaries of their topics and responded to editorial requests in a timely manner. This edition, with 8 volumes and 116 chapters, like the previous editions, is a reflection of the expanding complexity of medicinal chemistry and associated disciplines. Separate volumes have been added on anti-infectives, cancer, and the process of drug development. In addition, the coeditors elected to expand coverage of cardiovascular and metabolic disorders, aspects of CNS-related medicinal chemistry, and computational drug discovery. This provided the opportunity to delve into many subjects in greater detail and resulted in specific chapters on important subjects such as biologics and protein drug discovery, HIV, new diabetes drug targets, amyloid-based targets for treatment of Alzheimer's disease, high-throughput and other screening methods, and the key role played by metabolism and other pharmacokinetic properties in drug development.

The following individuals merit special thanks for their contributions to this complex endeavor: Surlan Alexander of John Wiley & Sons for her organizational skills and attention to detail, Sanchari Sil of Thomson Digital for processing the galley proofs, Jonathan Mason of Lundbeck, Andrea Mozzarelli of the University of Parma, Alex Tropsha of the University of North Carolina, John Block of Oregon State University, Paul Reider of Princeton University, William (Rick) Ewing of Bristol-Myers Squibb, William Hagmann of Merck, John Primeau and Rob Bradbury of AstraZeneca, Bryan Norman of Eli Lilly, Al Robichaud of Wyeth, and John Lowe for their input on topics and potential authors. The many reviewers for these chapters deserve special thanks for the constructive comments they provided to authors. Finally, we must express gratitude to our lovely, devoted wives, Nancy and Mary Beth, for their tolerance as we spent time with this task, rather than with them.

As coeditors, we sincerely hope that this edition meets the high expectations of the scientific community. We assembled this edition with the guiding vision of its namesake in mind and would like to dedicate it to Professor H.C. Brown and Professor Donald T. Witiak. Don collaborated with Dr. Witiak in the early days of his research in sickle cell drug discovery. Professor Witiak was Dave's doctoral advisor at Ohio State University and provided essential guidance to a young scientist. Professor Brown, whose love for chemistry infected all organic graduate students at Purdue University, arranged for Don to become a medicinal chemist by securing a postdoctoral position for him with Professor Alfred Burger.

It has been a real pleasure to work with all concerned to assemble an outstanding and up-to-date edition in this series.

DONALD J. ABRAHAM
DAVID P. ROTELLA

March 2010

CONTRIBUTORS

P. Angibaud, Ortho-Biotech Oncology Research & Development, Val de Reuil, France

J. Arts, Ortho-Biotech Oncology Research & Development, Turnhoutseweg, Beerse, Belgium

Surendar Reddy Bathula, University of North Carolina at Chapel Hill, Chapel Hill, NC

Gustave Bergnes, Cytokinetics, Inc., South San Francisco, CA

Robert H. Bradbury, AstraZeneca, Macclesfield, UK

Brian S. J. Blagg, University of Kansas, Lawrence, KS

Maureen G. Conlan, Cytokinetics, Inc., South San Francisco, CA

Gabriele Costantino, Università degli Studi di Parma, Via GP Usberti, Parma, Italy

Horace G. Cutler, Mercer University, Atlanta, GA

Stephen J. Cutler, University of Mississippi, University, MS

William A. Denny, University of Auckland, Auckland, New Zealand

Gerhard F. Ecker, University of Vienna, Vienna, Austria

K. Van Emelen, Tibotec-Virco, Gen. De Wittelaan, Mechelen, Belgium

Robertson Graeme, Siena Biotech SpA, Siena, Italy

M. Kyle Hadden, University of Connecticut, Storrs, CT

Leaf Huang, University of North Carolina at Chapel Hill, Chapel Hill, NC

Kyung Bo Kim, University of Kentucky, Lexington, KY

Steven D. Knight, GlaxoSmithKline, Collegeville, PA

Jeffrey Jie-Lou Liao, University of Science and Technology of China, Hefei, Anhui, China; Duke University, Durham, NC

Roberto Pellicciari, Università degli Studi di Perugia, Via del Liceo, Perugia, Italy

Marie Wehenkel, University of Kentucky, Lexington, KY

Barbara Zdrazil, University of Vienna, Vienna, Austria

NATURAL PRODUCTS AS CYTOTOXIC AGENTS

STEPHEN J. CUTLER[1]
HORACE G. CUTLER[2]
[1] Department of Medicinal Chemistry, University of Mississippi, University, MS
[2] Department of Medicinal Chemistry, Mercer University, Atlanta, GA

1. INTRODUCTION

In Sthe first decade of the twenty-first century, the cassette of cancers has increased not only in First World countries but also in those countries considered to have a lower standard of living, according to a world cancer report issued in 2008 for the IARC Nonserial Publication by Boyle and Levin. So widespread is cancer, in its various forms, that most adults have either had family members or had friends who have been afflicted by the disease. Projected figures for cancers indicate, in the aforementioned publication, that 12 million cancer cases were diagnosed in 2008 alone and that the number of cancer victims has doubled since 1978. Coupled to the psychological impact that a diagnosis has on an individual is the cost of treatment that must also be considered. Again, according to the National Cancer Institute (December 2008) these costs are estimated to be US$147 billion by 2020. An article, published in *USA Today*, claims that new cancer cases in 2008 were projected to be 40,480. And the cost for elderly Medicare patients will be US$21.1 billion for 5 years of treatment for the period 2004–2009. Sadly, one in eight patients with advanced stages of cancer will turn down the offer of recommended care because of costs. While cases may have been misdiagnosed in the twentieth century, the sophisticated techniques available today give a clearer graph of the incidence of the disease. With advanced diagnostics, there has been a concomitant advance in anticancer chemotherapy.

Many, but not all, of the cytotoxic agents employed to combat various cancers are secondary metabolites, or natural products, and may be divided into two categories: those agents that have cleared all clinical trials and others that are still undergoing extensive testing. The origins of natural products include those obtained from plants, sponges (marine), and microorganisms [1–5]. Some have their genesis in folk medicine, but most are discovered by bioassay-directed separation, giving rise to large databases of chemical structures which, in turn, may be synthetically modified to give second-generation derivatives, or which may be used as templates on which to pattern further bioactive compounds.

The subject of secondary metabolites has been a topic for some heated discussions among scientists. One point on which all agree is that nature makes molecules that have not yet entered the minds of synthetic chemists. Furthermore, nature whimsically builds structures from C, H, O, N and, in some cases, additions of halogens and other unlikely atoms. Beyond that, some believe that bioactive natural products are intended as defense mechanisms against several types of predators, including pathogens, others that they are end products of metabolism and, therefore, serve no useful purpose. Certainly, some have been shown to control mitosis *in vitro* and may serve to do so *in vivo* acting at the hormonal level *in situ*. Whatever the correct outcome may be, they offer unique products that may successfully conquer certain cancers. As the aphorism states, "Never look a gift horse in the mouth."

Fortunately, it has been estimated by natural product chemists that probably only 10–12% of bioactive secondary metabolites have so far been discovered. With the advent of organisms being found in odd places, for example, in thermal vents on the seabed where metallic temperature probes melt, and the cryogenic organisms that reside in ice sheets, plus the many terrestrial microecosystems that have yet to be explored, it is anticipated that we shall have a staggering number of bioactive compounds by the end of the twenty-first century. Indeed, uncultured microorganisms represent another reservoir for chemical diversity. Current estimates suggest that less than 1% of bacteria have been cultured, yet greater than 70% of antineoplastic drugs have been derived from this small population of cultured microbes.

The number of cancer bioassays has also increased in the past few years. Initially, screens were limited to P388 and L1210 (murine leukemias) but screens now include solid tumors, which are slow growing. Having discovered the bioactivity of a compound in several types of cancer models, the question next arises as to the toxicity of the material. Obviously, the amount that induces inhibition of cell growth in an *in vitro* system may, or may not, be relevant. And one is looking for cytotoxins or cytostatins that specifically act on mitosis during cell division, but which are not toxic to healthy, resting cells in an individual. While this seems to be a paradox, which it essentially is, the beauty of chemotherapeutic agents is that they can be delivered in such a way as to selectively attack the specific cancer, for the most part. It becomes self evident that the drug must be titrated against the patient. But the initial bioassay depends on demonstrating the total destruction of the cells before further development can proceed [6–8].

Cells undergoing rapid mitosis are excellent targets for cytotoxins, as opposed to cells that are in the resting stage and undergoing slow mitotic division. The former cells need extra supplies of chemical energy to continue replication and will readily latch onto molecules that can be catabolized to produce elemental building blocks for new cells. So, antineoplastic natural products are, essentially, Trojan horses. Unfortunately, this applies only to rapidly dividing cancers and the problem with solid tumors is that they are more like healthy cells in that they are resting. The problem is not easily resolved and there is always the unintended effect when using chemotherapeutics. These side effects are caused by damage to healthy rapidly dividing cells and include alopecia, gastrointestinal (GI) disturbance, ulcers, rashes, fertility inhibition, immune suppression, blood dyscrasis and general nausea.

We are only at the beginning of our journey to control and, perhaps, eradicate cancers. Natural products are, unfortunately, labeled secondary metabolites, but are of primary importance in nature. Perhaps, by the end of the 21st century we shall look on cancer as a disease that can be controlled in much the same way as we now look at diphtheria, measles, smallpox and a host of other diseases that ravaged families at the beginning of the 20th century.

The results of one of the principal screening methods presently used are stored in the National Cancer Institute database that was established in 1990 [6]. This is based on comparative potency against 60 different human cancer cell lines grown in tissue culture. More than 70,000 compounds have been put through this screen and the data for each are presented below. From this, insights into mechanism of action and mode of resistance can be deduced [9]. Many other tests are in present use, including screens for signal transduction inhibitors, antiangiogenesis, cell-cycle inhibition, exploitation of functional genomics, immunotherapeutics, vaccines, and chemoprevention. Much inventive biology is coming forward and exciting days appear to lie ahead.

Those natural agents presently in use can be conveniently classified according to their molecular modes of action as follows:

(1) Drugs attacking DNA
- Dactinomycin
- Bleomycin
- Mitomycin
- Plicamycin (mithramycin)

(2) Drugs inhibiting enzymes that process DNA
- Anthracyclines (daunorubicin, doxorubicin, epirubicin, idarubicin, and valrubicin)
- Camptothecins (topotecan and irinotecan)
- Isopodophyllotoxins (etoposide and tenipocide)

(3) Drugs interfering with tubulin polymerization/depolymerization
- Taxus diterpenes (docetaxel and paclitaxel/taxol)
- Vinca dimeric alkaloids (vinblastine, vincristine, and vinorelbine)

Figure 1 illustrates, in summary form, the various points of attack of prominent natural antitumor agents on growing cells. One notes that DNA or tubulin in one way or another (either by direct attack or by interference with enzymes processing these important cellular macromolecules) is the primary target of all of these agents and that most phases of the cell cycle are involved, especially when mixtures ("cocktails") are employed.

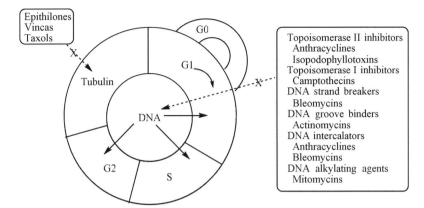

Figure 1. Synopsis of molecular modes of action of various prominent antitumor natural products.

Added biological detail of the properties and applications of these naturally occurring natural products can be found in *AHFS Drug Information 2001* [10] and in *Goodman and Gilman's Pharmacological Basis of Therapeutics* [11].

2. DRUGS ATTACKING DNA

2.1. Dactinomycin (Cosmegen)

2.1.1. Introduction The actinomycins are a family of yellow-red-peptides-containing antitumor antibiotics produced by fermentation of various *Streptomyces* and *Micromonospora* species. The first members of the actinomycin family were discovered in the early 1940s with the intention of finding nontoxic antibacterial antibiotics in fermentations of soil microorganisms [12], although in the actinomycin case this goal was scotched by high toxicity. Somewhat later (about 1958), this was compensated for by the discovery that the toxicity to rapidly growing cells could be useful in cancer chemotherapy. One should note, however, that later discoveries demonstrated that the potencies against microbes and against tumors do not correlate. Presently about seven different complexes of actinomycins have been identified, each differing from the others primarily by the various amino acids constituting the two cyclic depsipentapeptide side chains pendant from the common phenoxazinone chromophore, called actinosin (**1**). When the two cyclic peptide chains are identical, these agents are referred to as isoactinomycins ($R = R'$). When they are different from each other, they are known as anisoactinomycins ($R \neq R'$). Of the 20, or so, natural actinomycins plus a much larger number of synthetic and biosynthetic analogs, actinomycin D (**2**), from which the generic name dactinomycin is derived, is the most prominent medicinally. A useful trivial nomenclatural system has also developed. In this system, dactinomycin is referred to as Val-2-AM, and other analogs are named by the position and identity of the amino acids that are exchanged. Actinomycin C (cactinomycin-3) is thus known as Ile-2-AM.

2.1.2. Medicinal Uses As noted in the summarizing table, dactinomycin is used medicinally by intravenous (i.v.) injection for the treatment of Wilm's tumor, rhabdomyosarcoma, metastatic, and nonmetastatic choriocarcinoma, nonseminomatous testicular carcinoma, Ewing's sarcoma, nonmetastatic Ewing's sarcoma, and sarcoma botryoides. The usual dose is 10–15 µg/kg i.v. for 5 days. If no serious symptoms develop from this, additional treatments are given at 2–4-week intervals. Other treatment schedules have also been used. The drug is often combined with vincristine and

4 NATURAL PRODUCTS AS CYTOTOXIC AGENTS

cyclophosphamide in a cocktail to enhance the cure rate [13].

2.1.3. Contraindications and Side Effects
Dactinomycin is contraindicated in the presence of chicken pox or herpes zoster, wherein administration may result in severe exacerbation including, occasionally death. The drug is extremely corrosive to soft tissues, so extravasation can lead to severe tissue damage [14]. To avoid this, the drug is usually injected into infusion tubing rather than direct injection into veins. When combined with radiation therapy, exaggerated skin reactions can occur, as can an increase in GI toxicity and bone marrow problems. Secondary tumors can be observed in some cases that can be attributed to the drug. Dactinomycin is carcinogenic and mutagenic in animal studies and malformations in animal fetuses have also been observed. Nausea and vomiting are common along with renal, hepatic, and bone marrow function abnormalities. The usual alopecia, skin eruptions, GI ulcerations, proctitis, anemia, and other blood dyscrasias, esophagitis, anorexia, malaise, fatigue, and fever, for example, are also observed. Clearly, this is a very toxic drug.

2.1.4. Pharmacokinetic Features
Dactinomycin is not readily available after oral administration, so it is primarily administered by injection. About 2 h after i.v. administration, very little circulating dactinomycin can be detected in blood. It is primarily excreted in the bile and the urine. Hence, dactinomycin is only slightly metabolized. It does not pass the blood–brain barrier. Despite these factors it has a half-life of about 36 h. Persistence is largely accounted for by tight binding of the drug to DNA in nucleated cells [15–17].

2.1.5. Medicinal Chemical Transformations
Total synthesis of dactinomycin has been accomplished, but this has not proved to be of practical value as yet (see Fig. 2). Analogs can be assembled from appropriately substituted benzenoid analogs. The overall approach commonly involves construction of the external aromatic rings, attachment of the depsipeptide side-chain precursors, oxidative generation of the actinocin ring system, and functional group transformations to complete the synthesis [18–26].

Semisynthetic side-chain analogs of the actinomycins are prepared by removal of the depsipentapeptide side chains and their replacement by synthetic moieties. Analogs with altered peptide side chains are also prepared by directed biosynthetic manipulation of the fermentations. The synthetic replacement has been done in a combinatorial mode

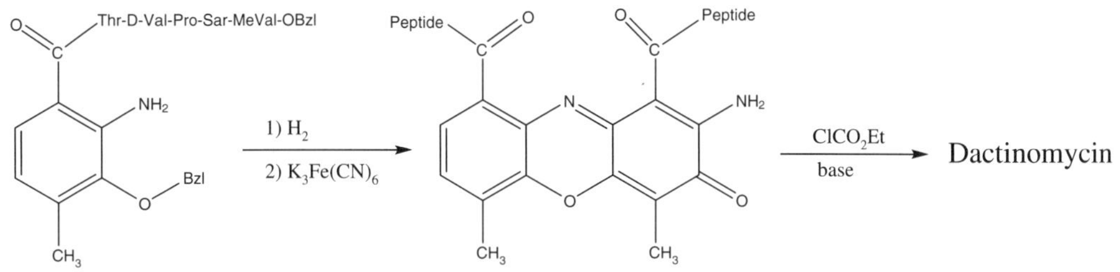

Figure 2. Synthesis of dactinomycin (actinomycin D).

as well [27]. Replacement of the normal side chains by simple amines leads to inactive products. Most of the other side-chain variations have led to compounds with reduced *in vivo* potency. None of those few analogs where this is not true have been commercialized.

Some chemical alterations in the chromophoric phenoxazinone moiety have also been examined. After considerable work, it has been observed that the C2- and the C7-positions can be substituted with retention of significant activity.

Among some of the useful reactions leading to viable analogs are a series of addition/elimination reactions, starting with careful alkali hydrolysis to produce the C2-OH analog. This can be converted by thionyl chloride treatment to the C2-Cl analog. This, in turn, can be reacted with a variety of amines to produce alkylated C2-amino substances [28]. Catalytic reduction of the 2-Cl analog results in the protio analog, which is inactive. The C2-chloro analog can be halogenated with chlorine, or bromine, to produce the C2-chloro–C7-chloro or bromo analogs. Subsequently, these can be solvolyzed to the C2-amino–C7-halo analogs [29–32]. Nitration and hydroxylation at the C7-position can be accomplished, but requires prior protection with pyruvate. After nitration, or oxidation, to the quinoneimine and reduction, careful alkaline hydrolysis of the blocking pyruvate moiety leads to the desired analogs [28,29,33]. The C7-OH analog can be converted to the allyl ether and this can be epoxidized to produce an analog that not only can intercalate by virtue of its aromatic rings but can also alkylate DNA. The nitrogen analog of the epoxide (aziridinylmethylene) can be prepared by a somewhat different route. Hydrogenation of this last product opens the aziridine ring to produce the primary amine [34–36] (See Fig. 3).

The central chromophoric ring can also be modified, for example, to the phenazine [37,38] analogs and to oxazinone and oxazole ring analogs [39,40]. These products appear not to have any importance [41].

In sum, these studies demonstrate that the side-chains are important determinants of activity, as is the basic chromophoric three-ring system. Peripheral adornments are tolerated, but not superior [42–48]. Considering the putative molecular mode of action described below, this definition of the pharmacophore is not surprising and is schematically represented in Fig. 4. Therein, the pharmacologically successful transformations that take place are represented by the boxes.

2.1.6. Molecular Mode of Action The flat three-ring fused aromatic portion of dactinomycin intercalates into double-helical DNA between the stacked bases (preferring guanine–cytosine pairs), whereas the attached cyclic peptide side chains of the drug bind into the minor grooves; thus, further anchoring the complex [49–58]. These combined interactions produce a tight and long-lasting binding. This model is supported by extensive X-ray studies with model nucleotides. As with other intercalating drugs, this interaction stretches the DNA and interferes with DNA transcription by RNA polymerase. The interference with the functioning of DNA-dependent RNA polymerase by dactinomycin is much stronger than the interference with DNA polymerases themselves. The consequences of intercalation are believed to be responsible for the antitumor action and most of the toxicity of dactinomycin. Some strand breaks are also reported. These broken products are believed to result from redox reactions of the quinine-like central chromophoric ring [57]. Although relatively noncell-cycle

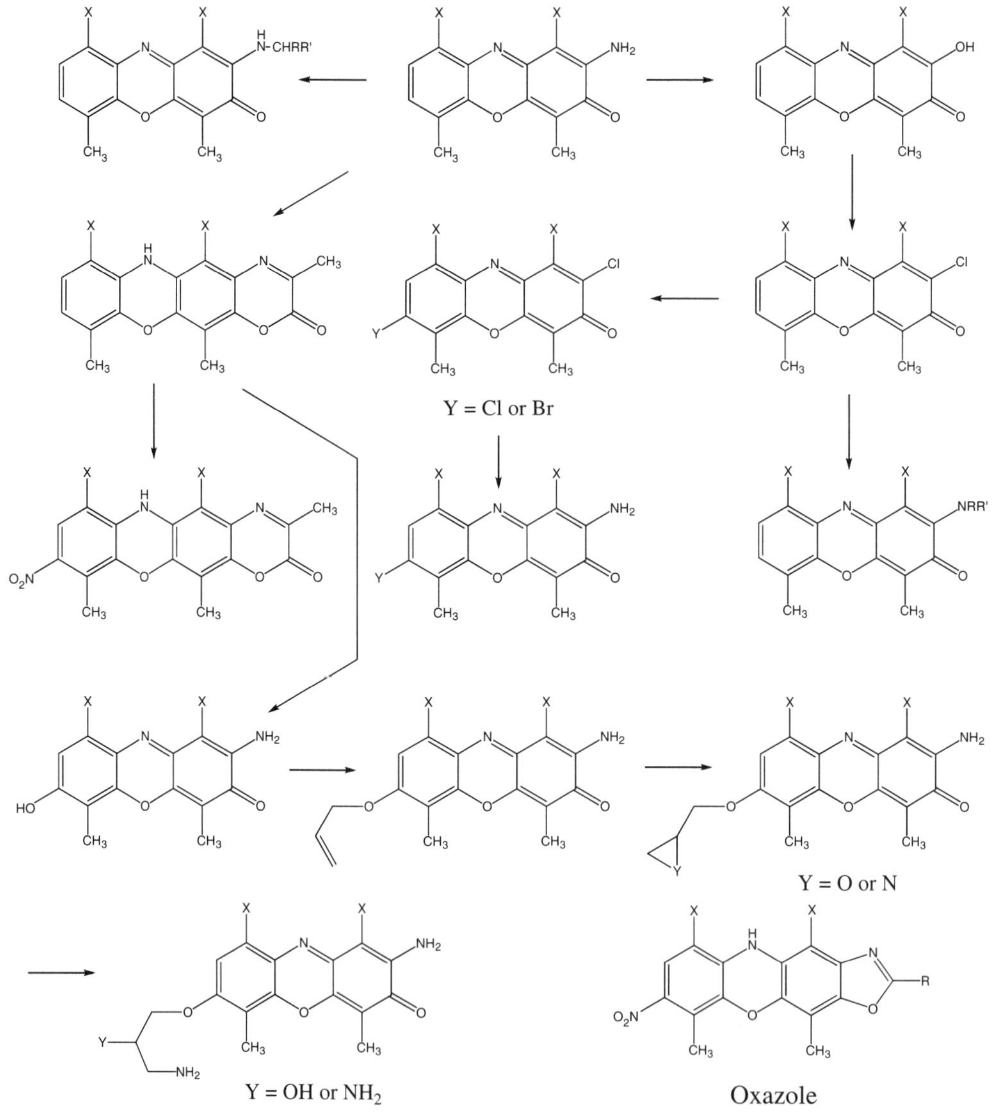

Figure 3. Synthesis of dactinomycin analogs.

specific, dactinomycin's action is particularly prominent in the G-1 phase. The cytotoxic action of dactinomycin on rapidly proliferating cells is pronounced, resulting not only in antitumor activity but also in severe toxicities to certain host organs. Figure 5 illustrates the intercalation and minor groove binding of dactinomycins.

Resistance to dactinomycin is primarily attributable to drug export through overexpression of P-glycoprotein and to alterations in tumor cell differentiation mechanisms [58–63].

2.1.7. Biosynthesis The actinomycins are biosynthesized starting with tryptamine (see Fig. 6). This passes through kynurenine to 3-hydroxyanthranilic acid then to 4-methyl-3-hydroxyanthranilic acid. Finally, to this the peptide side chains are added. Oxidative dimerization then results in completion of the phenoxazinone ring chromophore. This process is rather similar to that used in total chemical synthesis of dactinomycin. The unusual amino acids in the side chains are synthesized by a nonribosome peptide synthase.

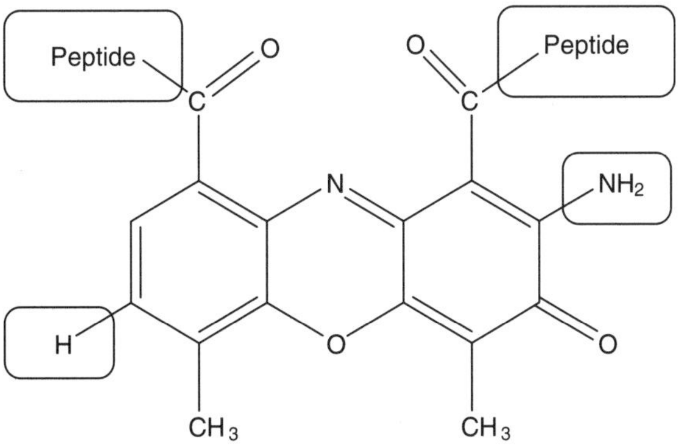

Figure 4. Synopsis of pharmacologically successful transformations of actinomycins. The "boxed" functional groups can be changed with retention of significant biological activity. Not all such changes, however, are successful.

The various D-amino acids are converted from the L-stereo-isomers and, in the case of dactinomycin, sarcosine is *N*-methylated [64]. By varying the amino acid composition of the medium, a variety of actinomycin analogs can be made by directed fermentation [65,66].

The chemistry of actinomycins has been the subject of a number of detailed reviews [67–71].

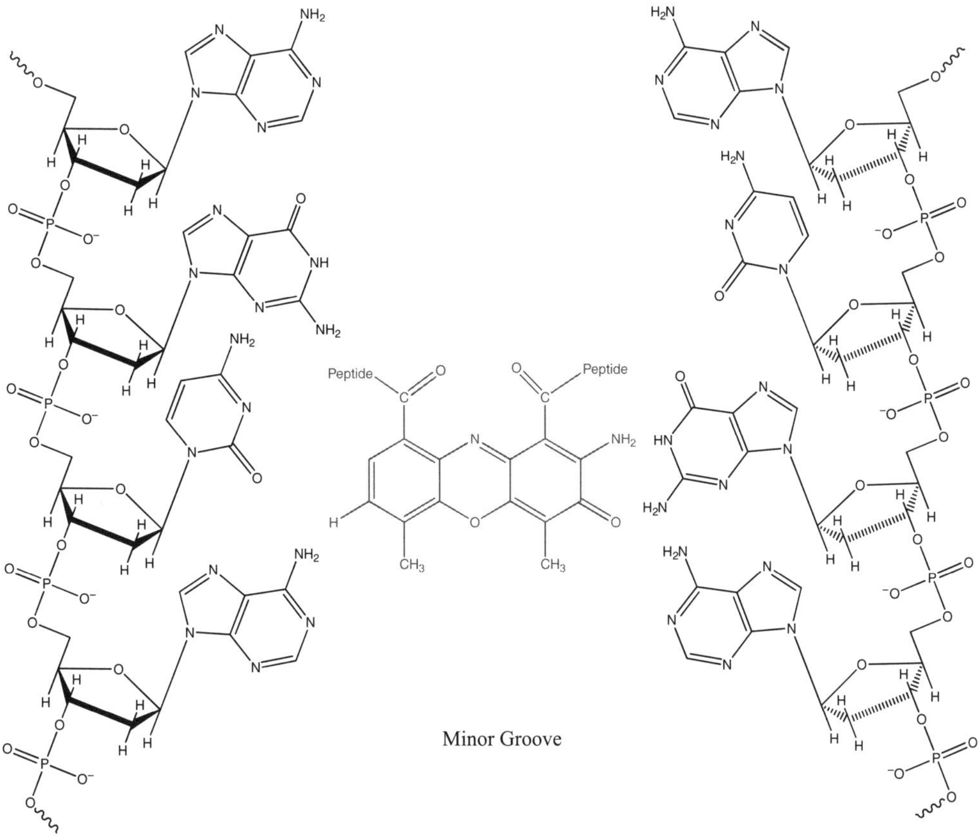

Figure 5. Schematic of intercalation and minor groove binding of dactinomycins. (This figure is available in full color at http://mrw.interscience.wiley.com/emrw/9780471266945/home.)

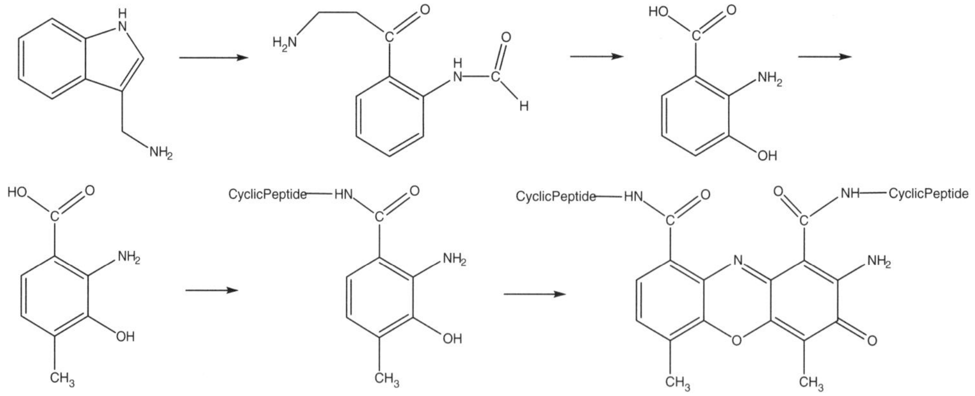

Figure 6. Biosynthesis of dactinomycin.

2.2. Bleomycin (Blenoxane)

2.2.1. Introduction
Bleomycin sulfate is a mixture of cytotoxic water-soluble basic glycopeptidic antibiotics isolated by the Umezawa group from fermentation broths of *Streptomyces verticillus*. The commercial form consists of cuprous chelates primarily of bleomycins A-2 (3) and B-2 (4). Subsequently, many analogs have been isolated by various groups and been given assorted names. Among these are the pepleomycins (5), phleomycins (11), cleomycins (12), tallysomycins (13), and zorbamycins (14).

2.2.2. Medicinal Uses
Bleomycin is used intramuscularly (i.m.), subcuteously (s.c.), intravenous (i.v.), or intrapleurally (i.p.), often in combination with other antibiotics, for the clinical treatment of squamous cell carcinomas, Hodgkin's disease, testicular and ovarian carcinoma, and malignant pleural effusion. It is also instilled into the bladder for bladder cancer so that less generalized side effects are obtained. It is often coadministered with a variety of other antitumor agents to enhance its antitumor efficacy. One advantage that bleomycin has in such combinations is that it possesses little bone marrow toxicity and is not very immune suppressant, so it is compatible, therapeutically, with other agents [72–75].

2.2.3. Contraindications and Side Effects
Bleomycin is contraindicated when idiosyncratic or hypersensitive reactions are observed. Immediate or delayed reactions resembling anaphylaxis occur in about 1% of lymphoma patients. Because of the possibility of anaphylaxis, it is wise to treat lymphoma patients with 2 units, or less, for the first 2 doses. If no acute reaction occurs, then the normal administration schedule can be followed.

The most severe toxicity of bleomycin is pulmonary fibrosis and it is more common with higher doses. This toxicity is observed in about 10% of patients and is difficult to anticipate, hard detecting in its early stages, and in about 10% of those affected it progresses to fatal lung compromise [76–78]. Renal damage occurs occasionally and further decreases the rate of excretion of the drug. In rats, bleomycin has been observed to be tumorigenic. In pregnant females, fetal damage can result.

Skin and mucous membrane damage, hair loss, rash, and itching, for example, are not uncommon and may require discontinuation of the drug. In addition, the common constellation of fever, nausea, chills, vomiting, anorexia, weight loss, pain at the tumor site, and phlebitis are seen.

Coadministration with digoxin and phenytoin may lead to a decrease in blood levels of these two drugs if used with bleomycin.

The side effects of bleomycin generally do not reinforce the toxicities of other antitumor agents, so it is often used in anticancer cocktails.

2.2.4. Pharmacokinetic Features
When injected i.v., bleomycin is rapidly distributed and has a half-life of 10–20 min. Intramuscular injections peak in 30–60 min, although the peak levels are less than about one-third those obtained i.v. The overall half-life of bleomycin is about 3 h. Skin and lungs accumulate particu-

larly high concentrations of the drug, in part because these are apparently the only tissues that do not rapidly deactivate it by enzymatic hydrolysis. It does not cross the blood–brain barrier efficiently because of its size and polarity. About 60–70% of the administered dose is recoverable as active bleomycin in the urine. Excretion is progressively delayed when the kidneys are damaged, so the doses are reduced by reference to creatinine levels [79].

2.2.5. Medicinal Chemistry The essential central core of bleomycin provides a chelating environment for transition metals, especially Cu (I) and Fe (II) [3]. The branched glycopeptide side chain is less essential for activity and appears to serve in facilitating passage across cell membranes and to assist in oxygen binding. Removal of the sugars and the oxygen to which they are attached produces molecules that are fully active, but distinct from bleomycin itself. The dipeptide unit is a linker arm, but contributes key hydrogen bonding and, perhaps, other binding interactions that intensify activity and produce degrees of base specificity to the cleavages. The bithiazole unit and its pendant terminal cation are important in DNA targeting of the drug. These contributions were uncovered by the chemical synthesis of analogs that could not readily have been prepared by degradation of bleomycin itself or by directed biosynthesis.

(3) A = $-NH(CH_2)_3S^+(CH_3)_2X-$
(4) A = $-NH(CH_2)_4NHC(NH)NH_2$
(5) A = $-NH(CH_2)_3NH(CH_2)_2$-2-pyridyl
(6) A = $-OH$
(7) A = $-NH(CH_2)_3S(O)CH_3$
(8) A = $-NH(CH_2)_3NH(CH_2)_4NH_2$
(9) A = $-NH(CH_2)_3NH(CH_2)_4NH(CH_2)_3NH_2$
(10) A = $-NH_2$
(11) A = as in (3); B = 44,45 Dihydrothiazole
(12) A = as in (3); C =
(13) A = as in (3); C = $-NHCH_2CH(OH)CH(CH_3)CO-$
    D = $-CH(-O$-4-amino-4,6-dideoxy-L-tallose$)-CH(OH)-$
(14) C_{50} = CH_3; A = as in (3); B = as in (11)
    C = $-NHCH_2(CH_2OH)CH(OH)CH(CH_3)CO-$

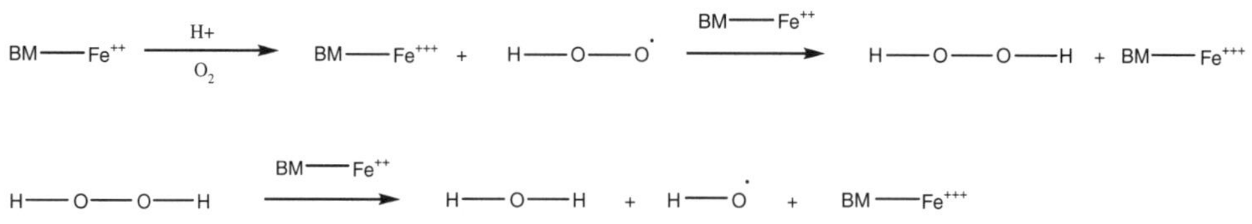

$$BM\text{—}Fe^{++} \xrightarrow[O_2]{H+} BM\text{—}Fe^{+++} + H\text{—}O\text{—}O^{\cdot} \xrightarrow{BM\text{—}Fe^{++}} H\text{—}O\text{—}O\text{—}H + BM\text{—}Fe^{+++}$$

$$H\text{—}O\text{—}O\text{—}H \xrightarrow{BM\text{—}Fe^{++}} H\text{—}O\text{—}H + H\text{—}O^{\cdot} + BM\text{—}Fe^{+++}$$

Figure 7. Generation of reactive oxygen species by transition metal chelates of bleomycins.

Partial chemical synthesis, with or without the aid of enzymes, has also produced a variety of analogs through modifications of this peripheral side-chain array [80–91]. Bleomycin is a conglomerate molecule built up from a collection of unusual subunits. Most of these were prepared independently by synthesis, in preparation for ultimate assembly into bleomycin itself or its analogs. The terminal bithiazole and its pendant amides are the portion of the molecule that binds to DNA. For the purpose of making analogs, the charged dimethylsulfonium group is monodemethylated by heat. The resulting compound is then cleaved to bleomycinic acid (**6**) using cyanogen bromide, followed by mild alkaline treatment. Some soil microorganisms possess acylagmatine amidohydrolase capable of converting bleomycin to bleomycinic acid. Bleomycinic acid is then converted to the desired amides by use of water-soluble carbodimide chemistry. Whereas the chemical method is capable of producing greater structural variation, in practice the semisynthetic method has proved more convenient.

Although bleomycin and its analogs have also been totally synthesized in various laboratories, the processes are too complex to yet be of commercial value [92–95].

The phleomycins [11] are related in that one of the thiazole rings has been reduced to its C-44,45-dihydro analog. The phleomycins have substantial antitumor activity, but are too nephrotoxic for clinical use. The cliomycins [12], tallysomycins [13], zorbamycins [14], zorbonamycins, platomycins, and victomycins are also structurally related to the bleomycins. None of these various alternative substances has displaced the bleomycin complex from the market, even though many possess significant antitumor properties. The specific potencies and toxicities vary widely with structural variations.

In the presence of a mild base, metal-free bleomycin isomerizes to isobleomycin through an O-to-O acyl migration of the carbamoyl moiety from position 22–23 of the mannosyl group. Copper (II) bleomycin, under the same conditions, slowly isomerizes at its masked aspartamine moiety attached to the pyrimidine substituent (at C6). This isomer is substantially less active than bleomycin itself.

Bleomycin chelates with various transition metals, the most relevant of which are iron (II) and copper (I), to form the corresponding complexes. The iron complex binds oxygen and becomes oxidized, producing the hydroxyl radical and the hydroperoxyl radical. This is illustrated in Fig. 7. The bithiazole moietly intercalates into DNA and the complex is stabilized by electrostatic attractions between the sulfonium or ammonium side chains with the phosphate backbone of DNA. This fixes the drug at DNA, whereupon the reactive oxygen species generated by its transition-metal complex breaks the DNA molecule at the sugar backbone, thus releasing purine and pyrimidine bases. This important reaction is illustrated in Fig. 8. Specific details of this complex interaction are still emerging.

Given that the biological action of bleomycin depends collectively on its ability to intercalate, to stabilize the intercalation complex by electrostatic forces, and to complex transition metals capable of generating oxygen radicals, the pharmacophore is distributed through the molecule. Acceptable variations involve substitution of various groups onto bleomycinic acid and a variety of other comparatively trivial changes, such as partial reduction of the thiazole moieties and alterations of the amino acids near the bleomycinic acid carboxyl group.

Recently, efforts have been directed to the synthesis of various macromolecular conjugates of bleomycin, in an attempt to produce tissue selectivity and, perhaps, reduce lung toxicity. Some of these agents retain very significant nucleic acid clastogenicity *in vitro* [96].

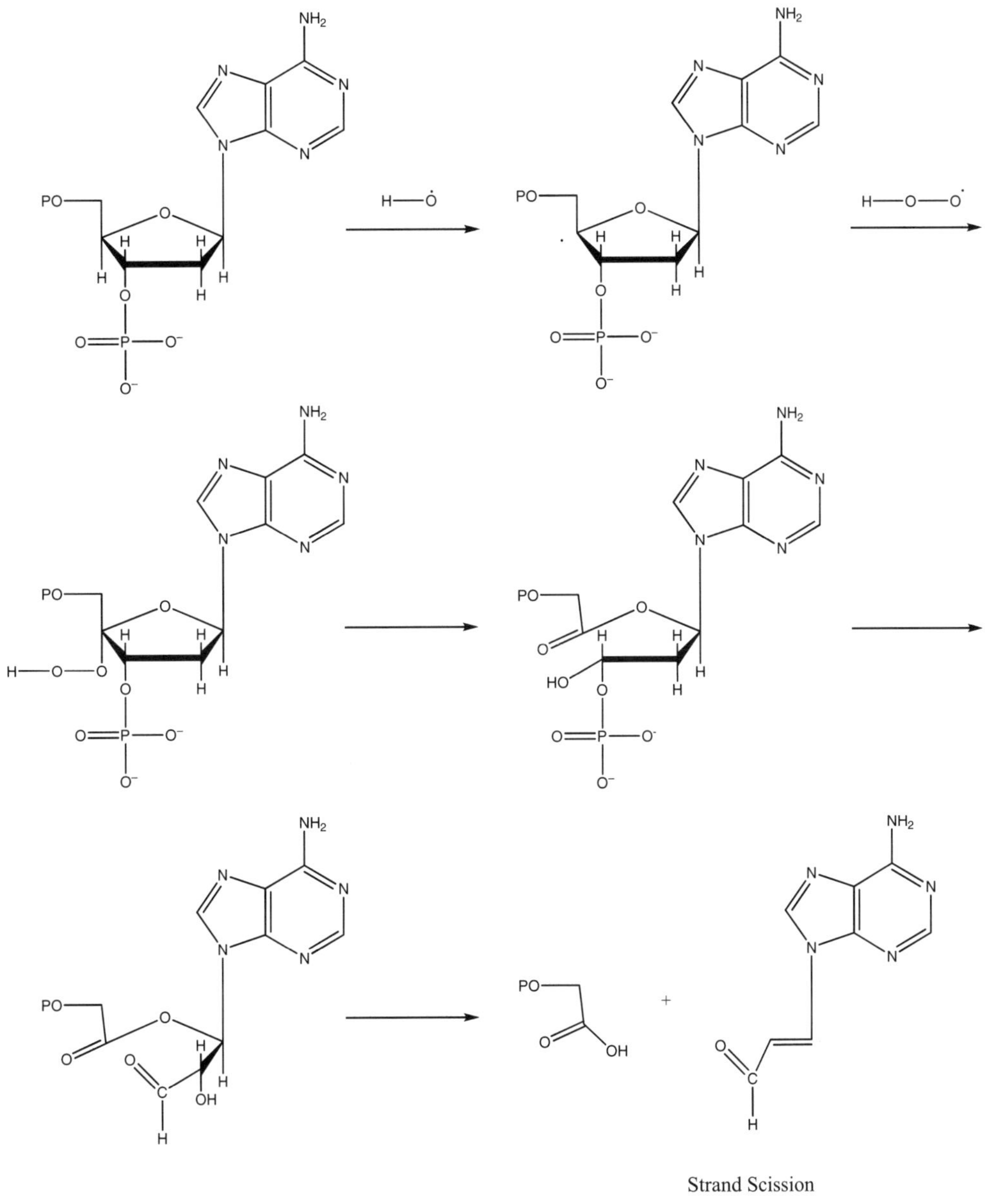

Figure 8. DNA backbone cleavage catalyzed by bleomycins.

2.2.6. Biosynthesis Many analogs of bleomycin have been prepared by directed biosynthesis through appropriate media supplementation [97–100]. Approximately 10 naturally occurring bleomycins have been reported (**3–4, 7–10**, etc.). These differ from one another by possessing a variety of different diamino analogs in place of the sulfoniumamino side chain attached to C49 of bleomycinic acid (**6**). In addition, directed biosynthetic methods involving media supplementation with suitable precursors have produced approximately 21 others, which also consist of a variety of diamino analogs in which the C49 moiety has been replaced. Thus, the biosynthesis of bleomycinic acid is relatively tightly controlled, although the amide synthase that puts on the various side chains is not very specific in its substrate tastes.

2.2.7. Molecular Mode of Action and Resistance

The precise molecular mode of action and bleomycin is incompletely understood because it has numerous actions in test systems. The bleomycins are known to bind preferentially to the minor groove of DNA, although the specific details of this host–guest interaction are still elusive. The cytotoxicity of the bleomycins is enhanced when a DNA-binding region is present and the specific nature of the DNA-binding moiety can convey sequence specificity. The nucleic acid-cleaving capacity is metal ion and oxygen dependent, and it is believed that the complexes generate reactive oxygen species that are responsible for the single- and double-strand nucleic acid cleavages observed (see Fig. 8). This DNA destruction is generally believed to account for its cytotoxicity [101–104]. Interestingly, in the absence of DNA, bleomycin is also capable of destroying itself instead, presumably through the action of the same reactive oxygen species [105]. A number of artificial analogs have been prepared to explore the contribution of various molecular features of these drugs and to exploit these features. Some of these products include agents that are inert by themselves, but they enhance the cytotoxicity of bleomycin fragments when attached craftily to them. These agents usually contain aromatic moieties and have the capacity to have a cationic moiety as well. Bleomycin is known to generate oxygen-based free radicals when chelated to certain metal ions, notably ferrous iron and copper. When chelated to ferric iron, a reducing agent adds an electron to convert the complex to ferrous iron. This, in turn, transfers an electron to oxygen, producing either the superoxide radical, or the hydroxide radical (see Fig. 7). These radicals attack ribosyl moieties in DNA and RNA, leading to nucleic acid fragmentation and subsequent interference with their biosynthesis. This action is believed to be cardinal in the cytotoxic action of bleomycin. Bleomycin's action is cell cycle specific, causing major damage in the G-2 and less in the M phase.

Resistance to bleomycin occurs primarily through the action of bleomycin hydrolase, which attacks metal-free bleomycin at the C4-carboxamine moiety to produce deamidobleomycin [106]. This last produces radicals at a much lower frequency than that of bleomycin itself. This causes a much lower cleavage of DNA and removes the majority of the antitumor action of bleomycin. In support of this idea, resistant cells usually possess a higher concentration of bleomycin hydrolase than do sensitive cells. The hydrolase is present in normal tissues, particularly in the liver. Interestingly, recent evidence implicates this enzyme in the formation of amyloid precursor protein characteristic of Alzheimer's disease [107]. Other experts implicate enhanced DNA repair capacity or decreased cellular uptake as contributory to resistance.

2.2.8. Recent Developments and Things to Come

Considering bleomycin's particular ability to destroy DNA and RNA molecules, there is comparatively little likelihood that molecular manipulation of bleomycin will soon produce a nontoxic version of the drug.

The chemical properties of the bleomycins have been recently reviewed [104,108–113].

2.3. Mitomycin (Mutamycin)

2.3.1. Introduction

Mitomycin C (15) was discovered initially at the Kitasato Institute [114] and at the Kyowa Hakko Kogyo laboratories in Japan, as a metabolite of *Streptomyces caespitosus* [115], and elsewhere [116]. A number of analogs have been discovered by several other groups. These drugs are a group of blue aziridine-containing quinines, of which mitomycin C is the most important from a clinical perspective. Mitomycin A and porphiromycin also belong to this group, but have not been marketed. Mitomycins, apparently, were the first of the useful bioreductively activated DNA alkylating agents to be discovered. Literally thousands of alkylating agents, notably the α,β-unsaturated sesquiterpene lactones of the Compositae, have been found in nature, and enormous effort has been expended in their synthesis and evaluation without notable success. The contrasting success of the mitomycins seems to derive from the finding that they are relatively inert until bioreductively activated, so they show greater biological selectivity compared with that of many other naturally occurring alkylating agents.

(15)

2.3.2. Clinical Use Mitomycin is administered i.v. in combinations with antitumor agents for treatment of disseminated adenocarcinoma of the stomach, colon, or pancreas, or for treatment of other tumors where other drugs have failed [117–120].

2.3.3. Contraindication and Side Effects It is contraindicated in cases of hypersensitivity, or idiosyncratic responses to the drug, or where there are preexisting blood dyscrasias. The drug can cause a serious cumulative bone marrow suppression, notably thrombocytopenia and leucopenia [121,122], that can contribute to the development of overwhelming infectious disease. This requires reducing dosages. Irreversible renal failure, as a consequence of hemolytic uremic syndrome is also possible [121]. Occasionally, adult respiratory distress syndrome has also been seen. When extravasation is seen during administration, cellulitis, ulceration, and sloughing of tissue may be the consequence [123,124]. The drug is known to be tumorigenic in rodents. Its safety in pregnancy is unclear and teratogenicity is seen in rodent studies. Other side effects include fever, anorexia, nausea, vomiting, headache, blurred vision, confusion, drowsiness, syncope, fatigue, edema, thrombophlebitis, hematemesis, diarrhea, and pain. It is not clear whether all of these symptoms are related to the use of mitomycin, or whether they are, at least partly, the consequence of other agents in antitumor cocktails.

2.3.4. Pharmacokinetics Mitomycin is poorly absorbed orally and is rapidly cleared when injected i.v., with a serum half-life of about 30–90 min after a bolus dose of 30 mg. Metabolism takes place primarily in the liver and is saturable. As a consequence, the amount of free drug in the urine increases proportionally. Only about 10% of an average administered dose is excreted unchanged in the urine and the bile because extensive metabolism takes place. The drug is distributed widely in the tissues, with the exception of the brain, where there is little penetration [125–128].

Because mitomycin C is activated as an antitumor agent by reduction, significant effort has been expended on trying to elucidate whether DT-diaphorase activity correlates well with antitumor activity *in vivo*. This is, as yet, imperfectly resolved but the correlation appears to be poor. Other studies suggest that NADPH:cytochrome P450 reductase (a quinone reductase) contributes strongly under some circumstances.

Inactivation, and activation, occurs by metabolism and/or by conjugation, and a number of metabolites, principally 2,7-diaminomitosene, have been identified [129–131]. The ratio between inactivation and activation is partially a function of whether DNA intercepts the reduced species before it is quenched by some other molecular species.

2.3.5. Medicinal Chemistry Much exploration of the chemistry of the mitomycins has been carried out accompanied by excellent reviews in the literature [132–134]. Total chemical syntheses of mitomycins A and C have been achieved, but these are not practical for production purposes [135–137]. More than a thousand analogs have been prepared by semisynthesis, but none of these agents has succeeded in replacing mitomycin C itself. Generally, it has been found that mitomycin C analogs are less toxic than mitomycins A derivatives. Most modifications have been achieved at the N-1a-, C7-, C6-, and C10-positions. The C7-position is particularly conveniently altered through addition/elimination sequences, and some of these agents have received extensive evaluation. It is noted that the C6- and C7-positions play only an indirect role in the activation of the ring system, so substitutions there might be regarded as primarily significant in altering the pharmacokinetic properties of the mitomycins. It has been found quite recently, however, that the participation of the C7-substituent in activation by thiols differs significantly when C7 bears a methoxyl group (the mitomycin A series) compared to the activation when C7 bears an amino group (the mitomycin C series). Indeed, thiols activate the methoxy analogs, but not the amino analogs. Mechanistically, both series arrive at the same bisalkylating species *in vivo*, but through different routes. This may help rationalize why mitomycin A is both more potent and more cardiotoxic than mitomycin C [138]. The results of a comparison of physicochemical properties and biological activity of the mitomycins led to the conclusion that potency correlates with uptake, as influenced primarily by log p, and also with the redox potential (E1/2) [139].

The metabolism of mitomycin C *in vivo* primarily leads through reduction and loss of methanol to a dihydromitosene end product. Interception by DNA, on the other hand, instead leads to alkylation of the latter [138,139].

2.3.6. Molecular Mode of Action and Resistance

Mitomycin C undergoes enzymatic reductive activation to produce reactive species capable of bisalkylation and cross-linking of DNA, resulting in inhibition of DNA biosynthesis [140–142]. This effect is particularly prominent at guanine-cytosine pairs. The reductive activation of mitomycin C makes it particularly useful in anaerobic portions of tumor masses that exhibit a generally reducing environment. Mitomycin is also capable of causing single-strand breaks in DNA molecules.

The apparent chemical mechanism by which mitomycin is reductively alkylated to a bisalkylating agent is illustrated in Fig. 9. The process is initiated by a quinine reduction followed by elimination of methanol, opening of the aziridine ring, conjugate addition of DNA, ejection of the carbamate function, and further addition of DNA.

The bisalkylation of DNA can be either intrastrand or interstrand, as illustrated in Fig. 10.

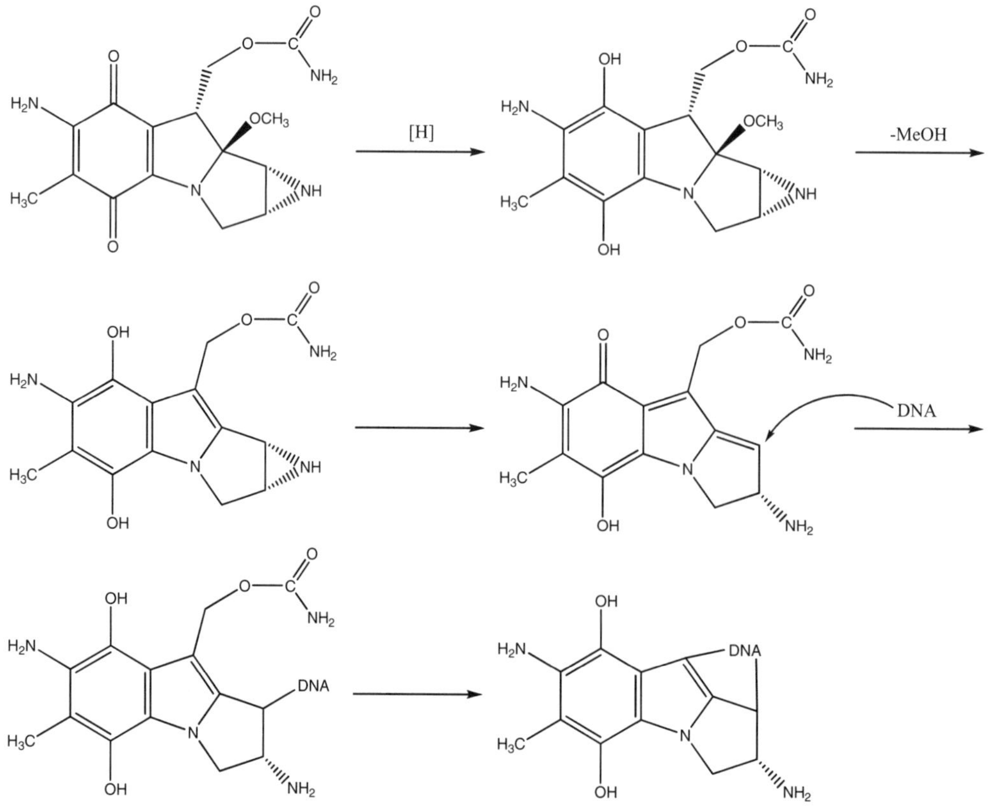

Figure 9. Reductive activation and bisalkylation of DNA by mitomycin C.

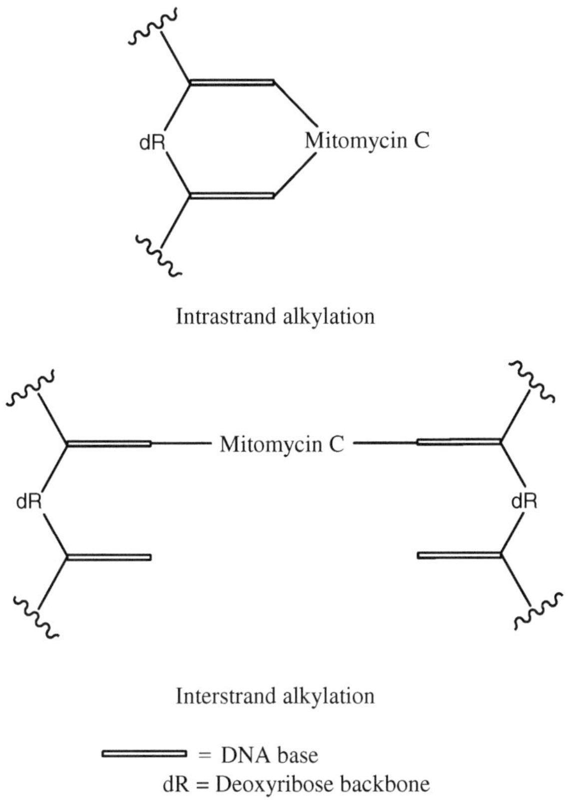

Figure 10. Interstrand and intrastrand alkylation of DNA by bioreductively activated mitomycin C.

Resistance is attributed to failure of reduction [143], to premature reoxidation [143,144], binding to a drug-intercepting protein that also has oxidase activity [145], and to P-glycoprotein-mediated efflux from cancer cells [146,147].

2.3.7. Medicinal Chemistry The pharmacologically successful chemical transformations of mitomycin are schematically summarized in Fig. 11.

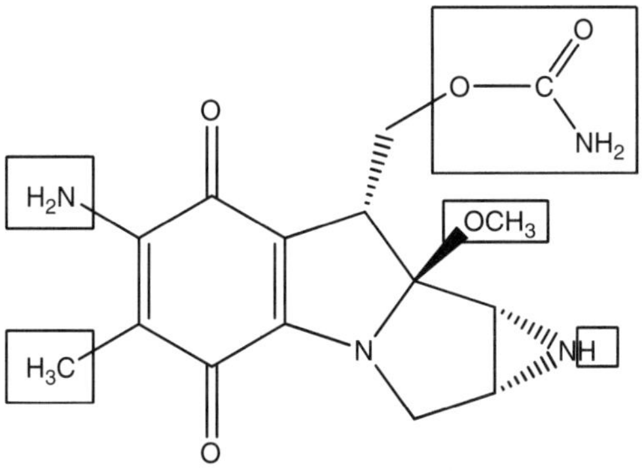

The "boxed functional groups can be changed with retention of significant biological activity. Not all changes, however, are successful.

Figure 11. Pharmacologically successful modifications of mitomycin C. The "boxed" functional groups can be changed with retention of significant biological activity. Not all changes, however, are successful.

The chemistry and pharmacological actions of the mitomycins have been reviewed in Refs [132–134,148].

2.4. Plicamycin (Formerly, Mithramycin and Mithracin)

2.4.1. Introduction Plicamycin (**16**), produced by fermentation of *Streptomyces plicatus* and *S. argillaceus*, was isolated in 1953 [149]. It is a member of the aureolic acid family of glycosylated polyketides, which also includes chromomycins, olivomycins, and UCH9. It was subsequently found to be identical to mithramycin, a fermentation product of *S. argillaceus* and *S. tanashiensis*.

(**16**)

2.4.2. Clinical Use Plicamycin is highly toxic, but is nevertheless administered i.v. for treatment of testicular tumors [150–153]. In lower doses, it is used for treatment of hypercalcemia and hypercalciuria associated with advanced cancer. It also finds use in Paget's bone disease [154–157].

2.4.3. Contraindications and Side Effects Severe thrombocytopenia, hemorrhagic tendency, and death can be encountered with the use of plicamycin [158,159]. Renal impairment, mutagenicity, and interference with fertility are also known to occur with its use. Anorexia, nausea, vomiting, diarrhea and stomatitis, fever, drowsiness, weakness, lethargy, malaise, headache, depression phlebitis, facial flushing, skin rash, hepatotoxicity, and electrolytic disturbances (decrease in serum calcium, potassium, and phosphate levels) are also encountered.

Plicamycin is contraindicated with coagulation disorders, thrombocytopenia, thrombocytopathy, impairment of bone marrow function, and in pregnancy.

The toxic reactions of plicamycin are much less severe and frequent in the lower dosages employed to lower calcium ion levels.

2.4.4. Pharmacokinetics Plicamycin is given i.v., whereupon a complex excretion pattern ensues, with a reported half-life of approximately 11 h [160].

2.4.5. Mode of Action and Resistance
The exact mechanism of action of plicamycin is elusive, but it is known to intercalate into DNA, favoring G-C base pairs, resulting in the inhibition of enzymes that process DNA [161–163]. Plicamycin also interferes in the biosynthesis of RNA [163]. The effect of plicamycin is enhanced in the presence of divalent metal ions such as magnesium (II). Its hypocalcemic action is unrelated to this, but is rather mediated by interference with the function of vitamin D in some unclear manner [164]. Plicamycin also acts on osteoclasts and blocks the action of parathyroid hormone [165,166]. Resistance to plicamycin involves efflux through the action of P-glycoprotein [167], although recent publications suggest that plicamycin has the capacity to suppress *MDR 1* gene expression *in vitro*, thereby modulating multidrug resistance [168].

2.4.6. Medicinal Chemistry
The chemistry of plicamycin and its analogs has been reviewed [169]. For a long time, there was considerable confusion about the precise chemical structure of plicamycin (mostly with respect to the number and arrangement of the sugars) but this has, apparently, now been resolved by detailed NMR studies [170].

The sugars must be present in plicamycin for successful DNA binding and magnesium ion also promotes the interaction.

2.4.7. Biosynthesis
Biosynthesis of the aureolic acid group of antitumor antibiotics begins with condensation of 10 acetyl units to produce a nascent polyketide that, on condensation, produces a tetracyclic intermediate whose structure, and that of the subsequent intermediates, is reminiscent of those involved in tetracycline biosynthesis [171]. After the formation of premithramycinone, a rather complex sequence of reactions ensues, as illustrated in Fig. 12. A sequence of methylations and glycosylations leads to premithramycin A3. Of particular interest in the remaining sequence is an oxidative ring scission and decarboxylation, which leads to the final tricyclic ring system. This is followed by oxidation level adjustment, producing plicamycin itself, or to one of the other members of this class, depending on the specifics of the biosynthetic intermediates [172,173]. Omission of the key C7-methylation step leads, for example, through a parallel pathway to the formation of 7-demethylmithramycin [174].

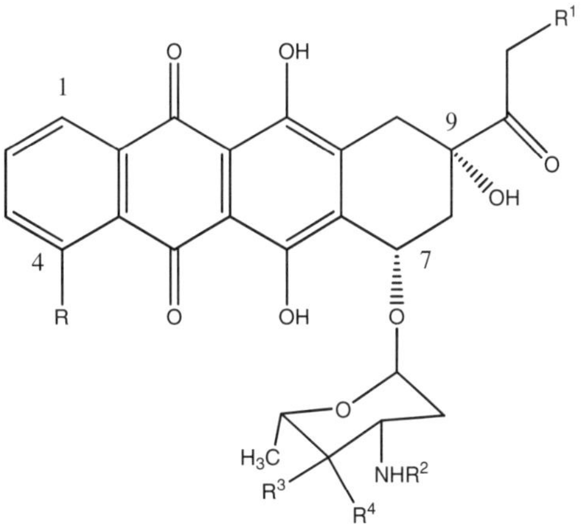

Compound	R	R_1	R_2	R_3	R_4
(17)	OCH$_3$	OH	H	H	OH
(18)	OCH$_3$	H	H	H	OH
(19)	OCH$_3$	OH	H	OH	H
(20)	H	H	H	H	OH
(21)	OCH$_3$	OCOBu	COCF$_3$	H	OH

Figure 12. Biosynthesis of plicamycin.

3. DRUGS INHIBITING ENZYMES THAT PROCESS DNA

3.1. Anthracyclines

The anthracyclines are an important class of streptomycete-derived tetracyclic glycosidic and intercalating red quinon-based drugs. None of the first generation of this widespread class of natural products became clinically prominent. The structures of some of these chemically interesting compounds, generally named as rhodomycins, including pyrromycin, musettamycin, and marcellomycin (whose names will please opera buffs), are given in Fig. 13. Those anthracyclines of clinical value were discovered, initially in the Pharmitalia Laboratories in Italy and subsequently in a number of other institutions [175,176]. The first of the clinically useful group was the *Streptomyces peucetius* metabolite, daunorubicin (**18**). This was followed by its hydroxy-

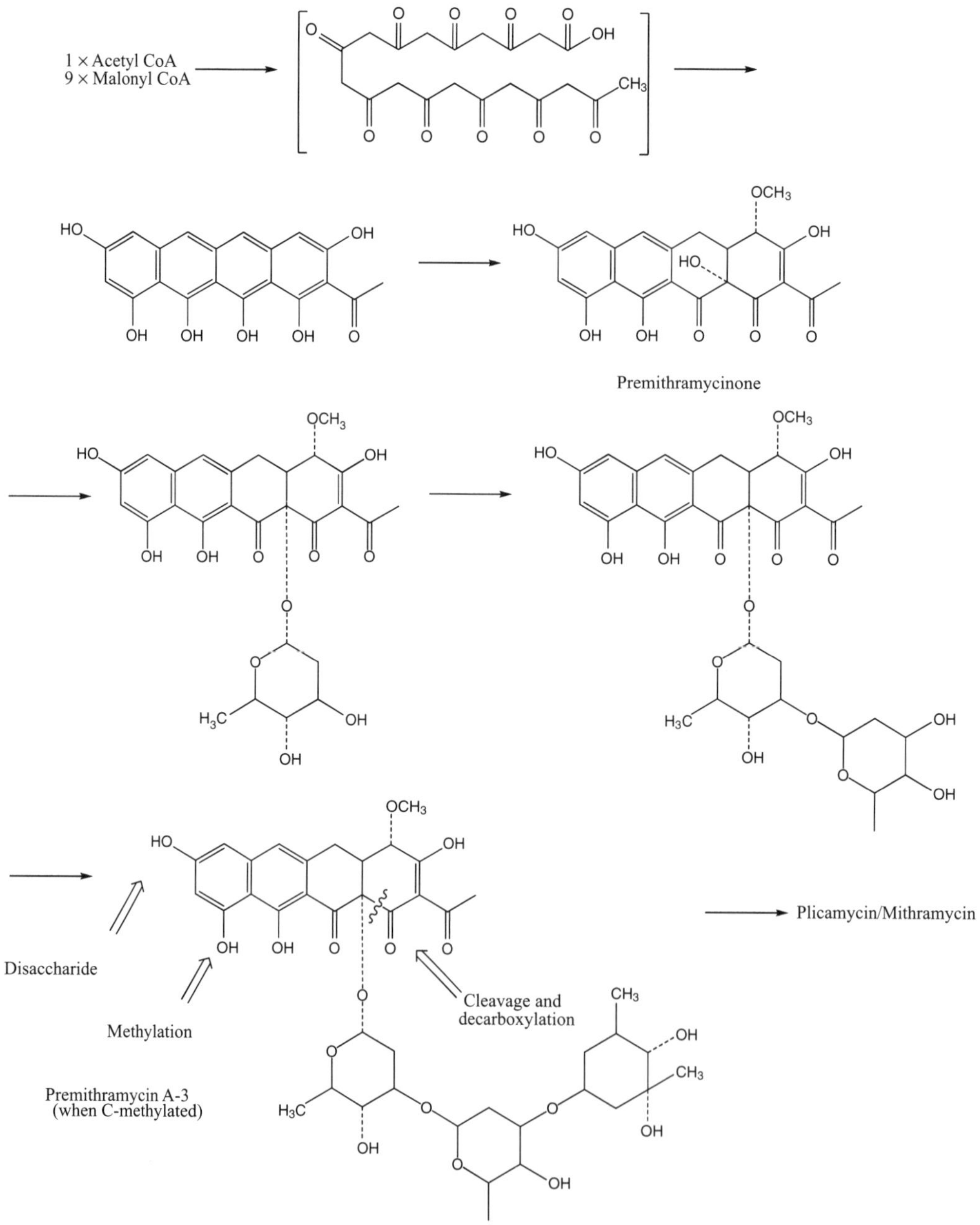

18 NATURAL PRODUCTS AS CYTOTOXIC AGENTS

Figure 12. (*Continued*)

lated analog doxorubicin (**17**), a metabolite of *S. peucetius* var. *caesius*. Many synthetic anthracyclines resulted from intense study in many laboratories. These synthetic methods led to a number of marketed products, including daunomycin's desmethoxy analog idarubicin (**20**) and doxorubicin's diastereomer epirubicin (**19**) [177, 178], and the bisacylated product of doxorubicin, valrubicin (**21**), Daunomycin and idarubicin are primarily used for the treatment of acute leukemia, and epirubicin is used for a much wider range of cancers.

Compound	R	R_1	R_2	R_3	R_4
(17)	OCH$_3$	OH	H	H	OH
(18)	OCH$_3$	H	H	H	OH
(19)	OCH$_3$	OH	H	OH	H
(20)	H	H	H	H	OH
(21)	OCH$_3$	OCOBu	COCF$_3$	H	OH

3.1.1. Daunorubicin (Daunomycin, Cerubidine, and Rubidomycin (18))

Therapeutic Uses Daunorubicin is used in combination with other agents by i.v. infusion for treatment of acute myelogenous and lymphocytic leukemias [179–181].

Side Effects and Contraindications It is not generally used i.m. or s.c. because of the severe tissue damage that may accompany extravasation [182]. It is contraindicated when hypersensitivity reactions are present. Among the side effects that are encountered are severe cumulative myocardial toxicity that can include acute congestive heart failure after cumulative doses above 400–500 mg/m^2 of body surface in adults and less in infants [183], severe myelosuppression (hemorrhage and superinfections), bone marrow suppression, secondary leukemia, renal/hepatic failure, carcinogenesis, mutagenesis, teratogenicity, and fertility impairment. The cardiomyopathy is characteristic of the anthracycline class can occur long after therapy is concluded [184,185]. The highly colored nature of the drug can lead to urine discoloration that alarms the patient because of drug excretion. In addition, alope-cia, rash, contact dermatitis, urticaria, nausea, vomiting, mucositis, diarrhea, abdominal pain, fever, chills, and (occasionally) anaphylaxis are observable. When given along with cyclophosphamide, its cardiotoxicity is enhanced, and enhanced toxicity is seen when given concurrently with methotrexate.

Pharmacokinetics On i.v. administration, the drug is rapidly distributed into tissues, but does not enter the central nervous system. Rapid liver reduction to daunomycinol is seen followed by hydrolytic or reductive loss of the sugar along with the ether atom with which it is attached to the ring system. These two reactions can also take place before reduction. Demethylation of the *O*-methyl ether moiety also occurs followed by sulfation, or glucuronidation of the resulting phenolic OH. These and other transformation products have lesser bioactivity [186]. Patients with decreased liver function should receive smaller doses because they are not able to detoxify the drug effectively. The half-life is about 8.5 h and about 25% of the active drug is found in the urine along with about 40% in the bile [187]. Liposomally encased daunorubicin citrate shows greater selectivity for solid tumors and is translocated in the lymph [180]. Nuclear binding of anthracyclines is sufficiently strong to complicate the excretion pattern and to determine the tissue distribution of these agents [188]. Different tissues bind doxorubicin in direct proportion to their DNA content. The metabolism of daunorubicin is illustrated in Fig. 14.

Mechanism of Action and Resistance The mode of action of daunorubicin, and the other clinically useful anthracyclines, is multiple. Authorities differ with respect to which is the most significant, but most attribute this to inhibition of the mammalian topoisomerase II, essential for shaping DNA, so that it can function and be processed [189]. The drug also intercalates into DNA, inhibits DNA and RNA polymerases, and also causes free-radical single- and double-strand damage to DNA [190]. These drugs are, therefore, also mutagenic and carcinogenic. Free-radical (reactive oxygen species) generation is promoted by the interaction of these drugs with P450 [191] and with iron, which they chelate [192]. The reactive oxygen species that they can generate also

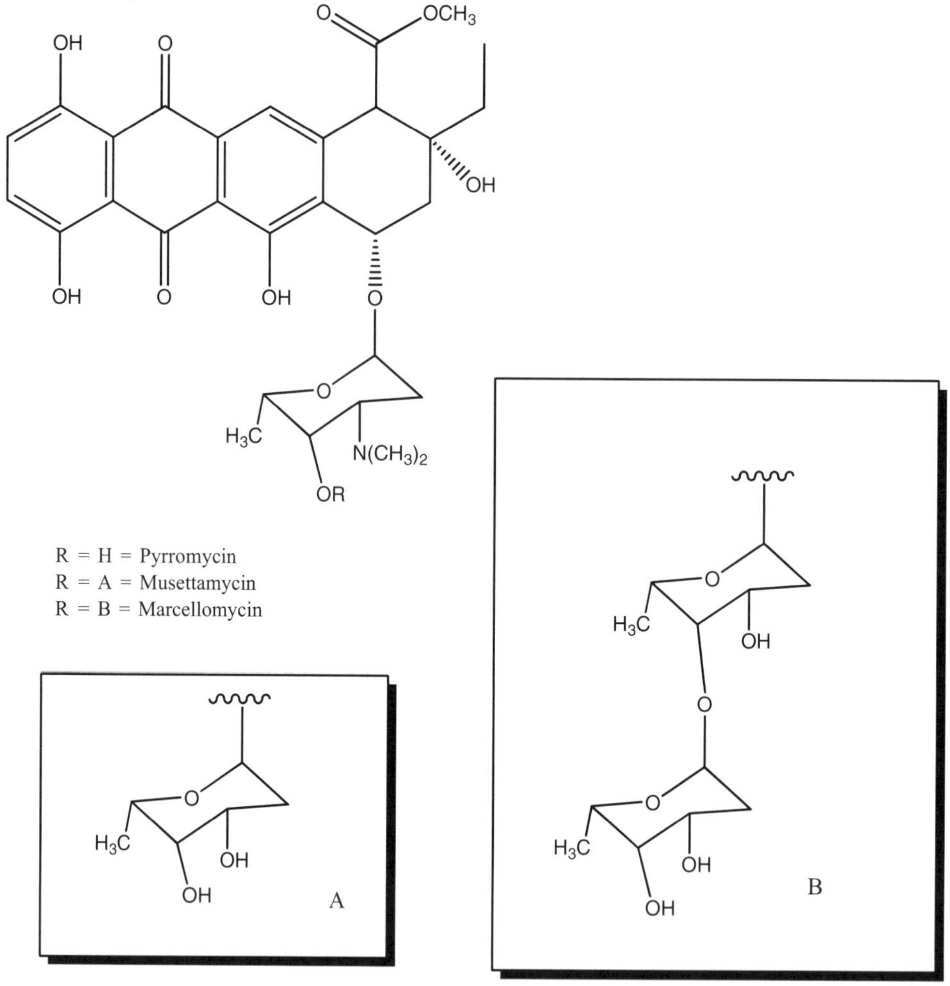

R = H = Pyrromycin
R = A = Musettamycin
R = B = Marcellomycin

Figure 13. Structures of some unmarketed anthracyclines.

causes severe damage to membranes and this may contribute not only to their antitumor efficacy but also to the cardiomyopathy that they cause [193].

Resistance to daunorubicin and the other anthracyclines is attributed to efflux mediated by P-glycoprotein, whose expression is amplified in response to their use [187,194–196]. A number of other mechanisms have been advanced as contributory such as use of other export mechanisms, and decreased action of mammalian topoisomerase II [197].

3.1.2. Doxorubicin Doxorubicin (Adriamycin and rubex (**17**)) is a hydroxylated analog of daunorubicin. It has a much wider anticancer use than daunorubicin.

Therapeutic Uses Doxorubicin is administered i.v. by rapid infusion for the treatment of disseminated neoplastic conditions such as acute lymphoblastic leukemia, acute myeloblastic leukemia, Wilms' tumor, neuroblastoma, soft tissue and bone sarcomas, breast carcinoma, ovarian carcinoma, transitional cell bladder carcinoma, thyroid carcinoma, Hodgkin's and non-Hodgkin's lymphomas, bronchogenic carcinoma, and gastric carcinoma are also target candidates. It has been reported that the effectiveness of doxorubicin against hepatocellular carcinoma is enhanced

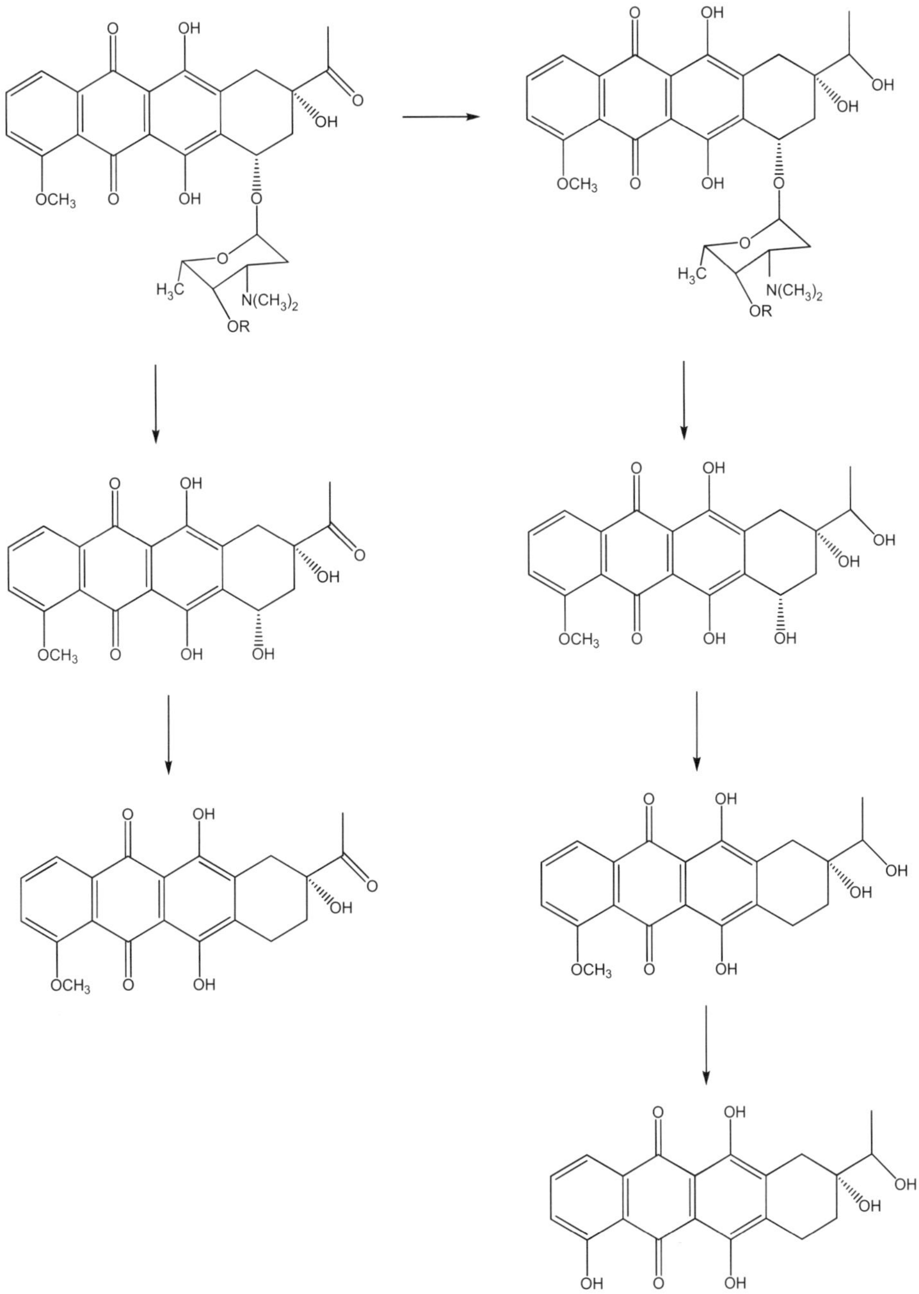

Figure 14. Metabolism of daunorubicin.

when conjugated with lactosaminated human albumin (L-HSA), a galactosyl-terminating neoglycoprotein.

Side Effects and Contraindications Doxorubicin is contraindicated in patients with preexisting severe myelosuppression consequent either to other antitumor treatments or to radiotherapy. It is also contraindicated when hypersensitivity to anthracyclines is present, or when significant previous doses of other anthracyclines have been administered, given that their doses coaccumulate toward congestive heart failure.

Side effects are generally similar to those seen with daunorubicin (see above), with particular reference to cumulative drug-related congestive heart failure, extravasation problems, myelosuppression, and hepatic damage.
Pharmacokinetics As with daunorubicin, the tissue distribution of doxorubicin is strongly influenced by the cellular content of DNA in various parts of the body [188]. Metabolites of doxorubicin are its aglycone, deoxyaglycone, doxorubibinol, and deoxyaglycone, and demethyldeoxyadriamycinol aglycone as its 4-O-β- glucuronide and O-sulfides. Thus, carbonyl reduction is the main metabolic reaction and this is followed by various hydrolytic and reductive losses of the sugar, O-demethylation, and various conjugative reactions [199]. These reactions directly parallel the findings with those of doxorubicin.
Molecular Mode of Action and Resistance The manifold cytotoxic actions of doxorubicin on cells are qualitatively the same as those of daunorubicin. Likewise, the resistance mechanisms, especially those involving P-glycoprotein expulsion, are closely similar. Interestingly, expulsion is significantly lessened by liposome encapsulation [200].

3.1.3. Epirubicin
Epirubicin (Ellence (**19**), a C4′-diastereoisomer of doxorubicin, is administered by i.v. infusion as an adjunct to the use of other agents for the treatment of breast cancer, when axillary node tumor involvement is seen after breast removal surgery. The toxicities of epirubicin are analogous to those described above for daunorubicin and doxorubicin (see above). Particular note should be paid to drug-related cumulative congestive heart failure, extravasation problems, myelosuppression, and hepatic damage.

3.1.4. Valrubicin
Valrubicin (**21**) and idarubicin (**20**) are anthracyclines that have significant use in the clinic [203]. Idarubicin differs from doxorubicin by lacking the methoxy group in the chromophore, and has an epimeric hydroxyl group in the sugar [204]. This compound is comparatively lipophilic, resulting in increased cellular uptake (the cellular concentrations exceed 100 times those achieved in plasma) [205] and strong serum protein binding [206]. The agent undergoes extensive extrahepatic metabolism to the 13-dihydro analog [207].

Valrubicin is the valeric ester trifluoroacetic amide of doxorubicin [208–210]. It is instilled into the bladder through a urethral catheter after bladder drainage and is voided after 2 h [211]. It is highly toxic on contact with tissues, but its means of administration limits systemic exposure. Its local adverse reactions are usually comparatively mild and resolve in about 24 h. Evidence indicates that the valeryl ester moiety is enzymatically cleaved *in vivo* before exerting its cytotoxic effect [212,213].
Biosynthesis The anthracyclines are polyketides, as can be readily discerned from their structures. Doxorubicin by a late hydroxylation step that is genetically unstable. As a consequence, it is apparently produced commercially by an efficient chemical transformation instead of through fermentation [214].
Medicinal Chemistry Several hundred analogs have been prepared either by chemical transformation of the natural products themselves or by total synthesis. As a result, there is a very good understanding of their structure–activity relationships [215–220].

The impressive anticancer activity and clinical potential of the anthracyclines resulted in intensive research toward the total synthesis and structural modification studies of these compounds [175,176,221,222]. From a structure–activity relationship (SAR) perspective, the anthracycline structural core can be divided into three major components: (1) Ring D, the alicyclic moiety bearing the two-carbon side-chain group and the tertiary hydroxyl group at C9, concomitantly having a chiral secondary hydroxyl group at C7, which, in turn, is connected to the aminosugar unit; (2) the aminosugar residue, attached to the C7-hydroxy group through an α-glycosidic linkage; and (3) the anthraquinone chromophore, consisting of a quinine and a hydroquinone moiety on adjacent rings. The C13- and C14-positions of the various anthracyclines are obvious functional sites for derivatization. Thus, the 13-keto functionality has been subjected to reduction, deoxygenation, hydrazide formation, and so forth, without adversely affecting the bioactivity. Similarly, incorporation of various ester and ether functionalities at C14, through initial halide formation and subsequent displacement of the halogen with nucleophiles, was found to be a useful approach in modulating the activity of the parent anthracyclines. However, homologation of the C9 alkyl chain, or introduction of amine functionalities at C14 is detrimental to the 9, 10-anhydro, or the 9-deoxy analogs and results in decreased activity. Interestingly, the natural stereochemical configurations at C7 and C9 were found to be an important contributor to bioactivity, wherein it has been proposed that H-bonding between the two *cis*-oxygen functionalities at these positions stabilizes the preferred half-chair conformation of the D ring.

The amino sugar residue of the various anthracyclines is an essential requirement for bioactivity. Among the various SAR studies involving the carbohydrate core, it has been seen that attachment of this moiety to the anthracycline nucleus through an α-anomeric bond is necessary for optimum activity. Conversion of the C3′-amine group to the corresponding dimethylamino or morpholino functionaliltes confers improved activity; however, acylation of the amine (the exception being trifluoroacetyl) or its replacement with a hydroxyl group results in loss of activity. Interestingly, conversion of the C4′-hydroxyl group to its corresponding methyl ether, C4′-epimerization, or deoxygenation, has a negligible effect on bioactivity. In more recent studies, novel disaccharide analogs of doxorubicin and idarubicin have been found to exhibit impressive antitumor activity [223].

The anthraquinone chromophore is an important structural feature of the anthracyclines. The various oxygenated functionalities present in this fragment have been the focus of considerable synthetic activity in search of analogs with improved activity. Thus, the phenolic hydroxyl groups present in this core were found to undergo ready acylation and alkylation under standard reaction conditions. It has been show that, *O*-methylation of the C6- or C11-phenolic groups results in analogs with markedly reduced activity, whereas C4 modifications, such as demethylation and deoxygenation, do not affect bioactivity. Interestingly, a serendipitous transformation of the C5-carbonyl to the corresponding imino functionality resulted in an analog that retained activity and was found to be significantly less cardiotoxic than the parent compound.

Biosynthesis The proposed biogenesis of the anthracyclines invokes the involvement of a polyketide synthon. In studies involving various blocked mutants of anthracycline-producing *Streptomyces* and utilization of 14C-labeled acetate and propionate precursors, it has been shown that there are two biosynthetic pathways responsible for the formation of the polyketide fragment. Daunomycinone, pyrromycinone, and related aglycones are derived from a polyketide synthon having one propionate and nine acetate units, whereas deviant members such as steffimycinone and nogalanol are obtained from a 10-acetate polyketide unit. Thus, a "head-to-tail" condensation of the decaketide chain forms the parent tetracycline core and the C9 quaternary center of the anthracyclines. A sequence of biotransformations involving C2- and C7-carbonyl reduction, dehydration (C-2/C-3), enolization/aromatization, and B ring oxidation leads to aklavinone. Further oxidation, decarboxylation (for some class of compounds), and glycosidation finally result in the corresponding bioactive glycosides. See Fig. 15 for a schematic illustration of the proposed biosynthetic pathway leading to anthracyclines.

Recent Developments and Things to Come Reviews of this topic are available in Refs [205,212,224].

3.2. Camptothecins

Camptothecin (**22**) was discovered almost at the same time (1966) as taxol by the same research group. It is present in extracts of the Chinese tree *Camptotheca acuminata* (growing in California) and has subsequently been found to be abundant in the extracts of *Mappia foetida*, a weed that grows prolifically in the Western Ghats of India. Despite its early promise in laboratory and rodent studies, it was disappointing in clinical studies because of severe toxicity. Subsequently, it has not found clinical use by itself, but serves as the template for the preparation of its clinical descendants prepared both by partial and by total

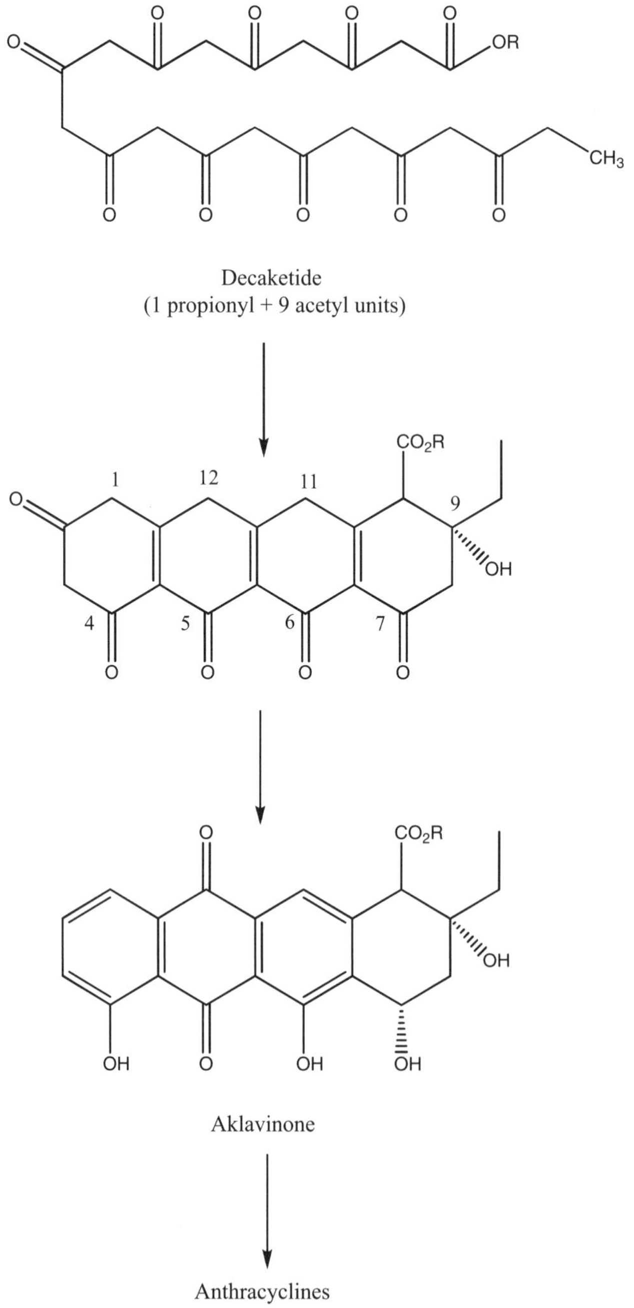

Figure 15. Proposed biosynthetic pathway leading to anthracyclines.

chemical syntheses. Camptothecin itself is very insoluble in polar and nonpolar solvents. This made early evaluation difficult. Tests were performed with its sodium salt (prepared by hydrolysis of the lactone ring) but clinical trials had to be discontinued because of severe, unpredictable hemorrhagic cystitis, even though some patients with gastric and colon cancers were responding. This area of research was very slow for a number of years.

Eventually, a resurgence of interest increased due to the discovery that the drug worked by inhibiting nuclear mammalian topoisomerase I, a novel mechanism of action among contemporary antitumor agents [227]. Topoisomerase I is a ubiquitous enzyme essential for changing the twisting number of DNA molecules (relaxing supercoils) so that they can be transcribed and repaired. The levels of topoisomerase I are often raised in tumor cells. The role of Topoisomerase I is to make transient single-strand breaks in duplex DNA, rotate the molecule, and reseal the strand. Camptothecin, and its analogs, form a ternary complex with the cut DNA and topoisomerase I, which prevents progression or regression. The cut DNA is unavailable to the cell, so that it is stranded in the S-phase of the cell cycle and the DNA is degraded, thereby leading to cell death.

Compound	R	R_1	R_2
(22)	H	H	H
(23)	H	$CH_2N(CH_3)_2$	OH
(24)	Et	H	O-C(=O)-N(piperidinyl)-N(piperidinyl)

Camptothecin is not easily formulated, because of its solubility characteristics, which impedes its use by injection. Furthermore, it is quite unstable in the body because of ease of hydrolysis of the lactone ring under physiological conditions, to produce the highly toxic acid analog (Fig. 16). The ring-opened form is also highly serum protein bound, helping to account for its comparatively poor activity *in vivo*. This high level of binding also displaces the equilibrium further in the direction of the undesirable ring-opened acid form. These factors apparently are less limiting in mice, producing a significant species difference in behavior. This raised the level of disappointment when, despite favorable animal studies, the drug performed poorly in clinical trials. Many analogs were subsequently prepared by total synthesis and by conversions of camptothecin itself. The more promising of these newer analogs are much more soluble in water and less serum protein bound, helping to overcome some of the defects of camptothecin itself.

3.2.1. Metabolism Hydrolysis to the less-active and toxic ring-opened lactone occurs readily *in vivo* under physiological conditions (Fig. 16). Furthermore, the lactone binds to serum proteins approximately 200 times more than does camptothecin itself. By mass action, this shifts the equilibrium toward ring opening. The lactone-opened analogs are significantly more water soluble than the lactone forms, but are generally rather less active.

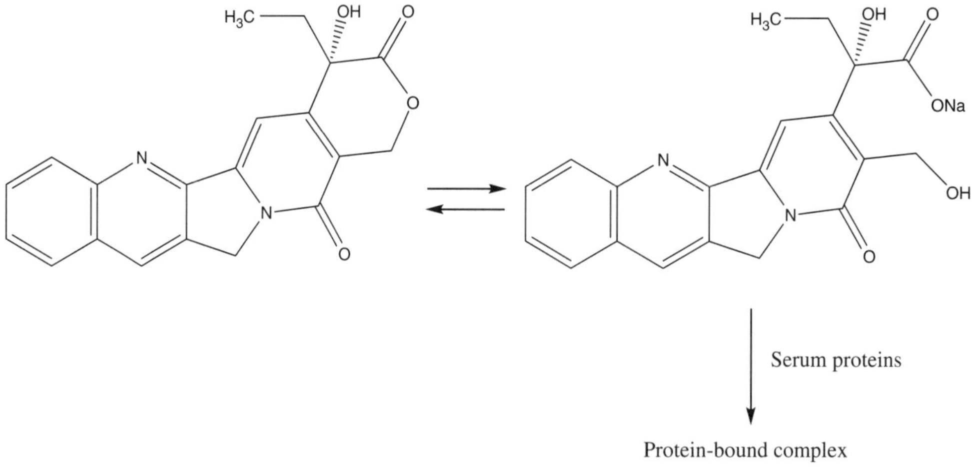

Figure 16. Hydrolysis of camptothecin analogs.

3.2.2. Irinotecan (CPT-11)

Clinical Use Irinotecan (**24**) is an analog hydroxylated in the quinoline ring and further converted to an amine-bearing prodrug carbamate linker. This represents the first water-soluble semisynthetic derivative of camptothecin to be used in the clinic [226]. It is given by i.v. infusion, often in combination with 5-FU and leucovorin (which combination is particularly toxic) for the treatment of metastatic carcinoma of the colon, or rectum [228]. Irinotecan and its metabolites are much less serum protein bound than topotecan and have a somewhat longer half-life in serum. Irinotecan, however, has poor orally bioavailability and is also subject to significant first-pass metabolism. In addition, the bulky dipiperidino side-chain imparts improved water solubility but leads to a substantial reduction in anticancer activity. Metabolic hydrolysis of this side-chain by hepatic carboxylesterases leads to a metabolite that is much more potent than irinotecan itself.

3.2.3. Topotecan

Clinical Uses Topotecan (**23**) is used for ovarian [229,230] and small-cell lung cancers [231–238]. It is rapidly metabolized by hydrolysis and the majority of the drug (75–80%) is hydrolyzed in the plasma, with a half-life of approximately 2 h. N-Demethyltopotecan and its glucuronide conjugate are metabolites that are found in smaller amounts.

Contraindications and Side Effects Extravasation of the camptothecin-derived drugs leads to tissue damage, and they are strongly emetic with neutropenia being common. Hypersensitivity to irinotecan is observed and is a contraindication. In addition it has a high incidence of diarrhea, which occurs through various mechanisms. Poor exposure to pelvic/abdominal irradiation enhances the risk of severe myelosuppression and deaths have been attributed to consequent infections [239]. Orthostatic dyspepsia, anemia, weight loss, dehydration, colitis/ileus, renal function, and fertility impairment are also seen with irinotecan, but are generally considered to be mild [240].

Pharmacokinetic Features The pharmacokinetic features of topotecan are very complex. The drug is subject to alteration by esterases and the products undergo glucuronidation at various points. Furthermore, it goes through oxidation by CYP 3A4, so it is subject to potentially numerous drug–drug interactions [241]. After oral administration, about 30–40% of the drug is bioavailable [242–244]. Following i.v. dosage of the prodrug irinotecan, rapid metabolic conversion by hydrolysis of the carbamoyl moiety to an active phenolic metabolite (SN-38) occurs as a result of the action of liver carboxylesterase; this is followed by glucuronidation to a metabolite that is much less potent. Metabolite SN-38 is about 1000-fold more active than irinotecan itself and accounts for the bulk of the antitumor activity of the drug. Fortunately, SN-38 has much less affinity toward serum proteins and this shifts the equilibrium toward retention of the active lactone form. Irinotecan is also converted in part to a metabolite in which the piperazine ring is oxidatively opened to produce an acid analog (presumably through its lactam) [245]. About 11–20% of active irinotecan is excreted in the urine, but the majority of the drug and its metabolites are excreted in the bile. There appears to be a significant patient-to-patient variation in ability to metabolize irinotecan [239].

Molecular Mode of Action and Resistance The camptothecins are inhibitors of the action of mammalian topoisomerase I. The normal function of this essential enzyme is to produce temporary single-strand breaks by which the topography of DNA can be altered, so that the molecule can be processed. In the presence of camptothecin and its analogs, a ternary complex forms (camptothecin analogs + DNA + enzyme) that results in single-strand breaks that cannot be resealed, and this leads to defective DNA. In particular, when the replicating fork of DNA reaches the cleavable complex generated by camptothecin derivatives, irreversible strand breaks result, causing a failure in DNA processing, thus causing the cells to die. The camptothecins are thus S-phase poisons (Fig. 17). The specific molecular details are still obscure, however [240].

Resistance to the camptothecins is believed to result in part from excretion mediated by P-glycoprotein and MRP-3 (multidrug resistance-associated protein) mechanisms [240,246], although a number of other biological effects are seen in *in vitro* studies [247–250]. Their meaning in clinical uses is, as yet, unclear. In some resistant cells, reduced levels of hydrolases capable of cleaving irinotecan to SN-38 seem to contribute. Another mode of resistance involves decreases in content and potency of topoisomerases I [240,251,252].

Medicinal Chemistry Several papers and reviews on the various total synthesis and analog studies of camptothecin and related molecules have been published [226,253–257].

The objectives of much of this work are clear. Drawbacks of camptothecin that have to be overcome are its poor water solubility,

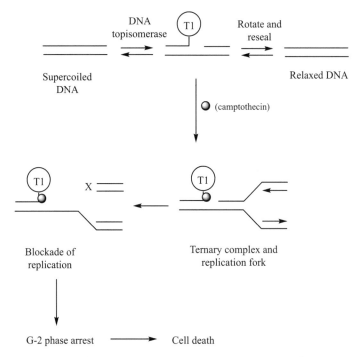

Figure 17. Schematic of camptothecin molecular mode of action.

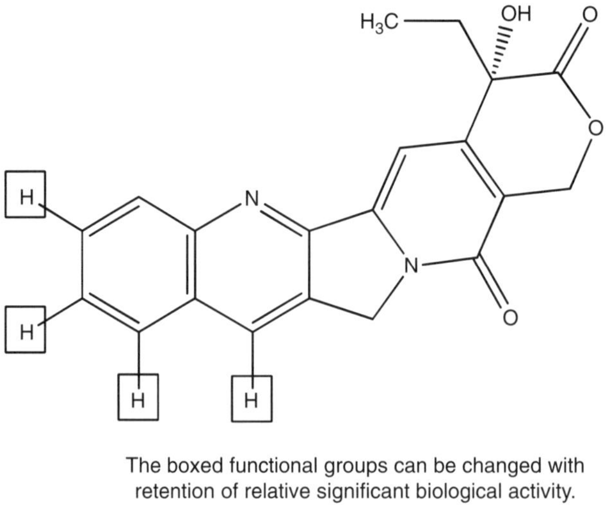

Figure 18. Summary of camptothecin structure–activity relationships. The boxed functional groups can be changed with retention of relative significant biological activity.

ease of hydrolytic lactone opening to the undesirable acid form, high serum protein binding, and the reversibility of its drug–target interaction. The solubility problem has been approached, interestingly, in quite opposite directions. Some groups have sought to increase water solubility and others to make the molecules even more lipophilic. Each approach has worked significantly.

After considerable effort it was discovered that placing substituents at the C9- and C11-positions considerably decreased serum protein binding, even with the lactone-ring opened analogs, and yet this did not interfere with antitumor activity. Among the analogs that have received clinical examination, but have not yet been marketed are lurtotecan (**26**), 9-nitro and 9-aminocamptothecin [258], and DX-8951f [259]. A number of other analogs stand out from the many that have been made. Among these are the hexacyclic 1, 4-oxazines [260], Ring E homocamptothecins, 7-cyanocamptothecins [261], and the silatecans (**25**) [262]. The latter are structurally unusual, in that very few candidate drugs contain silicon atoms. Furthermore, the best analogs are quite lipid soluble and, despite this, display superior stability in human blood and decrease albumin binding combined with significant potency. A summary of camptothecin SARs is illustrated in Fig. 18.

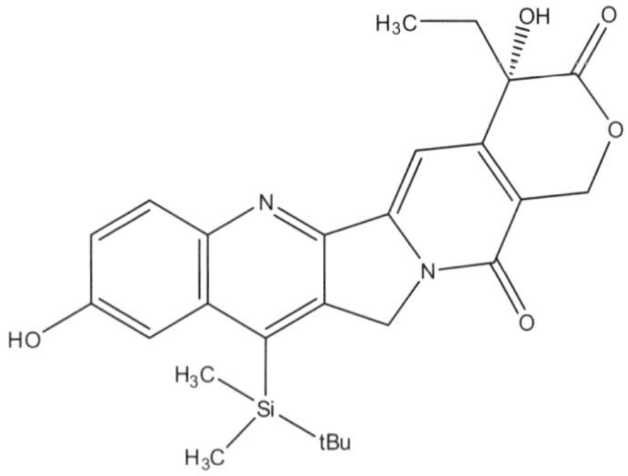

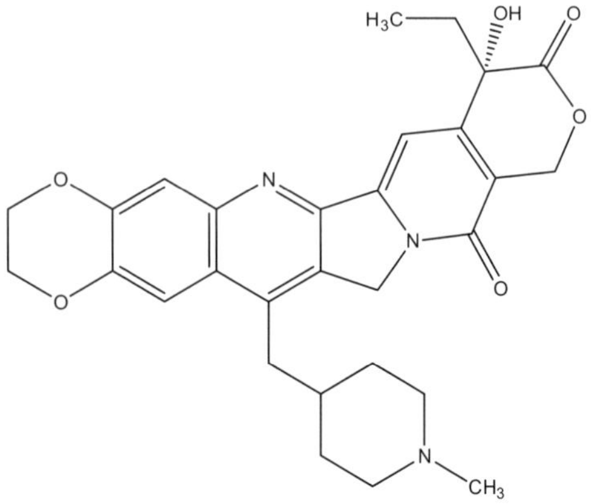

Much effort has been expended also to enhance the water solubility of the camptothecins by inventive formulations. A number of prodrugs have also been made in attempts to enhance stability and water solubility. Among these are the C16-esters, such as the butyrates and propionates, some sugar-containing molecules and the C11-carbamates, of which irinotecan is the most successful, to date.

Topotecan is, likewise, hydroxylated in the quinoline ring, but with a dimethylaminomethylene moiety adjacent. It is administered i.v. for the treatment of ovarian and small-cell lung cancer. As with the other camptothecins, topotecan undergoes a reversible pH-dependent hydrolysis of its lactone moiety and it is this form that is pharmacologically active. The drug has a complex excretion pattern, with a terminal half-life of about 2–3 h, and about 30% of the drug appears in the urine. Kidney damage decreases the excretion of the drug. Binding of topotecan to serum proteins is about 35%. The clinical side effects of topotecan are similar to those of irinotecan.

The chiral center of the camptothecins is S; the R-enantiomers are much less (10- to 100-fold) potent.

Quantitative Structure–Activity Relationships Many synthetic camptothecin analogs have been prepared in attempts to stabilize the active lactone form and to enhance water solubility. A quantitative structure–activity relationship (QSAR) correlation has been published based on the NCI database information for 167 camptothecin analogs. The key functions that emerged from this are the presence and comparative positions of the E-ring hydroxyl and lactone carbonyl and the D-ring carbonyl [263].

Recent Developments and Things to Come Topoisomerase I inhibition is a popular area of contemporary research, and a number of analogs are in various stages of preclinical and clinical workup. It seems likely that the immediate future will see the emergence of additional agents in this class [264–267].

3.3. Isopodophyllotoxins

The lignin podophyllotoxin (**27**) is an ancient folk remedy (classically used for treatment of gout) found in the May apple, *Podophyllum peltatum* [268,269]. Interestingly podophyllotoxin binds to tubulin at a site distinct from that occupied by taxol and the vinca bases, although its molecular mode of action does not involve this in any obvious way, and modern clinical interest lies in its isomers instead. The isopodophyllotoxins are semisynthetic analogs resulting from acid-catalyzed reaction with suitably protected sugars followed by additional transformations. This results in attachment of the sugars to the ring system, with opposite stereochemistry to podophyllotoxin itself. Etoposide (**28**) and teniposide (**29**) are the most prominent analogs so produced and these possess a different mode of action to that of podophyllotoxin [270]. Another diastereoisomer, picropodophyllotoxin (**30**), is produced by epimerization of podophyllotoxin at the lactone ring, but it has not led to interesting analogs.

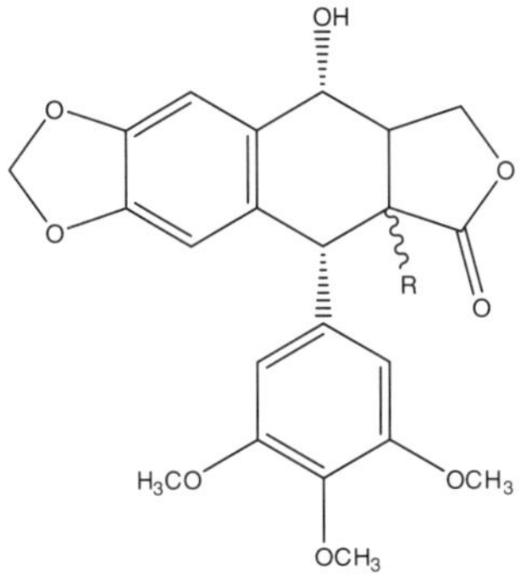

(**27**) R = -H
(**30**) R = -H

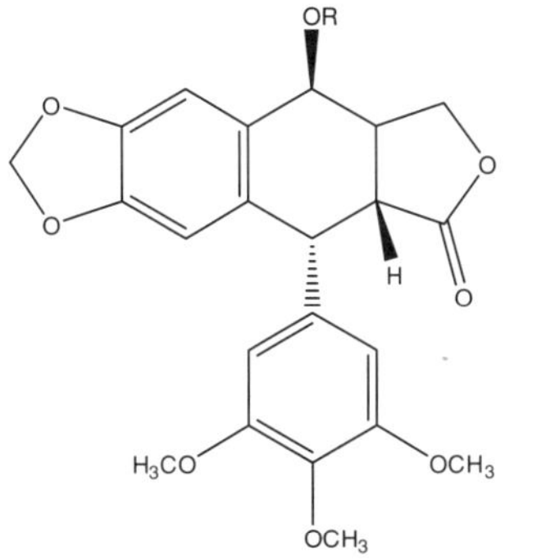

(**28**) R =

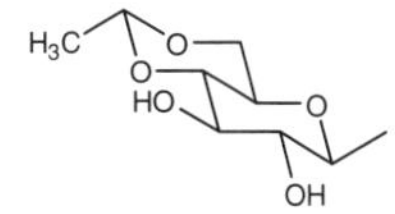

(**29**) R =

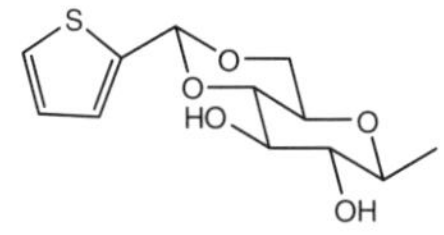

3.3.1. Etoposide

Therapeutic Uses Etoposide is injected for the treatment of refractory testicular tumors [271], small-cell lung cancer, and other less thoroughly established tumor regimes [272–274].

Side Effects and Contraindications Hypersensitivity to etoposide or to the cremophor EL vehicle is contraindication [275,276]. Myelosuppression, alopecia, nausea, vomiting, anorexia, diarrhea, hepatic damage, leucopenia, and thrombocytopenia are among other side effects. In a small number of patients treated with etoposide, a therapy-related leukemia results [277,278].

Pharmacokinetics The drug is given by slow i.v. infusion and may be given orally. About half of the administered dose is bioavailable and follows a biphasic elimination kinetic profile after infusion [279]. Oxidative de-O-methylation is seen as a result of the action of human cytochrome P450 3A4 [280]. Hydrolysis of the lactone is seen and conjugates are excreted in the urine [281]. With about half of the dose excreted, unchanged, in the urine. Etoposide binds 76–96% to serum proteins and is displaced therefrom by bilirubin, so liver damage may require dose reduction [282,283]. The drug does not effectively pass the blood–brain barrier [284]. The drug distributes best into small bowel, prostate, thyroid, bladder, spleen, and testicle, but does not stay in the body for extended times after treatment cessation [285].

Mode of Action and Resistance Etoposide administration causes DNA single- and double-strand breaks and DNA-protein links. This effect appears to be based on inhibition of topoisomerase II [286,287]. Furthermore, it is not an intercalator, nor does it bind directly to DNA in the absence of the enzyme. Its action is most prominent in the late S or early G-2 cell-cycle phases; thus, cells do not enter the M-phase. The details of the interaction between topoisomerase II, DNA, and the isopodophyllotoxins are still emerging. Human topoisomerase II is a homodimeric enzyme responsible for manipulating DNA supercoiling, chromosomal condensation/decondensation, and unlinking of intertwined daughter chromosomes. These steps require energy gained by hydrolysis of ATP. Etoposide (and teniposide) act by stabilizing the covalent topoisomerase II-DNA intermediate and this stabilized ternary complex containing enzyme-cleaved DNA acts as a cellular poison. Figure 19 illustrates the formation of a ternary complex between DNA, DNA topoisomerase II, and an epipodophyllotoxin glycoside. During one topoisomerase II catalytic cycle, two ATP atoms are hydrolyzed. Etoposide and teniposide inhibit release of the ADP resulting from hydrolysis of the first ATP in a manner yet to be precisely determined, although the net result is that the ATPase activity of the enzyme is inhibited and resealing is prevented [288]. Resistance takes the form of P-glyco-protein-related efflux [289], decreased

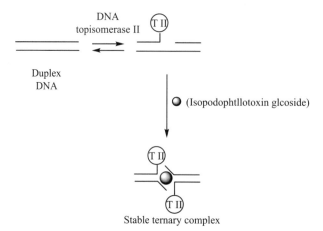

Figure 19. Illustration of the formation of a ternary complex between DNA, DNA topoisomerase II, and an isopodophyllotoxin glycoside.

expression and biosynthesis of topoisomerase II [290], or mutations in human topoisomerase IIα [291] or *p53* tumor-suppressor gene [292].
Medicinal Chemistry [293,294] Although etoposide is widely used, it is inconveniently water insoluble. A water-soluble prodrug, etopophos, has been introduced. This agent is rapidly and extensively converted back to etoposide after injection [270,295].

3.3.2. Teniposide

Therapeutic Uses Teniposide is injected for the treatment of acute nonlymphocytic leukemia, Hodgkin's disease and other lymphomas, Kaposi's sarcoma, neuroblastoma, and other less thoroughly validated tumor situations. Unfortunately, in a number of children treated for leukemia, later development of a teniposide-generated leukemia may occur [270].
Pharmacokinetics The drug is given by slow i. v. infusion and also can be given orally. About half of the administered dose is bioavailable and follows a biphasic elimination kinetic profile following infusion. The drug does not efficiently pass the blood–brain barrier and its tissue distribution and persistence are similar to those of etoposide [282]. Hydrolysis of the lactone is seen and conjugates are excreted in the urine after oxidative demethylation, more so than with etoposide [280].
Mode of Action and Resistance Teniposide administration causes DNA single- and double-strand breaks and DNA-protein links. This effect appears to be based on inhibition of topoisomerase II because the drug is not an intercalator, nor does it bind to DNA. Its action is most prominent in the late S or early G-2 cell-cycle phases; thus cells do not enter the M-phase. Resistance takes the form of P-glycoprotein-related efflux, decreased biosynthesis of topoisimerase II, or mutations in *p53* tumor-suppressor gene [296].
Structure–Activity Relationships Figure 20 illustrates the comparatively limited information relating to isopodophyllotoxin glycosides.

Detailed reviews of the properties of the isopodophyllotoxins are available [297–300].

4. DRUGS INTERFERING WITH TUBULIN POLYMERIZATION/DEPOLYMERIZATION

Microtubules provide a type of cytoskeleton for cells so that they may maintain their

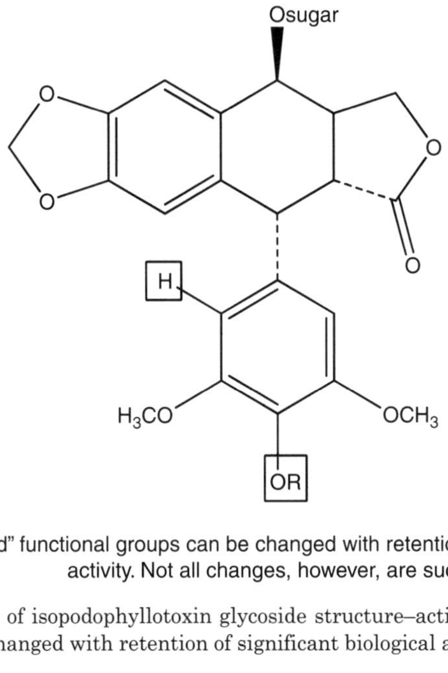

The "boxed" functional groups can be changed with retention of significant biological activity. Not all changes, however, are successful.

Figure 20. Summary of isopodophyllotoxin glycoside structure–activity relationships. The "boxed" functional groups can be changed with retention of significant biological activity. Not all changes, however, are successful.

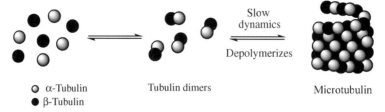

Figure 21. Tubulin polymerization to microtubules and their disassembly.

shapes. They also form a sort of "rails," along which the chromosomes move during mitosis. They are constructed by the controlled polymerization of monomeric tubulin proteins of which there are two types, α and β. Figure 21 illustrates this process. The dimeric vinca alkaloids interfere with polymerization, thus preventing cell division by preventing the formation of new microtubules. The taxus alkaloids, on the other hand, promote the polymerization into new microtubules but stabilize these and prevent their remodeling. This prevents cell growth and repair. These mechanisms are compatible with the modes of action of other antitumor agents, thereby allowing for synergy when combined with these drugs/chemotherapies in cocktails.

4.1. Taxus Diterpenes

4.1.1. Paclitaxel/Taxol

Taxol (**31**), a diterpene ester, was discovered initially as a minor component in the bark of the Pacific yew at approximately the same time as camptothecin was found by the same group in quite another source [225,301]. It took a great many years for taxol to come to the clinic, because its initial performance in tumor-bearing mice was comparatively unimpressive. The progress to market was accelerated materially by the discovery of a novel (at the time) molecular mode of action. Taxol stimulates the formation of microtubules from tubulin and stabilizes the polymer, which stops cells from dividing [302–304].

(**31**) R = Ph; R_1 = COCH$_3$
(**32**) R = tBuO; R_1 = H

After enormous effort, semisynthesis from 10-deacetylbaccatin II (**33**), itself available in quantity from the much more renewable needles of various abundant yew species, proved economical and also allowed the synthesis of many analogs, of which docetaxel (**32**) is the most prominent [305]. Many partial and total syntheses of taxol have been reported, but none of these has yet proved to be practical. Despite a long tradition allowing the discoverer to name an important compound, taxol was renamed paclitaxel for commercial purposes by the U.S. Federal Government CRADA winner, Bristol-Myers Squibb.

(33)

Clinical Uses Taxol is widely used in combinations for the therapy of refractory ovarian, breast, lung, esophageal, bladder, and head and neck cancers.

Side Effects and Contraindications Taxol is very toxic and it must be used with care. Bone marrow suppression (neutropenia) is a major dose-limiting side effect [306]. A few patients develop severe cardiac conduction abnormalities [307]. Patients with liver abnormalities may be especially sensitive to taxol. Fertility impairment and mutagenesis is seen in experimental animals, so taxol should only be given to pregnant patients with special care. Hypersensitivity is not uncommon and is often associated with the solvent in which this especially water insoluble drug must be administered (cremophor EL, a polyethoxylated castor oil) [308]. Peripheral neuropathy occurs frequently and requires reduced dosage. In contrast to many other antitumor agents, extravasation, although causing discomfort and local pathologies, does not generally lead to severe necrosis. Gastrointestinal distress (diarrhea, fever, anemia, mucositis, nausea, and vomiting), alopecia, edema, and opportunistic infections are also reported by many patients. Paclitaxel is metabolized by the P450 system, coadministration of drugs requiring processing by CYP2C8 and CYP3A4 requires caution. This statement also holds for docetaxol [309].

Pharmacokinetics The drug is generally given by long-time (3 or 24 h) infusions at 3-week intervals, or short-time (1 h) infusions at weekly intervals and is heavily protein bound (90–98%). Attempts to infuse the drug over very long times (96 h) have been made, but involve significant practical limitations. The drug is excreted after a biphasic mode, with an initial rapid serum decline as the drug is distributed to the tissues and the overflow excreted. Return from peripheral tissues is slow and accounts for the second part of the excretion curve. The excretion half-life is fairly long (~13–55 h) [310–312]. Extensive clearance, other than by urine, takes place given that only 1–13% of the drug is found in the urine. Metabolism is primarily oxidative, with the main metabolite being the 6-α-hydroxy analog and lesser amounts of the 3′-parahydroxybenzamide and the 6-α-hydroxy, 3′-parahydroxybenzamide analogs being detected [313–315].

Molecular Mode of Action and Resistance Taxol binds to the β-tubulin component and stimulates the formation of microtubules. These, however, do not break down, so the cell is unable to repair and undergo mitosis [304]. Resistant cells in culture are often seen to produce P-glycoprotein to excrete the drug [316], and also to have mutations in the β-tubulin component [317,318]. Whether this is responsible for clinical resistance is still being studied. Overexpression of the *ErbB2* gene occurs fairly often in breast tumors and this leads to overproduction of a transmembrane growth factor receptor belonging to the ErbB receptor tyrosine kinase subfamily. Cells with this characteristic have reduced responsiveness to taxol [319]. Other growth factor anomalies involving, for example, EGFRviii and HER-2 are also seen in some cell lines [320].

4.1.2. Docetaxel/Taxotere

Docetaxel (**32**) is a semisynthetic analog of taxol prepared by a variety of chemical means, starting with the more abundant 10-deacetyllbaccatin III (**33**) [305]. It has found a significant place in anticancer chemotherapy, but is still a significantly toxic drug that must be used with care.

Clinical Applications Docetaxel is administered by i.v. infusion for the treatment of breast cancer, nonsmall-cell lung cancer, and a variety of other less well established antitumor indications [321].

Side Effects and Contraindications Many of the adverse effects of docetaxel are similar to those of taxol itself. The drug, however, is administered in polysorbate 80 rather than cremophor, so allergy is more commonly to the drug itself and can be severe. Poor liver function greatly enhances patient sensitivity to docetaxel. Severe fluid retention can also be observed. Patients are often administered corticoids before being exposed to docetaxel to assist in their tolerance of the drug. Myelotoxicity is potentially severe, so blood cell counts should be monitored. The toxicity of docetaxel is exaggerated when liver disease is present [322].

Pharmacokinetics In contrast to paclitaxel, docetaxel has linear pharmacokinetics at the doses used in the clinic. As with paclitaxel, metabolism takes place in the liver through cytochrome P450 enzymic oxidation and the metabolites are excreted primarily in the bile. The involvement of P450 3A4 and 3A5 requires care in coadminstering drugs that are also metabolized by these common enzymes [323]. The metabolites are generally less toxic and less potent than docetaxel itself [324].

Mode of Action and Resistance See paclitaxel.

Chemical Transformations Although several total syntheses of taxol have been achieved during the last few years, low overall yields and high costs preclude them from being of commercial importance. Fortunately, isolation of the two taxol biosynthetic precursors, baccatin III and 10-deacetyl baccatin III, initially from the regenerable needles of the yew species *T. baccata*, and subsequent development of highly efficient semisynthesis of taxol and taxotere from the above precursors have, apparently, solved the present supply problem of these valuble drugs. Moreover, the semisynthetic routes have also provided the means to carry out extensive SAR studies and consequent access to a large number of taxol analogs. These SAR studies with taxol have demonstrated that the C3-phenylisoserinate side chain is an essential component for bioactivity, wherein limited modifications can be carried out at the 3′-phenyl and the 3′-*N*-benzoyl sites toward attenuating activity. Additionally, although the oxygen-bearing functionalities at C7, C9, and C10 allow various modifications, an acetoxy at C4 and an aroyloxy group at C2 are indispensable for optimum activity. Interestingly, A-ring-contracted taxol analogs (A-nortaxol) were found to retain tubulin assembly activity, albeit with significantly diminished cytotoxicity [325–328].

The SARs of the taxol series are summarized in Fig. 22. In addition, additional details may be found in the chapter of mictorubulin inhibitors in this series.

Biosynthesis Taxol is one of the structurally more complex members of the diterpene family, characterized by the presence of the unusual taxane ring system. The initial steps in taxol biosynthesis involve the cyclization of geranylgeranyl diphosphate to taxa-4 (5),11(12)-diene, forming the taxane core structure. Subsequent cytochrome P450-mediated hydroxylation at C5 of the olefin is followed by several other cytochrome P450-dependent oxygenations at C1, C2, C4, C7, C9, C10, and C13 (the precise order of these regiospecific oxidations, however, is not yet known) and Co-A-dependent acylations of the taxane core, en route to taxol [329]. Biosynthetically, the *N*-benzoyl phenylisoserine side chain has been shown to originate from phenylalanine and its further elaboration involves a late-stage esterification at C13 of an advanced baccatin III intermediate [330].

The biosynthetic pathway between geranylgeranyl diphosphate and taxol remains to be fully elucidated, but apparently passes through taxa-4(5),11(12)-diene and taxa-4 (20),11(12)-diene-5α-ol, as shown in Fig. 23.

Things to Come Recent interest has developed about the properties of the epothilones.

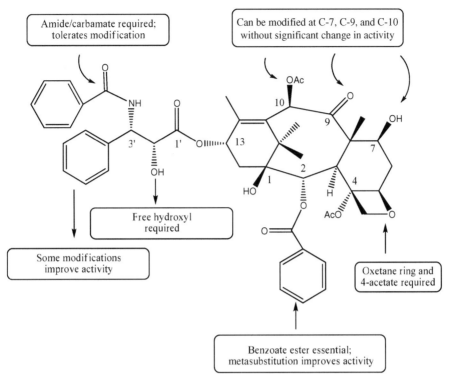

Figure 22. Structure–activity relationships of the taxol series.

These apparently bind to tubulin at approximately the taxol site but resistance by P-glycoprotein expulsion is apparently not significant with these fermentation products and they are active against a number of taxol-resistant cell lines. Elutherobin is another natural product binding to the taxane-binding site, but this agent is cross-resistant with taxol [331]. The clinical future of these agents is, as yet, uncertain and they have attracted much synthetic and biochemical attention. Please see the chapter on microtublin inhibitors in this series for additional information.

4.2. Dimeric Vinca Alkaloids

The dimeric indole-indoline alkaloids were initially isolated from the Madagascan periwinkle, *Catharanthus rosea* (formerly, *Vinca rosea*). The plant was originally investigated as a follow-up to folkloric reports of hypoglycemic activity, and it was hoped to be of value in treating diabetes mellitus. This did not prove to be the case but, during the investigation of extracts, certain fractions produced granulocytopenia and bone marrow suppression in animals. The active alkaloids were isolated from a matrix of indole compounds and were found to be active antileukemic agents against P-1534 cells. Development for human use followed after extensive experimentation. Four of these unsymmetrical dimeric alkaloids ultimately found use as antitumor agents. The best agents contain C-linked vindoline and 16β-carbomethoxy velbanamine units. Apparently minor structural differences between the alkaloids led to major differences in potency and utility [332]. Because of their relative scarcity and medicinal value, these dimmers have been attractive synthetic targets, and a rich synthetic and biosynthetic literature has grown up around them. Inspection of their structures readily

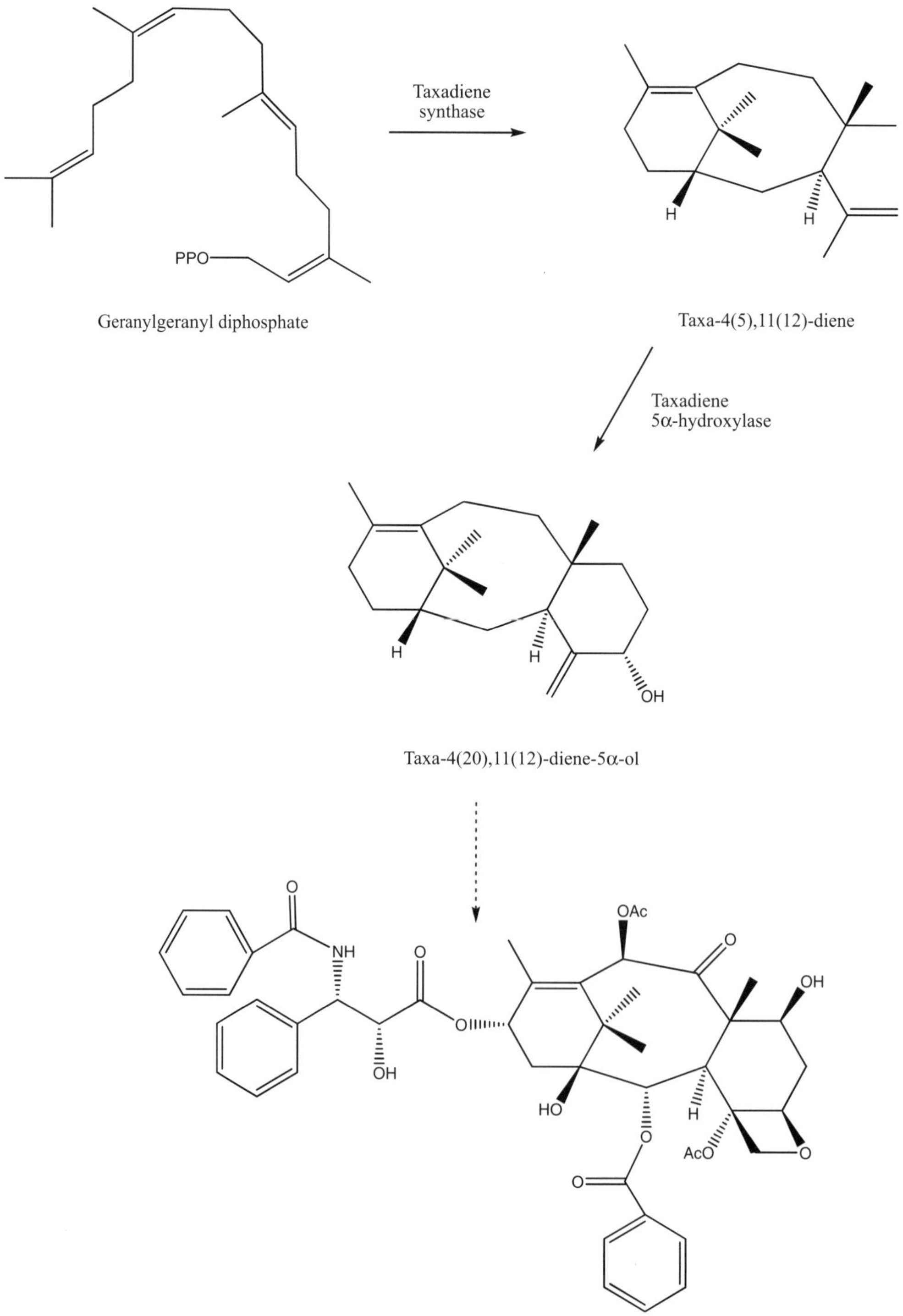

Figure 23. Biosynthesis of taxol.

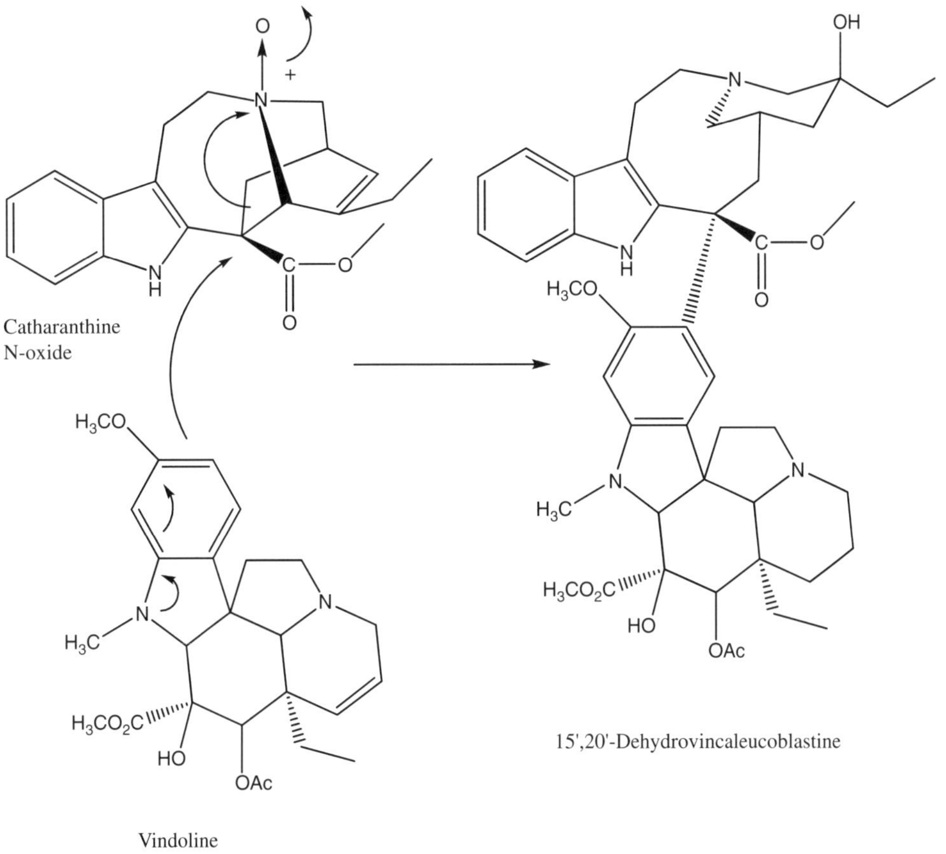

Figure 24. Partial chemical synthesis of vinca dimers with the natural stereochemistry.

leads to the inference that they are the products of asymmetrical free-radical coupling. After much work, two groups, those of Potier in France [333,334] and of Kutney in Canada [335], succeeded in stereoselective dimerization. Treatment of the abundant alkaloid catharanthine as its *N*-oxide with trifluoroacetic anhydride leads to fragmentation into an enamine that can be intercepted by vindoline, another comparatively abundant alkaloid, and the product reduced by sodium borohydride. Under low temperature the condensation is stereospecific in the desired manner. This is believed to reflect a concerted interaction. When the reaction is run at higher temperatures, a mixture of diastereomers is produced instead. This is believed to be the result of a stepwise condensation. Variation of this chemistry leads to the formation of useful synthetic analogs, and interconversions into natural analogs as well. Figure 24 illustrates the partial chemical synthesis of vinca dimers with the natural stereochemistry.

It is interesting to note that dolastatin 10, a marine natural product with exceptional antitumor properties, also binds near to the vinca alkaloid binding domain and inhibits tubulin polymerization [336,337].

4.2.1. Vinblastine (Velban)

Medicinal Uses Vinblastine sulfate (**34**) is given i.v. with great care, to avoid damaging extravasation [123], for the treatment of metastatic testicular tumors (usually in combination with bleomycin and cisplatin). Various lymphomas may also respond. It has only limited neurotoxicity, thus enhancing its utility.

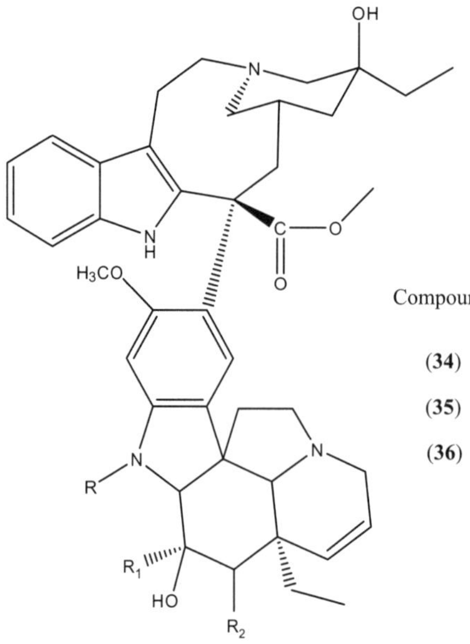

Compound	R	R_1	R_2
(34)	CH_3	CO_2CH_3	$OCOCH_3$
(35)	CHO	CO_2CH_3	$OCOCH_3$
(36)	CH_3	$CONH_2$	OH

Side Effects and Contraindications Vinblastine causes severe tissue necrosis upon extravasation. Mild neurotoxicity and myelosuppression occur and these effects should be monitored to prevent significant toxicity to the patient. The other side effects are in common with antitumor agents (alopecia, ulceration, nausea, etc.)

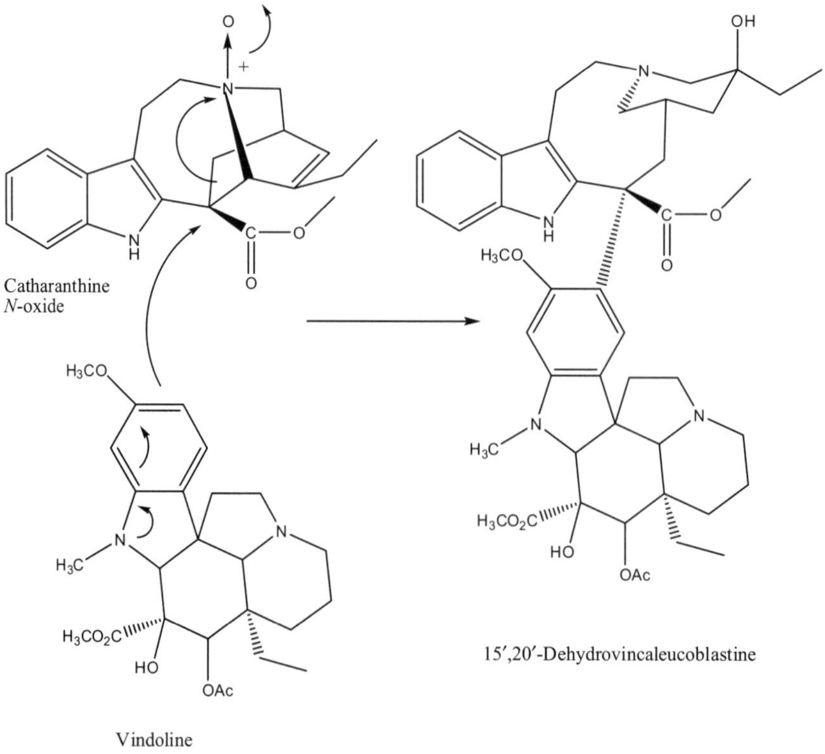

Catharanthine N-oxide

Vindoline

15′,20′-Dehydrovincaleucoblastine

Pharmacokinetic Features Vinblastine is extensively metabolized in the liver and the metabolites are excreted as conjugates in the bile. About 15% of the drug is found unchanged in the urine [338,339]. Oxidative degradation of the catharanthus alkaloids occurs, in part, by the action of myeloperoxidase. Cleavage occurs between C20' and C21' and is structurally facilitated by the presence of a C20'-hydroxyl moiety [340]. Peroxidase and ceruloplasmin also catalyze oxidative transformations of vinblastine [341,342].

Medicinal Chemical Transformations Hydrolysis of the acetyl group at C4 of vinblastine abolishes its antileukemic activity. Furthermore, acetylation of the free hydroxyl groups also inactivates the molecule. The dimeric structure is required, as is the stereochemistry of the point of attachment. Hydrogenation of the olefinic linkage and reduction to the carbinol also greatly diminish potency. Hence, the antileukemic activity is substantially dependent on the specific structural groups present in the molecule.

Molecular Mode of Action and Resistance Vinblastine blocks cells in the M-phase. It binds to the β-subunit of tubulin in its dimer in a one-to-one complex, thus preventing its polymerization into microtubules. The binding site is near to, but different from, that of colchicine but similar to that of maytansine and rhizoxin (although the consequences of binding of the latter are different from those of vinca binding). Nontubulin oligomers form from the component parts as a consequence, and preformed tubulin depolymerizes and the complex with vinblastine crystallizes [343]. Failure to produce functional microtubules prevents proper chromosome formation and so prevents cell division. The blocked cells then die (become apoptotic). Other cellular processes dependent on microtubules are also interfered with, although the blockade of chromosome formation is regarded as central to their action [344]. Resistance mainly takes the form of elaboration of P-glycoproteins that export vinblastine, and this cross-resistance is broad enough to include the other vinca alkaloids, and other antitumor agents as well [345]. Resistance is also attributed to alterations in the tubulin subunits [346].

4.2.2. Vincristine (Oncovin, Vincasar PFS)

Medicinal Uses Vincristine (**35**) is a common component of antitumor cocktails used in treating acute lymphoblastic leukemia and solid tumors of youngsters and adult lymphoma. It is commonly used with corticosteroids. Study of its use in the form of liposomes has also been carried out [347]. Its use produces limited myelosuppression, so it is an attractive component in cocktails. The reduced myelotoxicity may be attributable to oxidative degradation of the drug by myeloperoxidase, a heme-centered peroxidase enzyme present in acute myeloblastic leukemia, but not in acute lymphoblastic leukemia [348,349].

Pharmacokinetic Features Vincristine is extensively metabolized in the liver and the metabolites are excreted as conjugates in the bile. About 15% of the drug is found unchanged in the urine.

Side Effects and Contraindications Vincristine causes severe tissue necrosis upon extravasation [123]. Neurotoxicity is a significant potential problem with vincristine and is often treated in part by reducing the dose of the drug [350]. Myelosuppression also occurs, but to a lesser extent, and this effect should be monitored to prevent significant toxicity to the patient. Gout can occur with vincristine administration and can be controlled by use of allopurinol. The other side effects of vinblastine are common to antitumor agents (alopecia, ulceration, nausea, diarrhea, etc.)

Resistance Resistance to vincristine is mediated in part by export resulting from the multidrug resistance protein and, interestingly, is characterized by cotransport with reduced glutathione [351].

4.2.3. Vinorelbine (Navelbine)

Medicinal Uses Vinorelbine (**37**) is used against nonsmall-cell lung cancer, and against breast cancer [352–356]. It appears to be intermediate in its neurotoxicity and myelosuppression compared to that of the other vinca antitumor agents [357].

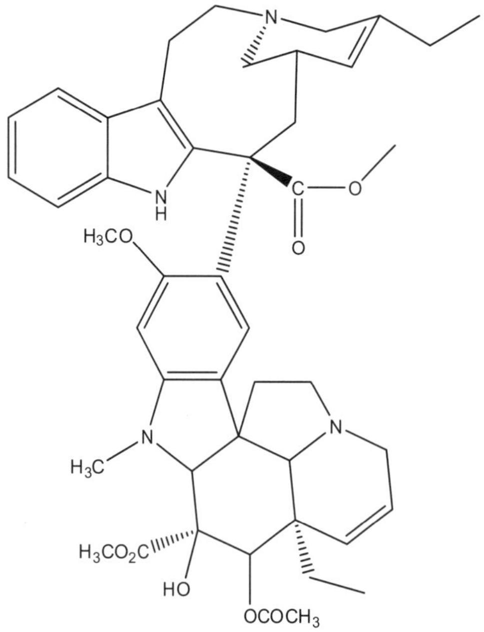

(37)

Pharmacokinetics Vinorelbine is extensively metabolized in the liver and the metabolites are excreted as conjugates in the bile. About 15% of the drug is found unchanged in the urine.

Side Effects and Contraindications Vinorelbine causes severe tissue necrosis upon extravasation as well as phlebitis [358]. Prior i.v. administration of cimetidine partially avoids this. Mild neurotoxicity and myelosuppression occur and these effects should be monitored to prevent significant toxicity to the patient. Its most notable toxic side effect appears to be granulocytopenia. The other side effects are common to antitumor agents (alopecia, ulceration, nausea, etc.).

Things to Come Vindisine (36) is an analog prepared from vinblastine (34). Its antitumor spectrum, however, is more closely similar to that of vincristine. Clinical studies show activity against acute leukemia; lung cancer; breast carcinoma; squamous cell carcinoma of the esophogous, head and neck; Hodgkin's disease; and non-Hodgkin's lymphomas. Its toxicities include myelosuppression and neurotoxicity. Despite these promising findings, it has yet to be introduced to the clinic [359].

Vinflunine (38) is a dimeric alkaloid, containing two gem-fluorine atoms, prepared by a mechanistically interesting process using super acidic reactants on vinorelbine. This compound has improved antitumor potency in a variety of model tumor systems, shows less drug resistance [360], and has entered clinical trials [361,362].

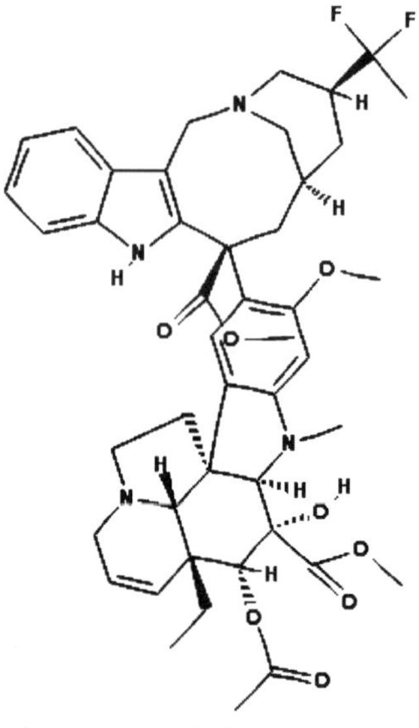

(38)

Some additional reviews of this topic are available in Refs [341,363–365].

SUMMARY

Nature has proved to be a wellspring of important antineoplastic agents. For the past 15, or so, years, the emphasis in the discovery of new therapeutic agents has been by more modern approaches such as high-throughput screening, parallel synthesis, combinatorial chemistry, and other programs. But recently, the search for novel natural products in the treatment of cancer has converted to secondary metabolites from nature. The literature coming into focus indicates that sources not previously considered as bioactive molecules have in fact, become potential valuable treasure-troves of novel chemical structures. For example, marine products from diverse sources such as thermal vents and sponges and terrestrial thermal vents now give rise to a vast assortment of microorganisms. Additionally, we now find cryogenic organisms in glacial systems and xerophytic organisms and plants in other hostile environments. It comes as no surprise that nature, for reasons that are not presently obvious, biosynthesize chemical compounds that chemists have not considered even in their wildest imaginations. Although this chapter provides a few of these examples, it is certain that more will be present in the literature as the intellectual property in these discoveries are protected.

ACKNOWLEDGMENTS

The authors wish to acknowledge the contributions of authors from previous edition of this chapter, which helped facilitate the writing of this chapter. The authors are thankful to Professor Lester A. Mitscher for his work in the sixth edition.

REFERENCES

1. daRocha AB, Lopes RM, Scwartsmann G. Curr Opin Pharmacol 2001;1:364.
2. Mukherjee AK, Basu S, Sarkar N, Ghosh AC. Curr Med Chem 2001;8:1467.
3. Pezzuto JM. Biochem Pharmacol 1997;53:121.
4. Pettit GR, Cragg GM, Herald CL. Biosynthetic Products for Cancer Chemotherapy. Vols 1–4. New York: Elsevier; 1984, 1985, 1989, and earlier.
5. Pettit GR, Pierson FH, Herald CL. Anticancer Drugs from Animals, Plants and Microorganisms. New York: Wiley-Interscience; 1994.
6. Cragg GM. Med Res Rev 1998;18:315.
7. NIH Workshop. Bioassays for Discovery of Antitumor and Antiviral Agents from Natural Sources, October 18–19 1988.
8. Valeriote FA, Corbett TH, Baker LH,editors. Cytotoxic Anticancer Drugs: Models and Concepts for Drug Discovery and Development. Dordrecht, The Netherlands: Kluwer Academic; 1992.
9. Cragg GM, Newman DJ. Cancer Invest 1999;17:153.
10. G.V. McEvoy,editor. AHFS Drug Information 2001. Bethesda, MD: American Society of Health System Pharmacists; 2001.
11. Brunton LL, Lazo JS, Parker KL.editors. Goodman and Gilman's Pharmacological Basis of Therapeutics. 11th ed. New York: McGraw-Hill; 2006.
12. Waksman SA, Woodruff HB. Proc Soc Exp Biol Med 1940;45:609.
13. Chabner B, Amrein PC, Druker B, Michaelson MD, Mitsiades CS, Godd PE, Ryan DP, Ramachandra S, Richardson PG, Supko JG, Wilson WH.In: Brunton LL, Lazo JS, Parker KL,editors. Goodman and Gilman's Pharmacological Basis of Therapeutics. 10th ed. New York: McGraw-Hill; 2006. p 1315.
14. Coppes MJ, Jorgenson K, Arlette JP. Med Pediatr Oncol 1997;31:128.
15. Juliano RL, Stamp D. Biochem Pharmacol 1978;27:21.
16. Benjamin RS, Hall SW, Burgess MA, et al. Cancer Treat Rep 1976;60:289.
17. Tattersall MH, Sodergren JE, Dengupta SK, et al. Clin Pharmacol Ther 1975;17:701.
18. Nakajima K, Tanaka T, Neya M, Okawa K. Pept Chem 1982;19:143.
19. Okawa K, Nakajima K, Tanaka T. J Heterocycl Chem 1980;17:1815.
20. Okawa K, Nakajima K, Tanaka T. Pept Chem 1977;15:131.
21. Tanaka T, Nakajima K, Okawa K. Bull Chem Soc Jpn 1980;53:1352.
22. Vlasov GP, Lashkov VN, Kulikov SV, Ginzburg OF. Zh Org Khim 1978;14:1961.
23. Meienhofer J. J Am Chem Soc 1970;92:3771.
24. Meienhofer J. Experientia 1968;24:776.
25. Brockmann H, Lackner H. Chem Ber 1968; 101:1312.
26. Brockmann H, Lackner H. Chem Ber 1968; 101:2231.
27. Tong G, Nielsen J. Bioorg Med Chem 1996;4:693.
28. Brockmann H, Pampus G, Mecke R. Chem Ber 1959;92:3082.
29. Brockmann H, Ammann J, Mueller W. Tetrahedron Lett 1966; 3595.
30. Meienhofer J. Cancer Chemother Rep 1974;58:21.
31. Sengupta SK, Anderson JE, Yury K, et al. J Med Chem 1981;24:1052.
32. Madhavarao MS, Chaykovsky M, Sengupta SK. J Med Chem 1978;21:958.
33. Brockmann H, Mueller W, Peterssen-Borstel H. Tetrahedron Lett 1966; 3531.
34. Sengupta SK, Rosenbaum DP, Sehgal RK, et al. J Med Chem 1988;31:1540.

35. Sehgal RK, Almassian B, Rosenbaum DP, et al. J Med Chem 1987;30:1626.
36. Sengupta SK, Anderson JE, Kelley C. J Med Chem 1982;25:1214.
37. Mosher CW, Lee DY, Enanoza RM, et al. J Med Chem 1979;22:1051.
38. Sengupta SK, Trites DH, Madhavarao MS, Beltz WR. J Med Chem 1979;22:297.
39. Sengupta SK, Tinter SK, Modest EJ. J Heterocycl Chem 1978;15:129.
40. Sengupta SK, Madhavarao MS, Kelly C, Blondin J. J Med Chem 1983;26:1631.
41. Sengupta SK, Kelly C, Sehgal RK. J Med Chem 1985;28:620.
42. Chowdhury AKA, Brown J, Longmore RB. J Med Chem 1978;21:607.
43. Moore S, Kondo M, Copeland M, et al. J Med Chem 1975;18:1098.
44. Sehgal RK, Almassian B, Rosenbaum DP, et al. J Med Chem 1988;31:790.
45. Seela F. J Med Chem 1972;15:684.
46. Brockmann H, Seela F. Chem Ber 1971;104:2751.
47. Lifferth A, Bahner I, Lackner H, Schafer M. Z Naturforsch B Chem Sci 1999;54:681.
48. Glibin EN, Plekhanova NG, Ovchinnikov DV, Korshunova ZI. Zh Org Khim 1996;32:406.
49. Kamitori S, Takusagawa F. J Mol Biol 1992;225:445.
50. Liu X, Chen H, Patel DJ. J Biomol NMR 1991;1:323.
51. Wadkins RM, Jares-Erijman EA, Klement R, et al. J Mol Biol 1996;262:53.
52. Chen H, Liu X, Patel DJ. J Mol Biol 1996;258:457.
53. Rill RL, Marsch GA, Graves DE. J Biomol Struct Dyn 1989;7:591.
54. Chen FM. Biochemistry 1988;27:6393.
55. Jones RL, Scott EV, Zon G, et al. Biochemistry 1988;27:6021.
56. Sobell HM. Mol Biol 1973;13:153.
57. Sehgal RK, Sengupta SK, Waxman DJ, Tauber AI. Anticancer Drug Des 1985;1:13.
58. Knutsen T, Mickley LA, Ried T, et al. Genes Chromosomes Cancer 1998;23:44.
59. Prados J, Melguizo C, Fernandez A, et al. J Pathol 1996;180:85.
60. Prados J, Melguizo C, Fernandez A, et al. Cell Mol Biol 1994;40:137.
61. Devine SE, Ling V, Melera PW. Proc Natl Acad Sci USA 1992;89:4564.
62. Selassie CD, Hansch C, Khwaja TA. J Med Chem 1990;33:1914.
63. Inaba M, Johnson RK. Cancer Res 1977;37:4629.
64. Jones GH. Antimicrob Agents Chemother 2000;44:1322.
65. Crooke ST.In: Chabner BA, Pinedo HM,editors. The Cancer Pharmacology Annual. Amsterdam: Excerpta Medica; 1983. p 69.
66. Katz E, Lloyd HA, Mauger AB. J Antibiot 1990;43:731.
67. Waksman SA,editor. Actinomycin: Nature, Formation, and Activities, New York: Wiley-Interscience; 1968.
68. Brockmann H. Progress in the Chemistry of Organic Natural Products. Vol. 18.New York: Springer; 1960. p 1.
69. Meienhofer J, Atherton E.In: Perlman D,editor. Structure–Activity Relationships among the Semisynthetic Antibiotics. New York: Academic Press; 1977. p 427.
70. Mauger AB.In: Sames PG,editor. Topics in Antibiotic Chemistry. Chichester, UK: Ellis Horwood; 1980. p 224.
71. Mauger AB.In: Wilman SV,editor. The Chemistry of Antitumor Agents. Glasgow, UK: Blackie; 1990. p 403–409.
72. Friedman MA. Recent Results Cancer Res 1978;63:152.
73. Evans WE, Yee GC, Coom WR, Pratt CB, Green AA. Drug Intell Clin Pharm 1982;16:448.
74. Muraoka Y, Takita T. Cancer Chemother Biol Response Modif 1988;10:40.
75. Lazo JS. Cancer Chemother Biol Response Modif 1999;18:39.
76. Jules-Elysee K, White DA. Clin Chest Med 1990;11:1.
77. Chisholm RA, Dixon AK, Williams MV, Oliver RT. Cancer Chemother Pharmacol 1992;30:158.
78. Hay J, Shahzeidi S, Laurent G. Arch Toxicol 1991;65:81.
79. Dorr RT. Semin Oncol 1992;19:3.
80. Rishel MJ, Hecht SM. Org Lett 2001;3:2867.
81. Highfield JA, Mehta LK, Parrick J, Wardman P. Bioorg Med Chem 2000;8:1065.
82. Huang L, Quada JC, Lown JW. Heterocycl Commun 1995;1:335.
83. Levin MD, Subrahamanian K, Katz H, et al. J Am Chem Soc 1980;102:1452.
84. Tsuchiva T, Miyake T, Kageyama S, et al. Tetrahedron Lett 1981;22:1413.
85. Ohgi T, Hecht SM. J Org Chem 1981;46:3761.
86. Leitheiser C, Rishel MJ, Wu X, Hecht SM. Org Lett 2000;2:3397.

87. Boger DL, Teramoto S, Cai H. Bioorg Med Chem 1997;5:1577.
88. Boger DL, Teramoto S, Cai H. Bioorg Med Chem 1996;4:179.
89. Boger DL, Colletti SL, Teramoto S, et al. Bioorg Med Chem 1995;3:1281.
90. Vloon WJ, Kruk C, Pandit UK, et al. J Med Chem 1987;30:20.
91. Takita T, Fujii A, Fukuoka T, Umezawa H. J Antibiot 1973;26:252.
92. Saito S, Umezawa Y, yoshioka T, et al. J Antibiot 1983;36:92.
93. Takita T, Umezawa Y, Saito S, et al. Proc Am Pept Symp 1981;7:29.
94. Aoyagi Y, Katano K, Sugana H, et al. J Am Chem Soc 1982;104:5537.
95. Saito S, Umezawa Y, Yoshioka T, et al. J Antibiot 1983;36:95.
96. Choudhury AK, Tao ZF, Hecht SM. Org Lett 2001;3:1291.
97. Shen B, Du L, Sanchez C, et al. Biiorg Chem 1999;27:155.
98. Takita T, Muraoka Y.In: Kleinkauf H, Von Dohren H,editors. Biochemistry of Peptide Antibiotics. Berlin: de Gruyter; 1990. p 289–309.
99. Nakatani T, Fujii A, Naganawa H, et al. J Antibiot 1980;33:717.
100. Fujii A, Takita T, Shimada N, Umezawa H. J Antibiot 1974;27:73.
101. Hecht SM. Fed Proc 1986;45:2784.
102. Umezawa H. Lloydia 1977;40:67.
103. Twentyman PR. Pharmacol Ther 1983;23:417.
104. Hecht SM. J Nat Prod 2000;63:158.
105. Nakamura M, Peisach J. J Antibiot 1988;41:638.
106. Sebti SM, Lazo JS. Pharmacol Ther 1988;38:321.
107. Lefterov LM, Koldamova RP, Lazo JS. FASEB J 2000;14:1837.
108. Caputo A,editor. Biological Basis for the Clinical Effect of Bleomycin. Basel: S. Karger AG; 1976.
109. Sikic BI, Rozensweig M, Carter S,editors. Bleomycin Chemotherapy. San Diego, CA: Academic Press; 1985.
110. Carter SK, Crooke ST, Umezawa H, editors. Bleomycin: Current Status and New Developments. San Diego, CA: Academic Press; 1978.
111. Meunier B,editor. Metal-Oxo and Metal-Peroxo Species in Catalytic Oxidations. New York: Springer-Verlag; 2000.
112. Umezawa H. Pure Appl Chem 1971;28:665.
113. Umezawa H. Rev Infect Dis 1987;9:147.
114. Hata T, Sano Y, Sugawara R, et al. J Antibiot 1956;9:141.
115. Wakaki S, Marumo H, Tomioka K, et al. Antibiot Chemother 1959;8:228.
116. DeBoer C, Dietz A, Lummus NE, Savage GM. Antimicrob Agents Chemother Ann 17: 1961;.
117. Sculier JP, Ghisdal L, Berghmans T, et al. Br J Cancer 2001;84:1150.
118. Bradner WT. Cancer Treat Rev 2001;27:35.
119. Verwij J, Sparreboom A, Nooter K. Cancer Chemother Biol Response Modif 1999;18:46.
120. Verwij J. Cancer Chemother Biol Response Modif 1997;17:46.
121. Medina PJ, Sipols JM, George JN. Curr Opin Hematol 2001;8:286.
122. Nishiyama Y, Komaba Y, Kitamura H, Katayama Y. Intern Med 2001;40:237.
123. Kassner E. J Pediatr Oncol Nurs 2000;17:135.
124. Patel JS, Krusa M. Pharmacotherapy 1999;19:1002.
125. Spanswick VJ, Cummings J, Ritchie AA, Smyth JF. Biochem Pharmacol 1998;56:1497.
126. Phillips RM, Burger AM, Loadman PM, et al. Cancer Res 2000;60:6384.
127. Spanswick VJ, Cummings J, Smyth JF. Gen Pharmacol 1998;31:539.
128. Theisohn M, Fischbach R, Joseph R, et al. Int J Clin Pharmacol Ther 1997;35:72.
129. Kumar GS, Lipman R, Cummings J, Tomasz M. Biochemistry 1997;36:14128.
130. Tomasz M, Lipman R. Biochemistry 1981;20:5056.
131. Iyengar BS, Dorr RT, Shipp NG, Remers WA. J Med Chem 1990;33:253.
132. Remers WA, Iyengar BS.In: Lukacs G, Ohno M,editors. Recent Progress in the Chemical Synthesis of Antibiotics. Berlin: Springer-Verlag; 1990.
133. Remers WA, Dorr RT.In: Pelletier SW,editor. Alkaloids: Chemical and Biological Properties. New York: John Wiley & Sons; 1988.
134. Frank RW, Tomasz M.In: Wilman E,editor. The Chemistry of Antitumor Agents. Glasgow, UK: Blackie; 1989.
135. Nakatsubo F, Cocuzza AJ, Keeley DE, Kishi Y. J Am Chem Soc 1977;99:4835.
136. Fukuyama T, Yang I. J Am Chem Soc 1987;109:7881.
137. Fukuyama T, Yang I. J Am Chem Soc 1989;111:8303.

138. Paz MM, Das A, Palom Y, He QY, Tomasz M. J Med Chem 2001;44:2834.
139. Kunz KR, Iyengar BS, Dorr RT, et al. J Med Chem 1991;34:2286.
140. Tomasz M, Palom Y. Pharmacol Ther 1997;76:73.
141. Boyer MJ. Oncol Res 1997;9:391.
142. Hargreaves RH, Hartley JA, Butler J. Front Biosci 2000;5:E172.
143. Baumann RP, Hodnick WF, Seow HA, et al. Cancer Res 2001;61:7770.
144. Cummings J. Drug Resist Updat 2000;3:143.
145. He M, Sheldon PJ, Sherman DH. Proc Natl Acad Sci USA 2001;98:926.
146. Belcourt MF, Penketh PG, Hodnick WF, et al. Proc Natl Acad Sci USA 1999;96:10489.
147. Dorr RT. Semin Oncol 1988;32:1584.
148. Carter SK, Cook ST, editors. Mitomycin C: Current Status and New Developments. San Diego, CA: Academic Press; 1980.
149. Grundy WE, Goldstein AW, Rickher CJ, et al. Antibiot Chemother 1953;3:1215.
150. Kennedy BJ, Torkelson JL. Med Pediatr Oncol 1995;24:327.
151. Hill GJ II Sedransk N, Rochlin D, et al. Cancer 1972;30:900.
152. Kennedy BJ. J Urol 1972;107:429.
153. Ream NW, Perlia CP, Wolter J, Taylor SG 3rd JAMA 1968;204:1030.
154. Zojer N, Keck AV, Pecherstorfer M. Drug Saf 1999;21:389.
155. Ralston SH. Cancer Surv 1994;21:179.
156. Bilezikian JP. N Engl J Med 1992;326:1196.
157. Harrington G, Olson KB, Horton J, et al. N Engl J Med 1970;283:1172.
158. Singh B, Gupta RS. Cancer Res 1985;45:2813.
159. Ahr DJ, Scialla SJ, Kimbali DB Jr. Cancer 1978;41:448.
160. Fang K, Koller CA, Brown N, et al. Ther Drug Monit 1992;14:255.
161. Chakrabarti S, Mir MA, Dasgupta D. Biopolymers 2001;62:131.
162. Dasgupta D, Shashiprabha BK, Podder SK. Indian J Biochem Biophys 1979;16:18.
163. Kiseleva OA, Volkova NG, Stukacheva EA, et al. Mol Biol 1974;7:741.
164. Wimalawansa SJ. Semin Arthritis Rheum 1994;23:267.
165. Cortes EP, Holland JF, Moskowitz R, Depoli E. Cancer Res 1972;32:74.
166. Minkin C. Calcif Tissue Res 1973;13:249.
167. Carulli G, Petrini M, Marinia A, et al. Haematologica 1990;75:516.
168. Tagashira M, Kitagawa T, Isonishi S, et al. Biol Pharm Bull 2000;23:926.
169. Umezawa H. In: DiVita VT Jr, Busch H, editors. Methods in Cancer Research. Vol. 16, Part A. New York: Academic Press; 1979. p 43–72.
170. Wohlert SE, Kuenzel E, Machinek R, et al. J Nat Prod 1999;62:119.
171. Lombo F, Kunzel E, Prado L, et al. Angew Chem Int Ed Engl 2000;39:796.
172. Lozano MJ, Remseng LL, Quiros LM, et al. J Biol Chem 2000;275:3065.
173. Blanco G, Fu H, Mendez C, et al. Chem Biol 1996;3:193.
174. Prado L, Fernandez E, Weissbach U, et al. Chem Biol 1999;6:19.
175. Lown JW. Chem Soc Rev 1993;22:165.
176. Arcamone F. Doxorubicin Anticancer Antibiotics. Medicinal Chemistry Monograph. Vol. 17. New York: Academic Press; 1981.
177. Silber R, Liu LF, Israel M, et al. NCI Monogr 1987; 111.
178. Horton D, Priebe W, Varela OJ. Antibiotics 1984;37:1635.
179. Young RC, Ozols RF, Myers CE. N Engl J Med 1981;305:139.
180. Cortes J, Estrey E, O'Brien S, et al. Cancer 2001;92:7.
181. Hortobagyi BN. Drugs 1997;54S4:1.
182. Langer SW, Sehested M, Jensen PB. Clin Cancer Res 2000;6:3680.
183. Singal PK, Li T, Kumar D, et al. Mol Cell Biochem 2000;207:77.
184. Semenov DE, Lushnikova EL, Nepomnyashchikh LM. Bull Exp Biol Med 2001;131:505.
185. Keefe DL. Semin Oncol 2001;28S12:2.
186. Takanashi S, Bachur NR. J Pharmacol Exp Ther 1975;195:41.
187. Galettis P, Boutagy J, Ma DD. Br J Cancer 1994;70:324.
188. Terasaki T, Iga T, Sugiyama Y, et al. J Pharmacobiodyn 1984;7:269.
189. Beck WT, Danks MK. Semin Cancer Biol 1991;2:235.
190. Kiyomiya K, Matsuo S, Kurebe M. Cancer Chemother Pharmacol 2001;47:51.
191. Serrano J, Palmeira CM, Kuchi DW, Wallace KB. Biochim Biophys Acta 1999;1411:201.
192. Rajagopalan S, Politi PM, Sinha BK, Myers CE. Cancer Res 1988;48:4766.

193. Tritton TR, Murphree SA, Sartorelli AC. Biochem Biophys Res Commun 1978; 84:802.
194. Endicott JA, Ling V. Annu Rev Biochem 1989;58:137.
195. Wielinga PR, Westerhoff HV, Lankelma J. Eur J Biochem 2000;267:649.
196. Michieli M, Damiani D, Ermacora A, et al. Br J Haematol 1999;104:328.
197. Ax W, Soldan M, Koch L, Maser E. Biochem Pharmacol 2000;59:293.
198. Fiume L, Baglioni M, Bolondi L, Farina C, Di Stefano G. Drug Discov Today 13:2008; (21–22): 1002.
199. Takanashi S, Bachur NR. Drug Metab Dispos 1976;4:79.
200. Kruh GD, Goldstein LJ. Curr Opin Oncol 1993;5:1029.
201. Plosker GL, Faulds D. Drugs 1993;45:788.
202. Vari FM, Lindemalm C, Choudhury A, Granstam-Björneklett H, Lekander M, Nilsson B, Ojutkangas M-L, Österborg A, Bergkvist L, Mellstedt H. Cancer Immunol Immunother 2009;58:111.
203. Grandi M, Pezzoni G, Ballinari D, et al. Cancer Treat Rev 1990;17:133.
204. Cabri W. Chim Ind 1993;75:314.
205. Arcamone F, Animati F, Bigioni M, et al. Biochem Pharmacol 1999;57:1133.
206. Toffoli G, Sorio R, Aita P, et al. Clin Cancer Res 2000;6:2279.
207. Ross DD, Doyle LA, Yang W, et al. Biochem Pharmacol 1995;50:1673.
208. Israel M, Potti PG, Seshadri R. J Med Chem 1985;28:1223.
209. Ma C, Wang Y-G, Shi J-F, et al. Synth Commun 1999;29:3581.
210. Onrust SV, Lamb HM. Drugs Aging 1999;15:69.
211. Kuznetsov DD, Aliskafi NF, O'Connor RC, Steinberg GD. Epert Opin Pharmacother 2001;2:1009.
212. Arcamone F, Animati F, Capranico G, et al. Pharmacol Ther 1997;76:117.
213. Abbruzzi R, Rizzardini M, Benigni A, et al. Cancer Treat Rep 1980;64:873.
214. Arcamone F, Cassinellit G. Curr Med Chem 1998;5:391.
215. Farquhar D, Cherif A, Bakina E, Nelson JA. J Med Chem 1998;41:965.
216. Binashi M, Bigioni M, Cipollone A, et al. Curr Med Chem 2001;1:113.
217. Suarato A, Angelucci F, Geroni C. Curr Pharm Des 1999;5:217.
218. Cameron DW, Feutrill GI, Griffiths PG. Aust J Chem 2000;53:25.
219. Nafziger J, Averland G, Bertounesque E, et al. J Antibiot 1995;48:1185.
220. Penco S. Chim Ind 1993;75:369.
221. Oki T.In: El-Khadem HS,editor. Anthracycline Antibiotics. New York: Academic Press; 1982. p 75.
222. Priebe W,editor. Anthracycline Antibiotics: New Analogues, Methods of Delivery, and Mechanisms of Action. ACS Symposium Series 574. Washington, DC: American Chemical Society; 1995.
223. Zunino F, Pratesi G, Perego P. Biochem Pharmacol 2001;61:933.
224. Crooke ST, Reich SD,editors. Anthracyclines: Current Status and New Developments. San Diego, CA: Academic Press; 1980.
225. Wall ME. Med Res Rev 1998;18:299.
226. Pizzolato JF, Saltz LB. Lancet 2003;361: 2235.
227. Hsiang YH, Hertzberg R, Hecht S, Liu LF. J Biol Chem 1985;260:14873.
228. Saijo N. Ann NY Acad Sci 2000;922:92.
229. Ozols RF. Semin Oncol 1999;6S18:34.
230. Sandler AB. Oncology 2001;1S1:11.
231. Adeji AA. Curr Opin Pulm Med 2000;6:384.
232. Forbes C, Shirran L, Bagnall AM, Duffy S, et al. Health Technol Assess 2001;5:1.
233. Stewart CF. Caner Chemother Biol Response Modif 2001;19:85.
234. Huang CH, Treat J. Oncology 2001;61S1:14.
235. Schiller JH. Oncology 2001;61S1:1.
236. Eltabbakh GH, Awtrey CS. Expert Onin Pharmacother 2001;2:109.
237. Kehrer DF, Soepenberg O, Loos WJ, et al. Anticancer Drugs 2001;12:89.
238. Schuette W. Lung Cancer 2001;33S1:S107.
239. Arun B, Frenkel EP. Expert Opin Phrarmacother 2001;2:491.
240. Rivory P, Robert J. Bull Cancer 1995;82:265.
241. Mathijssen RH, van Alphen RJ, Verweij J, et al. Clin Cancer Res 2001; 2182.
242. Iyer L, Ratain MJ. Cancer Chemother Pharmacol 1998;42S:S31.
243. Boucaud M, Pinguet F, Poujol S, et al. Eur J Cancer 2001;37:2357.
244. Kollmannsberger C, Mross K, Jakob A, et al. Oncology 1999;56:1.

245. Bourzat J-D, Vuilhorgne M, Rivory LP, et al. Tetrahedron Lett 1996;37:6327.
246. Fedier A, Schwarz VA, Walt H, et al. Int J Cancer 2001;93:571.
247. Popanda O, Flohr C, Dai JC, et al. Mol Carcinog 2001; 171.
248. Fedier A, Schwarz VA, Walt H, et al. Int J Cancer 2001;93:571.
249. Urasaki Y, Laco GS, Pourquier P, et al. Cancer Res 2001;61:1964.
250. Schellens JH, Maliepaard M, Scheper RJ, et al. Ann NY Acad Sci 2000;922:188.
251. Saleem A, Edwards TK, Rasheed Z, Rubin EH. Ann NY Acad Sci 2000;922:46.
252. Desai SD, Li TK, Rodriguez-Bauman A, et al. Cancer Res 2001;61:5926.
253. Fortunak JMD, Kitteringham J, Mastrocola AR, et al. Tetrahedron Lett 1996;37:5683.
254. Curran DP, Ko S-B, Josien H. Angew Chem Int Ed Engl 2683 1996;.
255. Sawada S, Yokokura T, Miyasaka T. Curr Pharm Des 1995;1:113.
256. Sawada S, Yokokura T. Ann NY Acad Sci 1996;803:13.
257. Henegar KE, Ashford SW, Baughman TA, et al. J Org Chem 1997;62:6588.
258. Takimoto CH, Thomas R. Ann NY Acad Sci 2000;922:224.
259. Ishii M, Iwahana M, Mitsui I, et al. Anticancer Drugs 2000;11:353.
260. Kim DK, Ryu DH, Lee JY, et al. J Med Chem 2001;44:1594.
261. Dellavalle S, Delsoldato T, Ferrari A, et al. J Med Chem 2000;43:3963.
262. Bom D, Curran DP, Kruszewski S, et al. J Med Chem 2000;43:3970.
263. Fan Y, Shi LM, Kohn KW, et al. J Med Chem 2001;44:3254.
264. Lavergne O, Bigg DC. Bull Cancer 1998; Spec. No. 51.
265. Boven E, Van Hattam AH, Hoogsteen I, et al. Ann NY Acad Sci 2001;922:175.
266. Demarquay D, Huchet M, Coulomb H, et al. Anticancer Drugs 2001;12:9.
267. Larsen AK, Gilbert C, Chyzak G, et al. Cancer Res 2001;61:2961.
268. Canel C, Moraes RM, Dayan FE, Ferreira D. Phytochemistry 2000;54:115.
269. Imbert TF. Biochemie 1998;80:207.
270. Hande KR. Eur J Cancer 1998;34:1514.
271. Oliver RT. Curr Opin Oncol 2001;13:191.
272. Schuette W. Lung Cancer 2001;33S1:S99.
273. Mavroudis D, Papadakis E, Veslemes M, et al. Ann Oncol 2001;12:463.
274. Hande KR. Biochem Biophys Acta 1998; 1400:173.
275. Hoetelmans RM, Schornagel JH, ten Bokkel Huinink WW, Beijnen JH. Ann Pharmacother 1996;30:367.
276. Bernstein BJ, Troner MB. Pharmacotheraphy 1999;19:989.
277. Felix CA. Biochem Biophys Acta 1998;1400:233.
278. Rubnitz JE, Look AT. J Pediatr Hematol Oncol 1998;20:1.
279. McLeod HL, Evans WE. Cancer Surv 1993;17:253.
280. Relling MV, Nemec J, Schuetz EG, et al. Mol Pharmacol 1994;45:352.
281. Kitamura R, Bandoh T, Tsuda M, Satoh T. J Chromatogr B Biomed Sci Appl 1997;690:283.
282. Nguyen L, Chatelut E, Chevreau C, et al. Cancer Chemother Pharmacol 1998;41:125.
283. Liu B, Earl HM, Poole CJ, et al. Cancer Chemother Pharmacol 1995;36:506.
284. Chen CL, Rawwas J, Sorrell A, et al. Leuk Lymphoma 2001;42:317.
285. Stewart DJ, Grewaal D, Redmond MD, et al. Cancer Chemother Pharmacol 1993;32:368.
286. Dolden JA. Ann Clin Lab Sci 1997;27:402.
287. Byl JA, Cline SD, Utsugi T, et al. Biochemistry 2001;40:712.
288. Morris SK, Lindsley JE. J Biol Chem 1999;274:30690.
289. Chiou WL, Chung SM, Wu TC, Ma C. Int J Clin Pharmacol Ther 2001;39:93.
290. Koshiyama M, Fujii H, Kinezaki M, Yoshida M. Anticancer Res 2001;21:905.
291. Mao Y, Yu C, Hsieh TS, et al. Biochemistry 1999;38:10793.
292. Matsumoto Y, Takano H, Kunishio K, et al. Jpn J Cancer Res 2001;92:1133.
293. Middel O, Woerdenbag HJ, van Uden W, et al. J Med Chem 1995;38:2112.
294. Stahelin HF, von Wartburg A. Cancer Res 1991;51:5.
295. Schacter L. Semin Oncol 1996;6S13:1.
296. Jain N, Lam YM, Pym J, Campling BG. Cancer 1996;77:1797.
297. Issel BF, Muggia FM, Carter SK. Etoposide (VP-16): Current Status and New Developments. San Diego, CA: Academic Press; 1984.
298. Botta B, Delle Monache G, Misiti D, et al. Curr Med Chem 2001;8:1363.

299. Kellner U, Rudolph P, Parwaresch R. Onkologie 2000;23:424.
300. Gordaliza M, Miguel del Corral JM, Angeles Castro M, et al. Farmaco 2001;56:297.
301. Kingston DGI. Chem Commun 2001; 867.
302. Schiff PB, Fant J, Horwitz SB. Nature 1979; 277:665.
303. Manfredi JJ, Horwitz SB. Pharmacol Ther 1984;25:83.
304. Horwitz SB. Ann Oncol 1994;5S6:S3.
305. Potier P. C R Acad Agric Fr 2000;86:179.
306. Wu CH, Yang CH, Lee JN, et al. Int J Gynecol Cancer 2001;11:295.
307. Minotti G, Saponiero A, Licata S, et al. Clin Cancer Res 2001;7:1511.
308. Gelderblom H, Verweij J, Nooter K, Sparreboom A. Eur J Cancer 2001;37:1590.
309. Kivisto KT, Kroemer HK, Eichelbaum M. Br J Clin Pharmacol 1995;40:523.
310. Karlsson MO, Molnar V, Freijs A, et al. Drug Metab Dispos 1999;27:1220.
311. Maier-Lenz H, Hauns B, Haering B, et al. Semin Oncol 1997;24S6:S19–S16.
312. Kearns CM, Gianni L, Egorin MJ. Semin Oncol 1995;22S6:S16–S23.
313. Monsarrat B, Royer I, Wright M, Cresteil T. Bull Cancer 1997;84:125.
314. Rahman A, Korzekwa KR, Grogan J, et al. Cancer Res 1994;54:5543.
315. Sparreboom A, Van Tellingen O, Scherenburg EJ, et al. Drug Metab Dispos 1996;24:655.
316. Jang SH, Wientjes MG, Au JL-S. J Pharmacol Exp Ther 2001;298:1236.
317. Giannakakou P, Sackett DL, Ksang Y-K, et al. J Biol Chem 1997;272:17118.
318. Kavallaris M, Kuo DY, Burkhart CA, et al. J Clin Invest 1997;100:1282.
319. Yu D. Semin Oncol 2001;28-5S16:S12.
320. Montgomery RB, Guzman J, O'Rourke DM, Stahl WL. J Biol Chem 2000;275:17358.
321. Lynch TJ Jr Semin Oncol 2001;3S9:5.
322. Bruno R, Vivier N, Veyrat-Follet C, et al. Invest New Drugs 2001;19:163.
323. Shou M, Martinet M, Korzekwa KR, et al. Pharmacogenetics 1998;8:391.
324. Vaishampayan U, Parchment RE, Jasti BR, Hussain M. Urology 1999;54S6A:22.
325. Georg GI, Boge TC, Cheruvallath ZS, et al. In: Sufness M,editor. Taxol: Science and Applications. Boca Raton, FL: CRC Press; 1995. p 317.
326. Ter hart E. Expert Opin Ther Patents 1998;8:57.
327. Lin S, Ojima I. Expert Opin Ther Patents 2000;10:1.
328. Gueritte F. Curr Pharm Des 2001;7:1229.
329. Jennewein S, Rithner CD, Williams RM, Croteau RB. Proc Natl Acad Sci USA 2001;98:13595.
330. Floss HG, Mocek U.In: Sufness M,editor. Taxol: Science and Applications. Boca Raton, FL: CRC Press; 1995. p 191.
331. Li Q, Sham HL, Rosenberg SH. Annu Rep Med Chem 1999;34:139.
332. Fahy J. Curr Pharm Des 2001;7:1181.
333. Potier P, Langlois N, Langlois Y, Gueritte F. Chem Commun 1975; 670.
334. Mangeney P, Andriamialisoa RZ, Langlois N, et al. J Am Chem Soc 1979;101:2243.
335. Kutney JP, Hibino T, Jahngen E, et al. Helv Chim Acta 1976;59:2858.
336. Bai RL, Pettit GR, Hamel E. Biochem Pharacol 1990;39:1941.
337. Bai RL, Pettit GR, Hamel E. Biochem Pharacol 1990;265:17141.
338. Zhou XJ, Rahmani R. Drugs 1992;44S4:1.
339. Van Tellingen O, Beijnen JH, Nooijen WJ, Bult A. Cancer Chemother Pharmacol 1993;32:286.
340. Schlaifer D, Cooper MR, Attal M, et al. Leukemia 1994;8:668.
341. Elmarakby SA, Duffel MW, Rosazza JP. J Med Chem 1989;32:2158.
342. Elmarakby SA, Duffel MW, Goswami A, et al. J Med Chem 1989;32:674.
343. D, Sackett l. Biochemistry 1995;34:7010.
344. Prakash V, Timasheff SN. Biochemistry 1985;24:5004.
345. Luker KE, Pica CM, Schreiber RD, Piwnica-Worms D. Cancer Res 2001;61:6540.
346. Kavallaris M, Tait AS, Walsh BJ, et al. Cancer Res 2001;61:5803.
347. Krishna R, Webb MS, St. Onge G, Mayer LD. J Pharmacol Exp Ther 2001;298:1206.
348. Schlaifer D, Cooper MR, Attal M, et al. Blood 1993;81:482.
349. Schlaifer D, Duchayne E, Demur C, et al. Leuk Lymphoma 1996; 441.
350. Naumann R, Mohm J, Reuner U, Kroschinsky FR. Br J Haematol 2001;115:323.
351. Loe DW, Deeley RG, Cole SP. Cancer Res 1998;58:5130.
352. Montalar J, Morales S, Maestu I, et al. Lung Cancer 2001;34:305.
353. Namer M, Soler-Michel P, Turpin F, et al. Eur J Cancer 2001;37:1132.

354. Lilenbaum R. Oncologist 2001;6S1:16.
355. Masters G. Oncologist 2001;6S1:12.
356. Burstein HJ, Bunnell CA, Winer EP. Semin Oncol 2001;28:344.
357. Sonoda K, Ohshiro T, Ohta M, et al. Gan To Kagaku Ryoho 2001;28:1397.
358. Vassilomanolakis M, Koumakis G, Barbounis V, et al. Support Care Cancer 2001;9:108.
359. Cersosimo RJ, Bromer R, Licciardello JT, Hong WK. Pharmacotherapy 1983;3:259.
360. Etievant C, Kruczynski A, Barret JM, et al. Cancer Chemother Pharmacol 2001;48:62.
361. Hill BT. Cyrr Pharm Des 2001;7:1199.
362. Kruczynski A, Hill BT. Crit Rev Oncol Hematol 2001;40:159.
363. Taylor WI. The Catharanthus Alkaloids: Botany, Chemistry, Pharmacology, and Clinical Use. New York: Marcel Dekker; 1975.
364. Rowinsky EK, Donehower RC.In: DeVita VT Jr, Hellman S, Rosenberg SA,editors. Cancer: Principle and Practice of Oncology. 5th ed. Philadelphia: Lippencott-Raven; 1997. p 467–483.
365. Shi Q, Chen K, Morris-Natchke SL, Lee KH. Curr Pharm Des 4:1998; 219.
366. Remers WA,editor. Antineoplastic Agents. New York: Wiley-Interscience; 1984.
367. Suffness M,editor. Taxol: Science and Applications. Boca Raton, FL: CRC Press; 1995.
368. Potmesil M, Pinedo H,editors. Camptothecins: New Anticancer Agents. Boca Raton, FL: CRC Press; 1995.
369. Kingston DGI, Molinero AA, Rimoldi JM. Prog Chem Org Nat Prod 1993;61:1.
370. Kingston DGI, Jagtap PG, Yuan H, Samala L. Prog Chem Org Nat Prod 2002;84:53.
371. Newman DJ, Craigg GM. J Nat Prod 2007;70:461.
372. Kingston DGI. J Nat Prod 2009;72:5–7.

HISTONE DEACETYLASE INHIBITORS: A BRIEF OVERVIEW OF THEIR ROLE AND MEDICINAL CHEMISTRY

P. Angibaud[1]
K. Van Emelen[2]
J. Arts[3]

[1] Department of Medicinal Chemistry, Ortho-Biotech Oncology Research & Development, campus de maigremont, Val de Reuil, France
[2] Medicinal Chemistry, Tibotec-Virco, Gen. De Wittelaan, Mechelen, Belgium
[3] Biology Department, Ortho-Biotech Oncology Research & Development, Turnhoutseweg, Beerse, Belgium

1. INTRODUCTION

Despite substantial efforts, cancer remains one of the major health concerns at the beginning of this century and is therefore the focus of intense research at specialized organizations and pharmaceutical companies. Cancers may result from genetic alterations but can also have an epigenetic origin that result in changes in the expression of oncogenes or tumor suppressors due to alterations of specific chromatin domains [1]. DNA is wrapped around nucleosomes, the principal constituents of chromatin fibers. Each nucleosome consists of a histone octamer core constituted by pairs of histones (H2A, H2B, H3, and H4) [2]. Histones are small proteins (12–16 kDa) with positively charged lysine-rich N-termini. Interactions of these protonated amino groups with negatively charged DNA help in compacting the DNA around the nucleosomes. When the amino groups of these lysines are acetylated this charged based interaction are annihilated and the DNA become accessible for transcription. The balance of histone acetylation is regulated by the activity of two enzymes histones acetyltransferases (HATs) and histone deacetylases (HDACs) (Fig. 1) that, respectively, catalyze the acetylation and the removal of an acetyl group on ε-amino groups of these lysine residues [3]. Inhibition HDACs results in increased chromatin acetylation and changes in transcription that have been linked to cell cycle arrest and apoptosis as well as down-regulation of genes involved in tumor progression, invasion, and angiogenesis [3–6]. This is highlighted by the ability of most HDAC inhibitors (HDACi) to cause induction of epigenetically silenced tumor suppressor genes like $p21^{WAF1/CYP1}$, a cyclin-dependent kinase inhibitor [7,8].

Based on these initial observations, HDAC inhibition has attracted a great deal of interest in the last 10 years as a potential way to add new weapons to the armentarium against cancer and set the basis of what can be called the "initial hypothesis." However, an increasing amount of data has unveiled part of the mist and revealed that HDACi also act through deacetylation of other protein substrates, resulting in the modulation of several signalling pathways that are key in tumorigenesis. Class I HDACs act on several substrates that are involved in proliferation and differentiation, including p53, NF-κB, and Bcl-6 [9–11]. Class II HDAC substrates, such as tubulin and HSP90, are involved in various cellular processes including protein trafficking, migration and oncogene degradation.

Moreover, through intense research in this domain, our understanding of HDAC biology has grown rapidly and novel therapeutic applications for HDACi, such as neurodegenerative diseases and inflammation [12,13], have been discovered. With the recent approval of the first histone deacetylase inhibitor, vorinostat (SAHA) [14], to treat patients diagnosed for cutaneous T-cell lymphoma, we are witnessing the emergence of a new class of promising therapeutic agents.

The pharmacology, patent literature and medicinal chemistry aspects of HDAC inhibition have been regularly reviewed. We therefore propose to focus instead on synthesis and structure–activity relationships (SARs) of HDACi compounds having reached clinical phase, highlighting some key features or new concepts, with limited consideration regarding clinical results. Our aim is not to be exhaustive but to provide elements that help to understand the challenges of medicinal

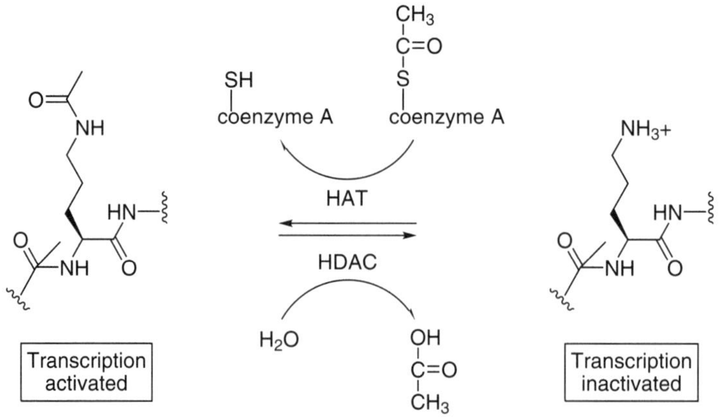

Figure 1. The balance of histone acetylation and deacetylation.

chemistry of HDAC inhibitors as well as providing key references to further improve comprehension of the HDAC field.

2. BIOLOGY AND ROLES OF HISTONE DEACETYLASES

The 18 HDACs in humans are usually divided in four structurally and functionally different classes [15–17]. HDAC-1,2,3, and 8 share homology with yeast RPD3 and constitute Class I HDACs. They are ubiquitously expressed and are predominatly located in the nucleus. Class II HDACs share homology with yeast HDA1 and can be divided into two subclasses. Class IIa comprises HDAC-4,5,7, and 9 and the HDAC9 splice variant MEF2-interacting transcription repressor (MITR). The Class IIb enzymes (HDAC-6 and HDAC-10) differ from other HDACs in that they contain two independent catalytic sites. Class II HDACs are large proteins (1000 amino-acids) approximately twice the size of Class I HDACs (350–500 amino acids) that are able to shuttle between the nucleus and the cytoplasm. HDAC-11 exhibits properties of both Class I and Class II and is often classified as Class IV. Whereas Class I and II HDACs are inhibited by HDAC inhibitors such as trichostatin A (TSA) or Zolinza® (vorinostat, suberoylanilide hydroxamic acid, or SAHA) through chelation of the Zn^{2+} cation, which is essential for their catalytic deacetylating activity, Class III HDACs are involved in the NAD^+-dependent acetylation of nicotinamide-deprived ADP-ribose groups. Class III HDACs are also named as sirtuins based on their sequence similarity with Sir2 (silent information regulator 2), a yeast transcription repressor. Seven different sirtuins have been identified in mammals [18]. However, deacetylase activity has only been reported for situins-1, 2, 3, and 5 and Sir1 has been recently linked to prostate cancer via FoxO1 activation [19]. Sirtuins inhibition has only recently received attention and therefore the number of sirtuin inhibitors is currently limited [20,21] with no compounds having reached clinical phases yet.

The Class I HDACs have been shown to be crucial for tumor cell proliferation. Knockdown of HDAC1 and HDAC3 using siRNA techniques caused inhibition of proliferation accompanied by distinct morphological changes [22]. Lagger et al. [23] showed that disruption of HDAC1 in mouse embryonic stem cells resulted in an increase in H3 and H4 acetylation and gene induction, thereby linking histone deacetylation and the subsequent transcriptional modulation to the enzymatic activity of the Class I HDACs. Another key example further emphasizing the key role of Class I HDACs is the HDAC1-dependent silencing of the cyclin-dependent kinase inhibitor $p21^{waf1,cip1}$ [8]. HDAC2 modulates transcriptional activity through regulation of p53 binding activity and its knockdown inhibits cellular proliferation [24]. Using a selective HDAC8 inhibitor, Balasubramanian et al. demonstrated that HDAC8 inhibition is involving phospholipase C-γ1 activation and calcium-induced apoptosis in cells derived from T-cell malignancies [25].

Class IIa HDACs lack intrinsic histone deacetylase enzymatic activity but regulate gene expression by functioning as bridging factors since they associate both with Class I HDAC complexes and with transcription factor/DNA complexes. Downregulating HDAC4 expression by small interfering RNA (siRNA) in HCT116 cells induced growth inhibition and apoptosis in vitro, reduced xenograft tumor growth, and increased p21 transcription [26]. HDAC5 downregulation has been linked to treatment of depression [27]. HDAC7 has been identified as a key modulator of endothelial cell migration and linked to angiogenesis and vascular integrity [28,29]. HDAC6, a member of Class IIb, has received attention due to its identification as an Hsp90 deacetylase. Hsp90 deacetylation results in degradation of Hsp-90 associated prosurvival and proproliferative client proteins. These include Her-2, Bcr-Abl, c-Raf, and Akt [30]. In addition to Hsp90, HDAC6 also mediates tubulin deacetylation, which results in microtubule destabilization [31]. The biological role of HDAC6 was further confirmed by a recent report showing that a small molecule inhibitor of HDAC6, tubacin, caused α-tubulin hyperacetylation and decreased cell motility without affecting cell cycle progression [32]. Moreover, Lee et al. have recently shown that deacetylase activity of HDAC6 is required to support malignant growth of cancer cells in vitro [33]. HDAC6 appears more and more as a pivotal regulatory molecule in many key biological processes [34]. Very little is known about HDAC11.

In summary, due to the large number of cell cycle regulatory proteins modulated by HDACs at the level of either their expression or their activity, the antiproliferative effect of HDAC inhibitors cannot be linked to a single mechanism of action and the relative importance of the different proteins affected by HDACs is likely to vary between tumors.

In agreement with the aberrant HDAC/HAT activity equilibrium in cancer, HDAC inhibitors have been shown to induce cell-cycle arrest, terminal differentiation, and/or apoptosis in a broad spectrum of human tumor cell lines in vitro, to inhibit angiogenesis and to exhibit in vivo antitumor activity in human xenograft models in nude mice. Several HDAC inhibitors are in advanced stages of clinical development and antitumor activity has been observed in hematological malignancies at doses that were well tolerated.

3. STRUCTURAL ELEMENTS

In 1999, Finnin et al. [35] published the X-ray crystal structure of an HDAC-like protein (HDLP) from bacterial origin (Aquifex aelicus) complexed with two inhibitors SAHA and TSA. The crystal structure revealed that the catalytic site where lysine deacetylation occurs is constituted of a deep, narrow pocket with a tube-like shape with a depth of 11 Å and an internal cavity adjacent to the pocket approximately 14 Å wide (Fig. 2a). The zinc atom is located at the bottom of the tube and is coordinated by two Asp (168 and 258), one His (170), and a water molecule. This water molecule is displaced by the NH–OH moiety of crystal bound hydroxamic acids SAHA and TSA, which then coordinates the zinc in a bidentate fashion through carbonyl and hydroxyl groups (Fig. 2c). The hydroxamic acid moiety also makes interaction with charge–relay histidines (131, 132) and the tyrosine 297. The tube walls are defined by hydrophobic and aromatic residues among which two Phe (141 and 198) facing each other are of particular importance. The aromatic moieties in SAHA and TSA make contacts at the pocket entrance and with the near surface groove capping the pocket.

This pioneer work shed light on the binding mode of HDAC inhibitors and explained why HDACi share the so-called HDAC pharmacophore (Fig. 2b), which consists of a metal binding domain, a linker domain which mimics the lysine side-chain and a hydrophobic "surface recognition motif" also named the capping group that interact with the pocket entrance.

In spite of the existence of a large number of different HDAC proteins, it took 5 years for the next HDAC crystal structure to be reported, which may be due to the tendency for HDACs to form multiprotein complexes. Two groups independently solved the crystal

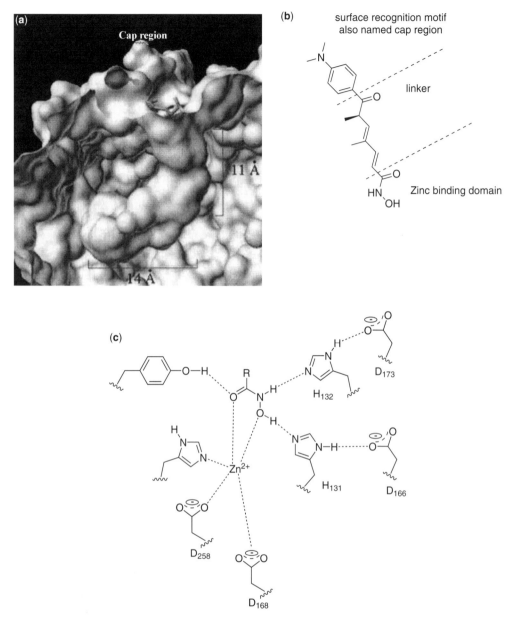

Figure 2. (a) Snapshot of trichostatin A bound into the catalytic site of HDLP (PDB code 1C3R). (b) Structure of trichostatin A and HDAC pharmacophore model. (c) Schematic interactions of the hydroxamic acid moiety with Zinc cation and aminoacid at the bottom of the tube-like shape pocket.

structure of human HDAC8 [36,37]. The active site architecture is conserved between HDLP, HDAC1, and HDAC8. This is probably also true for the other HDAC enzymes as sequence homology is high. SAHA and TSA showed similar mode of binding in HDAC8 and HDLP. The structural diversity is higher in the protein surface and this has led to the suggestion that medicinal chemistry focus on that region would identify more selective HDACi [38]. Very recently, Schuetz et al. disclosed the first crystal structure of HDAC7 adding valuable knowledge to the design of specific inhibitors [39].

4. HYDROXAMIC ACID-BASED HDAC INHIBITORS

The number of medicinal chemistry publications describing new molecules as HDAC inhibitors is extensive and generally classified based on the zinc binding group (ZBG). It is not the intention of this review to thoroughly report every effort made to design potent and efficacious HDACi, nor to go deeper into structure–activity relationships, but rather to focus on a brief description of HDACi in clinical development as they cover most of the different classes. We will attempt to provide a recent description of chemical synthesis and brief elements of SARs for each compound. For further in depth analysis of the recent literature, we would recommend, among others, the two reviews from Miller and Fattori, which together cover these last 8 years [40,41].

The most common group (Fig. 3) of HDACi employs hydroxamic acid (HA) to target the zinc cation. It includes SAHA, the archetype inhibitor that uses a linear chain as linker, but also LBH-589, PXD-101, CRA-024781, or SB939 where cinnamoyl moieties are connected to the hydroxamic acid. Moreover, phenyl or heterocycle-linked hydroxamics acids such as ITF-2357 or R306465 and JNJ-26481585 have also reached clinical phase evaluation stage.

In a search for alternative ZBG, orthophenylene diamines have been identified as promising HDACi. These have a different selectivity profile compared to HA, and are represented in clinical phases by CI-994, MS-275, or MGCD-0103. Finally, naturally occurring depsipeptides, such as FK-228, represent the last main class of HDACi in clinical phases.

4.1. SAHA

Suberoyl hydroxamic acid (SAHA, vorinostat, Zolinza, MERCK) was first discovered by Breslow et al. as cytodifferentiating agent in leukemia cells [42] and reported as an HDAC inhibitor ($IC_{50} \sim 50$ nM) 3 years later [43]. Kg scale synthesis of SAHA has recently been reported [44] and depicted in Scheme 1. In a first step, aniline is mixed with melt suberic acid **1** and the mixture added to potassium hydroxide in water and stirred 20 min. Filtration, acidification and several sequences of triturations provided acid **2** in 32.7% yield. Heating **2** in methanol in the presence of Dowex 50WX2-400 resin provided the methyl ester with 84% yield after recrystallization in MeOH/water. Crude SAHA was then obtained in a high yield (95.6%) by reacting ester **3** with hydroxylamine hydrochloride and sodium methoxide in methanol at room temperature, followed by acidification and filtration of the precipitate.

The length of the linker between hydroxamate and N-phenylamide is important (see Table 1) and clearly 6-methylene linker is needed for cellular potency [45].

SAHA similarly inhibits HDAC1 ($IC_{50} = 48$ nM), 3 ($IC_{50} = 52$ nM), and 6 ($IC_{50} = 62$ nM). SAHA was the first HDACi to be approved by Food and Drug Administration (FDA) in patients diagnosed with CTCL on or following two systematic therapies [14]. Vorinostat is given orally at doses of 400 mg q.d. and is well tolerated, main side effects being fatigue, diarrhea, and nausea [46,47].

Vorinostat is 71% bound to plasma proteins over a range of 0.5–50 µg/mL [48]. It is metabolized by glucuronidation and hydrolysis followed by oxidation. The mean half-life is about 2 h for vorinostat and the O-glucuronide metabolite. Vorinostat is not an inhibitor of CYP drug-metabolizing enzymes, but there is potential for suppression of CYP2C9 and CYP3A at concentrations above 10 µM [49]. Vorinostat is currently being investigated in more than 40 clinical studies both as a single agent and in combination.

The *in vitro* potency of SAHA has been greatly increased by Merck scientists who discovered that adding amino-substitution (amides, sulfonamides, or ureas) alpha to the phenyl amide bond better mimic lysine residues **7** (Scheme 2) [45,50]. Compound **9**, for example, where a 2-naphtylsulfonamide moiety has been added on SAHA backbone displayed subnanomolar potency in cell antiproliferation.

A low pharmacokinetic profile attributed to the hydroxamic acid function has triggered

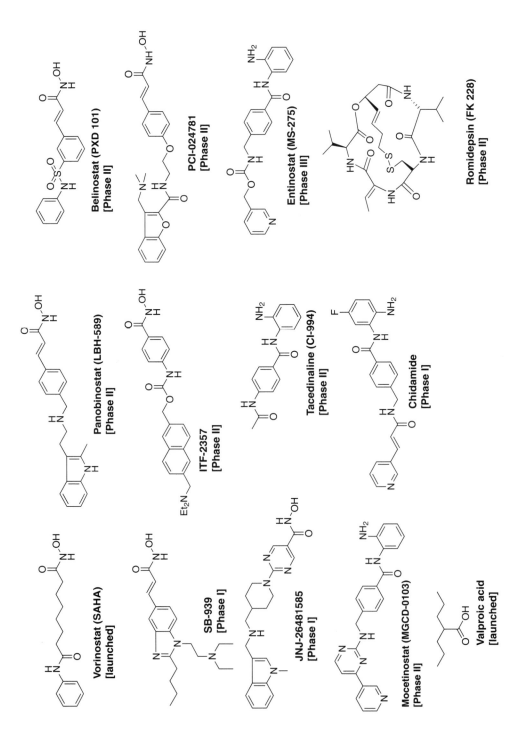

Figure 3. HDAC inhibitors in clinical phases and development status.

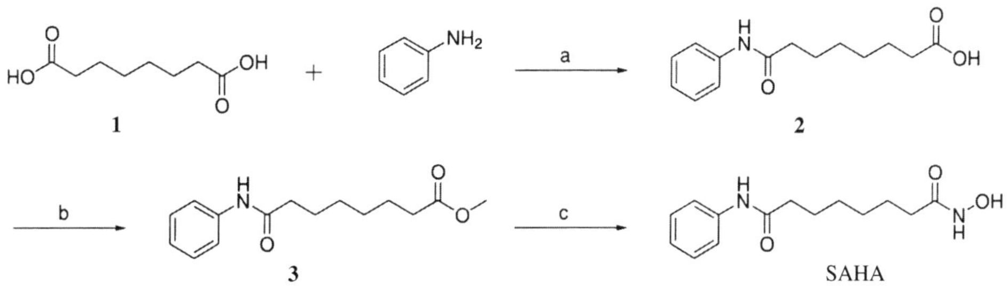

Scheme 1. Reagents and conditions: (a) 175–190 °C then KOH/H$_2$O and trituration 32.7%; (b) DOWEX 50WX2-400, MeOH, reflux, 84%; (c) NH$_2$OH•HCl, MeONa/CH$_3$OH, RT, 16 h, 95.6%.

Table 1. Influence of the Linker Length on HDAC Activity of SAHA Analogs

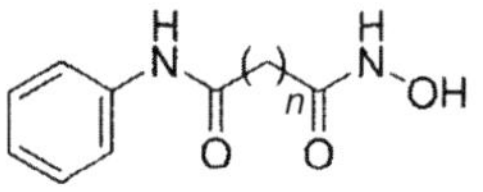

Compound	n	HDAC Enzyme IC$_{50}$ (nM)	SC9 IC$_{50}$ (µM)
4	3	2,100	>10,000
5	5	355	11,000
SAHA	6	40	600
6	7	100	18,000

significant efforts to replace hydroxamic acids. SAHA is probably the HDACi where this effort has been attempted most thoroughly. Table 2 gives an example of successful identification of some ZBG as HA surrogate.

Dehmel et al. have recently developed trithiocarbonates as a novel class of HDACi, first by preparing SAHA analogues [53] and then by further extending the concept to a variety of capping moieties [54]. Suzuki et al. have studied the impact of replacing hydroxamic acid by a thioacetyl moiety (**11** or its prodrug **12**), a thioacetamide (**13**), or an alkyl thiol group (**14**) [55,56]. It should be kept in mind that the apparent drop in enzymatic potency sometimes obtained while replacing the hydroxamic acid ZBG in SAHA by trifluoromethylketone (**15**) [57], or ortho-aminoanilide moiety (**16**) can be compensated by modifications in the cap region and or gains in selectivity on HDAC isoforms. Although not tested on SAHA skeleton, alpha-ketoamides such as **19** (IC$_{50}$ = 3.1 nM) (Fig. 4) illustrated another attempt of replacing hydroxamic moiety [58]. Without being exhaustive, successful efforts to identify various types of ZBG led mainly to compounds bearing sulfur atoms. KD5170 (EC$_{50}$ = 140 nM HCT116 cell line) is one example (Fig. 4) that may soon reach clinical phase [59]. Replacing HA is still an actively pursued area as demonstrated by recent disclosure of silanediol compounds [60] and sulfamides [61].

4.2. NVP-LAQ-824 and LBH-589

Novartis first HDACi were identified in a cell-based screen for compounds that induced the expression of the cyclin-dependent kinase inhibitor p21^{waf1} [60]. Although not suitable for

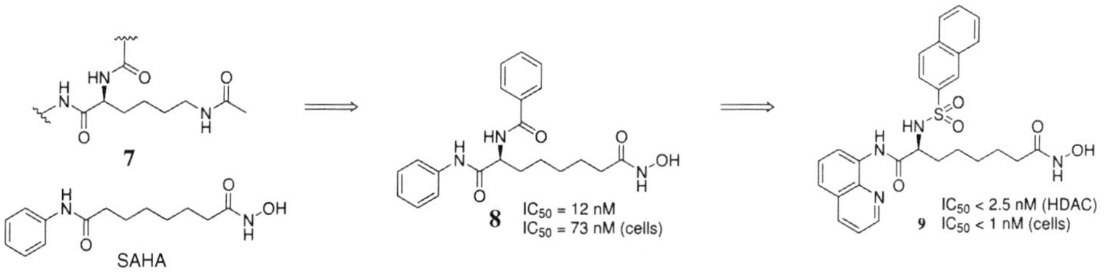

8 IC$_{50}$ = 12 nM
IC$_{50}$ = 73 nM (cells)

9 IC$_{50}$ < 2.5 nM (HDAC)
IC$_{50}$ < 1 nM (cells)

Scheme 2. Mimicking lysine residues with SAHA analogs.

Table 2. HDAC Enzymatic Inhibition (HeLa Cells Nuclear Extract) for Nonhydroxamates SAHA Analogs

Compound	IC$_{50}$(μM)	reference	Compound	IC$_{50}$(μM)	reference
SAHA	0.080/0.37/0.28	[51–53]	14	0.21	[57]
10	0.050	[53]	15	6.7[a]	[58]
11	0.15	[52]	16	120	[57]
12	0.081	[52]	17	1.5[b]	[61]
13	2.4/0.39	[55]	18	no data	[60]

[a] Inhibition of HDAC1, [b] Inhibition of HDAC6.

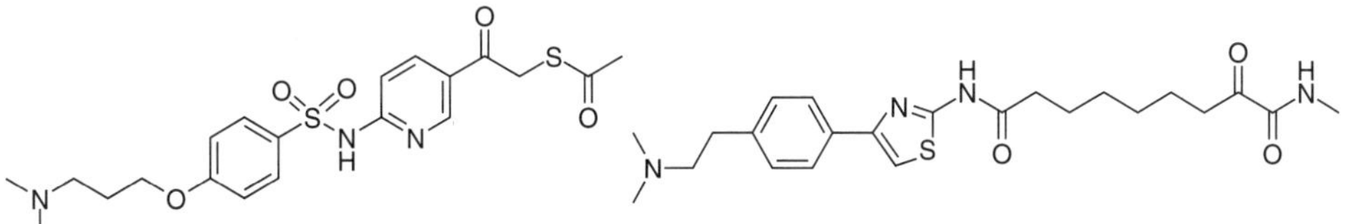

HYDROXAMIC ACID-BASED HDAC INHIBITORS 57

Figure 4. Structures of KD5170 and alpha-ketoamide 19.

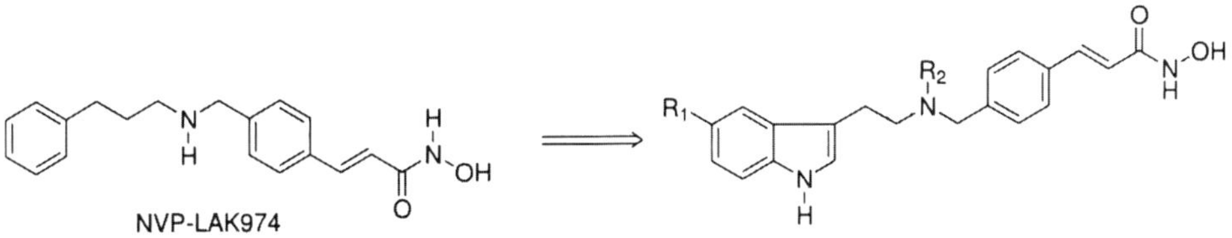

Scheme 3. Structure of NVP-LAK974, NVP-LAQ-824, and related compounds.

drug development, these hits have paved the way for Novartis HDAC program. An enzymatic screen revealed NVP-LAK974 that, despite good *in vitro* potency, showed poor efficacy *in vivo* but already possessed the cinnamoyl hydroxamic moiety that characterized Novartis' HDACi in clinic (Scheme 3). Because NVP-LAK974 was stable in plasma and as the benzylamino moiety was a requisite for solubility, main efforts were directed toward substituting this aminogroup. Several indole derivatives were found to be efficacious in xenograft studies, but LAQ824 **21d** distinguished itself due to its tolerability (MTD 200 mg/kg, less body weight loss at all doses, see Table 3), higher cellular activity and was selected for Phase I clinical studies [62,63].

Novartis is now developing a close analog, LBH-589 where a methyl substituent has been introduced on carbon-2 of the indole ring and the hydroxyethyl substituent has been removed. LBH-589 is slightly more potent *in vitro* (see Table 4) and currently is undergoing Phase II/III clinical studies.

A short synthesis scheme for LBH-589 has been recently patented (Scheme 4) [65]. The first step described the condensation of 5-chloropentan-2-one **23** on phenyl hydrazine **22** to provide 2-methyltryptamine **24**. Reaction of 2-methyl-tryptamine with (*E*)-3-(4-formylphenyl)acrylicmethyl ester **25** in MeOH at room temperature provide the imine that is reduced by NaBH$_4$ while keeping temperature below −10 °C. Introduction of the hydroxamic moiety is obtained by reacting the ester **26** with hydroxylamine in water in basic medium while controlling temperature. Seeding and careful pH adjustments enabled LBH-589 (free base) to crystallize in a nearly quantitative yield.

Table 3. Some *in vivo* Evaluation of Novartis Cinamoyl HDACi

Compound	HDAC IC$_{50}$ (nM)	H1299 IC$_{50}$ (nM)	% T/C at 10 mg/kg	% Body Weight Change
21a	63	400	56	−10.2 ± 0.47
21b	14	580	64	−4.6 ± 0.51
21c	14	310	40	−6.4 ± 0.39
21d	32	150	55	2.7 ± 0.33

Table 4. Comparison of *In Vitro* Profile of LAQ-824 and LBH-589 [64]

Compound	HDAC IC50 (nM)	p21 Activation AC50 (μM)	Inhibition of Monolayer Cell Proliferation IC50 (nM)			
			H1299	HCT116	DU145	PC3
LAQ 824	40	300	150	10	18	23
LBH 589	30	69	60	5	30	25

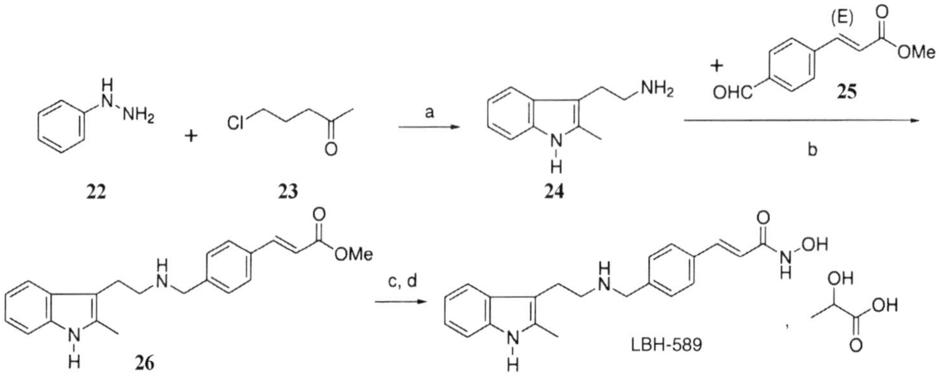

Scheme 4. Reagents and conditions: (a) EtOH, N_2, 40 °C, then reflux 50 min, 49.2%; (b) (i) MeOH 1 h, RT then (ii) NaBH4 < −10 °C (iii) HCl 0–5 °C; (c) NaOH, NH_2OH 50% in water, −15 °C, 7 h, 99%; (d) DL-Lactic acid, water 65 °C (dissolution) then crystallization at 33 °C upon seeding.

Panobinostat is 90% bound to plasma proteins and is a competitive inhibitor of CYP2D6 ($IC_{50} \sim 2\,\mu M$) drug-metabolizing enzyme. It also inhibits the hERG channel with an IC_{50} of 3.9 μM (compared to 0.03 μM for HDAC inhibition), indicating the possibility of cardiac arrhythmia as a complication of drug administration. A Phase I clinical study has demonstrated activity in patients with CTCL and Phase II/III for this indication as well as other hematological malignancies are ongoing [66]. Novartis is also evaluating panobinostat, using IV administration, for the treatment of metastatic hormone refractory prostate cancer (Phase I first line, Phase II second line) [67].

4.3. PXD-101 (Belinostat)

Another member of the cinnamoyl-hydroxamic acid family currently investigated intravenously and orally in Phase I and II clinical trials is PXD-101 (belinostat, Curagen/Topotarget). Belinostat inhibits HDAC activity in HeLa cell nuclear extracts with an IC_{50} of

Table 5. Enzymatic Inhibitory Activity of PXD-101 on HDAC Isoforms [68]

	Class I				Class II			
HDAC isoform	1	2	3	8	4	6	7	9
EC_{50} (nM)	41	125	30	216	115	81	67	128

27 nM [68] and recombinant Class I and Class II HDAC isoforms in the nanomolar range as can be seen in Table 5 below. This characterizes PXD-101 as a pan-HDAC inhibitor.

Synthesis of *N*-hydroxy-3-(3-phenylsulfamoylphenyl)acrylamide has been reported by Watkins et al. in a patent published in 2002 (Scheme 5) [69]. Sulfonylation of benzaldehyde in position *meta* constituted the first step of this synthesis. Acrylic acid methylester was introduced by Wittig reaction and chlorination of compound **28** gave a sulfonyl chloride intermediate, which was then reacted with aniline to provide ester **29**. Last steps illustrate a slightly different way than the one used

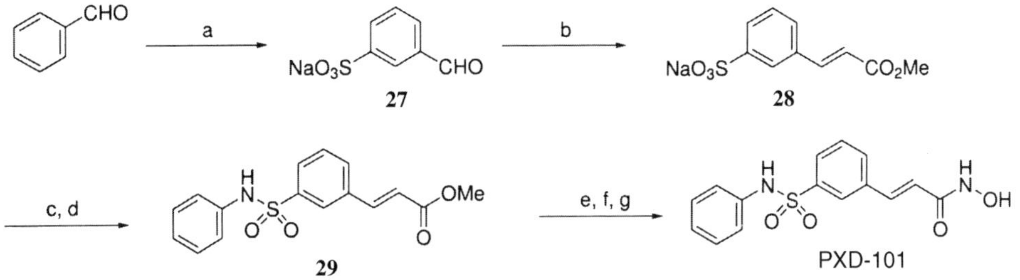

Scheme 5. Reagents and conditions: (a) Oleum, 40 °C, 10 h, 51%; (b) K$_2$CO$_3$, dimethylphosphoroacetate, H$_2$O, RT, 30 min, 55%; (c) SOCl$_2$, benzene; (d) Aniline, pyridine, DCM, 50 °C, 1 h, 29%; (e) NaOH, MeOH, RT, overnight, 82%; (f) Oxalylchloride, DMF, DCM, 40 °C, 1 h; (g) NH$_2$OH, HCl, NaHCO$_3$, THF, RT, 1 h, 36%.

in SAHA and LBH-589 synthesis to introduce the hydroxamic acid moiety. Ester **29** is saponified and the resulting acid converted in acid chloride, which was reacted with hydroxylamine to obtain PXD-101 with a moderate yield of 36%.

Belinostat has also demonstrated additivity or synergism when combined with the cytotoxic agents carboplatin and paclitaxel [70] and been subsequently evaluated in a Phase II relapsed ovarian cancer studies in combination with these agents at a dose of 1000 mg/m^2/day [71]. The most commonly reported adverse events were nausea, fatigue, vomiting, and diarrhea. Belinostat is currently studied as single agent or in combination in several Phase I/II trials, either with oral or with i.v. administration [72].

4.4. SB939

Benzimidazole cinnamoylhydroxamic acids have recently been reported by S*BIO Pte, [73] a company which initially developed SB639 a potent HDACi (Fig. 5) [74].

The *in vitro* metabolism data suggested that SB639 was unstable in mouse and rat liver microsomes, and further efforts led to the identification SB939, a potent HDACi (HDAC1 IC$_{50}$ = 52 nM, GI$_{50}$ = 560 nM (colo205 cells)) [75] with improved metabolic stability. Scheme 6 describes chemical access to SB939. Starting with the condensation of *N,N*-diethylethylene diamine onto *trans*-4-chloro-3-nitrocinnamic acid at high temperature, the benzimidazole ring was prepared, after esterification of the acid function, by one pot reduction of the nitro moiety followed by reductive amination using butyraldehyde and cyclization in acidic medium. Hydroxamic acid function is introduced by careful addition of sodium methylate at low temperature on a mixture of the ester and an excess of hydroxylamine hydrochloride followed by a slow warm up to room temperature.

First results of a Phase I study evaluating SB939 in patients with advanced solid tumors have recently been presented [76] showing that the drug was rapidly absorbed and had a mean elimination half-life value of 8 h. Common adverse events in the 20 patients participating in the dose escalation were nausea, vomiting, and diarrhea with dose limiting toxicities of grade 3 being reported as fatigue, asymptomatic QT prolongation, and troponin T elevation (40 and 80 mg doses).

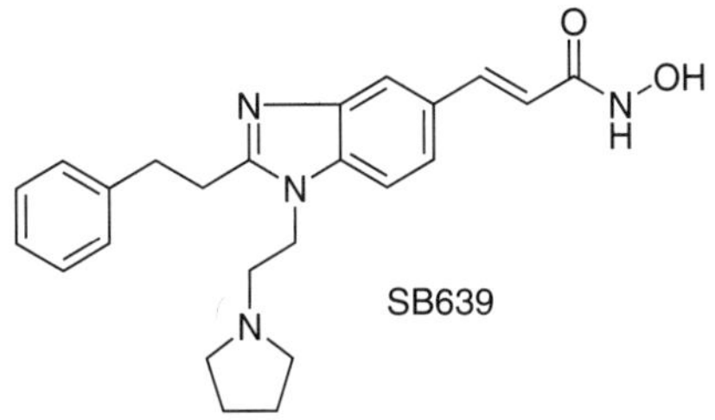

IC$_{50}$ = 52 nM (HDAC1)
GI$_{50}$ = 180 nM (colo 205)
$t_{1/2}$ = 3 min. (Mouse Liver Microsomes (5 µM))
$t_{1/2}$ = 1.7 h (nude mice, 10 mg/kg i.v.)
Bioavailability = 13% (nude mice)

Figure 5. Structure and biological activity of SB639.

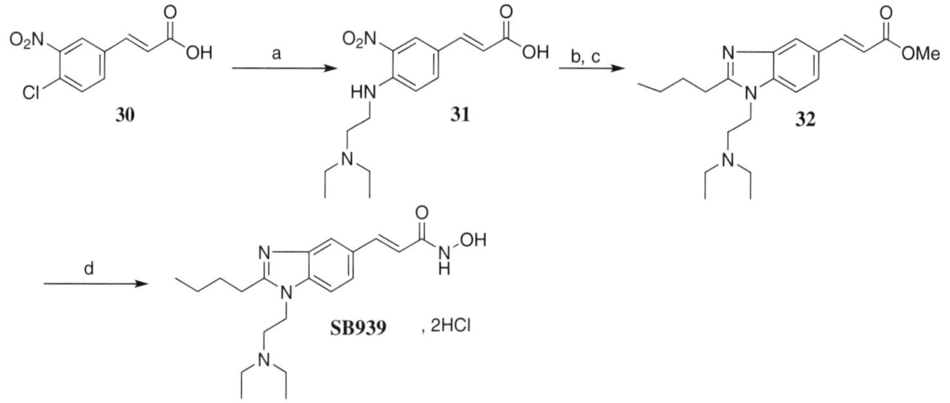

Scheme 6. Reagents and conditions: (a) H$_2$N–(CH$_2$)$_2$–N-(CH$_2$CH$_3$)$_2$, NEt$_3$, dioxane, 100 °C, 1–2 days; (b) H$_2$SO$_4$, CH$_3$OH, 80 °C; (c) nC$_4$H$_9$CHO, SnCl$_2$•2H$_2$O, AcOH, MeOH, 45 °C; (d) NH$_2$OH•HCl, NaOMe, MeOH, −78 °C to RT (no yield reported).

4.5. ITF-2357

Initially developed as anti-inflammatory and immunosuppressive agents [77], benzohydroxamic acids analogs of ITF-2357 have also shown HDAC inhibiting properties [78] with a nanomolar range potency on recombinant hHDAC isoforms (see Table 6) [79]. ITF-2357 inhibits cell growth in the same range (IC$_{50}$ = 495 nM in A549 and 73 nM in MDA-MB485).

ITF-2357 can be obtained in a five-steps synthesis starting from 2,6-naphtalene dicarboxylic acid 33 (Scheme 7).

Peptidic coupling of 33 using 3 equivalents of amine and only 1 equivalent of coupling reagents afforded monoamide 34 in 60% yield. Reduction of both acid and amide functions by LiAlH$_4$ provided primary alcohol 35 which was reacted with N,N'-disuccinimidylcarbonate and 4-aminobenzoic acid to obtain carbamate 36. Action of thionylchloride on 36 gave the corresponding acid chloride, which was not isolated but reacted *in situ* with hydroxylamine to provide ITF-2357 after acidification, as a chlorhydrate, with a moderate yield of 41%.

ITF-2357 has demonstrated activity on multiple myeloma (MM) and acute myelogenous leukemia (AML) cells *in vitro* and *in vivo* and apoptosis induction in these cell lines. It is being evaluated in MM and AML Phase I trials [80]. A Phase II study in Crohn disease showed that the compound is well tolerated in human.

4.6. PCI-024781

Pharmacyclics is developing CRA-024781, a broad spectrum HDACi initially from Celera. It is currently in Phase I/II clinical trials for solid tumors and lymphoma. PCI-024781 is a potent broad spectrum HDACi [81] (Table 7) and inhibits cell growth in the submicromolar range (IC$_{50}$ = 495 nM in A549 and 73 nM in MDA-MB485).

A convergent synthesis has been described in 8 steps in which intermediates 39 and 42 are built before being assembled by classical amide coupling (Scheme 8). Introduction of the dimethylamino group onto the 3-methylbenzofuryl-2-carboxylic acid is obtained following a sequence of esterification and

Table 6. Enzymatic Inhibitory Activity of ITF-2357 on HDAC Isoforms

	Class I			Class II		
rhHDAC isoform	1	2	3	4	6	7
EC$_{50}$ (nM)	28	56	21	52	27	173

Table 7. Enzymatic Inhibitory Activity of PCI-024781 on HDAC Isoforms

	Class I				Class II	
rhHDAC isoform	1	2	3 SMRT	8	6	10
K_i (nM)	7	19	8	82	17	24

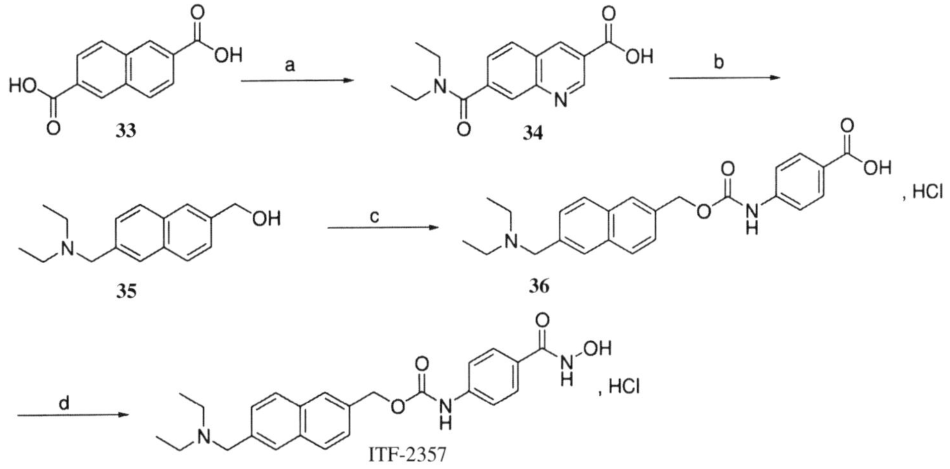

Scheme 7. Reagents and conditions: (a) Diethylamine, EDC, HOBT, DMF, RT, 2 h, 60%; (b) LiAlH₄, THF, reflux, 1 h, 79%; (c) 4-Aminobenzoic acid, N,N′-disuccinimidylcarbonate, CH₃CN, RT, overnight then acidification by HCl 64%; (d) SOCl₂, CHCl₃, reflux, 4 h then H₂NOH, HCl, Na₂CO₃, NaOH 1N, THF, H₂O, RT, overnight, acidification by HCl/Et₂O 41%.

radical bromination of the methyl group, and nucleophilic substitution of the bromine atom by the dimethylamino group. Saponification of **38** then regenerates the carboxylic acid bond to give **39**. Synthesis of compound **42** had been described in the literature [82,83] and starts by alkylation of the phenol **40** with bromomethylcyanide followed by reduction of

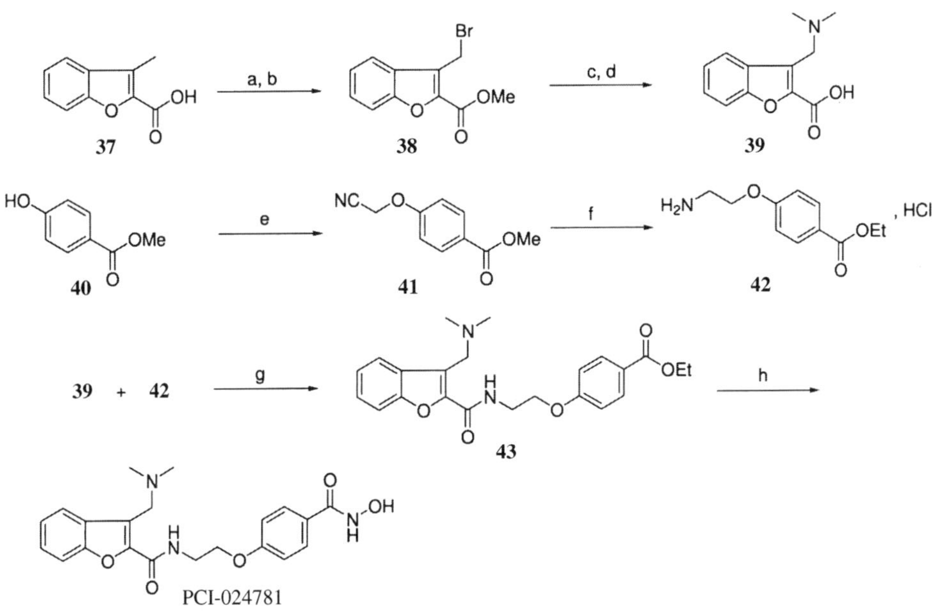

Scheme 8. Reagents and conditions: (a) DMF, oxalyl chloride, RT, 1 h, then MeOH, TEA, overnight; (b) NBS, AIBN, CCl₄, reflux, 3 h; (c) Dimethylamine, DMF, RT, 2 h, 56%; (d) MeOH, NaOH, 1 h 30 min then HCl; (e) BrCH₂CN, K₂CO₃, acetone, RT; (f) H₂, Pd/C, HCl, EtOH, Et₂O; (g) EDCI, HOBt, DMF, TEA; (h) MeOH, NaOH, NH₂OH, RT, overnight 48%.

the nitrile moiety in ethanol which results in both transesterification and formation of the amino moiety. Synthesis ends with the introduction of the hydroxamic moiety by action of hydroxylamine onto ester **43** in basic conditions as already reported for SAHA or LBH-589.

When PCI-024781 was evaluated in mice harboring HCT116 colon xenografts (i.v. administration at 80 mg/kg q.d. × 4 per week), an 82% reduction of tumor growth was observed. Preclinical studies have shown that PCI-024781 is synergistic in combination with the PARP inhibitor PJ34 and radiation in HCT116 cells. An oral capsule formulation has been developed and an oral bioavailability of 25% reported in a Phase I trial [84]. Pharmacyclics has also shown that this compound decreases RAD51 expression and is investigating this property as a marker of clinical efficacy [85].

4.7. R306465 and JNJ-26481585

4.7.1. R306465 the First-Generation HDACi

In the course of the search for novel, orally active HDAC inhibitors, Ortho-Biotech Oncology, Johnson & Johnson identified *N*-hydroxybenzamide **44** as an early lead compound, exhibiting micromolar activity on HDAC activity in HeLa cell nuclear extracts. In addition to submicromolar HDAC activity, the more favorable des-nitro analog **45** displayed moderate antiproliferative activity against human A2780 ovarian carcinoma cells, with an IC_{50} value of 1.9 µM.

Variation of the aromatic group, either through phenyl substitution or through replacement by various heterocycles, as shown in Table 8, suggests that an unsubstituted aromatic ring is required to position the hydroxamic acid into the active-site pocket. In their quest for R306365, chemists from Ortho-Biotech Oncology R&D speculated that the 1,4-disubstituted phenyl moieties in their hit was partly flanked by the two hydrophobic phenylalanine residues that constitute the wall of the narrow tubular pocket. In that respect, the degree of coplanarity of the aromatic moiety and the piperazine ring seemed to be a determining factor. The use of a 2,6-substituted pyridine ring in order to internalize the electron-withdrawing group by means of the aromatic nitrogen while removing any substitution that may disrupt the coplanarity of the bicyclic system, enhanced the activity to 0.01 µM. However, incorporation of a 3,6-pyridazine **47** or a 2,5-pyrazine **48** caused a drop in activity, in comparison to **46** (pyridine). Conversely, in the 2,5-pyrimidine **49**, the two electron-withdrawing aromatic nitrogen atoms are present *meta* to the hydroxamic acid, carrying lone pairs that are thought to allow for a minimal interplane angle. This compound possessed nanomolar inhibitory activity, which was believed to result from an optimal fit in the tubular pocket, possibly assisted by a π-stacking interaction with the phenylalanine side chains, although additional interactions between the two pyrimidine nitrogens and the peptide backbone or residues cannot be excluded. Importantly, only submicromolar concentrations of **49** are required to significantly induce the cyclin-dependent kinase inhibitor $p21^{waf1,cip1}$, a key denominator of the antiproliferative activity of HDAC inhibitors in tumor cells, with a MAD of 0.1 µM [86].

Variation of the diamine spacer in **49** revealed that this feature was amenable to variation of both length and directionality. While the inhibitory activity in the biochemical assay was invariably in the nanomolar range, none of these derivatives exhibited significantly improved antiproliferative activity, as illustrated by the micromolar IC_{50} values against the A2780 tumor cell line in Table 8.

Synthesis of **49** (R306465) is reported in Scheme 9 and most of the compounds depicted in Table 8 can be prepared in a similar way [87]. It starts from commercially available 4-chloro-2-methylthiopyrimidyl-5-carboxylic acid methyl ester that is dehalogenated by hydrogenation. Oxidation of the sulfur atom into **56** to the methylsulfone, a better leaving group prepared the field for condensation of benzyl-protected piperazine. Removal of protection by hydrogenation followed by reaction of the free aminogroup with 2-naphtylsulfonyle chloride afforded ester **59**. It is interesting to note that the strategy employed at Ortho-Biotech Oncology R&D differs from many other groups in that it introduces the hydroxamic acid function using

Table 8. Impact of Modification of the Aryl and the Linker Moiety on HDAC Enzymatic Inhibition

Compound	—Ar—	HDAC IC$_{50}$ (µM)	A2780 IC$_{50}$ (µM)	Compound	—Y—	HDAC IC$_{50}$ (nM)	A2780 IC$_{50}$ (µM)
44	2,4-disubstituted NO$_2$-phenyl	0.49	-	50	piperazine	6	0.07
45	1,4-phenyl	0.071	1.9	51	pyrrolidine-NH	14	5.96
46	2,5-pyridyl	0.01	0.29	52	piperidine-CH$_2$NH	8	3.34
47	3,6-pyridazyl	0.32	1.0	53	piperazine-NH	29	3.36
48	2,5-pyrazyl	0.069	0.86	54	piperazine-CH$_2$NH	9	2.04
49	2,5-pyrimidyl	0.006	0.07	55	hydroxypiperidine-CH$_2$NH	28	>10

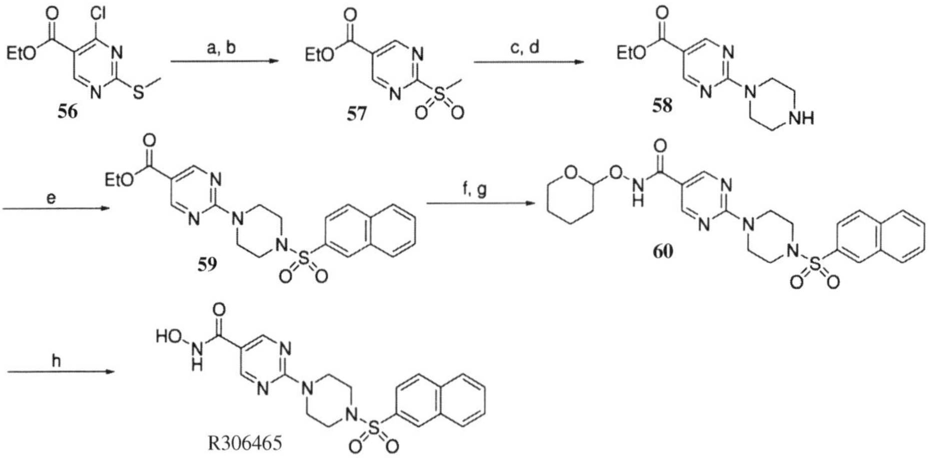

Scheme 9. Reagents and conditions: (a) H$_2$, NEt$_3$, Pd/C, THF, RT; (b) m-CPBA, DCM, 0 °C, 85% (2 steps); (c) N-Benzylpiperidine, K$_2$CO$_3$, CH$_3$CN, RT, 82;5%; (d) H$_2$, Pd/C 10%, EtOH, 50 °C; (e) 2-Naphtylsulfonyl chloride, TEA, DCM, RT, 30 min, 65%; (f) NaOH, MeOH/THF, RT, the acidification with HCl, 89%; (g) O-(Tetrahydro-2H-pyran-2-yl)hydroxylamine, EDC, HOBT, THF, CH$_2$Cl$_2$, RT, 82%; (h) TFA, DCM/MeOH, RT, 94%.

tetrahydropyranyl-protected hydroxylamine. Indeed, amide **60** and analogs are stable and can be easily purified on silica gel contrary to the final hydroxamic acid compounds. Removal of the THP protection in acidic media followed by crystallization constituted a clean method to obtain R306465.

In order to improve solubility, the naphtyl moiety was replaced by a series of functionalized di-aryl derivatives bearing solubility enhancing substituents. Although some examples showed considerable improvement of thermodynamic solubility over the naphtyl compound [88], upon oral administration of a single dose to rats, plasma concentrations for all these analogs were below the limit of quantification ($\leq$5 ng/mL), indicative of poor oral absorption. R306465, in contrast, displayed excellent systemic availability with an AUC of 2745 ng·h/mL. R306465 had an oral bioavailability of 20%, with an acceptable half-life of 3.4 h. The relatively short T_{max} of 0.5 h is indicative of dissolution rate controlled absorption. The transepithelial transport of the compound was later on confirmed in the CaCo-2 assay. The apparent permeability coefficients (P_{app}) of apical to basal = 9.42 × 10^{-6} cm/s and basal to apical = 8.11 × 10^{-6} cm/s, suggest a medium intestinal absorption. R306465 inhibited cell proliferation in all tumor lines tested, with IG$_{50}$ values ranging from 52 to 303 nM [86]. This antiproliferative effect was not affected by the genotypic status of these tumor cells, such as mutations in the p53 or Rb proteins.

R306465 induced potent dose-dependent inhibition of tumor growth in several models in immunodeficient mice including H460 lung and HCT116 colon carcinoma, resulting in maximal inhibition of tumor volume at 40 mpk (p.o., q.d.) of 69% and 64%, respectively, at the end of the study. Following successful preclinical evaluation, R306465 has been evaluated in a Phase I clinical trial in advanced cancer patients with solid malignancies.

4.7.2. JNJ-26481585, a Second-Generation HDACi

The discovery of JNJ-26481585 is very interesting from a strategic point of view. It shed light on a successful, nonclassical, way to identify new drug candidate. In the late optimization stage, medicinal chemists are often confronted with having to select a candidate to evaluate in xenograft antitumor studies *in vivo*, from a pool of compounds with very similar profiles. It often happens that these candidates that arise from the same chemical series, have *in vitro* profiles that are so similar, that selection is based on results in other assays (pharmacokinetic data, pharmacodynamic studies, and so on). These usually have a low throughput and if performed sequentially require a significant amount of time.

R306465, like most HDAC inhibitors in clinical development, induces histone acetylation only transiently in tumor tissue. In order to identify novel HDAC inhibitors with superior pharmacodynamic properties, Arts and colleagues have developed an animal model allowing noninvasive real-time evaluation of the response to HDAC inhibitors [89]. Human A2780 ovarian carcinoma cells were engineered to express ZsGreen fluorescent protein under control of the $p21^{waf1,cip1}$ promoter. Induction of fluorescence *in vivo* was found to accurately predict long-term antitumor activity of HDAC inhibitors. This model allowed Ortho-Biotech R&D to screen 140 potent pyrimidyl-hydroxamic acid analogs *in vivo*, and allowed real time analysis of the kinetic effect of drug substances *in vivo* (Fig. 6). This rapid *in vivo* evaluation of efficacy greatly reduced time-cycles and amount of compound needed: 50 mg per cpd (p.o. + i.v.)/5 days compared to 2 g per cpd/28 days. However, this strategy does aid in understanding which parameter (bioavailability, metabolism, and so on) must be modified in order to improve compound performance. Therefore, it is better suited for projects where a reference compound (internal or external) is available and then becomes a very powerful tool for comparison of very close analogs where the impact of small structural modifications can be detected.

These efforts resulted in the identification of JNJ-26481585, a novel "second generation" oral pan-HDAC inhibitor.

An improved synthesis of JNJ-26481585, suitable for large-scale production, has been recently reported [90] (Scheme 10). It also starts from commercially available 4-chloro-2-methylthiopyrimidyl-5-carboxylic acid methyl ester that is dehalogenated by hydro-

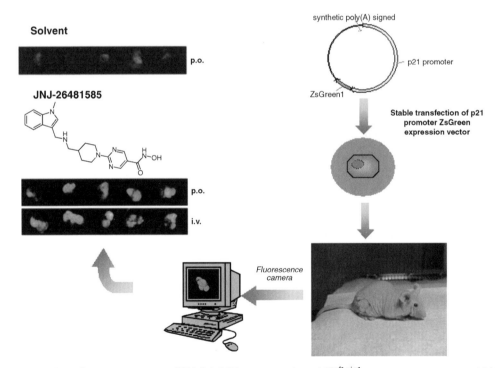

Figure 6. In order to generate an HDAC inhibitor-responsive $p21^{waf1,cip1}$ promotor construct, a 1.3 kb fragment containing the −1300 to +88 $p21^{waf1,cip1}$ promoter region was cloned into pGL3-basic-pZsGreen. Human A2780 ovarian carcinoma cells were stably transfected, and clones were selected and characterized. Mice were treated p.o. or i.v. for 5 days and fluorescence was evaluated on day 6 [85].

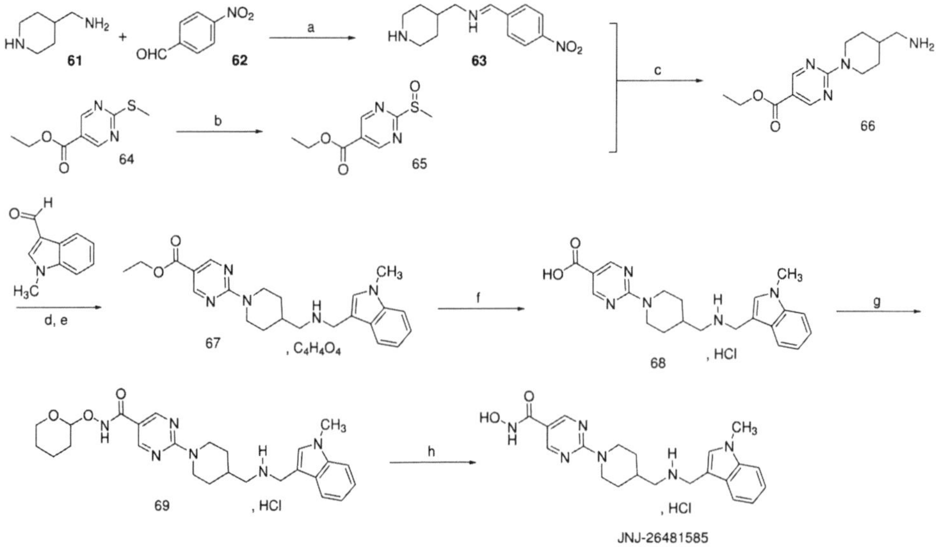

Scheme 10. Reagents and conditions: (a) EtOAc, 45 °C, 2 h; (b) mCPBA, EtOAc, 0 °C; (c) RT, overnight then acidification with HCl 2N, 74% (3 steps); (d) 1-Methyl-1H-indole-3-carboxaldehyde, reflux, overnight then NaBH$_4$, toluene, 10 °C, RT, 6 h; (e) Fumaric acid, EtOH, 50 °C then RT overnight, 73%; (f) NaOH, 2-methyltetrahydrofuran, H$_2$O, 1 h, 90 °C, 16 h, then acidification, 97%; (g) O-(Tetrahydro-2H-pyran-2-yl)-hydroxylamine, EDC, EtOAc, 5 h, 71%; (h) HCl, EtOH, 60 °C, 3 h, 77%.

genation to provide **64**. Then, **64** is not oxidized to the sulfone as for synthesis of R306465 but to the sulfoxide **65**. Crude **65** was then smoothly reacted with 4-piperidine methanamine **63** protected by the imine formed with *para*-nitrobenzaldehyde to provide **66** with good yields after acidic cleavage. Reductive amination of **66** with 1-methyl-1H-indole-3-carboxaldehyde provided ester **67** after isolation via fumaric acid salt formation. After saponification of ester **67**, as for R306465, JNJ-26481585 was obtained as an HCl salt, using THP-protected hydroxylamine to introduce a hydroxamic acid moiety.

JNJ-26481585 was found to be a potent pan-HDAC inhibitor (HDAC1 IC$_{50}$ = 0.16 nM) and demonstrated both acetylation of HDAC1 substrates (Histones) and HDAC6 substrates (Tubulin) in A2780 ovarian tumor cells. JNJ-26481585 inhibited tumor cell proliferation (IC$_{50}$ values 2–144 nM) and induced massive apoptosis in all tumor cell lines tested, irrespective of p53 or Ras mutation status [91]. Patient derived adriamycin-resistant breast carcinoma and taxol resistant nonsmall cell lung carcinoma (NSCLC) were also highly sensitive to JNJ-26481585. This potent single agent preclinical antitumoral activity combined with is favorable pharmacodynamic profile, makes JNJ-26481585 a promising "second-generation" HDAC inhibitor with potential applicability in a broad spectrum of both solid and hematological human malignancies. Developed as an HCl salt, JNJ-26481585 is currently undergoing Phase I clinical studies.

5. ANILIDES HDACi

5.1. CI-994 (Tacedinaline)

CI-994 or tacedinaline, initially identified as a metabolite of dinaline *in vivo* has been one of the first ortho-aminoanilide HDACi to be tested in clinical phases. It has the advantage of potentially being prepared with an excellent yield in a very short synthesis scheme as depicted below (Scheme 11) [92]. Commercially available *para*-acetamidobenzoic acid **70** is converted intermediately into the corresponding imidazole amide **71** that was not isolated but further reacted with orthoaminoaniline to provide CI-994 with a very good yield after crystallization.

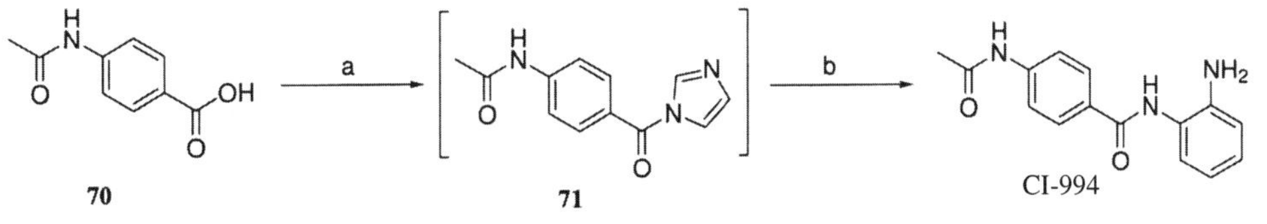

Scheme 11. Reagents and conditions: (a) Carbonyldiimidazole (CDI), THF, RT, 1 h; (b) orthoaminoaniline, TFA, RT, 16 h.

Phase II studies have been conducted in patients with metastatic renal cell carcinomas or advanced NSCLC with stable disease reported. However, no development has been reported from some time.

5.2. MS-275 (Entinostat)

SNDX-275 (entinostat), formely MS-275, is now developed by Syndax Pharmaceuticals under license from Bayer Schering Pharma for the potential treatment of Myeloid leukemia or solid tumors such as melanoma or breast cancer.

An improved synthesis of MS-275 has recently been published by Gediya et al. [92] in a two steps procedure (Scheme 12). Condensation of commercially available 3-(hydroxymethyl)pyridine **72** with 4-aminomethylbenzoic acid **73** at room temperature using CDI as source of carbonyl group provided **74** with a high yield of 91%. Treatment of **74** with CDI at reflux in THF gave a reactive imidazolide intermediate that is condensed *in situ* with 1,2-phenylenediamine to give MS-275 with 80% yield.

SNDX-275 inhibits Class I HDAC1 (HDAC1 $IC_{50} = 0.37\,\mu M$), HDAC3 (HDAC3 $IC_{50} = 0.5\,\mu M$) and only weakly HDAC8 ($IC_{50} = 63.4\,\mu M$) as well as Class II HDACs (HDAC4 $IC_{50} = 10.7\,\mu M$; HDAC10 $IC_{50} = 50.1\,\mu M$), but is inactive ($IC_{50} > 100\,\mu M$) on HDAC6. This compound has shown *in vivo* antitumor activity (tumor growth inhibition from 55% to 90%) in at least 13 different xenografts models when administered orally at doses ranging from 40 to 50 mg/kg [93]. The SNDX-275 core scaffold has also been used widely to identify other potential ZBG. One example worth highlighting describes efforts to replace the ortho-aminoaniline by a 3-diaminopyrazole moiety. Fifty-one examples were described in a recent patent although no quantitative data were provided [94].

5.3. MGCD-0103 (Mocetinostat)

In September 2007 and February 2008, the EMEA granted MGCD-0103 Orphan status for Hodgkins lymphoma and AML, respectively. MGCD-0103 inhibits Class I and IV HDACs (HDAC1 $IC_{50} = 0.15\,\mu M$; HDAC2 $IC_{50} = 0.29\,\mu M$; HDAC3 $IC_{50} = 1.76\,\mu M$; HDAC11 $IC_{50} = 0.59\,\mu M$) but is inactive against HDAC4 and HDAC8 ($IC_{50} > 20\,\mu M$) [95]. Mocetinostat, developed as the di-hydrobromide salt demonstrated significant antitumoral activity in several xenografts models. Tumor growth inhibition ranged from 64% to 93% when given orally at doses from 60 to 120 mg/kg.

MGCD-0103 can be obtained in a five-steps sequence (Scheme 13) [96] starting, respectively, from 4-acetylpyridine **75** that was submitted to Mannich type reaction to provide **76**, and 4-aminomethylbenzoic acid methylester **77** that was converted into

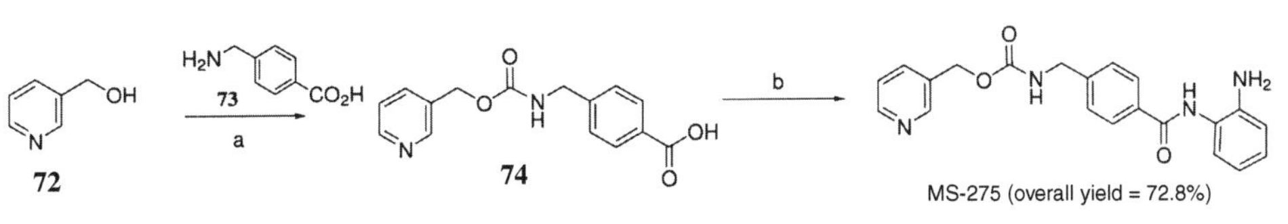

Scheme 12. Reagents and conditions: (a) CDI, THF, RT, 1 h then 4-aminomethyl benzoic acid DBU, Et3N, 91%; (b) CDI, THF, reflux, 1 h then orthoaminoaniline, TFA, RT, 16 h, 80%.

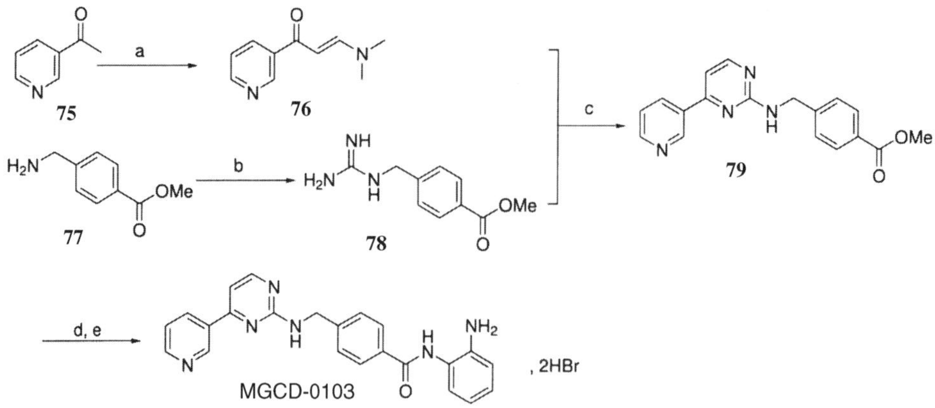

Scheme 13. Reagents and conditions: (a) Me$_2$NCH(CO$_2$Me)$_2$, neat, reflux, 85%; (b) 1-Pyrazole-carboxamidine, DMF, DIPEA, 77%; (c) i-PrOH, Molecular sieves 4A, 52%; (d) LiOH, THF, MeOH, H$_2$O, 40 °C, 45%; (e) Orthophenylene diamine, BOP, NEt$_3$, DMF, 66% then HBr.

4-guanidine **78**. Reaction of **76** with **78** in isopropanol in the presence of a dehydrating agent enabled to build the cap moiety of MGCD-0103 with a moderate yield. Saponification of the ester **79** followed by amide coupling with orthophenylene diamine ended the synthesis.

Methylgene scientists have extensively studied structure–activity relationships of N-(2-aminophenyl)-benzamides and reported their findings in recent publications [96–101]. Interestingly, the 4-aminomethylbenzamide core spacer, a feature also found in chidamide and MS-275 does not easily tolerate substitution to fit in the tube-like part of the HDAC catalytic site (Fig. 7).

On the other hand, the cap region accomodate a wide range of substituants, including incorporation of the nitrogen into bicycles, as can be shown by the nanomolar activity on HDAC1 of a selected panel of examples (Table 9, compounds **80** to **85**).

Although the hydroxamic acid moiety does not tolerate substitution on either the nitrogen or the oxygen atom, as demonstrated by the drastic decrease in potency on HDAC enzymatic activity, Moradei et al. [100] have shown that the orthophenylenediamine binding group can be substituted in para-position of the aniline, by phenyl, 2, or 3-thienyl, or 2-furyl moieties giving compounds with potency on HDAC1 ranging from 50 to 60 nM. It seems also possible to replace the NH$_2$ moiety of the aniline by a hydroxyl group although no such compound has been advanced to clinical trials.

MGCD-0103 has been tested as a single agent and in combination with other chemotherapeutics such as gemcitabine or azacitidine in various Phase I or Phase II clinical trials [72]. This includes myelodysplastic syndrome (MDS), AML, non-Hodgkin's lymphoma (NHL), chronic lymphocytic leukemia (CLL), B-cell lymphoma (BCL) as well as pancreatic tumor [71]. It was generally well tolerated, with the most common adverse events being fatigue, nausea, and vomiting. Pericarditis was also recently reported.

5.4. Chidamide

Shenzhen Chipscreen Biosciences is developing CS-055, also named chidamide, another 2-aminophenyl amide HDACi with the particularity of possessing a fluor atom on the phenyl ZBG. This compound has shown enzymatic and cellular activity in the same range as MS-275. Synthesis of CS-055 was reported in 2004, and starts with a Knoevenagel reaction to prepare the pyridinylacrylic acid **86** from 3-pyridylaldehyde (Scheme 14). The two amide bonds were then subsequently introduced using the same amide coupling reaction with carbonyldiimide to activate the acid bond [102,103].

Chidamide is currently undergoing Phase I clinical trials in China.

Compound	rHDAC1 IC$_{50}$ (µM)	MTT HCT 116 IC$_{50}$ (µM)
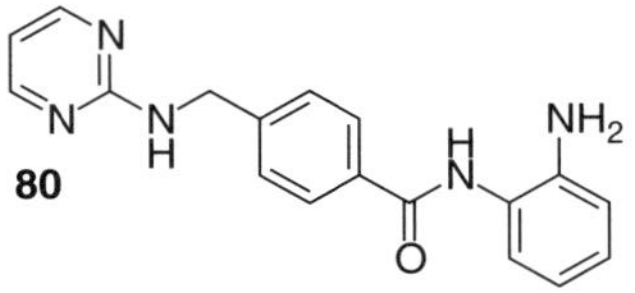 80	0.07	2
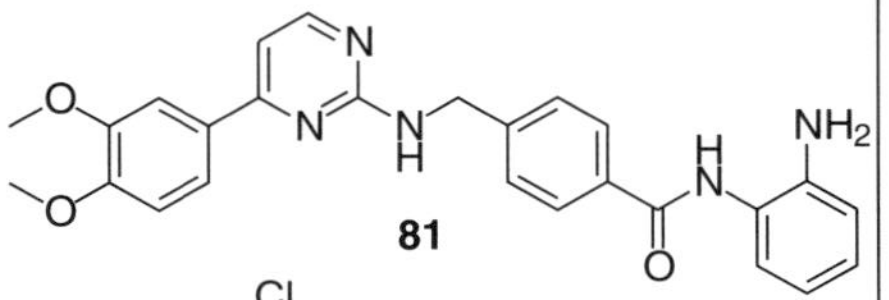 81	0.05	0.1
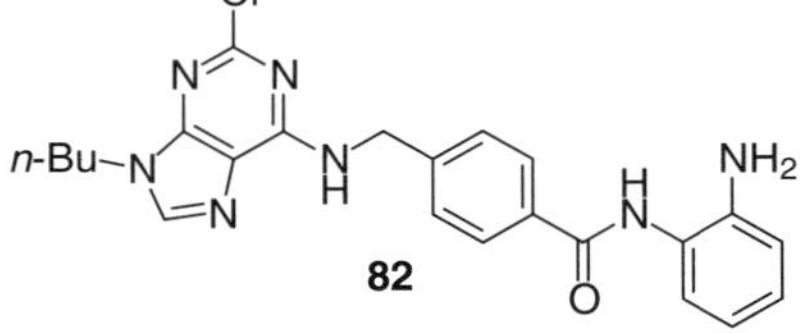 82	0.05	0.1

Compound	~HDAC1 IC$_{50}$ (µM)	MTT HCT 116 IC$_{50}$ (µM)
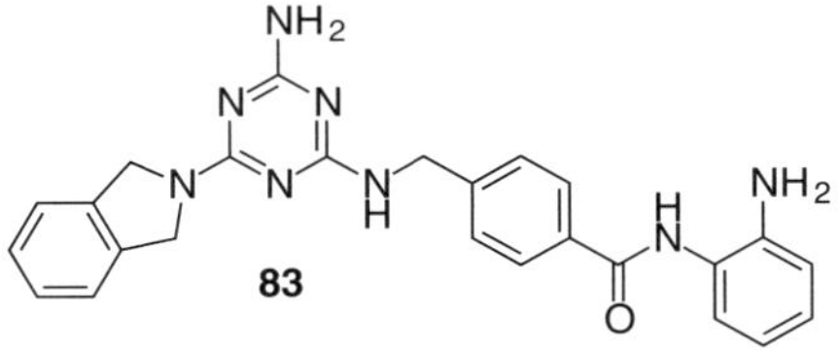 83	0.04	0.5
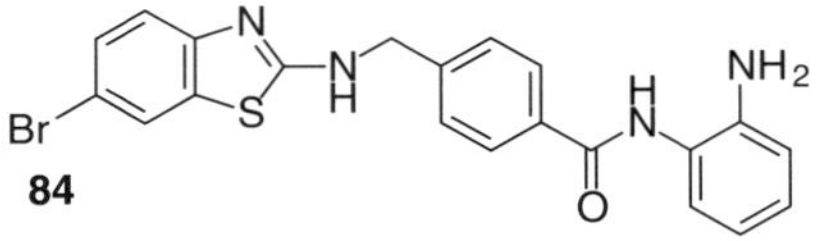 84	0.05	0.4
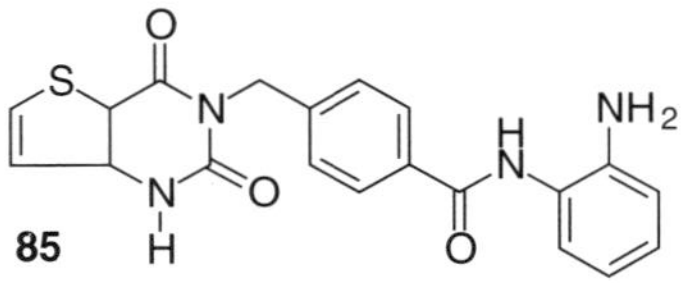 85	0.06	0.4

Figure 7. Some core features of MGCD-0103.

Table 9. Various Examples of Capping Groups Around MGCD-0103 Core Scaffold

Compound	r HDAC 1 IC50 (μM)	MTT HCT 116 IC50 (μM)
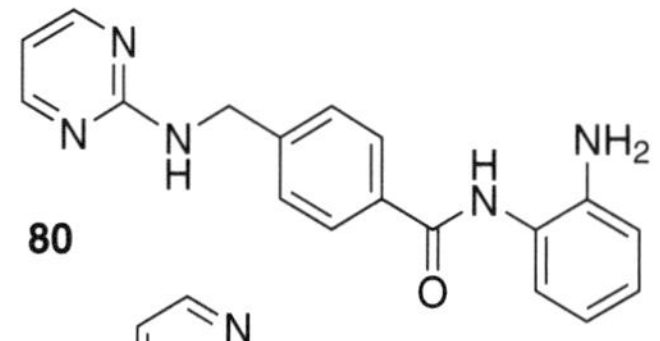 80	0.07	2
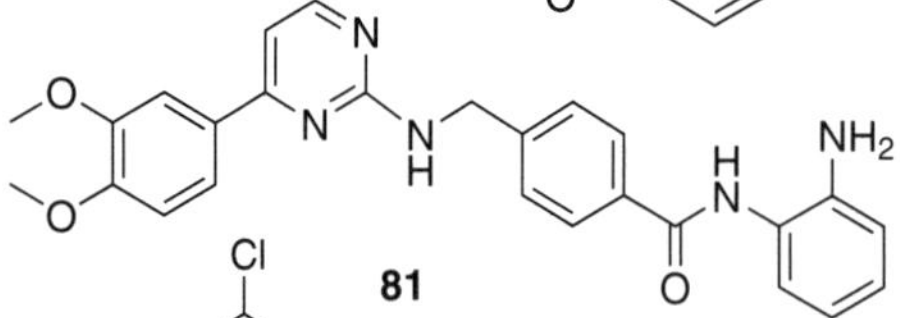 81	0.05	0.1
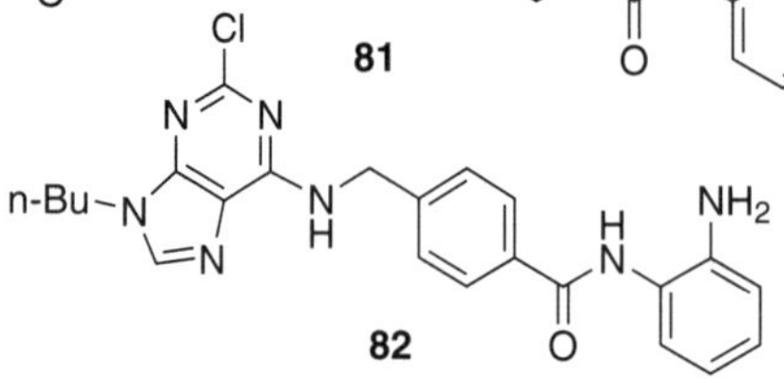 82	0.05	0.1
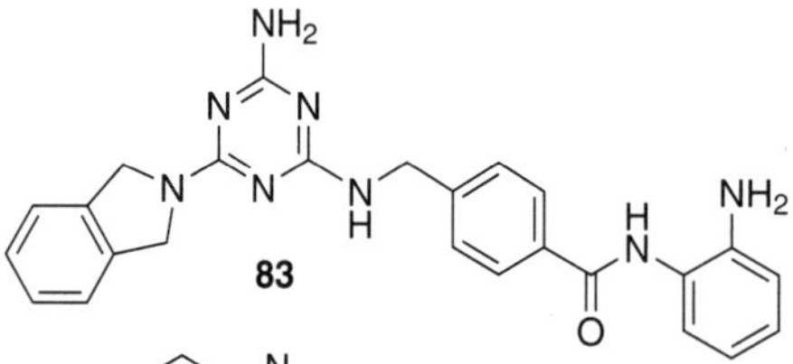 83	0.04	0.5
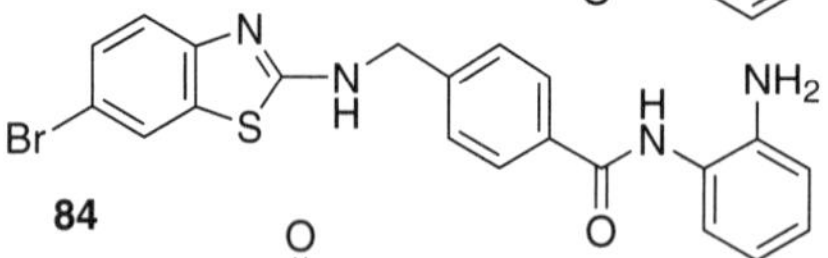 84	0.05	0.4
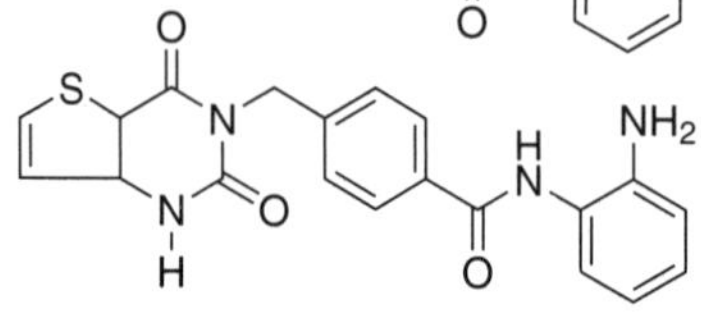 85	0.06	0.4

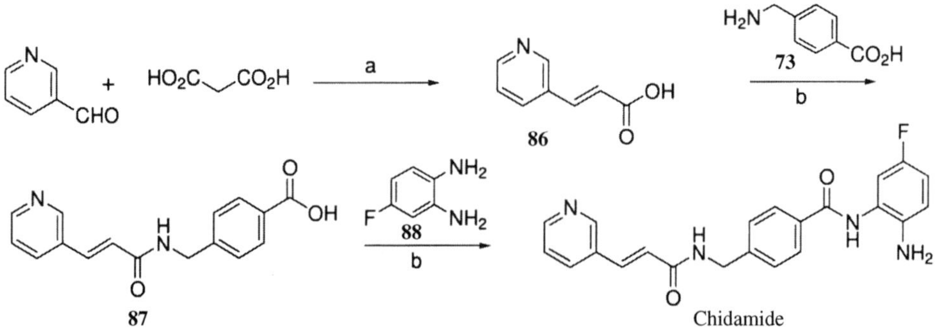

Scheme 14. Reagents and conditions: (a) DMAP, NEt$_3$, 1 h, reflux; (b) CDI, THF, 3 h, then NaOH, 4-aminomethyl-benzoic acid, RT, 8 h, 82%; (c) CDI, THF, 45 °C, 1 h, then 4-fluoro-1,2-phenylene diamine, CF$_3$CO$_2$H, RT, 24 h, 57%.

6. HDACi FROM NATURAL ORIGIN

6.1. FK-228 (Romidepsin)

FK-228 (Romidepsin, Gloucester Pharmaceuticals) is a Class I selective and potent bicyclic depsipeptide isolated from Chromobacterium Violaceum [104] with the particularity to be a natural prodrug [105] obtained by fermentation. In cells, the disulfur bond is reduced into a short methylthiol group and a longer 7-mercapto-4-heptenoic tail, a long chain able to insert its thiol moiety into the tube like pocket of HDACs catalytic site to interact with the zinc ion located at the bottom. Thiol binding to the zinc ion can be considered as relatively weak as compared to the tight chelation that has been characterized with some hydroxamic acids. However, the larger capping group is probably engaged in more significant binding interactions with the surface recognition motif than some small HA, as demonstrated by its high potency *in vitro* ($IC_{50} = 1.1$ nM) and in cell lines from a broad range of tumor types (mean $IC_{50} = 0.31$ nM) [106].

The first total synthesis of FK-228 was reported by Li et al. [107] but it is only 10 years later, in 2008, that the team of Williams and colleagues [108] improved key steps of this pathway based on a macrolactonization reaction to build the cyclic moiety. The same year, Ganesan and colleagues [109] published a slightly different disconnection approach (Fig. 8) using a macrolactamization to perform the cyclization. Both syntheses are completed by the disulfide formation using similar conditions (Fig. 8).

Ganesan and colleagues reported a convergent route using depsipeptide **89** prepared in 6 steps from acrolein and peptide **90** obtained in 6 steps. Scheme 15 depicts the final steps.

This methodology was also applied to prepare the first biologically active analogs of FK-228, thus broadening the SAR knowledge of this series [110]. Di Maro et al. replaced the double bond of the most synthetically challenging unit, the (3S,4E)-3-hydroxy-7-mercaptoheptenoic acid, with an amide bond [111] and employed solid phase synthesis to design

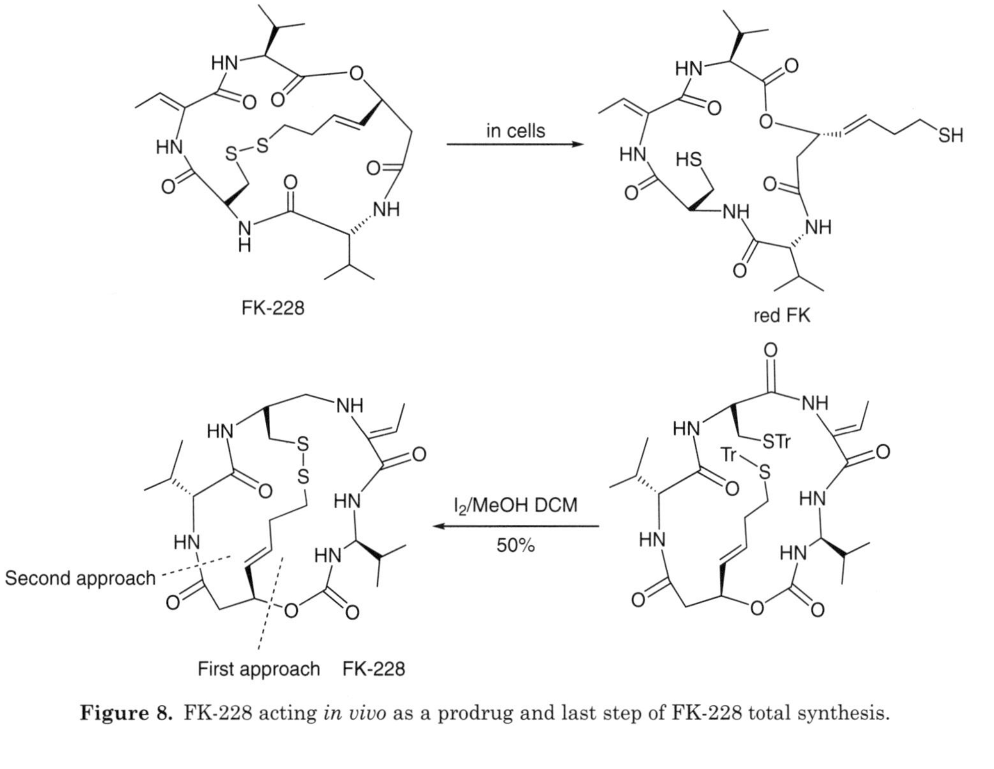

Figure 8. FK-228 acting *in vivo* as a prodrug and last step of FK-228 total synthesis.

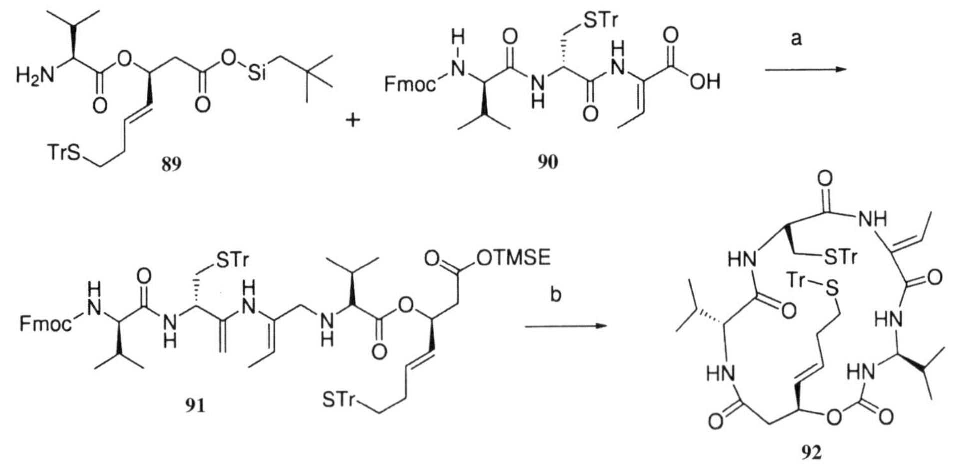

Scheme 15. Reagents and conditions: (a) PyBOP, i-Pr$_2$NEt, RT, overnight, 84%; (b) (i) Piperidine, DCM, RT, 1 h; (ii) Bu$_4$NBF, THF, 1 h 40 min, RT; (iii) HATU, i-Pr$_2$NEt, DCM, RT, overnight, 54%.

analogs showing antitumoral activity on various human cancer cells (Fig. 9).

Numerous Phase I and II trials have been conducted with Romidepsin. The agent demonstrated significant activity in a Phase I with CTCL to warrant further evaluation of its efficacy in CTCL in Phase II [72]. In June 2009, a New Drug Application submission for romidepsin in CTCL was accepted by the FDA for review. Romidepsin was also associated with a high number of potentially serious cardiac adverse events in patients with metastatic neuroendocrine tumor [112], which was another incentive to synthesize analogs of FK-228, but also increase interest in other structurally related despsipeptides of natural origin potentially inhibiting HDACs such as spiruchostatins [113]. Hopefully, this will result in identification of potent and selective new HDACi.

6.2. Largazole

Largazole, a depsipeptide recently isolated from the marine cyanobacterium *Symploca* sp. [114] has attracted attention due to its antiproliferative activities as demonstrated by a GI$_{50}$ of 7.7 nM in growth inhibition of the transformed mammary epithelial cell line MDA-MB-231 [115]. It consists (Fig. 10) of a macrocycle unit incorporating two, linked five-membered heterocycles, namely, a thiazole and a 4-methylthiazoline moiety, which is not often seen in natural products. It has been

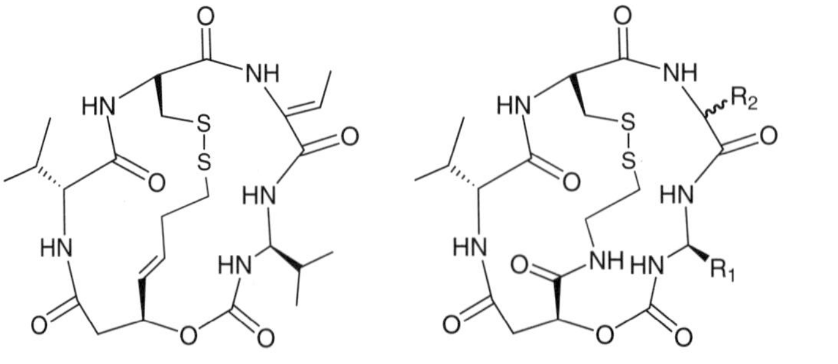

FK-228

93 R$_1$=R$_2$= Phe IC$_{50}$=0.26 µM (T24 bladder cell line)
94 R$_1$=R$_2$= 1-Nal IC$_{50}$=0.06 µM

Figure 9. FK-228 and amides analogs **93** and **94** [107].

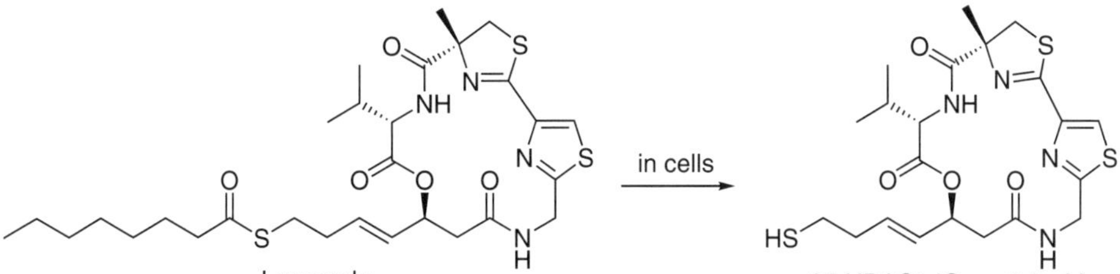

Figure 10. Structure of largazole and HDAC inhibition of deactylated product **95**.

Largazole

95 HDAC1 IC$_{50}$ = 2.5 nM
HDAC6 IC$_{50}$ = 380 nM

postulated that the 7-heptanoylsulfanyl-3-hydroxy-7-hept-4-eneoic acid unit might be responsible for the cellular activity through metabolization of the thioester into the thiol **95**, depicted in Fig. 10. A similar mercapto unit is the key-binding moiety of FK-228 that drives HDAC inhibition and potency in cells. Ying et al. [116] reported in 2008 the first total synthesis of largazole in eight steps with an overall yield of 19%. They also prepared the putative metabolite and confirmed its potency against HDAC1 in the nanomolar range.

The desire to further establish SAR and to provide novel structural variations of these unusual macrocycle inhibitors has recently been the focus of several groups with no less than six synthesis of largazole or largazole analogs reported in 2008 [116–121]. High overall total synthesis yields could be a real driver for this scaffold and it will be interesting to follow both medicinal chemistry efforts and further profiling of related analogs in the near future.

7. FATTY ACIDS

This review would not be complete without mentioning valproic acid (VPA), a molecule already on the market for a long time for the treatment of epilepsy, bipolar disorders, and migraine [122]. Anticancer activity of VPA is currently being investigated by TopoTarget Germany AG in particular for the potential treatment of familial adenomatous polyposis (FAP), and has been postulated to occur through HDAC inhibition. VPA is inhibiting rhHDACs 1, 2, 3, and 8 in the millimolar range contrasting with the other HDACi in clinic which are several order of magnitude more potent *in vitro* [79]. However, this millimolar range also corresponds well with the clinical dose range. Valproic acid is also reported to interfere with multiple processes in cancer cells [123], which probably contributes to its therapeutic potential.

8. CONCLUSION AND PERSPECTIVES

As can be deduced from the examples above, one additional interesting property of the HDACi is their synthetic accessibility based on the relatively short number of steps required for their synthesis. Furthermore, the synthesis of HDACi from natural origin has been recently improved, thus enabling further exploration of the depsipeptide scaffold.

So far, HDAC inhibitors have shown clear potency in CTCL, which remains a rare disease. However, this has generated high hopes for use in a wide range of other cancer indications as single agent therapy. Most of the inhibitors presented above have also been evaluated in combination with other therapeutic agents to optimize or maximize their efficacy. There is increasing preclinical evidence that HDACi can synergize with cytotoxic agents (topoisomerase inhibitors, DNA-damaging agents), tyrosine kinase inhibitors (erlotinib, gefitinib, imatinib) inhibitors of the proteasome, and DNA-hypomethylating agents or ionizing radiations, for example [124]. Some of these preclinical studies have led to the investigation of SAHA, PXD-101, LBH-589, MGCD-0103, and FK-228 in combination with well-known anticancer agents, such as 5-fluorouracil, paclitaxel, erlotinib, bortezomib, gemcitabine, and others, in Phase II clinical trials.

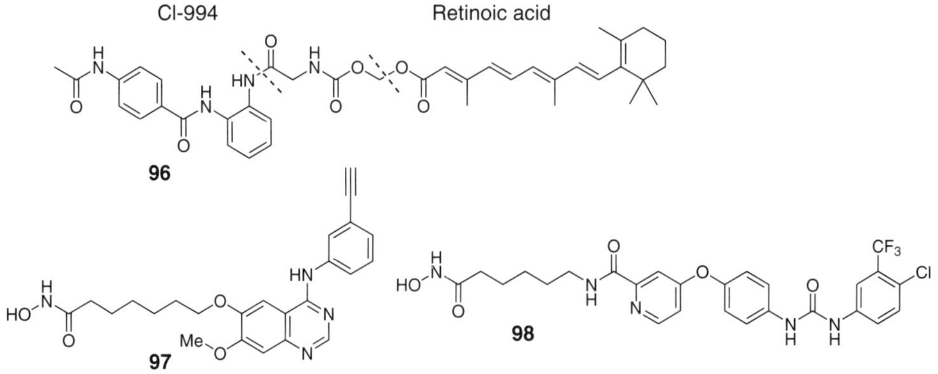

Figure 11. Examples of combining HDAC inhibition with another therapeutic target into a single molecule.

It is also interesting to mention the development of innovative combination strategies. For example, combining an HDAC inhibitor with all-*trans* retinoic acid (ATRA) in a single molecule linked using a prodrug moiety (**96**) (Fig. 11). This concept aims at delivering the two agents to the site of action simultaneously in order to synergize to overcome resistance to retinoids [125]. Recent patents from Curis, Inc. described, them, the synthesis of single agents combining both features of HDAC and EGFR (**97**), or VEGFR2 inhibitors (**98**) [126,127].

It will be interesting to see if the reported enzymatic activity could be translated into *in vivo* efficacy.

As it happens many times in the scientific world, the quest for HDACi was initially based on an appealing hypothesis. This triggered a multitude of studies that increased our understanding to a level that revealed that the picture is not so simple.

Today, the HDACi on the market or in late clinical phases are pan-HDACi (they inhibit all HDACs nonspecifically, e.g., SAHA, LBH-589, and PXD-101) or are only broadly selective for Class I HDACs. For example, depsipeptide preferentially inhibits HDACs 1 and 2, MS-275 inhibits HDACs 1, 2, 3, and Class IIa HDAC9, and MGCD-0103 inhibits HDAC1, 2, 3, and 11. Therefore, information collected from the numerous clinical trials performed with these agents, although already very useful, draw a partial picture only as they may also disrupt additional multiple cellular processes that depend on protein acetylation [128]. The design of isoform selective HDAC inhibitors has faced two main hurdles. First, structural information is incomplete. Only structural information for human HDAC8, 7 and very recently HDAC4 structure have been reported [129]. This has posed a major challenge to rational drug design. Second, data relative to the isoform selectivity profile of the many HDACi described in the literature is limited. This is probably due to lack of commercial availability or difficulties in expressing human HDAC isoforms. Nevertheless, our understanding of the biological role of the different HDACs is increasing aided by medicinal chemistry efforts to design selective inhibitors as can be seen from the following, nonexhaustive, selection of examples. The discovery of tubacin, (Fig. 12) [130,131], the first HDAC6 inhibitor was completed in 2007 by the design of a novel series of HDACi (**99**) showing selective HDAC6 inhibition as compared to HDAC1 and 4 [132]. Moreover, Kozikowski et al. [133], used nitrile oxide cycloaddition chemistry to prepare hydroxamic acid inhibitors containing a phenylisoxazole moiety as the capping group (**100**) with potency in the picomolar range toward HDAC6 and excellent selectivity toward other HDACs.

Recently, Balasubramanian et al. [25] described PCI-34051, a specific HDAC-8 inhibitor (HDAC8 $IC_{50} = 10$ nM, HDAC1, 2, 3, 6, 10 $IC_{50} > 2\,\mu M$)). A group from Merck Research Laboratories at Rome, Italy identified 5-(trifluoroacetyl)thiophene-2-carboxamides derivatives as selective HDAC4 and 6 inhibitors, thus providing tools to further study the

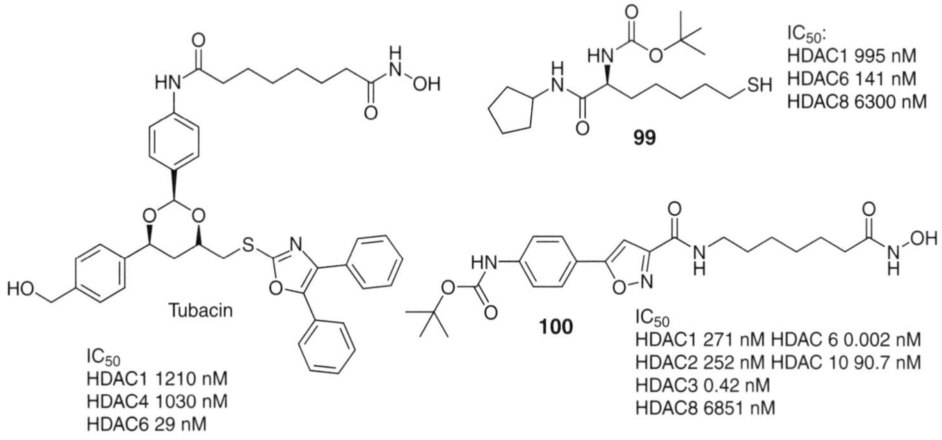

Figure 12. Examples of HDAC6 selective inhibitors.

therapeutic impact of Class II HDAC inhibition [134]. We are currently witnessing the use of rational design of increasing number of chemical features dedicated to increase HDAC isoform selectivity by targeting either the ZBG or the cap region as reported by two recent reviews [135,136].

Hydroxamic acids have been the pioneers and their general potency as ZBG guaranties that their contribution to the discovery of new HDAC inhibitors will remain relevant in the coming years. Indeed, as the hydroxamic acid function is responsible for tight binding interactions with the zinc ion and the protein, it drives potency and allows more space for chemical variations of the cap region. We will likely also witness the emergence in clinical trials of agents bearing new zinc binding moieties aimed at selective inhibition of specific HDACs, much like the benzamide-type compounds have demonstrated selectivity for Class I HDACs. It will be very interesting to correlate their overall deacetylation inhibition profile with the therapeutic advantages generated to patients. There is great hope that a new generation of HDACi showing increased isoform selectivity compared to the agents known today may have more limited side effects and ultimately improve patient life. The HDAC inhibition field is reaching its maturity and chances are high that we will soon see patients benefiting from several new HDACi. These will not only be anticancer agents but also be probably target novel therapeutic indications.

9. ACKNOWLEDGMENTS

The authors would like to thank everyone within Johnson & Johnson group of companies who have contributed to increase scientific knowledge on HDACi over the years. We would also like to thank Christèle Vignon for her great support in editing this manuscript.

REFERENCES

1. Mottet D, Castronovo V. Les histones deacetylases: nouvelles cibles pour les thérapies anticancereuses. M S Med Sci 2008;24:742–6.

2. Remiszewski SW. Recent advances in the discovery of small molecule histone deacetylase inhibitors. Curr Opin Drug Discov Dev 2002;5(4):487–499.

3. Johnstone RW. Histone-deacetylase inhibitors: novel drugs for the treatment of cancer. Nat Rev 2002;1:287–299, DOI: 10.1038/nrd772.

4. Marks PA, Richton VM, Rifkind RA. Histone deacetylase inhibitors: inducers of differentiation or apoptosis of transformed cells. J Natl Cancer Inst 2000;92:1210–1216.

5. Bolden JE, Peart MJ, Johnstone RW. Anticancer activities of histone deacetylase inhibitors. Nat Rev Drug Discov 2006;5(9):769–784, DOI: 10.1038/nrd2133.

6. Timmermann S, Lehrmann H, Polesskaya A, Harel-Bellan A. Histone acetylation and disease. Cell Mol Life Sci 2001;58(5/6):728–736.
7. ten Holte P, Van Emelen K, Janicot M, Fong PC, de Bono JS, Arts J. HDAC inhibition in cancer therapy: an increasingly intriguing tale of chemistry, biology and clinical benefit. Top Med Chem 2007;1:293–331.
8. Ocker M, Schneider-Stock R. Histone deacetylase inhibitors: signalling towards p21cip1/waf1. Int J Biochem Cell Biol 2007;39(7–8):1367–1374, DOI: 10.1016/j.biocel.2007.03.001.
9. Meunier D, Seiser C. Histone deacetylase 1. In: Verdin E, Editor. Histone deacetylases: transcriptional regulation and other cellular functions. Totowa, NJ: Humana Press, Inc; 2006.
10. Senese S, Zaragoza K, Minardi S, Muradore I, Ronzoni S, Passafaro A, Bernard L, Draetta GF, Alcalay M, Seiser C, Chiocca S. Role for histone deacetylase 1 in human tumor cell proliferation. Mol Cell Biol 2007;27(13):4784–4795.
11. Zupkovitz G, Tischler J, Posch M, Sadzak I, Ramsauer K, Egger G, Grausenburger R, Schweifer N, Chiocca S, Decker T, Seiser C. Negative and positive regulation of gene expression by mouse histone deacetylase 1. Mol Cell Biol 2006;26(21):7913–7928.
12. Kazantsev AG, Thompson LM. Therapeutic application of histone deacetylase inhibitors for central nervous system disorders. Nat Rev Drug Discov 2008;7:854–868, DOI: 10.1038/nrd2681.
13. Sun Y, Reddy P. Histone deacetylase inhibitors: novel immunomodulators. Curr Enzyme Inhib 2007;3(3):207–215.
14. Marks PA. Discovery and development of SAHA as an anticancer agent. Oncogene 2007; 1351–1356, DOI: 10.1038/sj.onc.1210204.
15. Dokmanovic M, Clarke C, Marks PA. Histone deacetylase inhibitors: overview and perspectives. Mol Cancer Res 2007;5(10):981–989, DOI: 10.1158/1541-7786.
16. De Ruijter AJ.M, van Gennip AH, Caron HN, Kemp S, van Kuilenburg AB.P. Histone deacetylases (HDACs): characterization of the classical HDAC family. Biochem J 2003;370(3):737–749.
17. Gregoretti IV, Lee YM, Goodson HV. Molecular Evolution of the histone deacetylase family: functional implications of phylogenetic analysis. J Mol Biol 2004;338(1):17–31.
18. Michan S, Sinclair D. Sirtuins in mammals: insights into their biological function. Biochem J 2007;404:1–13, DOI: 10.1042/BJ20070140.
19. Jung-Hynes B, Nihal M, Zhong W, Ahmad N. Role of sirtuin histone deacetylase Sirt1 in prostate cancer: a target for prostate cancer management via its inhibition. J Biol Chem 2009;284:3823–3832, DOI: 10.1074/jbc.M807869200.
20. Neugebauer RC, Sippl W, Jung M. Inhibitors of NAD^+ dependent histone deacetylases (sirtuins). Curr Pharm Design 2008;14:562–573, DOI: 10.2174/138161208783885380.
21. Napper AD, Hixon J, McDonagh T, Keavey K, Pons JF, Barker J, Yau WT, Armouzegh P, Flegg A, Hamelin E, Thomas RJ, Kates M, Jones S, Navia MA, Saunders JO, Distefano PS, Curtis R. Discovery of indoles as potent and selective inhibitors of the deacetylase SIRT1. J Med Chem 2005;48:8045–8054, DOI: 10.1021/jm050522v.
22. Foglietti C, Filocamo G, De Rinaldis E, Lahm A, Cortese R, Steinkuhler C. Dissecting the biological function of *Drosophila* histone deacetylase by RNA interference and transcriptional profiling. J Biol Chem 2006;281:17968–17976.
23. Lagger G, O'Carrol D, Rembold M, Khier H, Tishler J, Weitzer G, Schuettengruber B, Hauser C, Brunmeir R, Jenuwein T, Seiser C. Essential function of histone deacetylase 1 in proliferation control and CDK inhibitor repression. EMBO J 2002;21:2672–2681, DOI: 10.1093/emboj/21.11.2672.
24. Harms KL, Chen X. Histone deacetylase 2 modulates p53 transcriptional activities through regulation of p53-DNA binding activity. Cancer Res 2007;67(7):3145–3152, DOI: 10.1158/0008-5472.
25. Balasubramanian S, Ramos J, Luo W, Sirisawad M, Verner E, Buggy JJ. A novel histone deacetylase 8 (HDAC8)-specific inhibitor PCI-34051 induces apoptosis in T-cell lymphomas. Leukemia 2008;22:1026–1034, DOI: 10.1038/leu.2008.9.
26. Wilson AJ, Byun DS, Nasser S, Murray LB, Ayyanar K, Arango D, Figueroa M, Melnick A, Kao GD, Augenlicht LH, Mariadason JM. HDAC4 promotes growth of colon cancer cells via repression of p21. Mol Biol Cell 2008;19(10):4062–4075.
27. Iga J, Ueno S, Yamauchi K, Numata S, Kinouchi S, Tayoshi-Shibuya S, Song H, Ohmori T. Altered HDAC5 and CREB mRNA expressions in the peripheral leukocytes of major depression. Prog Neuropsychopharmacol Biol Psychiatry 2007;31(3):628–632, DOI: 10.1016/j.pnpbp.2006.12.014.

28. Mottet D, Bellahcene A, Pirotte S, Waltregny D, Deroanne C, Lamour V, Lidereau R, Castronovo V. Histone deacetylase 7 silencing alters endothelial cell migration, a key step in angiogenesis. Circ Res 2007;101(12): 1237–1246.
29. Chang S, Young BD, Li S, Qi X, Richardson JA, Olson EN. Histone deacetylase 7 maintains vascular integrity by repressing matrix metalloproteinase 10. Cell 2006;126(2):321–334, DOI: 10.1016/j.cell.2006.05.040.
30. Bali P, Pranpat M, Bradner J, Balasis M, Fiskus W, Guo F, Rocha K, Kumaraswamy S, Boyapalle S, Atadja P, Seto E, Bhalla K. Inhibition of histone deacetylase 6 acetylates and disrupts the chaperone function of heat shock protein 90: a novel basis for antileukemia activity of histone deacetylase inhibitors. J Biol Chem 2005;280(29):26729–26734.
31. Zhang Y, Li N, Caron C, Matthias G, Hess D, Khochbin S, Matthias P. HDAC 6 interacts with and deacetylates tubulin and microtubules in vivo. EMBO J 2003;22:1168–1179, DOI: 10.1093/emboj/cdg115.
32. Tran AD.A, Marmo TP, Salam AA, Che S, Finkelstein E, Kabarriti R, Xenias HS, Mazitschek R, Hubbert C, Kawaguchi Y, Sheetz MP, Yao TP, Bulinski JC. HDAC6 deacetylation of tubulin modulates dynamics of cellular adhesions. J Cell Sci 2007;120(8):1469–1479.
33. Lee YS, Lim KH, Guo X, Kawaguchi Y, Gao Y, Barrientos T, Ordentlich P, Wang XF, Counter CM, Yao TP. The cytoplasmic deacetylase HDAC6 is required for efficient oncogenic tumorigenesis. Cancer Res 2008;68(18): 7561–7569.
34. Valenzuela-Fernandez A, Cabrero JR, Serrador JM, Sanchez-Madrid F. HDAC6: a key regulator of cytoskeleton, cell migration and cell-cell interactions. Trends Cell Biol 2008; 18(6):291–297, DOI: 10.1016/j.tcb.2008.04.003.
35. Finnin MS, Donigian JR, Cohen A, Richon VM, Rifkind RA, Marks PA, Breslow R, Pavletich NP. Structures of a histone deacetylase homologue bound to the TSA and SAHA inhibitors. Nature 1999;401:188–193, DOI: 10.1038/43710.
36. Somoza JR, Skene RJ, Katz BA, Mol C, Ho JD, Jennings AJ, Luong C, Arvai A, Buggy JJ, Chi E, Tang J, Sang BC, Verner E, Wynands R, Leahy EM, Dougan DR, Snell G, Navre M, Knuth MW, Swanson RV, McRee DE, Tari LW. Structural snapshots of human HDAC8 provide insights into the Class I histone deacetylases. Structure 2004;12(7):1325–1334, DOI: 10.1016/j.str.2004.04.012.
37. Vannini A, Volpari C, Filocamo G, Casavola EC, Brunetti M, Renzoni D, Chakravarty P, Paolini C, De Francesco R, Gallinari P, Steinkuehler C, Di Marco S. Crystal structure of a eukaryotic zinc-dependent histone deacetylase, human HDAC8, complexed with a hydroxamic acid inhibitor. Proc Natl Acad Sci USA 2004;101(42):15064–15069.
38. KrennHrubec K, Marshall BL, Hedglin M, Verdin E, Ulrich SM. Design and evaluation of "Linkerless" hydroxamic acids as selective HDAC8 inhibitors. Bioorg Med Chem Lett 2007;17:2874–2878, DOI: 10.1016/j.bmcl.2007.02.064.
39. Schuetz A, Min J, Allali-Hassani A, Schapira M, Shuen M, Loppnau P, Mazitschek R, Kwiatkowski NP, Lewis TA, Maglathin RL, McLean TH, Bochkarev A, Plotnikov AN, Vedadi M, Arrowsmith CH. Human HDAC7 harbors a Class IIa histone deacetylase-specific zinc binding motif and cryptic deacetylase activity. J Biol Chem 2008;283(17): 11355–11363.
40. Miller TA, Witter DJ, Belvedere S. Histone deacetylase inhibitors. J Med Chem 2003;46 (24):5097–5116, DOI: 10.1021/jm0303094.
41. Paris M, Porcelloni M, Binaschi M, Fattori D. Histone deacetylase inhibitors: from bench to clinic. J Med Chem 2008;51(6):1505–1529, DOI: 10.1021/jm7011408.
42. Breslow R, Marks PA, Rifkind RA, Jursic B, Novel potent inducers of terminal differentiation and methods of use thereof. 1993; WO93107148.
43. Richon VM, Emiliani S, Verdin E, Webb Y, Breslow R, Rifkind RA, Marks PA. A class of hybrid polar inducers of transformed cell differentiation inhibits histone deacetylases. Proc Natl Acad Sci USA 1998;95 (6):3003–3007.
44. Wong JC, Cote AS, Dienemann EA, Gallagher K, Ikeda C, Moser J, Rajniak P, Reed RA, Starbuck C, Tung HH, Wang Q, Cohen BM, Capodanno VR, Sell B, Miller TA, Formulations of suberoylanilide hydroxamic acid and methods for producing same. 2006; WO 2006127319.
45. Miller T. Histone deacetylase inhibitors: bench to bedside, data presented at international conference, CHI's first annual HDAC inhibitors, discovery on target, Oct 16, 2007. Boston MA.
46. Duvic M, Vu J. Vorinostat in cutaneous T-cell lymphoma. Drugs Today 2007;43(9):585–599 DOI: 10.1358/dot.2007.43.9.1112980.

47. Duvic M, Vu J. Vorinostat: A new oral histone deacetylase inhibitor approved for cutaneous T-cell lymphoma. Expert Opin Investig Drugs 2007;16(7):1111–1120, DOI: 10.1517/13543784.16.7.1111.
48. Kelly WK, Marks PA. Drug insight: Histone deacetylase inhibitors-development of the new targeted anticancer agent suberoylanilide hydroxamic acid. Nat Clin Paract Oncol 2005;2:150–7, DOI: 10.1038 ncponc0106.
49. Desai D, Das A, Cohen L, el-Bayoumy K, Amin S. Chemopreventive efficacy of suberoylanilide hydroxamic acid (SAHA) against 4-(methylnitrosamino)-1-(3-pyridyl)-1-butanone (NNK)-induced lung tumorigenesis in female A/J mice. Anticancer Res 2003;23:499–503.
50. Belvedere S, Witter DJ, Yan J, Secrist JP, Richon V, Miller TA. Aminosuberoyl hydroxamic acids (ASHAs): a potent new class of HDAC inhibitors. Bioorg Med Chem Lett 2007;17(14):3969–3971, DOI: 10.1016/j.bmcl.2007.04.089.
51. Suzuki T, Nagano Y, Kouketsu A, Matsuura A, Maruyama S, Kurotaki M, Nakagawa H, Miyata N. Novel inhibitors of human histone deacetylases: design, synthesis, enzyme inhibition, and cancer cell growth inhibition of SAHA-based non-hydroxamates. J Med Chem 2005;48(4):1019–1032, DOI: 10.1021/jm049207j.
52. Gu W, Nusinzon I, Smith RD, Horvath CM, Silverman RB. Carbonyl-and sulfur-containing analogs of suberoylanilide hydroxamic acid: potent inhibition of histone deacetylases. Bioorg Med Chem 2006;14(10): 3320–3329.
53. Dehmel F, Ciossek T, Maier T, Weinbrenner S, Schmidt B, Zoche M, Beckers T. Trithiocarbonates-exploration of a new head group for HDAC inhibitors. Bioorg Med Chem Lett 2007;17(17):4746–4752, DOI: 10.1016/j.bmcl.2007.06.063.
54. Dehmel F, Weinbrenner S, Julius H, Ciossek T, Maier T, Stengel T, Fettis K, Burkhardt C, Wieland H, Beckers T. Trithiocarbonates as a novel class of HDAC inhibitors: SAR studies, isoenzyme selectivity, and pharmacological profiles. J Med Chem 2008;51(13): 3985–4001, DOI: 10.1021/jm800093c.
55. Suzuki T, Kouketsu A, Matsuura A, Kohara A, Ninomiya SI, Kohda K, Miyata N. Thiol-based SAHA analogues as potent histone deacetylase inhibitors. Bioorg Med Chem Lett 2004;14(12):3313–3317, DOI: 10.1016/j.bmcl.2004.03.063.
56. Suzuki Takayoshi Miyata Naoki Non-hydroxamate histone deacetylase inhibitors. Curr Med Chem 2005;12(24):2867–2880.
57. Frey RR, Wada CK, Garland RB, Curtin ML, Michaelides MR, Li J, Pease LJ, Glaser KB, Marcotte PA, Bouska JJ, Murphy SS, Davidsen SK. Trifluoromethyl ketones as inhibitors of histone deacetylase. Bioorg Med Chem Lett 2002;12(23):3443–3447.
58. Wada CK, Frey RR, Ji Z, Curtin ML, Garland RB, Holms JH, Li J, Pease LJ, Guo J, Glaser KB, Marcotte PA, Richardson PL, Murphy SS, Bouska JJ, Tapang P, Magoc TJ, Albert DH, Davidsen SK, Michaelides MR. α -Keto amides as inhibitors of histone deacetylase. Bioorg Med Chem Lett 2003;13(19):3331–3335, DOI: 10.1016/S0960-894X(03)00685-1.
59. Payne JE, Bonnefous C, Hassig CA, Symons KT, Guo X, Nguyen PM, Annable T, Wash PL, Hoffman TZ, Rao TS, Shiau AK, Malecha JW, Noble SA, Hager JH, Smith ND. Identification of KD5170: a novel mercaptoketone-based histone deacetylase inhibitor. Bioorg Med Chem Lett 2008;18(23):6093–6096, DOI: 10.1016/j.bmcl.2008.10.029.
60. Davis D, Kim HM, Ramphal JY, Spencer JR, Tai VW.F, Verner EJ,Silanol derivatives as inhibitors of histone deacetylase. 2006; WO2006069096.
61. Wahhab A, Smil D, Ajamian A, Allan M, Chantigny Y, Therrien E, Nguyen N, Manku S, Leit S, Rahil J, Petschner AJ, Lu AH, Nicolescu A, Lefebvre S, Montcalm S, Fournel M, Yan TP, Li Z, Besterman JM, Déziel R, Bioorg Med Chem Lett 2009;19:336–340, DOI: 10.1016/j.bmcl.2008.11.081.
62. Remiszewski SW. The discovery of NVP-LAQ824: From concept to clinic. Curr Med Chem 2003;10:2393–2402, DOI: 10.2174/0929867033456675.
63. Remiszewski SW, Sambucetti LC, Bair KW, Bontempo J, Cesarz D, Chandramouli N, Chen R, Cheung M, Cornell-Kennon S, Dean K, Diamantidis G, France D, Green MA, Howell KL, Kashi R, Kwon P, Lassota P, Martin MS, Mou Y, Perez LB, Sharma S, Smith T, Sorensen E, Taplin F, Trogani N, Versace R, Walker H, Weltchek-Engler S, Wood A, Wu A, Atadja P. N-Hydroxy-3-phenyl-2-propenamides as novel inhibitors of human histone deacetylase with in vivo antitumor activity: discovery of (2E)-N-hydroxy-3-[4-[[(2-hydroxyethyl)[2-(1H-indol-3-yl)ethyl]amino]methyl]phenyl]-2-propenamide (NVP-LAQ824). J Med Chem 2003;46: 4609–4624, DOI: 10.1021/jm030235w.

64. Atadja P.Histone deacetylase inhibition, a promising anticancer strategy, data presented at international conference, EORTC-NCI-AACR molecular targets and cancer therapeutics, Geneva, 2004.
65. Cemoglu M, Bajwa JS, Parker DJ, Slade J, Process for preparation of (2E)-N-hydroxy-3-[4-[[[2-(2-methyl-1H-indol-3-yl)ethyl]amino]methyl]phenyl]-2-propenamide. 2007; WO2007/146718.
66. Revill P, Mealy N, Seradell N, Bolos J, Rosa E. Antimitotic drug, microtubule-stabilizing agent. Oncolytic. Drugs Fut 2007;32(4): 315–322, DOI: 10.1358/dof.2007.032.04.1088164.
67. A list of on-going clinical studies with LBH-589 can be found on www.clinicaltrials.gov.
68. Plumb JA, Finn PW, Williams RJ, Bandara MJ, Romero RM, Watkins CJ, La Thangue NB, Brown B. Pharmacodynamic response and inhibition of growth of human tumor xenografts by the novel histone deacetylase inhibitor PXD101. Mol Cancer Ther 2003;2: 721–749.
69. Watkins CJ, Romero-Martin MR, Moore KG, Ritchie J, Finn PW, Kalvinsh I, Loza E, Dikovska K, Gailite V, Vorona M, Piskunova I, Starchenkov I, Adrianov V, Harris CJ, Duffy JE.S,Preparation of aryl-substituted N-hydroxy amides with sulfonamide linkages as HDAC inhibitors for treatment of proliferative conditions. 2002; WO2002030879.
70. Qian X, LaRochelle WJ, Ara G, Wu F, Petersen KD, Thougaard A, Sehested M, Lichenstein HS, Jeffers M. Activity of PXD101, a histone deacetylase inhibitor, in preclinical ovarian cancer studies. Mol Cancer Ther 2006; 5(8):2086–2095, DOI: 10.1158/1535-7163.
71. Lee MJ, Kim YS, Kummar S, Giaccone G, Trepel JB. Histone deacetylase inhibitors in cancer therapy. Curr Opin Oncol 2008; 20(6): 639–649.
72. Glaser KB. HDAC inhibitors: clinical update and mechanism-based potential. Biochem Pharmacol 2007;74(5):659–671, DOI: 10.1016/j.bcp. 2007.04.007.
73. Chen D, Deng W, Sangthongpitag K, Song HY, Sun ET, Yu N, Zou Y,Preparation of benzimidazolyl-hydroxamates as inhibitors of histone deacetylase (HDAC). 2005; WO2005028447.
74. Venkatesh PR, Goh E, Zeng P, New LS, Xin L, Pasha MK, Sangthongpitag K, Yeo P, Khantaraj E. *In vitro* phase I cytochrome P450 metabolism, permeability and pharmacokinetics of SB639, a novel histone deacetylase inhibitor in preclinical species. Biol Pharm Bull 2007;30(5):1021–1024, DOI: 10.1248/bpb.30.1021.
75. Chen D, Deng W, Lee KC.L, Lye PL, Sun ET, Wang H, Yu N,Heterocyclic compounds as HDAC inhibitors and their preparation, pharmaceutical compositions and use in the treatment of diseases associated with enzymes that have HDAC activity. 2007; WO2007030080.
76. Yong WP, Goh BC, Ethirajulu K, Yeo P, Otheris O, Chao SM, Soo R, Yeo WL, Seah E, Zhu J. A phase I dose escalation study of oral SB939 when administered thrice weekly (every other day) for 3 weeks in a 4-week cycle in patients with advanced solid malignancies. Eur J Cancer Suppl 2008;6(12):130.
77. Bertolini G, Biffi M, Leoni F, Mizrahi J, Pavich G, Mascagni P,Preparation of benzohydroxamic acids as antiinflammatory and immunosuppressive agents. 1997; WO9743251.
78. Leoni F, Mascagni P,Use of hydroxamic acid derivatives for antitumor medicaments. 2004; WO2004064824.
79. Khan N, Jeffers M, Kumar S, Hackett C, Boldog F, Khramtsov N, Qian X, Mills E, Berghs SC, Carey N, Finn PW, Collins LS, Tumber A, Ritchie JW, Jensen PB, Lichenstein HS, Sehested M. Determination of the class and isoform selectivity of small-molecule histone deacetylase inhibitors. Biochem J 2008;409: 581–589, DOI: 10.1042/BJ20070779.
80. Golay J, Cuppini L, Leoni F, Mico C, Barbui V, Domenghini M, Lombardi L, Neri A, Barbui AM, Salvi A, Pozzi P, Porro G, Pagani P, Fossati G, Mascagni P, Introna M, Rambaldi A. The histone deacetylase inhibitor ITF2357 has anti-leukemic activity *in vitro* and *in vivo* and inhibits IL-6 and VEGF production by stromal cells. Leukemia 2007;21(9):1892–1900.
81. Buggy JJ, Cao ZA, Bass KE, Verner E, Balasubramanian S, Liu L, Schultz BE, Young PR, Dalrymple SA. RA-024781: a novel synthetic inhibitor of histone deacetylase enzymes with antitumor activity *in vitro* and *in vivo*. Mol Cancer Ther 2006;5(5):1309–1317, DOI: 10.1158/1535-7163.MCT-05-0442.
82. Benarab A, Boye S, Savelon L, Guillaumet G. Utilisation du groupement cyanométhyle comme motif protecteur des phénols, amines et carbamates. Tetrahedron Lett 1993;34 (47):7567–7568, DOI: 10.1016/S0040-4039 (00)60401-X.
83. Himmelsbach F, Linz G, Pieper H, Austel V, Mueller T, Weisenberger J, Guth B.Preparation of amide group-containing compounds as antithrombotics. 1995; Ger Offen 4326344.

84. Balasubramanian S.Biomarkers for broad-spectrum and isoform semective HDAC inhibitors, data presented at international conference, CHI's first annual HDAC inhibitors, Discovery on Target, Oct 16, 2007, Boston MA.
85. Adimoolam S, Sirisawad M, Chen J, Thieman P, Ford JM, Buggy JJ. HDAC inhibitor PCI-24781 decreases RAD51 expression and inhibits homologous recombination. PNAS 2007; 19482–19487.
86. Arts J, Angibaud P, Marien A, Floren W, Janssens B, King P, van Dun J, Janssen L, Geerts T, Tuman RW, Johnson DL, Andries L, Jung M, Janicot M, van Emelen K. R306465 is a novel potent inhibitor of class I histone deacetylases with broad-spectrum antitumoral activity against solid and hematological malignancies. Br J Cancer 2007;97(10):1344–1353, DOI: 10.1038/sj.bjc.6604025.
87. Van Emelen K, Arts J, Backx LJ.J, De Winter HL.J, Van Brandt SF.A, Verdonck MG.C, Meerpoel L, Pilatte IN.C, Poncelet VS, Dyatkin AB,Preparation of sulfonyl-derivatives as novel inhibitors of histone deacetylase. 2003; WO2003076422.
88. Van Emelen K.The identification of a novel series of N-hydroxybenzamides as potent HDAC inhibitors: synthesis, biological evaluation and structure–activity relationships, CNIO Cancer Conference, Medicinal chemistry in Oncology, October 2–4, 2006, Madrid, Spain.
89. Belien A, De Schepper S, Floren W, Janssens B, Mariën A, King P, Van Dun J, Andries L, Voeten J, Bijnens L, Janicot M, Arts J. Realtime gene expression analysis in human xenografts for evaluation of histone deacetylase inhibitors. Mol Cancer Ther 2006; 5(9):2317–2323, DOI: 10.1158/1535-7163.MCT-06-0112.
90. Dickens JWJ, Houpis IN, Lang YL, Leys C, Stokbroekx SCM, Weerts JEE.Monohydrochloric salt of N-hydroxy-2-[4-({[(1-methyl-1H-indol-3-yl)methyl]amino}methyl)-1-piperidinyl]-5-pyrimidinecarboxamide as HDAC inhibitors and its preparation, pharmaceutical compositions and use in the treatment of cancer. 2008; WO2008138918.
91. Arts J.JNJ-26481585-a novel "second generation" pan-Histone Deacetylase (HDAC) inhibitor-showed broad-spectrum antotumoral activity against solid and haematological malignacies, Discovery on Target, Oct 16–17, 2007. Boston MA.
92. Kgediya LK, Belosay A, Khandelwal A, Purushottamachar P, Njar VC.O. Improved synthesis of histone deacetylase inhibitors (HDIs) (MS-275 and CI-994) and inhibitory effects of HDIs alone or in combination with RAMBAs or retinoids on growth of human LNCaP prostate cancer cells and tumor xenografts. Bioorg Med Chem 2008;16:3352–3360, DOI: 10.1016/j.bmc.2007.12.007.
93. Hess-Stumpp H, Bracker TU, Henderson D, Politz O. MS-275, a potent orally available inhibitor of histone deacetylases: the development of an anticancer agent. Int J Biochem Cell Biol 2007;39:1388–1405, DOI: 10.1016/j.biocel.2007.02.009.
94. Close J, Heidebrecht RW, Kattar S, Miller TA, Sloman D, Stanton MG, Tempest P, Witter DJ, Preparation of pyrazoles and related compounds as histone deacetylase inhibitors for the treatment of cancer. 2007; WO2007055941.
95. Fournel M, Bonfils C, Hou Y, Yan PT, Trachy-Bourget MC, Kalita A, Liu J, Lu AH, Zhou NZ, Robert MF, Gillespie J, Wang JJ, Ste-Croix H, Rahil J, Lefebvre S, Moradei O, Delorme D, MacLeod AR, Besterman JM, Li Z. MGCD-0103, a novel isotype-selective histone deacetylase inhibitor, has broad spectrum antitumor activity in vitro and in vivo. Mol Cancer Ther 2008;7(4):759–768, DOI: 10.1158/1535-7163.MCT-07-2026.
96. Zhou N, Moradei O, Raeppel S, Leit S, Frechette S, Gaudette F, Paquin I, Bernstein N, Bouchain G, Vaisburg A, Jin Z, Gillespie J, Wang J, Fournel M, Yan PT, Trachy-Bourget MC, Kalita A, Lu A, Rahil J, MacLeod AR, Li Z, Besterman JM, Delorme D. Discovery of N-(2-Aminophenyl)-4-[(4-pyridin-3-ylpyrimidin-2-ylamino)methyl] benzamide (MGCD-0103), an orally active histone deacetylase inhibitor. J Med Chem 2008;51(14):4072–4075, DOI: 10.1021/jm800251w.
97. Frechette S, Leit S, Woo SH, Lapointe G, Jeannotte G, Moradei O, Paquin I, Bouchain G, Raeppel S, Gaudette F, Zhou N, Vaisburg A, Fournel M, Yan PT, Trachy-Bourget MC, Kalita A, Robert MF, Lu A, Rahil J, MacLeod AR, Besterman JM, Li Z, Delorme D. 4-(Heteroarylaminomethyl)-N-(2-aminophenyl)-benzamides and their analogs as a novel class of histone deacetylase inhibitors. Bioorg Med Chem Lett 2008;18(4):1502–1506, DOI: 10.1016/j.bmcl.2007.12.057.
98. Paquin I, Raeppel S, Leit S, Gaudette F, Zhou N, Moradei O, Saavedra O, Bernstein N, Raeppel F, Bouchain G, Frechette S, Woo SH, Vaisburg A, Fournel M, Kalita A, Robert MF, Lu A, Trachy-Bourget MC, Yan PT, Liu J, Rahil J,

MacLeod AR, Besterman JM, Li Z, Delorme D. Design and synthesis of 4-[(s-triazin-2-ylamino)methyl]-N-(2-aminophenyl)-benzamides and their analogues as a novel class of histone deacetylase inhibitors. Bioorg Med Chem Lett 2008;18(3):1067–1071, DOI: 10.1016/j.bmcl.2007.12.009.

99. Vaisburg A, Paquin I, Bernstein N, Frechette S, Gaudette F, Leit S, Moradei O, Raeppel S, Zhou N, Bouchain G, Woo SH, Jin Z, Gillespie J, Wang J, Fournel M, Yan PT, Trachy-Bourget MC, Robert MF, Lu A, Yuk J, Rahil J, MacLeod AR, Besterman JM, Li Z, Delorme D. N-(2-Amino-phenyl)-4-(heteroarylmethyl)-benzamides as new histone deacetylase inhibitors. Bioorg Med Chem Lett 2007;17(24):6729–6733, DOI: 10.1016/j.bmcl.2007.10.050.

100. Moradei Oscar M, Mallais Tammy C, Frechette Sylvie, Paquin Isabelle, Tessier Pierre E, Leit Silvana M, Fournel Marielle, Bonfils Claire, Trachy-Bourget Marie-Claude, Liu-Jianhong, Yan Theresa P, Lu Ai-Hua, Rahil Jubrail, Wang James, Lefebvre Sylvain, Li Zuomei, Vaisburg Arkadii F, Besterman Jeffrey M. Novel aminophenyl benzamide-type histone deacetylase inhibitors with enhanced potency and selectivity. J Med Chem 2007; 50(23):5543–5546, DOI: 10.1021/jm701079h.

101. Moradei O, Leit S, Zhou N, Frechette S, Paquin I, Raeppel S, Gaudette F, Bouchain G, Woo SH, Vaisburg A, Fournel M, Kalita A, Lu A, Trachy-Bourget MC, Yan PT, Liu J, Li Z, Rahil J, MacLeod AR, Besterman JM, Delorme D. Substituted N-(2-aminophenyl)-benzamides, (E)-N-(2-aminophenyl)-acrylamides and their analogues: novel classes of histone deacetylase inhibitors. Bioorg Med Chem Lett 2006;16(15):4048–4052, DOI: 10.1016/j.bmcl.2006.05.005.

102. Yin ZH, Wu ZW, Lan YK, Liao CZ, Shan S, Li ZL, Ning ZQ, Lu XP, Li ZB, Synthesis of chidamide, a new histone deacetylase (HDAC) inhibitor Zhongguo Xinyao Zazhi 2004;13(6):536–538.

103. Lu XP, Li ZP, Xie A, Li B, Ning ZQ, Shan S, Deng T, Hu W,Histone deacetylase inhibitors of novel benzamide derivatives with potent differentiation and anti-proliferation activity. 2004; WO2004/071400.

104. Fujisawa Pharmaceutical Co., Ltd, Jpn. Kokai Tokkyo Koho. Antitumor antibiotic FR901375 manufacture with Pseudomonas, 1991; JP, 03141296.

105. Furumai R, Matsuyama A, Kobashi N, Lee KH, Nishiyama M, Nakajima H, Tanaka A, Komatsu Y, Nishino N, Yoshida M, Horinouchi S. FK228 (depsipeptide) as a natural prodrug that inhibits class I histone deacetylases. Cancer Res 2002;62(17):4916–4921.

106. Vigushin David M. FR-901228 (Fujisawa/National Cancer Institute). Curr Opin Investig Drugs 2002;3(9):1396–1402.

107. Li KW, Wu J, Xing W, Simon JA. Total synthesis of the antitumor depsipeptide FR-901,228. J Am Chem Soc 1996;118:7237–7238, DOI: 10.1021/ja9613724.

108. Greshock TJ, Johns DM, Noguchi Y, Williams RM. Improved total synthesis of the potent HDAC inhibitor FK228 (FR-901228). Org Lett 2008;10(4):613–616, DOI: 10.1021/ol702957z.

109. Wen S, Packham G, Ganesan A. Macrolactamization versus macrolactonization: total synthesis of FK228, the depsipeptide histone deacetylase inhibitor. J Org Chem 2008;73: 9353–9361, DOI: 10.1021/jo801866z.

110. Yurek-George A, Cecil AR.L, Mo AH.K, Wen S, Rogers H, Habens F, Maeda S, Yoshida M, Packham G, Ganesan A. The first biologically active synthetic analogues of FK228, the depsipeptide histone deacetylase inhibitor. J Med Chem 2007;50(23):5720–5726; DOI: 10.1021/jm0703800.

111. Di Maro S, Pong RC, Hsieh JT, Ahn JM. Efficient solid-phase synthesis of FK228 analogues as potent antitumoral agents. J Med Chem 2008;51(21):6639–6641, DOI: 10.1021/jm800959f.

112. Shah MH, Binkley P, Chan K, Xiao J, Arbogast D, Collamore M, Farra Y, Young D, Grever M. Cardiotoxicity of histone deacetylase inhibitor depsipeptide in patients with metastatic neuroendocrine tumors. Clin Cancer Res 2006;12 (13):3997–4003, DOI: 10.1158/1078-0432.CCR-05-2689.

113. Townsend PA, Crabb SJ, Davidson SM, Johnson PW.M, Packham G, Ganesan A. The bicyclic depsipeptide family of histone deacetylase inhibitors. Chem Biol 2007;2: 693–720.

114. Taori K, Paul VJ, Luesch H. Structure and activity of largazole, a potent antiproliferative agent from the Floridian marine cyanobacterium *Symploca* sp. J Am Chem Soc 2008;130:1806–1807, : 10.1021/ja7110064.

115. Ying Y, Taori K, Kim H, Hong J, Luesch H. Total synthesis and molecular target of largazole, a histone deacetylase inhibitor. J Am Chem Soc 2008;130:8455–8459, DOI: 10.1021/ja8013727.

116. Numajiri Y, Takahashi T, Takagi M, Shin-ya K, Doi T. Total synthesis of largazole and its biological evaluation. Synlett 2008; (16):2483–2486, DOI: 10.1055/s-2008-1078263.
117. Ren Q, Dai L, Zhang H, Tan W, Xu Z, Ye T. Total synthesis of largazole. Synlett 2008;15:2379–2383, DOI: 10.1055/s-2008-1078270.
118. Seiser T, Kamena F, Cramer N. Synthesis and biological activity of largazole and derivatives. Angew Chem Int Ed 2008;47(34):6483–6485, DOI: 10.1002/anie.200802043.
119. Ying Y, Liu Y, Byeon SR, Kim H, Luesch H, Hong J. Synthesis and activity of largazole analogues with linker and macrocycle modification. Org Lett 2008;10(18):4021–4024, DOI: 10.1021/ol801532s.
120. Ghosh AK, Kulkarni S. Enantioselective total synthesis of (+)-largazole, a potent inhibitor of histone deacetylase. Org Lett 2008;10 (17):3907–3909, DOI: 10.1021/ol8014623.
121. Bowers A, West N, Taunton J, Schreiber SL, Bradner JE, Williams RM. Total synthesis and biological mode of action of largazole: a potent Class I histone deacetylase inhibitor. J Am Chem Soc 2008;130(33):11219–11222, DOI: 10.1021/ja8033763.
122. Lagace DC, O'Brien WT, Gurvich N, Nachtigal MW, Klein PS. Valproic acid: how it works. Or not. Clin Neurosci Res 2004;4:215–225, DOI: 10.1016/j.cnr.2004.09.013.
123. Michaelis M, Doerr HW, Cinatl J Jr. Valproic acid as anti-cancer drug. Curr Pharm Design 2007;13:3378–3393.
124. Nolan L, Johnson PWM, Ganesan A, Pacham G, Crabb SJ. Will histone deacetylase inhibitors require combination with other agents to fulfil their therapeutic potential?. Br J Cancer 2008;99:689–694, DOI: 10.1038/sj.bjc.6604557.
125. Gediya LK, Khandelwal A, Patel J, Belosay A, Sabnis G, Mehta J, Purushottamachar P, Njar VC.O. Design, synthesis, and evaluation of novel mutual prodrugs (hybrid drugs) of all-*trans*-retinoic acid and histone deacetylase inhibitors with enhanced anticancer activities in breast and prostate cancer cells *in vitro*. J Med Chem 2008;51(13):3895–3904, DOI: 10.1021/jm8001839.
126. Cai X, Qian C, Gould S, Zhai H.Preparation of pyridine derivatives as Raf kinase inhibitors. WO 2008115263 Cai, Xiong; Qian, Changgeng; Gould, Stephen. Preparation of 2-indolinone derivatives as protein tyrosine kinase (PTK) inhibitors containing a zinc binding moiety. 2008; WO2008033743.
127. Cai X, Qian C, Gould S, Zhai H,Multi-functional small molecules as anti-proliferative agents and their preparation. 2008; WO2008033747.
128. Balasubramanian S, Verner E, Buggy JJ. Isoform-specific histone deacetylase inhibitors: the next step? Cancer Lett 2009;280: 211–221, DOI: 10.1016/j.canlet.2009.02.013.
129. Wang D. Computational studies on the histone deacetylases and the design of selective histone deacetylase inhibitors. Curr Top Med Chem 2009;9:241–256, DOI: 10.2174/156802609788085287.
130. Sternson SM, Wong JC, Grozinger CM, Schreiber SL. Synthesis of 7200 Small molecules based on a substructural analysis of the histone deacetylase inhibitors trichostatin and trapoxin. Org Lett 2001;3: 4239–4242, DOI: 10.1021/ol016915f.
131. Haggarty SJ, Koeller KM, Wong JC, Grozinger CM, Schreiber SL. Domain-selective small-molecule inhibitor of histone deacetylase 6 (HDAC6)-mediated tubulin deacetylation. Proc Natl Acad Sci USA 2003;100(8): 4389–4394.
132. Itoh Y, Suzuki T. Design, synthesis, structure–selectivity relationship, and effect on human cancer cells of a novel series of histone deacetylase 6-selective inhibitors. J Med Chem 2007;50 (22):5425–5438, DOI: 10.1021/jm7009217.
133. Kozikowski AP, Tapadar S, Luchini DN, Kim KH, Billadeau DD. Use of the nitrile oxide cycloaddition (NOC) reaction for molecular probe generation: a new class of enzyme selective histone deacetylase inhibitors (HDACIs) showing picomolar activity at HDAC6. J Med Chem 2008;51(15):4370–4373, DOI: 10.1021/jm8002894.
134. Bottomley MJ, Lo Surdo P, Di Giovine P, Cirillo A, Scarpelli R, Ferrigno F, Jones P, Neddermann P, De Francesco R, Steinkuehler C, Gallinari P, Carfi A. Structural and functional analysis of the human HDAC4 catalytic domain reveals a regulatory structural zinc-binding domain. J Biol Chem 2008;283(39):26694–26704, DOI: 10.1074/jbc.M803514200.
135. Itoh Y, Suzuki T, Miyata N. Isoform-selective histone decacetylase inhibitors. Curr Pharm Des 2008;14:529–544, DOI: 10.2174/138161208783885335.
136. Bieliauskas AV, Pflum MK.H. Isoform-selective histone deacetylase inhibitors. Chem Soc Rev 2008;37:1402–1413, DOI: 10.1039/b703830p.

SYNTHETIC DNA-TARGETED CHEMOTHERAPEUTIC AGENTS AND RELATED TUMOR-ACTIVATED PRODRUGS

WILLIAM A. DENNY
Auckland Cancer Society Research Centre,
University of Auckland,
Auckland, New Zealand

1. INTRODUCTION

Synthetic drugs have always played an important role in cancer therapy. In fact, systemic chemotherapy for cancer began in the 1940s and 1950s with the nitrogen mustards developed from war gases [1], and with antimetabolites developed from early knowledge about DNA metabolism [2]. Large-scale random screening programs over the next 25 years (mainly by the U.S. National Cancer Institute) [3] seeking cytotoxic agents resulted in the identification of a number of cytotoxic natural products that target DNA. Many of these (e.g., anthracyclines, epipodophylloxins, and vinca alkaloids) became very useful drugs that are still widely used today. Most of the natural products were so complex that neither they nor close analogs could be economically produced by synthesis, limiting the role of synthetic chemistry in optimizing their potencies or pharmacokinetic properties. However, the discovery of their activity and mechanism of action sparked much work on simpler synthetic analogs. One result was the development of the large class of synthetic topoisomerase inhibitors that are now an important group of drugs. More recently, the increasing power of organic synthesis has greatly improved chances that quite complex natural product leads can be synthesized economically, and therefore that close analogs can be made to try and optimize physicochemical properties; recent examples are the cyclopropylindolines and the epothilones. However, the primary focus of this chapter is synthetic compounds that have not been derived from a natural product lead. Finally, our increasing understanding of tumor physiology and genetics has allowed the development of a new class of synthetic agents, tumor-activated prodrugs. The latter compounds attempt to exploit tumor-specific phenomena such as unique antigen expression, low pH and hypoxia, in order to achieve tumor-selective activation of prodrugs of the more classical cytotoxins and so increase their therapeutic range.

2. ALKYLATING AGENTS

2.1. Introduction

Compounds that alkylate DNA have long been of interest as anticancer drugs. Alkylating agents can be strictly defined as electrophiles that can replace a hydrogen atom by an alkyl group under physiological conditions, but the term is usually more broadly interpreted to include any compound that can replace hydrogen under these conditions, including metal complexes forming coordinate bonds. Many different types of chemicals are able to alkylate DNA, and several are used as anticancer drugs, but the most important classes of such agents in clinical use are the nitrogen mustards and the platinum complexes, the nitrosoureas and the triazene-based DNA-methylating agents. The DNA minor groove alkylating cyclopropylindoles are also a fascinating group of compounds that are finding a niche in cancer therapy as components of immunotoxins. Other important classes of DNA alkylating agents, the pyrrolobenzodiazepines and the mitosenes, are covered elsewhere in this series, although the bioreductive aspects of the mitosenes are mentioned in Section 6.3.

2.2. Clinical Examples of Alkylating Agents

The most commonly used mustards and platinum complexes are listed in Table 1, along with other recent DNA-alkylating agents that have received clinical trial. These compounds are invariably used in combination with other agents in multidrug therapy regimens.

2.3. Mustards

2.3.1. History As noted above, the mustards were among the very earliest class of anticancer agents developed, and they have been extensively reviewed. Mechlorethamine (**1**) [4]

Table 1. Alkylating Agents Used in Cancer Chemotherapy

Generic Name (Structure)	Trade Name	Marketer	Chemical class
Mustards			
Mechlorethamine (**1**)	Mustargen	Merck	Aliphatic mustard
Chlorambucil (**2**)	Leukeran	GlaxoSmithKlein	Aromatic mustard
Melphalan (**3**)	Alkeran		Aromatic mustard
Bendamustine (**4**)	Treanda	Cephalon	Aromatic mustard
Cyclophosphamide (**5**)	Cytoxan	Bristol-Myers	Phosphoramide mustard
Ifosfamide (**6**)	Ifex	Bristol-Myers	Phosphoramide mustard
Platinum complexes			
Cisplatin (**15**)	Cisplatin	Bristol-Myers	Platinum complex
Carboplatin (**16**)	Paraplatin	Bristol-Myers	Platinum complex
Tetraplatin (**17**)	Ormaplatin		Platinum complex
Oxaliplatin (**18**)	Eloxantin	Sanofi	Platinum complex
ZD-0473 (**19**)	Picoplatin	AstraZeneca	Platinum complex
JM-216 (**20**)	Satraplatin	Johnson-Matthey	Platinum complex
BBR 3464 (**21**)		Cell Therapeutics	Triplatinum complex
Cyclopropylindoles			
Adozelesin (**23**)		Upjohn	Cyclopropylindole
Carzelesin (**24**)		Upjohn	Cyclopropylindole
KW 2189 (**25**)		Kyowa Hakko	Cyclopropylindole
Nitrosoureas			
BCNU (**30**)	Carmustine	Bristol-Myers	Nitrosourea
CCNU (**31**)	Lomustine	Bristol-Myers	Nitrosourea
Streptozotocin (**32**)	Zanosar	Upjohn	Nitrosourea
Cloretazine (**33**)		Vion	
Methylating agents			
Dacarbazine (**33**)	DTIC	Bayer	Triazene
Mitozolomide (**34**)	Azolastone		Triazene
Temozolomide (**35**)	Temodar	Schering-Plough	Triazene

was the first systemic agent approved for use in cancer therapy in 1949. Chlorambucil (**2**) [5] was approved in 1957, melphalan (**3**) [6] in 1964. Bendamustine (Treanda; **4**) is a benzimidazole mustard structurally related to chlorambucil, which also is containing some "purine-like" features. In was originally prepared in the 1960s, making it broadly contemporary with other mustards, and was used extensively and marketed (as Ribomustin) in Germany in the 1970s [7]. The phosphoramide mustard cyclophosphamide (**5**) [8] was introduced in 1959, and the analog ifosfamide (**6**) [9] in 1988. Cyclophosphamide is currently the most widely used mustard, while chlorambucil and melphalan are still in use as components of many combination chemotherapy regimens.

1: R=Me
2: R=pPh(CH$_2$)$_3$CO$_2$H
3: R=pPhCH(NH$_2$)CO$_2$H

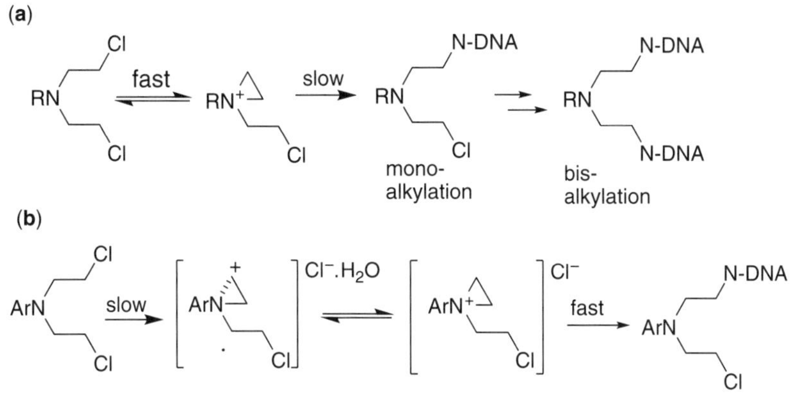

Figure 1. Mechanism of alkylation by mustards. (a) Aliphatic mustards. (B) Aromatic mustards.

2.3.2. Mechanism and SAR The biologically important initial lesion formed by mustards in cells is interstrand cross-links between different DNA bases [10], although there is also evidence that they cause termination of transcription [11]. The overall process of alkylation is a two-step sequence involving formation of a cyclic cationic intermediate, followed by nucleophilic attack on that intermediate by DNA (Fig. 1a). Mustards can be divided into two broad classes, depending on the mechanism of the rate-determining step in this process. The less basic compounds aromatic mustards have formation of the solvated cyclic carbocation (which is in equilibrium with the aziridinium cation) as the rate-determining step, following first-order kinetics [12] (Fig. 1b). Nucleophilic attack on this is then rapid, so that the cyclic form does not accumulate, and the overall reaction is first order (S_N1), with the rate depending only on the concentration of the mustard (Eq. 1).

$$R-X \rightarrow R^+ \rightarrow R-DNA \quad (1)$$

For the more basic aliphatic mustards, the first step (formation of the aziridinium cation) is rapid and the rate-determining step is a second-order nucleophilic substitution on this by DNA [12]. In these cases, the aziridinium cation can be detected as an intermediate, and the overall reaction is second order (S_N2), with the rate depending on the concentrations of both the mustard and the DNA (Eq. 2).

$$DNA + R-X \rightarrow DNA\ldots R\ldots H \rightarrow R-DNA + X \quad (2)$$

This kinetic classification is only broad, but it useful as a rough predictor of the spectrum of adducts formed. Generally, S_N1-type compounds are expected to be less discriminating in their pattern of alkylation (reaction at N, P, and O sites on DNA), while most S_N2 type compounds tend to alkylate only at N sites on the DNA bases [13].

The primary site of DNA alkylation by mustards is at the N7 position of guanine, particularly at guanines in contiguous runs of guanines [14], which have the lowest molecular electrostatic potentials [15]. However, the level of selectivity of the initial attack by mustards (to form monoadducts) is quite low, with evidence [16] that most guanines are attacked. Studies with alkyl mustards have also shown significant levels of alkylation at the N3 position of adenine [17,18]. However, the sequence selectivity of cross-link formation by mustards is necessarily higher, because of the requirement to have two suitable sites juxtaposed. Early work [19] on the interaction of mechlorethamine (**1**) with DNA resulted in isolation of the 7-linked bis-guanine adduct, and it has been widely assumed that the cross-links were between adjacent

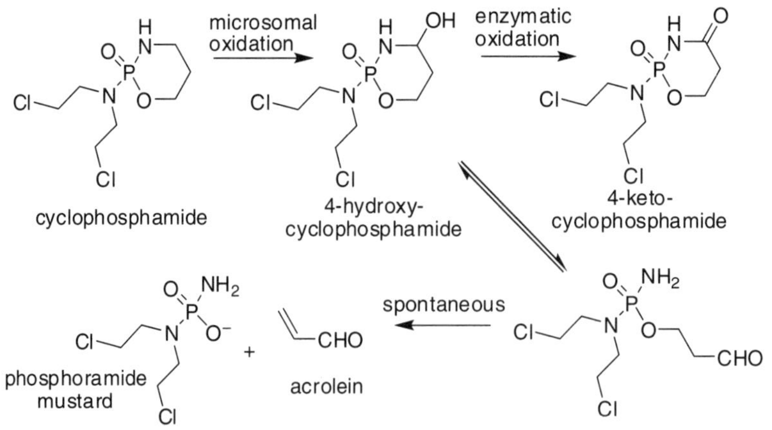

Figure 2.

guanines (i.e., at 5'-GC or 5'-CG sites). However, later results [20] showed that the preferred cross-links are between nonadjacent guanines (i.e., at 5'-GNC sites). Adenine-containing cross-links have also been reported [21].

Cyclophosphamide (5) is a nonspecific prodrug of the active metabolite phosphoramide mustard, requiring enzymic activation by cellular mixed function oxidases (primarily in the liver) to 4-hydroxycyclophosphamide, which is in equilibrium with the open-chain aldophosphamide. This species then spontaneously eliminates to form acrolein and phosphoramide mustard (Fig. 2). The isomeric ifosfamide (6) is activated more slowly, but in a broadly similar fashion, to give the analogous isophosphoramide mustard as the active species [22]. A significant difference in the metabolism between the two isomers is a higher level of dechloroethylation with ifosfamide, which may account for its greater neurotoxicity [23].

The rates of the various reactions of aromatic nitrogen mustards (hydrolysis, alkylation of DNA) can be correlated closely with the basicity of the nitrogen that in turn can be systematically altered by ring substituents. The rates of hydrolysis (K_H) of a series of substituted aromatic nitrogen mustards in aqueous acetone can be described [24] by Equation 3, where σ is the Hammett electronic parameter.

$$\log K_H = -1.84\sigma - 4.02 \qquad (3)$$

The negative slope is evidence for an S_N1 mechanism, indicating that electron-releasing substituents (negative σ values) increase the rate of hydrolysis by accelerating formation of the carbocation. The same broad correlations hold for how well the compounds alkylate DNA, with a similar equation (Eq. 4) describing the rates of alkylation (K) of 4-(4-nitrobenzyl)pyridine (a nucleophile similar to DNA nucleophilic sites) by substituted aromatic nitrogen mustards [24], where σ^- is an electronic parameter closely related to σ.

$$\log K = -1.92\sigma - 1.17 \qquad (4)$$

The cytotoxicities of the above compounds ($1/IC_{50}$ values) also correlate well with substituent σ values, with the more reactive compounds (bearing electron-donating substituents) being the more cytotoxic, as in Equation 5 [25].

$$\log\left(\frac{1}{IC_{50}}\right) = -2.46\sigma + 0.53 \qquad (5)$$

The cytotoxicity of aromatic mustards can thus be predictably varied over a very wide range by controlling the basicity of the mustard nitrogen via ring substitution or other means. Bendamustine (4) appears to act somewhat differently to other nitrogen mustards, in particular inducing a much stronger p53-mediated stress response. It may also activate different DNA repair pathways to traditional mustards, and be less susceptible to drug resistance based on alkylguanyl transferase expression [26].

2.3.3. Biological Activity and Side Effects

The (necessarily) high chemical reactivity of mustards leads to rapid loss of drug by interaction with other cellular nucleophiles, particularly proteins and low molecular weight thiols. This results in the development of cellular resistance by increases in the levels of low molecular weight thiols (particularly glutathione) [27,28]. Of equal importance for efficacy, much of the drug can reach the DNA with only one alkylating moiety intact, leading to monoalkylation events that are considered to be genotoxic rather than cytotoxic [29]. The fact that cross-linking is a two-step process adds to the proportion of (genotoxic) monoalkylation events, since the second step is very dependent on spatial availability of a second nucleophilic DNA site. Mustards have no intrinsic biochemical or pharmacological selectivity for cancer cells, and act as classical antiproliferative drugs, whose therapeutic effects are primarily cytokinetic. They target rapidly dividing cells rather than cancer cells, and this, together with their generally systemic distribution, causes killing of rapidly dividing normal cell populations in the bone marrow and gut, resulting in myelosupression as usually the dose-limiting side effect. Because of their genotoxicity, there is a risk of the development of second cancers from their mutagenic effects [30]. The most frequent alkylator-induced malignancy is acute leukemia, usually occurring a long period (3–7 years) after treatment. These usually demonstrate deletions of chromosome 13 and loss of parts or all of chromosomes 5 or 7 (loss of the coding regions for tumor suppressor genes). The induced tumors are typically myelodysplasias [31]. In one study [32], 6% of all myeloid leukemias were therapy related, with mustards, nitrosoureas, and procarbazine producing the greatest levels of induction. This has sparked a trend toward the discontinuance of alkylating agents in some combination therapies [33], although there are indications where they remain very useful [34]. Early clinical studies with bendamustine, confirmed in more recent trials, show it is active in combination therapy in a number of cancer types, particularly chronic lymphocytic leukemia [35], non-Hodgkin's lymphoma [36] and multiple myeloma [37]. Bendamustine was approved by the FDA early in 2008 for the treatment of chronic lymphocytic leukemia.

2.3.4. Recent Developments: Minor Groove Targeting

Many of the limitations noted above could in principle be ameliorated by targeting the mustard moiety more specifically to DNA-affinic carrier molecule. This has resulted in much work where mustards have been attached to DNA-affinic compounds [38–40]. This could mean less chance of losing active drug by reaction with other cell components, rendering less effective the development of cellular resistance by elevation of thiol levels. A higher proportion of bifunctional alkylating agent delivered intact to the DNA would also contribute to a higher proportion of cross-links over monoalkylation events. The use of a carrier with sequence-specific reversible binding ability should also result in greater specificity of alkylation; both sequence specificity (at the favored reversible binding site of the carrier) and region specificity (at particular atoms on the DNA bases).

Attachment to DNA-intercalating carriers goes back to the work of Creech et al. [41], who originally suggested that the attachment to acridine carriers might serve to target the reactive center to DNA. They showed that such "targeted mustards" such as (7) were more potent than the corresponding untargeted moiety against ascitic tumors in vivo, but these proved to be exceptionally potent frameshift mutagens in bacteria, and this property has tended to dominate the perception of these compounds. Later work showed that such targeting by an intercalator could also drastically modify the pattern of DNA alkylation by the mustard. Thus, while untargeted mustards react largely at the N7 of guanines in runs of guanines, quinacrine mustard (8) also alkylates at guanines in 5'-GT sites [14]. Isolation and identification of DNA adducts showed that while the acridine-linked mustard (9) formed primarily guanine N7 adducts, the similar analog (10) formed exclusively adenine N1 adducts [42] showing the extent to which DNA targeting by attachment to carrier molecules can alter the usual pattern of DNA alkylation by mustards.

However, most DNA-targeting of mustards has been done using minor groove binding carriers. These ligands offer much larger binding site sizes (up to 5–6 base pairs) than intercalators, together with a highly defined binding orientation. While several other minor groove binding carriers have been used [43–45], most work has employed polypyrrole and related ligands. These compounds have been well documented as reversible AT-specific minor groove binders [46], and early work using a variety of alkylating units (e.g., bromoacetyl) showed highly specific alkylation at adenines in runs of adenines [47]. The benzoic acid mustard derivative tallimustine (11; FCE 24517) was selected for further development on the basis of its broad-spectrum solid tumor activity [48]. This is a difunctional alkylator, yet it only monoalkylates DNA at the N3 of adenine in the minor groove, almost exclusively at the sequence 5'-TTTTGA [49,50]. A single base modification in the hexamer completely abolishes its ability to alkylate [51]. The number of pyrroleamide units also affected the pattern of DNA alkylation, with a monopyrrole analog showing mainly guanine-N7 alkylation similar to that of the untargeted mustard, but with additional adenine-N3 lesions [52]. Di- and tripyrrole conjugates alkylated only in AT tracts, with increasing specificity for alkylation at the 3'-terminal units in two 5'-TTTTGG and 5'-TTTTGA sequences (guanine N3 and adenine N3 lesions, respectively).

Tallimustine showed biological effects somewhat different to those of mustards like melphalan. It induces blockage of the cell cycle in G2 but without the delay through S-phase normally seen with untargeted mustards, suggesting a different mechanism of cytotoxicity via monoadduct formation [53,54]. As a highly sequence-specific alkylator, it selectively blocks the binding of transcription binding protein and complexes to a TATA box to their AT-rich cognate sequences [55]. Clinical trials of tallimustine [56,57] reported severe myelosuppression as the dose-limiting toxicity. Recent work with halogenoacrylic derivatives such as brostacillin (12) show these may work differently, possibly via Michael-type reactions involving reaction with thiols such as glutathione [58], with much better cytotoxicity/myelotoxicity indices [59]. Brostacillin is reported to be in clinical trial [60].

Perhaps, the ultimate in targeting mustards to specific DNA sites has been achieved by Dervan and coworkers, who have developed the "hairpin polyamide" concept where poly(pyrrole/imidazole) compounds bind in a side-by-side" manner in the minor groove [61]. These compounds can bind tightly and selectively to individual designated sequences of up to 12 base pairs long [62]. As an example polyamide 13, with an incipient mustard side chain attached, binds to its designated sites 5'-AGCTGCT and 5'-TGCAGCA with equilibrium association constants K of 1.6 and $1.3 \times 10^{10} \, M^{-1}$, respectively, and binds >100-fold less strongly to double mismatch sites [63]. The corresponding mustard 14 alkylated at adenine N3 sited in target 5'-(A/T)GC(A/T)GC(A/T) sequences on a 241-bp HIV-1 promotor sequence in high yield and approximately 20-fold selectively over double mismatch sites [63]. Compound 14 was shown to growth-arrest a wide range of human cancer cell lines at the G2/M phase of the cell cycle, downregulating histone H4c gene expression [64].

2.4. Platinum Complexes

2.4.1. History The complex *cis*-diamminodichloroplatinum (II) (cisplatin; **15**) was first described in 1845, but it was not until 1969 that it was reported to have antitumor activity. These studies were sparked by experiments by Rosenberg on the effects of electric fields on bacteria, when the peculiar effects seen with *Escherichia coli* cells were shown to be caused by the electrochemical synthesis of cisplatin from the ammonium chloride electrolyte and the platinum electrodes [65]. Clinical trials began in 1972, and after slow progress due to high toxicity, cisplatin became one of the most widely used anticancer drugs; it is the main reason for the spectacular successes in drug treatment of testicular and ovarian cancer. Thousands of analogs of cisplatin have been made and evaluated, with two major driving forces.

The first has been to seek compounds with lower neurotoxicity than cisplatin. While better clinical management has improved things, one of the main drivers of analog development has been agents with less neurotoxicity. Carboplatin (**16**) has carboxylate instead of chloride leaving groups. These hydrolyze much less rapidly, resulting in lower nephro- and neurotoxicity (the dose-limiting toxicity of carboplatin is myelosuppression), while retaining the broad spectrum of activity of cisplatin [66].

The second impetus to analog development has been to seek agents active in cell lines that become resistant to cisplatin. One mechanism of resistance to cisplatin is an increased ability to repair the DNA adducts formed [67], and analogs such tetraplatin (**17**; ormaplatin) and oxaliplatin (**18**), with trans-1,2-diaminocyclohexane (DACH) ligands, were shown to be more effective against such resistant cell lines [68]. These compounds proved to be neurotoxic and tetraplatin was difficult to formulate, but oxaliplatin has shown promise in colorectal cancer, where it is synergistic with 5-fluorouracil [69]. A second significant mechanism of resistance is elevation of thiol levels (primarily glutathione) in cells [70]. Picoplatin (**19**; JM-473) is more resistant than cisplatin to thiols, possibly because of steric hindrance by the pyridine ligand [71], and has shown clinical activity in small cell lung cancer [72].

Another Pt(IV) complex, satraplatin (**20**; JM216) was developed as an orally available platinum agent [73]. It has potent *in vitro* cytotoxicity against a variety of tumor cell lines, and also had oral antitumor activity against a variety of murine and human subcutaneous tumor models *in vivo*, broadly comparable to the level of activity obtainable with parenterally administered cisplatin. Satraplatin has shown activity in Phase I trials in lung cancer, with no neurotoxicity or nephrotoxicity [73], and responses have also been seen in small cell lung cancer and hormone refractory prostate cancer. An ongoing Phase III trial suggests satraplatin and prednisone may be beneficial in terms of progression-free survival compared to prednisone alone as a second-line treatment for hormone-refractory metastatic prostate cancer [74].

2.4.2. Mechanism and SAR As with the nitrogen mustards, the mechanism of action of the platinum complexes involves formation of DNA cross-links. In the platinum complexes, the chloride or carboxylato ligands are the leaving groups, with the ammine ligands being substitutionally inert and serving to modulate other properties. The bonds formed and broken in this case are coordinate metal–ligand bonds that are not permanent, but have characteristic half-lives (although these may be very long), making the chemistry quite different to that of the mustards. Thus, the Pt-Cl bonds in cisplatin (**15**) are more stable in the relatively high chloride conditions in plasma than they are in the lower chloride conditions inside cells, where the reaction with water to form aquo species is more facile [75]. The cationically charged aquo species have higher affinity for DNA, and react primarily at guanine N7 sites in the major groove to form long-lived ammine complexes (Fig. 3).

Cisplatin reacts with DNA to form a number of different adducts. However, by far the commonest are intrastrand guanine N7-guanine N7 adducts between adjacent guanines on the same strand (about 65%), followed by similar intrastrand guanineN7-adenineN7

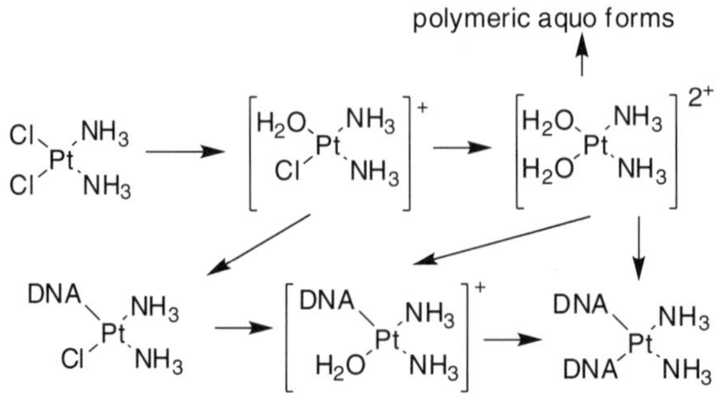

Figure 3. Reaction of platinum species with DNA.

adducts (about 25%), with DNA–protein cross-links and monofunctional adducts making up less than 10%, and DNA interstrand adducts less than 1% of the total adducts. A major difference between mustards and platinum complexes is that while hydrolysis in the former case is a deactivating event, leading to loss of bifunctionality (and thus cross-linking ability), with platinum complexes formation of the aquo species is a necessary activating process. Thus, there is a much higher proportion of cross-links to monoadducts formed with platinum complexes than with mustards. The use of [^{1}H, ^{15}N] HSQC 2D NMR has recently allowed a better understanding of the kinetics of the multiple processes involved in the reaction of platinum drugs with DNA [76].

2.4.3. Biological Activity and Side Effects

While cisplatin is an extremely useful drug, it has many side effects. In addition to the myelosuppressive activity typical of a DNA alkylating agent, it also showed severe renal and neural toxicity. Nevertheless, its effectiveness has meant it has become one of the most widely used cancer drugs. A recent analysis of the clinical literature [77] showed that in about two-thirds of all cell models, cisplatin-resistant cells were sensitive to the taxane paclitaxel and vice versa. This striking inverse relationship is mirrored in the clinical responses to these agents in ovarian cancer. The analog development work described above has been aimed primarily at overcoming some of these side effects. Thus carboplatin is less nephro- and neurotoxic, tetraplatin, oxaliplatin, and picoplatin are more effective against various types of resistance mechanisms, and satraplatin is orally effective. However, none of these compounds show major differences in their interaction with DNA.

2.4.4. Recent Developments: Increased Interstrand Cross-linking

The tetracationic triplatinum complex BBR 3464 (**21**) appears to represent a new structural class of DNA-modifying anticancer agents [78]. It reacts with DNA faster than cisplatin, suggesting rapid cellular uptake and nuclear access [79] to give a different profile of adducts than cisplatin, with approximately 20% being interstrand GG cross-links. DNA modified by BBR3464 cross-reacted with antibodies to transplatin-adducted DNA but not to antibodies to cisplatin-adducted DNA [78]. Crystal structure studies of a noncovalent analog of BBR3464 (triplatinNC) showed that the Pt(II) centers formed extensive H-bond interactions with the phosphate oxygens termed "phosphate clamping" [80]. It was suggested that similar preassociation of BBR3464 with DNA may be important in the efficiency and specificity of covalent cross-linking of this compound. BBR 3464 was 30-fold more cytotoxic than cisplatin in L1210 cells, and showed no cross-resistance in sublines resistant to cisplatin because of impaired accumulation and lower DNA binding [81]. Consistent with this, it was also highly active in a panel of cisplatin-resistant xenografts, giving longer growth delays [82]. Unlike cisplatin, BBR 3464 was able to induce the p53/p21 pathway to a similar extent in both cisplatin-sensitive and -resistant cells [83], and had a quite different sensitivity profile to cisplatin in the U.S. National Cancer Institute's 60-cell-line screening panel [82]. However, no significant activity was seen in Phase II clinical trials in small cell lung [84] and gastric [85] cancer.

2.5. Cyclopropylindoles

2.5.1. History

Interest in DNA minor groove alkylating agents was stimulated by the discovery [86] of the natural product CC-1065

(22) from *Streptomyces zelensis* [87], which showed extraordinary potency in a number of animal tumor models [88], but with concomitant fatal delayed hepatotoxicity at therapeutic doses [89]. An extensive synthesis program at Upjohn prepared a large number of analogs in an attempt to understand structure–activity relationships for the class, and succeeded in developing the structurally simpler agents adozelesin (23) and the open-chain form carzelesin (24), that did not show the delayed hepatotoxicity of CC-1065 [90,91]. The related semisynthetic duocarmycin analog KW 2189 (25) is a carbamate prodrug, releasing the active moiety DU-86 (26) by esterase hydrolysis [92]. KW 2189 has been prepared on a large scale by a three-step synthesis in overall 55% yield from natural duocarmycin B2 [93]. Although it is less potent than duocarmycin in cell culture assays, it has high activity in a wide range of human solid tumor xenografts in mice, and lacks the delayed lethal toxicity seen with some other cyclopropylindoles [94].

2.5.2. Mechanism and SAR These compounds bind initially reversibly in the minor groove of DNA with minimal structural distortion, and subsequently alkylate specifically at the N3 position of adenine [95]. This provides further evidence that targeting alkylating functionality to the DNA minor groove can provide compounds of very high cytotoxic potency. While the lead compound is a natural product, it has sparked a vast amount of synthetic chemistry, and the analogs developed for clinical studies are synthetic. It has been proposed [96] that binding of these compounds in the minor groove of DNA requires a propellor twist of the cyclopropyldienone and indole subunits around the amide bond, and that this interrupts the vinylogous amide stabilization of the cyclopropyldienone, activating the conjugated cyclopropane electrophile (Fig. 4). Changes in the DNA binding side chain have only minor effects on the sequence selectivity of alkylation; both adozelesin and carzelesin alkylate DNA at the consensus sequences 5'-(A/T)(A/T)$\underline{A}$ and 5'-(A/T)(G/C)(A/T)$\underline{A}$, broadly similar [97] to the consensus sequence 5'-(A/T)(A/T)$\underline{A}$ for CC-1065. A series of analogs of KW 2189 with water-solubilizing cinnamate side chains were reported to have potent *in vivo* antitumor activity and low peripheral blood toxicity compared with the trimethoxyindole congeners [98], and more potent ring A dialkylaminoalkyl derivatives have also been reported [99].

2.5.3. Biological Activity and Side Effects Many of the synthetic compounds developed from the original natural product lead were also extremely potent, and showed broad-spectrum activity in human tumor colony-forming assays [100], and both adozelesin and carzelesin proceeded to clinical trial. However, adozelsin had only marginal efficacy in a Phase II trial of untreated metastatic breast carcinoma [101]. Similarly, carzelesin showed no activity in a Phase II trial in patients with a variety of advanced solid tumors [102]. A Phase I trial of KW 2189 established the maximum tolerated dose at 0.04 mg/m^2/day when given daily for 5 days, with leukopenia, neutropenia, and thrombocytopenia as dose-limiting toxicities [103]. A Phase II pilot study in metastatic renal cell carcinoma showed a good safety profile but no activity [104].

2.5.4. Recent Developments Amino analogs (e.g., 27) of the corresponding phenolic open-chain forms (e.g., 28) were reported to have comparable cytotoxicity [105,106] and similar patterns of DNA interaction, alkylating preferentially at 5'-A(A/T)$\underline{A}$N sequences [107]. An efficient synthesis of 27 and analogs has been reported [108].

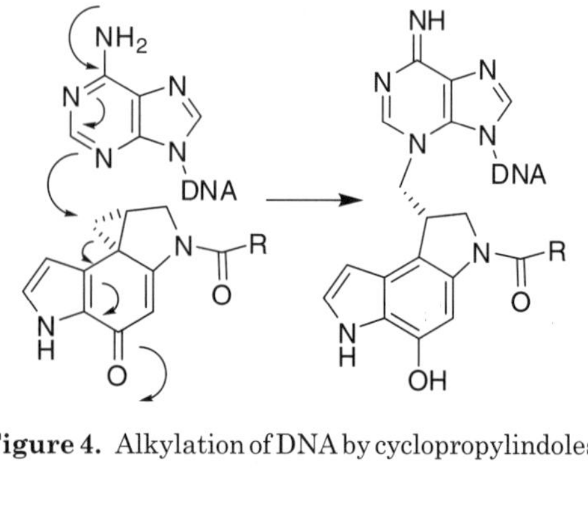

Figure 4. Alkylation of DNA by cyclopropylindoles.

2.6. Nitrosoureas

2.6.1. History This class of compounds has a long history, and extensive reviews exist on all aspects of their chemistry and biology [109]. The initial impetus for their development came from screening at the U.S. National Cancer Institute, where 1-methyl-1-nitrosourea (**29**) showed some activity in the *in vivo* leukemia screen [110]. Development of this lead resulted in the urea-based clinical agents BCNU (**30**; carmustine) and CCNU (**31**; lomustine). These reactive compounds have very short half-lives (a few minutes) [111], but their very lipophilic nature suggested they might cross the blood–brain barrier and be useful in brain tumors [112]. This has been borne out, with both compounds finding some use in various combination therapies for gliomas [113]. The more hydrophilic streptozotocin (**32**) is a natural product isolated from *Streptomyces* species. It was evaluated initially as an antibacterial agent but proved too toxic [114]. The bis(sulfonyl)hydrazine derivative Cloretazine (**33**) was designed as a compound able to generate several of the electrophilic species believed to be responsible for the alkylating and/or carbamoylating activities of the chloroethylnitrosoureas (BCNU and CCNU) [115].

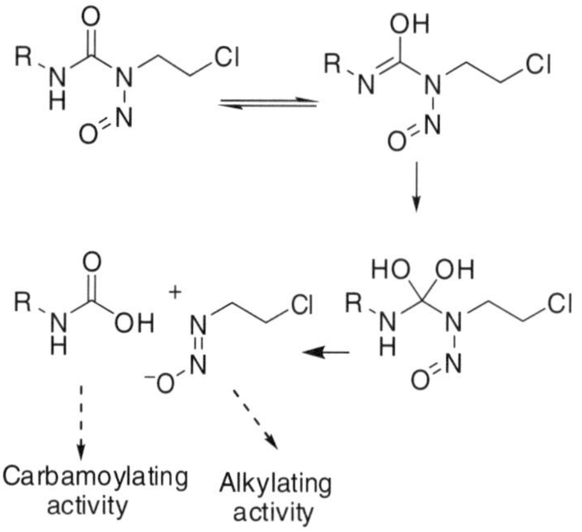

Figure 5. Mechanism of action of nitrosoureas.

2.6.2. Mechanism and SAR
The mechanism of the nitrosoureas is complex. They possess both alkylating and carbamoylating activities [116]. Decomposition occurs spontaneously in aqueous media by cleavage of the N–CO bond to give a diazoacetate (alkylating agent) and an isocyanic acid (carbamoylating agent) [117] (Fig. 5). The bis(sulfonyl)hydrazines produced a greater degree of DNA cross-linking and a lower level of single-strand breaks than the nitrosoureas. This was attributed to lower levels of guanine N7 alkylation by 2-chloroethylamine, generated by hydrolysis of nitrosoureas but not by bis(sulfonyl) hydrazines [118]. The prodrug form Cloretazine, on activation, generates chloroethylating species that alkylate DNA at the O6-position of guanine, and methyl isocyanate which inhibits O6-alkylguanine-DNA alkyltransferase (AGT), enhancing the yield of G–C interstrand cross-links responsible for the antitumor activity [119].

2.6.3. Biological Activity and Side Effects
Streptozotocin has been used as a component of multidrug protocols for the treatment of Hodgkin's disease [120], and for pancreatic [121] and colorectal [122] carcinomas, where some responses were seen, but the drug is not now widely used. Streptozocin-based therapy has recently been suggested to have some clinical utility in patients with pancreatic neuroendocrine tumors [123]. The bis(sulfonyl)hydrazine Cloretazine (33) showed broad-spectrum activity in a range of tumors, including xenografts of the human LX-1 lung carcinoma, and was extremely effective in human glioblastoma U251 xenografts [124]. Recent Phase II trials of single-agent Cloretazine showed substantial activity in patients with both previously untreated [125] or relapsed [126] acute myeloid leukemia, but was not active in patients with recurrent glioblastoma [127].

2.7. Triazenes

2.7.1. History
Dacarbazine (34; DTIC) came from studies by Shealy and coworkers, and has been well reviewed [128,129]. Mitozolomide (35) was developed by Stevens and coworkers as a potential prodrug of linear triazenes such as DTIC [130]. The same workers later followed up with the development of the related temozolomide (36), which lacks the 2-chloroethyl group [131]. The general class has been reviewed recently [132].

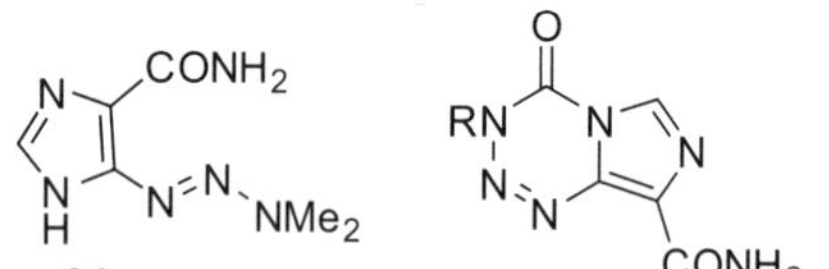

35: R=CH$_2$CH$_2$Cl
36: R=Me

2.7.2. Mechanism and SAR
The cyclic triazenes undergo base-catalyzed ring opening, followed by spontaneous decarboxylation. Thus temozolomide forms the open-chain triazene MTIC, which then fragments to a methyldiazonium species, the DNA-methylating agent (Fig. 6) [133]. Temozolomide alkylates DNA primarily at the N7 of guanine, but O6 guanine alkylation also occurs. The rate of conversion to the alkylating species is not influenced by the presence of DNA, suggesting no or very weak binding of the prodrug [133]. As with mitozolomide, cytotoxicity correlates with the alkylation of the O6-position of guanine [134]. L1210 cells treated with mitozolomide form DNA interstrand cross-links, presumably via the 2-chloroethyldiazo metabolite, suggesting this is a major mechanism of cytotoxicity [135]. Mitozolomide preferen-

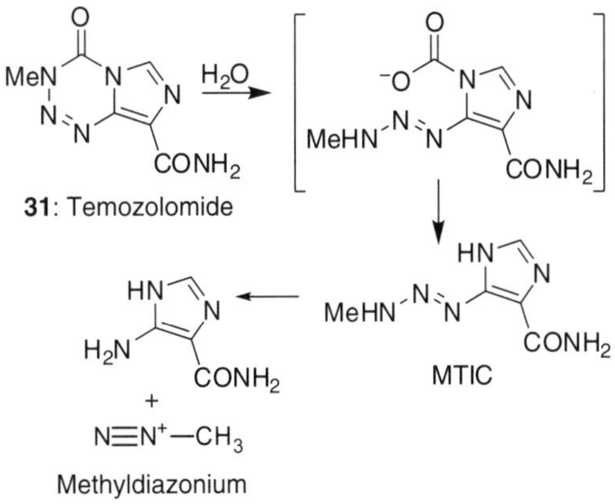

Figure 6. Mechanism of DNA alkylation by temozolomide.

tially alkylates DNA at guanines in runs of guanines, forming 7-hydroxyethyl and 7-chloroethyl adducts [136].

2.7.3. Biological Activity and Side Effects
Dacarbazine has been widely used for many years, and in particular has been the cornerstone of drug therapy for malignant melanoma [137,138]. It is metabolized by N-hydroxylation, followed by N-demethylation, to give a monomethyltriazene that then methylates DNA [139]. Not surprisingly, dacarbazine is strongly carcinogenic in animal models [140], suggesting it may also be a human carcinogen. Single-administration dacarbazine, coupled with modern antiemetic therapy, is the standard therapy for patients with advanced melanoma [141].

Mitozolomide proved curative against a broad range of murine tumor models in vivo [142], and showed very pronounced antitumor effects in a range of human tumor xenografts [143]. Cell lines with constitutive levels of O6-methylguanine-DNA methyltransferase (Mer+ phenotype) were less sensitive to the cytotoxic effects of mitozolomide, consistent with alkylation of the O6-position of guanine being the cytotoxic event [134]. In 1998, mitozolomide entered Phase I clinical trials [144], but despite demonstrable activity in SCLC and melanoma [145] unpredictable myelosuppression precluded further development [146]. However, recent work on the successful transduction of human hematopoietic progenitor cells with variants of this enzyme has led to suggestions that this could be used clinically to protect against myelosuppression, to allow safer use of agents such as mitozolomide and temozolomide in conjunction with O6-benzylguanine [147].

Temozolomide also demonstrated good in vivo activity against a variety of mouse tumor models, including the TLX5 lymphoma [148], and excellent antitumor activity, including cures, on oral administration to athymic mice bearing both subcutaneous and intracerebral human brain tumor xenografts [149]. Many later studies confirmed the good activity in brain tumor models and this, together with the lesser myelosupression seen in toxicology screens, led to Phase I trials [150]. Trials in radiotherapy-resistant astrocytomas confirmed animal data suggesting that temozolomide efficiently passes the blood-brain barrier [151]. Recent reviews show that oral temozolomide has almost 100% bioavailability and acceptable noncumulative myelosuppression, and is clinically useful in the treatment of gliomas [152,153] and brain metastases in advanced melanoma [154].

3. SYNTHETIC TOPOISOMERASE INHIBITORS

3.1. Introduction

The topoisomerases are enzymes that regulate DNA topology by successive cleavage-religation reactions, and are a major target for anticancer drugs. Topoisomerase (topo) II is a homodimeric protein, associated with the mitotic chromosome scaffold. It initially binds to DNA reversibly, and then executes a series of concerted strand-breaking and religation processes to relieve torsional stresses generated during DNA replication [155]. Topoisomerase inhibitors interfere to some extent with the normal function of the enzymes, resulting in the accumulation of DNA breaks. The observation of these breaks, characterized by protein covalently attached to the 5′-ends, first led to suggestions [156] that a topoisomerase enzyme was involved. Drugs that inhibit the function of the topoisomerase enzymes do so primarily through the formation of a ternary enzyme/DNA/drug complex, which interferes with the DNA religation step of the enzyme [157,158]. DNA intercalating agents in particular can unwind the double helix and distort DNA structure at the enzyme cleavage sites, hindering registration between the cleaved ends of the DNA during religation, leading to accumulation of the cleavable complex and an increase in strand breaks. Such drugs essentially convert the topo enzyme into a DNA damaging agent, and are also termed topo poisons [157]. The selective killing of tumor cells by topo inhibitors arises because many tumor cells overexpress these enzymes to enhance cellular proliferation, and the degree of poisoning is a function of the amount of the enzyme present.

The gene for the topo I enzyme gene is located on chromosome 20, with the gene copy number varying from 2 to 8 in a panel of colorectal cancer cell lines [159]. The topo II enzyme has major isozymes coded by two separate genes [160]. The IIα isozyme (170 kDa) maps to chromosome 17 [161], is regulated during the cell cycle, and is the target of virtually all of the DNA intercalators. The IIβ isozyme (180 kDa) maps [162] to chromosome 3 and becomes the predominant isozyme in both noncycling cells and in cells resistant to "classical" topo II agents [157,162]. Topoisomerase inhibitors are classified primarily by end function, as inhibitors of topo I, topo II, or as dual topo I/II inhibitors [163]. The largest class of such drugs (the majority of the topo II and dual topo I/II inhibitors) are DNA intercalating agents.

Intercalation of ligands into DNA was first described by Lerman [164] for the acridine proflavine (**37**), and involves insertion of the chromophore between the base pairs. This is now understood to be the major DNA binding mode of virtually any flat polyaromatic ligand of sufficiently large surface area and suitable steric properties. Intercalative binding is driven primarily by stacking (charge-transfer and dipole-induced dipole) and electrostatic interactions, with entropy (dislodgement of ordered water around the DNA) of lesser and variable importance [165]. A great deal of work has been done delineating the ligand structural properties which favor intercalation, the geometry, kinetics and DNA sequence-selectivity of the binding process, and the effect of such binding on the structure of the DNA substrate [166]. A large number of compounds have been shown to be DNA intercalating agents, and many of these show cytotoxic activity by the mechanism outlined above [167].

H_2N — acridine — NH_2

37

3.2. Clinical Use of Agents

The major original topo I inhibitor was the natural product camptothecin (**38**), which is outside the scope of this chapter (and in any case has been widely reviewed). However, most of the more recent topo I agents are synthetic or semisynthetic analogs, and will be briefly dealt with here, along with other synthetic topoisomerase inhibitors (Table 2).

3.3. Topo I inhibitors

3.3.1. History Camptothecin itself (**38**) was first isolated from the plant *Camptotheca acuminate* [168]. Clinical trials were delayed due to problems with insolubility and instability, particularly of the ring E lactone, which is required for activity but is unstable in aqueous solution. The analogs topotecan (**39**) and the carbamate-based prodrug irinotecan (**40**) have become established clinically [169], but several more recent analogs are now in trial. These feature further substituents at positions 9–11 of ring A or 7 of ring B to increase water solubility and bioavailability, together with changes to stabilize the ring E lactone.

3.3.2. Mechanism of Action and SAR Human topo I relaxes superhelical tension generated during DNA replication and transcription by reversibly cutting one strand of duplex DNA via a covalent 3′-phosphotyrosine link. Crystal structure [170] and modeling [171] studies of the topo I/DNA/camptothecin (and analogs) ternary complex have provided valuable insights into the mechanism of this process. Camptothecin binds to the topo I/DNA

Table 2. Synthetic Topoisomerase (Topo) Inhibitors Used in Cancer Chemotherapy

Generic Name (Structure)	Trade Name	Marketer	Chemical class
Topo I inhibitors			
Rubitecan (**41**)	Orathecin	Supergen	Camptothecin
Belotecan (**42**)	Camtobell	Chong Keun Dang	Camptothecin
Exatecan (**43**)			Camptothecin
Gimatecan (**44**)		Novartis	Camptothecin
Karenitecin (**45**)		Bionumerik	Camptothecin
Diflomotecan (**46**)		Ipsen/Roche	Camptothecin
+			
Topo II inhibitors			
Amrubicin (**48**)	Amrubicin	Pharmion	Anthracycline
MEN-10755 (**49**)	Sabarubicin	Menarini	Anthracycline
Amsacrine (**50**)	Amsidyl	Warner-Lambert	9-Anilinoacridine
Asulacrine (**51**)		Sparta	9-Anilinoacridine
Mitoxantrone (**52**)	Novantrone	Wyeth	Anthracenedione
Pixantrone (**54**)		Boehringer	Anthracenedione
Losoxantrone (**56**)	Biantrazole	Warner-Lambert	Anthrapyrazolone
Piroxantrone (**57**)	Oxantrazole	Warner-Lambert	Anthrapyrazolone
C-1311 (**58**)	Symadex	Xanthus	Imidazoacridinone
XL-119 (**59**)	Becatecarin	Helsinn	Pyrrolocarbazole
Dual topo I/II inhibitors			
DACA (**62**)	XR-5000	Xenova	Acridine
Pyrazoloacridine (**63**)			Acridine
Intoplicine (**64**)		Ilex	Pyridoindole
TAS 103 (**65**)		Taisho	Indenoquinolone
Tafluposide (**66**)			
LU 79553 (**71**)	Elinafide		Bis(naphthalimide)
MLN-944 (**72**)		Millennium	bis(phenazine)

covalent complex with the A, B and E rings of the drug stacked in the DNA on place of a flipped-out guanine base, with additional stacking interactions with this base and with Asn722. The lactone and 20(S)-hydroxy group make additional H-bond contacts to Arg364 and Asp533. This model is consistent with the fact that mutations in these residues confer resistance to camptothecins [172].

3.3.3. Biological Activity and Side Effects

Rubitecan (9-nitrocamptothecin) (**41**) is the most advanced of the new camptothecin analogs. It undergoes rapid *in vivo* reduction to the corresponding 9-amino analog [173], which is also an active drug that has been evaluated clinically; thus, it could be considered a type of prodrug. Unlike camptothecin, rubitecan is little affected by cellular efflux pumps [174]. A number of Phase II trials showed very limited activity, but a recent trial in previously treated refractory pancreatic cancer showed 3/43 partial responses and 7/43 stable disease [175], and Phase III trials are in progress. Rubitecan is very insoluble, and there is interest in liposomal formulations to improve delivery [176].

Belotecan (**42**) is a more soluble analog, with an N-aminoethyl group at the 7-position of ring B where there is some bulk tolerance [177]. It is equipotent with camptothecin as an inhibitor of topo I *in vitro*, but is susceptible (like camptothecin) to P-glycoprotein-mediated efflux, and was comparable to topotecan in a range of human tumor xenografts in nude mice (e.g., SKOV-3 ovarian, MX-1, LX-1, HT-29, WIDR, and CX-1 colon) [178]. In Phase II trials, belotecan showed high response rates (65%) in topo I inhibitor-naïve patients [179].

Exatecan (**43**) is a more extensively modified camptothecin, with an additional alicyclic ring fused to rings A and B bearing a solubilizing primary amine. Lipophilic substituents at positions 10- and 11- on ring A may help to enhance membrane permeability [180]. It is fully active in Pgp overexpressing cell lines [181] possibly due to its enhanced cell penetration ability [182]. Extensive Phase II trial studies suggest the drug has little activity as a single agent in gastric, biliary tract and colon cancer, but may have some utility in the treatment of refractory ovarian and pancreatic cancer [183,184].

Gimatecan (**44**) is a more lipophilic analog, following a QSAR study that showed overall lipophilicity correlated best with both cytotoxicity and the potency of topo I inhibition [185]. Fluorescence studies suggest lysosomal rather than mitochondrial disposition of gimatecan compared with topotecan [186]. It showed lower levels of intracellular accumulation than other camptothecin analogs, but was the most active against neuroblastoma cell lines, suggested due to an ability to cause high levels of DNA breaks [187]. *In vitro* it is a modest substrate for the efflux transporter BCRP, but not for MDR1/2 [188]. Preliminary results from a Phase II trial in recurring ovarian cancer suggest oral gimatecan administered as a single agent is active [189].

Karenitecin (**45**) is also a lipophilic camptothecin derivative, designed for higher oral bioavailability. Cell culture studies in a variety of tumor cell lines showed that karenitecin was not affected by overexpression of P-gp [190], and was a much poorer substrate than topotecan for BCRP [191]. Studies in karenitecin-resistant cell lines suggest that resistance is related to upregulation of the chk1 signaling pathways that mediate the G2/M cell cycle checkpoint [192]. A Phase II trial of karenitecin in patients with metastatic melanoma (which expresses high levels of topo I), using an intravenous infusion of 1 mg/m^2 karenitecin daily for 5 days every 3 weeks, showed 1/43 complete responses, and 33% patients with some degree of stable disease. The major toxicity was reversible noncumulative myelosuppression [193].

Diflomotecan (**46**) is a homocamptothecin possessing a 7-membered hydroxylactone ring [194], and was the first homocamptothecin taken to clinical trial, on the basis of its high activity and the improved stability of the lactone ring (half-life of about 2 h in human plasma at 37 °C, compared with about 5 min for camptothecin) [195]. It has higher overall antiproliferative activity than camptothecin, topotecan or SN-38 in a series of 43 early passage human colon cancer cell lines in culture [196], and broad activity in human xenograft models [195,197]. Phase I trials of both oral and intravenous diflomotecan in solid tumors in adults have been reported [198,199]. These showed linear pharmacokinetics, and thrombocytopenia/neutropenia as dose-limiting toxicities.

3.3.4. Recent Developments Included above.

3.4. Topo II inhibitors

3.4.1. History The natural product doxorubicin (**47**) is still the most important single topo II inhibitor clinically, and several semisynthetic analogs (e.g., epirubicin, idarubicin, and esorubicin) have been well reviewed and are not reported on here, but the newer synthetic analogs amrubicin (**48**) and sabarubicin (MEN-10755; **49**) are discussed. The other two main classes of synthetic topo II inhibitors are the 9-anilinoacridines and the anthracenediones. The 9-anilinoacridine amsacrine (*m*-AMSA; **50**) evolved from work carried out by Cain and associates on antileukemic quinolinium-type agents [200,201]. Encouraging results in both leukemias and lymphomas, including an apparent lack of cross-resistance to doxorubicin [202], resulted in amsacrine becoming the first synthetic DNA intercalator to show clinical efficacy [203]. The success of amsacrine led to a search for analogs with a broader spectrum of action. Because the high pK_a (8.02) of amsacrine was thought to play a part in limiting its distribution, analogs with a lower pK_a that still retained high DNA binding and had improved aqueous solubility were sought. QSAR studies [204] suggested that the anilino side chain was close to optimal, and focused attention on the 4- and 5-positions of the acridine as being the most suitable for modification. The 4-methyl-5-methylcarboxamide asulacrine (**51**) was the most active analog against both a human solid tumor cell line panel [205] and a wide range of murine solid tumors *in vivo* [206].

Mitoxantrone (**52**) was discovered through screening of industrial dye compounds [207]. Both mitoxantrone (**52**) and the des-hydroxy analog ametantrone (**53**) showed broad-spectrum activity in animal tumor models. Mitoxantrone has become perhaps the most widely used synthetic DNA intercalating agent, and has been well reviewed [208].

3.4.2. Mechanism of Action and SAR

Much of the early SAR work on the DNA intercalator class of drugs focused around their interaction with DNA, delineating the requirements for successful intercalation and tight binding [166]. Early SAR studies for several classes of intercalators showed positive relationships between cytotoxic potency and strength of DNA binding [209,210] and long residence times for the intercalators at individual DNA sites [211]. However, the discovery that the primary mode of cytotoxicity of these compounds was inhibition of topo II via formation of a ternary drug/DNA/protein complex [157,158] made it clear that drug design through modeling of DNA binding properties alone could be misleading. A new model of a ternary topo II enzyme/DNA/anthracycline complex [212] suggests the anthracycline molecule intercalates within the "gate"-DNA at the cleavage site, stabilized by stacking interactions between the anthracycline and the cleaved DNA and H-bond interactions with both the DNA and the enzyme. The exact hydrophilic interactions depend on the structure of the anthracycline, and involve the sugar of the drug. The authors suggest the anthracycline acts as a mechanical obstacle to the DNA religation reaction, due to the larger distance between the cleaved-DNA bases in the presence of an intercalation site.

Amsacrine binds to DNA by reversible, enthalpy-driven [213] intercalation of the acridine chromophore, with an association constant of $1.8 \times 10^5 \, M^{-1}$ for calf thymus DNA in 0.01 M salt [214]. It was postulated to bind with the anilino ring lodged in the minor groove, with the 1'-substituent pointing tangentially away from the helix with the possibility of it interacting with another (protein) macromolecule to form a ternary complex [214]. In such a complex, DNA intercalation of the acridine occurs, with the anilino side chain making additional protein contacts [215,216].

In a panel of human breast cancer cell lines, sensitivity to amsacrine was shown to correlate with the level of expression of the topo IIα protein (although not with the level of topo IIα mRNA) [217], suggesting that the former is the most important mechanism of resistance to these topo II inhibitors. A human SCLC line with acquired resistance to amsacrine did not overexpress P-glycoprotein but had an 82% decrease in topo IIβ protein level [218]. A classification of antitumor drugs by their topo II-induced DNA cleavage activity and sequence preference placed amsacrine in the class which enhanced the stabilization of cleavable complexes at a single major site, acting upstream of the DNA cleavage step [219]). Amsacrine appears unique among topo II poisons in that its ability to trap both topo IIα and topo IIβ-induced lesions is only modestly reduced in ATP-depleted cells, suggesting it produces mainly prestrand and not poststrand passage DNA lesions in intact cells [220].

Mitoxantrone also binds reversibly to DNA by intercalation [221], with an unwinding angle of 23°, probably with the chromophore inserted perpendicular to the base pair axis with the side chains lying in the major groove [222], although this has not been rigorously proven. Footprinting studies show the preferred intercalation site for mitoxantrone to be 5'-(A/T)CG or 5'-(A/T)CA sites [223]. Mitoxantrone and the related ametantrone bind tightly and about equally well to DNA with association constants of approximately $5 \times 10^6 \, M^{-1}$ at physiological salt concentrations [224], but mitoxantrone has approximately fourfold slower dissociation kinetics [221,225]. The higher cytotoxicity of mitoxantrone compared with ametantrone correlated with its higher capacity to induce topo II-mediated cleavable complexes, suggested because of greater stability of the ternary complex [226]. Mitoxantrone induces DNA fragmentation and activates caspases, demonstrating that the ultimate cytotoxic effect is induction of apoptosis [227]. It is readily oxidized (for example, by human myeloperoxidase) to metabolites that covalently bind to DNA [228] and also, like the anthracyclines, forms formaldehye-induced adducts that function as virtual interstrand cross-links [229].

3.4.3. Biological Activity and Side Effects
Amrubicin (**48**) was designed as a less cardiotoxic anthracycline [230] and is a prodrug, with the major metabolite (the 13-alcohol amrubicinol) being more cytotoxic in cell culture than amrubicin itself [231]. Both compounds act via topo II-generated double-strand DNA breaks [232], and are subject to P-glycoprotein-mediated efflux [233]. In human tumor xenografts the metabolite accumulates to higher levels in the tumors but at lower levels in normal tissues than does doxorubicin, suggesting that its selective distribution plays a role in the observed therapeutic activity [234]. A number of Phase II studies suggest that amrubicin, likely in combination with cisplatin, will become an important therapy for both small cell and nonsmall cell lung cancer [235,236].

Sabarubicin (**49**) was the first anthracycline disaccharide to be used clinically. A crystal structure of sabarubicin bound to the oligodeoxynucleotide d(CGATGG)$_2$ shows the anthracycline binding similarly to other analogs, while both sugar rings lie in the minor groove [237]. Sabarubicin is more effective than doxorubicin in human tumor xenografts in nude mice [238], but was susceptible to cell efflux mechanisms [239]. A human ovarian tumor cell line resistant to sabarubicin showed lower expression of the transporter protein MRP, lower topo II content and reduced activation of the NF-κB pathway [240]. A phase II study of sabarubicin in hormone refractory prostate cancer gave a 26% response rate (PSA reduction), with the authors concluding that larger trials were warranted [241].

The 9-anilinoacridine amsacrine (51) is used clinically mainly in acute myeloid leukemia (AML). Amsacrine/etoposide therapy with or without azacytidine in relapsed childhood acute myeloid leukemia was effective (34% complete responses), with azacytidine not improving response rates [242]. Recent successful use in various adult leukemias has also been reported [243,244]. A trial in elderly patients with AML showed that amsacrine could replace daunorubicin in multidrug regimes with equal efficacy and lowered cardiac toxicity [245]. Amsacrine has generally not been successful in the treatment of solid tumors, except for some responses in head-and-neck cancer, where high-dose amsacrine was a toxic but very effective drug for first-line treatment [246], with a response rate of 65%. While much less cardiotoxic than the anthracyclines, preexposure to amsacrine is a risk factor for cardiotoxicity following subsequent anthracycline treatment for childhood cancer [247].

Asulacrine (52) showed a similar mechanism of action to amsacrine, generating DNA protein cross-links and DNA breaks via inhibition of topo II [248]. In initial Phase II trials, some drug-induced remissions were seen in nonsmall cell lung cancer and breast cancer, but not in colorectal and gastric cancer [249,250]. A pilot study of a trial of oral administration was carried out [251], but there are no reports of asulacrine being used clinically in combination therapy.

Mitoxantrone (53) is used in first-line therapy for AML [252], and with cytosine arabinoside is suggested as salvage therapy in AML and CML [253]. It is also an effective treatment for secondary progressive multiple sclerosis, but the duration of treatment is limited by cumulative cardiotoxicity [254]. Again, while less cardiotoxic than the anthracyclines, mitoxantrone has been shown to have cumulative cardiotoxic effects [255]. Mitoxantrone is genotoxic in the *in vitro* micronucleus test and in mutation assays [256], and has been reported to induce secondary cancers after use in the treatment of breast cancer [257]. Resistance to mitoxantrone can develop in a number of ways; by lower expression of topo II [258], by expression of a topo II with altered DNA cleavage activity [259], by decreased drug uptake even in the absence of elevated levels of P-glycoprotein [260], and by inherent resistance to the induction of apoptosis [261]. Many cells develop multifactorial resistance to mitoxantrone [262].

Topoisomerase inhibitors are also known to be tumorigenic, related to the formation of multiple DNA strand breaks. A frequent chromosomal translocation is at 11q23, where the myeloid-lymphoid leukemia (MLL) gene is located [31], but other translocations are also seen. The onset of induction of AML is shorter than with alkylating agents, with the average of approximately 2 years and an incidence of 2–12% [263]. Anthracyclines, mitoxantrone, and epipodophyllotoxins have all been shown to induce AML [264].

3.4.4. Recent Developments: Compounds with Lower Cardiotoxicity

The development of analogs of mitoxantrone has been driven largely by the requirement for lower cardiotoxicity. Two broad classes of analogs can be distinguished, and much work has been done on both. The first are close analogs of mitoxantrone, where the tricyclic chromophore has been maintained and variations occur in the side chains or the chromophore atoms. The positioning of the aza group appears to be important, with the 2-aza derivative pixantrone (54) being the most potent [265]. This compound bound less tightly to DNA, but induced topo II-mediated DNA cleavage [266], and was less cardiotoxic than both mitoxantrone and doxorubicin in mouse studies [267]. As for mitoxantrone, pixantrone forms cross-links with DNA in the presence of formaldehyde, but at much lower formaldehyde levels [268]. Pixantrone showed broadly similar effectiveness when replacing doxorubicin in a number of multidrug treatment protocols for relapsed aggressive non-Hodgkin's lymphoma [269]. A phase III trial of pixantrone as a single agent in patients with relapsed non-Hodgkin's lymphoma is in progress [270].

The second broad class is tetracyclic compounds, primarily the imidazoacridinones and the anthrapyrazoles. The initial SAR studies [271] showed that biological activity was maximal with alkylamino side chains at the N-2 and C-5 positions with two to three carbon spacers between proximal and distal nitrogens, and also showed they induced less oxygen consumption than doxorubicin in the rat liver microsomal system. This class of compounds bound very tightly to DNA by intercalation, with association constants of approximately $2 \times 10^8 M^{-1}$ [272], and were highly active in murine leukemias and a range of human tumor xenografts [273]. Three analogs were taken to clinical trial. Teloxantrone (**55**) has not been reported on following an initial clinical trial [274], but losoxantrone (**56**) and piroxantrone (**57**) have been more widely studied. Losoxantrone showed classical topo II inhibition [275]. The major metabolites detected in humans resulted from oxidation of the hydroxymethylene side chains to either mono- or dicarboxylic acid derivatives [276]. A Phase II trial of losoxantrone in hormone-refractory metastatic prostate cancer showed improvement of clinical symptoms in one-third of patients [277]. Piroxantrone showed some responses in Phase I trials [278], but this was not borne out in Phase II trials [279,280].

Structure–activity studies on imidazoacridinones [281] identified Symadex (C-1311; **58**) as a potential anticancer drug that intercalates DNA and inhibits the catalytic activity of topo II [282]. It shows potent growth inhibition in a wide range of human tumor cell lines, independent of p53 status [283]. Symadex undergoes metabolism under either oxidative [284] or electrochemical [285] conditions via intermediates that may be involved in its demonstrated ability to form covalent adducts with DNA in cells. The drug is reported to be in Phase II trial in metastatic colorectal cancer [286].

Becatecarin (XL-119; **59**) is a more water-soluble semisynthetic [287] derivative of the natural antitumor antibiotic rebeccamycin [288]. It showed IC_{50} values between 0.5-1 µM against a series of established and early passage cell lines derived from pediatric solid tumors; this is similar to the plasma levels achieved during adult phase I clinical trials [289]. Phase II studies in renal [290] and refractory breast cancer [291] showed modest activity, but advanced colorectal [292] and non-small cell lung cancer [293] were not responsive. The major toxicity was myelosuppression.

3.5. Dual Topo I/II Inhibitors

3.5.1. History
The topo I and topo II enzymes are expressed at different absolute levels in different cell types. Topo II levels are reported to be high in many breast and ovarian lines [294], while topo I levels are reported to be high in many colon cancer lines [295]; the good clinical activity of camptothecin analogs against colon tumors has been suggested due in part to this high level of topo I expression [296]. The time course of expression of topo I and topo II also differs markedly, with topo II levels at their highest during S-phase, while levels of topo I remain relatively constant through the cell cycle [297]. Since expression of either enzyme appears to be sufficient to support cell division, the development of resistance to topo I inhibitors is often accompanied by a concomitant rise in the level of topo II and vice versa [298,299].

Thus, one of the recent interests in topo inhibitors has been in agents capable of simultaneous inhibition of both enzymes, although relatively few compounds have been reported as dual topo I/II inhibitors [163]. The quaternary alkaloid nitidine (**60**) is reported [300] to be a dual poison, although more active against topo I. The related quaternary salt NK 109 (**61**) is described as a topo II poison, but etoposide-resistant lines with reduced topo II levels are still sensitive [301], suggesting a dual activity. Most work has been focused on the DNA intercalators DACA (XR-5000; **62**), pyrazoloacridine (**63**), intoplicine (**64**), TAS 103 (**65**), and the epipodophyllotoxin analog tafluposide (**66**).

The 9-aminoacridine-4-carboxamides are DNA-intercalating agents [302] with well-defined structure–activity relationships for both the chromophore and side chain [303]. The derived acridine-4-carboxamide analog DACA (**61**) also binds to DNA by intercalation, and induces DNA cleavage in the presence of either topo I or topo II enzymes, being unaffected by either P-glycoprotein-mediated multidrug resistance or "atypical" multidrug resistant due to low topo II activity [304]. DACA showed remarkable activity against multidrug resistant cells [305] and *in vivo* activity against the Lewis lung carcinoma [306].

Pyrazoloacridine (**63**) is a DNA intercalator selected for advancement due to its outstanding activity in solid tumor compared with leukemia models [307]. It is a potent inhibitor of the catalytic activity of both topo I and topo II enzymes [308], but without stabilization of the topo-DNA cleavable complexes [309] suggesting an unusual mechanism of action. It is also a hypoxia-selective agent [310], with bioreduction of the 5-nitro group likely to form transient alkylating species; the corresponding 5-amino derivative is a major metabolite in mice [311].

The DNA-intercalating [312] pyridoindole intoplicine (**64**) is reported to also be a dual topo I/II poison [312,313]. Analogs of intoplicine that were only topo I or topo II poisons were less cytotoxic [312], suggesting the possible utility of a dual poisoning ability. Intoplicine showed activity in a variety of human tumor explants in a soft agar cloning assay [314] and in transplantable mouse tumors *in vivo* [315].

The indenoquinolone TAS-103 (**65**) is also reported to be a DNA intercalating agent [316] and to enhance both topo I and topo II-mediated DNA cleavage in treated cells [317], but it is now considered that topo II is the primary cellular target [318]. An affinity bead approach detected a 54 kDa subunit (SRP54) of the signal recognition particle (SRP) as a TAS-103 binding protein and cellular studies confirmed that TAS-103 disrupts SRP complex formation [319]. TAS-103 showed broad-spectrum activity against a number of cell lines, with no cross-resistance in cells with lower topo I expression and only slight cross-resistance in those where topo II was downregulated [320].

Finally, tafluposide (**66**) is a novel phosphate prodrug of a lipophilic, perfluorinated epipodophyllotoxin, and is an example where significant modification of a topo II inhibitor has altered its spectrum of enzyme activity. It is a potent inhibitor of the catalytic activities of both enzymes, inhibiting the binding of the enzymes to DNA in a drug- and enzyme-dependent manner [321] It has superior antitumor activity *in vivo* compared to etoposide [322,323] and may represent a new class of topoisomerase agent.

3.5.2. Mechanism and SAR There appears to be no clear structural features predisposing to dual topo I/II activity. Raman and CD studies of intoplicine analogs suggest that the dual poisoning abilities of intoplicine are due to its ability to simultaneously form two types of DNA complexes; a "deep intercalation mode" responsible for topo I-mediated cleavage and an "outside binding mode" responsible for topo II-mediated cleavage [324].

3.5.3. Biological Activity and Side Effects DACA (**62**) undergoes rapid oxidative metabolism by aldehyde oxidase to give the acridone (**67**), and oxidative demethylation of the side chain dimethylamino group has also been observed [325]. Pharmacological studies showed high binding to human α1-acid glycoprotein, followed by albumin [326]. In Phase I clinical trials, the major urinary metabolite was the *N*-oxide (**68**) [327]. The dose-limiting toxicity in the Phase I trials was arm pain of unknown cause at the infusion site [328]. This was avoided in Phase II trials by using a 5-day infusion, but the compound was not active in glioblastoma or ovarian cancer [329,330].

67: R = NMe$_2$
68: R = N(O)Me$_2$

Pyrazoloacridine has received extensive clinical trial because of its broad-spectrum activity in animal solid tumor models and its novel mechanism of action, but was inactive in many Phase II trials [331]. More recently, it was shown to be highly cytotoxic in a series of multidrug-resistant neuroblastoma cell lines, suggesting it may be effective clinically against neuroblastoma [332]. A Phase II study of a PZA/carboplatin combination in patients with recurrent glioma showed only modest activity, with short-term disease stabilization in approximately 38% of the patients [333].

Phase I trials of intoplicine have been conducted [334,335] and noted hepatotoxicity rather than myelosuppression as the major dose-limiting toxicity [336]. A Phase I clinical trial of TAS-103 recommended a dose of 130–160 mg/m^2 for Phase II trials [337], but these have not yet been reported. A P388 leukemia subline resistant to tafluposide showed decreased nucleotide excision repair activity, suggesting that both topoisomerase IIα and DNA repair enzymes are major targets [338]. Tafluposide showed synergistic effects in the A549 human nonsmall cell lung cancer line in culture with cisplatin, etoposide, doxorubicin, and mitomycin C [339]. The drug has been advanced to clinical trial [340], but results have not yet been reported.

3.5.4. Recent Developments: Bis Analogs as Dual Topo I/II Inhibitors

Because of the early SAR suggesting a positive correlation between cytotoxic potency and the strength of DNA binding, and because bis-intercalation would theoretically greatly increase DNA binding, many dimeric compounds designed as bis-intercalators were evaluated as anticancer drugs [166,341]. However, the biological activities of these compounds were generally disappointing. The bis(acridine) (**69**) was considered for clinical trial [342] but had significant CNS toxicity, and the bis(ellipticine) analog ditercalinium (**70**) had unacceptable mitochondrial toxicity [343].

However, several examples of dimers of more lipophilic, neutral chromophores have shown potent and broad-spectrum activity against a variety of human solid tumor cell lines, both in culture and as xenografts in nude mice. Elinafide (**71**) is a symmetric dimeric bis(naphthalimide) [344] reported as a DNA bis-intercalator binding in the major groove [345]. An NMR structure of elinafide complexed with d(ATGCAT)$_2$ confirmed this, showing that the two naphthalimide chromophores bis-intercalate at the TpG and CpA steps of the DNA, with the linker chain lies in the major groove. The naphthalimide rings exchange by rotational ring flipping (at 1800 s^{-1} at 36 °C), without affecting the binding of the linker chain region [346]. Elinafide proved highly cytotoxic *in vitro* (IC$_{50}$ values as low as 0.2 nM), and active in a series of human xenograft models in nude mice [347]. Phase I trials showed a dose-limiting neuromuscular toxicity [348,349], which may limit further development.

The dicationic bis(phenazine) (**72**; MLN-944) is in the same broad structural class as elinafide, binding tightly ($K_a = 1.6 \times 10^9 \text{M}^{-1}$) to DNA by intercalation, preferentially to GC-rich regions, suggesting a binding mode with the linker chain lying in the major groove [350]. This was again confirmed by an NMR structure of a complex with the DNA duplex d(ATGCAT)$_2$, showing the two phenazine rings intercalated at 5′-TpG sites, with the linker chain lying in the major groove [351]. MLN-944 was a very potent cytotoxin (IC$_{50}$ of 0.08 nM in human leukemia cells) [350], stabilizing cleavable complex formation between DNA and both topo I and human topo IIα, with fragmentation patterns different to that generated by the specific inhibitors etoposide and camptothecin [352]. Studies in synchronized human HCT 116 cells showed MLN-944-induced G1 and G2 arrest, unlike the typical G2-M arrest noted with known topoisomerase poisons. Transcriptional profiling analysis of treated xenografts showed clusters of regulated genes distinct from those observed in irinotecan-treated tumors, suggesting that the primary mechanism does involve DNA binding, but not topoisomerase inhibition [353]. MLN-944 inhibited transcription initiation of all RNA polymerases, as well as inhibiting transcription elongation at higher concentrations, suggesting that transcription is the primary target, and the reason for its high cytotoxicity [354]. A later study [355] showed it specifically inhibited the binding of the estrogen receptor to its consensus DNA sequence, and suggested that this novel mechanism may overcome resistance to current antiestrogens. *In vivo*, MLN-944-induced regressions of both HT29 and H69/P xenografts, inducing 100% cures in the latter model at an intravenous dose of 5 mg/kg on a daily schedule [352]. A Phase I clinical trial of MLN-944 [356] showed 1/27 partial responses, but noted that the lack of correlation between toxicity and pharmacokinetics needed more work before a recommendation for Phase II studies.

(Fig. 7). They are also key intermediates in the GAR-formyltransferase- and AICR-formyltransferase-mediated construction of the purines [357]. The first antifolate used clinically was aminopterin (**73**), rapidly followed by methotrexate (**74**), which was registered for clinical use in 1953. These "classical" (glutamate-containing) antifolates bind tightly to the enzyme dihydrofolate reductase (DHFR) (Fig. 7). Methotrexate has been very widely used, and has been extensively reviewed [358].

4. ANTIMETABOLITES

4.1. Introduction

The class of compounds known broadly as antimetabolites interfere in varying ways with the synthesis of DNA. Along with the alkylating agents, antimetabolites such as methotrexate, 5-fluorouracil, cytosine arabinoside, 6-mercaptopurine, and 6-thioguanine were some of the earliest drugs used in cancer chemotherapy.

4.2. Clinical Use of Agents

The most commonly used antimetabolites are listed in Table 3. These compounds are invariably used in combination with other agents in multidrug therapy regimens.

4.3. Antifolates

4.3.1. History Antifolates interfere at various points in the process (folic acid metabolism) that provides the one-carbon unit required to convert deoxyuridine monophosphate to thymidylic acid for synthesis of the pyrimidines

A more recent classical antifolate is the 10-ethyl analog edatrexate (**75**). This was developed following observations that 1-alkyl analogs showed better relative uptake into tumor tissue, and edatrexate does show enhanced uptake, retention and polyglutamate formation in tumor cells [359]. While edatrexate binds to DHFR similarly to methotrexate, it showed better activity in animal tumor models, including models resistant to methotrexate [360]. In a Phase II trial in pretreated metastatic breast cancer, edatrexate was well tolerated but showed limited effectiveness as a single agent [361]. Resistance to methotrexate and edatrexate arises in several ways, the most important of which are elevation of DHFR levels and lowering of both folate transport and polyglutamylation activities [362]. A recent study [363] identified a mutation (Cys346Phe) mapping to the active site of folylpoly-γ-glutamate synthetase as responsible for the decrease in catalytic activity of the enzyme, by lowering its affinity L-glutamate.

The enzyme thymidylate synthase (TS) is also intimately involved in folate metabolism, catalysing the reductive methylation of deox-

Table 3. Antimetabolites Used in Cancer Chemotherapy

Generic Name (Structure)	Trade Name	Marketer	Chemical Class
Folic acid analogs			
Methotrexate (**74**)		Cyanamid	Folate analog
Edatrexate (**75**)			Folate analog
Raltitrexed (**77**)	Tomudex	Lilly	Folate analog
Pemetrexed (**78**)	Alimta	Lilly	Folate analog
Trimetrexate (**79**)	NeuTrexin	US BioScience	Folate analog
Piritrexim (**80**)		Burroughs-Wellcome	Folate analog
Nolatrexted (**81**)	Thymitaq	Zarix	Folate analog
Pyrimidine analogs			
5-Fluorouracil (**82**)	Adrucil	Roche	Pyrimidine
Cytosine arabinoside (**83**)	Cytosar	Pharmacia & Upjohn	Pyrimidine
Gemcitabine (**84**)	Gemzar	Lilly	Pyrimidine
Capecitabine (**85**)	Xeloda		Pyrimidine
Decitabine (**88**)	Dacogen	Supergen/MGI	Triazinone
Azacitidine (**89**)	Vidaza	Pharmion	Triazinone
Purine analogs			
6-Mercaptopurine (**90**)	Purinethol	Burroughs-Wellcome	Urine analog
6-Thioguanine (**91**)	Lanvis	Glaxo-Wellcome	Purine analog
Fludarabine (**92**)	Fludara	Berlex Laboratories	Purine analog
2′-Deoxycoformycin (**93**)	Pentostatin	Supergen	Purine analog
2-Chloro-2′deoxyadenosine (**94**)	Cladribine	Bedford Laboratories	Purine analog
Forodesine (**95**)			

yuridine monophosphate (dUMP) to thymidylate (dTMP), a reaction in which N^5,N^{10}-methylenetetrahydrofolate is a cofactor (Fig. 7). While the pyrimidine binding site on the TS enzyme has been a major target for anticancer drugs such as 5-fluorouracil (see Section 4.4), it also has a folate binding site that has been a target for drug development. Methotrexate itself binds weakly to this site, and can exercise cytotoxicity via TS inhibition in cells that highly overexpress DHFR [364]. The design of highly specific inhibitors of the folate binding site of TS led initially to the quinazoline derivative CB 3717 (**76**) [365]. This proved a tight binding inhibitor of TS ($K_i = 4.5$ nM), with 10-fold selectivity over DHFR, and able to undergo polyglutamylation in cells to metabolites that are more potent and more selective for TS over DHFR [366]. CB 3717 showed some activity in a number of Phase I/II clinical trials, but severe nephrotoxicity, due probably to precipitation of drug in the kidneys [367], led to its withdrawal [368].

Raltitrexed (Tomudex; **77**) is another "classical" folic acid derivative that exerts its therapeutic effect through inhibition of the folate site of TS [369]. It is polyglutamylated in cells into metabolites are more potent inhibitors of TS than the parent drug, and are retained in cells. Raltitrexed showed activity in a number of tumor types in Phase I/II trials, including colon cancer [370] and mesothelioma, where cisplatin/raltitrexed combinations significantly improve median survival compared to single-agent cisplatin [371].

Pemetrexed (**78**; MTA) also has TS as a major target, with DHFR and glycinamide ribonucleotide formyltransferase (GARFT) being important secondary sites of action [372]. Pemetrexed is an excellent substrate for FPGS, and it and its polyglutamylated metabolites are potent inhibitors for all of the above enzymes [373]. Pemetrexed performed well in the human tumor cloning assay against colorectal (32% of cell lines inhibited) and nonsmall cell lung cancer (25% of cell lines inhibited) [374]. It showed broad antitumor activity in phase II trials with breast, colon, pancreatic, bladder, head and neck, and cervical carcinomas, and nonsmall cell lung can-

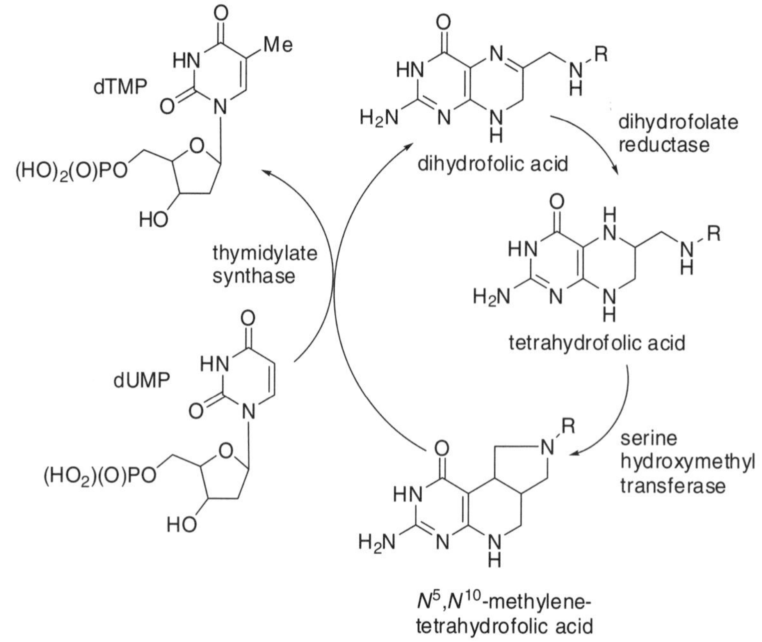

Figure 7. Folate biosynthesis.

cer [375]. The pemetrexed-cisplatin combination has become the new standard of care for patients with mesothelioma [376], significantly improving response rates over cisplatin alone (41% versus 17%) [371].

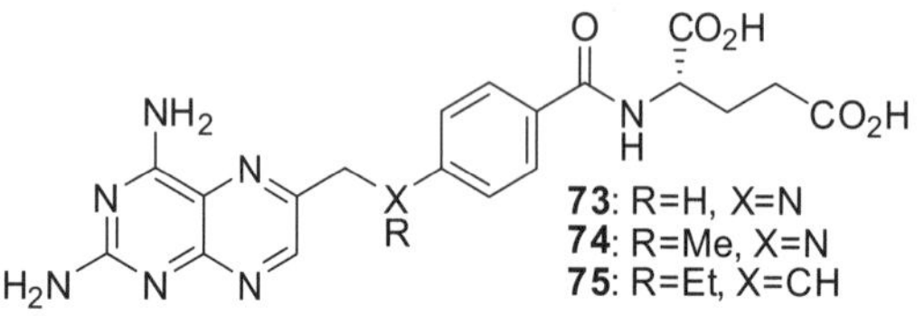

4.3.2. Mechanism and SAR Methotrexate, introduced in 1953, and the related aminopterin bind to DHFR, prevent transfer of the one-carbon unit from dihydrofolic acid to methylenetetrahydrofolic acid and ultimately to thymidine (Fig. 7). Methotrexate is taken into cells by the folate transporter, and converted in cells to active polyglutamate metabolites by folylpolyglutamate synthase [377]; this also has the effect of trapping the drug in cells [359]. A large amount of work has been done to delineate the SAR for 2,4-diaminopteridines binding to DHFR, but no clinical successor to methotrexate as a DHFR inhibitor has yet been found among the "classical" antifolates [367], although edatrexate is still in development. Since there is also a folate site on TS, these compounds have some level of binding to this as well. CB 3717 and raltitrexed were designed specifically to target TS rather than DHFR, while pemetrexed is closer to a general folate pathway inhibitor.

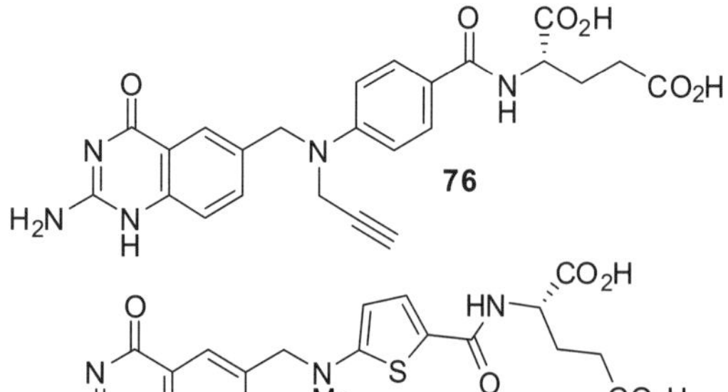

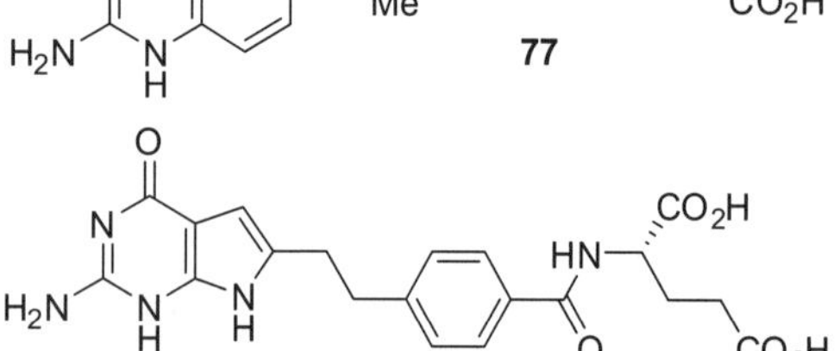

4.3.3. Biological Activity and Side Effects Methotrexate has broad-spectrum clinical activity and is still the most widely used antifolate, despite high myelosupressive activity and frequent development of resistance by various mechanisms. The newer antifolates have broadly similar toxicity profiles.

4.3.4. Recent Developments: Lipophilic Antifolates These compounds were designed to circumvent resistance to methotrexate that arises by reduced folate uptake or reduced polyglutamylation. They are relatively lipophilic compounds, lacking a glutamate residue, that get into cells by passive diffusion. The first examples to receive clinical evaluation were trimetrexate (**79**) and piritrexim (**80**). Trimetrexate was superior to methotrexate in animal models, with activity in methotrexate-resistant lines [378] but (unlike methotrexate) is susceptible to P-glycoprotein-mediated multidrug resistance [379]. Trimetrexate has had extensive clinical trials, and has shown activity in a number of tumors, including breast, nonsmall cell lung, head and neck and prostate [380], and particularly in colon cancer in conjunction with 5-fluorouracil/leucovorin [381]. Piritrexim was chosen for development from a range of lipid-soluble diaminoheterocyclic compounds on the basis of potent DHFR inhibition and minimal effects on histamine metabolism [382]. Piritrexim is approximately 75% bioavailable when given orally [383], and in Phase II trials showed some activity using oral dosing in bladder cancer [384]. It is also more effective than methotrexate in severe psoriasis, since its lack of polyglutamylated metabolites makes it less hepatotoxic in long-term dosing [385].

Nolatrexed (Thymitaq; **81**) is a lipophilic folate analog designed as a TS inhibitor, using structure-based methods to maximize binding at the folate site [386]. It is a potent ($K_i = 11$ nM), noncompetitive inhibitor of human TS, with modest growth-inhibitory effects (IC_{50} values 0.4–7 µM) against a wide variety of murine and human cell lines. Nolatrexed does not enter cells by the reduced folate carrier, is not polyglutamylated, and does not inhibit DHFR. The activity of the drug is abrogated by thymidine (but not hypoxanthine), and TS overexpressing cells are strongly resistant, demonstrating that the primary target is TS [387]. Oral bioavailability in rats was 30–50%, and oral nolatrexed showed curative activity against both i.p.- and i.m.-implanted thymidine kinase-deficient murine L5178Y/ TK- lymphomas [386]. Combinations of nolatrexed and cisplatin showed synergistic activity in both 5-FU- and cisplatin-resistant ovarian and colon cancer cells. A single-agent Phase II study of nolatrexed (i.v. infusion of 725 mg/m^2/day for 5 days) in patients with advanced hepatocellular carcinoma showed minimal activity [388], as did a Phase III single-agent study in the same indication [389].

4.4. Pyrimidine Analogs

4.4.1. History The pyrimidine analogs 5-fluorouracil (**82**) and cytosine arabinoside (ara-C; **83**) were developed from a knowledge of DNA metabolism [2], and were registered for clinical use in 1962 and 1969, respectively. A huge amount of work has gone into developing further analogs, and this has recently begun to pay off with the more recent introduction of gemcitabine (Gemzar; **84**) and capecitabine (Xeloda; **85**).

4.4.2. Mechanism and SAR The mechanisms by which 5-fluorouracil exerts its cytotoxicity have been extensively reported [390,391]. It is converted in cells to the monophosphate 5-FdUMP, which binds initially reversibly at the dUMP site of the enzyme thymidylate synthetase (Fig. 7). This is followed by Michael-type attack of an SH group on the enzyme to give an enolate-type intermediate, which reacts at the methylene moiety of N^5, N^{10}-methylenetetrahydrofolate to form a

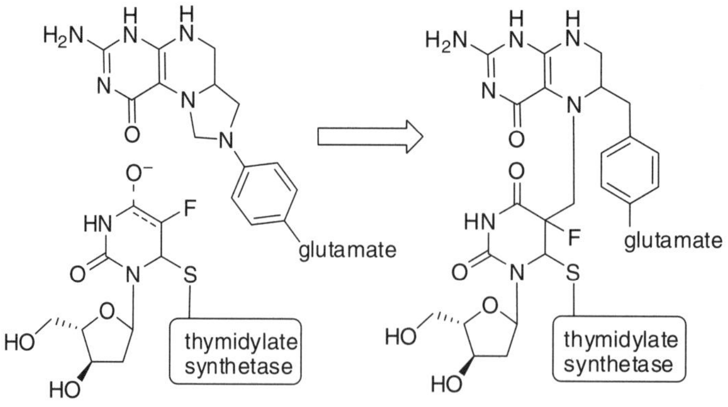

Figure 8. Mechanism of action of 5-florouracil inhibition of thymidylate synthetase.

covalent drug–enzyme–cofactor ternary complex (Fig. 8). Because the fluorine cannot be displaced, as with the natural (nonfluorinated) substrates, this results in permanent poisoning of the enzyme with 1:1 stoichiometry. 5-Fluorouracil is also converted into the triphosphate 5-FdUTP, which incorporated into both RNA and DNA.

The mechanism of action of cytosine arabinoside has also been well reviewed [392]. It acts primarily as a chain terminator during the elongation phase of DNA synthesis, incorporating into the growing chain and preventing the action of DNA polymerases [393]. Gemcitabine (**84**) also acts primarily as a chain terminator, but has additional effects, through rapid phosphorylation by deoxycytidine kinase to di- and triphosphate metabolites. The diphosphate inhibits ribonucleotide reductase (RR), the enzyme responsible for producing the deoxynucleotides required for DNA synthesis and repair, and the subsequent depletion of cellular deoxynucleotides favors gemcitabine triphosphate incorporation into DNA over the normal dCTP, in a "self-potentiating" mechanism [394]. Incorporation of gemcitabine into the elongating DNA strand halts the DNA polymerase after the addition of one more additional deoxynucleotide, in a "masked chain termination" event that appears to lock the drug into DNA, preventing proofreading exonucleases from removing it. Gemcitabine is synergistic with cisplatin due to the triphosphate preventing chain elongation during the DNA resynthesis process following nucleotide excision repair of the lesions [395]. The mechanism of RR inhibition by gemcitabine has been studied in *E. coli*, and appears to be different to that of other 2′-substituted nucleotide inhibitors, involving inactivation of the R1 subunit [396], and overexpression of RR is a resistance mechanism for gemcitabine [397]. It is also an effective radiation sensitizer, probably via depletion of dATP pools in cells [398], and can increase cellular apoptosis in irradiated cells [399].

Capecitabine (**85**) is an oral prodrug of 5-fluorouracil (Fig. 9) [400]. It is converted by carboxylesterase to 5′-deoxy-5-fluorocytidine (**86**), which is then converted to doxifluridine (**87**) by cytidine deaminase, which is elevated in many tumors [401]. Doxifluridine has also been used clinically as a prodrug of 5-fluorouracil (**82**), but shows gastrointestinal toxicity [402]; replacing uridine with carbamoylated cytosine protects it from prerelease by intestinal thymidine phosphorylase [400,403]. Both cytidine deaminase and thymidine phosphorylase are present at generally higher levels in tumor cells than in most normal tissues, which allows for some selectivity of activation of the drug and lesser systemic toxicity.

4.4.3. Biological Activity and Side Effects
Both 5-fluorouracil and cytosine arabinoside remain widely used in combination cancer

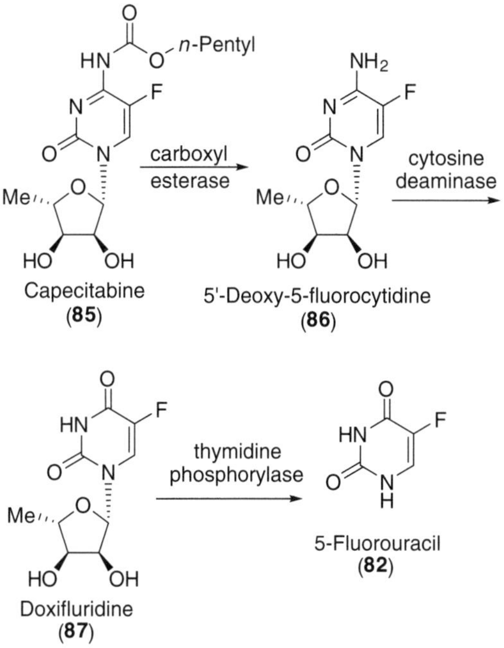

Figure 9. Activation of capecitabine.

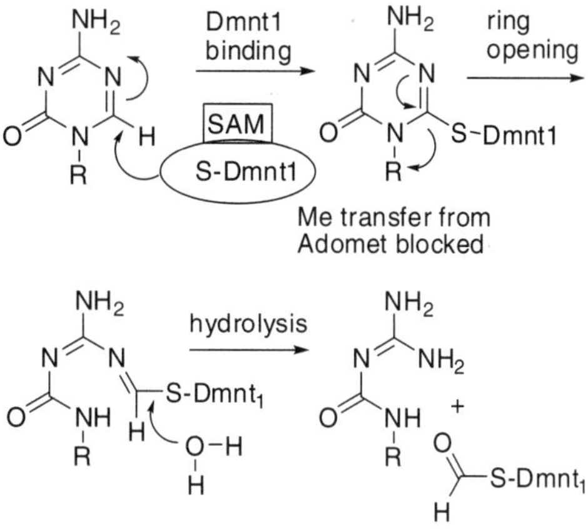

Figure 10. Suicide inhibition of Dmnt1.

chemotherapy. 5-Fluorouracil is one of the most effective drugs against colon cancer [390]. Cytosine arabinoside is effective in leukemias and lymphomas but has a very short half-life (about 12 min in man), due to catabolism by cytidine deaminase [404], and various nonspecific prodrug forms are used [405]. Gemcitabine was shown in Phase I trials to be active in a number of cancers, especially in nonsmall cell lung cancer (NSCLC), where it showed >20% responses as a single agent and up to 54% in combination with cisplatin [406]. In Phase II trials, it has proved active in a wide range of tumors, including NSCLC (>60% responses in combination with cisplatin) [407], urothelial (22–28% responses as monotherapy, 42–66% in combination with cisplatin) [408], advanced breast cancer (25.0% responses as monotherapy) [409] and metastatic bladder cancer (42–66% responses in combination with cisplatin) [410]. The main adverse effects were hematological, but were generally mild. A number of large Phase III trials are in progress. Capecitabine is becoming increasingly popular and has largely replaced 5-fluorouracil in several indications [411].

4.4.4. Recent Developments: Inhibitors of DNA Methylation
Hypermethylation of cytosines in short CpG sequences (islands) is common in cancer, and is a major mechanism of gene silencing [412]. The enzyme DNMT1, the "maintenance methylase" that keeps cytosines in these CpG islands 5-methylated, is overexpressed in many human cancers [413], and DNA methylation status is a useful prognostic cancer marker [414]. Recently, the azapyrimidines Dacogen (decitabine; **88**) and Vidaza (5-azacitidine; **89**) were approved for use in myelodysplastic syndrome (MDS) [415], a group of hematopoietic disorders whose pathogenesis is poorly understood. They act as suicide inhibitors, being first incorporated into the DNA in place of cytosine, then acting to covalently trap the enzyme as it alkylates (Fig. 10). Dacogen showed response rates of 17–49% in MDS in multiple phase II and III studies and also activity in acute and chronic myelogenous leukemia [416]. The closely related Vidaza has a very similar activity profile [417]. A recent review of clinical trials of both agents concluded that both were effective in MDS, but that it was too early to make a meaningful comparison of the two [418].

4.5. Purine analogs

4.5.1. History
The purine analogs 6-mercaptopurine (**90**) and 6-thioguanine (**91**) were among the first anticancer drugs to be used, registered in 1953 and 1966, respectively. Later, the purine nucleoside analogs fludarabine (**92**) and pentostatin (2′-deoxycoformycin; **93**) were registered in 1991, and cladribine (2-chloro-2′-deoxyadenosine; **94**) in 1992.

4.5.2. Mechanism and SAR
Cytosine arabinoside, fludarabine and cladribine are taken into cells via a specific nucleoside transporter protein, and are phosphorylated to the mono-, di-, and triphosphates, with the first phosphorylation mainly by deoxycytidine kinase [419]. The active triphosphate derivatives are incorporated into DNA, blocking polymerase function and thus DNA synthesis. Cladribine is resistant to degradation by adenosine deaminase [420], and induces apoptosis in leukemia cell lines through the Fas/Fas ligand pathway [421]. It also interrupts deoxyadenosine metabolism, blocking both phosphorylation and deamination [420]. Pentostatin is also converted to the triphosphate and incorporated into DNA, where it blocks polymerase function [422], but is also an extremely potent inhibitor of adenosine deaminase ($K_i = 2.5 \times 10^{12}\,M^{-1}$) [423].

4.5.3. Biological Activity and Side Effects
These three adenosine analogs, which are cytotoxic to both dividing and resting lymphocytes, have revolutionized the treatment of indolent lymphoid malignancies such as chronic lymphocytic leukemia, non-Hodgkin's lymphoma, cutaneous T-cell lymphoma and hairy cell leukemia. Both fludarabine and cladribine showed similar good response rates, but were cross-resistant, in refractory

non-Hodgkin's lymphoma [424]. Cladribine is active in hairy cell leukemia (>80% complete responses) [425], non-Hodgkin's lymphoma (89% responses) [426], refractory chronic lymphocytic leukemia (44% responses) [427], untreated chronic lymphocytic leukemia (85% response rate) [428], and cutaneous T-cell lymphomas (28% responses) [429], but showed little activity in solid tumors. Clinical trials of parenteral cladribine in patients with progressive forms of multiple sclerosis showed that that it can be effective in slowing relapse rates and disability progression. Oral cladribine is currently being assessed as monotherapy and in addition to interferon-β1 [430]. The purine nucleosides have been reviewed recently [431].

4.5.4. Recent Developments A recent development in purine nucleoside-based drugs has been as transition-state mimics of the enzyme purine nucleoside phosphorylase (PNP), which catalyzes the reversible phosphorolysis of 6-amino-2′-deoxyribonucleosides to the free base and 2′-deoxyribose-1-phosphates [432]. This class of compounds, known collectively as the immunocillins, and represented by forodesine (**95**), are D-ribofuranosyl C-glycosides [433], designed to structurally resemble the transition state of PNP as the base-glycoside bond is cleaved (Fig. 11).

Transition state inhibitors, by conforming to a low-energy state in the reaction profile, should bind much more tightly than the substrate [434], and this is illustrated by **95**, which has a K_D of 56 pM for human PNP [435]. This inhibition of PNP results in the accumulation of deoxyguanosine, which is then processed intracellularly into GTP. This leads to inhibition of ribonucleotide reductase, imbalance in the dNTP pool and inhibition of DNA synthesis/repair [433]. In a Phase IIa trial in patients with acute T-cell leukemia, forodesine at 40 mg/m^2 for 5 days weekly gave an overall response rate of 32% and a complete remission rate of 20% [433]. A recent review [436] suggests it may also be of value in the therapy of some B-cell malignancies.

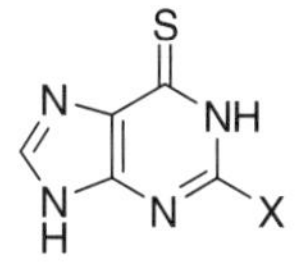

90: X = H
91: X = NH$_2$

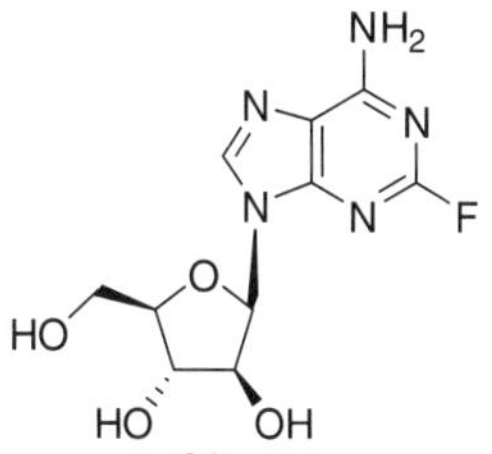

92

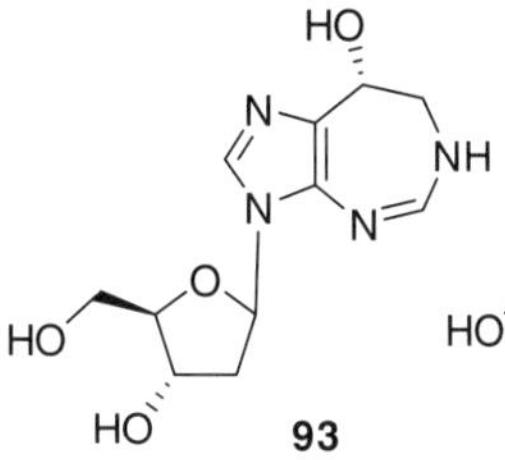

93

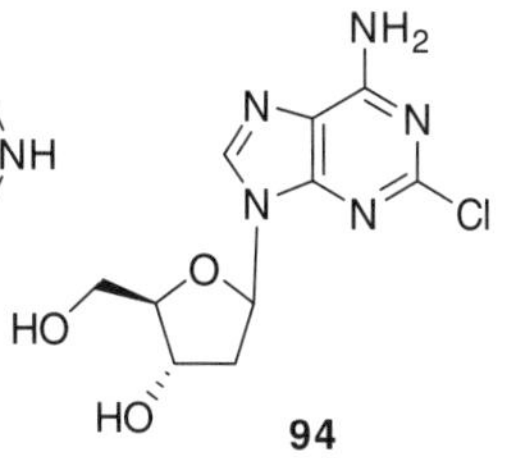

94

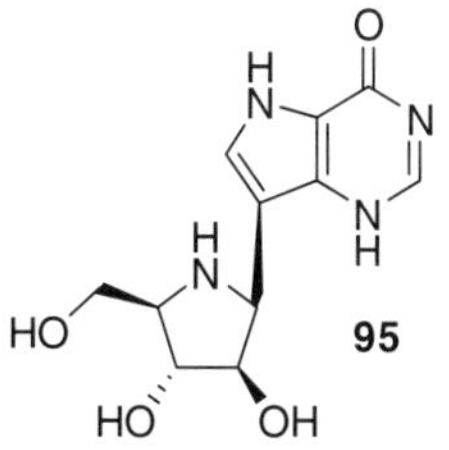

95

5. TUMOR-ACTIVATED PRODRUGS

5.1. Introduction

As noted above, the majority of clinically used anticancer drugs are systemic antiproliferative agents (cytotoxins). These kill cells by a

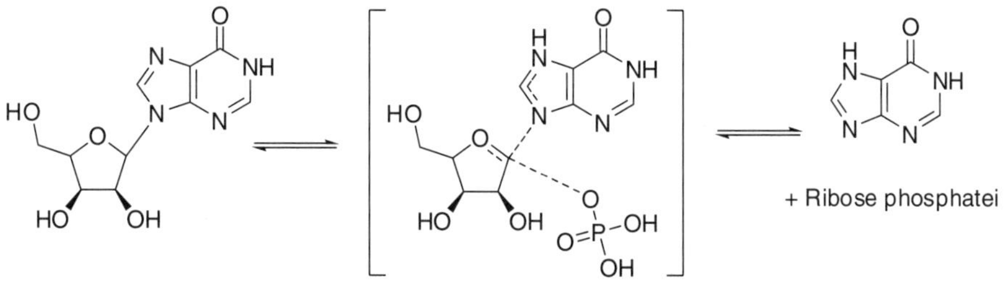

Figure 11. PNP transition state.

variety of mechanisms, but primarily by attacking their DNA at some level (synthesis, replication, or processing). However, a large part of their selectivity for cancer cells is based on cytokinetics, in that they (to varying extents) are preferentially toxic to cycling cells. For this reason, their therapeutic efficacy is limited by the damage they also cause to proliferating normal cells such as those in the bone marrow and gut epithelia. This is especially true in the treatment of solid tumors, where cell doubling times may be very long. While efforts to physically target cytotoxins to tumor tissue has not been very successful, the development of relatively nontoxic prodrug forms of cytotoxins, that can be selectively activated in tumor tissue, is beginning to achieve some success, and in the future may become a major strategy.

Prodrugs can be defined broadly as agents that are transformed after administration, either by metabolism or by spontaneous chemical breakdown, to form a pharmacologically active species. Strictly speaking, agents such as cyclophosphamide (4) are prodrugs, but these undergo nonspecific activation in all tissues. Of more interest are tumor-activated prodrugs that exploit various aspects of tumor physiology and other techniques to become selectively activated in tumor tissue to toxic species [437]. The field is quite active and many compounds have been studied, but relatively few have proceeded to clinical trial.

The multiple criteria required of a tumor-activated prodrug has meant that these compounds, while sometimes using natural products such as doxorubicin as the toxins, are primarily synthetic agents. They fall into four broad categories, defined by their mode of selective activation; hypoxia-selective prodrugs (bioreductives; reviewed in Refs [438,439]), prodrugs for antibody-directed enzyme–prodrug therapy (ADEPT prodrugs; reviewed in Ref. [440]), prodrugs for gene-directed enzyme–prodrug therapy (GDEPT prodrugs; reviewed in Ref. [441]), and antibody-toxin conjugates ("armed" antibodies). Only the first three categories are covered in this chapter, which is focused on synthetic small-molecule therapeutics.

5.2. Clinical use of Tumor-Activated Prodrugs

Because interest in tumor-activated prodrugs is relatively recent, only the hypoxia-selective agent tirapazamine has had extensive clinical use, and even this is still in development. The limited clinical experience with these various drugs (Table 4) is discussed below in each subclass.

5.3. Hypoxia-Activated Bioreductive Prodrugs

5.3.1. History The imperfect neovascularization that develops in growing solid tumors

Table 4. Tumor-Activated Prodrugs in Clinical Trial for Cancer Chemotherapy

Generic Name (Structure)	Trade Name	Supplier	Chemical class
Hypoxia-activated prodrugs			
Mitomycin C (**96**)			Aziridinylquinone
Porfiromycin (**97**)		Vion	Aziridinylquinone
EO9 (**98**)	Apaziquone	Spectrum	Benzoquinone
Tirapazamine (**99**)	Tirazone	Sanofi-Aventis	Benzotriazine-di-N-oxide
Banoxantrone (AQ4N) (**101**)		British Technology Group	Aliphatic N-oxide
PR-104 (**102**)		Proacta	Aromatic mustard
TH-302 (**103**)		Threshold	Heteroaromatic mustard
ADEPT prodrugs			
ZD 2767P (**93**)		AstraZeneca	Aromatic mustard
GDEPT prodrugs			
Ganciclovir (**99**)	Cytovene	Roche	
CB 1954 (**102**)		Cobra Therapeutics	Dinitrophenylaziridine

results in limited and inefficient blood vessel networks, and restricted and often chaotic blood flow [442]. This generates chronic or diffusion hypoxia, where cells sufficiently distant from the nearest blood capillary are hypoxic for long periods, due to the steep diffusion gradient of oxygen in tissue. The high and variable interstitial pressures caused by the growing tumor [443] can also result in transient or perfusion hypoxia, resulting from the temporary shut down of blood vessels placing sections of tissue under hypoxia for shorter periods [444]. Because severe hypoxia is a common and unique property of cells in solid tumors, it is an important potential mechanism for the tumor-specific activation of prodrugs. This concept grew initially out of the development of radiosensitizers, drugs designed to take the place of oxygen in hypoxic tissue by oxidatively "fixing" the initial DNA radicals formed by ionizing radiation to generate cytotoxic strand breaks [445]. Such compounds tended to be easily reduced electron-deficient species such as misonidazole (**81**). In addition to their radiosensitizing properties as "oxygen mimetics," many of these compounds were also found to have modestly higher levels (about 10-fold) of cytotoxicity in hypoxic compared to oxygenated cells in culture [446]. The mechanism of such hypoxia-selective cytotoxicity is the ability of the prodrug to be metabolized by reductive enzymes such as cytochrome P450 reductase and xanthine oxidase [447] to a transient one-electron intermediate. In normal oxygenated tissue this is efficiently back-oxidized by molecular oxygen to the parent compound, but in hypoxic cells it is further metabolized or spontaneously breaks down to more cytotoxic species [448] (Fig. 12).

Quinones were the first class of compounds explored as hypoxia-activated prodrugs. The widely used clinical agent mitomycin C (**96**) shows reasonable hypoxic selectivity in cell culture (the ratio of aerobic and hypoxic IC_{50} values) [449] due to complex but well-understood fragmentation to alkylating species that cross-link DNA [450]. The analog porfiromycin (**97**) has greater selectivity and was developed primarily as a hypoxia-activated prodrug [451]. It underwent Phase I clinical trials for head and neck cancer in combination with

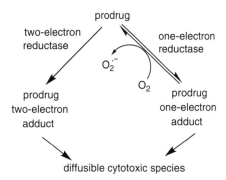

Figure 12. Hypoxia-activated prodrugs.

radiotherapy but was inferior to mitomycin C [452], and was not developed further. Apaziquone (EO9; **98**) was developed originally as a hypoxia-activated prodrug but was found to be activated (to a DNA cross-linking agent) primarily by the two-electron (and thus oxygen-insensitive) reductase NQO1 (DT-diaphorase) [453]. Early clinical results were disappointing [454] and were attributed to this and a short drug half-life [455]. More recently, these characteristics have been exploited in the use of the compound for bladder cancer by intravesical instillation, where it has been shown active in two Phase II trials [456,457].

The most well-studied hypoxia-activated prodrug is the di-*N*-oxide tirapazamine (**99**). This drug was originally evaluated as an antimicrobial agent but was discovered to have hypoxia-selective cytotoxicity in a screening program. Tirapazamine has been well reviewed [458,459] and has undergone several Phase III studies, some of which are still in progress [460].

Aliphatic *N*-oxides of DNA binding agents have also been explored as synthetic hypoxia-selective prodrugs. This was first demonstrated by the drug nitracrine-*N*-oxide (**100**), which is much more hypoxia-selective (>1000-fold in cell culture) [461] than the free amine itself, although this also shows moderate hypoxic selectivity through reductive activation of the nitro group [462]. The most advanced drug of this type in development is the bis-*N*-oxide banoxantrone (AQ4N; **101**) [463], which is in Phase II trial.

The final major class, nitroaromatics, are at first sight the most logical approach, since reduction of aromatic nitro groups proceeds in a series of one-electron steps to hydroxylamines and ultimately amines. This converts the group from one of the most powerful electron withdrawing functions (NO$_2$; Hammett σ_p +0.78) to one of the most powerful electron donors (NH$_2$; Hammett σ_p -0.66), via an oxygen-scavengeable nitro radical anion [464]. This system is a potentially powerful "trigger" for hypoxia activation via either electron redistribution or induced fragmentation, but has been one of the last approaches to be utilized. Two recent compounds now in trial exemplify these two approaches; PR-104 (**102**) [465] and TH-302 (**103**) [466], respectively. The synthesis of PR-104 includes a novel one-pot formation of the asymmetric halomesylate mustard using aziridineeethanol [467].

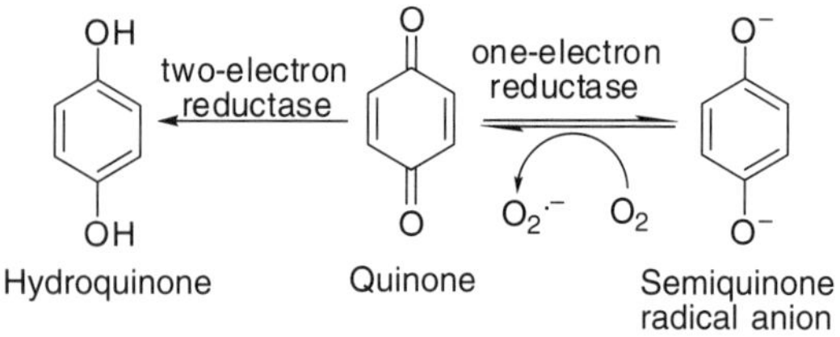

Figure 13. Bioreactive metabolism of quinones.

The quinones mitomycin C (**96**) and porfiromycin (**97**) undergo ready one-electron reduction, primarily by NADPH:cytochrome C (P-450) reductase [468], to the corresponding semiquinone radical anion that is capable of back-oxidation by molecular oxygen (Fig. 13). Following this, mitomycin C undergoes fragmentation to DNA cross-linking agents that form guanine–guanine cross-links in the major groove [450]. One potential drawback to quinone-based compounds as hypoxia-activated prodrugs is that they are also often good

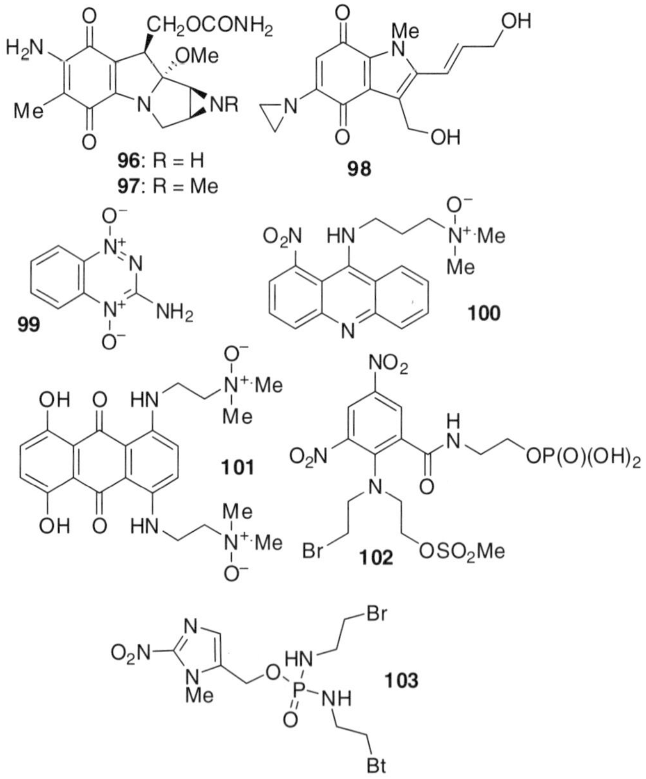

5.3.2. Mechanism and SAR The classes of hypoxia-activated prodrugs discussed above work by a variety of different mechanisms.

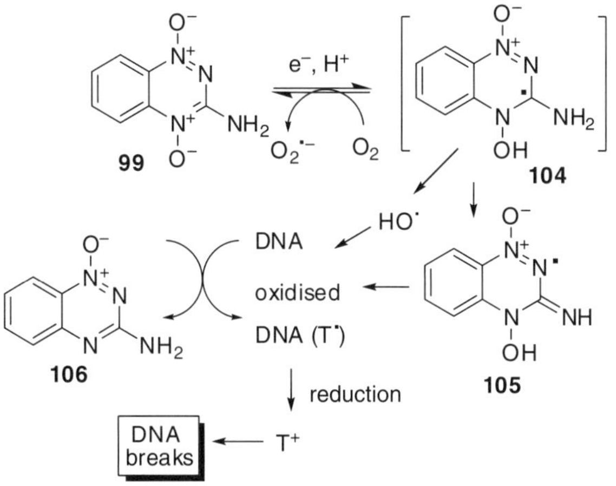

Figure 14. Metabolism of tirapazamine.

substrates for two-electron reductases, particularly NQO1 (DTD-diaphorase) [469].

Tirapazamine (**99**) undergoes ready one-electron reduction, primarily by cytochrome P450 and cytochrome P450 reductase [447], to initially form the reducing radical (**104**), which in the absence of oxygen dehydrates form the oxidizing benzotriazinyl radical (**105**) (Fig. 14) [470]. This is the active species, reacting with DNA to generate primarily C-4′ ribose radicals [471] and the two-electron reduction product (**106**). Oxidative "fixing" of these radicals under hypoxia by (among other things) tirapazamine or **106** [472] results in cytotoxic DNA strand breaks. The two-electron reduction product (**106**) is not toxic, resulting in TPZ showing good hypoxic selectivity (100–200-fold) in cell culture. Although only forming a monofunctional radical, tirapazamine generates a high proportion of double-strand DNA breaks, suggested due to high local radical concentrations generated by an undefined intranuclear reductase associated with DNA [473]. Although TPZ has no bystander effect, due to the reactivity of the radical metabolites, its hypoxic killing does extend over a wider range of oxygen concentrations than for most bioreductive drugs (as high as 20 μM O_2) [474]. This means its activation is not restricted to completely anoxic tissues. Related quinoxalinecarbonitrile-1,4-di-N-oxides (e.g., **107**), where the 2-nitrogen in the benzotriazine unit of tirapazamine is replaced with a C-CN unit, is also potent and highly selective hypoxia-selective drugs [475]. Structure–activity studies with these compounds show that hypoxic selectivity is retained when H or NHR replaces the 3-amino group.

The aliphatic tertiary amine N-oxides of the bis-bioreductive prodrug banoxantrone (**101**) are also reduced (to the free amines) largely by the CYP3A isozyme of NADPH:cytochrome C (P-450) reductase [476]. Although this is not a one-electron process it is still oxygen inhibited, being a direct competition between oxygen and the drug at the enzyme site. Regeneration of the cationic side chains allows tight binding to DNA and an ability to function as a topo II poison, similarly to the closely related drug mitoxantrone (**53**) [477]. This approach seems quite general, with compounds such as the mono-N-oxide DACA N-oxide (**108**) [478] and the nitracrine N-oxide (**100**) [461] also showing significant hypoxic selectivity in cell culture. The latter is an interesting example of a bis-bioreductive prodrug with two different reductive centers (nitro and N-oxide); both centers need to be reduced for full activation, with the N-oxide demasking occurring before nitro reduction [479].

The dinitrobenzamide mustard prodrugs such as PR-104 (**102**) work on the concept [480] that aromatic nitrogen mustards show a high dependence of cytotoxicity on the electron density at the mustard nitrogen, so that enzymic reduction of an electron-withdrawing nitro group to an electron-donating hydroxylamine or amine can provide a large increase

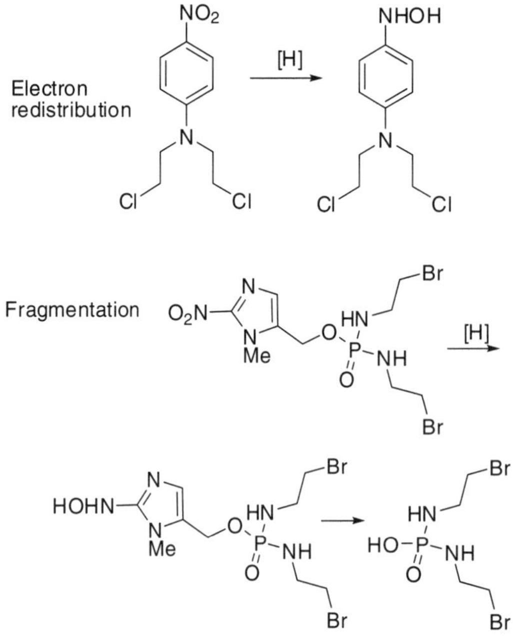

Figure 15. Bioreactive activation of nitroaromatics.

in the cytotoxicity of a prepositioned mustard [25] (Fig. 15). Nitroheterocyclic prodrugs such TH-302 (**103**) [481] and the related nitroimidazole mustard **109** [482] rely on nitro reduction causing fragmentation of the molecule, thus releasing an activated mustard (Fig. 15).

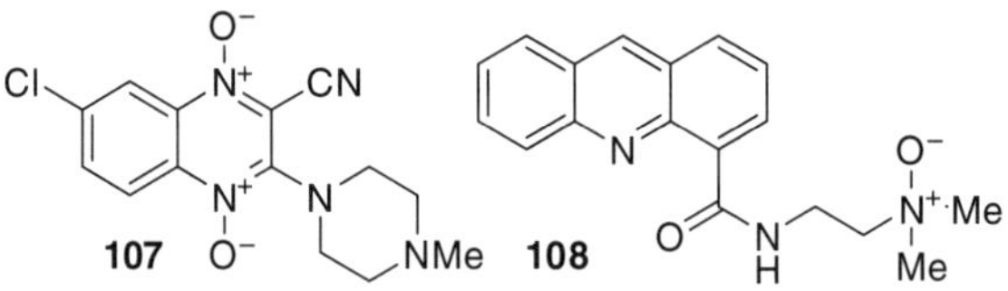

5.3.3. Biological Activity and Side Effects
Tirapazamine shows high selective toxicity (100–200-fold) toward hypoxic cells in culture, but its diffusion through tissue is limited by its metabolism to the (nondiffusible) radical species [483]. In animal studies, tirapazamine enhanced the effect of both single-dose or fractionated radiation [484]. Combinations of tirapazamine with cytotoxic agents, including etoposide, bleomycin, cisplatin [485] and paclitaxel [486] showed additive or greater than additive effects on both tumor cell killing and tumor growth delay. In the case of cisplatin, the effects were suggested due to delaying the repair of cisplatin-induced DNA cross-links in hypoxic cells [485]. Combinations with hypoxia-expanding agents such as the blood flow inhibitor DMXAA (ASA404) showed marked increases in activity in a variety of tumor models [487]. Phase II trials in cervical cancer (with cisplatin) [488] and in NSCLC (with cisplatin and vinorelbine) [489] were encouraging, but a Phase III trial in NSCLC (with paclitaxel and carboplatin) [490] showed the addition of tirapazamine increased toxicity without improving survival. A Phase II trial of tirapazamine in advanced head and neck cancer (with radiation and cisplatin) showed superior loco-regional control rates compared with fluorouracil, cisplatin, and radiation [491], and is being followed up with a Phase III trial. A laboratory study showed that tirapazamine caused time- and dose-dependent retinal damage in mice [492], but this does not appear to be a clinical issue.

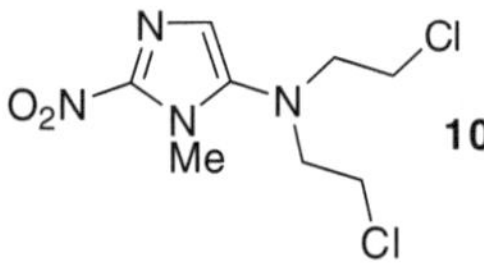

Quantitation by mass spectrometry of banoxantrone and its metabolites in nude mice bearing s.c. or orthotopically implanted human BxPC-3 or Panc-1 tumors tumor-bearing mice showed that banoxantrone readily penetrated the tumors and that the metabolite AQ4 rapidly accumulated in tumor tissues at high levels in a dose-dependent fashion [493].

A Phase I clinical study showed that banoxantrone was well tolerated when administered on a weekly schedule, and that sufficient plasma levels were achieved with a dose of 768 mg/m^2 [494].

The nitrophenyl mustard PR-104 (**102**) has a lower K-value (the oxygen concentration that halves cytotoxic potency) value than tirapazamine (0.13 compared with 1.3 µM), which suggests activation only in very hypoxic regions. However, PK modeling predicted better penetration to hypoxic cells in tumors because of its slower metabolism [495]. PR-104 is not a substrate for the two-electron reductase NQO1; the major activating reductase appears to be the one-electron cytochrome P450 reductase [496]. In mice, PR-104 is rapidly converted to the corresponding alcohol, with reduction of this to the major intracellular hydroxylamine metabolite resulting in DNA cross-linking selectively under hypoxia. PR-104 showed single-agent activity in six of eight xenograft models and greater than additive antitumor activity in combination with gemcitabine and docetaxel [465]. PR-104 is in Phase II clinical trial.

TH-302 (**103**) showed hypoxia selectivities (ratio of aerobic/hypoxic IC$_{50}$ values) of up to 300-fold in a range of cancer cell lines (PC-3, MIA PaCa-2, ACHN, and HCT 116), and showed no difference in potency or selectivity in lines overexpressing various efflux pumps. The compound was more potent in cell lines deficient in homologous recombination, base or nucleotide excision repair, consistent with DNA cross-linking as the mechanism of action. In combination studies in orthotopic xenograft models in nude mice, TH-302 in combination with gemcitabine in MiaPaca-2 pancreatic tumors, and with taxotere in GFP-PC3 prostate tumors, showed significantly better results in each case than either drug alone; tumor growth delays (TH-302:other drug:combination) of 4, 12, and 26 days in pancreatic and 16, 31 days and cures in prostate tumors [466]. TH-302 is in Phase I trial.

5.3.4. Recent Developments Studies on the mechanism of action of tirapazamine continue. Evidence has been put forward for a benzotriazinyl radical (Fig. 14) [470,471] and the hydroxyl radical [497,498] as the DNA-reacting species. The mechanism by which the initial DNA radicals are then oxidized to generate strand breaks in the absence of oxygen has been shown due to both tirapazamine [499] and its two-electron reduced metabolite SR 4317 [500] acting as oxidants for this process. It is suggested that a combination DNA single-strand breaks, base damage, and DNA-protein cross-links result in stalling and collapse of replication forks, resulting in double-strand breaks which invoke homologous recombination for their repair [501]. Studies of tirapazamine transport through multi-cellular layers of tumor cells under hyperoxic conditions (to suppress reductive metabolism) determined a diffusion coefficient of 0.40×10^{-6} cm^2 s^{-1}, with this being strongly suppressed under hypoxia. The results suggested that inefficient transport is a limitation for tirapazamine, and that analog development should focus on this, by increase in the rate of diffusion and/or decreases in the rate of reductive metabolism [502].

In a search for improved analogs, a detailed study of ring-substituted analogs of tirapazamine showed that these could be used to predictably vary one-electron reduction potentials (and thus rates of bioreduction). Hypoxic cytotoxicity generally increased with increasing reduction potential, which had an optimum range for desirable biological activity profile of between −370 and −400 mV [503]. To improve potency, tirapazamine was linked to a variety of DNA-affinic carriers, which greatly increased *in vitro* potency and showed that activity depends on close association with the DNA [504]. Such targeted analogs (e.g., **110**) [505] showed increases of 12–18-fold in single-strand break formation and 60–110-fold in double-strand breaks, compared with tirapazamine [506]. The extravascular diffusion coefficient of tirapazamine analogs was shown to depend primarily on their lipophilicity, and to lesser extents on the number of H-bond donors/acceptors and the overall molecular weight. The extravascular diffusion coefficient of tirapazamine analogs was shown to depend primarily on their lipophilicity, and to lesser extents on the number of H-bond donors/acceptors and the overall molecular weight. Hypoxic cell kill was shown to be in turn very sensitive to changes in the extravascular diffusion coefficient, by using a pharmacokinetic/pharmacodynamic model in which diffusion in the extravascular compartment of tumors was explicitly considered [507]. Analogs bearing 3-aminoalkylamino weak bases (e.g., **111**) were prepared to improve aqueous solubility and extravascular transport properties. Several of these were superior to tirapazamine in xenograft models, showing improved killing of hypoxic cells and better transport and/or plasma pharmacokinetics [508]. A series of highly active tricyclic analogs of TPZ (e.g., **112**) has also been reported recently [509].

5.4. Prodrugs for ADEPT

5.4.1. History

ADEPT is an adaption of the earlier immunotoxin concept [510,511] (not covered in this chapter). The difference is that instead of the toxin being attached to the antibody for localization on tumors, an enzyme (usually nonhuman) is attached and thus localized instead [512,513] (Fig. 16). A prodrug that can be activated efficiently and selectively by the enzyme is then administered, and is catalytically activated by the localized enzyme only in the vicinity of the tumor cells. The advantage of using nonhuman enzymes is the enhanced ability to find prodrugs that can be selectively activated. ADEPT shares with the original immunotoxin concept the problems of limited access of the (large) antibody–enzyme conjugate to tumors, and the usually heterogeneous expression of the target antigen on tumor cells. However, provided the released cytotoxin has the appropriate properties (high potency and an efficient bystander effect), it can ameliorate these problems by diffusing from the cells where it is generated to enter and kill surrounding tumor cells that may not possess prodrug activating ability. A further increase in efficacy can be achieved if the prodrug is designed to be excluded from cells until it is activated [514].

5.4.2. Mechanism and SAR

The specific mechanism of action depends on the type of enzyme used to activate the prodrug. Particular requirements of the prodrug include being a selective and efficient substrate for the enzyme used. General requirements are an ability to be excluded from cells (usually achieved by high hydrophilicity and/or possession of a negative charge) until activation, and the capability to then release a potent and diffusible toxin with a substantial bystander effect.

Prodrugs for Phosphatase Enzymes
Phosphates have been employed as ADEPT prodrugs, since both aromatic (e.g., etoposide phosphate; **113**) [515] and aliphatic (e.g., mitomycin phosphate; **114**) [516] examples are efficiently cleaved by alkaline phosphatases, and are substantially cell excluded. However it proved difficult to achieve selectivity since there is an abundance of such phosphatase enzymes in human serum and other tissues, and phosphates are primarily now used directly as nonspecific prodrugs; the antivascular agent **115** (combretastatin phosphate) is an example [517].

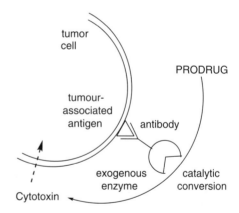

Figure 16. Antibody-directed enzyme–prodrug therapy.

Prodrugs for Peptidase Enzymes Glutamate-type prodrugs of mustards are effectively excluded from cells by the diacid side chain, and can be cleaved by the *Pseudomonas*-derived enzyme carboxypeptidase G2 [518]. Cleavage of the amide or carbamate releases more lipophilic agent that is also activated by electron release through the aromatic ring to the mustard. The amide CMDA (**116**) was the first ADEPT prodrug evaluated clinically [519], and the carbamate prodrug ZD 2767P (**117**), releasing a more cytotoxic phenol iodomustard, is currently in Phase I clinical trial [520].

to treatment with 96.5/bL, a mAb/β-lactamase conjugate that binds to specific surface antigens on these cells [522]. A cephem derivative of doxorubicin (**119**) showed higher intratumoral levels of doxorubicin after treatment with the conjugate than with doxorubicin alone [523]. However, the differential cytotoxicities between drug and prodrug in this approach are only moderate, and no such prodrugs have yet proceeded to clinical trial.

Prodrugs for Glucuronidase Enzymes Because serum levels of β-glucuronidase (GUS) are

Prodrugs for β-Lactamase Enzymes These enzymes from *Enterobacter* species can selectively hydrolyze the four-membered β-lactam ring of penicillins and cephalosporins, and have been used in a variety of prodrugs [521]. Hydrolysis is followed by spontaneous fragmentation of the carbamate side chain and release of a variety of toxic amines (Fig. 17). Carboxy and sulfoxide groups on the cephem nucleus assist with cell exclusion. Several nitrogen mustard prodrugs for β-lactamase have been evaluated. The cephem analog (**118**) effected cures in mice bearing xenografts of human melanoma cells if given subsequent

very low (it is largely confined to lysozymes in cells), it is possible to use the human version as an activating enzyme in ADEPT, avoiding potential immunogenicity problems [524]. Work with this enzyme has tended to focus on anthracycline effectors ([525]). The epirubicin *O*-glucuronide prodrug (**120**) was 100–1000-fold less cytotoxic than epirubicin itself *in vitro* [526], but pretreatment of antigen-positive cells with a 323/A3-GUS-*E. coli* immunoconjugate gave equivalent cytotoxicity to that of the free drug. The daunomycin prodrug (**121**) used an immolative spacer unit [525] and although only 10-fold less

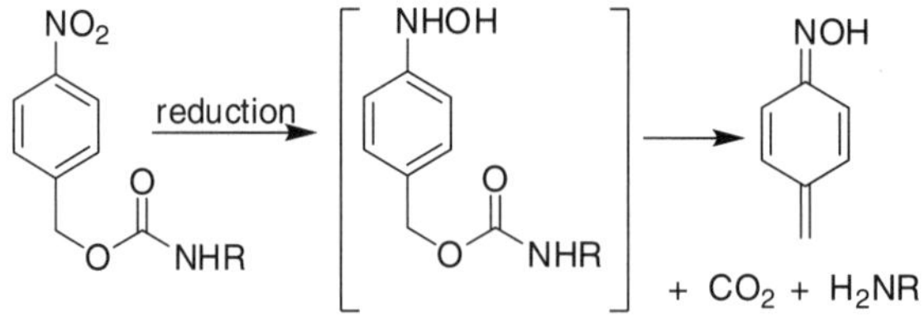

Figure 17. Carbamate fragmentation.

cytotoxic than free doxorubicin was a better substrate for the enzyme [527]. The doxorubicin prodrug DOX-GA3 (**122**) was 12-fold less toxic than doxorubicin in cells, in the human ovarian cancer cell line FMa, and was somewhat superior to doxorubicin itself against FMa xenografts in mice in conjunction with 323-A3/human β-glucuronidase conjugate [528]. However, the utility of such tightly DNA binding, cell cycle specific (topo II inhibitor) effectors is not yet known. Again, no clinical trials have been reported.

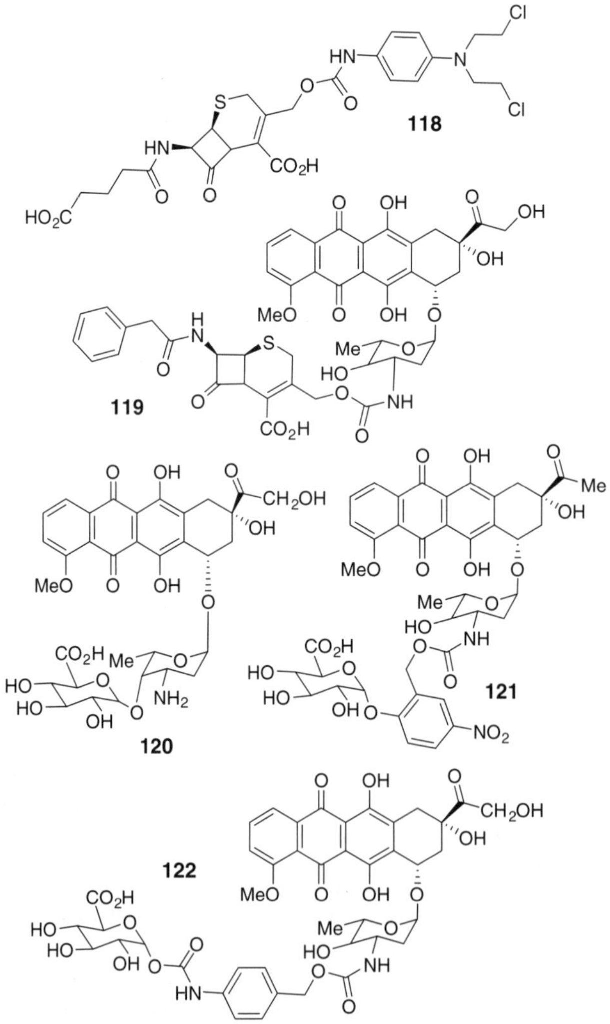

118

119

120

121

122

5.4.3. Biological Activity and Side Effects

Etoposide phosphate and etoposide appear to be equivalent in many settings [529], but a recent report [530] suggests the phosphate can usefully be used in cases of hypersensitivity to etoposide. Two Phase I trials of combretastatin A4 phosphate have been reported [531,532]. The drug was well tolerated, with the dose-limiting toxicities being reversible ataxia [531], and Phase II trials are in progress. No clinical trials of β-lactamase or β-glucuronidase prodrugs have been reported. A Phase I study of the prodrug ZD2767P was conducted, using a single administration of MFECP1 (a recombinant antibody–enzyme fusion protein of an anticarcinoembryonic antigen single-chain Fv antibody) and the bacterial enzyme carboxypeptidase G2, followed by the prodrug. Optimal conditions for effective therapy were established 521].

5.5. Prodrugs for GDEPT

5.5.1. History

The ADEPT approach is generally limited to the use of enzymes that do not require energy-producing cofactors, and also has the likelihood of generating immune responses to the foreign proteins used. In GDEPT, the enzyme is targeted to tumor cells by integrating the gene that produces it into the genome of the tumor cells, followed by administration of the prodrug. A small but growing proportion of the large number of "gene therapy" trials now in progress is for gene-directed enzyme–prodrug therapy, or "suicide gene therapy," although problems with systemic gene delivery remain [533]. This concept theoretically retains the advantages of ADEPT in terms of selective and sufficient access of the activated drug to tumor cells, and expands the class of available enzymes to those that require endogenous cofactors. However, one approach to design in selectivity between prodrug and toxin is lost compared with ADEPT, since the prodrugs must be able to enter cells freely.

5.5.2. Mechanism and SAR

Similarly for ADEPT prodrugs, the specific mechanism of action depends on the type of enzyme used to activate the prodrug. GDEPT offers a wider choice of enzymes, since those with cofactors not readily available outside cells can also be used. The protocol is also generally less immunogenic than ADEPT.

Prodrugs for Kinase Enzymes The most widely used prodrug in GDEPT protocols is the antifungal agent ganciclovir (**123**), which is activated by the thymidine kinase enzyme from Herpes simplex virus converting it into the monophosphate (**124**). This can then be converted by cellular enzymes into the toxic triphosphate, which acts as an antimetabolite. This combination has been evaluated in numerous clinical trials, primarily in gliomas by intratumoral injection [534]. A limitation of the approach is the poor bystander properties of the active drug, which cannot enter cells by passive diffusion, but uses gap junction connections [535] that are not well developed in many types of tumors [536].

Prodrugs for Cytosine Deaminase The yeast enzyme cytosine deaminase [537] has also been widely studied as a GDEPT system in conjunction with 5-fluorocytosine (**125**), which it converts to the thymidylate synthetase inhibitor 5-fluorouracil (**82**) (see Section 4.4). This has good diffusion properties, and shows better bystander effects [538]. Experimental studies have focused mainly on colon cancer models for the use of this combination, because clinically 5-fluorouracil is one of the most effective drugs for colon cancer. Possible drawbacks include the relatively low potency of **82**, coupled with its pronounced cell cycle selectivity [539], and no clinical trials of the protocol have yet been reported.

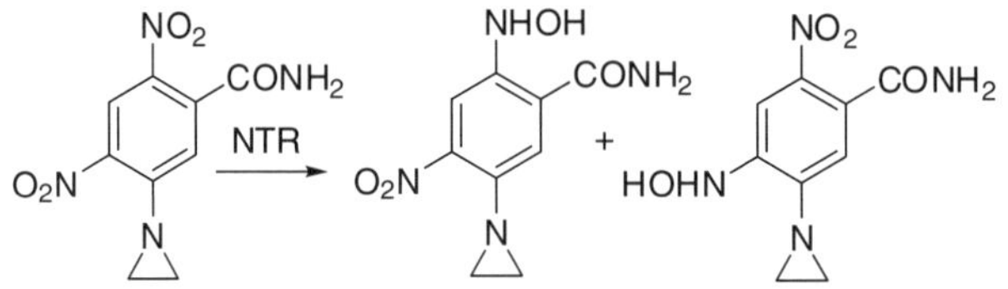

Figure 18. Metabolism of CB 1954 by *E. coli* NTR.

Prodrugs for Oxidative Enzymes The cytochrome P450 enzymes that nonspecifically activate the clinical agent cyclophosphamide (4) in the liver to the active species phosphoramide mustard (see Section 2.3.2 and Fig. 2) have also been employed in a GDEPT protocol with cyclophosphamide. Treatment of sc 9L gliosarcoma tumors transduced with various isozymes, especially CYP2B6 or CYP2C18-Met) with 4 gave large enhancements over the normal liver P450-dependent antitumor effect seen with control 9L tumors (growth delays of 25–50 days compared with 5–6 days), with no apparent increase in host toxicity [540].

Prodrugs for Reductase Enzymes The dinitrophenylaziridine CB 1954 (**126**) is activated by the aerobic nitroreductase (NTR) from *E. coli* [541], in conjunction with NADH or NADPH, to a mixture of hydroxylamines (Fig. 18). The 4-hydroxylamine (**127**) is further metabolized by cellular enzymes to DNA cross-linking species. CB 1954 shows high selectivity (up to 1000-fold) for a variety of cell lines transduced with the enzyme over the corresponding wild-type cell lines [542].

5.5.3. Biological Activity and Side Effects

A recent review of ganciclovir as a prodrug in suicide gene therapy [543] concluded that despite much work and good activity in a number of preclinical models, relatively poor results have been obtained so far in clinical trials. In studies with the HT-29 colorectal tumor cell line, maximum *in vitro* bystander cell death was always obtained when TK-transduced cells were also transduced with connexin-43, to enhance gap junction protein expression [544]. Modest levels of response have been seen in recent clinical trials for prostate cancer [545] and mesothelioma [546]. There have been very few recent clinical studies of 5-fluorocytosine/cytosine deaminase gene therapy. A trial was proposed using as vector an attenuated strain of *S. typhimurium* with chromosomal insertion of an *E. coli* cytosine deaminase gene, designated VNP20009 [547], but results have not been reported.

Work on enhancing the intratumoral levels of the active 4-hydroxy metabolite of cyclophosphamide in mice bearing 9L gliosarcomas expressing the CPA 4-hydroxylase P450 2B6 showed higher levels in the tumors than in liver and plasma [548]. A Phase I trial of oral cyclophosphamide with directly injected MetXia-P450, a recombinant retroviral vector encoding the human cytochrome P450 type 2B6 gene (CYP2B6) in patients with advanced breast cancer or melanoma showed the treatment was well tolerated. Gene transfer (evaluated by β-galactosidase activity) was seen at all dose levels of MetXia-P450, and there was a modest level of therapeutic response (1/12 PR and 4/12 SD [549]. This and other clinical trials have been reviewed recently [550].

A Phase I safety study of a replication-defective adenovirus encoding nitroreductase noted minimal side effects, given by direct intratumoral inoculation, noted a short half-life in the circulation, a robust antibody response and a dose-related increases in tumoral nitroreductase expression; further studies with CTL102 and CB 1954 are planned [551].

REFERENCES

1. Goodman LM, Wintrobe MM, Dameshek W, Goodman MJ, Gilman A, McLennan MT. Nitrogen mustard therapy. Use of methylbis(2-chloroethyl)amine hydrochloride and tris(2-chloroethyl)amine hydrochloride for Hodgkin's disease, lymphosarcoma, leukemia and certain allied and miscellaneous disorders. J Am Med Assoc 1946;132:126–132.

2. Elion GB. The purine path to chemotherapy. Biosci Rep 1989;9:509–529.

3. Zee-Cheng RK-Y, Cheng CC. Screening and evaluation of anticancer agents. Methods Find Exp Clin Pharmacol 1988;10:67–101.
4. Brookes P. The early history of the biological alkylating agents, 1918–1968. Mutat Res Fund Mol Mech Mutagen 1990;233(1-2):3–14.
5. Begleiter A, Mowat M, Israels LG, Johnston JB. Chlorambucil in chronic lymphocytic leukemia: mechanism of action. Leuk Lymphoma 1996;23:187–201.
6. Furner RL, Brown RK. L-phenylalanine mustard (L-PAM): the first 25 years. Cancer Treat Rep 1980;64:559–574.
7. Keating MJ, Bach C, Yasothan U, Kirkpatrick P. Bendamustine. Nat Rev Drug Discov 2008;7:473–474.
8. Colvin OM. An overview of cyclophosphamide development and clinical applications. Curr Pharm Des 1999;5:555–560.
9. Boddy AV, Yule SM. Metabolism and pharmacokinetics of oxazaphosphorines. Clin Pharm 2000;38:291–304.
10. Hansson J, Lewensohn R, Ringborg U, Nilsson B. Formation and removal of DNA cross-links induced by melphalan and nitrogen mustard in relation to drug-induced cytotoxicity in human melanoma cells. Cancer Res 1987;47: 2631–2637.
11. Pieper RO, Futscher BW, Erickson LC. Transcription-terminating lesions induced by bifunctional alkylating agents *in vitro*. Carcinogenesis 1989;10:1307–1314.
12. Niculescu-Duvaz I, Baracu I, Balaban AT, Alkylating agents. In: Wilman DEV, editor. Chemistry of Antitumour Agents. London: Blackie; 1989. p 63–130.
13. Wilman DEV, Connors TA, Molecular structure and antitumour activity of alkylating agents. In: Neidle S, Waring MJ, editors. Molecular Aspects of Anticancer Drug Action. London: MacMillan; 1983. p 233–282.
14. Kohn KW, Hartley JA, Mattes WB. Mechanisms of DNA sequence selective alkylation of guanine-N7 positions by nitrogen mustards. Nucleic Acids Res 1987;15:10531–10549.
15. Pullman A, Pullman B. Molecular electrostatic potential of the nucleic acids. Q Rev Biophys 1981;14:289–380.
16. Prakash AS, Denny WA, Gourdie TA, Valu KK, Woodgate PD, Wakelin LPG. DNA-directed alkylating ligands as potential antitumor agents: sequence specificity of alkylation by intercalating aniline mustards. Biochemistry 1990;29:9799–9807.
17. Osborne MR, Wilman DEV, Lawley PD. Alkylation of DNA by the nitrogen mustard bis(2-chloroethyl)methylamine. Chem Res Toxicol 1995;8:316–320.
18. Osborne MR, Lawley PD. Alkylation of DNA by melphalan with special reference to adenine derivatives and adenine-guanine crosslinking. Chem. Biol. Interact. 1993;89:49–60.
19. Lawley PD, Brookes P. Interstrand cross-linking of DNA by difunctional alkylating agents. J Mol Biol 1967;25:143–160.
20. Millard JT, Raucher S, Hopkins PB. Mechlorethamine cross-links deoxyguanosine residues at 5′-GNC sequences in duplex DNA fragments. J Am Chem Soc 1990;112:2459–2460.
21. Balcome S, Park S, Dorr DRQ, Hafner L, Phillips L, Tretyakova N. Adenine-containing DNA-DNA cross-links of antitumor nitrogen mustards. Chem Res Toxicol 2004;17 (7):950–962.
22. Sladek NE. Metabolism of oxazaphosphorines. Pharmacol Ther 1988;37:301–355.
23. Goren MP, Wright RK, Pratt CB, Pell FE. Dechloroethylation of ifosfamide and neurotoxicity. Lancet 1986;2(8157):1219–1220.
24. Panthananickal A, Hansch C, Leo A, Quinn FR. Structure–activity relationships in antitumor aniline mustards. J Med Chem 1977;21:16–26.
25. Palmer BD, Wilson WR, Pullen SM, Denny WA. Hypoxia-selective antitumor agents. 3. Relationships between structure and cytotoxicity against cultured tumor cells for substituted *N,N*-bis(2-chloroethyl)anilines. J Med Chem 1990;33:112–121.
26. Leoni LM, Bailey B, Reifert J, Bendall HH, Zeller RW, Corbeil J, Elliott G, Niemeyer CC. Bendamustine (Treanda) displays a distinct pattern of cytotoxicity and unique mechanistic features compared with other alkylating agents. Clin Cancer Res 2008;14:309–317.
27. Wang AL, Tew KD. Increased glutathione-S-transferase activity in a cell line with acquired resistance to nitrogen mustards. Cancer Treat. Rep. 1985;69:677–682.
28. Suzukake K, Vistica BP, Vistica DT. Dechlorination of L-phenylalanine mustard by sensitive and resistant tumor cells and its relationship to intracellular glutathione content. Biochem Pharmacol 1983;32:165–167.
29. Brendel M, Ruhland A. Relationships between functionality and genetic toxicology of selected DNA-damaging agents. Mutat Res Rev Genet Toxicol 1984;133:51–85.

30. Palmer RG, Denman AM. Malignancies induced by chlorambucil. Cancer Treat Rev 1984;2:121–129.
31. Leone G, Voso MT, Sica S, Morosetti R. Therapy related leukemias: susceptibility, prevention and treatment. Leuk Lymphoma 2001;41:255–276.
32. Leone G, Mele L, Pulsoni A, Equitani F, Pagano L. The incidence of secondary leukemias. Haematologica 1999;84:937–945.
33. Chronowski GM, Wilder RB, Levy LB, Atkinson EN, Ha CS, Hagemeister FB, Barista I, Rodriguez MA, Sarris AH, Hess MA, Cabanillas F, Cox JD. Second malignancies after chemotherapy and radiotherapy for Hodgkin's disease. Am J Clin Oncol 2004;27(1):73–80.
34. Nicolle A, Proctor SJ, Summerfield GP. High dose chlorambucil in the treatment of lymphoid malignancies. Leuk Lymphoma. 2004;45(2):271–275.
35. Bergmann MA, Goebeler ME, Herold M, Emmerich B, Wilhelm M, Ruelfs C, Boening L, Hallek M, Efficacy of bendamustine in patients with relapsed or refractory chronic lymphocytic leukemia: results of a phase I/II study of the German CLL Study Group J Haematologica 2005;90(10):1357–1364.
36. Herold M, Schulze A, Niederwieser D, Franke A, Fricke HJ, Richter P, Freund M, Ismer B, Dachselt K, Boewer C, Schirmer V, Weniger J, Pasold R, Winkelmann C, Klinkenstein C, Schulze M, Arzberger H, Bremer K, Hahnfeld S, Schwarzer A, Muller C, Muller Ch. Bendamustine, vincristine and prednisone (BOP) versus cyclophosphamide, vincristine and prednisone (COP) in advanced indolent non Hodgkin's lymphoma and mantle cell lymphoma. J Cancer Res Clin Oncol 2006;132(2):105–212.
37. Poenisch W, Mitrou PS, Merkle K, Herold M, Assmann M, Wilhelm G, Dachselt K, Richter P, Schirmer V, Schulze A, Subert R, Harksel B, Grobe N, Stelzer E, Schulze M, Bittrich A, Freund M, Pasold R, Friedrich T, Helbig W, Niederwieser D. Treatment of Bendamustine and Prednisone in patients with newly diagnosed multiple myeloma results in superior complete response rate, prolonged time to treatment failure and improved quality of life compared to treatment with Melphalan and Prednisone. J Cancer Res Clin Oncol 2006;132(4):205–212.
38. Denny WA. DNA minor groove alkylating agents. Curr Med Chem 2001;8:533–554.
39. Gourdie TA, Valu KK, Gravatt GL, Boritzki TJ, Baguley BC, Wilson WR, Woodgate PD, Denny WA. DNA-directed alkylating agents. 1. Structure–activity relationships for acridine-linked aniline mustards: consequences of varying the reactivity of the mustard. J Med Chem 1990;33:1177–1186.
40. Mattes WB, Hartley JA, Kohn KW. DNA sequence selectivity of guanine-N7 alkylation by nitrogen mustards. Nucleic Acids Res 1986;14:2971–2987.
41. Creech HJ, Preston RK, Peck RM, O'Connell AS, Ames BN. Antitumor and mutagenic properties of a variety of heterocyclic nitrogen and sulfur mustards. J Med Chem 1972;15:739–746.
42. Boritzki TJ, Palmer BD, Coddington JM, Denny WA. Identification of the major lesion from the reaction of an acridine-targeted aniline mustard with DNA as an adenine N1 adduct. Chem Res Toxicol 1994;7:41–46.
43. Lee M, Rhodes AL, Wyatt MD, D'Incalci M, Forrow S, Hartley JA. *In vitro* cytotoxicity of GC sequence directed alkylating agents related to distamycin. J Med Chem 1993;36:863–870.
44. Gravatt GL, Baguley BC, Wilson WR, Denny WA. DNA-directed alkylating agents. 6. Synthesis and antitumor activity of DNA minor groove targeted aniline mustard analogues of pibenzimol (Hoechst 33258). J Med Chem 1994;37:4338–4345.
45. Atwell GJ, Yaghi BM, Turner PR, Boyd M, O'Connor CJ, Ferguson LR, Baguley BC, Denny WA, Synthesis DNA interactions and biological activity of DNA minor groove targeted polybenzamide-linked nitrogen mustards. Bioorg Med Chem 1995;6:679–691.
46. Pelton JG, Wemmer DE. Binding modes of distamycin A with d(CGCAAATTTGCG)$_2$ determined by two-dimensional NMR. J Am Chem Soc 1990;112:1393–1399.
47. Baker BF, Dervan PB. Sequence-specific cleavage of double-helical DNA, *N*-Bromoacetyldistamycin. J Am Chem Soc 1985;107:8266–8268.
48. Arcamone FM, Animati F, Barbieri B, Configliacchi E, D'Alessio R, Geroni C, Giuliani FC, Lazzari E, Menozzi M, Mongelli N, Penco S, Verini MA, Synthesis DNA-binding properties, and antitumor activity of novel distamycin derivatives. J Med Chem 1989;32:774–778.
49. Broggini M, Erba E, Ponti M, Ballinari D, Geroni C, Spreafico F, D'Incalci M. Selective DNA interaction of the novel distamycin

derivative FCE 24517. Cancer Res 1991; 51:199–204.
50. Sunavala-Dossabhoy G, Van Dyke MW. Combinatorial identification of a novel consensus sequence for the covalent DNA-binding polyamide tallimustine. Biochemistry 2005;44 (7):2510–2522.
51. Broggini M, Coley HM, Mongelli N, Pesenti E, Wyatt MD, Hartley JA, D'Incalci M. DNA sequence-specific adenine alkylation by the novel antitumor drug tallimustine (FCE 24517), a benzoyl nitrogen mustard derivative of distamycin. Nucleic Acids Res 1995;23: 81–87.
52. Wyatt MD, Lee M, Garbiras BJ, Souhami RL, Hartley JA. Sequence specificity of alkylation for a series of nitrogen mustard-containing analogs of distamycin of increasing binding site size: evidence for increased cytotoxicity with enhanced sequence specificity. Biochemistry 1995;34:13034–13041.
53. Cozzi P, Mongelli N. Cytotoxics derived from distamycin A and congeners. Curr Pharm Des 1998;4:181–201.
54. Erba E, Mascellani E, Pifferi E, D'Incalci M. Comparison of cell-cycle phase perturbations induced by the DNA-minor-groove alkylator tallimustine and by melphalan in the SW626 cell line. Int J Cancer 1995;62:170–175.
55. Bellorini M, Moncollin V, D'Incalci M, Mongelli N, Mantovani R. Distamycin A and tallimustine inhibit TBP binding and basal in vitro transcription. Nucleic Acids Res 1995;23: 1657–1653.
56. Weiss GR, Poggesi I, Rocchetti M, Demaria D, Mooneyham T, Reilly D, Vitek LV, Whaley F, Patricia E, von Hoff DD, O'Dwyer PA. A phase I and pharmacokinetic study of tallimustine [PNU 152241 (FCE 24517)] in patients with advanced cancer. Clin Cancer Res 1998;4: 53–59.
57. Punt CJA, Humblet Y, Roca E, Dirix LY, Wainstein R, Polli A, Corradino I. Tallimustine in advanced previously untreated colorectal cancer, a phase II study. Br J Cancer 1996;73: 803–804.
58. Cozzi P, Beria I, Caldarelli M, Capolongo L, Geroni C, Mongelli N. Cytotoxic halogenoacrylic derivatives of distamycin A. Bioorg Med Chem Lett 2000;10:1269–1272.
59. Cozzi P. Recent outcome in the field of distamycin-derived minor groove binders. Farmaco 2000;55:168–173.
60. Beria I, Baraldi PG, Cozzi P, Caldarelli M, Geroni C, Marchini S, Mongelli N, Romagnoli R. Cytotoxic α-halogenoacrylic derivatives of distamycin A and ongeners. J Med Chem 2004;47(10):2611–2623.
61. Trauger JW, Baird EE, Dervan PB. Recognition of DNA by designed ligands at subnanomolar concentrations. Nature 1996;382: 559–561.
62. Trauger JW, Baird EE, Eldon E, Dervan PB. Recognition of 16 base pairs in the minor groove of DNA by a pyrrole-imidazole polyamide dimer. J Am Chem Soc 1998;120: 3534–3535.
63. Wurtz NR, Dervan PB. Sequence specific alkylation of DNA by hairpin pyrrole-imidazole polyamide conjugates. Chem Biol 2000;7: 153–161.
64. Alvarez David Chou CJ, Latella L, Zeitlin SG, Ku S, Puri PL, Dervan PB, Gottesfeld JM. A two-hit mechanism for pre-mitotic arrest of cancer cell proliferation by a polyamide-alkylator conjugate. Cell Cycle 2006;5 (14):1537–1548.
65. Rosenberg B, Van Camp L, Grimley EB, Thomson AJ. The inhibition of growth or cell division in Escherichia coli by different ionic species of platinum(IV) complexes. Biol Chem 1967;242: 1347–1352.
66. Weiss RB, Christian MC. New cisplatin analogues in development. A review. Drugs 1993; 46:360–377.
67. Scanlon KJ, Kashani-Sabet M, Tone T, Funato T. Cisplatin resistance in human cancers. Pharmacol Ther 1991;52:385–406.
68. Burchenal JH, Kalaher K, O'Toole T, Chisholm J. Lack of cross-resistance between certain platinum coordination compounds in mouse leukemia. Cancer Res 1977;37: 3455–3457.
69. Pelley RJ. Oxaliplatin: a new agent for colorectal cancer. Curr Oncol Rep 2001;3:147–155.
70. Behrens BC, Hamilton TC, Masuda H, Grotzinger KR, Whang-Peng J, Louie KG, Knutsen T, McKoy WM, Young RC, Ozols RF. Characterization of a cis-diamminedichloroplatinum (II)-resistant human ovarian cancer cell line and its use in evaluation of platinum analogues. Cancer Res 1987;47:414–418.
71. Kelland LR, Sharp SY, O'Neill CF, Raynaud IF, Beale PJ, Judson IR. Discovery and development of platinum complexes designed to circumvent cisplatin resistance. J Inorg Biochem 1999;77:111–115.
72. Kelland L. Broadening the clinical use of platinum drug-based chemotherapy with new

analogues: satraplatin and picoplatin. Expert Opin Investig Drugs 2007;16(7):1009–1021.
73. Kelland LR. An update on satraplat the first orally available platinum anticancer drug. Expert Opin Investig Drugs 2000;29:1373–1382.
74. Armstrong AJ, George DJ. Satraplatin in the treatment of hormone-refractory metastatic prostate cancer. Ther Clin Risk Manag 2007;3(5):877–883.
75. Berners-Price SJ, Appleton TG, The chemistry of cisplatin em aqueous solution. In: Kelland LR, Farrell NP, editors. Platinum-Based Drugs in Cancer Therapy. Totowa, NJ: Humana Press; 2000. p 3–35.
76. Davies MS, Berners-Price SJ, Hambley TW. Rates of platination of -AG- and -GA- containing double-stranded oligonucleotides: effect of chloride concentration. J Inorg Biochem 2000; 79:167–172.
77. Stordal B, Pavlakis N, Davey R. A systematic review of platinum and taxane resistance from bench to clinic: an inverse relationship. Cancer Treat Rev 2007;33(8):688–703.
78. Brabec V, Kasparkova J, Vrana O, Novakova O, Cox JW, Qu Y, Farrell N. DNA modifications by a novel bifunctional trinuclear platinum phase I anticancer agent. Biochemistry 1999; 38:6781–6790.
79. Komeda S, Moulaei T, Woods KK, Chikuma M, Farrell NP, Williams LD. A third mode of DNA binding: phosphate clamps by a polynuclear platinum complex. J Am Chem Soc 2006;128 (50):16092–16103.
80. Roberts JD, Peroutka J, Farrell N. Cellular pharmacology of polynuclear platinum anticancer agents. J Inorg Biochem 1999;77:51–57.
81. Di Blasi P, Bernareggi A, Beggiolin G, Piazzoni L, Menta E, Formento ML. Cytotoxicity, cellular uptake and DNA binding of the novel trinuclear platinum complex BBR 3464 in sensitive and cisplatin resistant murine leukemia cells. Anticancer Res 1998;18:3113–3117.
82. Manzotti C, Pratesi G, Menta E, Di Domenico R, Cavalletti E, Fiebig HH, Kelland LR, Farrell N, Polizzi D, Supino R, Pezzoni G, Zunino F, BBR 3464: a novel triplatinum complex, exhibiting a preclinical profile of antitumor efficacy different from cisplatin. Clin Cancer Res 2000;6:2626–2634.
83. Servidei T, Ferlini C, Riccardi A, Meco D, Scambia G, Segni G, Manzotti C, Riccardi R. The novel trinuclear platinum complex BBR 3464 induces a cellular response different from cisplatin. Eur J Cancer 2001;37:930–938.
84. Hensing TA, Hanna NH, Gillenwater HH, Camboni MG, Allievi C, Socinski MA. Phase II study of BBR 3464 as treatment in patients with sensitive or refractory small cell lung cancer. Anticancer Drugs 2006;17(6):697–704.
85. Jodrell DI, Evans TRJ, Steward W, Cameron D, Prendiville J, Aschele C, Noberasco C, Lind M, Carmichael J, Dobbs N, Camboni G, Gatti B, De Braud F. Phase II studies of BBR3464, a novel tri-nuclear platinum complex, in patients with gastric or gastro-oesophageal adenocarcinoma. Eur J Cancer 2004;40 (12):1872–1877.
86. Hurley LH, Lee C-S, McGovren JP, Warpehoski MA, Mitchell MA, Kelly RC, Aristoff PA. Molecular basis for sequence-specific DNA alkylation by CC-1065. Biochemistry 1988; 27:3886–3892.
87. Hanka LJ, Dietz A, Gerpheide SA, Kuentzel SL, Martin DG. CC-1065 (NSC-298223), a new antitumor antibiotic. Production, in vitro biological activity, microbiological assays and taxonomy of the producing microorganism. J Antibiot 1978;31:1211–1217.
88. Martin DG, Biles C, Gerpheide SA, Hanka LJ, Krueger WC, McGovren JP, Mizak SA, Neil GL, Stewart JC, Visser J. C-1065 (NSC 298223), a potent new antitumor agent improved production and isolation, characterization and antitumor activity. J Antibiot 1981; 34:1119–1126.
89. McGovren JP, Clarke GL, Pratt EA, DeKoning TF. Preliminary toxicity studies with the DNA-binding antibiotic, CC-1065. J Antibiot 1984;37:63–70.
90. Li LH, Kelly RC, Warpehoski MA, McGovren JP, Gebhard I, DeKoning TF. Adozelesin, a selected lead among cyclopropylpyrroloindole analogs of the DNA-binding antibiotic, CC-1065. Invsest New Drugs 1991;9:137–148.
91. Li LH, DeKoning TF, Kelly RC, Krueger WC, McGovren JP, Padbury GE, Petzold GL, Wallace TL, Ouding RJ, Prairie MD, Gebhard I. Cytotoxicity and antitumor activity of carzelesin, a prodrug cyclopropylpyrroloindole analog. Cancer Res 1992;52:4904–4913.
92. Ogasawara H, Nishio K, Takeda Y, Ohmori T, Kubota N, Funayama Y, Ohira T, Kuraishi Y, Isogai Y, Saijo N. A novel antitumor antibiotic KW-2189 is activated by carboxyl esterase and induces DNA strand breaks in human small cell lung cancer cells. J Cancer Res 1994;85: 418–425.
93. Kinugawa M, Nakamura S, Sakaguchi A, Masuda Y, Saito H, Ogasa T, Kasai M. Practical

Synthesis of the high-quality antitumor agent KW-2189 from Duocarmycin B2 using a facile one-pot synthesis of an intermediate. Org Process Res Dev 1998;2:344–350.

94. Kobayashi E, Okamoto A, Asada M, Okabe M, Nagamura S, Asai A, Saito H, Gomi K, Hirata T. Characteristics of antitumor activity of KW-2189, a novel water-soluble derivative of duocarmycin, against murine and human tumors. Cancer Res 1994;54:2404–2410.

95. Boger DL, Johnson DS. CC-1065 and the duocarmycins: understanding their biological function through mechanistic studies. Angew Chem Int Ed Engl 1996;35:1438–1474.

96. Boger DL, Bollinger B, Hertzog DL, Johnson DS, Cai H, Mesini P, Garbaccio RM, Jin Q, Kitos PA. Reversed and sandwiched analogs of Duocarmycin SA: establishment of the origin of the sequence-selective alkylation of DNA and new insights into the source of catalysis. J Am Chem Soc 1997;119:4987–4998.

97. Yoon JH, Lee CS. Sequence selectivity of DNA alkylation by adozelesin and carzelesin. Arch Pharm Res 1998;21:385–390.

98. Amishiro N, Nagamura S, Kobayashi E, Gomi K, Sato H. New water-soluble duocarmycin derivatives: synthesis and antitumor activity of A-ring pyrrole compounds bearing β-heteroarylacryloyl groups. J Med Chem 1999;42:669–676.

99. Amishiro N, Okamoto A, Murakata C, Tamaoki T, Okabe M, Saito H. Synthesis and antitumor activity of duocarmycin derivatives: modification of segment-A of A-ring pyrrole compounds. J Med Chem 1999;42:2946–2960.

100. Hidalgo M, Izbicka E, Cerna C, Gomez L, Rowinsky EK, Weitman SD, von Hoff DD. Comparative activity of the cyclopropylpyrroloindole compounds adozelesin, bizelesin and carzelesin in a human tumor colony-forming assay. Anticancer Drugs 1999;10:295–305.

101. Cristofanilli M, Bryan WJ, Miller LL, Chang AYC, Gradishar WJ, Kufe DW, Hortobagyi GN. Phase II study of adozelesin in untreated metastatic breast cancer. Anticancer Drugs 1998;9:779–782.

102. Pavlidis N, Aamdal S, Awada A, Calvert AH, Fumoleau P, Sorio R, Punt C, Verweij J, van Oosterom A, Morant R, Wanders J, Hanauske AR. Carzelesin phase II study in advanced breast, ovarian, colorectal, gastric, head and neck cancer, non-Hodgkin's lymphoma and malignant melanoma: a study of the EORTC early clinical studies group (ECSG). Cancer Chemother Pharmacol 2000;46:167–171.

103. Alberts SR, Erlichman C, Reid JM, Sloan JA, Ames MM, Richardson RL, Goldberg RM. Phase I study of the duocarmycin semisynthetic derivative KW-2189 given daily for five days every six weeks. Clin Cancer Res 1998;4:2111–2117.

104. Small EJ, Figlin R, Petrylak D, Vaughn DJ, Sartor O, Horak I, Pincus R, Kremer A, Bowden C. A phase II pilot study of KW-2189 in patients with advanced renal cell carcinoma. Invest New Drugs 2000;18:193–197.

105. Atwell GJ, Tercel M, Boyd M, Wilson WR, Denny WA. Synthesis and cytotoxicity of 5-amino-1-(chloromethyl)-3-[(5,6,7-trimethoxyindol-2-yl)carbonyl]-1,2-dihydro-3H-benz[e]indole (amino-seco-CBI-TMI) and related 5-alkylamino analogues: new DNA minor groove alkylating agents. J Org Chem 1998;63:9414–9420.

106. Atwell GJ, Milbank JJB, Wilson WR, Hogg A, Denny WA. 5- Amino -1-(chloromethyl)-1,2-dihydro-3H-benz[e]indoles: relationships between structure and cytotoxicity for analogues bearing different DNA minor groove binding subunits. J Med Chem 1999;42(17): 3400–3411.

107. Gieseg MA, Matejovic J, Denny WA. Comparison of the patterns of DNA alkylation by phenol and amino seco-CBI-TMI compounds: use of a PCR method for the facile preparation of single end-labelled double stranded DNA. Anticancer Drug Des 1999;14:77–84.

108. Yang S, Denny WA. A new short synthesis of 3-substituted-1-(chloromethyl)-2,3-dihydro-(1H)-benzo[e]indol-5-ylamines (amino-CBIs). J Org Chem 2002;67:8958–8961.

109. Montgomery JA, Johnson TP, Nitrosoureas. In: Wilman DEV, editor. Chemistry of Antitumour Agents London: Blackie; 1989. p 131–158.

110. Skipper HE, Schabel FM, Trader MW, Thomson JR. Experimental evaluation of potential anticancer agents. VI. Anatomical distribution of leukemic cells and failure of chemotherapy. Cancer Res 1961;21:1154–1164.

111. DeVita VT, Denham C, Davidson JD, Olivero VT. Physiological disposition of the carcinostatic 1,3-bis(2-chloroethyl)-1-nitrosourea in man and animals. Clin Pharm Ther 1967;8:566–577.

112. Salmon SE. Nitrosoureas in multiple myeloma. Cancer Treat Rep 1976;60:789–794.

113. Brandsma D, van den Bent MJ. Molecular targeted therapies and chemotherapy in malignant gliomas Curr Opin Oncol 2007;19(6):598–605.

114. Weiss RB. Streptozocin: a review of its pharmacology, efficacy, and toxicity. Cancer Treat Rep 1982;66:427–438.
115. Shyam K, Penketh PG, Loomis RH, Rose WC, Sartorelli AC. Antitumor 2-(aminocarbonyl)-1,2-bis(methylsulfonyl)-1-(2-chloroethyl)- hydrazines. J Med Chem 1996;39(3):796–801.
116. Montgomery JA, James R, McCaleb GS, Kirk MC, Johnston TP. Decomposition of N-(2-chloroethyl)-N-nitrosoureas in aqueous media. J. Med. Chem. 1975;18:568–571.
117. Buckley N. Structure–activity relations of (2-chloroethyl)nitrosoureas. 1. Deuterium isotope effects in the hydrolysis of 1-(2-chloroethyl)-1-nitrosoureas: evidence for the rate-limiting step. J Org Chem 1987;52:484–488.
118. Penketh PG, Shyam K, Sartorelli AC. Comparison of DNA lesions produced by tumor-inhibitory 1,2-bis(sulfonyl)hydrazines and chloroethylnitrosoureas. Biochem Pharmacol 1999;59(3):283–291.
119. Baumann RP, Shyam K, Penketh PG, Remack JS, Brent TP, Sartorelli AC. 1,2-Bis(methylsulfonyl)-1-(2-chloroethyl)-2-[(methylamino)carbonyl]hydrazine (VNP40101M): II. Role of O6-alkylguanine-DNA alkyltransferase in cytotoxicity. Cancer Chemother Pharmacol 2004;53(4):288–295.
120. Schein PS, O'Connell MJ, Blom J, Hubbard S, Magrath IT, Bergevin PA, Wiernik PH, Zeigler JL, DeVita VT. Clinical antitumor activity and toxicity of streptozotocin (NSC-85998). Cancer 1974;34:993–1000.
121. Bukowski RM, Abderhalden RT, Hewlett JS, Weick JK, Groppe CW. Phase II trial of streptozotocin, mitomycin C, and 5-fluorouracil in adenocarcinoma of the pancreas. Cancer Clin Trials 1980;3:321–324.
122. Kemeny N, Yagoda A, Braun D, Golbey R. Therapy for metastatic colorectal carcinoma with a combination of methyl-CCNU, 5-fluorouracil vincristine and streptozotocin (MOF-Strep). Cancer 1980;45:876–891.
123. Chan JA, Kulke MH. Emerging therapies for the treatment of patients with advanced neuroendocrine tumors. Expert Opin Emerg Drugs 2007;12(2):253–270.
124. Finch RA, Shyam K, Penketh PG, Sartorelli AC. 1,2-Bis(methylsulfonyl)-1-(2-chloroethyl)-2-(methylamino)carbonylhydrazine (101M): a novel sulfonylhydrazine prodrug with broad-spectrum antineoplastic activity. Cancer Res 2001;61(7):3033–3038.
125. Giles F, Rizzieri D, Karp J, Vey N, Ravandi F, Faderl S, Khan KD, Verhoef G, Wijermans P, Advani A, Roboz G, Kantarjian H, Bilgrami SFA, Ferrant A, Daenen S, Karsten V, Cahill A, Albitar M, Mufti G, O'Brien S. Cloretazine (VNP40101M), a novel sulfonylhydrazine alkylating agent, in patients age 60 years or older with previously untreated acute myeloid leukemia. J Clin Oncol 2007;25(1):25–31.
126. Giles F, Verstovsek S, Faderl S, Vey N, Karp J, Roboz G, Khan KD, Cooper M, Bilgrami SFA, Ferrant A, Daenen S, Karsten V, Cahill A, Albitar M, Kantarjian H, O'Brien S, Feldman E. A phase II study of cloretazine (VNP40101M), a novel sulfonylhydrazine alkylating agent, in patients with very high risk relapsed acute myeloid leukemia. Leukemia Res 2006;30(12):1591–1595.
127. Badruddoja MA, Penne K, Desjardins A, Reardon DA, Rich JN, Quinn JA, Sathornsumetee S, Friedman AH, Bigner DD, Herndon JE, Cahill A, Friedman HS, Vredenburgh JJ. Phase II study of cloretazine for the treatment of adults with recurrent glioblastoma multiforme. Neuro-Oncol. 2006;9(1):70–74.
128. Shealy YF, Krauth CA, Montgomery JA, Imidazoles I. Coupling reactions of 5 diazoimidazole-4-carboxamide. J Org Chem 1962;27:2150–2154.
129. Shealy YF, Montgomery JA, Laster WR. Antitumor activity of triazenoimidazoles. Biochem Pharmacol 1962;11:674–676.
130. Stevens MGF, Hickman JA, Stone R, Gibson NW, Baig GU, Lunt E, Newton CG. Synthesis and chemistry of 8-carbamoyl-3-(2-chloroethyl)imidazo[5,1-d]-1,2,3,5-tetrazin-4($3H$)-one, a novel broad-spectrum antitumor agent. J Med Chem 1984;27:196–201.
131. Newlands ES, Stevens MGF, Wedge SR, Wheelhouse RT, Brock C. Temozolomide: a review of its discovery, chemical properties, pre-clinical development and clinical trials. Cancer Treat Rev 1997;23:35–61.
132. Marchesi F, Turriziani M, Tortorelli G, Avvisati G, Torino F, De Vecchis L. Triazene compounds: mechanism of action and related DNA repair systems. Pharmacol Res 2007;56:275–287.
133. Clark AS, Deans B, Stevens MGF, Tisdale MJ, Wheelhouse RT, Denny BJ, Hartley JA. Antitumor imidazotetrazines. 32. 1 Synthesis of novel imidazotetrazinones and related bicyclic heterocycles to probe the mode of action of the antitumor drug Temozolomide. J Med Chem 1995;38:1493–1504.
134. Tisdale MJ. Antitumor imidazotetrazines. XV. Role of guanine O6 alkylation in the mechan-

ism of cytotoxicity of imidazotetrazinones. Biochem Pharmacol 1987;36:457–462.
135. Gibson NW, Erickson LC, Hickman JA. Effects of the antitumor agent 8-carbamoyl-3-(2-chloroethyl)imidazo[5,1-d]1,2,3, 5-tetrazin-4(3H)-one on the DNA of mouse L1210 cells. Cancer Res 1984;44:1767–1771.
136. Hartley JA, Gibson NW, Kohn KW, Mattes WB. DNA sequence selectivity of guanine-N7 alkylation by three antitumor chloroethylating agents. Cancer Res 1986;46:1943–1947.
137. Cohen GL, Falkson CI. Current treatment options for malignant melanoma. Drugs 1998;55:791–799.
138. Dreiling L, Hoffman S, Robinson WA. Melanoma: epidemiology, pathogenesis, and new modes of treatment. Adv Intern Med 1996;41:553–604.
139. Connors TA, Goddard PM, Merai K, Ross WCJ, Wilman DEV. Tumor inhibitory triazenes: structural requirements for an active metabolite. Biochem Pharmacol 1976;25:241–246.
140. Meer L, Janzer RC, Kleihues P, Kolar GF. In vivo metabolism and reaction with DNA of the cytostatic agent, 5-(3,3-dimethyl-1-triazeno)imidazole-4-carboxamide (DTIC). Biochem Pharmacol 1986;35:3243–3247.
141. Eggermont AMM, Kirkwood JM. Re-evaluating the role of dacarbazine in metastatic melanoma: what have we learned in 30 years? Eur J Cancer 2004;40(12):1825–1836.
142. Hickman JA, Stevens MFG, Gibson NW, Langdon SP, Fizames C, Lavelle F, Atassi G, Lunt E, Tilson RM. Experimental antitumor activity against murine tumor model systems of 8-carbamoyl-3-(2-chloroethyl)imidazo[5,1-d]-1,2,3,5-tetrazin-4(3H)-one (mitozolomide), a novel broad-spectrum agent. Cancer Res 1985;45:3008–3013.
143. Fodstad O, Aamdal S, Pihl A, Boyd MR. Activity of mitozolomide (NSC 353451), a new imidazotetrazine, against xenografts from human melanomas, sarcomas, and lung and colon carcinomas. Cancer Res 1985;45:1778–1786.
144. Newlands ES, Blackledge G, Slack JA, Goddard C, Brindley CJ, Holden L, Stevens MF. Phase I clinical trial of mitozolomide. Cancer Treat Rep 1985;69:801–805.
145. Harding M, Docherty V, Mackie R, Dorward A, Kaye S. Phase II studies of mitozolomide in melanoma, lung and ovarian cancer. Eur J Cancer Clin Oncol 1989;25:785–788.
146. Somers R, Santoro A, Verweij J, Lucas P, Rouesse J, Kok T, Casali A, Seynaeve C, Thomas D. Phase II study of mitozolomide in advanced soft tissue sarcoma of adults: the EORTC Soft Tissue and Bone Sarcoma Group. Eur J Cancer 1992;28A:855–857.
147. Hickson I, Fairbairn LJ, Chinnasamy N, Lashford LS, Thatcher N, Margison GP, Dexter TM, Rafferty JA. Chemoprotective gene transfer. I. Transduction of human hemopoietic progenitors with O6-benzylguanine-resistant O6-alkylguanine-DNA alkyltransferase attenuates the toxic effects of O6-alkylating agents in vitro. Gene Ther 1998;5:835–841.
148. Stevens MF, Hickman JA, Langdon SP, Chubb D, Vickers L, Stone R, Baig G, Goddard C, Gibson NW, Slack JA. Antitumor activity and pharmacokinetics in mice of 8-carbamoyl-3-methyl-imidazo[5,1-d]-1,2,3,5-tetrazin-4(3H)-one (CCRG 81045; M & B 39831), a novel drug with potential as an alternative to dacarbazine. Cancer Res 1987;47:5846–5852.
149. Plowman J, Waud WR, Koutsoukos AD, Rubinstein LV, Moore TD, Grever MR. Preclinical antitumor activity of temozolomide in mice: efficacy against human brain tumor xenografts and synergism with 1,3-bis(2-chloroethyl)-1-nitrosourea. Cancer Res 1994;54: 3793–3799.
150. Newlands ES, Blackledge GR, Slack JA, Rustin GJ, Smith DB, Stuart NS, Quarterman CP, Hoffman R, Stevens MF, Brampton MH. Phase I trial of temozolomide (CCRG 81045: M&B 39831: NSC 362856). Br J Cancer 1992;65: 287–291.
151. O'Reilly SM, Newlands ES, Glaser MG, Brampton M, Rice-Edwards JM, Illingworth RD, Richards PG, Kennard C, Colquhoun IR, Lewis P. Temozolomide: a new oral cytotoxic chemotherapeutic agent with promising activity against primary brain tumours. Eur J Cancer 1993;29A:940–942.
152. Friedman HS, Kerby T, Calvert H. Temozolomide and treatment of malignant glioma. Clin Cancer Res 2000;6:2585–2597.
153. Prados MD. Future directions in the treatment of malignant gliomas with temozolomide. Semin Oncol 2000;27(Suppl 6):41–46.
154. Agarwala SS, Kirkwood JM. Temozolomide, a novel alkylating agent with activity in the central nervous system, may improve the treatment of advanced metastatic melanoma. Oncologist 2000;5:144–151.
155. De Isabella P, Capranico G, Zunino F. The role of topoisomerase II in drug resistance. Life Sci 1991;48:2195–2205.
156. Zwelling LA, Michaels S, Erickson LC, Ungerleider RS, Nichols M, Kohn KW. Protein-asso-

ciated deoxyribonucleic acid strand breaks in L1210 cells treated with the deoxyribonucleic acid intercalating agents 4'-(9-acridinylamino)methanesulfon-*m*-anisidide and adriamycin. Biochemistry 1981;20:6553–6563.

157. Liu LF. DNA topoisomerase poisons as antitumor drugs. Ann Rev Biochem 1989;58: 351–375.

158. Robinson MJ, Osheroff N. Effects of antineoplastic drugs on the post-strand-passage DNA cleavage/religation equilibrium of topoisomerase II. Biochemistry 1991;30:1807–1813.

159. Boonsong A, Marsh S, Rooney PH, Stevenson DAJ, Cassidy J, McLeod HL. Characterization of the topoisomerase I locus in human colorectal cancer. Cancer Genet Cytogenet 2000;12: 56–60.

160. Chung TDY, Drake FH, Tan SR, Per M, Crooke ST, Mirabelli CK. haracterization and immunological identification of cDNA clones encoding two human DNA topoisomerase II isozymes. Proc Natl Acad Sci USA 1989;86: 9431–9435.

161. Tan KB, Dorman TF, Falls KM, Chung TDY, Mirabelli CK, Crooke ST, Mao J. Topoisomerase IIα and topoisomerase IIβ genes: characterization and mapping to human chromosomes 17 and 3, respectively. Cancer Res 1992;52:231–234.

162. Drake FH, Hofmann GA, Bartus HF, Mattern MR, Crooke ST, Mirabelli CK. Biochemical and pharmacological properties of p170 and p180 forms of topoisomerase II. Biochemistry 1989;28:8154–8160.

163. Denny WA, Baguley BC. Dual topoisomerase I/II inhibitors. Curr Top Med Chem 2003;3: 339–353.

164. Lerman LS. Structural considerations in the interaction of DNA and acridines. J Mol Biol 1961;3:18–30.

165. Wadkins RM, Graves DE. Interactions of anilinoacridines with nucleic acids: effects of substituent modifications on DNA-binding properties. Biochemistry 1991;30:4277–4283.

166. Denny WA. DNA-Intercalating agents as antitumour drugs: prospects for future design. Anticancer Drug Des 1989;4:241–263.

167. Nelson EM, Tewey KM, Liu LF. Mechanism of antitumor drug action: poisoning of mammalian DNA topoisomerase II on DNA by 4'-(9-acridinylamino)methanesulfon-m-anisidide. Proc Natl Acad Sci USA 1984;181:1361–1365.

168. Wall ME. Camptothecin and taxol: discovery to clinic. Med Res Rev 1998;18:299–314.

169. Pizzolato JF, Saltz LB. The camptothecins. Lancet 2003;361:2235–2242.

170. Chrencik JE, Staker BL, Burgin AB, Pourquier P, Pommier Y, Stewart Lance Redinbo MR. Mechanisms of camptothecin resistance by human topoisomerase I mutations. J Mol Biol 2004;339(4):773–784.

171. Laco GS, Collins JR, Luke BT, Kroth H, Sayer JM, Jerina DM, Pommier Y. Human topoisomerase I inhibition: docking camptothecin and derivatives into a structure-based active site model. Biochemistry 2002;41(5):1428–1435.

172. Redinbo MR, Champoux JJ, Hol WGJ. Structural insights into the function of type IB topoisomerases. Curr Opin Struct Biol 1999;9(1):29–36.

173. Jung LL, Ramanathan RK, Egorin MJ, Jin R, Belani CP, Potter DM, Strychor S, Trump DL, Walko C, Fakih M, Zamboni WC. Pharmacokinetic studies of 9-nitrocamptothecin on intermittent and continuous schedules of administration in patients with solid tumors. Cancer Chemother Pharmacol 2004;54:487–496.

174. Rajendra R, Gounder MK, Saleem A, Schellens JHM, Ross DD, Bates SE, Sinko P, Rubin EH. Differential effects of the breast cancer resistance protein on the cellular accumulation and cytotoxicity of 9-aminocamptothecin and 9-nitrocamptothecin. Cancer Res 2003;63: 3228–3233.

175. Burris HA, Rivkin S, Reynolds R, Harris J, Wax A, Gerstein H, Mettinger KL, Staddon A. Phase II trial of oral rubitecan in previously treated pancreatic cancer patients. Oncologist 2005;10:183–190.

176. Verschraegen CF, Gilbert BE, Loyer E, Huaringa A, Walsh G, Newman RA, Knight V. Clinical evaluation of the delivery and safety of aerosolized liposomal 9-nitro-20(S)-camptothecin in patients with advanced pulmonary malignancies. Clin Cancer Res 2004;10: 2319–2326.

177. Jew SS, Kim MG, Kim HJ, Rho EY, Park HG, KimJK, Han HJ, Lee H. Synthesis and *in vitro* cytotoxicity of C(20)(RS)-camptothecin analogs modified at both B (or A) and E ring. Bioorg Med Chem Lett 1998;8:1797–1800.

178. Lee JH, Lee JM, Kim JK, Ahn SK, Lee SJ, Kim MY, Jew SS, Park JG, Hong CI. Antitumor activity of 7-[2-(*N*-isopropylamino)ethyl]-(20S)-camptothecin, CKD602, as a potent DNA topoisomerase I inhibitor. Arch Pharm Res 1998;21:581–590.

179. Lee DH, Kim SW, Suh C, Lee JS, Lee JH, Lee SJ, Ryoo BY, Park K, Kim JS, Heo DS, Kim

NK. Belotecan, new camptothecin analogue, is active in patients with small-cell lung cancer: results of a multicenter early phase II study. Ann Oncol 2008;19(1):123–127.
180. Bom D, Curran DP, Chavan AJ, Kruszewski S, Zimmer SG, Fraley KA, Burke TG. Novel A,B, E-ring-modified camptothecins displaying high lipophilicity and markedly improved human blood stabilities. J Med Chem 1999;42: 3018–3022.
181. Kumazawa E, Jimbo T, Ochi Y, Tohgo A. Potent and broad antitumor effects of DX-8951f, a water-soluble camptothecin derivative, against various human tumors xenografted in nude mice. Cancer Chemother Pharmacol 1998;42:210–220.
182. Joto N, Ishii M, Minami M, Kuga H, Mitsui I, Tohgo A. DX-8951f, a water-soluble camptothecin analog, exhibits potent antitumor activity against a human lung cancer cell line and its SN-38-resistant variant. Int J Cancer 1997;72:680–686.
183. Chilman-Blair K, Mealy NE, Castaner J, Bayes M. Exatecan mesylate: anticancer agent DNA topoisomerase I inhibitor. Drugs Future 2004;29:9–22.
184. Reichardt P, Nielsen OS, Bauer S, Hartmann JT, Schoeffski P, Christensen TB, Pink D, Daugaard S, Marreaud S, Van Glabbeke M, Blay JY. Exatecan in pretreated adult patients with advanced soft tissue sarcoma: results of a phase II - Study of the EORTC soft tissue and bone sarcoma group. Eur J Cancer 2007;43 (6):1017–1022.
185. Dallavalle S, Ferrari A, Merlini L, Penco S, Carenini N, De Cesare M, Perego P, Pratesi G, Zunino F. Novel cytotoxic 7-iminomethyl and 7-aminomethyl derivatives of camptothecin. Bioorg Med Chem Lett 2001;11:291–294.
186. Croce AC, Bottiroli G, Supino R, Favini E, Zuco V, Zunino F. Subcellular localization of the camptothecin analogues, topotecan and gimatecan. Biochem Pharm 2004;67:1035–1045.
187. Di Francesco AM, Riccardi AS, Barone G, Rutella S, Meco D, Frapolli R, Zucchetti M, D'Incalci M, Pisano C, Carminati P, Riccardi R, Biochem Pharmacol 2005;70:1125–1136.
188. Marchetti S, Oostendorp RL, Pluim D, van Eijndhoven M, van Tellingen O, Schinkel AH, Versace R, Beijnen JH, Mazzanti R, Schellens JH. In vitro transport of gimatecan (7-t-butoxyiminomethylcamptothecin) by breast cancer resistance protein, P-glycoprotein, and multidrug resistance protein 2. Mol Cancer Ther 2007;6:3307–3313.

189. Pecorelli S, Ray-Coquard I, Colombo N, Katsaros D, Lhomme C, Lissoni A, Vermorken JB, Du Bois A, Poveda A, Frigerio L. A phase II study of oral gimatecan (ST1481) in women with progressing or recurring advanced epithelial ovarian, fallopian tube and peritoneal cancers. J Clin Oncol 2006;24(Suppl 18S):5088.
190. Van Hattum AH, Pinedo HM, Schluper HM, Hausheer FH, Boven E. New highly lipophilic captothecin BNP1350 is an effective drug in experimental human cancer. Int J Cancer 2000;88:260–266.
191. Van Hattum AH, Schluper HM, Hausheer FH, Pinedo HM, Boven E. Novel camptothecin derivative BNP1350 in experimental human ovarian cancer: determination of efficacy and possible mechanisms of resistance. Int J Cancer 2002;100:22–29.
192. Yin MB, Hapke G, Wu J, Azrak R, Frank C, Wrzosek C, Rustum YM. Chk1 signaling pathways that mediated G2M checkpoint in relation to the cellular resistance to the novel topoisomerase I poison BNP1350. Biochem Biophys Res Commun 2002;295:435–444.
193. Daud A, Valkov N, Centeno B, Derderian J, Sullivan P, Munster P, Urbas P, DeConti RC, Berghorn E, Liu Z, Hausheer F, Sullivan D. Phase II trial of karenitecin in patients with malignant melanoma: clinical and translational study. Clin Cancer Res 2005;11: 3009–3016.
194. Lavergne O, Demarquay D, Bailly C, Lanco C, Rolland A, Huchet M, Coulomb H, Muller N, Baroggi N, Camara J, Le Breton C, Manginot E, Cazaux J, -B Bigg DCH. Topoisomerase I-mediated antiproliferative activity of enantiomerically pure fluorinated homocamptothecins. J Med Chem 2000;43:2285–2289.
195. Lesueur-Ginot L, Demarquay D, Kiss R, Kaspryzk PG, Lavergne O. Homocamptothecin, an E-ring modified camptothecin with enhanced lactone stability, retains topoisomerase I-targeted activity and antitumor properties. Cancer Res 1999;59:2939–2943.
196. Philippart P, Harper L, Chaboteaux C, Decaestecker C, Bronckart Y, Gordover L, Lesueur-Ginot L, Malonne H, Lavergne O, Bigg DCH, Da Costa PM, Kiss R. Homocamptothecin, an E-ring-modified camptothecin, exerts more potent antiproliferative activity than other topoisomerase I inhibitors in human colon cancers obtained from surgery and maintained in vitro under histotypical culture conditions. Clin Cancer Res 2000;6:1557–1562.

197. Demarquay D, Huchet M, Coulomb H, Lesueur-Ginot L, Lavergne O, Kasprzyk PG, Bailly C, Camara J, Bigg DCH. The homocamptothecin BN 80915 is a highly potent orally active topoisomerase I poison. Anticancer Drugs 2001;12:9–19.

198. Gelderblom H, Salazar R, Verweij J, Pentheroudakis G, de Jonge MJA, Devlin M, van Hooije C, Seguy F, Obach R, Prunonosa J, Principe P, Twelves C. Phase I Pharmacological and bioavailability study of oral Diflomotecan (BN80915), a novel E-ring-modified camptothecin analogue in adults with solid tumors. Clin Cancer Res 2003;9:4101–4107.

199. Troconiz IF, Garrido MJ, Segura C, Cendros J-M, Principe P, Peraire C, Obach R. Phase I dose-finding study and a pharmacokinetic/pharmacodynamic analysis of the neutropenic response of intravenous diflomotecan in patients with advanced malignant tumours. Cancer Chemother Pharmacol 2006;57(6):727–735.

200. Denny WA, Atwell GJ, Baguley BC, Cain BF. Potential antitumor agents. Part 29. QSAR for the antileukemic bisquaternary ammonium heterocycles. J Med Chem 1979;22:134–151.

201. Atwell GJ, Cain BF, Seelye RN. Potential antitumor agents. 12. 9-Anilinoacridines. J Med Chem 1972;15:611–615.

202. Lawrence HJ, Ries CA, Reynolds BD, Lewis JP, Koretz MM, Torti FM. AMSA-a promising new agent in refractory acute leukemia. Cancer Treat Rep 1982;66:1475–1478.

203. Denny WA, Amsacrine. In: Lednicer D, Bindra J, editors. Chronicles of Drug Discovery. Vol. 3. Washington, DC: ACS Publications; 1993. p 381–404.

204. Denny WA, Atwell GJ, Cain BF, Leo A, Panthananickal A, Hansch C. Potential antitumor agents. Part 36. Quantitative relationships between antitumor potency, toxicity and structure for the general class of 9-anilinoacridine antitumor agents. J Med Chem 1982;25:276–316.

205. Baguley BC, Denny WA, Atwell GJ, Finlay GF, Rewcastle GW, Twigden SJ, Wilson WR. Synthesis, antitumor activity and DNA-binding properties of a new derivative of amsacrine (CI-921: NSC 343499). Cancer Res 1984;44:3245–3252.

206. Leopold WR, Corbett TH, Griswold DP, Plowman J, Baguley BC. Experimental antitumor activity of the amsacrine analog CI-921. J Natl Cancer Inst 1987;79:343–349.

207. Zee-Cheng RKY, Cheng CC. Antineoplastic agents. Structure–activity relationship study of bis(substituted aminoalkylamino)anthraquinones. J Med Chem 1978;21:291–294.

208. Faulds D, Balfour JA, Chrisp P, Langtry HD. Mitoxantrone. A review of its pharmacodynamic and pharmacokinetic properties, and therapeutic potential in the chemotherapy of cancer. Drugs 1991;41(3):400–449.

209. Baguley BC, Denny WA, Atwell GJ, Cain BF. Potential Antitumor Agents Part. 35. Quantitative relationships between antitumor (L1210) potency and DNA binding for 4'-(9-acridinylamino)methanesulfon-m-anisidide analogues. J Med Chem 1981;24:520–525.

210. Valentini L, Nicolella V, Vannini E, Menozzi M, Penco S, Arcamone FM. Association of anthracycline derivatives with DNA: a fluorescence study. Farmaco 1985;40:377–390.

211. Feigon J, Denny WA, Leupin W, Kearns DR. The interactions of antitumor drugs with natural DNA: a ^{1}H NMR study of binding mode and kinetics. J Med Chem 1984;23:450–465.

212. Dal Ben D, Palumbo M, Zagotto G, Capranico G, Moro S. DNA topoisomerase II structures and anthracycline activity: insights into ternary complex formation. Curr Pharm Des 2007;13:2766–2780.

213. Wadkins RM, DEGraves DE. Thermodynamics of the interactions of m-AMSA and o-AMSA with nucleic acids: influence of ionic strength and DNA base composition. Nucleic Acids Res 1989;17:9933–9946.

214. Wilson WR, Baguley BC, Wakelin LPG, Waring MJ. Interaction of the antitumor drug 4'-(9-acridinylamino)methanesulfon-m-anisidide and related acridines with nucleic acids. Mol Pharmacol 1981;20:404–414.

215. Chourpa I, Morjani H, Riou J-F, Manfait M. Intracellular molecular interactions of antitumor drug amsacrine (m-AMSA) as revealed by surface-enhanced Raman spectroscopy. FEBS Lett 1996;397:61–64.

216. Finlay GJ, Atwell GJ, Baguley BC. Inhibition of the action of the topoisomerase II poison amsacrine by simple aniline derivatives: evidence for drug-protein interactions. Oncol Res 1999;11:249–254.

217. Houlbrook S, Addison CM, Davies SL, Carmichael J, Stratford IJ, Harris AL, Hickson ID. Relationship between expression of topoisomerase II isoforms and intrinsic sensitivity to topoisomerase II inhibitors in breast cancer cell lines. Br J Cancer 1995;72:1454–1461.

218. Withoff S, de Vries EGE, Keith WN, Nienhuis EF, van der Graaf WTA, Uges DRA, Mulder NH. Differential expression of DNA topoisomerase IIα and -β in P-gp and MRP-negative VM26, mAMSA and mitoxantrone-resistant sublines of the human SCLC cell line GLC4. Br J Cancer 1996;74:1869–1876.

219. Van Hille B, Perrin D, Hill BT. Differential *in vitro* interactions of a series of clinically useful topoisomerase-interacting compounds with the cleavage/religation activity of the human topoisomerase IIα and II-β isoforms. Anticancer Drugs 1999;10:551–560.

220. Sorensen M, Sehested M, Jensen PB. Effect of cellular ATP depletion on topoisomerase II poisons. Abrogation of cleavable-complex formation by etoposide but not by amsacrine. Mol Pharmacol 1999;55:424–431.

221. Denny WA, Wakelin LPG. Kinetics of the binding of mitoxantrone, ametantrone and analogues to DNA: relationship to binding mode and antitumour activity. Anticancer Drug Des 1990;5:189–200.

222. Lown JW, Hanstock CC. High field proton NMR analysis of the 1:1 intercalation complex of the antitumor agent mitoxantrone and the DNA duplex [d(CpGpCpG)]2. J Biomol Struct Dyn 1985;2:1097–1106.

223. Bailly C, Routier S, Bernier JL, Waring MJ. DNA recognition by two mitoxantrone analogues: influence of the hydroxyl groups. FEBS Lett 1996;379:269–272.

224. Kapuscinski J, Darzynkiewicz Z. Interactions of antitumor agents Ametantrone and mitoxantrone (Novatrone) with double-stranded DNA. Biochem Pharmacol 1985;34: 4203–4213.

225. Krishnamoorthy CR, Yen SF, Smith JC, Lown JW, Wilson WD. Stopped-flow kinetic analysis of the interaction of anthraquinone anticancer drugs with calf thymus DNA, poly[d(G-C)].poly[d(G-C)], and poly[d(A-T)].poly[d(A-T)]. Biochemistry 1986;25:5933–5940.

226. De Isabella P, Capranico G, Palumbo M, Sissi C, Krapcho AP, Zunino F. Sequence selectivity of topoisomerase II DNA cleavage stimulated by mitoxantrone derivatives: relationships to drug DNA binding and cellular effects. Mol Pharmacol 1993;43:715–721.

227. Bellosillo B, Colomer D, Pons G, Gil J. Mitoxantrone, a topoisomerase II inhibitor, induces apoptosis of B-chronic lymphocytic leukaemia cells. Br J Haematol 1998;100:142–146.

228. Panousis C, Kettle AJ, Phillips DR. Myeloperoxidase oxidizes mitoxantrone to metabolites which bind covalently to DNA and RNA. Anticancer Drug Des 1995;10:593–605.

229. Parker BS, Cullinane C, Phillips DR. Formation of DNA adducts by formaldehyde-activated mitoxantrone. Nucleic Acids Res 1999;27:2918–2923.

230. Tsujimoto S, Satoh E, Sugimoto S, Katoh T, Kaneko M, Takada H, Katoh T. General pharmacological studies of a novel anthracycline derivative, (7S, 9S)-9-acetyl-9-amino-7-[(2-deoxy-β-D-erythro-pentopyranosyl)oxy]-7,8,9,10-tetrahydro-6,11-dihydroxy-5,12-naphthacenedione hydrochloride (SM-5887), with antitumor activity. Pharmacometrics 1996;52:351–370.

231. Yamaoka T, Hanada M, Ichii S, Morisada S, Noguchi T, Yanagi Y. Cytotoxicity of amrubicin, a novel 9-aminoanthracycline, and its active metabolite amrubicinol on human tumor cells. Jpn J Cancer Res 1998;89:1067–1073.

232. Hanada M, Mizuno S, Fukushima A, Saito Y, Noguchi T, Yamaoka T. A new antitumor agent amrubicin induces cell growth inhibition by stabilizing topoisomerase II-DNA complex. Jpn J Cancer Res 1998;89:1229–1238.

233. Hira A, Watanabe H, Maeda Y, Yokoo K, Sanematsu E, Fujii J, Sasaki J, Hamada A, Saito H. Role of P-glycoprotein in accumulation and cytotoxicity of amrubicin and amrubicinol in MDR1 gene-transfected LLC-PK1 cells and human A549 lung adenocarcinoma cells. Biochem Pharmacol 2008;75(4):973–980.

234. Hanada M, Noguchi T, Murayama T. Profile of the anti-tumor effects of amrubicin, a completely synthetic anthracycline. Nippon Yakurigaku Zasshi 2003;122:141–150.

235. Kurata T, Okamoto I, Tamura Kenji Fukuoka M. Amrubicin for non-small-cell lung cancer and small-cell lung cancer. Invest New Drugs 2007;25(5):499–504.

236. Ferraldeschi R, Baka S, Jyoti B, Faivre-Finn C, Thatcher N, Lorigan P. Modern management of small-cell lung cancer. Drugs 2007;67 (15):2135–2152.

237. Temperini C, Messori L, Orioli P, Di Bugno C, Animati F, Ughetto G. The crystal structure of the complex between a disaccharide anthracycline and the DNA hexamer d(CGATCG) reveals two different binding sites involving two DNA duplexes. Nucleic Acids Res 2003;31: 1464–1469.

238. Bos AME, de Vries EGE, Dombernovsky P, Aamdal S, Uges DRA, Schrijvers D, Wanders J, Roelvink MWJ, Hanauske AR, Bortini S, Capriati A, Crea AEG, Vermorken JB. Pharma-

cokinetics of MEN-10755, a novel anthracycline disaccharide analog, in two phase I studies in adults with advanced solid tumors. Cancer Chemother Pharmacol 2001;48: 361–369.

239. Pratesi G, De Cesare M, Caserini C, Perego P, Dal Bo L, Polizzi D, Supino R, Bigioni M, Manzini S, Iafrate E, Salvatore C, Casazza A, Arcamone F, Zunino F. Improved efficacy and enlarged spectrum of activity of a novel anthracycline disaccharide analog of doxorubicin against human tumor xenografts. Clinical Cancer Res 1998;4:2833–2839.

240. Salvatore C, Camarda G, Maggi CA, Goso C, Manzini S, Binaschi M. NF-κB activation contributes to anthracycline resistance pathway in human ovarian carcinoma cell line A2780. Int J Oncol 2005;27(3):799–806.

241. Fiedler W, Tchen N, Bloch J, Fargeot P, Sorio R, Vermorken JB, Collette L, Lacombe D, Twelves C. A study from the EORTC new drug development group: open label phase II study of sabarubicin (MEN-10755) in patients with progressive hormone refractory prostate cancer. Eur J Cancer 2006;42(2):200–204.

242. Steuber CP, Krischer J, Holbrook T, Camitta B, Land V, Sexauer C, Mahoney D, Weinstein H. Therapy of refractory or recurrent childhood acute myeloid leukemia using amsacrine and etoposide with or without azacitidine: a pediatric oncology group randomized phase II study. J Clin Oncol 1996;14:1521–1525.

243. Reman O, Buzyn A, Lheritier V, Huguet F, Kuentz M, Stamatoullas A, Delannoy A, Fegueux N, Miclea J-M, Boiron J-M, Vernant JP, Gardin C, Hacini M, Georges M, Fiere D, Thomas X. Rescue therapy combining intermediate-dose cytarabine with amsacrine and etoposide in relapsed adult acute lymphoblastic leukemia. Hematol J 2004;5(2):123–129.

244. Sung WJ, Kim DH, Sohn SK, Kim JG, Baek JH, Jeon SB, Moon JH, Ahn BM, Lee KB. Phase II trial of amsacrine plus intermediate-dose Ara-C (IDAC) with or without etoposide as salvage therapy for refractory or relapsed acute leukemia. Jpn J Clin Oncol 2005;35(10):612–616.

245. Kessler T, Mohr M, Muller-Tidow C, Krug U, Brunnberg U, Mohr B, Schliemann C, Sauerland C, Serve H, Buchner T, Berdel WE, Mesters RM. Amsacrine containing induction therapy in elderly AML patients: comparison to standard induction regimens in a matched-pair analysis. Leukemia Res 2008;32:491–494.

246. Jelic S, Nikolic-Tomasevic Z, Kovcin V, Milanovic N, Tomasevic Z, Jovanovic V, Vlajic MJ. A two-step reevaluation of high-dose amsacrine for advanced carcinoma of the upper aerodigestive tract: a pilot phase II study. Chemother (Firenze) 1997;9:364–370.

247. Krischer JP, Epstein S, Cuthbertson DD, Goorin AM, Epstein ML, Lipshultz SE. Clinical cardiotoxicity following anthracycline treatment for childhood cancer: the pediatric oncology group experience. J Clin Oncol 1997;15:1544–1552.

248. Covey JM, Kohn KW, Kerrigan D, Tilchen EJ, Pommier Y. Topoisomerase II-mediated DNA damage produced by 4′-(9-acridinylamino) methanesulfon-m-anisidide and related acridines in L1210 cells and isolated nuclei: relation to cytotoxicity. Cancer Res 1988;48: 860–865.

249. Hardy JR, Harvey VJ, Paxton JW, Evans P, Smith S, Grove W, Grillo-Lopez AJ, Baguley BC. Phase I trial of the amsacrine analogue 9-[(2-methoxy-4-[(methylsulfonyl)amino]-phenyl]amino)-N,5-dimethyl-4- acridinecarboxamide (CI-921). Cancer Res 1988;48: 6593–6596.

250. Harvey VJ, Hardy JR, Smith S, Grove W, Baguley BC. Phase II study of the amsacrine analogue CI-921 (NSC 343499) in non-small cell lung cancer. Eur J Cancer 1991; 27:1617–1620.

251. Fyfe D, Raynaud F, Langley RE, Newell DR, Halbert G, Gardner C, Clayton K, Woll PJ, Judson I, Carmichael J. A study of amsalog (CI-921) administered orally on a 5-day schedule, with bioavailability and pharmacokinetically guided dose escalation. Cancer Chemother Pharmacol 2002;49(1):1–6.

252. Koller CA, Kantarjian HM, Feldman EJ, O'Brien S, Rios MB, Estey E, Keating M. A phase I-II trial of escalating doses of mitoxantrone with fixed doses of cytarabine plus fludarabine as salvage therapy for patients with acute leukemia and the blastic phase of chronic myelogenous leukemia. Cancer 1999;86: 2246–2451.

253. Thomas X, Archimbaud E. Mitoxantrone in the treatment of acute myelogenous leukemia: a review. Hematol Cell Ther 1997;39:163–174.

254. Jain KK. Evaluation of mitoxantrone for the treatment of multiple sclerosis. Expert Opin Investig Drugs 2000;9:1139–1149.

255. Dunn CJ, Goa KL. Mitoxantrone: a review of its pharmacological properties and use in acute nonlymphoblastic leukaemia. Drugs Aging 1996;9:122–147.

256. Boos G, Stopper H. Genotoxicity of several clinically used topoisomerase II inhibitors. Toxicol Lett 2000;116:7–16.

257. Carli PM, Sgro C, Parchin-Geneste N, Isambert N, Mugneret F, Girodon F, Maynadie M. Increase therapy-related leukemia secondary to breast cancer. Leukemia 2000;14: 1014–1017.
258. Nielsen D, Eriksen J, Maare C, Litman T, Kjaersgaard E, Plesner T, Friche E, Skovsgaard T. Characterization of non-P-Glycoprotein multidrug-resistant Ehrlich ascites tumor cells selected for resistance to mitoxantrone. Biochem Pharmacol 2000;60:363–370.
259. Sullivan DM, Feldhoff PW, Lock RB, Smith NB, Pierce WM. Characterization of an altered DNA topoisomerase IIα from a mitoxantrone resistant mammalian cell line hypersensitive to DNA crosslinking agents. Int J Oncol 1995;7:1383–1393.
260. Kellner U, Hutchinson L, Seidel A, Lage H, Hermann D, Danks MK, Dietel M, Kaufmann SH. Decreased drug accumulation in a mitoxantrone-resistant gastric carcinoma cell line in the absence of P-glycoprotein. Int J Cancer 1997;71:817–824.
261. Bailly JD, Skladanowski A, Bettaieb A, Mansat V, Larsen AK, Laurent G. Natural resistance of acute myeloid leukemia cell lines to mitoxantrone is associated with lack of apoptosis. Leukemia 1997;11:1523–1532.
262. Hazlehurst LA, Foley NE, Gleason-Guzman MC, Hacker MP, Cress AE, Greenberger LW, De Jong MC, Dalton WS. Multiple mechanisms confer drug resistance to mitoxantrone in the human 8226 myeloma cell line. Cancer Res 1999;59:1021–1028.
263. Felix CA. Secondary leukemias induced by topoisomerase-targeted drugs. Biochimica et Biophysica Acta 1998;1400:233–255.
264. Seiter K, Feldman EJ, Sreekantaiah C, Pozzuoli M, Weisberger L, Liu D, Papageorgio C, Weiss M, Kancherla R, Ahmed T. Intravenous bolus topotecan in patients with myelodysplastic syndrome. Leukemia 2001;15:963–968.
265. Krapcho AP, Petry ME, Getahun Z, Landi JJ, Stallman J, Polsenberg JF, Gallagher CE, Maresch MJ, Hacker MP, Giuliani FC, Beggiolin G, Pezzioni G, Menta E, Manzotti C, Oliva A, Spinelli S, Tognnella S. 6,9-Bis[(aminoalkyl)amino]benzo[g]isoquinoline-5,10-diones. A novel class of chromophore-modified antitumor anthracene-9,10-diones: synthesis and antitumor evaluations. J Med Chem 1994;37:828–837.
266. De Isabella P, Palumbo M, Sissi C, Capranico G, Carenini N, Menta E, Oliva A, Spinelli S, Krapcho AP, Giuliani FC, Zunino F. Topoisomerase II DNA cleavage stimulation, DNA binding activity, cytotoxicity, and physicochemical properties of 2-aza- and 2-aza-oxide-anthracenedione derivatives. Mol Pharmacol 1995;48:30–38.
267. Cavalletti E, Crippa L, Mainardi P, Oggioni N, Cavagnoli R, Bellini O, Sala F. Pixantrone (BBR 2778) has reduced cardiotoxic potential in mice pretreated with doxorubicin: comparative studies against doxorubicin and mitoxantrone. Invest New Drugs 2007;25(3):187–195.
268. Evison BJ, Mansour OC, Menta Ernesto Phillips DR, Cutts SM. Pixantrone can be activated by formaldehyde to generate a potent DNA adduct forming agent. Nucleic Acids Res 2007;35(11):3581–3589.
269. El-Helw LM, Hancock BW. Pixantrone: a novel aza-anthracenedione in the treatment of non-Hodgkin's lymphomas. Expert Opin Invest Drugs 2007;16(10):1683–1691.
270. Engert A, Herbrecht R, Santoro A, Zinzani PL, Gorbatchevsky I. EXTEND PIX301: a phase III randomized trial of pixantrone versus other chemotherapeutic agents as third-line monotherapy in patients with relapsed, aggressive non-Hodgkin's lymphoma. Clin Lymphoma Myeloma 2006;7(2): 152–154.
271. Showalter HD, Johnson JL, Hoftiezer JM, Turner WR, Werbel LM, Leopold WR, Shillis JL, Jackson RC, Elslager EF. Anthrapyrazole anticancer agents. Synthesis and structure-activity relationships against murine leukemias. J Med Chem 1987;30:121–131.
272. Hartley JA, Reszka K, Zuo ET, Wilson WD, Morgan A, Lown JW. Characteristics of the interaction of anthrapyrazole anticancer agents with deoxyribonucleic acids: structural requirements for DNA binding, intercalation, and photosensitization. Mol Pharmacol 1988;33:265–271.
273. Leopold WR, Nelson JM, Plowman J, Jackson RC. Anthrapyrazoles, a new class of intercalating agents with high-level, broad spectrum activity against murine tumors. Cancer Res 1985;45:5532–5539.
274. Belanger K, Jolivet J, Maroun J, Stewart D, Grillo-Lopez A, Whitfield L, Wainman N, Eisenhauer E. Phase I pharmacokinetic study of DUP-937, a new anthrapyrazole. Invest New Drugs 1993;11:301–308.
275. Leteurtre F, Kohlhagen G, Paull KD, Pommier Y. Topoisomerase II inhibition and cytotoxicity of the anthrapyrazoles DuP 937 and DuP 941 (losoxantrone) in the National Cancer Insti-

tute preclinical antitumor drug discovery screen. J Natl Cancer Inst 1994;86:1239–1244.
276. Joshi AS, Pieniaszek HJ, Vokes EE, Vogelzang NJ, Davidson AF, Richards LE, Chai MF, Finizio M, Ratain MJ. Elimination pathways of [14C]losoxantrone in four cancer patients. Drug Metab Dispos 2001;29:96–99.
277. Huan SD, Natale RB, Stewart DJ, Sartiano GP, Stella PJ, Roberts JD, Symes AL, Finizio M. A multicenter phase II trial of losoxantrone (DuP-941) in hormone-refractory metastatic prostate cancer. Clin Cancer Res 2000;6: 1333–1336.
278. Ames MM, Loprinzi CL, Collins JM, van Haelst-Pisani C, Richardson RL, Rubin J, Moertel CG. Phase I and clinical pharmacological evaluation of pirozantrone hydrochloride (oxantrazole). Cancer Res 1990;50:3905–3909.
279. Albain KS, Liu PY, Hantel A, Poplin EA, O'Toole RV, Wade JL, Maddox AM, Alberts DS. A phase II trial of piroxantrone in advanced ovarian carcinoma after failure of platinum-based chemotherapy. Gyn Oncol 1995;57:407–411.
280. Pazdur R, Bready B, Scalzo AJ, Brandof JE, Close DR, Kolbye S, Winn RJ. Phase II trial of piroxantrone in metastatic gastric adenocarcinoma. Invest New Drugs 1994;12:263–265.
281. Cholody WM, Martelli S, Paradziej-Lukowicz J, Konopa J. 5-[(Aminoalkyl)amino]imidazo[4,5,1-de]acridin-6-ones as a novel class of antineoplastic agents. Synthesis and biological activity. J Med Chem 1990;33:49–52.
282. Burger AM, Jenkins TC, Double JA, Bibby MC. Cellular uptake, cytotoxicity and DNA-binding studies of the novel imidazoacridinone antineoplastic agent C1311. Br J Cancer 1999;81:367–375.
283. De Marco C, Zaffaroni N, Comijn E, Tesei Anna Zoli W, Peters G. Comparative evaluation of C1311 cytotoxic activity and interference with cell cycle progression in a panel of human solid tumour and leukaemia cell lines. Int J Oncol 2007;31:907–913.
284. Mazerska Z, Sowinski P, Konopa J. Molecular mechanism of the enzymatic oxidation investigated for imidazoacridinone antitumor drug, C-1311. Biochem Pharmacol 2003;66(9): 1727–1736.
285. Mazerska Z, Zon A, Stojek Z. Electrochemical formation of the adduct between antitumor agent C-1311 and DNA nucleoside dG. Electrochem Commun 2003;5(9):770–775.
286. Skwarska A, Augustin E, Konopa J. Sequential induction of mitotic catastrophe followed by apoptosis in human leukemia MOLT4 cells by imidazoacridinone C-1311. Apoptosis 2007;12(12):2245–2257.
287. Kaneko T, Wong H, Utzig J, Schuring J, Doyle TW. Water soluble derivatives of rebeccamycin. J Antibiot 1990;43:125–127.
288. Prudhomme M. Recent developments of rebeccamycin analogues as topoisomerase I inhibitors and antitumor agents. Curr Med Chem 2000;7:1189–1212.
289. Weitman S, Moore R, Barrera H, Cheung NK, Izbicka E, Von Hoff DD. In vitro antitumor activity of rebeccamycin analog (NSC# 655649) against pediatric solid tumors. J Pediatr Hematol Oncol 1998;20:136–139.
290. Hussain M, Vaishampayan U, Heilbrun LK, Jain V, LoRusso PM, Ivy P, Flaherty L. A phase II study of rebeccamycin analog (NSC-655649) in metastatic renal cell cancer. Invest New Drugs 2003;21:465–471.
291. Burstein HJ, Overmoyer B, Gelman R, Silverman P, Savoie J, Clarke K, Dumadag L, Younger J, Ivy P, Winer EP. Rebeccamycin analog for refractory breast cancer: a randomized phase II trial of dosing schedules. Invest New Drugs 2007;25(2):161–164.
292. Goel S, Wadler S, Hoffman A, Volterra F, Baker C, Nazario E, Ivy P, Silverman A, Mani S. A Phase II study of rebeccamycin analog NSC 655649 in patients with metastatic colorectal cancer. Invest New Drugs 2003;21: 103–107.
293. Dowlati A, Chapman R, Subbiah S, Fu P, Ness A, Cortas T, Patrick L, Reynolds S, Xu N, Levitan N, Ivy P, Remick SC. Randomized phase II trial of different schedules of administration of rebeccamycin analogue as second line therapy in non-small cell lung cancer. Invest New Drugs 2005;23(6):563–567.
294. Holden JA, Rolfson DH, Wittwer CT. Human DNA topoisomerase II: evaluation of enzyme activity in normal and neoplastic tissues. Biochemistry 1990;29:2127–2134.
295. Husain I, Mohler JL, Seigler HF, Easterman JM. Elevation of topoisomerase I messenger RNA, protein, and catalytic activity in human tumors: demonstration of tumor-type specificity and implications for cancer chemotherapy. Cancer Res 1994;54:539–546.
296. Giovanella B, Stehlin JS, Wall ME, Wani MC, Nicholas AW, Liu LF, Silber R, Potmesil M. DNA topoisomerase I-targeted chemotherapy of human colon cancer in xenografts. Science 1989;246:1046–1048.
297. Heck MMS, Hittelman WN, Earnshaw WC. Differential expression of DNA topoisome-

rases I and II during the eukaryotic cell cycle. Proc Natl Acad Sci USA 1988;85:1086–1090.
298. Tan KB, Mattern MR, Eng WK, McCabe FL, Johnson RK. Nonproductive rearrangement of DNA topoisomerase I and II genes: correlation with resistance to topoisomerase inhibitors. J Natl Cancer Inst 1989;81:1732–1735.
299. Whitacre CM, Zborowska E, Gordon NH, MacKay W, Berger NA. Topotecan increases topoisomerase IIalpha levels and sensitivity to treatment with etoposide in schedule-dependent process. Cancer Res 1997;57:1425–1428.
300. Wang LK, Johnson RK, Hecht SM. Inhibition of topoisomerase I function by nitidine and fagaronine. Chem Res Toxicol 1993;6:813–818.
301. Kanzawa F, Nishio K, Ishida T, Fukuda M, Kurokawa H, Fukumoto H, Nomoto Y, Fukuoka K, Bojanowski K, Saijo N. Antitumor activities of a new benzo[c]phenanthridine agent, 2,3-(methylenedioxy)-5-methyl-7-hydroxy-8-methoxybenzo[c]phenanthridinium hydrogen sulfate dihydrate (NK109), against several drug-resistant human tumor cell lines. Br J Cancer 1997;76:571–581.
302. Atwell GJ, Cain BF, Baguley BC, Finlay JG, Denny WA. Potential antitumor agents. 43. Synthesis and biological activity of dibasic 9-aminoacridine-4-carboxamides: a new class of antitumor agent. J Med Chem 1984;27:1481–1485.
303. Denny WA, Atwell GJ, Rewcastle GW, Baguley BC. Potential antitumor agents. 49. 5-Substituted derivatives of N-[2-(dimethylamino) ethyl]-9-aminoacridine-4-carboxamide with in vivo solid tumor activity. J Med Chem 1987;30: 658–663.
304. Finlay GJ, Riou JF, Baguley BC. From amsacrine to DACA (N-[2-(dimethylamino)ethyl]acridine-4-carboxamide): selectivity for topoisomerases I and II among acridine derivatives. Eur J Cancer 1996;32A:708–714.
305. Finlay GJ, Marshall ES, Matthews JHL, Paull KD, Baguley BC. In vitro assessment of N-[2-(dimethylamino)ethyl]acridine-4-carboxamide (DACA), a DNA-intercalating antitumor drug with reduced sensitivity to multidrug resistance. Cancer Chemother Pharmacol 1993;31:401–406.
306. Finlay GJ, Baguley BC. Selectivity of N-[2-(dimethylamino)ethyl]acridine-4-carboxamide towards Lewis lung carcinoma and human tumor cell lines in vitro. Eur J Cancer Clin Oncol 1989;25:271–277.
307. Sebolt JS, Scavone SV, Pinter CD, Hamelehle KL, Von Hoff DD, Jackson RC. Pyrazoloacridines, a new class of anticancer agents with selectivity against solid tumors in vitro. Cancer Res 1987;47:4299–4304.
308. Adjei AA, Charron M, Rowinsky EK, Svingen PA, Miller J, Reid JM, Sebolt-Leopold J, Ames MM, Kaufmann SH. Effect of pyrazoloacridine (NSC 366140) on DNA topoisomerases I and II. Clin Cancer Res 1998;4:683–691.
309. Grem JL, Politi PM, Balis FM, Sinha BK, Dahut W, Allegra CJ. Cytotoxicity and DNA damage associated with pyrazoloacridine in MCF-7 breast cancer cells. Biochem Pharmacol 1996;51:1649–1659.
310. Capps DB, Dunbar J, Kesten SR, Shillis J, Werbel LM, Plowman J, Ward DL. 5-(Aminoalkyl)-5-nitropyrazolo[3,4,5-kl]acridines, a new class of anticancer agents. J Med Chem 1992;35:4770–4778.
311. Palomino E, Foster B, Kempff M, Corbett T, Wiegand R, Horwitz J, Baker L. Identification and antitumor activity of a reduction product in the murine metabolism of pyrazoloacridine (NSC-366140). Cancer Chemother Pharmacol 1996;38:453–458.
312. Riou JF, Fosse P, Nguyen CH, Larsen AK, Bissery MC, Grondard L, Saucier JM, Bisagni E, Lavelle F. Intoplicine (RP 60475) and its derivatives, a new class of antitumor agents inhibiting both topoisomerase I and II activities. Cancer Res 1993;53:5987–5993.
313. Poddevin B, Riou JK, Lavelle F, Pommier Y. Dual topoisomerase I and II inhibition by intoplicine (RP-60475), a new antitumor agent in early clinical trials. Mol Pharmacol 1993;44: 767–774.
314. Eckhardt JR, Burris HA, Kuhn JG, Bissery MC, Klink-Alaki M, Clark GM, von Hoff DD. Activity of intoplicine (RP60475), a new DNA topoisomerase I and II inhibitor, against human tumor colony-forming units in vitro. J Natl Cancer Inst 1994;86:30–33.
315. Bissery MC, Nguyen CH, Bisagni E, Vrignaud P, Lavelle F, Antitumor activity of intoplicine (RP 60475, NSC 645008), a new benzo-pyridoindole: evaluation against solid tumors and leukemias in mice. Invest New Drugs 1993;11:263–277.
316. Fortune JM, Velea L, Graves DE, Utsugi T, Yamada Y, Osheroff N. DNA Topoisomerases as targets for the anticancer drug TAS-103: DNA interactions and topoisomerase catalytic inhibition. Biochemistry 1999;38: 15580–15586.
317. Utsugi T, Aoyagi K, Asao T, Okazaki S, Aoyagi Y, Sano M, Wierzba K, Yamada Y. Antitumor

317. activity of a novel quinoline derivative, TAS-103, with inhibitory effects on topoisomerases I and II. Jpn J Cancer Res 1997;88:992–1002.
318. Byl JA, Fortune JM, Burden DA, Nitiss JL, Utsugi T, Yamada Y, Osheroff N. DNA topoisomerases as targets for the anticancer drug TAS-103: primary cellular target and DNA cleavage enhancement. Biochemistry 1999;38:15573–15579.
319. Yoshida M, Kabe Y, Wada T, Asai A, Handa H. A new mechanism of 6-((2-(dimethylamino)ethyl)amino)-3-hydroxy-7H-indeno(2,1-c)quinolin-7-one dihydrochloride (TAS-103) action discovered by target screening with drug-immobilized affinity beads. Mol Pharmacol 2008;73(3):987–994.
320. Aoyagi Y, Kobunai T, Utsugi T, Oh-hara T, Yamada Y. *In vitro* antitumor activity of TAS-103, a novel quinoline derivative that targets topoisomerases I and II. Jpn J Cancer Res 1999;90:578–587.
321. Perrin D, van Hille B, Barret J-M, Kruczynski A, Etievant C, Imbert T, Hill BT. F 11782, a novel epipodophylloid non-intercalating dual catalytic inhibitor of topoisomerases I and II with an original mechanism of action. Biochem Pharmacol 2000;59:807–819.
322. Barret J-M, Montaudon D, Etievant C, Perrin D, Kruczynski A, Robert J, Hill BT. Detection of DNA-strand breaks in cells treated with F 11782, a catalytic inhibitor of topoisomerases I and II. Anticancer Res 2000;20:4557–4562.
323. Kruczynski A, Etievant C, Perrin D, Imbert T, Colpaert F, Hill BT. Preclinical antitumour activity of F 11782, a novel dual catalytic inhibitor of topoisomerases. Br J Cancer 2000;83:1516–1524.
324. Nabiev I, Chourpa I, Riou JF, Nguyen CH, Lavelle F, Manfait M. Molecular interactions of DNA-topoisomerase I and II inhibitor with DNA and topoisomerases and in ternary complexes: binding modes and biological effects for intoplicine derivatives. Biochemistry 1994;33:9013–9023.
325. Robertson IGC, Palmer BD, Officer M, Siegers DJ, Paxton JW, Shaw GJ. Cytosol mediated metabolism of the experimental antitumor agent acridine carboxamide to the 9-acridone derivative. Biochem Pharmacol 1991;42:1879–1884.
326. Evans SMH, Robertson IGC, Paxton JW. Plasma protein binding of the experimental antitumor agent acridine-4-carboxamide in man, dog, rat and rabbit. J Pharm Pharmacol 1994;46:63–67.
327. Schofield PC, Robertson IGC, Paxton JW, McCrystal MR, Evans BD, Kestell P, Baguley BC. Metabolism of N-[2-(dimethylamino)ethyl]acridine-4-carboxamide in cancer patients undergoing a phase I clinical trial. Cancer Chemother Pharmacol 1999;44:51–58.
328. McCrystal MR, Evans BD, Harvey VJ, Thompson PL, Porter DJ, Baguley BC. Phase I study of the cytotoxic agent N-[2-(dimethylamino)ethyl]acridine-4-carboxamide. Cancer Chemother Pharmacol 1999;44:39–44.
329. Twelves C, Campone M, Coudert B, Van den Bent M, de Jonge M, Dittrich C, Rampling R, Sorio R, Lacombe D, de Balincourt C, Fumoleau P. Phase II study of XR5000 (DACA) administered as a 120-h infusion in patients with recurrent glioblastoma multiforme. Ann Oncol 2002;13:777–780.
330. Dittrich C, Dieras V, Kerbrat P, Punt C, Sorio R, Caponigro F, Paoletti X, de Balincourt C, Lacombe D, Fumoleau P. Phase II study of XR5000 (DACA), an inhibitor of topoisomerase I and II, administered as a 120-h infusion in patients with advanced ovarian cancer. Invest New Drugs 2003;21:47–352.
331. Berg SL, Blaney SM, Sullivan J, Bernstein M, Dubowy R, Harris MJ. Phase II trial of pyrazoloacridine in children with solid tumors: a pediatric oncology group phase II study. Pediatr Hemat Oncol 2000;22:506–509.
332. Keshelava N, Tsao-Wei D, Reynolds CP. Pyrazoloacridine is active in multidrug-resistant neuroblastoma cell lines with nonfunctional p53. Clin Cancer Res. 2003;9:3492–3502.
333. Galanis E, Buckner JC, Maurer MJ, Reid JM, Kuffel MJ, Ames MM, Scheithauer BW, Hammack JE, Pipoly G, Kuross SA. Phase I/II trial of pyrazoloacridine and carboplatin in patients with recurrent glioma. Invest New Drugs 2005;23:495–503.
334. van Gijn R, ten Bokkel Huinink WW, Rodenhuis S, Vermorken JB, van Tellingen O, Rosing H, van Warmerdam LJ, Beijnen JH. Topoisomerase I/II inhibitor intoplicine administered as a 24 h infusion: phase I and pharmacologic study. Anticancer Drugs 1999;10: 17–23.
335. Newman RA, Kim J, Newman BM, Bruno R, Bayssas M, Klink-Alaki M, Pazdur R. Phase I trial of intoplicine (RP 60475) administered as a 72 h infusion every 3 weeks in patients with solid tumors. Anticancer Drugs 1999;10: 889–897.
336. Abigerges D, Armand JP, Chabot GG, Bruno R, Bissery MC, Bayssas M, Klink-Alaki M, Clavel M, Catimel G. Phase I and pharmacology study of intoplicine (RP 60475; NSC 645008), a novel

topoisomerase I and II inhibitor, in cancer patients. Anticancer Drugs 1996;7:166–174.

337. Ewesuedo RB, Iyer L, Das S, Koenig A, Mani S, Vogelzang NJ, Schilsky RL, Brenckman W, Ratain MJ. Phase I clinical and pharmacogenetic study of weekly TAS-103 in patients with advanced cancer. J Clin Oncol 2001;19: 2084–2090.

338. Kruczynski A, Barret J-M, Van Hille B, Chansard N, Astruc J, Menon Y, Duchier C, Creancier L, Hill BT. Decreased nucleotide excision repair activity and alterations of topoisomerase IIβ are associated with the *in vivo* resistance of a P388 leukemia subline to F11782, a novel catalytic inhibitor of topoisomerases I and II. Clin Cancer Res 2004;10(9):3156–3168.

339. Barret J-M, Kruczynski A, Etievant C, Hill BT. Synergistic effects of F 11782, a novel dual inhibitor of topoisomerases I and II, in combination with other anticancer agents. Cancer Chemother Pharmacol 2002;49(6):479–486.

340. Kluza J, Mazinghien R, Irwin H, Hartley JA, Bailly C. Relationships between DNA strand breakage and apoptotic progression upon treatment of HL-60 leukemia cells with tafluposide or etoposide. Anticancer Drugs 2006;17 (2):155–164.

341. Wakelin LPG. Polyfunctional DNA intercalating agents. Med Res Rev 1986;6:275–340.

342. Goldin A, Vendetti JM, McDonald JS, Muggia FM, Henney JE, DeVita VT. Current results of the screening program at the Division of Cancer Treatment, National Cancer Institute. Eur J Cancer 1981;17:129–142.

343. Segal-Bendirjian E, Coulaud D, Roques BP, Le Pecq JB. Selective loss of mitochondrial DNA after treatment of cells with ditercalinium (NSC 335153), an antitumor bis-intercalating agent. Cancer Res 1988;48:4982–4992.

344. Brana MF, Ramos A. Naphthalimides as anticancer agents: synthesis and biological activity. Curr Med Chem Anticancer Agents 2001;1:237–255.

345. Bailly C, Brana MF, Waring MJ. Sequenceselective intercalation of antitumour bisnaphthalimides into DNA Evidence for an approach via the major groove. Eur J Biochem 1996;240:195–208.

346. Gallego J, Reid BR. Solution structure and dynamics of a complex between DNA and the antitumor bisnaphthalimide LU-79553: intercalated ring flipping on the millisecond time scale. Biochemistry 1999;38:15104–15115.

347. Bousquet PF, Brana MF, Conlon D, Fitzgerald KM, Perron D, Cocchiaro C, Miller R, Moran M, George J, Preclinical evaluation of LU 79553: a novel bis-naphthalimide with potent antitumor activity. Cancer Res 1995;55: 1176–1180.

348. Villalona-Calero MA, Eder JP, Toppmeyer DL, Allen LF, Fram R, Velagapudi R, Myers M, Amato A, Kagen-Hallet K, Razvillas B, Kufe DW, Von Hoff DD, Rowinsky EK. Phase I and pharmacokinetic study of LU79553, a DNA intercalating bisnaphthalimide, in patients with solid malignancies. J Clin Oncol 2001; 19:857–869.

349. Awada A, Thoedtmann R, Piccart MJ, Wanders J, Schrijvers AHGJ, Von Broen I-M, Hanauske AR. An EORTC-ECSG phase I study of LU 79553 administered every 21 or 42 days in patients with solid tumors. Eur J Cancer 2003;39:742–747.

350. Gamage SA, Spicer JA, Finlay GJ, Stewart AJ, Charlton P, Baguley BC, Denny WA. Dicationic bis(9-methylphenazine-1-carboxamides): relationships between biological activity and linker chain structure for a series of potent topoisomerase-targeted anticancer drugs. J Med Chem 2001;44: 1407–1415.

351. Dai J, Punchihewa C, Mistry P, Ooi AT, Yang D. Novel DNA bis-intercalation by MLN944, a potent clinical bisphenazine anticancer drug. J Biol Chem 2004;279:46096–46103.

352. Stewart AJ, Mistry P, Dangerfield W, Bootle M, Baker D, Kofler B, Okiji S, Baguley BC, Denny WA, Charlton P. Antitumour activity of XR5944, a novel and potent topoisomerase poison. Anticancer Drugs 2001;12: 359–367.

353. Sappal DS, McClendon AK, Fleming JA, Thoroddsen V, Connolly K, Reimer C, Blackman RK, Bulawa CE, Osheroff N, Charlton P, Rudolph-Owen LA. Biological characterization of MLN944: a potent DNA binding agent. Mol Cancer Ther 2004;3:47–58.

354. Byers SA, Schafer B, Sappal DS, Brown J, Price DH. The antiproliferative agent MLN944 preferentially inhibits transcription. Mol Cancer Ther 2005;4:1260–1267.

355. Punchihewa C, De Alba A, Sidell N, Yang D. XR5944: a potent inhibitor of estrogen receptors. Mol Cancer Ther 2007;6(1):213–219.

356. Verborg W, Thomas H, Bissett D, Waterfall J, Steiner J, Cooper M, Rankin EM. First-intoman phase I and pharmacokinetic study of XR5944.14, a novel agent with a unique mechanism of action. Br J Cancer 2007;97 (7):844–850.

357. Nair AG, Chemistry of antifolates. In: Wilman DEV, editor. Chemistry of Antitumour Agents. London: Blackie; 1989; p 202–233.
358. Huennekens FM. The methotrexate story: a paradigm for development of cancer chemotherapeutic agents. Adv Enzyme Regul 1994;34:397–419.
359. Sirotnak FM, DeGraw JI, Moccio DM, Samuels LL, Goutas LJ. New folate analogs of the 10-deazaaminopterin series. Basis for structural design and biochemical and pharmacologic properties. Cancer Chemother Pharmacol 1984;12:18–25.
360. Brown DH, Braakhuis BJ, van Dongen GA, van Walsum M, Bagnay M, Snow GB. Activity of the folate analog 10-ethyl, 10-deaza-aminopterin (10-EdAM) against human head and neck cancer xenografts. Anticancer Res 1989;9:1549–1552.
361. Beinart GA, Gonzalez-Angulo AM, Broglio K, Frye D, Walters R, Holmes FA, Gunale S, Booser D, Rosenthal J, Dhingra K, Young JA, Hortobagyi GN. Phase II trial of 10-EDAM in the treatment of metastatic breast cancer. Cancer Chemother Pharmacol 2007;60(1):61–67.
362. van der Laan BF, Jansen G, Kathmann I, Schornagel JH, Hordijk GJ. Mechanisms of acquired resistance to methotrexate in a human squamous carcinoma cell line of the head and neck, exposed to different treatment schedules. Eur J Cancer 1991;27:1274–1280.
363. Liani E, Rothem L, Bunni MA, Smith CA, Jansen G, Assaraf YG. Loss of folylpoly-γ-glutamate synthetase activity is a dominant mechanism of resistance to polyglutamylation-dependent novel antifolates in multiple human leukemia sublines. Int J Cancer 2003;103(5):587–599.
364. Jackson RC, Niethammer D. Acquired methotrexate resistance in lymphoblasts resulting from altered kinetic properties of dihydrofolate reductase. Eur J Cancer 1977;13:567–575.
365. Jones TR, Smithers MJ, Taylor MA, Jackman AL, Calvert AH, Harland SJ, Harrap KR. Quinazoline antifolates inhibiting thymidylate synthase: benzoyl ring modifications. J Med Chem 1986;29:468–472.
366. Sikora E, Jackman AL, Newell DR, Calvert AH. Formation and retention and biological activity of N10-propargyl-5,8-dideazafolic acid (CB3717) polyglutamates in L1210 cells in vitro. Biochem Pharmacol 1988;37:4047–4054.
367. Berman EM, Werbel LM. The renewed potential for folate antagonists in contemporary cancer chemotherapy. J Med Chem 1991;34:479–485.
368. Jackman AL, Calvert AH. Folate-based thymidylate synthase inhibitors as anticancer drugs. Ann Oncol 1995;6:871–881.
369. Judson IR. 'Tomudex' (raltitrexed) development: preclinical, phase I and II studies. Anticancer Drugs 1997;8(Suppl 2):5–9.
370. Cao S, Bhattacharya A, Durrani FA, Fakih M. Irinotecan, oxaliplatin and raltitrexed for the treatment of advanced colorectal cancer. Expert Opin Pharmacother 2006;7(6):687–703.
371. Ellis P, Davies AM, Evans WK, Haynes E, Lloyd NS. The use of chemotherapy in patients with advanced malignant pleural mesothelioma: a systematic review and practice guideline. J Thoracic Oncol 2006;1(6):591–601.
372. Schultz RM, Patel VF, Worzalla JF, Shih C. Role of thymidylate synthase in the antitumor activity of the multitargeted antifolate, LY231514. Anticancer Res 1999;19:437–443.
373. Mendelsohn LG, Shih C, Chen VJ, Habeck LL, Gates SB, Shackelford KA. Enzyme inhibition, polyglutamation, and the effect of LY231514 (MTA) on purine biosynthesis. Semin Oncol 1999;26(Suppl 6):42–47.
374. Britten CD, Izbicka E, Hilsenbeck S, Lawrence R, Davidson K, Cerna C, Gomez L, Rowinsky EK, Weitman S, Von Hoff DD. Activity of the multitargeted antifolate LY231514 in the human tumor cloning assay. Cancer Chemother Pharmacol 1999;44:105–110.
375. Adjei AA. Pemetrexed: a multitargeted antifolate agent with promising activity in solid tumors. Ann Oncol 2000;11:1335–1341.
376. Garcia-Carbonero R, Paz-Ares L. Systemic chemotherapy in the management of malignant peritoneal mesothelioma. Eur J Surgical Oncol 2006;32(6):676–681.
377. Kumar P, Kisliuk L, Guamont Y, Nair MG, Baugh CM, Kaufmann BT. Interaction of polyglutamyl derivatives of methotrexate, 10-deazaaminopterin, and dihydrofolate with dihydrofolate reductase. Cancer Res 1986;46:5020–5023.
378. O'Dwyer PJ, Shoemaker DD, Plowman J, Cradock J, Grillo-Lopez A, Leyland-Jones B. Trimetrexate: a new antifol entering clinical trials. Invest New Drugs 1985;3:71–75.
379. Klohs WD, Steinkampf RW, Besserer JA, Fry DW. Cross resistance of pleiotropically drug resistant P338 leukemia cells to the lipophilic antifolates trimetrexate and BW 301U. Cancer Lett 1986;31:256–260.

380. Haller DG. Trimetrexate: experience with solid tumors. Semin Oncol 1997;24(Suppl 18):71–76.
381. Blanke CD, Messenger M, Taplin SC. Trimetrexate: review and current clinical experience in advanced colorectal cancer. Semin Oncol 1997;24(Suppl 18):57–63.
382. Duch DS, Edelstein MP, Bowers SW, Nichol CA. Biochemical and chemotherapeutic studies on 2,4-diamino-6-(2,5-dimethoxybenzyl)-5-methylpyrido[2,3-d]pyrimidine (BW 301U), a novel lipid-soluble inhibitor of dihydrofolate reductase. Cancer Res 1982;42:3987–3994.
383. de Wit R, Kaye SB, Roberts JT, Stoter G, Scott J, Verweij J. Oral piritrexim, an effective treatment for metastatic urothelial cancer. Br J Cancer 1993;67:388–390.
384. Khorsand M, Lange J, Feun L, Clendeninn NJ, Collier M, Wilding G. Phase II trial of oral piritrexim in advanced, previously treated transitional cell cancer of bladder. Invest New Drugs 1997;15:157–163.
385. Perkins W, Williams RE, Vestey JP, Tidman MJ, Layton AM, Cunliffe WJ, Saihan EM, Klaber MR, Manna VK, Baker H. A multicentre 12-week open study of a lipid-soluble folate antagonist, piritrexim in severe psoriasis. Br J Dermatol 1993;129:584–589.
386. Webber S, Bartlett C, Boritzki TJ, Hilliard JA, Howland EF, Johnston AL, Kosa M, Margosiak SA, Morse CA, Shetty BV. AG337, a novel lipophilic thymidylate synthase inhibitor: *In vitro* and *in vivo* preclinical studies. Cancer Chemother Pharmacol 1996;37:509–517.
387. Rhee MS, Webber S, Galivan J. Evaluation of the mechanism of growth inhibition of AG337. Cell Pharmacol 1995;2:97–101.
388. Jhawer M, Rosen L, Dancey J, Hochster H, Hamburg S, Tempero M, Clendeninn N, Mani S. Phase II trial of nolatrexed dihydrochloride [Thymitaq, AG 337] in patients with advanced hepatocellular carcinoma. Invest New Drugs 2007;25(1):85–94.
389. Gish RG, Porta C, Lazar L, Ruff P, Feld R, Croitoru A, Feun L, Jeziorski K, Leighton J, Gallo J, Kennealey GT. Phase III randomized controlled trial comparing the survival of patients with unresectable hepatocellular carcinoma treated with nolatrexed or doxorubicin. J Clin Oncol 2007;25(21):3069–3075.
390. Parker WB, Cheng YC. Metabolism and mechanism of action of 5-fluorouracil. Pharmacol Ther 1990;48:381–395.
391. Broom AD. Rational design of enzyme inhibitors: multisubstrate analogue inhibitors. J Med Chem 1989;32:2–7.
392. Pallavicini MG. Cytosine arabinoside: molecular, pharmacokinetic and cytokinetic considerations. Pharmacol Ther 1984;25:207–238.
393. Cohen SS. The lethality of aranucleotides. Med Biol 1976;54:299–326.
394. Plunkett W, Huang P, Gandhi V. Preclinical characteristics of gemcitabine. Anticancer Drugs 1995;6(Suppl 6):7–13.
395. Yang LY, Li L, Jiang H, Shen Y, Plunkett W. Expression of ERCC1 antisense RNA abrogates gemcitabine-mediated cytotoxic synergism with cisplatin in human colon tumor cells defective in mismatch repair but proficient in nucleotide excision repair. Clin Cancer Res 2000;6:773–781.
396. van der Donk WA, Yu G, Perez I, Sanchez RJ, Stubbe J, Samano V, Robins MJ. Detection of a new substrate-derived radical during inactivation of ribonucleotide reductase from *Escherichia coli* by gemcitabine 5'-diphosphate. Biochemistry 1998;37:6419–6426.
397. Goan YG, Zhou B, Hu E, Mi S, Yen Y. Overexpression of ribonucleotide reductase as a mechanism of resistance to 2, 2-difluorodeoxycytidine in the human KB cancer cell line. Cancer Res 1999;59:4204–4207.
398. Lawrence TS, Chang EY, Hahn TM, Hertel LW, Shewach DS. Radiosensitization of pancreatic cancer cells by 2', 2'-difluoro-2'-deoxycytidine. Int J Radiat Oncol Biol Phys 1996;34:867–872.
399. Lawrence TS, Davis MA, Hough A, Rehemtulla A. The role of apoptosis in 2', 2'-difluoro-2'-deoxycytidine (Gemcitabine)-mediated radiosensitization. Clin Cancer Res 2001;7: 314–319.
400. Walko CM, Lindley C. Capecitabine: a review. Clin Ther 2005;27:23–44.
401. Miwa M, Eda H, Ura M, Ouchi KF, Keith DD, Foley LH, Ishitsuka H. High susceptibility of human cancer xenografts with higher levels of cytidine deaminase to a 2'-deoxycytidine antimetabolite, 2'-deoxy-2'-methylidenecytidine. Clin Cancer Res 1998;4:493–497.
402. Bajetta E, Colleoni M, Rosso R, Sobrero A, Amadori D, Comella G, Marangolo M, Scanni A, Lorusso V, Calabresi F. Prospective randomised trial comparing fluorouracil versus doxifluridine for the treatment of advanced colorectal cancer. Eur J Cancer 1993;29A: 1658–1663.
403. Miwa M, Ura M, Nishida M, Sawada N, Ishikawa T, Mori K, Shimma N, Umeda I, Ishitsuka H. Design of a novel oral fluoropyrimidine carbamate, capecitabine, which generates 5-fluorouracil selectively in tumours by enzymes

concentrated in human liver and cancer tissue. Eur J Cancer 1998;34:1274–1281.
404. Ho DHW, Frei E. Clinical pharmacology of 1-β-D-arabinofuranosylcytosine. Clin Pharm Ther 1971;12:944–954.
405. MacCoss M, Robins MJ, Anticancer pyrimidines, pyrimidine nucleosides and prodrugs. In: Wilman DEV, editor. Chemistry of Antitumour Agents. London: Blackie; 1989. p 261–298.
406. Cartei G, Sacco C, Sibau A, Pella N, Iop A, Tabaro G. Cisplatin and gemcitabine in non-small-cell lung cancer. Ann Oncol 1999;10 (Suppl 5):57–62.
407. Scagliotti GV. Gemcitabine/cisplatin as induction therapy for stage IIIA N2 non-small-cell lung cancer. Oncology 2000;14(Suppl 4):15–19.
408. Culine S. Gemcitabine and urothelial tumours. Anticancer Drugs 2000;11(Suppl 1):9–13.
409. Carmichael J, Possinger K, Phillip P, Beykirch M, Kerr H, Walling J, Harris AL. Advanced breast cancer: a phase II trial with gemcitabine. J Clin Oncol 1995;13:2731–2736.
410. von der Maase H. Gemcitabine and cisplatin in locally advanced and/or metastatic bladder cancer. Eur J Cancer 2000;36(Suppl 2):13–16.
411. Schellens JHM. Capecitabine. Oncologist 2007;12(2):152–155.
412. Smiraglia DJ, Plass C. The development of CpG island methylation biomarkers using restriction landmark genomic scanning. Ann NY Acad Sci 2003;983:110–119.
413. Schmidt WM, Sedivy R, Forstner B, Steger GG, Zoechbauer-Mueller S, Mader Robert M. Progressive up-regulation of genes encoding DNA methyltransferases in the colorectal adenoma–carcinoma sequence. Mol Carcinog 2007;46(9):766–772.
414. Brena RM, Plass C, Costello JF. Mining methylation for early detection of common cancers. PLoS Med 2006;3(12):2184–2185.
415. Sekeres MA. The myelodysplastic syndromes. Expert. Opin Biol Ther 2007;7(3):369–377.
416. Issa J-PJ, Gharibyan V, Cortes J, Jelinek J, Morris G, Verstovsek S, Talpaz M, Garcia-Manero G, Kantarjian HM. Phase II study of low-dose decitabine in patients with chronic myelogenous leukemia resistant to imatinib mesylate. J Clin Oncol 2005;23(17): 3948–3956.
417. Oki Y, Issa J-PJ. Review: recent clinical trials in epigenetic therapy. Rev Rec Clin Trials 2006;1(2):169–182.
418. Kuykendall JR. 5-Azacytidine and decitabine monotherapies of myelodysplastic disorders. Ann Pharmacother 2005;39 (10):1700–1709.
419. Sasvari-Szekely M, Spasokoukotskaja T, Szoke M, Csapo Z, Turi A, Szanto I, Eriksson S, Staub M. Activation of deoxycytidine kinase during inhibition of DNA synthesis by 2-chloro-2′-deoxyadenosine (cladribine) in human lymphocytes. Biochem Pharmacol 1998; 56:1175–1179.
420. Fabianowska-Majewska K, Tybor K, Duley J, Simmonds A. The influence of 2-chloro-2′-deoxyadenosine on metabolism of deoxyadenosine in human primary CNS lymphoma. Biochem Pharmacol 1995;50:1379–1383.
421. Nomura Y, Inanami O, Takahashi K, Matsuda A, Kuwabara M. 2-Chloro-2′-deoxyadenosine induces apoptosis through the Fas/Fas ligand pathway in human leukemia cell line MOLT-4. Leukemia 2000;14:299–306.
422. Siaw MFE, Coleman MS. In vitro metabolism of deoxycoformycin in human T lymphoblastoid cells. Phosphorylation of deoxycoformycin and incorporation into cellular DNA. J Biol Chem 1984;259:9426–9433.
423. Agarwal RP. Inhibitors of adenosine deaminase. Pharmacol Ther 1982;17:399–429.
424. Tondini C, Balzarotti M, Rampinelli I, Valagussa P, Luoni M, De Paoli A, Santoro A, Bonadonna G. Fludarabine and cladribine in relapsed/refractory low-grade non-Hodgkin's lymphoma: a phase II randomized study. Ann Oncol 2000;11:231–233.
425. Cheson BD, Sorensen JM, Vena DA, Montello MJ, Barrett JA, Damasio E, Tallman M, Annino L, Connors J, Coiffier B, Lauria F. Treatment of hairy cell leukemia with 2-chlorodeoxyadenosine via the Group C protocol mechanism of the National Cancer Institute: a report of 979 patients. Clin Oncol 1998;16: 3007–3015.
426. Saven A, Emanuele S, Kosty M, Koziol J, Ellison D, Piro LD. 2-Chlorodeoxyadenosine activity in patients with untreated, indolent non-Hodgkin's lymphoma. Blood 1995;86: 1710–1716.
427. Saven A, Piro LD. 2-Chlorodeoxyadenosine: a newer purine analog active in the treatment of indolent lymphoid malignancies. Ann Int Med 1994;20:784–791.
428. Saven A, Lemon RH, Kosty M, Beutler E, Piro LD. 2-Chlorodeoxyadenosine activity in patients with untreated chronic lymphocytic leukemia. J Clin Oncol 1995;13:570–574.

429. Kuzel TM, Hurria A, Samuelson E, Tallman MS, Roenigk HH, Rademaker AW, Rosen ST. Phase II trial of 2-chlorodeoxyadenosine for the treatment of cutaneous T-cell lymphoma. Blood 1996;87:906–911.

430. Leist TP, Vermersch P. The potential role for cladribine in the treatment of multi ple sclerosis: clinical experience and development of an oral tablet formulation. Curr Med Res Opin 2007;23(11):2667–2676.

431. Robak T, Lech-Maranda E, Korycka A, Robak E. Purine nucleoside analogs as immunosuppressive and antineoplastic agents: mechanism of action and clinical activity. Curr Med Chem 2006;13(26):3165–3189.

432. Mao C, Cook WJ, Zhou M, Koszalka GW, Krenitsky TA, Ealick SE. The crystal structure of *Escherichia coli* purine nucleoside phosphorylase: a comparison with the human enzyme reveals a conserved topology. Structure 1997;5:1373–1383.

433. Korycka A, Blonski JZ, Robak T. Forodesine (BCX-1777, Immucillin H): a new purine nucleoside analogue: mechanism of action and potential clinical application. Mini Rev Med Chem 2007;7:976–983.

434. Wolfenden R. Transition state analogues for enzyme catalysis. Nature 1969;223:704–705.

435. Lewandowicz A, Tyler PC, Evans GB, Furneaux RH, Schramm VL. Achieving the ultimate physiological goal in transition state analogue inhibitors for purine nucleoside phosphorylase. J Biol Chem 2003;278: 31465–31468.

436. Furman RR, Hoelzer D. Purine nucleoside phosphorylase inhibition as a novel therapeutic approach for B-cell lymphoid malignancies. Semin Oncol 2007;34(Suppl 5): S29–S34

437. Denny WA. Prodrug strategies in cancer therapy. Eur J Med Chem 2001;36:577–595.

438. Boyle RG, Travers S. Hypoxia: targeting the tumour. Anticancer Agents Med Chem 2006;6 (4):281–286.

439. Denny WA. Prospects for hypoxia-activated anticancer drugs. Curr Med Chem Anticancer Agents 2004;4:395–399.

440. Tietze LF, Feuerstein T. Review: highly selective compounds for the antibody-directed enzyme prodrug therapy of cancer. Aust J Chem 2003;56(9):841–854.

441. Denny WA. Prodrugs for gene-directed enzyme-prodrug therapy (suicide gene therapy). J Biomed Biotech 2003;3:49–70.

442. Brown JM, Wilson WR. Exploiting tumor hypoxia in cancer treatment. Nature Rev Cancer 2004;4:437–447.

443. Baish JW, Netti PA, Jain RL. Transmural coupling of fluid flow in microcirculatory network and interstitium in tumors. Microvascular Res 1997;53:128–141.

444. Brown JM, Giaccia AJ. The unique physiology of solid tumors: opportunities (and problems) for cancer therapy. Cancer Res 1998;58: 1405–1416.

445. Narayanan VL, Lee WW. Development of radiosensitizers: a medicinal chemistry perspective. Adv Pharmacol Chemother 1982;19: 155–205.

446. Brown JM. The mechanisms of cytotoxicity and chemosensitization by misonidazole and other nitroimidazoles. Int J Radiat Oncol Biol Phys 1982;8:675–682.

447. Patterson AV, Saunders MP, Chinje EC, Patterson LH, Stratford IJ. Enzymology of tirapazamine metabolism: a review. Anticancer Drug Design 1998;13:541–573.

448. Denny WA, Wilson WR, Hay MP. Recent developments in the design of bioreductive drugs. Br J Cancer 1996;74(Suppl 27):32–38.

449. Rockwell S, Sartorelli AC, Tomasz M, Kennedy KA. Cellular pharmacology of quinone bioreductive alkylating agents. Cancer Met Rev 1993;12:165–176.

450. Tomasz M, Palom Y. The mitomycin bioreductive antitumor agents: cross-linking and alkylation of DNA as the molecular basis of their activity. Pharmacol Ther 1997;76:73–87.

451. Belcourt MF, Hodnick WF, Rockwell S, Sartorelli AC. Exploring the mechanistic aspects of mitomycin antibiotic bioactivation in Chinese hamster ovary cells overexpressing NADPH: cytochrome C (P-450) reductase and DT-diaphorase. Adv Enzyme Regul 1998;38: 111–133.

452. Haffty BG, Wilson LD, Son YH, Cho EI, Papac RJ, Fischer DB, Rockwell S, Sartorelli AC, Ross DA, Sasaki CT, Fischer JJ. Concurrent chemo-radiotherapy with mitomycin C compared with porfiromycin in squamous cell cancer of the head and neck: final results of a randomized clinical trial. Int J Radiat Oncol Biol Phys 2005;61(1):119–128.

453. Bailey SM, Wyatt MD, Friedlos F, Hartley JA, Knox RJ, Lewis AD, Workman P. Involvement of DT-diaphorase (EC 1. 6. 99. 2) in the DNA crosslinking and sequence selectivity of the bioreductive antitumor agent EO9. Br J Cancer 1997;76:1596–1603.

454. Dirix LY, Tonnesen F, Cassidy J, Epelbaum R, ten Bokkel Huinink WW, Pavlidis N, Sorio R, Gamucci T, Wolff I, Te Velde A, Lan J, Verweij J. EO9 phase II study in advanced breast, gastric, pancreatic and colorectal carcinoma by the EORTC early clinical studies group. Eur J Cancer 1996;32A:2019–2022.

455. McLeod HL, Graham MA, Aamdal S, Setanoians A, Groot Y, Lund B. Phase I pharmacokinetics and limited sampling strategies for the bioreductive alkylating drug EO9. Eur J Cancer 1996;32A:1518–1522.

456. Puri R, Palit V, Loadman PM, Flannigan M, Shah T, Choudry GA, Basu S, Double JA, Lenaz G, Chawla S, Beer M, Van Kalken C, de Boer R, Beijnen JH, Twelves CJ, Phillips RM. Phase I/II pilot study of intravesical apaziquone (EO9) for superficial bladder cancer. J Urol 2006;176(4 Pt 1):1344–1348.

457. van der Heijden AG, Moonen PMJ, Cornel EB, Vergunst H, de Reijke TM, van Boven E, Barten EJ, Puri R, van Kalken CK, Witjes JA. Phase II marker lesion study with instillation of apaziquone for superficial bladder cancer: toxicity and marker response. J Urol 2006;176 (4 Pt 1):1349–1353.

458. Marcu L, Olver I. Tirapazamine: from bench to clinical trials. Curr Clin Pharm 2006;1 (1):71–79.

459. Denny WA, Wilson WR. Tirapazamine: a bioreductive anticancer drug which exploits tumour hypoxia. Expert Opin Investig Drugs 2000;9:2889–2901.

460. Rischin D, Peters L, O'Sullivan B, Giralt J, Yuen K, Trotti A, Bernier J, Bourhis J, Henke M, Fisher R. Phase III study of tirapazamine, cisplatin and radiation versus cisplatin and radiation for advanced squamous cell carcinoma of the head and neck. J Clin Oncol 2008;26 (Suppl): Abstract LBA6008.

461. Wilson WR, van Zijl P, Denny WA. Bis-bioreductive agents as hypoxia-selective cytotoxins: N-oxides of nitracrine. Int J Radiat Oncol Biol Phys 1992;22:693–697.

462. Wilson WR, Denny WA, Stewart GM, Fenn A, Probert JC. Reductive metabolism and hypoxia-selective cytotoxicity of nitracrine. Int J Radiat Oncol Biol Phys 1986;12:1235–1238.

463. Smith PJ, Blunt NJ, Desnoyers R, Giles Y, Patterson LH. DNA topoisomerase II-dependent cytotoxicity of alkylaminoanthraquinones and their N-oxides. Cancer Chemother Pharmacol 1997;39:455–4651.

464. Rickert DE. Metabolism of nitroaromatic compounds. Drug Met Rev 1987;18:23–53.

465. Patterson AV, Ferry DM, Edmunds SJ, Gu Y, Singleton RS, Patel K, Pullen SM, Hicks KO, Syddall SP, Atwell GJ, Yang S, Denny WA, Wilson WR. Mechanism of action and preclinical antitumor activity of the novel hypoxia-activated DNA cross-linking agent PR-104. Clin Cancer Res 2007;13(13):3922–3932.

466. Hart CP, Meng F, Banica M, Evans JW, Lan L, Lorente G, Duan JX, Matteucci M. In vitro activity profile of the novel hypoxia-activated cytotoxic prodrug TH-302. Proc Am Assoc Cancer Res 2008; Abstract 1441.

467. Yang S, Atwell GJ, Denny WA. A new synthesis of asymmetric dinitrobenzamide mustards. Tetrahedron 2007;63:5470–5476.

468. Belcourt MF, Hodnick WF, Rockwell S, Sartorelli AC. Differential toxicity of mitomycin C and porfiromycin to aerobic and hypoxic Chinese hamster ovary cells overexpressing human NADPH:cytochrome c (P-450) reductase. Proc Natl Acad Sci USA 1996;93:456–460.

469. Beall HD, Winski SL. Mechanisms of action of quinone-containing alkylating agents I. NQO1-directed drug development. Front Biosci 2000;5:D639–D648

470. Anderson RF, Shinde SS, Hay MP, Gamage SA, Denny WA. Activation of 3-amino-1,2, 4-benzotriazine 1,4-dioxide antitumor agents to oxidizing species following their one-electron reduction. J Am Chem Soc 2003;125:748–756.

471. Shinde SS, Anderson RF, Hay MP, Gamage SA, Denny WA. Oxidation of 2-deoxyribose by benzotriazinyl radicals of antitumor 3-aminobenzotriazine 1,4-dioxides. J Am Chem Soc 2004;126:7865–7874.

472. Hwang JT, Greenberg MM, Fuchs T, Gates KS. Reaction of the hypoxia-selective antitumor agent tirapazamine with a C1′-radical in single-stranded and double-stranded DNA: the drug and its metabolites can serve as surrogates for molecular oxygen in radical-mediated DNA damage reactions. Biochemistry 1999;38: 14248–14255.

473. Brown JM. The hypoxic cell: a target for selective cancer therapy. Cancer Res 1999;59: 5863–5870.

474. Koch CJ. Unusual concentration dependence of toxicity of SR-4233, a hypoxic cell toxin. Cancer Res 1993;53:3992–3997.

475. Monge A, Martinez-Crespo FJ, Lopez de Cerain A, Palop JA, Narro S, Senador V, Marin A, Sainz Y, Gonzalez M, Hamilton EJ. Hypoxia-selective agents derived from 2-quinoxalinecarbonitrile 1,4-di-n-oxides. 2. J Med Chem 1995;38:4488–4494.

476. Raleigh SM, Wanogho E, Burke MD, McKeown SR, Patterson LH. Involvement of human cytochromes P450 (CYP) in the reductive metabolism of AQ4N, a hypoxia activated anthraquinone di-*N*-oxide prodrug. Int J Radiat Oncol Biol Phys 1998;42:763–767.

477. Patterson LH. Rationale for the use of aliphatic *N*-oxides of cytotoxic anthraquinones as prodrug DNA binding agents: a new class of bioreductive agent. Cancer Met Rev 1993;12:119–134.

478. Wilson WR, Denny WA, Pullen SM, Thompson KM, Li AE, Patterson LH, Lee HH. Tertiary amine *N*-oxides as bioreductive drugs: DACA *N*-oxide, nitracrine *N*-oxide and AQ4N. Br J Cancer 1996;74(Suppl 27):43–47.

479. Siim BG, Hicks KO, Pullen SM, van Zijl PL, Denny WA, Wilson WR. Comparison of aromatic and aliphatic *N*-oxides of acridine DNA intercalators as bioreductive drugs: cytotoxicity, DNA binding, cellular uptake and metabolism. Biochem Pharmacol 2000;60:969–978.

480. Denny WA, Wilson WR. Considerations for the design of nitrophenyl mustards as drugs selectively toxic for hypoxic mammalian cells. J Med Chem 1986;29:879–887.

481. Duan J-X, Jiao H, Kaizerman J, Stanton T, Evans JW, Lan L, Lorente G, Banica M, Jung D, Wang J, Ma H, Li X, Yang Z, Hoffman RM, Ammons WS, Hart CP, Matteucci M. Potent and highly selective hypoxia-activated achiral phosphoramidate mustards as anticancer drugs. J Med Chem 2008;51:2412–2420.

482. Lee HH, Palmer BD, Wilson WR, Denny WA. Synthesis and hypoxia-selective cytotoxicity of a 2-nitroimidazole mustard. Bioorg Med Chem Lett 1998;8:1741–1744.

483. Hicks KO, Fleming Y, Siim BG, Koch CJ, Wilson WR. Extravascular diffusion of tirapazamine: effect of metabolic consumption assessed using the multicellular layer model. Int J Radiat Oncol Biol Phys 1998;42:641–649.

484. Dorie MJ, Menke D, Brown JM. Comparison of the enhancement of tumor responses to fractionated irradiation by SR 4233 (tirapazamine) and by nicotinamide with carbogen. Int J Radiat Oncol Biol Phys 1994;28:145–150.

485. Kovacs MS, Hocking DJ, Evans JW, Siim BG, Wouters BG, Brown JM. Cisplatin anti-tumour potentiation by tirapazamine results from a hypoxia-dependent cellular sensitization to cisplatin. Br J Cancer 1999;80:1245–1251.

486. Dorie MJ, Brown JM. Modification of the antitumor activity of chemotherapeutic drugs by the hypoxic cytotoxic agent tirapazamine. Cancer Chemother Pharmacol 1997;39:361–366.

487. Cliffe S, Taylor ML, Rutland M, Baguley BC, Hill RP, Wilson WR. Combining bioreductive drugs (SR 4233 or SN 23862) with the vasoactive agents flavone acetic acid or 5,6-dimethylxanthenone acetic acid. Int J Radiat Oncol Biol Phys 1994;29:373–377.

488. Maluf FC, Leiser AL, Aghajanian C, Sabbatini P, Pezzulli S, Chi DS, Wolf JK, Levenback C, Loh E, Spriggs DR. Phase II study of tirapazamine plus cisplatin in patients with advanced or recurrent cervical cancer. Int. J. Gyn. Cancer 2006;16(3):1165–1171.

489. Gatineau M, Rixe O, Le Chevalier T. Tirapazamine with cisplatin and vinorelbine in patients with advanced non-small-cell lung cancer: a phase I/II study. Clin Lung Cancer 2005;6(5):293–298.

490. Williamson SK, Crowley JJ, Lara PN, McCoy J, Lau DHM, Tucker RW, Mills GM, Gandara DR. Phase III trial of paclitaxel plus carboplatin with or without tirapazamine in advanced non-small-cell lung cancer. J Clin Oncol 2005;23(36):9097–9104.

491. Rischin D, Peters L, Fisher R, Macann A, Denham J, Poulsen M, Jackson M, Kenny L, Penniment M, Corry June Lamb D, McClure B. Tirapazamine, cisplatin, and radiation versus fluorouracil, cisplatin, and radiation in patients with locally advanced head and neck cancer: a randomized phase II trial of the Trans-Tasman Radiation Oncology Group (TROG 98.02). J Clin Oncol 2005;23(1):79–87.

492. Lee AE, Wilson WR. Hypoxia-dependent retinal toxicity of bioreductive anticancer prodrugs in mice. Toxicol Appl Pharmacol 2000;163:50–59.

493. Lalani AS, Alters SE, Wong A, Albertella MR, Cleland JL, Henner WD. Selective tumor targeting by the hypoxia-activated prodrug AQ4N blocks tumor growth and metastasis in preclinical models of pancreatic cancer. Clin Cancer Res 2007;13(7):2216–2225.

494. Sarantopoulos J, Tolcher AW, Wong A, Goel S, Beeram M, Lam G, Desai K, Woody K, Mani S, Papadopoulos KP. Banoxantrone (AQ4N), tissue CYP 450 targeted prodrug: the results of a phase I study using an accelerated dose escalation. Proc Am Soc Clin Oncol 2006; Abstract 2011

495. Hicks KO, Myint H, Patterson AV, Pruijn FB, Siim BG, Patel K, Wilson WR. Oxygen dependence and extravascular transport of hypoxia-

activated prodrugs: comparison of the dinitrobenzamide mustard PR-104A and tirapazamine. Int J Radiat Oncol Biol Phys 2007;69 (2):560–571.

496. Guise CP, Wang AT, Theil A, Bridewell DJ, Wilson WR, Patterson AV. Identification of human reductases that activate the dinitrobenzamide mustard prodrug PR-104A: a role for NADPH:cytochrome P450 oxidoreductase under hypoxia. Biochem Pharmacol 2007;74: (6):810–820.

497. Chowdhury G, Junnotula V, Daniels JS, Greenberg MM, Gates KS. DNA strand damage product analysis provides evidence that the tumor cell-specific cytotoxin tirapazamine produces hydroxyl radical and acts as a surrogate for O_2. J Am Chem Soc 2007;129 (42):12870–12877.

498. Shi X, Mandel SM, Platz MS. On the mechanism of reaction of radicals with tirapazamine. J Am Chem Soc 2007;129(15):4542–4550.

499. Anderson RF, Shinde SS, Hay MP, Denny WA. Potentiation of the cytotoxicity of the anticancer agent tirapazamine by benzotriazine N-oxides: the role of redox equilibria. J Am Chem Soc 2006;128(1):245–249.

500. Siim BG, Pruijn FB, Sturman JR, Hogg A, Hay MP, Brown JM, Wilson WR. Selective potentiation of the hypoxic cytotoxicity of tirapazamine by its 1-N-oxide metabolite SR 4317. Cancer Res 2004;64(2):736–742.

501. Evans JW, Chernikova SB, Kachnic LA, Banath JP, Sordet O, Delahoussaye YM, Treszezamsky A, Chon BH, Feng Z, Gu Y, Wilson WR, Pommier Y, Olive PL, Powell SN, Brown JM. Homologous recombination is the principal pathway for the repair of DNA damage induced by tirapazamine in mammalian cells. Cancer Res 2008;68(1):257–265.

502. Hicks KO, Pruijn FB, Sturman JR, Denny WA, Wilson WR. Multicellular resistance to tirapazamine is due to restricted extravascular transport: a pharmacokinetic/pharmacodynamic study in HT29 multicellular layer cultures. Cancer Res 2003;63 (18):5970–5977.

503. Hay MP, Gamage SA, Kovacs MS, Pruijn FB, Anderson RF, Patterson AV, Wilson WR, Brown JM, Denny WA. Structure–activity relationships of 1,2,4-benzotriazine 1,4-dioxides as hypoxia-selective analogues of tirapazamine. J Med Chem 2003;46(1):169–182.

504. Delahoussaye YM, Hay MP, Pruijn FB, Denny WA, Brown JM. Improved potency of the hypoxic cytotoxin tirapazamine by DNA-targeting. Biochem Pharmacol 2003;65 (11):1807–1815.

505. Hay MP, Pruijn FB, Gamage SA, Liyanage HDS, Kovacs MS, Patterson AV, Wilson WR, Brown JM, Denny WA. DNA-targeted 1,2,4-benzotriazine 1,4-dioxides: potent analogues of the hypoxia-selective cytotoxin tirapazamine. J Med Chem 2004;47:475–488.

506. Anderson RF, Harris TA, Hay MP, Denny WA. Enhanced conversion of DNA radical damage to double strand breaks by 1,2,4-benzotriazine 1,4-dioxides linked to a DNA binder compared to tirapazamine. Chem Res Toxicol 2003;16 (11):1477–1483.

507. Pruijn FB, Sturman JR, Liyanage HDS, Hicks KO, Hay MP, Wilson WR. Extravascular transport of drugs in tumor tissue: effect of lipophilicity on diffusion of tirapazamine analogs in multicellular layer cultures. J Med Chem 2005;48(4):1079–1087.

508. Hay MP, Hicks KO, Pruijn FB, Pchalek K, Siim BG, Wilson WR, Denny WA. Pharmacokinetic/pharmacodynamic model-guided identification of hypoxia-selective 1,2,4-benzotriazine 1, 4-dioxides with antitumor activity: the role of extravascular transport. J Med Chem 2007;50(25):6392–6404.

509. Hay MP, Blaser A, Denny WA, Hicks KO, Lee HH, Pchalek K, Pruijn FB, Siim BG, Wilson WR, Yang S. Preparation of tricyclic 1,2,4-triazine oxides as cancer treatments. PCT Int Appl WO2006104406A1 (204 pp). Published 5 October 2006.

510. Green MC, Murray JL, Hortobagyi GN. Monoclonal antibody therapy for solid tumors. Cancer Treat Rev 2000;26:269–286.

511. Dubowchik GM, Walker MA. Receptor-mediated and enzyme-dependent targeting of cytotoxic anticancer drugs. Pharm Ther 1999;83:67–123.

512. Deonarain MP, Epenetos AA. Targeting enzymes for cancer therapy: old enzymes in new roles. Br J Cancer 1994;70:786–794.

513. Niculescu-Duvaz I, Springer CJ. Antibody-directed enzyme prodrug therapy (ADEPT): a review. Adv Drug Deliv Rev 1997;26:151–172.

514. Hay MP, Denny WA. Antibody-directed enzyme prodrug therapy (ADEPT). Drugs Future 1996;12:917–931.

515. Senter PD, Saulnier MG, Schreiber GJ, Hirschberg DL, Brown JL, Hellstrom I, Hellstrom KE. Anti-tumor effects of antibody-alkaline phosphatase conjugates in combination with etoposide phosphate. Proc Natl Acad Sci USA 1988;85:4842–4846.

516. Senter PD, Schreiber GD, Hirschberg DL, Ashe SA, Hellstrom CE, Hellstrom I. Enhancement of the in vitro and in vivo antitumor activities of phosphorylated mitomycin C and etoposide derivatives by monoclonal antibody-alkaline phosphatase conjugates. Cancer Res 1989;49:5789–5792.

517. Griggs J, Hesketh R, Smith GA, Brindle KM, Metcalfe JC, Thomas GA, Williams ED. Combretastatin-A4 disrupts neovascular development in non-neoplastic tissue. Br J Cancer 2001;84:832–835.

518. Niculescu-Duvaz I, Friedlos F, Niculescu-Duvaz D, Davies L, Springer CJ. Prodrugs for antibody- and gene-directed enzyme prodrug therapies (ADEPT and GDEPT). Anticancer Drug Des 1999;14:517–538.

519. Bagshawe KD, Sharma SK, Springer CJ, Rogers G. Antibody directed enzyme prodrug therapy (ADEPT). A review of some theoretical, experimental and clinical aspects. Anal Oncol 1994;5:879–891.

520. Mayer A, Francis RJ, Sharma SK, Tolner B, Springer CJ, Martin J, Boxer GM, Bell J, Green AJ, Hartley JA, Cruickshank C, Wren J, Chester KA, Begent RHJ. A Phase I study of single administration of antibody-directed enzyme prodrug therapy with the recombinant anti-carcinoembryonic antigen antibody-enzyme fusion protein MFECP1 and a bis-iodo phenol mustard prodrug. Clin Cancer Res 2006;12(21):6509–6516.

521. Smyth TP, O'Donnell ME, O'Connor MJ, St Ledger JO. β-Lactamase-dependent prodrugs-recent developments. Tetrahedron 2000;56: 5699–5707.

522. Kerr DE, Schreiber GJ, Vrudhula VM, Svensson HP, Hellstrom I, Hellstrom KE, Senter PD. Regressions and cures of melanoma xenografts following treatment with monoclonal antibody β-lactamase conjugates in combination with anticancer prodrugs. Cancer Res 1995;55: 3558–3563.

523. Svensson HP, Vrudhula VM, Emswiler JE, MacMaster JF, Cosand WL, Senter PD, Wallace PM. In vitro and in vivo activities of a doxorubicin prodrug in combination with monoclonal antibody β-lactamase conjugates. Cancer Res 1995;55:2357–2365.

524. Wang SM, Chern JW, Yeh MY, Ng JC, Tung E, Roffler SR. Specific activation of glucuronide prodrugs by antibody-targeted enzyme conjugates for cancer therapy. Cancer Res 1992;52:4484–4491.

525. Leenders RGG, Damen EWP, Bijsterveld EJA, Houba PHJ, van der Meulen-Muilman IH, Boven Haisma HJ. Novel anthracycline-spacer-β-glucuronide, -β-glucoside, and -β-galactoside prodrugs for application in selective chemotherapy. Bioorg Med Chem 1999;7: 1597–1610.

526. Haisma HJ, Boven E, van Muijen M, De Jong L, Van der Vigh WJ, Pinedo HM. A monoclonal antibody-β-glucuronidase conjugate as activator of the prodrug epirubicin-glucuronide for specific treatment of cancer. Br J Cancer 1992;66:474–478.

527. Haisma HJ, Van Muijen M, Pinedo HM, E Boven E. Comparison of two anthracycline-based prodrugs for activation by a monoclonal antibody-β-glucuronidase conjugate in the specific treatment of cancer. Cell Biophys 1994;24/25:185–192.

528. Houba PHJ, Boven E, Van Der Meulen-Muileman IH, Leenders RGG, Scheeren JW, Pinedo HM, Haisma HJ. Pronounced antitumor efficacy of doxorubicin when given as the prodrug DOX-GA3 in combination with a monoclonal antibody β-glucuronidase conjugate. Int J Cancer 2001;91:550–554.

529. Dorr RT, Briggs A, Kintzel P, Meyers R, Chow HHS, List A. Comparative pharmacokinetic study of high-dose etoposide and etoposide phosphate in patients with lymphoid malignancy receiving autologous stem cell transplantation. Bone Marrow Transpl 2003;31(8):643–649.

530. Collier K, Schink C, Young AM, How K, Seckl M, Savage P. Successful treatment with etoposide phosphate in patients with previous etoposide hypersensitivity. J Onc Pharm Pract 2008;14(1):51–55.

531. Rustin GJS, Galbraith SM, Anderson H, Stratford M, Folkes LK, Sena L, Gumbrell L, Price PM. Phase I clinical trial of weekly combretastatin A4 phosphate: clinical and pharmacokinetic results. J Clin Oncol 2003;21 (15):2815–2822.

532. Stevenson JP, Rosen M, Sun W, Gallagher M, Haller DG, Vaughn D, Giantonio B, Zimmer R, Petros WP, Stratford M, Chaplin D, Young SL, Schnall M, O'Dwyer PJ. Phase I trial of the antivascular agent combretastatin A4 phosphate on a 5-day schedule to patients with cancer: magnetic resonance imaging evidence for altered tumor blood flow. J Clin Oncol 2003;21(23):4428–4438.

533. Cao M-Y, Benatar T, Lee Y, Young AH, Yen Y, Wright JA. Recent progress in cancer gene therapy. Horiz. Cancer Res. 2005;20:27–73.

534. Wildner O. *In situ* use of suicide genes for therapy of brain tumours. Ann Med 1999;31:421–429.
535. Ishii-Morita H, Agbaria R, Mullen CA, Hirano H, Koeplin DA, Ram Z, Oldfield EH, Johns CG, Blaese RM. Mechanism of 'bystander effect' killing in the herpes simplex thymidine kinase gene therapy model of cancer treatment. Gene Ther 1997;4:244–251.
536. Mesnil M, Yamasaki H. Bystander effect in herpes simplex virus-thymidine kinase/ganciclovir cancer gene therapy: role of gap-junctional intercellular communication. Cancer Res 2000;60:3989–3999.
537. Kievit E, Bershad E, Ng E, Sethna P, Dev I, Lawrence TS, Rehemtulla A. Superiority of yeast over bacterial cytosine deaminase for enzyme/prodrug gene therapy in colon cancer xenografts. Cancer Res 1999;59:1417–1421.
538. Huber B, Austin E, Richards C, Davis S, Good S. Metabolism of 5-fluorocytosine to 5-fluorouracil in human colorectal tumor cells transduced with the cytosine deaminase gene: significant antitumor effects when only a small percentage of tumor cells express cytosine deaminase. Proc Natl Acad Sci USA 1994;91:8302–8306.
539. Leichman CG. Schedule dependency of 5-fluorouracil. Oncology 1999;13:26–32.
540. Jounaidi Y, Hecht JE, Waxman DJ. Retroviral transfer of human cytochrome P450 genes for oxazaphosphorine-based cancer gene therapy. Cancer Res 1998;58:4391–4401.
541. Anlezark GM, Melton RG, Sherwood RF, Coles B, Friedlos F, Knox RJ. The bioactivation of 5-(aziridin-1-yl)-2, 4-dinitrobenzamide (CB 1954). I. Purification and properties of a nitroreductase enzyme from *Escherichia coli*. A potential enzyme for antibody-directed enzyme prodrug therapy (ADEPT). Biochem Pharmacol 1992;44:2289–2295.
542. Grove JI, Searle PF, Weedon SJ, Green NK, McNeish LA, Kerr DJ. Virus-directed enzyme prodrug therapy using CB1954. Anticancer Drug Des 1999;14:461–472.
543. Fillat C, Carrio M, Cascante A, Sangro B. Suicide gene therapy mediated by the Herpes Simplex virus thymidine kinase gene/Ganciclovir system: fifteen years of application. Curr Gene Ther 2003;3(1):13–26.
544. Nicholas TW, Read SB, Burrows FJ, Kruse CA. Suicide gene therapy with herpes simplex virus thymidine kinase and ganciclovir is enhanced with connexins to improve gap junctions and bystander effects. Histol Histopath 2003;18(2):495–507.
545. Nasu Y, Saika T, Ebara S, Kusaka N, Kaku H, Abarzua F, Manabe D, Thompson TC, Kumon H. Suicide gene therapy with adenoviral delivery of HSV-tK gene for patients with local recurrence of prostate cancer after hormonal therapy. Molec Ther 2007;15(4):834–840.
546. Sterman DH, Recio A, Vachani A, Sun J, Cheung L, DeLong P, Amin KM, Litzky LA, Wilson JM, Kaiser LR, Albelda SM. Long-term follow-up of patients with malignant pleural mesothelioma receiving high-dose adenovirus herpes simplex thymidine kinase/ganciclovir suicide gene therapy. Clin Cancer Res 2005;11(20):7444–7453.
547. Cunningham C, Nemunaitis J. A phase I trial of genetically modified Salmonella typhimurium expressing cytosine deaminase (TAPET-CD, VNP20029) administered by intratumoral injection in combination with 5-fluorocytosine for patients with advanced or metastatic cancer. Hum Gene Ther 2001;12(12):1594–1596.
548. Chen CS, Jounaidi Y, Su T, Waxman DJ. Enhancement of intratumoral cyclophosphamide pharmacokinetics and antitumor activity in a P450 2B11-based cancer gene therapy model. Cancer Gene Ther 2007;14(12):935–944.
549. Braybrooke JP, Slade A, Deplanque G, Harrop R, Madhusudan S, Forster MD, Gibson R, Makris A, Talbot CD, Steiner J, White L, Kan O, Naylor S, Carroll MW, Kingsman SM, Harris AL. Phase I study of MetXia-P450 gene therapy and oral cyclophosphamide for patients with advanced breast cancer or melanoma. Clin Cancer Res 2005;11:1512–1520.
550. Roy P, Waxman DJ. Activation of oxazaphosphorines by cytochrome P450: application to gene-directed enzyme prodrug therapy for cancer. Toxicol *In Vitro* 2006;20:176–186.
551. Palmer DH, Mautner V, Mirza D, Oliff S, Gerritsen W, van der Sijp JRM, Hubscher S, Reynolds G, Bonney S, Rajaratnam R, Hull D, Horne M, Ellis J, Mountain A, Hill S, Harris PA, Searle PF, Young LS, James ND, Kerr DJ. Virus-directed enzyme prodrug therapy: intratumoral administration of a replication-deficient adenovirus encoding nitroreductase to patients with resectable liver cancer. J Clin Oncol 2004;22(9):1546–1552.

PARP INHIBITORS AS ANTICANCER AGENTS

GABRIELE COSTANTINO[1]
ROBERTO PELLICCIARI[2]

[1] Dipartimento Farmaceutico,
Università degli Studi di Parma,
Via GP Usberti, Parma, Italy

[2] Dipartimento di Chimica e
Tecnologia del Farmaco,
Università degli Studi di Perugia,
Via del Liceo, Perugia, Italy

1. INTRODUCTION

Poly(ADP-ribosyl)ation is a posttranslational modification resulting in a unique pattern of covalent alterations which significantly impact the functions of target proteins, among which several nuclear proteins deeply involved in the control of genomic integrity and in gene transcription.

The first evidence on a poly(ADP-ribosyl)ating activity in cells date back to 1963, when the existence of a DNA-dependent polyadenylic acid-synthesizing enzyme was reported [1]. Subsequently, the structure of the product (poly(ADP-ribose) (PAR)) of such enzymatic activity was elucidated. The discovery of an enzyme, known today as poly(ADP-ribose)glycohydrolase (PARG), involved in the degradation of poly(ADP-ribose) brought additional interest in the study of the metabolism of poly(ADP-ribose), until the mid-1980s, when, also with the availability of the first small molecule inhibitors, the involvement of poly(ADP-ribosyl)ation in the mechanisms of DNA repair was identified, thus opening the way to a potential application into cancer (as well as other diseases) therapy.

The enzyme responsible for poly(ADP-ribosyl)ation in eukaryotes, identified as poly(ADP-ribose)polymerase (PARP, PARP-1 in the current nomenclature), was first biochemically characterized and, then, cloned in 1987 [2]. The availability of the recombinant protein from various species has allowed, in the 1990s, to gain insights into the three-dimensional structure of the PARP-1, thus helping the structure-based design of chemically diverse classes of inhibitors. Coupled with the information gathered from studies on transgenic animals, this availability has allowed to define more precisely the druggability of PARP and to extend the scope of its modulation in cancer as well as other pathological conditions. The last decade, finally, has seen the growth of the so-called PARP family, as up to 17 proteins have been identified to be endowed with a poly(ADP-ribose)polymerase activity [3]. Some of these proteins, including the most abundant and most studied PARP-1, are activated by DNA break, some others are induced by different stimuli, thus leaving still open the nature of their precise involvement in the control of genomic integrity.

In this chapter, we introduce the heterogeneity of the PARP family, discuss the biochemical grounds of PAR metabolism, and then we focus on the potential use of PARP inhibitors in anticancer therapy.

2. BIOCHEMISTRY OF POLY(ADP-RIBOSYL)ATION: NAD^+, PARPs, AND PARG

PARP-1, one of the key components of the enzymatic machinery involved in DNA repair, is activated by DNA-strand breaks and participates in the base-excision repair (BER) pathway [4]. As a result of the DNA damage, chromatin-bound PARP-1 starts transferring multiple units of ADP-ribose (ADR), up to 200, from NAD^+ to several acceptor proteins, including DNA polymerases, endonucleases, histones, and PARP itself, to generate a poly(ADP-ribosyl)ated protein (Fig. 1).

The mechanistic meaning of such an intensive covalent modification resides in the fact that poly(ADP-ribosyl)ation brings in physical proximity an amount of negatively charged residues (the phosphate groups of ADR). This causes electrostatic repulsion between the poly(ADR-ribosyl)ated nuclear proteins and the phosphate groups of DNA, an event which significantly alters the overall architecture of the DNA-nuclear protein complex thus inducing the opening of the condensed form of chromatin, and facilitating the intervention of DNA-repairing enzymes (Fig. 2).

The generation of poly(ADP-ribosyl) conjugates is a biochemical event, complex in nat-

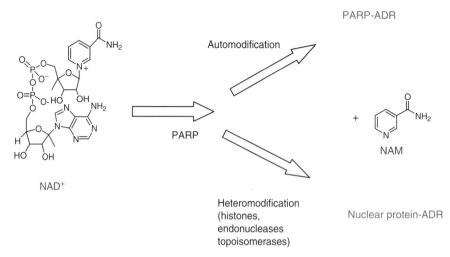

Figure 1. PARPs transfer multiple units of ADP-ribose to acceptor proteins, including PARP itself, and a number of chromatin bound protein. The covalent modification releases nicotinamide (NAM).

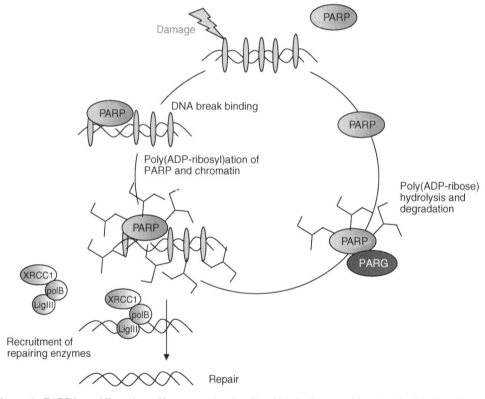

Figure 2. PARP is rapidly activated by genotoxic stimuli and binds the strand break. After binding, it starts poly(ADP-ribosyl)ating chromatin bound proteins, including PARP itself. The massive accumulation of negative charges causes chromatin relaxation, and allows for recruitment of repairing enzymes, such as XRCC1, Polβ, and LigIII. After repairing, poly(ADP-ribose) polymers are metabolized by PARG.

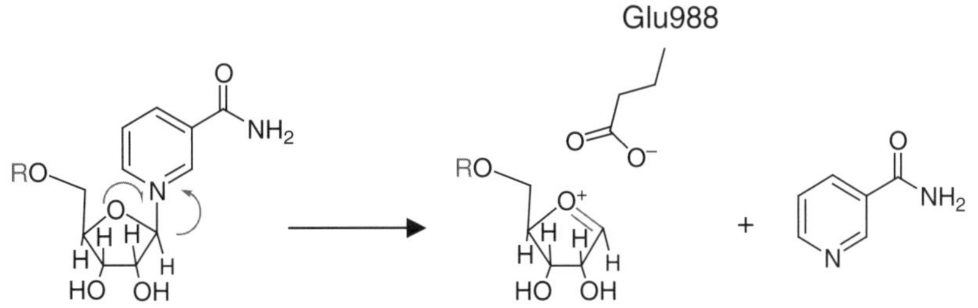

Figure 3. Catalytic mechanism of PARP. Also with the help of Glu988 as general base, there is the formation of the strong electrophile oxacarbenium ion intermediate that is readily transferred to a nucleophile of the acceptor protein.

ure, whose anabolic and metabolic steps should be examined in details for a comprehension of the role of PARPs (and PARP inhibitors) in cell death or survival.

The covalent poly(ADP-ribosyl)ation of nuclear proteins requires in fact three formally different steps, which can be enunciated as (i) hydrolysis of NAD^+ and transferring of a unit of ADP-ribose to the target acceptor; (ii) elongation of the poly(APD-ribose) chain; and (iii) branching of the poly(ADP-ribose) chain [5]. PARP-1 owes all the three enzymatic activities, while the ability of other enzymes of the PARP family to accomplish the whole reaction is not yet fully verified.

In the first step of the process, PARP-1 recognizes NAD^+, which binds the PARP-1 catalytic site, comprised of a nicotinamide binding pocket, a phosphate binding channel, and a nucleoside binding pocket (see below). The PARP-bound NAD^+ adopts a 3'-endo conformation, which is supposed to help the subsequent reaction by relieving steric stress. The catalytic processing of NAD^+ by PARP-1 releases nicotinamide (NAM) and generates a reactive ADP-ribose oxacarbenium ion intermediate, which is highly susceptible of nucleophilic attack. The oxacarbenium ion is transiently stabilized by Glu988, a catalytically relevant residue in the PARP catalytic site (Fig. 3).

The most common nucleophilic residue that quenches the oxacarbenium ion is a glutamic acid residue, but it can also be an aspartic acid residue. In general, the nucleophilic glutamic acid can be located in the so-called PARP automodification domain (to give automodified poly(ADP-ribosyl)ated PARP) or on the surface of the acceptor protein (Fig. 4).

The second step of the reaction is the elongation of the poly(ADP-ribose) polymer. This is accomplished by the formation of an α glicosidic linkage between the mono(ADP-ribose)conjugate and an incoming NAD^+ molecule. The reaction is depicted in Fig. 5, and involves the nucleophilic attack of the 2'-hydroxy group of the ribose moiety of the acceptor unity to the α position of the ribose moiety of the donor NAD^+ molecule. The 2'-hydroxy group is significantly less nucleophilic of the carboxylate group of a glutamic (or aspartic) acid residue, hence the elongation reaction requires a deprotonating base, which is again identified in the Glu988 residue. The same residue also promotes the reaction by polarizing the donor molecule through a hydrogen bonding to the 2'-hydroxy group of NAD^+.

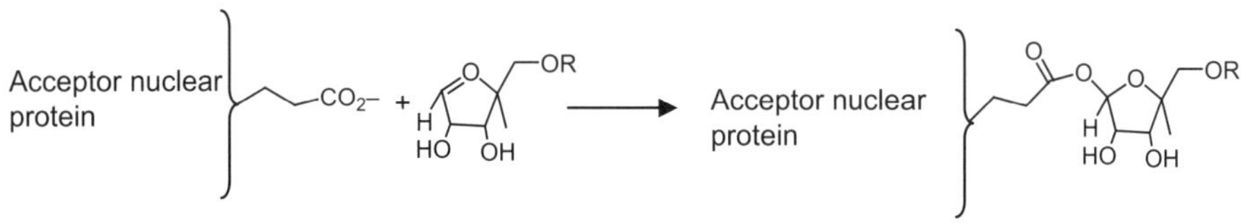

Figure 4. First step of poly(ADP-ribosyl)ation. The transfer of the oxacarbenium ion to a glutamic acid residues of the acceptor protein.

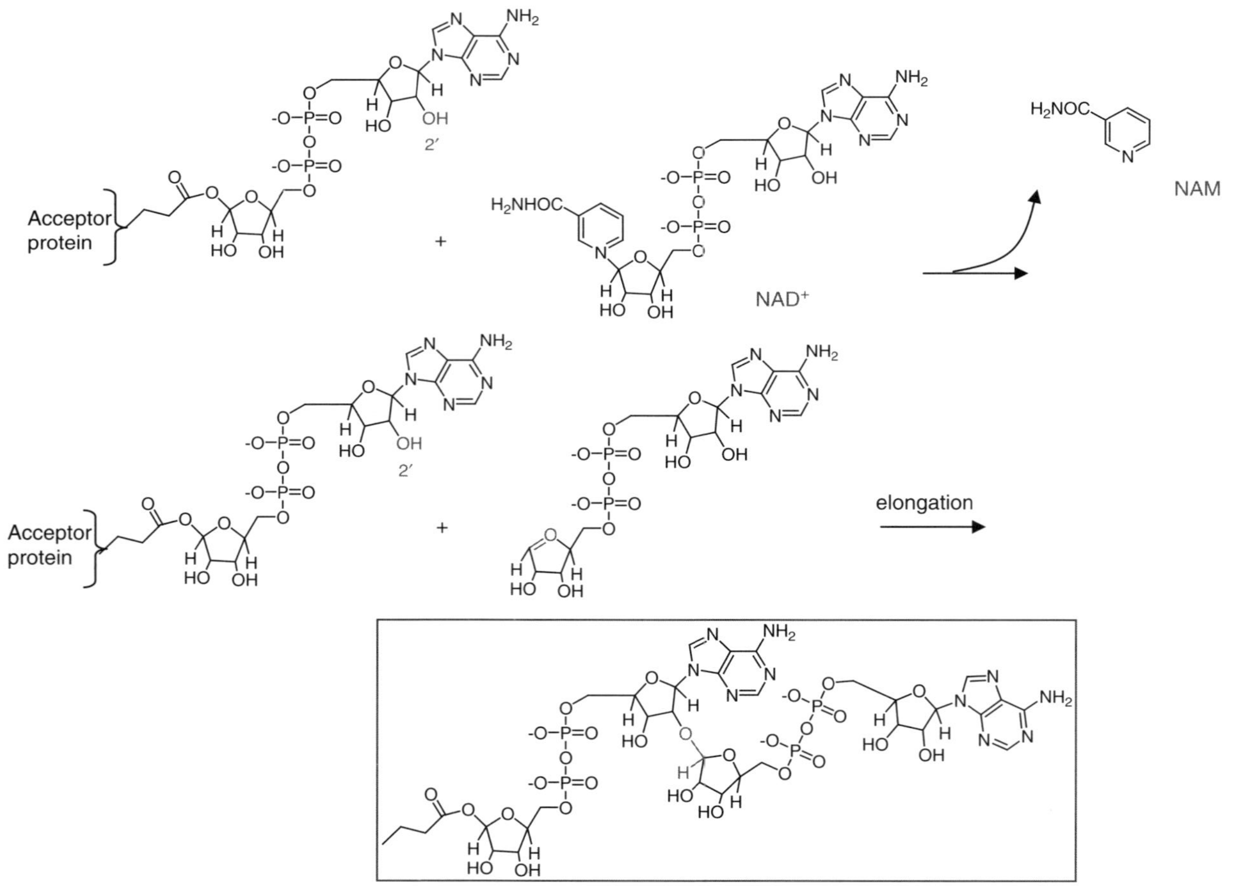

Figure 5. Second step of poly(ADP-ribosyl)ation, the elongation. An α-glicosidic bond is formed between two ribose units, releasing one molecule of nicotinamide (NAM).

The third step of the anabolic reaction of poly(ADP-ribosyl)ation of the acceptor protein involves the branching of the polymer. The branching reaction is thought to result from an inverted (with respect to elongation) orientation of the acceptor protein into the PARP catalytic site, which is broad enough to accommodate the substrate in two 180°-rotated orientations. The inverted orientation causes the α glicosidic linkage to occur at nicotinamide ribose, rather than at the adenine ribose, thus producing an $\alpha(1 \rightarrow 2)$ glycosidic linkage (Fig. 6).

The formation of poly(ADP-ribosyl)ated proteins is strongly induced by genotoxic stimuli, which increase the PARP activity and the amount of synthesized polymers of up to 500-fold over basal [6]. Nevertheless, poly (ADP-ribosyl)ated polymers are only transiently detectable in living cells under genotoxic stimuli, as their half life is below 1 min [7]. The degradation of the polymers is accomplished by two activities. The first one is a PARG activity [8,9], the second one is an ADP-ribosyl protein lyase activity [10]. The PARG activity is responsible for the cleavage of the glycosidic bonds between ADP-ribose units, while the lyase activity cleaves the "first" unit from the protein.

One of the most biochemically relevant consequences of PARP activation is the depletion of the cellular store of NAD^+ [11]. Indeed, NAD^+, a crucial cofactor in the energy metabolism, is involved in the ATP synthesis and in the regulation of the redox potential of cells. Poly(ADP-ribosyl)ation is the major determinant for NAD^+ catabolism, and indeed the NAD^+ levels in a cellular context are regulated by PARP activity. It has largely been demonstrated that genotoxic damage, and the consequent PARP activation, rapidly and massively depletes NAD^+ *in vivo* [12,13]. The NAD^+ depletion is associated with a reduction of the ATP levels. Two factors lead to ATP consumption. The first one is related to the stoichiometry of NAD^+ resynthesis, which requires at least 2 mol of ATP per mol of NAD^+ [14,15]. Thus, in conditions of genomic stress, the cell starts consuming ATP in the attempt of resynthesizing the NAD^+ processed by poly(ADP-ribosyl)ation. The second path leading to ATP consumption is related to the impairment of the ATP-synthesizing branch of the anaerobic glycolysis, which ultimately leads to piruvate synthesis [16] (Fig. 7).

The ATP depletion after genotoxic stimuli and subsequent PARP activation has important consequence for the cellular vitality, and it has been instrumental to depict the so called "suicide hypothesis" of PARP activation [17,18], which has driven much of the research in the field of PARP inhibitors over the last decade (Fig. 8).

In its simplest representation, the suicide hypothesis of PARP activation is related to the experimental evidence that the amount of consumed NAD^+ (and hence ATP) is proportional to the extent of the genomic damage. Thus, the hypothesis involves a scenario in which the extent of genotoxic damage defines the fate of the cell. If the damage is mild and thus recoverable, the reduction in the NAD^+ levels reaches about 70% allowing the preservation of the cell energetics and, at the same time, a level of poly(ADP-ribosyl)ation of nuclear proteins which is compatible with chromatin opening and intervention of the DNA repairing enzymes [19].

If the damage is too high, and hence the risk of unrecoverable genomic instability unacceptable, the depletion of NAD^+ is complete, and the cell enters into an energy crisis related to the consumption of ATP [20]. In this context, PARP can be seen as a housekeeper of the genomic stability: if the damage is recoverable, its activation coordinates the intervention of the cell's repairing machinery; in the case of unrecoverable DNA stress, the PARP helps the cell to commit suicide, thus avoiding the risk of hand down an unstable genome. The suicide hypothesis of PARP activation displays some inconsistencies, which are mainly related to the existence of a family of poly(ADP-ribosyl)ating proteins, and the hypothesis has been revisited and flanked by complementary models. But at its most fundamental terms, it elegantly provides a theoretical and experimental grounds to discuss the potential therapeutic relevance of PARP inhibition.

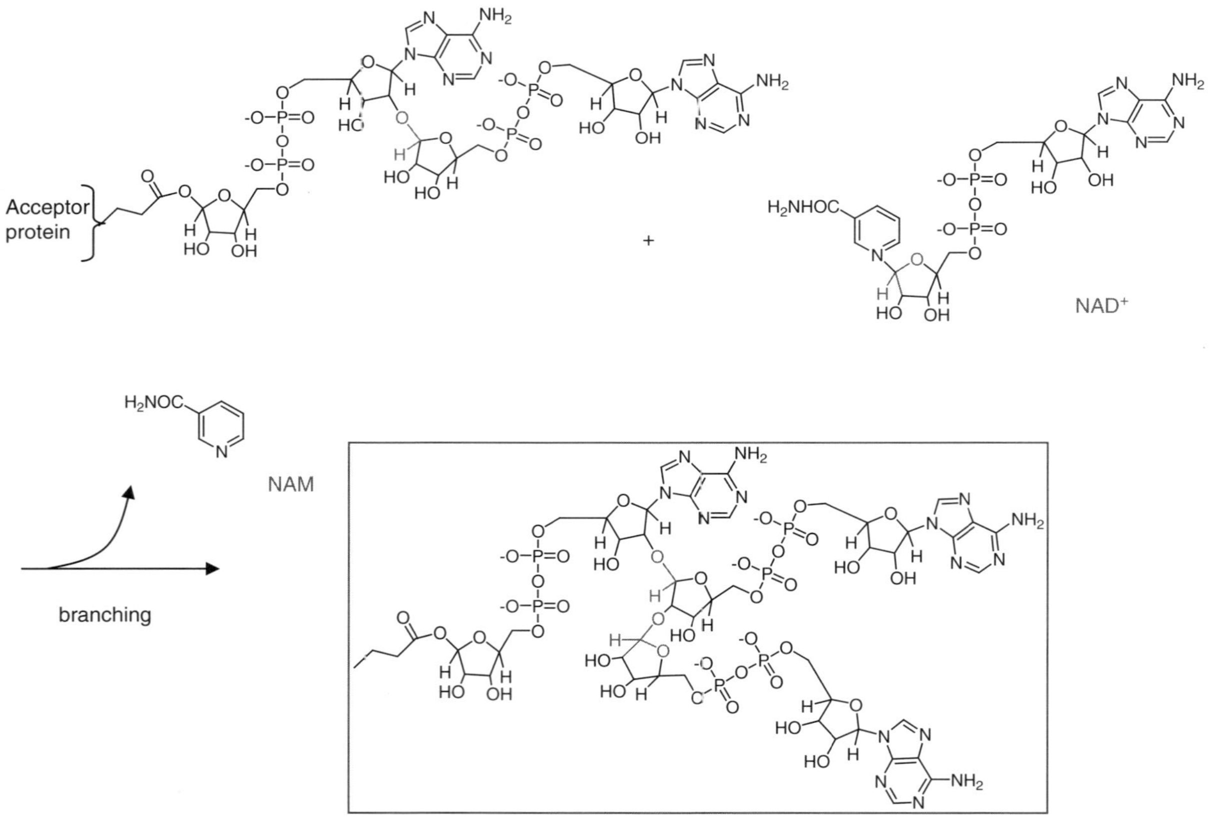

Figure 6. Third step of poly(ADP-ribosyl)ation, the branching.

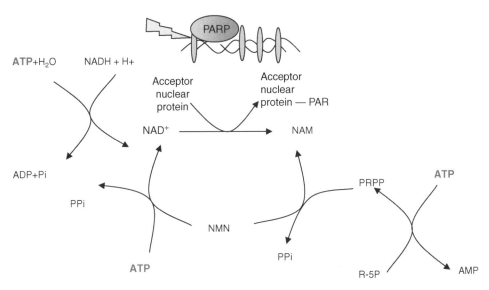

Figure 7. Activation of PARP causes depletion of the ATP stores.

3. MOLECULAR BIOLOGY OF THE PARP FAMILY

Poly(ADP-ribosyl)ation is a posttranslation modification employed by all the eukaryote organisms, with the notable exception of yeasts. A portion of the PARP catalytic fragment is entirely conserved among species [21], and has therefore been defined as the "PARP-signature." The PARP signature has been

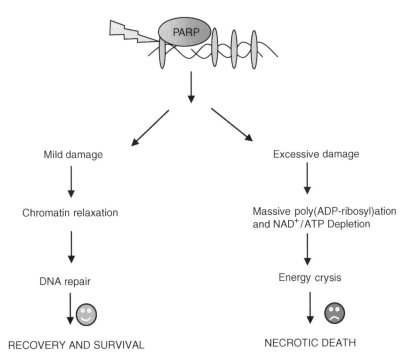

Figure 8. Schematic representation of the "suicide hypothesis" of the PARP activation.

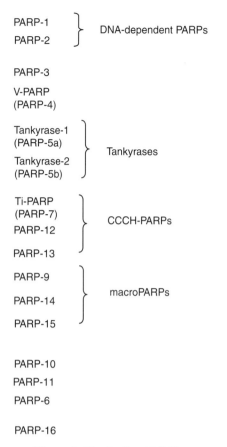

Figure 9. The family of PARPs.

employed to extensively screen the nonredundant NCBI database thus unveiling up to 17 genes that may be endowed with poly(ADP-ribosyl)ating activity [22]. Some of them have been cloned and expressed, and their potential physiological role investigated. For some others, the poly(ADP-ribosyl)ating activity, if present, is still at speculative levels.

In addition to the founding member of the superfamily (Fig. 9), now referred to as PARP-1 [23–25], a second PARP enzyme (called PARP-2) was identified after the unexpected observation of residual poly(ADP-ribosyl)ating activity in embryonic fibroblasts from PARP-1 knocked out mice [26]. PARP-2 shares a high similarity with PARP-1 in the catalytic domain, a feature that is reflected by the very similar three-dimensional structure of the two domains. PARP-2 has also a DNA binding domain and retains the ability to modify chromatin upon DNA damage, although with a different substrate specificity of PARP-1 [27,28]. The definition of the precise physiological role of PARP-2 still awaits for further clarification although it can be anticipated that PARP-1 and PARP-2 share complementary but partially overlapping function, a feature that must be carefully taken into account when interpreting data from genetically modified animals, or from enzyme inhibition studies with small molecule inhibitors.

A third PARP protein, termed PARP-3 [29], was identified in the core of the centrosome. PARP-3 shows sequence identity with PARP-1 and PARP-2 at the catalytic fragment level, and, thus, owes poly(ADP-ribosyl)ating properties that can be blocked by unselective PARP inhibitors as well as by selective PARP-3 ligands. The nuclear localization and the associated poly(ADP-ribose)polymerase activity suggests an involvement of PARP-3 in transcriptional silencing and in the cellular response after DNA damage.

V-PARP (or PARP-4) stands for vault-PARP, the largest protein in the superfamily, that has been identified in the vault particles, a cytoplasmic ribonucleoprotein complex of still unknown function [30]. Despite a peculiar domain arrangement, VPARP has a PARP domain, which retains poly(ADP-ribose)polymerase activity.

Other two members of the superfamily are tankyrases (tankyrase and tankyrase-2). Tankyrase stands for TRF1-interacting, ankyrin-related ADP-ribose polymerase and, as above, it contains a PARP domain with catalytic activity [31,32]). The role of tankyrases is to promote the segregation of cohesion complexes of telomeres during anaphase. Indeed, silencing of tankyrase by siRNA blocks the cell cycle in anaphase.

In addition to the above-described PARP family members, whose functions have been partially elucidated in vitro and in vivo, there are up to other 11 proteins that owe a PARP signature domain [33]. These proteins, termed from PARP-7 to PARP-15, are of still unknown function, and is intriguing to observe that their PARP domain is coupled to very different other domains, which include for example the zinc finger domain, the sterile alpha motif

(SAM) domain, the BRCT domain and many others. This puts forward the intriguing possibility that these domains target the poly (ADP-ribose)polymerase activity to a variety of functionally diverse acceptors, thus opening the way for an extension of the scope of poly (ADP-ribosyl)ation in the cellular context.

4. FOCUS ON PARP-1: DOMAIN ORGANIZATION AND STRUCTURE OF THE CATALYTIC SITE

PARP-1 is a modular enzyme containing, in the human ortholog, 1014 residues (Fig. 10). The N-terminal portion of the sequence (residues 1–206) contains the DNA binding domain (DBD) which, in turn, is composed of a zinc finger module, and a nuclear localization signal (NLS). The central domain (residues 383–476) is the automodification domain which contains the evolutionary conserved BCRT module (carboxyl-terminal domain of the breast cancer gene) [34,35]. The carboxy terminus (residues 524–1014) of the sequence contains the catalytic domain, which encompasses the so-called minimal catalytic fragment and the PARP-signature (residues 859–908), 100% conserved in all the orthologs [36].

The catalytic fragment of PARP-1 is very well characterized from a structural point of view, as several X-ray structures of mammalian and chicken PARPs, also in complex with inhibitors, are available in the RSCB database [5,37–40]. The catalytic fragment is composed by two parts, the purely α-helix N-terminal portion (residues 662–784) and the C-terminal portion (785–1014), which is composed by a five-stranded antiparallel β-sheet and a four-stranded mixed β-sheet. These two sheets are connected via a single pair of hydrogen bonds between strands c and d. The nicotinamide binding pocket of NAD^+ is located in the C-terminal module, and discussing the way in which nicotinamide is bound to the enzyme is of particular relevance for the discussion of the structure–activity relationships of PARP inhibitors (see below). Thus, the inspection of inhibitor-bound structures of the catalytic fragment of PARP-1 indicates the

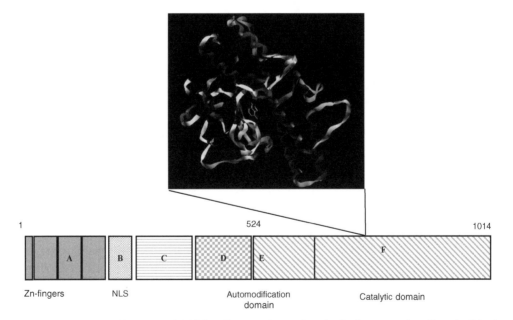

Figure 10. Modular architecture of PARP-1. Domain A contains the Zn-finger portion. Domain B is the nuclear localization signal; domain C is a link segment; domain D is the automodification domain, which includes the BRCT subdomain. Domain E and F constitute the catalytic fragment, containing the PARP signature (residues 859–908). Several X-ray structures of the catalytic fragment are available.

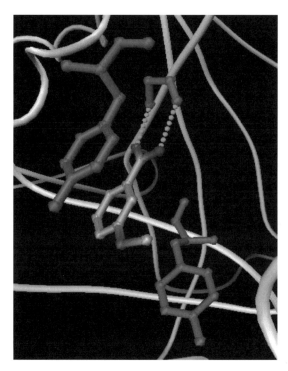

Figure 11. Nicotinamide binding pocket of the catalytic fragment of PARP-1, complexed with 3-methoxy benzamide. The primary interactions are shown. The amido moiety is hydrogen bonded to Glu863, and the aromatic ring is sandwiched by π–π interaction between Tyr907 and Tyr896.

binding is stabilized by two hydrogen bonds, involving the backbone atoms of Gly863. An additional hydrogen bonding can be formed between the amide oxygen of the inhibitors and the side chain of Ser904. The aromatic ring of the nicotinamide moiety is, furthermore, sandwiched by Tyr907 and Tyr896 in a face-to-face and hedge-to-face π–π interaction, respectively. This mode of binding has been inferred by the analysis of several inhibitor-bound structures, with nicotinamide-based inhibitors (Fig. 11).

It should also be mentioned that, at the moment of writing this chapter, the 3D structure of the catalytic fragment of PARP-2, PARP-3, PARP-10, PARP-14, and PARP-15 were also available [40], displaying an overall fold and a specific recognition motif similar to that of PARP-1. The comparative analysis of these structures is expected to shed light on molecular requirements for selective isoform inhibition.

5. PARP AND CANCER

PARP-1 is one of the major components of the constitutive DNA repairing machinery in eukaryotic cells. PARP-1 is activated by, and binds to, DNA-strand breaks thus initiating the cascade of events leading to DNA repairing or, alternatively, necrotic death of the injured cell [41,42].

This scenario has been validated by a variety of experiments *in vitro* and *in vivo* demonstrating the increased susceptibility to DNA damaging agents in PARP inhibited or shutdown systems. In particular, PARP-1$^{-/-}$ mouse fibroblast embryonic cells and PARP-1$^{-/-}$ mice show markedly increased response to ionizing radiations as well as to alkylating chemotherapeutics [43,44]. Incidentally, the PARP-1$^{-/-}$ mice also showed an unexpected resistance to brain ischemia [45], an observation that opened the way to studies aimed at verifying the potential therapeutic significance of PARP

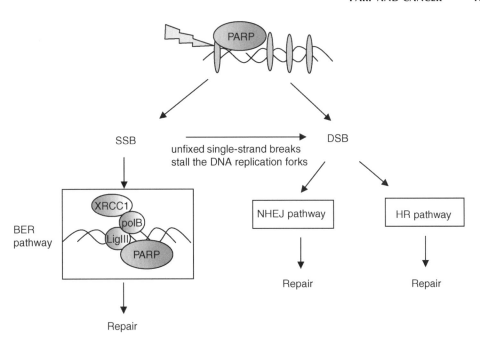

Figure 12. PARP-1 is involved in single-strand break (SSB) repair through the BER pathway. If unrepaired, SSB can degenerate into double-strand breaks (DSBs), which are repaired by the NHEJ or HR pathways.

inhibitors in brain (as well as other organs) ischemia [46].

The mechanism by which PARP-1 cooperates in DNA repairing is intimately connected to its ability to sense, and to bind to, DNA breaks thus operating chromatin poly (ADP-ribosyl)ation and promoting the recruitment of repairing enzymes Fig (12).

In the case of single-strand breaks, the involvement of PARP in the BER is well understood [47]. The transient poly(ADP-ribosyl)ation of histone proteins causes a local opening of the compacted structure of chromatin, thus allowing the recruitment of the complex cascade of repair proteins. The hydrolysis of poly(ADP-ribosyl)ated proteins by PARG and lyases allows for a fine tuned control of the repairing events. The presence of poly(ADP-ribose)polymers, including poly (ADP-ribosyl)ated PARPs is furthermore necessary to recruit XRCC1 (X-ray repairing cross-complement 1), DNA ligase III, and DNA polymerase β to the site of the single-strand break [48].

Double-strand DNA breaks originates from ionizing agents, chemotherapeutic treatments, and also by unfixed single-strand breaks that stall the DNA replication forks at the site of the break [49]. The nonhomologous end-joining (NHEJ) and homologous recombination (HR) pathways are involved in the repairing of double-strand breaks, and defects in these pathways cause genomic instability and carcinogenesis [50,51]. Of particular interest is the notion that specific genetic backgrounds are particularly sensitive to pharmacological or molecular PARP (PARP-1 and PARP-2) silencing [52]. Thus, PARP-1 inhibitors kill potently cells lacking the tumor suppressor proteins BRCA1 and BRCA2. PARP inhibition in these tumor cells with deficient homologous-recombination repair generates unrepaired DNA single-strand breaks that are likely to cause the accumulation of DNA double-strand breaks and collapsed replication forks [53–55]. The therapeutic relevance in a number of tumors, including breast cancer, ovarian cancer, and prostate cancer in men, will be discussed in Section 8.

It must be mentioned that the implication of poly(ADP-ribosyl)ation on tumorogenesis is likely to go beyond its role in DNA-strand repair. For instances, it has been suggested that PARP-1 and p53 may participate in some common pathways related to genomic stabilization. Thus, p53 operates cell cycle arrest and promote apoptosis in response to DNA damage, thereby preventing accumulation and propagation of altered DNA. The notion that p53 is poly(ADP-ribosyl)ated by PARP-1, and that PARP-1 inhibition or genetic shutdown impact the p53 pathway after DNA damage, puts forward the possibility that p53 and PARP-1 cooperate in suppressing tumorogenesis by genome stabilization.

Poly(ADP-ribosyl)ation is emerging as an important player in the control of protein expression at the transcriptional level, and PARP acts both as negative [56] or positive [57] regulators of the transcription. The first evidence of the link between PARP and transcription came from the observation, in 1998, that NF-κB is an activator of the expression of iNOS and that inhibition of the induction of iNOS by PARP is mediated by NF-κB [58].

Most of the data now available on the involvement of poly(ADP-ribosyl)ation in tumorogenesis and genome instability come from the observation of genetically modified animals [59]. Despite some initial controversy in the data interpretation, there is now a large consensus in recognizing an increased susceptibility to DNA damaging agents in PARP-1$^{-/-}$ modified animals. It should be noticed, however, that not all the carcinogens are responsive to the status of PARP. Susceptibility to carcinogenesis might thus be explained by the involvement of PARP-1 in the repair pathway for BER, but not for NER.

6. PARP INHIBITORS

As described above, PARP-1 is endowed with a modular architecture that, in principle, offers several sites for inhibition or modulation through small molecules. However, it is the C-terminal catalytic fragment of the protein, which has attracted most of interest in medicinal chemistry. The C-terminal catalytic fragment of PARP-1 recognizes NAD$^+$, which is therefore the obvious template for the design of competitive inhibitors. Although all the attempts to cocrystallize NAD$^+$ together with PARP have failed, the NAD$^+$ binding pocket could be inferred from homology with the NAD$^+$ binding site of diphtheria toxin [60], and subsequently validated by experiments of random point site mutagenesis and photo-labeling [61,62]. The crystal structure of PARP-1 cocrystallized with carbaNAD, a nonhydrolyzable analog of NAD$^+$ provided further evidence on the mode of binding of NAD$^+$ into the catalytic fragment of PARP-1 [39]. According to these information, NAD$^+$ interacts with the catalytic fragment by filling three subsites or pockets, which are conventionally defined as (i) nicotinamide binding pocket; (ii) ribose binding pocket; and (iii) nucleoside (or adenosine) binding pocket (Fig. 13).

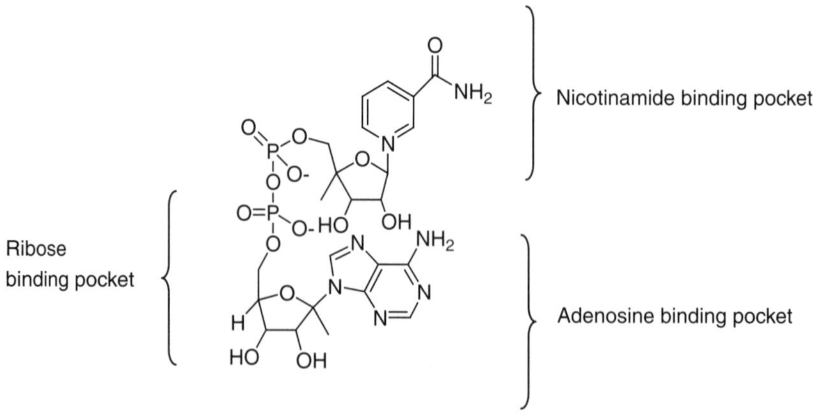

Figure 13. The three binding sites for NAD$^+$.

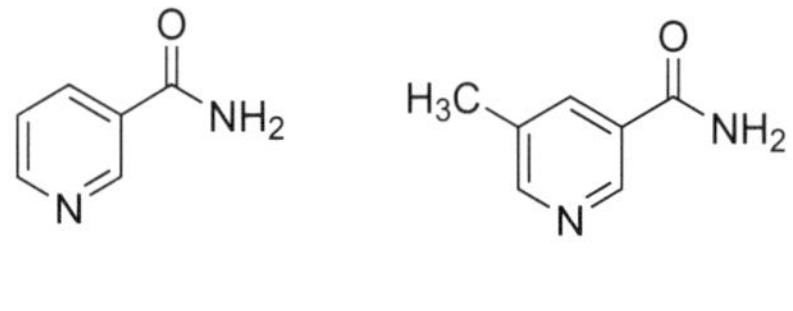

Scheme 1.

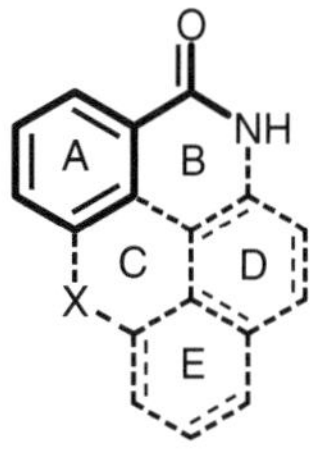

Figure 14. General pharmacophore model for PARP inhibition.

Inhibitors directed to one of the three pockets are expected to behave as NAD^+ competitive PARP inhibitors, and in fact examples of the three classes have been reported in the literature. Nevertheless, most of the work in the field of PARP competitive inhibitors has been devoted to the design and characterization of ligands directed to the so-called nicotinamide binding pocket.

In 1971, Clark et al. first described nicotinamide (1) (Scheme 1) itself, and its 5-methyl derivative (2) as inhibitors of poly(ADP-ribose) formation after DNA damage in cells [63].

While this is clearly consistent with the notion that the product of the enzymatic reaction (nicotinamide) may regulate through a negative feedback control the reaction itself, it is worth mentioning that this first observation opened the way to a systematic investigation of the structural elements required for the interaction with the nicotinamide pocket of the catalytic fragment of PARP, an effort that is culminated with the elaboration of a pharmacophore model of nicotinamide-based PARP inhibitors, which describes most of the structure–activity relationship up to now available.

Thus, also in agreement with available data derived from the inspection of enzyme–inhibitors X-ray complexes, a general pharmacophore for the nicotinamide binding pocket of PARP can be represented as composed by an aromatic ring connected by an amide group in an *anti* conformation [64,65] (Fig. 14).

The amide can be inserted into a saturated or unsaturated six- or five-membered ring, provided that the *anti* disposition is maintained. Additional ring systems can further decorate this scheme, giving rise to polycyclic inhibitors. Indeed, all the currently known nicotinamide-based inhibitors fit the above-described pharmacophore and can be classified as (i) primary amide derivatives and (ii) polycyclic amide derivatives.

Primary amides were the first to be investigated following the discovery of nicotinamide as a low-potency PARP inhibitors. Incidentally, it should be mentioned that in view of its action as vitamin (vitamin B3, niacin), nicotinamide (1) has gathered interest for its *in vivo* properties as PARP inhibitors [66]. The low potency (>30 µM) and the very low selectivity, however, have prevented its further development. A close bioisoster of nicotinamide, namely, benzamide (3) (Scheme 2) was described as a PARP inhibitor in 1975 [67]. Since then, the benzamide nucleus has been instrumental in the preparation of several chemotypes as potential PARP inhibitors, an effort culminated by the publication by Banasik et al., in 1992, of a throughout structure–activity relationship study [68]. In details, it turned out that the benzamide scaffolds can be substituted, and that the 3-position is the most

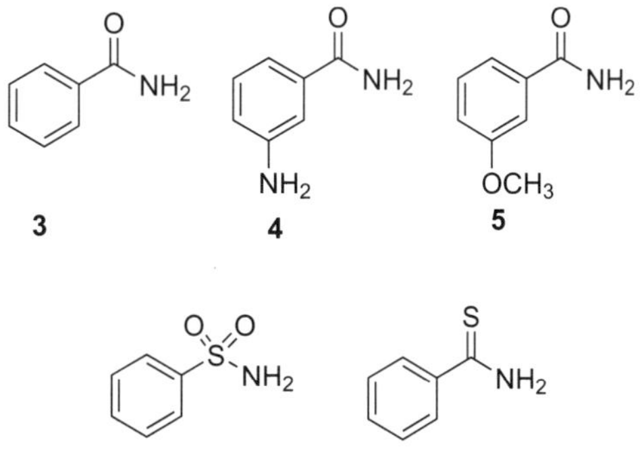

Scheme 2.

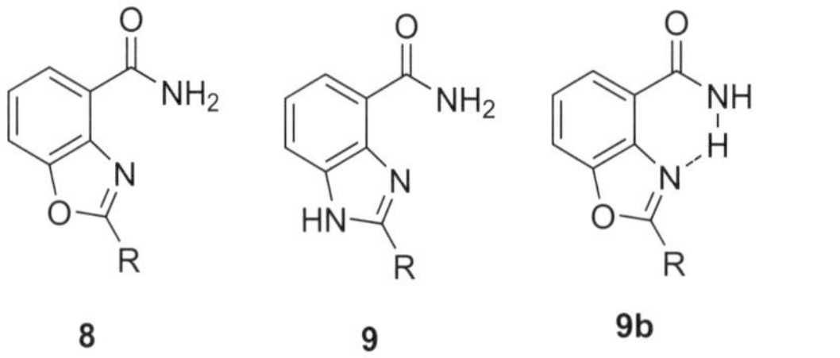

Scheme 3.

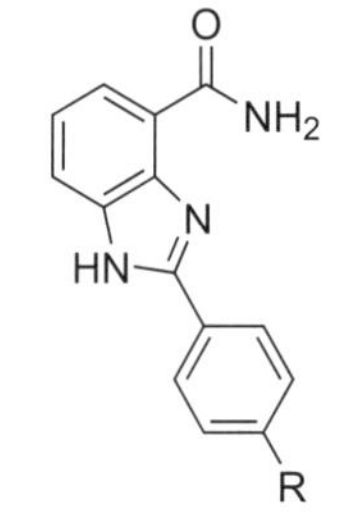

Scheme 4.

productive one, being 3-amino benzamide (**4**) and 3-methoxybenzamide (**5**) still used as pharmacological tools in *in vitro* experiments [69–72]. The amido moiety is less prone to modification, as the solfonamide (**6**) or the thioamide (**7**) are essentially inactive [73].

Retrospectively, this can be easily interpreted on the basis of the crystal structure of the catalytic fragment of PARP in complex with benzamide inhibitors, where it clearly appears that the amido moiety of the inhibitors is involved in a bidirectional hydrogen bonding with the backbone of Gly863. A particularly productive pattern of decoration on the benzamide scaffold turned out to be the 2,3-substitution with (substituted) heterocycle rings. Thus, benzoxazole or benzimidazole derivatives(**8**, **9**) (Scheme 3) were reported by Griffin et al. in 1995 [74] and have shown moderate to high potency as PARP-1 inhibitors.

A particularly interesting feature of these derivatives is that the heterocyclic iminonitrogen of the benzoxazole derivative or of the benzimidazole derivatives in the correct tautomeric form, may be involved in a intramolecular hydrogen bond with the amino group of the side-chain amide, thus forming a highly stabilized six-membered pseudo-ring (**9b**) [75]. This observation further confirm the need for an antidisposition of the side-chain amido group. Benzoxazole and, in particular, benzimidazoles offer additional potential for further decoration. Indeed, 2-arylbenzimidazole-4-carboxyamides (Scheme 4) are potent inhibitors of recombinant PARP-1, and the 2-aryl moiety can be further substituted thus allowing the achievement of a fine tuning of potency and pharmacokinetic properties. X-ray crystallography has allowed to elucidate the mode of binding of substituted 2-benzoimidazoles, where the 2-aryl system is project toward a solvent filled crevice of the enzyme, while the carboxamido group retains the interaction with Gly863 [76]. This mode of binding is shared by other chemotypes, as it is discussed below.

The second large groups of inhibitors falls into the class of cyclic amides. Thus, in 1991, Suto et al. were the first to investigate the relationship between conformation of the side-chain amide and PARP inhibition [76]. Thus, by using the rigid analog approach, they synthesized and tested 5- or 7-substituted dihydroisoquinolinones as potential PARP inhibitors (Scheme 5). As it is apparent, the 5-substituted derivatives are rigid analogs of 3-substituted benzamides in an *anti* conformation, while the 7-substituted compounds are analogs of the *syn*-oriented benzamides.

The activity is confined to the 5-substituted analogs, thus confirming again the need for an *anti* disposition of the side chain. The same

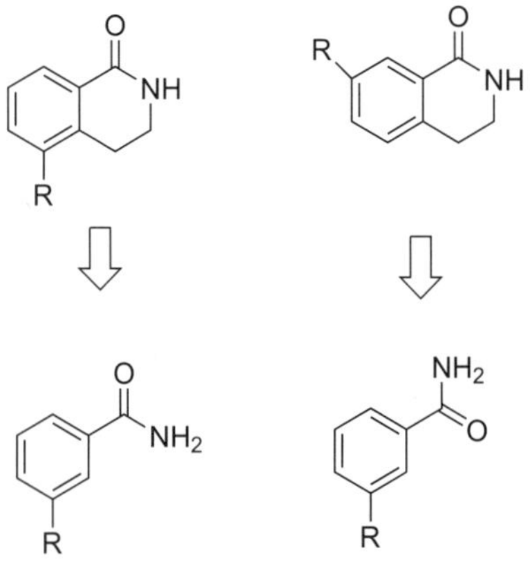

Scheme 5.

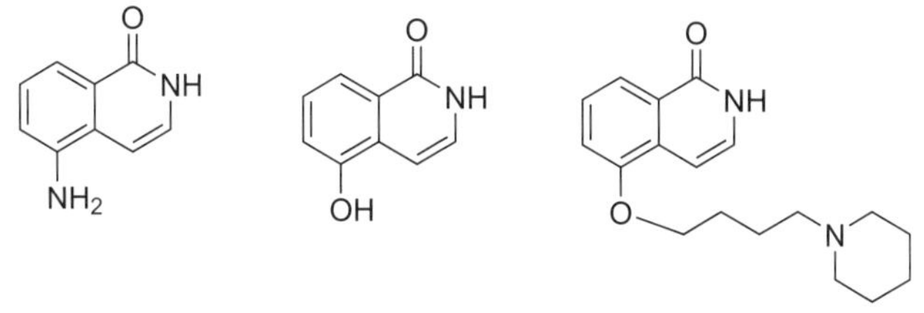

Scheme 6.

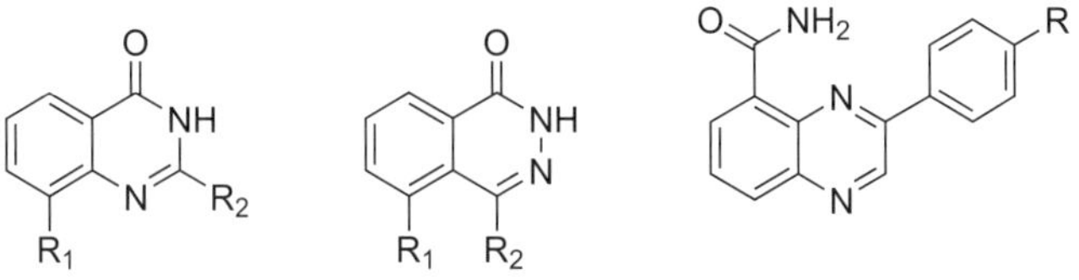

Scheme 7.

trend was obtained with a series of 5-, and 7-substituted isoquinolinones. The potential embedded into the dihydroisoquinolinone and the isoquinolinone scaffolds was further and productively exploited by decoration of both ring systems, with the aim of optimizing both the potency and the pharmacokinetic properties. Thus, the 5-amino-isoquinolinone (AIQ, **10**) (Scheme 6) derivative is endowed with high potency as PARP-1 inhibitor and achieves a reasonable water solubility [77]. The close analog 5-hydroxy-dihydroisoquinolinone (**11**) can be manipulated by introducing a 4-(1-piperidinyl)butoxy chain at the 5-position, and the corresponding derivative DPQ (**12**) is highly potent and water soluble [45].

A throughout structure–activity relationship on isoquinolinones and diydroisoquinolinens was reported by Chiarugi et al. in 2003 [78]. Scaffolds other than the (dihydro) isoquinolinone one can be employed to generate cyclic amides as PARP inhibitors. Thus,

quinazolinone derivatives (Scheme 7) are another class of conformationally constrained benzamide analogs, endowed with moderate to very high potency [79]. Furthermore, phtalazinones and quinoxalines have been demonstrated to be very productive scaffolds (Scheme 7) [68,80].

Attempts have been made to modify the cyclic amide ring. Thus, enlargement of the ring to a seven-membered resulted into [5,6,6,]-tricyclic indole lactams and [5,6,7]-ticyclic benzimidazole lactams which were designed to combine the favorable features of benzimidazoles with the requirement for an antioriented amido group (Scheme 8) [81]. The seven-membered lactam ring was anticipated, and then confirmed by X-ray crystallography, to optimize the bidirectional hydrogen bond with Gly863. Other seven-membered lactams, related to the dihydroisoquinolinone scaffold were reported to be, however, essentially inactive compounds [78]. Interestingly, incor-

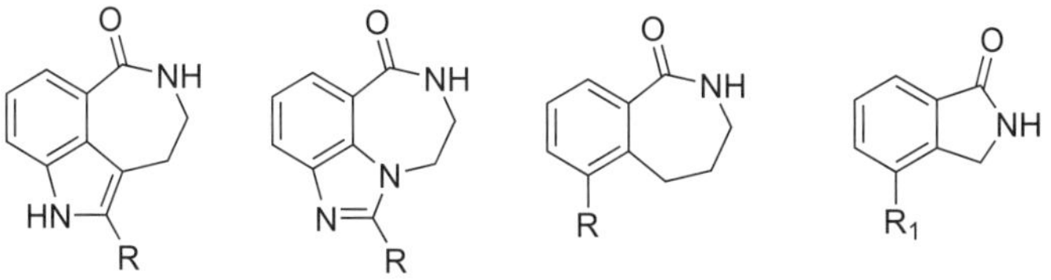

Scheme 8.

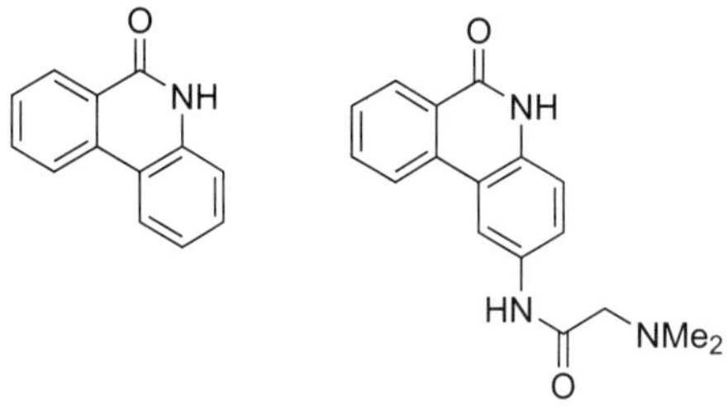

13

14

Scheme 9.

poration of the amide into the more strained five-membered ring preserved activity as PARP inhibitors (Scheme 8).

A very productive strategy toward PARP inhibitors has been the incorporation of the benzamide moiety into a polycyclic system. This approach started after the observation, again first reported by Banasik in 1992, that the tricyclic phenantridinone (PND, **13**) (Scheme 9) was a moderately potent PARP inhibitor [68]. Since then, this scaffold has been productively employed to generate series of PARP inhibitors claimed with superior properties in terms of potency and physico-chemical profile. For instances, introduction of an acidic or a basic side chain at 2- or 3-position of the PND scaffold yielded potent and water soluble derivatives. Among them, PJ34 (**14**), which was characterized by Inotek to be endowed with neuroprotective properties in both *in vivo* and *in vitro* model of cerebral ischemia [82,83].

Pellicciari et al. reported a bioisoster of PND, in which ring C is substituted by a thiophene ring. The resulting 4*H*-thieno[2,3-*c*]isoquinolinon-5-one (TIQ-A, **15**, Scheme 10) is a moderately potent PARP inhibitor, while the corresponding 5-amino (**16**), 5-hydroxy (**17**), or 5-methoxy (**18**) derivatives have IC_{50} in the low nanomolar range [78,84]. TIQ-A was widely characterized as neuroprotective in *in vitro* and *in vivo* model of cerebral ischemia.

Substituted pyrazolo[1,5-*a*]quinolin-5(4*H*)-one were also shown to be endowed with potent PARP inhibition profile, being compound **19** the most potent one, with a nanomolar IC_{50} [85]. Tricyclic quinaxolinones, exemplified by structure **20**, were also identified as submicromolar PARP-1 inhibitors, and the quinoxalinone scaffold appeared to be very sensitive to modifications to both the amine side chain and the tricyclic core [86].

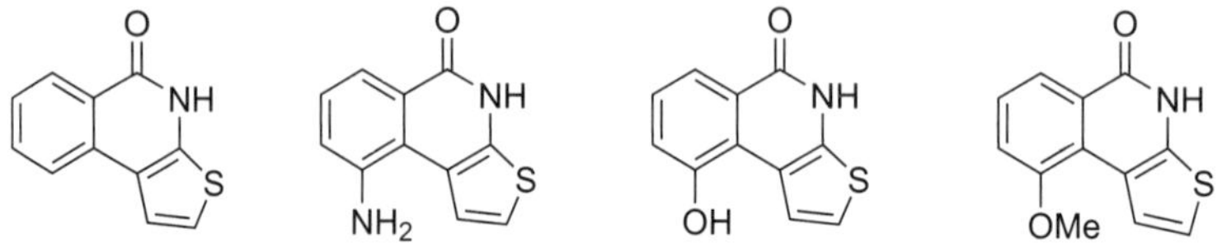

15   **16**   **17**   **18**

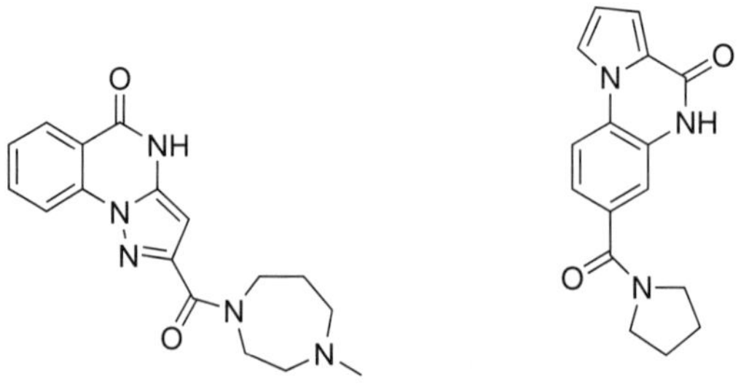

19   **20**

Scheme 10.

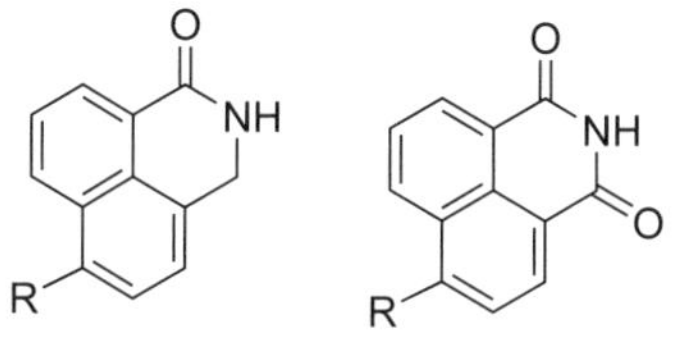

Scheme 11.

The 1,8-naphthalamide system was also investigated as a scaffold bearing the anticonformation of benzamide [68]. Again, introduction of polar substituents in the scaffold, like the amino group in the 4-position yielded potent and water soluble compounds. The 2,3, dihydrobenzo[*de*]isoquinolindiones (Scheme 11) were reported by Guilford pharmaceuticals to have submicromolar activity as PARP inhibitors [87].

Tetracyclic derivates incorporating the amido group were widely described as potent PARP inhibitors. The prototype is GPI6150 (**21**, Scheme 12), a tetracyclic derivative in which the dihydroisoquinolinone moiety is decorated by two additional rings, possibly exploiting the presence of hydrophobic clefts in the nicotinamide binding pocket of PARP-1 [88]. Analogs of GPI6150 incorporating the phtalazinone moiety have also been reported, belonging to the series of benzopyranophtalazinones (**22**) or indeno[1,2,3-*de*]-phtalazinones (**23**). Finally, indeno[1,2-*c*]isoquinolinones were reported, possessing an additional five-membered ring between the B and C rings of 6(5*H*)-phenanthridinone (**24**), and displaying very high potency as PARP inhibitors [89].

Another interesting example of incorporation of the benzamide scaffold into polycyclic structures is represented by the series of pyrrolo dihydroisoquinolinones, reported by Branca et al. [90], among which the urea derivatives **25** and **26** (Scheme 13) are the most potent representatives.

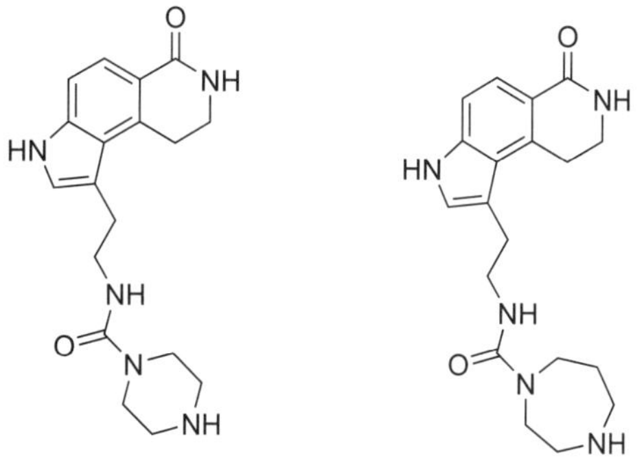

Scheme 13.

7. SELECTIVE PARP INHIBITORS

Most of the available literature on PARP inhibition refers to a generic "PARP" protein or to a generic poly(ADP-ribose)polymerase activity, meaning that the protein or the activity are those elicited by the product of the gene now called PARP-1. Indeed, PARP-1 is most abundantly expressed member of the PARP family, and most of activity is related to its activation. As new data accumulate, however, we know that the poly(ADP-ribose)polymerase activity is shared by up to 17 proteins, and that, at least in the case of PARP-2, redundant and overlapping activity with PARP-1 is pos-

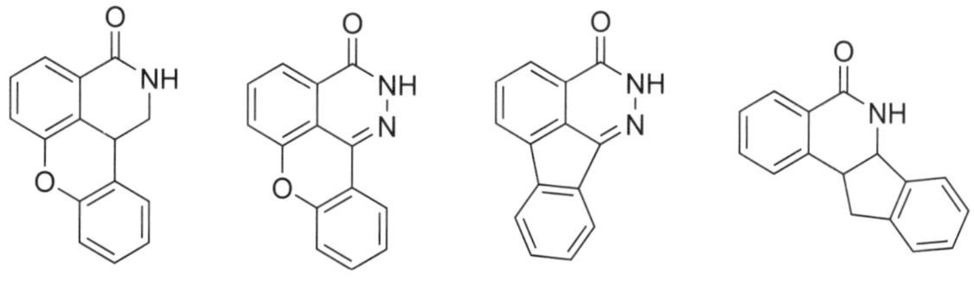

Scheme 12.

sible. From one point of view, this observation demands attention when interpreting data coming from experiments involving animals or whole cells. Indeed, the observed phenotype can be a consequence of the activation or inhibition of other PARP family members. On the other hand, experiments using purified or recombinant protein (mainly PARP-1) must be carefully judged since nothing can be foreseen on the activity towards other members of the family. In either case, there is a strong quest for selective PARP inhibitors, motivated by the need of clarifying the physiologic and the possible pathologic role of individual PARP proteins. Most of the problems of designing selective PARP isoform inhibitors reside in high similarity between the PARP catalytic domains. A step beyond is given by the growing availability (nine different members, as August 2009) of X-ray structures of the catalytic fragments of various proteins, which should facilitate the structure-based design of selective inhibitors. It should also be mentioned, on the other hand, that the claim of "selective" inhibition is only seldom based on evaluation on the whole PARP panel, and that in most cases, 'selectivity' means selectivity versus PARP-1. In this respect, PARP-2 selective have been described, among which quinoxaline derivatives [91,92] and, more recently, the 5-benzoyloxyisoquinolin-1(2H)-one, the most selective PARP-2 inhibitor reported so far, with a PARP-2/PARP-1 selectivity index greater than 60 [93]. More recently, the crystal structure of the catalytic fragment PARP-3 in complex with potent (but not selective over PARP-1) inhibitors has been reported thus opening the way for a better elucidation of the structural factors underlying specificity [94].

8. PARP INHIBITORS AS ANTICANCER AGENTS

Inability to properly repairing single- or double-strand breaks results into large chromosomal alterations, which include deletion, fusion, and translocation, ultimately determining genomic instability. Genomic instability is integral to cancerogenesis and is supposed to represent the way in which malignant cells acquire the ability to propagate and to metastasize. Members of the PARP family, and PARP-1 in particular, are important components of the nuclear machinery involved in single-strand break (mainly through the base-excision repair pathway) and in double-strand break repairing, a notion which is confirmed by the high vulnerability of PARP-1$^{-/-}$ mice to ionizing radiations or alkylating agents.

Given this context, it may appear counterintuitive to use PARP inhibitors, which silence poly(ADP-ribosyl)ation in a situation where it may contribute to preserve genomic stability. However, given the efficiency of the DNA repairing mechanisms, the protective effect of the repair pathways can be seen a disadvantage in case of DNA damaging chemo (or radio)therapy, since the protective pathways may well reduce the efficacy of cytotoxic agents and be causally related to resistance insurgence. Thus, following a seminal paper by Durkacz et al. [95], in which for the first time was envisaged the use of inhibitors of poly(ADP-ribose) synthesis as coadjutant for the cytotoxic treatment of leukemia, the use of PARP inhibitors in chemopotentiation and radiopotentition has received a lot of interest, and several candidates have proceeded in advanced clinical trials, as discussed below [96].

There is another important situation where PARP inhibition displayed therapeutic potential as monotherapy. This results from the notion that in the presence of specific genetic backgrounds, poly(ADP-ribose)polymerase activity promotes rather than protects from tumorogenesis. The most relevant case is represented by cell lines expressing mutated BRCA1 or BRCA2 genes, which have been involved in about 10% of all the breast cancer cases. Cell lines expressing mutated BRCA1 or BRCA2 possess an impaired ability to cope with double-strand breaks, due to a defective homologous repair (HR) pathway. These cells are normally viable, but tend to accumulate genomic defects. If these cells are treated with a PARP-1 inhibitor, they also fail to repair single-strand break lesions, which propagate and accumulate to unrepairable double-strand breaks, thus leading to cell death.

9. PARP INHIBITORS AS COADJUVANTS OF CHEMOTHERAPY AND RADIOTHERAPY

Several preclinical studies have confirmed the potential for cotherapy in which a PARP inhibitor is coadministered with a cytotoxic agent able to induce single-strand break damage. Thus, seven-membered lactams belonging to the family of [5,6,6,]-tricyclic indole lactams, such as AG-14361 (**27**) or AG-14699 (**28**) (Scheme 14) were evaluated (as chemosensitizers of temozolomide and topotecan using LoVo and SW620 human colorectal cells [97].

Temozolomide is an alkylating agent, while topotecan is topoisomerase I inhibitor. The water soluble AG-014699 (**28**) has been selected for clinical trials, and at the time of writing this chapter, is in phase II for BRCA1/BRCA2 positive breast and ovarian cancers [98]. 2-Substituted-(1*H*)-benzimidazole-4-carboxyamide derivatives were also evaluated for their effect in malignant cells treated with cytotoxic agents. In particular, ABT-888 (**29**) (Scheme 15) was assayed in several tumor cell lines, in combination with temozolomiode, cisplatin, carboplatin, irinotecan, and cyclophosphamide, resulting in either a synergic or enhancing effect. ABT-888 is entering phase I/II studies [99,100].

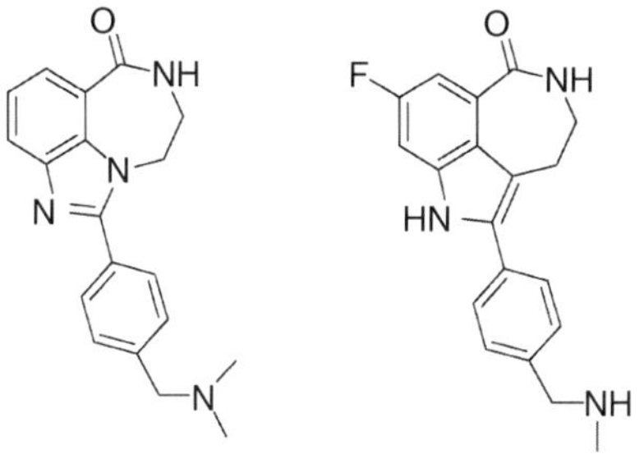

Olaparib (**30**), formerly known as AZD2281 or KU59436, belongs to the class of phtalazinones and is in phase II for ovarian cancer patients bearing BRCA1/2 mutations [101]. Olaparib (**30**) (Scheme 16) has been granted with the orphan drug status for the treatment of this kind of tumors.

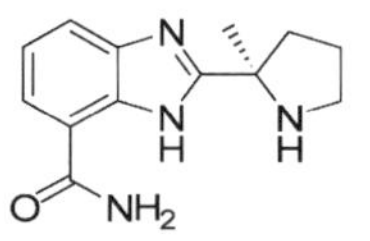

BSI-201, whose structure is not yet disclosed and developed by BiPar, is entering a phase III, multicenter, study in combination with Gemcitabine/Carboplatin in patients with ER-, PR-, and Her2-negative metastatic breast cancer [102,103].

CEP-9722, Cephalon pharma, not disclosed, is in phase I as single therapy or in combination with temozolomide in patients with advanced solid tumors [104].

The notion that PARP knockout mice are hypersensitive to radiation supports the potential use of PARP inhibitors as coadjuvants not only in chemotherapy but also in radiotherapy, although some studies report only a modest overall effect of PARP inhibitors as radiosensitizers in cell lines. Radiations cause both single and double-strand breaks, and PARP inhibition is likely to be effective only in single-strand breaks through the BER pathways. It can be anticipated that the effectiveness of PARP inhibitors will be maximized in replicating cells, where single-strand breaks produce collapsed replication forks which degenerate into double-strand breaks and cell death. Thus, PARP inhibitors should be particularly effective as radiosensitizers for the treatments of brain tumors, where the tissue surrounding the cancer cells is made of essentially nonreplicating cells. This concep-

30

Scheme 16.

27 **28**

Scheme 14.

29

Scheme 15.

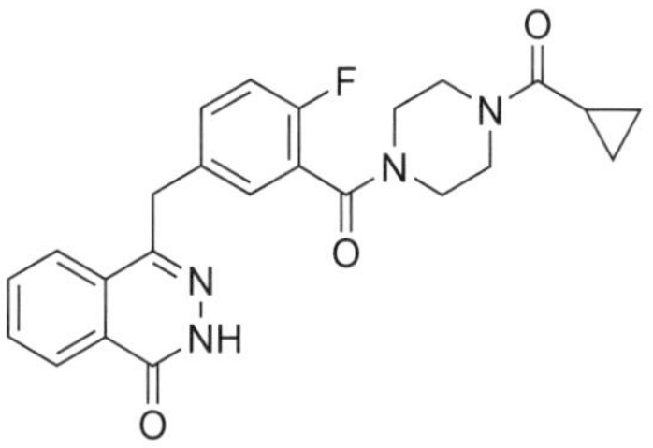

tual ground must be verified in clinical trials, which are particularly challenging owing the particular nature of radiotherapy settings, including the fine tuning of drug dose escalation and the assessment of cumulative and delayed effects [105].

10. PARP INHIBITORS AS MONOTHERAPY

As discussed in the above paragraphs, breast and ovarian cancer cells expressing mutated BRCA1 or BRCA2 are defective in HR pathway and thus tend to accumulate double-strand breaks that lead to an increased cell lethality. These cells are particularly sensitive to PARP inhibition, since PARP controls the repairing mechanism of single-strand breaks, and in its absence, they propagate into double-strand breaks upon replication fork collapse. PARP inhibitors such as AZD2281 (Olaparib, **30**) are currently in phase II as monotherapy for breast or ovarian cancer bearing BRCA1 or BRCA2 mutation [101].

11. PARG AND PARG INHIBITORS IN CANCER

Homeostasis of ploy(ADP-ribosyl)ation is assured not only by members of the PARP superfamily but also by degrading enzymes that hydrolyze the poly(ADP-ribose) polymers. Most of the hydrolyzing properties are owned by PARG, which is responsible for the cleavage of the glycosidic bonds between ADP-ribose units. Compared to the diversity of the PARP superfamily, so far only one gene has been identified to carry the poly(ADP-ribose)glycohydrolase activity in humans. The potential for therapeutic exploitation of PARG inhibitors in cancer resides in the fact that PARG inhibition disrupts the whole poly (ADP-ribose) metabolism thereby affecting the PARP signalling. Thus, genetic shutdown of a 110 KDa portion of PARG produced a viable phenotype particularly sensitive to DNA damage, likely due to a reduction of PARP-1 automodification, and thus PARP-1 activity impairments [106]. This opens the possibility for a therapeutic use of PARG inhibitors as chemo- (or radio-)sensitizers. This possibility, so far, remains at an inves-

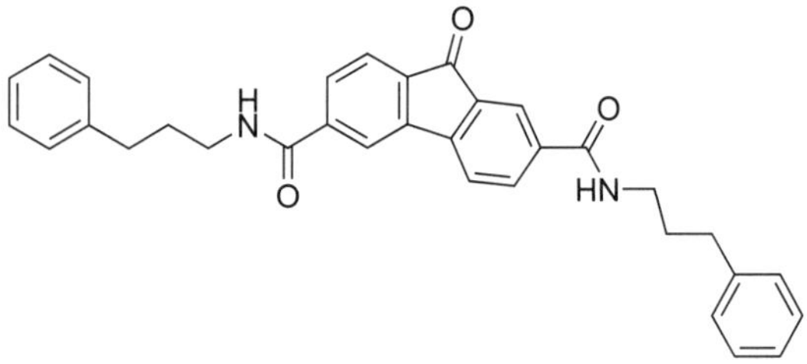

31

Scheme 17.

tigational level, also due to lack of a sufficiently varied array of PARG inhibitors. It is worth mentioning, in this context, that GPI 16552 (**31**) (Scheme 17), a PARG inhibitors, has been shown to enhance temozolomide citotoxicity in an in mice injected with B16 melanoma cells [107].

12. CONCLUDING REMARKS AND OUTLOOKS

Poly(ADP-ribosyl)ation of nuclear proteins is one of the fastest cellular response to DNA damage. DNA of every organism is prone to a massive attack by a variety of damaging agents, including radiations, chemicals, and endogenous reactive species, and instruments to repair the damage or to bring the injured cell to death must be effective. The family of PARPs, and PARP-1 in particular, operates covalent modifications of nuclear proteins, thereby allowing the relaxing of chromatin and the recruitment of repairing enzymes. Interestingly, the same mechanism is operative in attempting to reduce the DNA damage to cancer cells operated by radiation or alkylating agents, thus contributing to the developing of resistance to chemo- or radiotherapy. Thus, PARPs are becoming promising target for the developing of antitumor drugs, and several chemotypes are in advanced clinical trials at the time of writing. PARP inhibitors are usually endowed with a low acute toxicity and their clinically development is, in a sense, facilitated by the fact that many of these chemotypes are or have been under clinical evaluation, with the same mechanism, for uncorrelated pathologies such as stroke, cardiac ischemia, diabetes, inflammation.

Despite the hope for a rapid clinical exploitation, many aspects related to the use of PARP inhibitors in therapy await for further clarification, and a particularly important issue is that of selectivity among the 17 members of the superfamily. Most of still opened question, in this context, depends on the availability of selective inhibitors, possibly optimized for their physicochemical properties, a task which can and must properly be addressed by medicinal chemists.

REFERENCES

1. Chambon P, Weill JD, Mandel P. Biochem Biophys Res Commun 1963;11:39–43.
2. Alkhatib HM, Chen DF, Cherney B, Bhatia K, Notario V, Giri C, Stein G, Slattery E, Roeder RG, Smulson ME. Proc Natl Acad Sci USA 1987;84(5):1224–1228.
3. Schreiber V, Dantzer F, Ame JC, de Murcia G. Nat Rev Mol Cell Biol 2006;7(7):517–528.
4. Masson M, Niedergang C, Schreiber V, Muller S, Menissier-de Murcia J, de Murcia G. Mol Cell Biol 1998;18(6):3563–3571.
5. Ruf A, Rolli V, de Murcia G, Schulz GE. J Mol Biol 1998;278(1):57–65.
6. Alvarez-Gonzalez R, Althaus FR. Mutat Res 1989;218(2):67–74.
7. Wielckens K, George E, Pless T, Hilz H. J Biol Chem 1983;258(7):4098–4104.
8. Yamada M, Miwa M, Sugimura T. Arch Biochem Biophys 1971;146(2):579–586.
9. Min W, Wang ZQ. Front Biosci 2009;14: 1619–1626.
10. Oka J, Ueda K, Hayaishi O, Komura H, Nakanishi K. J Biol Chem 1984;259(2):986–995.
11. Berger NA. Radiat Res 1985;101(1):4–15.
12. Benjamin RC, Gill DM. J Biol Chem 1980; 255(21):10493–10501.
13. Ménissier-de Murcia J, Molinete M, Gradwohl G, Simonin F, de Murcia G. J Mol Biol 1989; 210(1):229–233.
14. Jacobson EL, Lange RA, Jacobson MK. J Cell Physiol 1979;99(3):417–425.
15. Bernofsky C. Mol Cell Biochem 1980;33 (3):135–143.
16. Goodwin PM, Lewis PJ, Davies MI, Skidmore CJ, Shall S. Biochim Biophys Acta 1978;543 (4):576–582.
17. Nagele A. Radiat Environ Biophys 1995;34 (4):251–254.
18. Chiarugi A. Trends Pharmacol Sci 2002;23 (3):122–129.
19. Yu SW, Wang H, Poitras MF, Coombs C, Bowers WJ, Federoff HJ, Poirier GG, Dawson TM, Dawson VL. Science 2002;297(5579):259–263.
20. Ha HC, Snyder SH. Proc Natl Acad Sci USA 1999;96(24):13978–13982.
21. Otto H, Reche PA, Bazan F, Dittmar K, Haag F, Koch-Nolte F. BMC Genomics 2005;6:139.
22. Amé JC, Spenlehauer C, de Murcia G. Bioessays. 2004;26(8):882–893.
23. de Murcia JM, Niedergang C, Trucco C, Ricoul M, Dutrillaux B, Mark M, Oliver FJ, Masson M, Dierich A, LeMeur M, Walztinger C, Chambon P, de Murcia G. Proc Natl Acad Sci USA 1997;94(14):7303–7307.
24. Wang ZQ, Stingl L, Morrison C, Jantsch M, Los M, Schulze-Osthoff K, Wagner EF. Genes Dev 1997;11(18):2347–2358.
25. Masutani M, Suzuki H, Kamada N, Watanabe M, Ueda O, Nozaki T, Jishage K, Watanabe T, Sugimoto T, Nakagama H, Ochiya T, Sugimura T. Proc Natl Acad Sci USA 1999;96 (5):2301–2304.
26. Shall S, de Murcia G. Mutat Res. 2000;460 (1):1–15.
27. Amé JC, Rolli V, Schreiber V, Niedergang C, Apiou F, Decker P, Muller S, Höger T, Ménissier-de Murcia J, de Murcia G. J Biol Chem 1999;274(25):17860–17868.
28. Schreiber V, Amé JC, Dollé P, Schultz I, Rinaldi B, Fraulob V, Ménissier-de Murcia J, de Murcia G. J Biol Chem 2002;277 (25):23028–23036.
29. Augustin A, Spenlehauer C, Dumond H, Ménissier-De Murcia J, Piel M, Schmit AC, Apiou F, Vonesch JL, Kock M, Bornens M, De Murcia G. J Cell Sci. 2003;116(Pt 8):1551–1562.
30. Kickhoefer VA, Siva AC, Kedersha NL, Inman EM, Ruland C, Streuli M, Rome LH. J Cell Biol 1999;146(5):917–928.
31. Smith S, Giriat I, Schmitt A, de Lange T. Science 1998;282(5393):1484–1487.
32. Sbodio JI, Lodish HF, Chi NW. Biochem J 2002;361(Pt 3):451–459.
33. Schreiber V, Dantzer F, Ame JC, de Murcia G. Nat Rev Mol Cell Biol 2006;7(7):517–528.
34. Lamarre D, Talbot B, de Murcia G, Laplante C, Leduc Y, Mazen A, Poirier GG. Biochim Biophys Acta 1988;950(2):147–160.

35. Mazen A, Menissier-de Murcia J, Molinete M, Simonin F, Gradwohl G, Poirier G, de Murcia G. Nucleic Acids Res 1989;17(12):4689–4698.
36. Uchida K, Uchida M, Hanai S, Ozawa Y, Ami Y, Kushida S, Miwa M. Gene 1993;137(2):293–297.
37. Jung S, Miranda EA, de Murcia JM, Niedergang C, Delarue M, Schulz GE, de Murcia GM. J Mol Biol 1994;244(1):114–116.
38. Ruf A, Mennissier de Murcia J, de Murcia G, Schulz GE. Proc Natl Acad Sci USA 1996;93(15):7481–7485.
39. Ruf A, de Murcia G, Schulz GE. Biochemistry 1998;37(11):3893–3900.
40. http://www.pdb.org/pdb/home/home.do.
41. Miwa M, Masutani M. Cancer Sci 2007;98(10):1528–1535.
42. Haince JF, Rouleau M, Hendzel MJ, Masson JY, Poirier GG. Trends Mol Med 2005;11(10):456–463.
43. Ménissier de Murcia J, Ricoul M, Tartier L, Niedergang C, Huber A, Dantzer F, Schreiber V, Amé JC, Dierich A, LeMeur M, Sabatier L, Chambon P, de Murcia G. EMBO J 2003;22(9):2255–2263.
44. Tsutsumi M, Masutani M, Nozaki T, Kusuoka O, Tsujiuchi T, Nakagama H, Suzuki H, Konishi Y, Sugimura T. Carcinogenesis. 2001;22(1):1–3.
45. Eliasson MJ, Sampei K, Mandir AS, Hurn PD, Traystman RJ, Bao J, Pieper A, Wang ZQ, Dawson TM, Snyder SH, Dawson VL. Nat Med 1997;3(10):1089–1095.
46. Virág L, Szabó C. Pharmacol Rev 2002;54(3):375–429.
47. Dantzer F, Schreiber V, Niedergang C, Trucco C, Flatter E, De La Rubia G, Oliver J, Rolli V, Ménissier-de Murcia J, de Murcia G. Biochimie 1999;81(1–2):69–75.
48. Horton JK, Watson M, Stefanick DF, Shaughnessy DT, Taylor JA, Wilson SH. Cell Res 2008;18(1):48–63.
49. Woodhouse BC, Dianova II, Parsons JL, Dianov GL. DNA Repair (Amst). 2008;7(6):932–940.
50. Weterings E, Chen DJ. Cell Res 2008;18(1):114–124.
51. Shrivastav M, De Haro LP, Nickoloff JA. Cell Res. 2008;18(1):134–147.
52. Lord CJ, Ashworth A. Curr Opin Pharmacol 2008;8(4):363–369.
53. Kyle S, Thomas HD, Mitchell J, Curtin NJ. Br J Radiol 2008;81(1):S6–S11.
54. Rottenberg S, Jaspers JE, Kersbergen A, van der Burg E, Nygren AO, Zander SA, Derksen PW, de Bruin M, Zevenhoven J, Lau A, Boulter R, Cranston A, O'Connor MJ, Martin NM, Borst P, Jonkers J. Proc Natl Acad Sci USA 2008;105(44):17079–17084.
55. Fong PC, Boss DS, Yap TA, Tutt A, Wu P, Mergui-Roelvink M, Mortimer P, Swaisland H, Lau A, O'Connor MJ, Ashworth A, Carmichael J, Kaye SB, Schellens JH, de Bono JS. N Engl J Med 2009;361(2):123–134.
56. Oei SL, Griesenbeck J, Ziegler M, Schweiger M. Biochemistry 1998;37(6):1465–1469.
57. Meisterernst M, Stelzer G, Roeder RG. Proc Natl Acad Sci USA 1997;94(6):2261–2265.
58. Le Page C, Sanceau J, Drapier JC, Wietzerbin J. Biochem Biophys Res Commun 1998;243(2):451–457.
59. Masutani M, Nakagama H, Sugimura T. Cell Mol Life Sci 2005;62(7–8):769–783.
60. Bell CE, Eisenberg D. Biochemistry 1997;36(3):481–488.
61. Rolli V, O'Farrell M, Ménissier-de Murcia J, de Murcia G. Biochemistry 1997;36(40):12147–12154.
62. Kim H, Jacobson MK, Rolli V, Ménissier-de Murcia J, Reinbolt J, Simonin F, Ruf A, Schulz G, de Murcia G. Biochem J 1997;322(Pt 2):469–475.
63. Clark JB, Ferris GM, Pinder S. Biochim Biophys Acta 1971;238(1):82–85.
64. Costantino G, Macchiarulo A, Camaioni E, Pellicciari R. J Med Chem 2001;44(23):3786–3794.
65. Bellocchi D, Macchiarulo A, Costantino G, Pellicciari R. Bioorg Med Chem 2005;13(4):1151–1157.
66. Kirkland JB. Nutr Cancer 2003;46(2):110–118.
67. Shall S. J Biochem 1975;77:1.
68. Banasik M, Komura H, Shimoyama M, Ueda K. J Biol Chem 1992;267(3):1569–1575.
69. Endres M, Wang ZQ, Namura S, Waeber C, Moskowitz MA. Cereb Blood Flow Metab 1997;17(11):1143–1151.
70. Tokime T, Nozaki K, Sugino T, Kikuchi H, Hashimoto N, Ueda K. J Cereb Blood Flow Metab. 1998;18(9):991–997.
71. Ducrocq S, Benjelloun N, Plotkine M, Ben-Ari Y, Charriaut-Marlangue C. J Neurochem 2000;74(6):2504–2511.
72. Ding Y, Zhou Y, Lai Q, Li J, Gordon V, Diaz FG. Brain Res 2001;915(2):210–217.
73. Cantoni O, Sestili P, Spadoni G, Balsamini C, Cucchiarini L, Cattabeni F. Biochem Int 1987;15(2):329–337.
74. Griffin RJ, Pemberton LC, Rhodes D, Bleasdale C, Bowman K, Calvert AH, Curtin NJ,

Durkacz BW, Newell DR, Porteous JK. Anticancer Drug Des 1995;10(6):507–514.
75. White AW, Almassy R, Calvert AH, Curtin NJ, Griffin RJ, Hostomsky Z, Maegley K, Newell DR, Srinivasan S, Golding BT. J Med Chem 2000;43(22):4084–4097.
76. Suto MJ, Turner WR, Arundel-Suto CM, Werbel LM, Sebolt-Leopold JS. Anticancer Drug Des 1991;6(2):107–117.
77. McDonald MC, Mota-Filipe H, Wright JA, Abdelrahman M, Threadgill MD, Thompson AS, Thiemermann C. Br J Pharmacol 2000;130(4):843–850.
78. Chiarugi A, Meli E, Calvani M, Picca R, Baronti R, Camaioni E, Costantino G, Marinozzi M, Pellegrini-Giampietro DE, Pellicciari R, Moroni F. J Pharmacol Exp Ther 2003;305(3):943–949.
79. Griffin RJ, Srinivasan S, Bowman K, Calvert AH, Curtin NJ, Newell DR, Pemberton LC, Golding BT. J Med Chem 1998;41(26):5247–5256.
80. Perkins E, Sun D, Nguyen A, Tulac S, Francesco M, Tavana H, Nguyen H, Tugendreich S, Barthmaier P, Couto J, Yeh E, Thode S, Jarnagin K, Jain A, Morgans D, Melese T. Cancer Res 2001;61(10):4175–4183.
81. Skalitzky DJ, Marakovits JT, Maegley KA, Ekker A, Yu XH, Hostomsky Z, Webber SE, Eastman BW, Almassy R, Li J, Curtin NJ, Newell DR, Calvert AH, Griffin RJ, Golding BT. J Med Chem 2003;46(2):210–213.
82. Jagtap P, Soriano FG, Virág L, Liaudet L, Mabley J, Szabó E, Haskó G, Marton A, Lorigados CB, Gallyas F Jr, Sümegi B, Hoyt DG, Baloglu E, VanDuzer J, Salzman AL, Southan GJ, Szabó C. Crit Care Med 2002;30(5):1071–1082.
83. Abdelkarim GE, Gertz K, Harms C, Katchanov J, Dirnagl U, Szabó C, Endres M. Int J Mol Med 2001;7(3):255–260.
84. Pellicciari R, Camaioni E, Costantino G, Marinozzi M, Macchiarulo A, Moroni F, Natalini B. Farmaco 2003;58(9):851–858.
85. Orvieto F, Branca D, Giomini C, Jones P, Koch U, Ontoria JM, Palumbi MC, Rowley M, Toniatti C, Muraglia E. Identification of substituted pyrazolo[1,5-a]quinazolin-5(4H)-one as potent poly(ADP-ribose)polymerase-1 (PARP-1) inhibitors. Bioorg Med Chem Lett 2009;19(15):4196–4200.
86. Miyashiro J, Woods KW, Park CH, Liu X, Shi Y, Johnson EF, Bouska JJ, Olson AM, Luo Y, Fry EH, Giranda VL, Penning TD. Synthesis and SAR of novel tricyclic quinoxalinone inhibitors of poly(ADP-ribose)polymerase-1 (PARP-1) Bioorg Med Chem Lett 2009;19(15):4050–4054.
87. Li JH, Zhang J, Ferraris DV. PCT Int Appl WO9959975. (1999) Chemical Abstract 136, 355165.
88. Zhang J, Lautar S, Huang S, Ramsey C, Cheung A, Li JH. Biochem Biophys Res Commun 2000;278(3):590–598.
89. Jagtap PG, Baloglu E, Southan GJ, Mabley JG, Li H, Zhou J, van Duzer J, Salzman AL, Szabó C. J Med Chem 2005;48(16):5100–5103.
90. Branca D, Cerretani M, Jones P, Koch U, Orvieto F, Palumbi MC, Rowley M, Toniatti C, Muraglia E. Identification of aminoethyl pyrrolo dihydroisoquinolinones as novel poly(ADP-ribose) polymerase-1 inhibitors Bioorg Med Chem Lett 2009;19(15):4042–4045.
91. Iwashita A, Hattori K, Yamamoto H, Ishida J, Kido Y, Kamijo K, Murano K, Miyake H, Kinoshita T, Warizaya M, Ohkubo M, Matsuoka N, Mutoh S. FEBS Lett 2005;579(6):1389–1393.
92. Ishida J, Yamamoto H, Kido Y, Kamijo K, Murano K, Miyake H, Ohkubo M, Kinoshita T, Warizaya M, Iwashita A, Mihara K, Matsuoka N, Hattori K. Bioorg Med Chem 2006;14(5):1378–1390.
93. Pellicciari R, Camaioni E, Costantino G, Formentini L, Sabbatini P, Venturoni F, Eren G, Bellocchi D, Chiarugi A, Moroni F. ChemMedChem 2008;3(6):914–923.
94. Lehtiö L, Jemth AS, Collins R, Loseva O, Johansson A, Markova N, Hammarström M, Flores A, Holmberg-Schiavone L, Weigelt J, Helleday T, Schüler H, Karlberg T. J Med Chem 2009;52(9):3108–3111.
95. Durkacz BW, Omidiji O, Gray DA, Shall S. Nature 1980;283(5747):593–596.
96. Rodon J, Iniesta MD, Papadopoulos K. Development of PARP inhibitors in oncology Expert Opin Investig Drugs 2009;18(1):31–43.
97. Thomas HD, Calabrese CR, Batey MA, Canan S, Hostomsky Z, Kyle S, Maegley KA, Newell DR, Skalitzky D, Wang LZ, Webber SE, Curtin NJ. Mol Cancer Ther 2007;6(3):945–956.
98. Wang Y, Castaner R, Bolos J. AG-14699. Drugs Fut 2009;34:177–182.
99. Penning TD, Zhu GD, Gandhi VB, Gong J, Liu X, Shi Y, Klinghofer V, Johnson EF, Donawho CK, Frost DJ, Bontcheva-Diaz V, Bouska JJ, Osterling DJ, Olson AM, Marsh KC, Luo Y, Giranda VL. J Med Chem 2009;52(2):514–523.
100. Horton TM, Jenkins G, Pati D, Zhang L, Dolan ME, Ribes-Zamora A, Bertuch AA, Blaney SM, Delaney SL, Hegde M, Berg SL. Mol Cancer Ther 2009;8(8):2232–2242.

101. Fong PC, Boss DS, Yap TA, Tutt A, Wu P, Mergui-Roelvink M, Mortimer P, Swaisland H, Lau A, O'Connor MJ, Ashworth A, Carmichael J, Kaye SB, Schellens JH, de Bono JS. N Engl J Med 2009;361(2):123–134.
102. Pal SK, Mortimer. J Maturitas 2009;63(4):269–274.
103. Kling J. PARP inhibitors blaze a trail in difficult-to-treat cancers Nat Biotechnol 2009;27(9):784–786.
104. Miknyoczki S, Chang H, Grobelny J, Pritchard S, Worrell C, McGann N, Ator M, Husten J, Deibold J, Hudkins R, Zulli A, Parchment R, Ruggeri B. Mol Cancer Ther. 2007;6(8):2290–2302.
105. Chalmers AJ. Br Med Bull 2009;89:23–40.
106. Cortes U, Tong WM, Coyle DL, Meyer-Ficca ML, Meyer RG, Petrilli V, Herceg Z, Jacobson EL, Jacobson MK, Wang ZQ. Mol Cell Biol 2004;24(16):7163–7178.
107. Tentori L, Leonetti C, Scarsella M, Muzi A, Vergati M, Forini O, Lacal PM, Ruffini F, Gold B, Li W, Zhang J, Graziani G. Eur J Cancer 2005;41(18):2948–2957.

PROTEASOME INHIBITORS

Marie Wehenkel
Kyung Bo Kim
Department of Pharmaceutical Sciences,
University of Kentucky, Lexington, KY

1. INTRODUCTION TO THE PROTEASOME

The ubiquitin-proteasome pathway (UPP) is the principal conduit for protein turnover in all eukaryotic cells and utilizes a 76-amino acid polypeptide, ubiquitin, to target proteins for degradation. Many essential cellular processes are regulated by the UPP, such as stress and immune responses, transcription, cell cycle progression, cell differentiation, and apoptosis. In eukaryotic cells, the UPP is the major protein degradation pathway, responsible for the destruction of more than 70% of intracellular proteins. These proteins are targeted for degradation by the conjugation of polyubiquitin chains to lysine residues on the protein. This ubiquitination process is carried out by ubiquitin-conjugating enzymes and regulated by a deubiquitinating metalloprotease [1]. The attachment of multiple ubiquitin chains is mediated by three such conjugating enzymes: a ubiquitin-activating enzyme (E1), a ubiquitin-conjugating enzyme (E2), and a ubiquitin ligase (E3), which confers substrate specificity (Fig. 1). Once proteins are ubiquitinated by these enzymes, the 26S proteasome rapidly recognizes and degrades targeted proteins in an ATP-dependent fashion (Fig. 2) [2,3]. The 26S proteasome is composed of the multisubunit 20S catalytic core complex and 19S regulatory complexes that assist in binding and unfolding ubiquitinated protein substrates. To prevent promiscuous protein degradation by the proteasome, proteolytic activity is confined to the 20S inner compartment that is only accessible to unfolded proteins. The 20S catalytic core, which is constitutively expressed in all eukaryotic cells, is a barrel-shaped ring structure with seven distinct α-subunits forming the two outer α-rings and seven distinct β-subunits forming the two inner catalytic β-rings. Each β-ring contains three different catalytic subunits displaying distinct proteolytic activities [4–7], which are referred to as the chymotrypsin-like (CT-L), trypsin-like (T-L), and caspase-like (C-L) activities [8,9]. Among these proteolytic activities, CT-L activity is thought to be the most physiologically relevant [10–13].

In addition to the constitutive proteasome, mammalian cells produce an alternative proteasome form called the "immunoproteasome" in response to stimuli such as interferon-γ [14]. Exposure of mammalian cells to these stimuli induces upregulation of certain catalytic subunits such as LMP2/β1i, MECL1/β2i, and LMP7/β5i, which replace their constitutive proteasome counterparts Y/β1, Z/β2, and X/β5, respectively, to create the immunoproteasome [14]. When the X-ray crystal structures of the bovine proteasomes were compared to that of the yeast proteasome, some alterations of the structures of the subunits α2, β1, β5, β6, and β7 were found to be necessary for the incorporation of either the constitutive or the inducible subunits [15]. As compared to the constitutive proteasome, the immunoproteasome has an enhanced capacity to generate peptide fragments bearing hydrophobic and basic amino acids at their C-termini and a reduced capacity to produce peptides bearing acidic residues at their C-termini. Thus, early research suggested that the primary function of the immunoproteasome was the generation of antigenic peptides.

To target either proteasome form specifically and effectively, inhibitors are needed that block the active site moiety of one or more of the beta-subunits. These beta-subunits are part of the family of N-terminal nucleophilic (NTN) hydrolases and each utilize a catalytic threonine to facilitate proteolytic cleavage of substrates. Additionally, the design of specific proteasome inhibitors necessarily takes into account the substrate binding pocket, which is unique to each catalytic subunit and has been defined for each subunit by structural and mutational studies [16]. The differences in the substrate binding pockets encourage the development of subunit-specific inhibitors, general inhibitors, and perhaps even inhibitors preferentially targeting prokaryotes' proteasomes to combat diseases such as tuberculosis.

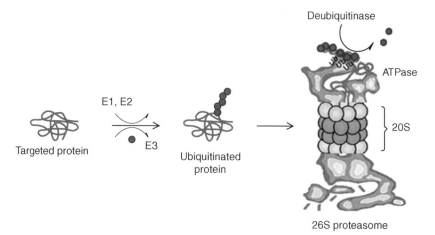

Figure 1. The ubiquitin-proteasome system. (This figure is available in full color at http://mrw.interscience.wiley.com/emrw/9780471266945/home.)

2. THE FUNDAMENTAL RATIONALE FOR TARGETING THE UPP: POTENTIAL APPLICATIONS IN CANCER AND OTHER DISEASES

It has been shown that regulation via ubiquitination is essential to many proteins involved in important cellular activities such as cell cycle progression, apoptosis, angiogenesis, and inflammation. Examples include cyclins, cyclin-dependent kinases, cyclin-dependent kinase inhibitors, cdc25 phosphatase, nuclear factor κB (NF-κB), C-fos, C-jun, N-MYC, p53, p21, Bax, β-catenin, and hypoxia-inducible factor 1-α (HIF-1α) [17,18]. Given that many cyclins (A, B, D, and E) and the cyclin-dependent protein kinases (CDK) regulate cell division, inhibition of the degradation of these proteins can cause a blockade of cell cycle progression. CDK inhibitors p21 and p27, which function to arrest the cell cycle by inhibiting cdk, are transiently degraded by the proteasome during cell division [19,20]. Thus, the levels of p21 and p27 are increased by treatment with proteasome inhibitors, blocking the cell cycle in G1/S and providing a rationale to use proteasome inhibitors against rapidly proliferating tumor cells. In this regard, it has been reported that a low level of p27 is associated with more aggressive tumors

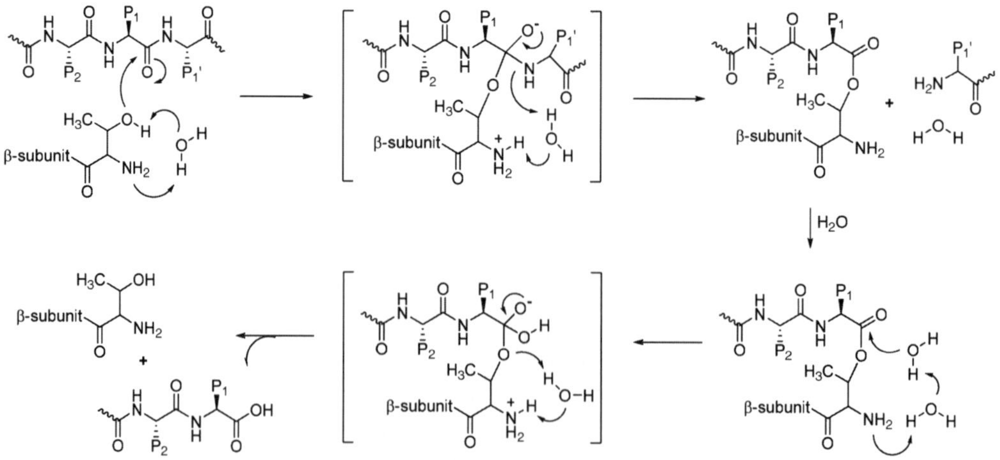

Figure 2. The proposed mechanism of proteasome-catalyzed hydrolysis.

and worse prognosis [20], further indicating the importance of controlling CDK level by the proteasome.

NF-κB is a proteasome-regulated transcription factor that activates genes involved in cell proliferation, cytokine synthesis, cell adhesion, cell survival, and angiogenesis. Under normal circumstances, NF-κB forms an inactive complex with IκBα in cytoplasm. Upon stimulation by cytokines, the IκB is polyubiquitinated and degraded by the proteasome, allowing nuclear translocation of NF-κB and transcriptional activation of downstream genes [21,22]. NF-κB-activated gene products promote cell survival and, thus, can potentiate tumor development and metastasis. In fact, NF-κB exhibits particularly high levels of activity in many tumor models (such as prostate cancer, melanomas, myeloma, lymphomas, and leukemias) [23]. Preclinical studies show that proteasome inhibitors have a higher cytotoxicity against tumor cells with high-level NF-κB activation than normal or cancer cells with lower expression of NF-κB [20,24]. Moreover, as compared to normal cells, tumor cells appear to be more sensitive to the proapoptotic effects of proteasome inhibitors. Many preclinical models have also shown that proteasome inhibition can restore tumor sensitivity to other cytotoxic agents and enhance apoptosis by preventing NF-κB activation [24]. As a result, inhibition of NF-κB by proteasome inhibitors has emerged as a new therapeutic strategy against tumor cells with high levels of NF-κB activity [25,26].

In the ubiquitin-proteasome pathway, E3 ubiquitin ligases play a key role in the selective targeting of substrate proteins [27–29]. Among the E3 ubiquitin ligases in cells, many play a major role in inducing ubiquitination and degradation of cellular oncogenic proteins, thus implicating these ligases in tumor growth [27,28]. For example, MDM2, which is a negative regulator of p53 and a target for ubiquitination and degradation by the proteasome, is overexpressed in many cancers where it may be targeted for cancer treatment [30]. E6-AP is another p53 regulator that associates with the human papilloma virus oncogene E6 and then functions as an E3 ubiquitin ligase targeting p53 [28]. Conversely, BRCA acts as a tumor suppressor and E3 ubiquitin ligase, and mutation of this protein confers an increased risk for breast and ovarian cancers. Meanwhile, the Von Hippel–Lindau (VHL) protein regulates the cellular oxygen stress response via HIF-1α stabilization. Another E3 ubiquitin ligase SKP2 leads to UPP-mediated removal of cell cycle regulators such as p21, and therefore is also linked to cancer. Currently, efforts are ongoing to develop inhibitors of these E3 ligases for cancer treatment [27,31–33].

The role of the ubiquitin-proteasome pathway in cell proliferation and inflammation clearly indicates that the proteasome is an attractive target for cancer therapeutics. Furthermore, the high metabolic rate in cancer cells in comparison to normal resting cells would provide a proteasome inhibitor with an opportunity to target cancer cells with relative selectivity. Additional inhibitors of tumor-associated components of the UPP could provide rational and selective therapies for diseases with dysregulation of these proteins.

3. THE RATIONAL DEVELOPMENT OF BORTEZOMIB AND OTHER PROTEASOME INHIBITORS IN CLINICAL TRIALS

Inhibition of proteasome function results in dysregulation of normal cellular activities, leading to cell growth inhibition and eventually cell death. As compared to normal resting cells, this effect is more pronounced in tumor cells, which require higher levels of UPP activity to maintain their high metabolic rate. This was a major rationale for the development of proteasome inhibitors as a cancer therapy. A highlight of this effort is the development of the "first-in-class" proteasome inhibitor drug bortezomib (Velcade®) [34], which is currently approved by the FDA for the treatment of multiple myeloma (MM) and relapsed mantle cell lymphoma.

Bortezomib (Fig. 3) is a dipeptide boronic acid that targets the active site of the catalytic β-subunits of the proteasome. Rapid but reversible inhibition of the proteasome's enzymatic activity is caused by the binding of bortezomib directly to the 20S proteasome active sites. Boron-pharmacophore-based peptides were originally developed to inhibit

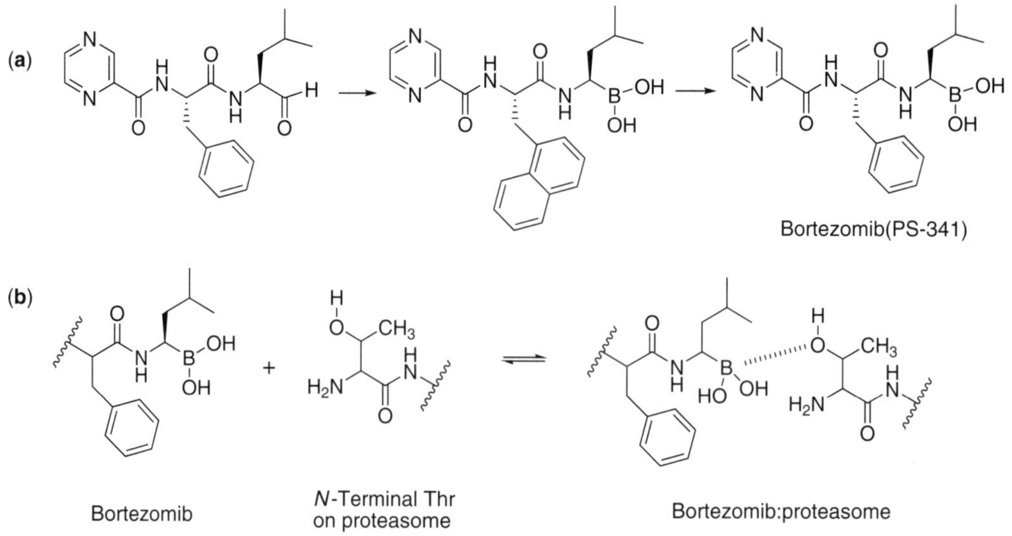

Figure 3. (a) Development of boronic acid pharmacophore-based proteasome inhibitors. (b) The proposed inhibitory mechanism of bortezomib.

the function of serine proteases [35], since boron is thought to form a complex with the hydroxyl group of serine proteases but not the thiol group of cysteine proteases. Given that the proteasome also contains a nucleophilic hydroxyl group in the active site of its catalytic β-subunits, Adams et al. used boronic acid as a pharmacophore in the design of proteasome inhibitors [36]. Based on information gathered from SAR studies on peptide aldehydes, they prepared peptide boronic acids of varying lengths (Fig. 3). In the end, through this medicinal chemistry approach, they eventually developed a potent dipeptidyl boronic acid proteasome inhibitor, bortezomib (PS-341), which has a comparable activity to tri- or tetrapeptide boronic acids [34].

In the NCI (National Cancer Institute) 60 cell lines, bortezomib showed significant growth inhibition and single-agent activity in several murine and human xenograft tumor models [37]. In several diverse preclinical models, bortezomib directly inhibited proliferation of MM cell lines and induced apoptosis in primary MM cells at clinically achievable concentrations of less than 10 nmol/L [38]. Mechanistically speaking, bortezomib induced apoptosis in a p53-independent manner in these tumor cell lines. It also sensitized MM cell lines to other antitumor agents, such as doxorubicin and melphalan. In a mouse xenograft model, overall survival was doubled in bortezomib-treated mice compared with controls, while the malignant cells showed decreased angiogenesis and increased apoptosis. Recently, bortezomib was shown to be effective in protecting against nephritis in a murine model of lupus, providing a proof-of-concept study for the application of bortezomib to autoimmune diseases characterized by aberrant antibody production [39].

Since bortezomib showed broad spectrum preclinical activity in *in vitro* and *in vivo* models, several phase I trials were initiated to evaluate bortezomib as a treatment for advanced solid tumors and refractory hematological malignancies [40]. When patients with advanced solid tumors were treated with bortezomib, antitumor efficacy was clearly observed. Increased toxicity was found with more frequent treatment, and the maximum tolerated dose was determined (1.56 mg/m^2). In patients with advanced hematological malignancies, bortezomib showed a dose- and time-dependent inhibition of the proteasome [38]. The maximum tolerated dose was slightly lower than that for solid tumors (1.04 mg/m^2). Bortezomib can be delivered to all human tissues except the brain, eyes, and testes.

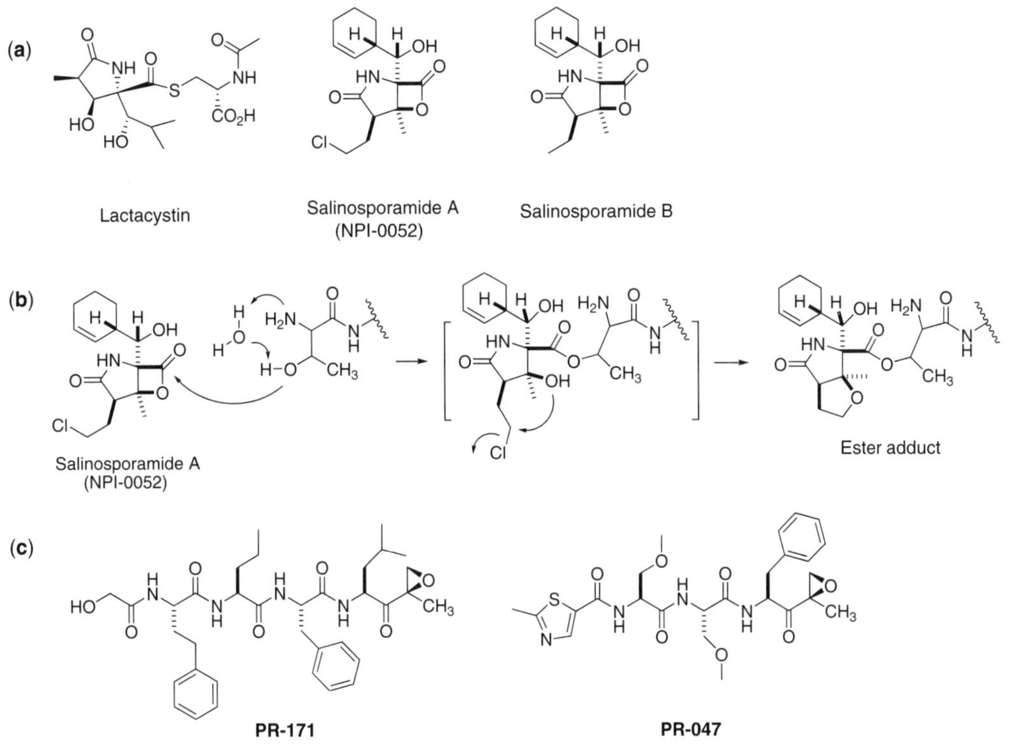

Figure 4. (a) Natural product (lactacystin and salinosporamide B) and natural product-derived proteasome inhibitor under clinical trials (NPI-0052). (b) The proposed proteasomal inhibitory mechanism of the β-lactone pharmacophore-containing NPI-0052. (c) PR-171 and its orally bioavailable analog (PR-047).

These phase I trial results, along with the data gathered from preclinical studies, provided the rationale for phase II clinical trials with bortezomib in MM [38]. In these trials, the combined complete and partial response rate was 27%, whereas 59% of patients had stable disease or better. This eventually led to the approval of bortezomib by the FDA as therapy for patients with multiple myeloma who had received at least two prior therapies. Single-agent or combination therapy trials with bortezomib targeting various nonhematologic malignancies are currently ongoing [41,42]. Bortezomib has also shown promising activity in mantle cell lymphoma and non-small cell lung cancer, but the clinical activity of bortezomib in other tumors has been less promising [43].

Although bortezomib has been widely used as a successful chemotherapeutic agent, the broad application of bortezomib is currently limited due to drug-associated side effects [44,45]. In addition, bortezomib is oxidatively metabolized via the cytochrome P450 enzymes and as a result, the boronic acid residue of bortezomib is released to yield inactive boron-less bortezomib [46]. This may be one of the mechanisms of bortezomib clearance in the body. Currently, due to the development of resistance to bortezomib by certain types of multiple myeloma, proteasome inhibitors having other pharmacophores are being investigated for their activity in clinical trials [47]. However, bortezomib will likely continue to be approved for additional indications as ongoing clinical trials, including combination therapy studies, conclude.

In addition to bortezomib, the development of new types of proteasome inhibitors with better efficacy has been actively pursued for the treatment of cancer. Salinosporamide A (NPI-0052) (Fig. 4) is an irreversible proteasome inhibitor developed by Nereus Pharmaceuticals (San Diego, CA) through natural product screening [48]. It is derived from marine *Actinomycete* and has the same β-lactone pharmacophore as the active form of lactacystin (clasto-lactacystin) [49]. Studies showed that NPI-0052 induces apoptosis in MM cells that are resistant to conventional and bortezomib therapies [48]. In a MM tumor xenograft mouse model, NPI-0052 inhibited tumor growth and prolonged survival, with no recurrence of tumor in 57% of the NPI-0052-treated mice at day 300. NPI-0052 is currently being tested in a phase I clinical trial for relapsed and relapsed/refractory MM patients [50]. Carfilzomib (PR-171) (Fig. 4), a proteasome inhibitor developed by Proteolix (South San Francisco, CA), is currently undergoing phase I clinical trials [51,52]. PR-171 is a peptide epoxyketone inhibitor derived from epoxomicin, a highly selective, potent proteasome inhibitor isolated as a microbial metabolite from *Actinomycete* [53]. The epoxyketone peptide inhibitor has been shown to form a six-membered morpholino ring between the amino terminal catalytic Thr-1 of the 20S proteasome and the α',β'-epoxy ketone pharmacophore, providing a unique specificity toward the proteasome [54]. Like NPI-0052, carfilzomib has been shown to be active against MM tumor cells resistant to bortezomib [51]. More recently, there has been an effort to prepare an orally bioavailable analog of carfilzomib, resulting in PR-047 [55]. It was shown that PR-047 tolerates repeated oral administration at doses resulting in >80% proteasome inhibition in most tissues and elicited an antitumor response equivalent to intravenously administered carfilzomib in multiple human tumor xenograft and mouse syngeneic models. This indicates that PR-047 may have potential for improved dosing flexibility and patient convenience over intravenously administered agents.

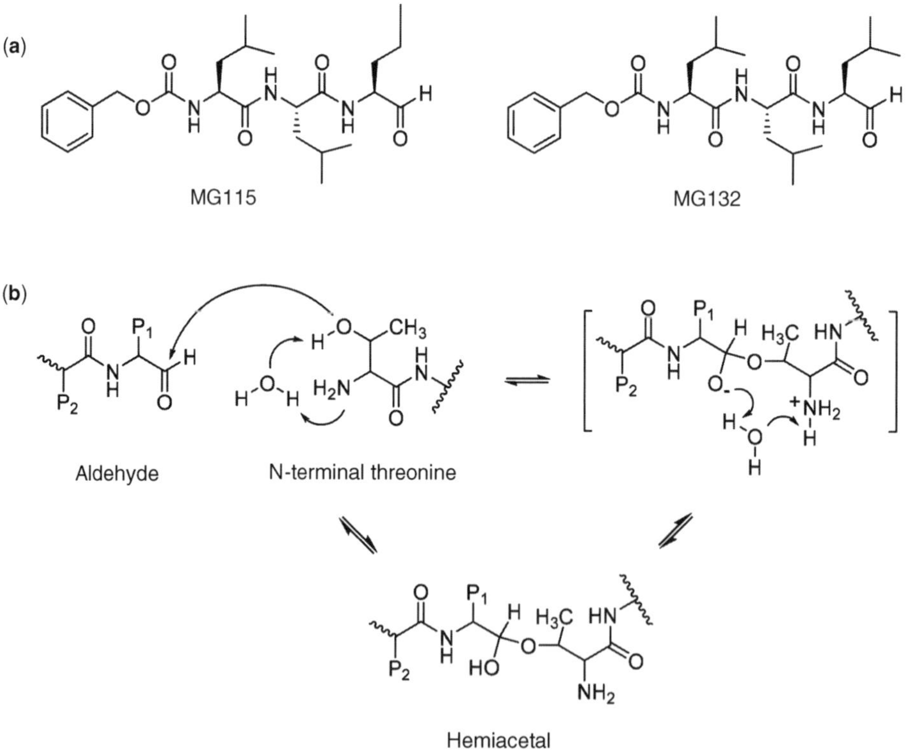

Figure 5. (a) Synthetic peptide aldehyde proteasome inhibitors. (b) The proposed inhibitory mechanism of the aldehyde pharmacophore.

4. MEDICINAL CHEMISTRY OF OTHER KNOWN PROTEASOME INHIBITORS

In addition to these proteasome inhibitors undergoing clinical trials, a number of proteasome inhibitors having other pharmacophores have been developed. While many of these compounds are potent proteasome inhibitors, the majority will not be developed into potential therapeutic agents. For these compounds, several concerns regarding clinical feasibility have been raised, such as inadequate specificity toward the proteasome. Meanwhile, new types of compounds targeting the immunoproteasome are currently being investigated as potential therapeutic agents [56–58].

Peptide aldehydes were the first compounds used as proteasome inhibitors (Fig. 5) [59]. Originally, these cell-permeable compounds were known to inhibit serine and cysteine proteases. Later, they were found to potently but reversibly inhibit the chymotrypsin-like activity of the 20S proteasome, and as such were the first example of substrate-based inhibitors [60]. Such inhibitors have been extensively studied and shown to block MHC class I antigen processing while also selectively targeting cancer cells for apoptosis [61–63] and are still widely used as molecular probes for the study of proteasome biology. MG132 and MG115, tripeptide aldehydes developed by Rock and colleagues, are the two most commonly used peptide aldehydes in proteasome biology [64]. Although these molecules are potent proteasome inhibitors, they show off-target activity by inhibiting other proteases. For this reason, the electrophilic aldehyde warhead has limited potential as a pharmaceutical agent.

Peptide vinylsulfones are another class of proteasome inhibitors (Fig. 6) [65,66]. The vinylsulfone pharmacophore acts as a Michael acceptor, thereby forming a covalent modification with the nucleophilic hydroxyl side

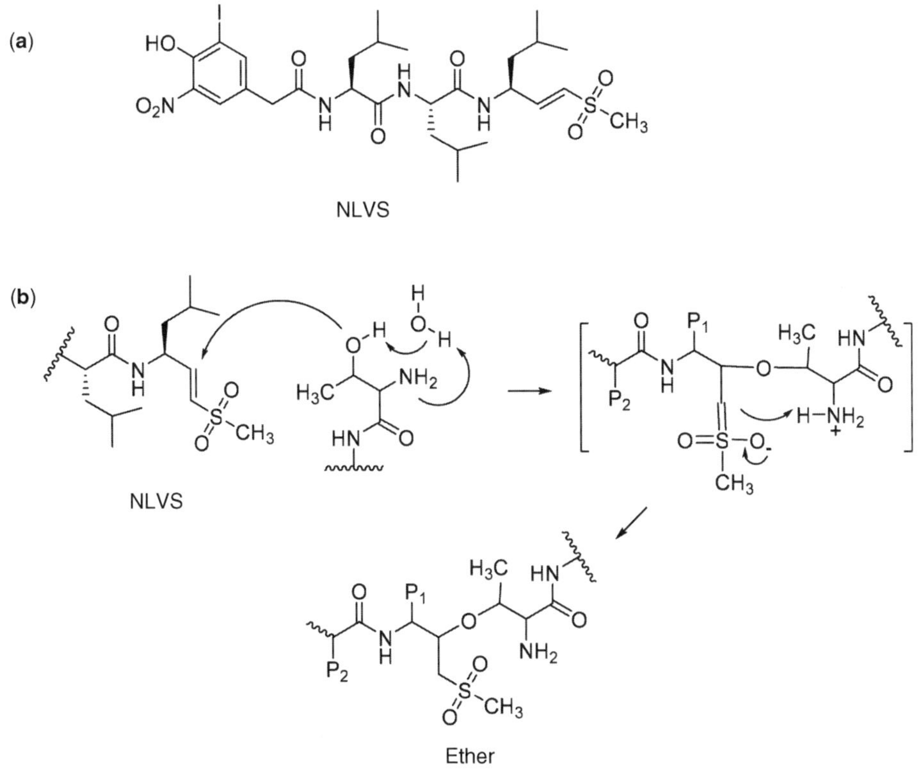

Figure 6. (a) A synthetic peptide vinylsulfone proteasome inhibitor. (b) The proposed inhibitory mechanism of the peptide vinylsulfone.

chain of the catalytic β-subunits of the proteasome [67,68]. However, the lack of specificity is again a major concern for this class of inhibitors, as peptide vinylsulfones inhibit both the proteasome and the cysteine proteases, restricting their further development as a therapeutic agent. Recently, analogs of vinylsulfone proteasome inhibitors have found increasing use as molecular probes in cellular models of disease states. For example, fluorescent-tagged peptide vinylsulfones have been developed to correlate the real-time activity of the proteasome with disease processes [69–71].

To overcome these specificity issues, an extensive medicinal chemistry approach has been undertaken to develop proteasome inhibitors with higher potency or efficacy, either by introducing new types of pharmacophores or by optimizing the backbone of known inhibitors. For example, researchers have introduced new types of pharmacophores that mimic the substrate–peptide bond. These new pharmacophores include β-lactam [72], semicarbazone [73], furan [74], α-ketocarbonyl [75,76], C-terminal constrained phenylalanine [77], and epoxypiperidine [78] (Fig. 7). Among these, the constrained C-terminal group, such as a tetrahydroisoquinoline (constrained version of phenylalanine), provided a good lead compound for the development of new analogs with improved pharmacokinetics. A structure-based optimization approach has also been widely applied to design better proteasome inhibitors [59]. For example, extended hydrophobic groups were introduced at the N-terminus of inhibitors to accommodate the hydrophobic pockets of the proteasome active site (Fig. 7) [79,80]. Meanwhile, the peptide backbone of inhibitors was derivatized to improve binding affinity. One such example is a cyclic peptide inhibitor (Fig. 7) [81], whose structure is derived from the backbone of TMC-95A, a macrocyclic natural product proteasome inhibitor. This approach is based on the hypothesis that a rigid

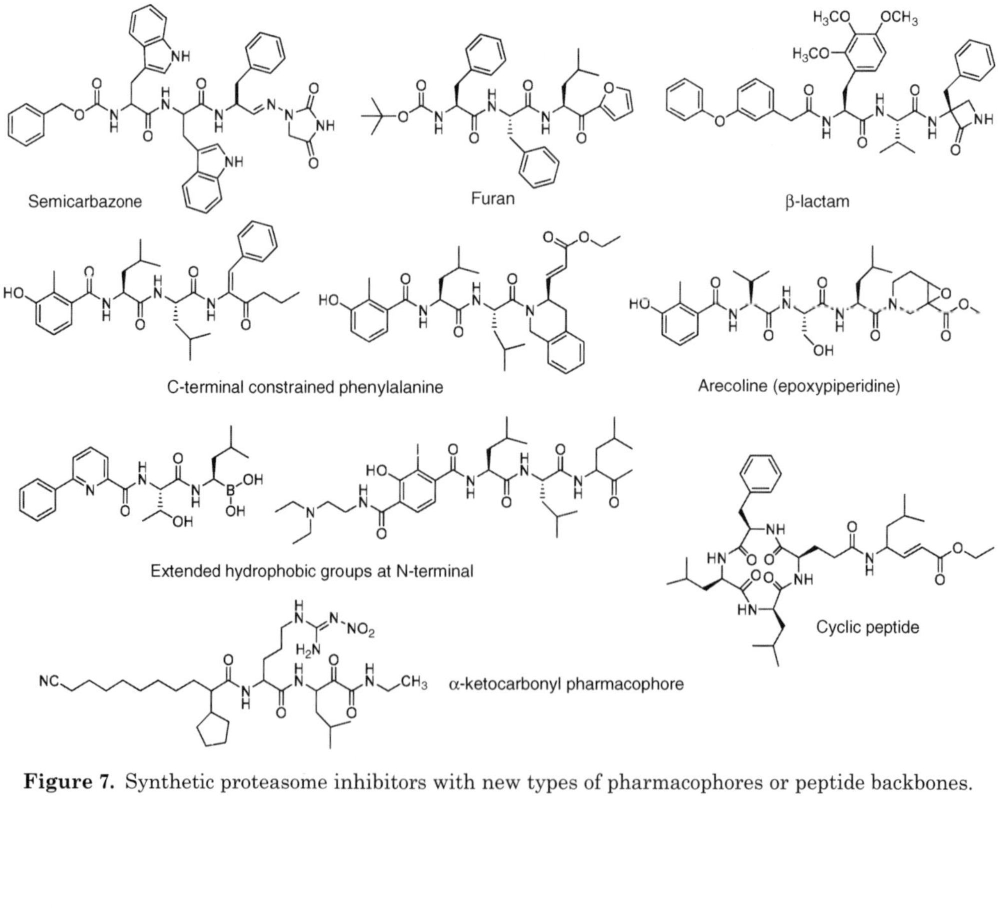

Figure 7. Synthetic proteasome inhibitors with new types of pharmacophores or peptide backbones.

conformation could provide much higher potency and selectivity [16]. On the other hand, some researchers have attempted to optimize proteasome inhibitors at the P2–P4 positions, either by constructing a small library of inhibitors [82] or by introducing a fluoro group or a long hydrocarbon chain [83].

In addition to synthetic approaches, natural products have been a rich source of proteasome inhibitors. Lactacystin, a metabolite of *Streptomyces lactacystinaeus*, is a natural product with a β-lactone pharmacophore that inhibits the proteasome [84]. Lactacystin was initially thought to be highly specific toward the proteasome; however, it was later revealed that it also inhibits other proteases [85,86]. Despite this concern, lactacystin is currently one of the most widely used proteasome inhibitors in laboratories worldwide for the study of proteasome biology.

As previously mentioned, a family of linear peptide epoxyketones is another class of proteasome inhibitors arising from natural products. The antitumor natural product epoxomicin (Fig. 8) is a widely studied peptide epoxyketone that targets the proteasome [53,54,87,88]. Epoxomicin is a highly specific proteasome inhibitor that preferentially targets the CT-L activity. Two of its derivatives, PR-171 and PR-047, have promising preclinical efficacy with improved delivery, illustrating the potential of natural product-based synthetic libraries. In addition to epoxomicin, natural products have provided a number of additional linear peptide α',β'-epoxyketone proteasome inhibitors [59,89].

A number of proteasome inhibitors have been discovered through systematic natural product screening. For example, TMC-95 (Fig. 8) is a macrocyclic molecule, screened

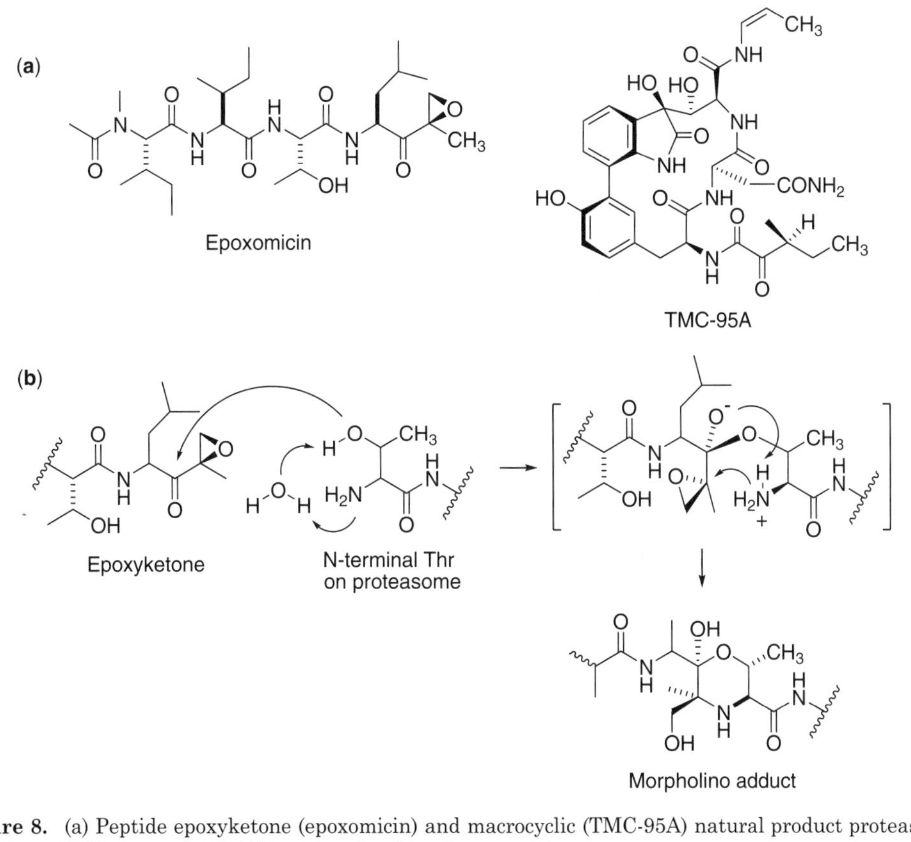

Figure 8. (a) Peptide epoxyketone (epoxomicin) and macrocyclic (TMC-95A) natural product proteasome inhibitors. (b) The proposed inhibitory mechanism of epoxomicin.

from the fermentation broth of *Apiospora montagnei* Sacc. (TC 1093) [90,91]. In addition, many traditional medicines have been shown to exert their activity through targeting of the proteasome [92]. Examples include epigallocatechin-3-gallate (EGCG), celastrol, genistein (Fig. 9). Researchers have successfully derivatized 18β-glycyrrhetinic acid to yield compounds with high potency and specificity toward the CT-L activity of the proteasome [93]. These kinds of nonpeptidic proteasome inhibitors may have some advantages, such as bioavailability and pharmacokinetics, over peptide-based inhibitors. Similarly, EGCG, a major component of green tea, has been a subject for derivatization [94–96]. These natural products may provide not only potential therapeutic agents but also tools for proteasome biology.

5. INHIBITORS OF ALTERNATIVE PROTEASOMES: IMMUNOPROTEASOME-SPECIFIC INHIBITORS

Recent studies indicate that the immunoproteasome may be implicated in a number of diseases, such as neurodegenerative diseases, inflammatory bowel diseases (IBD) [97,98], and cancers [56,99]. Currently, ongoing efforts are focused on the development of immunoproteasome-specific inhibitors for therapeutic applications. Thus far, only a limited number of immunoproteasome-specific inhibitors have been developed (Fig. 10) [56–58]. An important implication of immunoproteasome-specific inhibitors is that immunoproteasomes are not constitutively expressed in many human tissues. As compared to broadly acting proteasome inhibitors, which target essential biological processes in all eukaryotic cells, immunoproteasome-specific inhibitors provide an opportunity for therapeutic intervention with significantly lower toxicity.

6. CONCLUSION

The proteasome has been validated as a target for cancer treatment, and a proteasome inhibitor (bortezomib) has been approved by the FDA for the treatment of multiple myeloma and relapsed mantle cell lymphoma. However, drug-associated side effects still remain a major concern, as the proteasome is essential for normal cellular activities. Thus, continuous

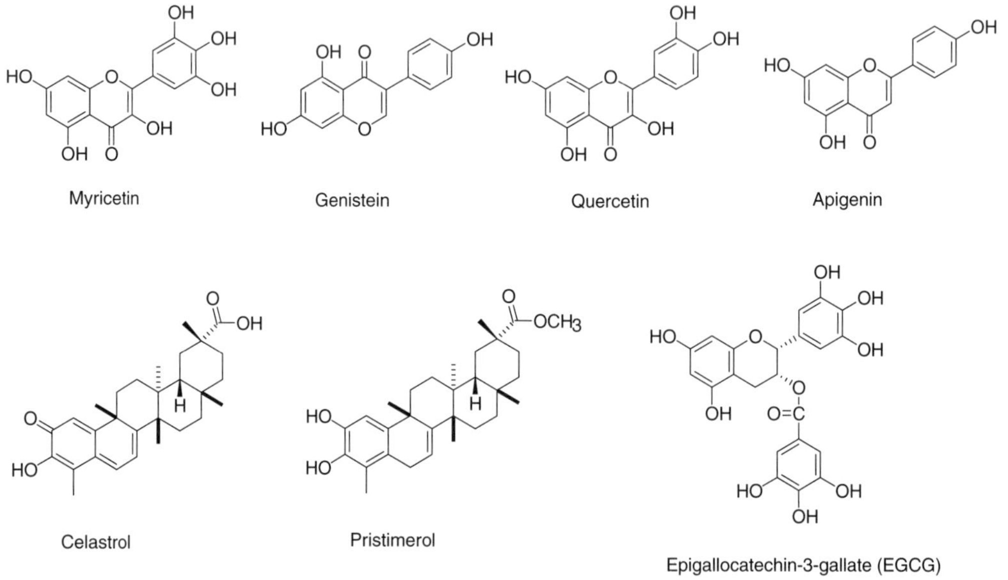

Figure 9. Dietary phytochemicals whose mode of action has been partially attributed to proteasome inhibition.

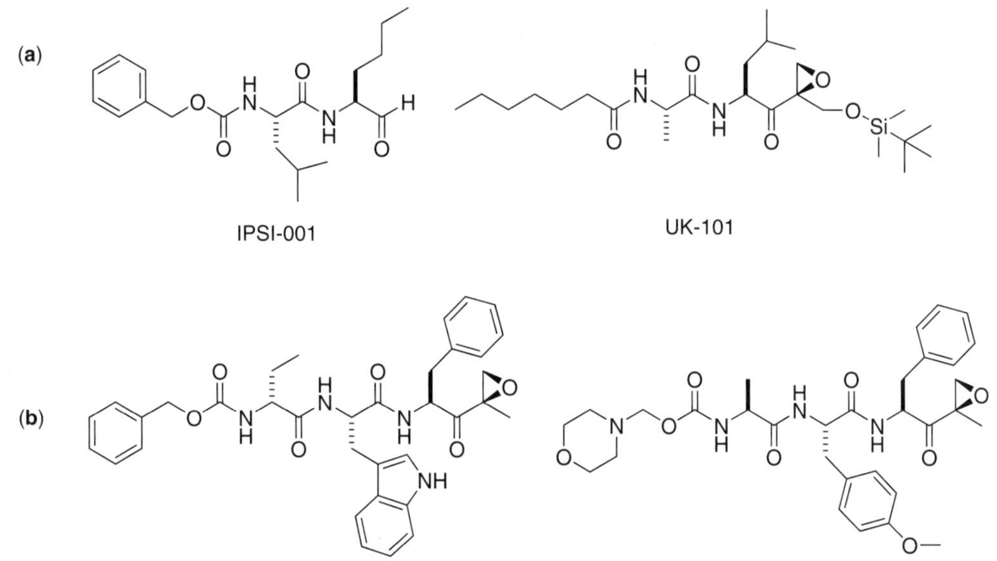

Figure 10. (a) Immunoproteasome LMP2 subunit-specific inhibitors. (b) Selective inhibitors of the immunoproteasome LMP7 catalytic subunit developed by Proteolix Inc.

efforts toward the development of new types of proteasome inhibitors with lower toxicity are actively ongoing.

REFERENCES

1. Verma R, Aravind L, Oania R, McDonald WH, Yates JR, 3rd, Koonin EV, Deshaies RJ. Role of Rpn11 metalloprotease in deubiquitination and degradation by the 26S proteasome. Science 2002;298:611–615.
2. Hershko A, Ciechanover A. The ubiquitin system. Annu Rev Biochem 1998;67:425–479.
3. Hershko A, Ciechanover A, Varshavsky A. Basic Medical Research Award. The ubiquitin system. Nat Med 2000;6:1073–1081.
4. Hilt W, Enenkel C, Gruhler A, Singer T, Wolf DH. The PRE4 gene codes for a subunit of the yeast proteasome necessary for peptidylglutamyl-peptide-hydrolyzing activity. Mutations link the proteasome to stress- and ubiquitin-dependent proteolysis. J Biol Chem 1993; 268:3479–3486.
5. Enenkel C, Lehmann H, Kipper J, Guckel R, Hilt W, Wolf DH. PRE3, highly homologous to the human major histocompatibility complex-linked LMP2 (RING12) gene, codes for a yeast proteasome subunit necessary for the peptidyl-glutamyl-peptide hydrolyzing activity. FEBS Lett 1994;341:193–196.
6. Heinemeyer W, Gruhler A, Mohrle V, Mahe Y, Wolf DH. PRE2, highly homologous to the human major histocompatibility complex- linked RING10 gene, codes for a yeast proteasome subunit necessary for chrymotryptic activity and degradation of ubiquitinated proteins. J Biol Chem 1993;268:5115–5120.
7. Salzmann U, Kral S, Braun B, Standera S, Schmidt M, Kloetzel PM, Sijts A. Mutational analysis of subunit I beta2 (MECL-1) demonstrates conservation of cleavage specificity between yeast and mammalian proteasomes. FEBS Lett 1999;454:11–15.
8. Cardozo C. Catalytic components of the bovine pituitary multicatalytic proteinase complex (proteasome). Enzyme Protein 1993;47: 296–305.
9. Orlowski M. The multicatalytic proteinase complex (proteasome) and intracellular protein degradation: diverse functions of an intracellular particle. J Lab Clin Med 1993;121:187–189.
10. Kisselev AF, Callard A, Goldberg AL. Importance of the different proteolytic sites of the proteasome and the efficacy of inhibitors varies with the protein substrate. J Biol Chem 2006;281:8582–8590.
11. Figueiredo-Pereira ME, Berg KA, Wilk S. A new inhibitor of the chymotrypsin-like activity of the multicatalytic proteinase complex (20S proteasome) induces accumulation of ubiquitin–protein conjugates in a neuronal cell. J Neurochem 1994;63:1578–1581.

12. Figueiredo-Pereira ME, Chen WE, Li J, Johdo O. The antitumor drug aclacinomycin A, which inhibits the degradation of ubiquitinated proteins, shows selectivity for the chymotrypsin-like activity of the bovine pituitary 20S proteasome. J Biol Chem 1996;271:16455–16459.
13. Myung J, Kim KB, Lindsten K, Dantuma NP, Crews CM. Lack of proteasome active site allostery as revealed by subunit-specific inhibitors. Mol Cell 2001;7:411–420.
14. Kloetzel PM. Antigen processing by the proteasome. Nat Rev Mol Cell Biol 2001;2:179–187.
15. Unno M, Mizushima T, Morimoto Y, Tomisugi Y, Tanaka K, Yasuoka N, Tsukihara T. The structure of the mammalian 20S proteasome at 2.75 Å resolution. Structure 2002;10:609–618.
16. Borissenko L, Groll M. 20S proteasome and its inhibitors: crystallographic knowledge for drug development. Chem Rev 2007;107:687–717.
17. Adams J. The proteasome: a suitable antineoplastic target. Nat Rev Cancer 2004;4:349–360.
18. Adams J. The development of proteasome inhibitors as anticancer drugs. Cancer Cell 2004;5:417–421.
19. Glotzer M, Murray AW, Kirschner MW. Cyclin is degraded by the ubiquitin pathway. Nature 1991;349:132–138.
20. Voorhees PM, Dees EC, O'Neil B, Orlowski RZ. The proteasome as a target for cancer therapy. Clin Cancer Res 2003;9:6316–6325.
21. Karin M, Ben-Neriah Y. Phosphorylation meets ubiquitination: the control of NF-κB activity. Annu Rev Immunol 2000;18:621–663.
22. Karin M, Delhase M. The I kappa B kinase (IKK) and NF-kappa B: key elements of proinflammatory signalling. Semin Immunol 2000;12:85–98.
23. Pikarsky E, Ben-Neriah Y. NF-kappaB inhibition: a double-edged sword in cancer? Eur J Cancer 2006;42:779–784.
24. Voorhees PM, Orlowski RZ. The proteasome and proteasome inhibitors in cancer therapy. Annu Rev Pharmacol Toxicol 2006;46:189–213.
25. Sethi G, Sung B, Aggarwal BB. Nuclear factor-kappaB activation: from bench to bedside. Exp Biol Med (Maywood) 2008;233:21–31.
26. Orlowski RZ, Baldwin AS, Jr. NF-kappaB as a therapeutic target in cancer. Trends Mol Med 2002;8:385–389.
27. Lakshmanan M, Bughani U, Duraisamy S, Diwan M, Dastidar S, Ray A. Molecular targeting of E3 ligases: a therapeutic approach for cancer. Expert Opin Ther Targets 2008;12:855–870.
28. Bernassola F, Karin M, Ciechanover A, Melino G. The HECT family of E3 ubiquitin ligases: multiple players in cancer development. Cancer Cell 2008;14:10–21.
29. Ardley HC, Robinson PA. E3 ubiquitin ligases. Essays Biochem 2005;41:15–30.
30. Vazquez A, Bond EE, Levine AJ, Bond GL. The genetics of the p53 pathway, apoptosis and cancer therapy. Nat Rev Drug Discov 2008;7:979–987.
31. Chan DA, Giaccia AJ. Targeting cancer cells by synthetic lethality: autophagy and VHL in cancer therapeutics. Cell Cycle 2008;7:2987–2990.
32. Shangary S, Wang S. Targeting the MDM2-p53 interaction for cancer therapy. Clin Cancer Res 2008;14:5318–5324.
33. Cardozo T, Pagano M. Wrenches in the works: drug discovery targeting the SCF ubiquitin ligase and APC/C complexes. BMC Biochem 2007;8(1 Suppl): S9.
34. Adams J. Development of the proteasome inhibitor PS-341. Oncologist 2002;7:9–16.
35. Fevig JM, Buriak J, Jr, Cacciola J, Alexander RS, Kettner CA, Knabb RM, Pruitt JR, Weber PC, Wexler RR. Rational design of boropeptide thrombin inhibitors: beta, beta-dialkyl- phenethylglycine P2 analogs of DuP 714 with greater selectivity over complement factor I and an improved safety profile. Bioorg Med Chem Lett 1998;8:301–306.
36. Adams J, Behnke M, Chen S, Cruickshank AA, Dick LR, Grenier L, Klunder JM, Ma YT, Plamondon L, Stein RL. Potent and selective inhibitors of the proteasome: dipeptidyl boronic acids. Bioorg Med Chem Lett 1998;8:333–338.
37. Adams J, Palombella VJ, Sausville EA, Johnson J, Destree A, Lazarus DD, Maas J, Pien CS, Prakash S, Elliott PJ. Proteasome inhibitors: a novel class of potent and effective antitumor agents. Cancer Res 1999;59:2615–2622.
38. Orlowski RZ, Stinchcombe TE, Mitchell BS, Shea TC, Baldwin AS, Stahl S, Adams J, Esseltine DL, Elliott PJ, Pien CS, Guerciolini R, Anderson JK, Depcik-Smith ND, Bhagat R, Lehman MJ, Novick SC, O'Connor OA, Soignet SL. Phase I trial of the proteasome inhibitor PS-341 in patients with refractory hematologic malignancies. J Clin Oncol 2002;20:4420–4427.
39. Neubert K, Meister S, Moser K, Weisel F, Maseda D, Amann K, Wiethe C, Winkler TH, Kalden JR, Manz RA, Voll RE. The proteasome inhibitor bortezomib depletes plasma cells and protects mice with lupus-like disease from nephritis. Nat Med 2008;14:748–755.
40. Aghajanian C, Soignet S, Dizon DS, Pien CS, Adams J, Elliott PJ, Sabbatini P, Miller V, Hensley ML, Pezzulli S, Canales C, Daud A,

Spriggs DR. A phase I trial of the novel proteasome inhibitor PS341 in advanced solid tumor malignancies. Clin Cancer Res 2002;8: 2505–2511.

41. Terpos E, Kastritis E, Roussou M, Heath D, Christoulas D, Anagnostopoulos N, Eleftherakis-Papaiakovou E, Tsionos K, Croucher P, Dimopoulos MA. The combination of bortezomib, melphalan, dexamethasone and intermittent thalidomide is an effective regimen for relapsed/refractory myeloma and is associated with improvement of abnormal bone metabolism and angiogenesis. Leukemia 2008; 22:2292.

42. Sterz J, von Metzler I, Hahne JC, Lamottke B, Rademacher J, Heider U, Terpos E, Sezer O. The potential of proteasome inhibitors in cancer therapy. Expert Opin Investig Drugs 2008;17: 879–895.

43. Davies AM, Chansky K, Lara PN, Jr, Gumerlock PH, Crowley J, Albain KS, Vogel SJ, Gandara DR. Bortezomib plus gemcitabine/carboplatin as first-line treatment of advanced non-small cell lung cancer: a phase II Southwest Oncology Group Study (S0339). J Thorac Oncol 2009;4:87–92.

44. Argyriou AA, Iconomou G, Kalofonos HP. Bortezomib-induced peripheral neuropathy in multiple myeloma: a comprehensive review of the literature. Blood 2008;112:1593–1599.

45. Roussou M, Kastritis E, Migkou M, Psimenou E, Grapsa I, Matsouka C, Barmparousi D, Terpos E, Dimopoulos MA. Treatment of patients with multiple myeloma complicated by renal failure with bortezomib-based regimens. Leuk Lymphoma 2008;49:890–895.

46. Lu C, Gallegos R, Li P, Xia CQ, Pusalkar S, Uttamsingh V, Nix D, Miwa GT, Gan LS. Investigation of drug–drug interaction potential of bortezomib in vivo in female Sprague–Dawley rats and in vitro in human liver microsomes. Drug Metab Dispos 2006;34:702–708.

47. Joazeiro CA, Anderson KC, Hunter T. Proteasome inhibitor drugs on the rise. Cancer Res 2006;66:7840–7842.

48. Chauhan D, Hideshima T, Anderson KC. A novel proteasome inhibitor NPI-0052 as an anticancer therapy. Br J Cancer 2006;95:961–965.

49. Ruiz S, Krupnik Y, Keating M, Chandra J, Palladino M, McConkey D. The proteasome inhibitor NPI-0052 is a more effective inducer of apoptosis than bortezomib in lymphocytes from patients with chronic lymphocytic leukemia. Mol Cancer Ther 2006;5:1836–1843.

50. Mitsiades CS, Hayden PJ, Anderson KC, Richardson PG. From the bench to the bedside: emerging new treatments in multiple myeloma. Best Pract Res Clin Haematol 2007;20:797–816.

51. Demo SD, Kirk CJ, Aujay MA, Buchholz TJ, Dajee M, Ho MN, Jiang J, Laidig GJ, Lewis ER, Parlati F, Shenk KD, Smyth MS, Sun CM, Vallone MK, Woo TM, Molineaux CJ, Bennett MK. Antitumor activity of PR-171, a novel irreversible inhibitor of the proteasome. Cancer Res 2007;67:6383–6391.

52. Stapnes C, Doskeland AP, Hatfield K, Ersvaer E, Ryningen A, Lorens JB, Gjertsen BT, Bruserud O. The proteasome inhibitors bortezomib and PR-171 have antiproliferative and proapoptotic effects on primary human acute myeloid leukaemia cells. Br J Haematol 2007;136: 814–828.

53. Meng L, Mohan R, Kwok BH, Elofsson M, Sin N, Crews CM. Epoxomicin, a potent and selective proteasome inhibitor, exhibits in vivo antiinflammatory activity. Proc Natl Acad Sci USA 1999;96:10403–10408.

54. Groll M, Kim KB, Kairies N, Huber R, Crews CM. Crystal structure of epoxomicin:20S proteasome reveals a molecular basis for selectivity of α',β'-epoxyketone proteasome inhibitors. J Am Chem Soc 2000;122:1237–1238.

55. Zhou HJ, Aujay MA, Bennett MK, Dajee M, Demo SD, Fang Y, Ho MN, Jiang J, Kirk CJ, Laidig GJ, Lewis ER, Lu Y, Muchamuel T, Parlati F, Ring E, Shenk KD, Shields J, Shwonek PJ, Stanton T, Sun CM, Sylvain C, Woo TM, Yang J. Design and synthesis of an orally bioavailable and selective peptide epoxyketone proteasome inhibitor (PR-047). J Med Chem 2009.

56. Ho YK, Bargagna-Mohan P, Mohan R, Kim KB. LMP2-specific inhibitors: novel chemical genetic tools for proteasome biology. Chem Biol 2007;14:419–430.

57. Orlowski R, Orlowski M, Potent and specific immunoproteasome inhibitors US Patent Application 20060241056, 2006.

58. Kuhn DJ, Hunsucker SA, Chen Q, Voorhees PM, Orlowski M, Orlowski RZ. Targeted inhibition of the immunoproteasome is a potent strategy against models of multiple myeloma that overcomes resistance to conventional drugs and non-specific proteasome inhibitors. Blood 2008.

59. Kim KB, Crews CM. Natural product and synthetic proteasome inhibitors. In: Adams J, editors. Cancer Drug Discovery and Development: Proteasome Inhibitors in Cancer Therapy. Totowa, NJ: Humana Press Inc.; 2003. p 47–63.

60. Iqbal M, Chatterjee S, Kauer JC, Das M, Messina P, Freed B, Biazzo W, Siman R. Potent inhibitors of proteasome. J Med Chem 1995;38: 2276–2277.

61. Harding CV, France J, Song R, Farah JM, Chatterjee S, Iqbal M, Siman R. Novel dipeptide aldehydes are proteasome inhibitors and block the MHC-I antigen-processing pathway. J Immunol 1995;155:1767–1775.
62. An B, Goldfarb RH, Siman R, Dou QP. Novel dipeptidyl proteasome inhibitors overcome Bcl-2 protective function and selectively accumulate the cyclin-dependent kinase inhibitor p27 and induce apoptosis in transformed, but not normal, human fibroblasts. Cell Death Differ 1998;5:1062–1075.
63. Sun J, Nam S, Lee CS, Li B, Coppola D, Hamilton AD, Dou QP, Sebti SM. CEP1612, a dipeptidyl proteasome inhibitor, induces p21WAF1 and p27KIP1 expression and apoptosis and inhibits the growth of the human lung adenocarcinoma A-549 in nude mice. Cancer Res 2001;61: 1280–1284.
64. Rock KL, Gramm C, Rothstein L, Clark K, Stein R, Dick L, Hwang D, Goldberg AL. Inhibitors of the proteasome block the degradation of most cell proteins and the generation of peptides presented on MHC class I molecules. Cell 1994;78:761–771.
65. Bogyo M, McMaster JS, Gaczynska M, Tortorella D, Goldberg AL, Ploegh H. Covalent modification of the active site threonine of proteasomal beta subunits and the *Escherichia coli* homolog HslV by a new class of inhibitors. Proc Natl Acad Sci USA 1997;94:6629–6634.
66. Bogyo M, Shin S, McMaster JS, Ploegh HL. Substrate binding and sequence preference of the proteasome revealed by active-site-directed affinity probes. Chem Biol 1998;5:307–320.
67. Bromme D, Klaus JL, Okamoto K, Rasnick D, Palmer JT. Peptidyl vinyl sulfones: a new class of potent and selective cysteine protease inhibitors. Biochem J 1996;315:85–89.
68. Palmer JT, Rasnick D, Klaus JL, Bromme D. Vinyl sulfones as mechanism-based cysteine protease inhibitors. J Med Chem 1995;38: 3193–3196.
69. Verdoes M, Berkers CR, Florea BI, van Swieten PF, Overkleeft HS, Ovaa H. Chemical proteomics profiling of proteasome activity. Methods Mol Biol 2006;328:51–69.
70. Verdoes M, Florea BI, Menendez-Benito V, Maynard CJ, Witte MD, van der Linden WA, van den Nieuwendijk AM, Hofmann T, Berkers CR, van Leeuwen FW, Groothuis TA, Leeuwenburgh MA, Ovaa H, Neefjes JJ, Filippov DV, van der Marel GA, Dantuma NP, Overkleeft HS. A fluorescent broad-spectrum proteasome inhibitor for labeling proteasomes *in vitro* and *in vivo*. Chem Biol 2006;13:1217–1226.
71. Berkers CR, van Leeuwen FW, Groothuis TA, Peperzak V, van Tilburg EW, Borst J, Neefjes JJ, Ovaa H. Profiling proteasome activity in tissue with fluorescent probes. Mol Pharm 2007.
72. Imbach P, Lang M, Garcia-Echeverria C, Guagnano V, Noorani M, Roesel J, Bitsch F, Rihs G, Furet P. Novel beta-lactam derivatives: potent and selective inhibitors of the chymotrypsin-like activity of the human 20S proteasome. Bioorg Med Chem Lett 2007;17:358–362.
73. Leban J, Blisse M, Krauss B, Rath S, Baumgartner R, Seifert MH. Proteasome inhibition by peptide-semicarbazones. Bioorg Med Chem 2008;16:4579–4588.
74. Fu Y, Xu B, Zou X, Ma C, Yang X, Mou K, Fu G, Lu Y, Xu P. Design and synthesis of a novel class of furan-based molecules as potential 20S proteasome inhibitors. Bioorg Med Chem Lett 2007;17:1102–1106.
75. Iqbal M, Chatterjee S, Kauer JC, Mallamo JP, Messina PA, Reiboldt A, Siman R. Potent α-ketocarbonyl and boronic ester derived inhibitors of proteasome. Bioorg Med Chem Lett 1996;6:287–290.
76. Chatterjee S, Dunn D, Mallya S, Ator MA. P'-extended alpha-ketoamide inhibitors of proteasome. Bioorg Med Chem Lett 1999;9: 2603–2606.
77. Baldisserotto A, Marastoni M, Lazzari I, Trapella C, Gavioli R, Tomatis R. C-terminal constrained phenylalanine as a pharmacophoric unit in peptide-based proteasome inhibitors. Eur J Med Chem 2008;43:1403–1411.
78. Marastoni M, McDonald J, Baldisserotto A, Canella A, De Risi C, Pollini GP, Tomatis R. Proteasome inhibitors; synthesis and activity of arecoline oxide tripeptide derivatives. Bioorg Med Chem Lett 2004;14:1965–1968.
79. Dorsey BD, Iqbal M, Chatterjee S, Menta E, Bernardini R, Bernareggi A, Cassara PG, D'Arasmo G, Ferretti E, De Munari S, Oliva A, Pezzoni G, Allievi C, Strepponi I, Ruggeri B, Ator MA, Williams M, Mallamo JP. Discovery of a potent, selective, and orally active proteasome inhibitor for the treatment of cancer. J Med Chem 2008;51:1068–1072.
80. Vivier M, Rapp M, Papon J, Labarre P, Galmier MJ, Sauziere J, Madelmont JC. Synthesis, radiosynthesis, and biological evaluation of new proteasome inhibitors in a tumor targeting approach. J Med Chem 2008;51: 1043–1047.
81. Baldisserotto A, Marastoni M, Fiorini S, Pretto L, Ferretti V, Gavioli R, Tomatis R. Vinyl ester-based cyclic peptide proteasome inhibitors. Bioorg Med Chem Lett 2008;18:1849–1854.

82. Marastoni M, Baldisserotto A, Trapella C, Gavioli R, Tomatis R. P3 and P4 position analysis of vinyl ester pseudopeptide proteasome inhibitors. Bioorg Med Chem Lett 2006;16: 3125–3130.
83. Formicola L, Marechal X, Basse N, Bouvier-Durand M, Bonnet-Delpon D, Milcent T, Reboud-Ravaux M, Ongeri S. Novel fluorinated pseudopeptides as proteasome inhibitors. Bioorg Med Chem Lett 2009;19:83–86.
84. Omura S, Fujimoto T, Otoguro K, Matsuzaki K, Moriguchi R, Tanaka H, Sasaki Y. Lactacystin, a novel microbial metabolite, induces neuritogenesis of neuroblastoma cells. J Antibiot 1991;44:113–116.
85. Ostrowska H, Wojcik C, Omura S, Worowski K. Lactacystin, a specific inhibitor of the proteasome, inhibits human platelet lysosomal cathepsin A-like enzyme. Biochem Biophys Res Commun 1997;234:729–732.
86. Ostrowska H, Wojcik C, Wilk S, Omura S, Kozlowski L, Stoklosa T, Worowski K, Radziwon P. Separation of cathepsin A-like enzyme and the proteasome: evidence that lactacystin/beta-lactone is not a specific inhibitor of the proteasome. Int J Biochem Cell Biol 2000;32:747–757.
87. Hanada M, Sugawara K, Kaneta K, Toda S, Nishiyama Y, Tomita K, Yamamoto H, Konishi M, Oki T. Epoxomicin, a new antitumor agent of microbial origin. J Antibiot 1992;45:1746–1752.
88. Sin N, Kim KB, Elofsson M, Meng L, Auth H, Kwok BH, Crews CM. Total synthesis of the potent proteasome inhibitor epoxomicin: a useful tool for understanding proteasome biology. Bioorg Med Chem Lett 1999;9:2283–2288.
89. Koguchi Y, Kohno J, Suzuki S, Nishio M, Takahashi K, Ohnuki T, Komatsubara S. TMC-86A, B and TMC-96, new proteasome inhibitors from *Streptomyces* sp. TC 1084 and *Saccharothrix* sp. TC 1094. II. Physico-chemical properties and structure determination. J Antibiot 53:2000; 63–65.
90. Koguchi Y, Kohno J, Nishio M, Takahashi K, Okuda T, Ohnuki T, Komatsubara S. TMC-95A, B, C, and D, novel proteasome inhibitors produced by *Apiospora montagnei* Sacc. TC 1093. Taxonomy, production, isolation, and biological activities. J Antibiot 53:2000; 105–109.
91. Kohno J, Koguchi Y, Nishio M, Nakao K, Kuroda M, Shimizu R, Ohnuki T, Komatsubara S. Structures of TMC-95A-D: novel proteasome inhibitors from *Apiospora montagnei* Sacc. TC 1093. J Org Chem 2000; 65:990–995.
92. Yang H, Landis-Piwowar KR, Chen D, Milacic V, Dou QP. Natural compounds with proteasome inhibitory activity for cancer prevention and treatment. Curr Protein Pept Sci 2008; 9:227–239.
93. Huang L, Yu D, Ho P, Qian K, Lee KH, Chen CH. Synthesis and proteasome inhibition of glycyrrhetinic acid derivatives. Bioorg Med Chem 2008;16:6696–6701.
94. Osanai K, Landis-Piwowar KR, Dou QP, Chan TH. A para-amino substituent on the D-ring of green tea polyphenol epigallocatechin-3-gallate as a novel proteasome inhibitor and cancer cell apoptosis inducer. Bioorg Med Chem 2007;15: 5076–5082.
95. Wan SB, Landis-Piwowar KR, Kuhn DJ, Chen D, Dou QP, Chan TH. Structure-activity study of epi-gallocatechin gallate (EGCG) analogs as proteasome inhibitors. Bioorg Med Chem 2005;13:2177–2185.
96. Wan SB, Chen D, Dou QP, Chan TH. Study of the green tea polyphenols catechin-3-gallate (CG) and epicatechin-3-gallate (ECG) as proteasome inhibitors. Bioorg Med Chem 2004; 12:3521–3527.
97. Fitzpatrick LR, Khare V, Small JS, Koltun WA. Dextran sulfate sodium-induced colitis is associated with enhanced low molecular mass polypeptide 2 (LMP2) expression and is attenuated in LMP2 knockout mice. Dig Dis Sci 2006; 51:1269–1276.
98. Fitzpatrick LR, Small JS, Poritz LS, McKenna KJ, Koltun WA. Enhanced intestinal expression of the proteasome subunit low molecular mass polypeptide 2 in patients with inflammatory bowel disease. Dis Colon Rectum 2007; 50:337–348; discussion 348–350.
99. Orlowski RZ. The ubiquitin proteasome pathway from bench to bedside. Hematology 2005; 220–225.

INHIBITORS OF KINESIN SPINDLE PROTEIN FOR THE TREATMENT OF CANCER

GUSTAVE BERGNES[1]
MAUREEN G. CONLAN[1]
STEVEN D. KNIGHT[2]

[1] Cytokinetics, Inc., South San Francisco, CA
[2] GlaxoSmithKline, Collegeville, PA

1. INTRODUCTION

Kinesin spindle protein (KSP/Eg5/kinesin 5) is a member of the superfamily of motor proteins known as kinesins [1]. Motor proteins are a class of enzymes that utilize ATP hydrolysis to power a cascade of large conformational changes that result in force generation against cytoskeletal structures, such as microtubules (MTs) [2]. Members of this family include not only kinesins but also myosins and dyneins. Within the kinesin superfamily are members that support cellular homeostasis and others that play roles in cell division. KSP is among those that act during mitosis and is responsible for driving the formation of a bipolar mitotic spindle (see Bergnes et al. for a review of mitotic kinesins) [3]. Inhibition of KSP, via genetic or chemical modulation, results in unseparated spindle poles, cell cycle arrest, and cell death.

By targeting mitosis, KSP-directed therapeutics would follow such anticancer agents as the taxanes and the vinca alkaloids. As successful as the latter two agent classes have been in the clinic, their targeting of MT polymerization dynamics has led to toxicities associated with perturbation of tubulin's wide range of cellular functions outside of cell division. In contrast, KSP acts by manipulating tubulin polymer during mitosis and as such is an accessory protein whose function is limited to dividing cells. In this way, inhibitors of KSP would be expected to exhibit many of the antitumor properties of the so-called "MT poisons" while avoiding many of the off-target effects associated with existing agents.

In this chapter, we discuss the challenges involved in targeting a structurally and functionally unique protein and the ways in which this problem was approached by various teams of scientists. The focus will be predominantly on how lead molecules for optimization were discovered and how those have led to proof of concept experiments preclinically and clinically.

2. MITOSIS AS A THERAPEUTIC TARGET

Proliferation of cancer cells, as with all dividing cells, requires progression through a cycle of events that includes duplication of the genetic material (interphase) and the separation of that material in equal parts into two daughter cells (M phase). The precise distribution of those two identical genetic blueprints is managed through a highly coordinated set of events, or phases, known as mitosis (Fig 1). Validation of mitosis as a target for chemotherapeutic intervention was established by natural products in the taxane and vinca alkaloid structural classes [4]. Even before their targets were known, their preclinical properties and visible perturbation of mitosis helped progress these agents into clinical trials. Interestingly, both taxanes and vinca alkaloids target tubulin, a protein of the cytoskeleton that plays essential roles in all phases of cellular life, but with opposing mechanisms of action [5]. While taxanes potently stabilize MTs, vinca alkaloids prevent tubulin polymerization. For example, cells treated with *paclitaxel* have much of their soluble tubulin depleted to form disorganized MTs that fail to form a mitotic spindle. Conversely, cells treated with vincristine or vinblastine have difficulty forming MTs in mitosis and hence also fail to form a viable mitotic spindle. Regardless of mechanism, both perturb tubulin polymerization dynamics at all stages in the cell cycle [6]. As a consequence of this lack of specificity for mitotic cells, these mechanisms can have deleterious effects on interphase cells, such as neuronal cells, resulting in dose-limiting peripheral neuropathies [7,8].

Although tubulin is the most abundant protein involved in mitosis, many other proteins play critical roles in the successful completion of cell division [9–11]. Kinesins are proteins that interact with MTs to perform functions that include cargo transport [12], spindle and chromosome movement [13,14],

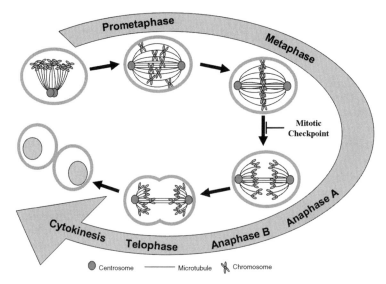

Figure 1. Illustration of the major stages of mitosis from the formation of the mitotic spindle in prometaphase to the generation of two fully formed daughter cells. The mitotic spindle, a structure that includes centrosomes, MTs and chromosomes, evolves into a bipolar structure during prometaphase, driven by the action of KSP. In this elliptical structure, the individual chromosomes become fully attached to MTs emanating from both centrosomes and migrate toward a central position in the spindle, which is known as the metaphase plate. When all the chromosomes have aligned in the center, the mitotic checkpoint disengages, chromosomes break apart into their sister chromatids, and MTs shorten, drawing the chromatids toward their respective spindle poles. Finally, by the process of cytokinesis, a contractile ring forms between the two nascent nuclei that pinches to form two daughter cells. (This figure is available in full color at http://mrw.interscience.wiley.com/emrw/9780471266945/home.)

and management of microtubule dynamics [15,16]. Mitotic kinesins are those that play unique roles exclusive to mitosis. Specifically, they provide the mechanochemical force required to manage the cell-wide movements that align, divide, and parse the cell's genetic materials. KSP has no effect on tubulin polymerization dynamics *per se*, but rather acts by manipulating existing polymer to drive mitotic spindle formation and function. In cells, KSP bundles microtubules and uses the energy derived from ATP hydrolysis to slide antiparallel MTs relative to one another. The consequence of this activity in prometaphase is the pushing apart of mitotic centrosomes to generate a bipolar mitotic spindle [17]. Prior to the identification of small-molecule affectors, inhibitory antibodies were used to demonstrate that KSP inhibition alone was sufficient to bring about mitotic arrest and cell death [18–20]. KSP is present in all dividing cells and has an exclusive role in M-phase [19]. Furthermore, it has been demonstrated that KSP is most abundant in tumor tissue relative to normal adjacent tissue [21]. As a result, its role as an accessory protein, its nonredundant function in mitosis, the requirement for activation, and higher levels of expression in tumor tissue contribute to the specificity of agents targeting this protein.

3. KSP STRUCTURE AND MECHANISM

In general, kinesins are composed of three structural domains: a motor (or head) domain, a rod (or linker) domain, and a tail (or cargo-binding) domain [22]. The relative positions of these domains along the polypeptide sequence can vary among kinesin subfamily members, but the intrinsic structure and function of each domain is maintained. For example, the tail domain specifically binds cargoes that range from cytoskeletal elements to whole organelles, while the motor domain always contains both the catalytic ATPase pocket and

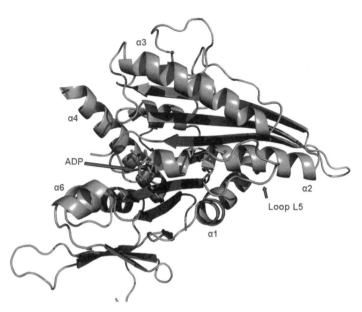

Figure 2. X-ray crystal structure of KSP-ADP complex in secondary structure rendering illustrating α-helical (blue), β-sheet (magenta), and unstructured (beige) regions. The bound molecule of ADP is shown (left of center) in stick form, colored by element [24]. (See color insert.)

MT binding region. In cells, KSP is found as a stable homotetramer with two motor domains displayed on each end, held together by a coiled coil linker domain mass and binding between the motor and the tail domains [23].

The engine that drives KSP is the motor domain, shown schematically in Fig 2 [24]. Through a complex hydrogen bonding communication network, the catalytic, and MT-binding pockets work together to orchestrate the conformational mobility and force generation that are a hallmark of this family [25]. Actions at the MT binding site that include MT association, force generation, and MT release are coupled to actions at the ATPase active site that include ATP binding, hydrolysis, and substrate release [26–29]. Beyond the intramolecular communication and function just described, a KSP homotetramer is also capable of intermolecular coordination between paired motor domains at each end of the homotetramer that work together to produce a unidirectional movement along the MT polymer [30–34]. In the context of early prometaphase, when KSP cross-links oppositely polarized MTs, it produces opposing forces on the MTs that serves to slide them in opposite directions (Fig 3). Because the processional movement is toward the (+) ends of MTs, the centrosome-containing (−) ends are thus driven apart to generate a bipolar mitotic spindle [35].

When viewed in this way, KSP presents an array of targets for small-molecule perturbation:

(1) Inhibition of ATP turnover
(2) Inhibition of MT binding
(3) Inhibition of conformational mobility
(4) Inhibition of cargo binding/tetramer assembly

4. KSP INHIBITORS

4.1. Monastrol: Discovery of a Unique "Induced Fit" Binding Pocket

The earliest report of a selective KSP inhibitor was by a group at Harvard of a molecule named "monastrol" [36], shown in Fig 4. Derived from a cell-based phenotypic screen, this molecule was found to cause monoastral mitotic figures that persisted for a time before leading to cell death (Fig 5). The monoastral

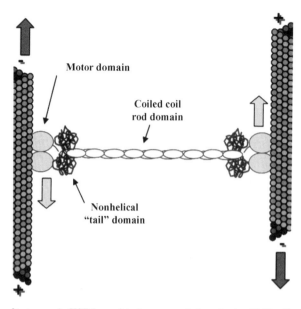

Figure 3. Schematic of tetrameric KSP bound to two oppositely polarized MTs. Four KSP motor domains (yellow) are paired at opposite ends of the tetramer and connected to each other via coiled coil and nonhelical (gray) domains. KSP motor domains process along MTs (green/blue) toward plus (+) ends (yellow arrows) and, as they do so, cause MTs to slide in opposite directions (teal arrows) forcing minus (−) ends to move apart. (See color insert.)

units were shown to consist of unseparated centrosomes with MTs radiating outward to a ring of condensed chromatin. Since this was consistent with expectations for KSP inhibition, KSP motor domain activity was assayed in the presence of this compound and found to indeed be inhibited. Subsequent enzymatic characterization showed that this compound acted specifically at the motor domain, and was ATP-uncompetitive and MT-noncompetitive [37]. The resulting conclusion was that this molecule bound to an allosteric pocket on the motor domain.

Although monastrol has not yet led to a clinical agent, it was an invaluable tool for exploration of KSP inhibition at the biochemical and cellular level. For example, its continued characterization shows that at steady state

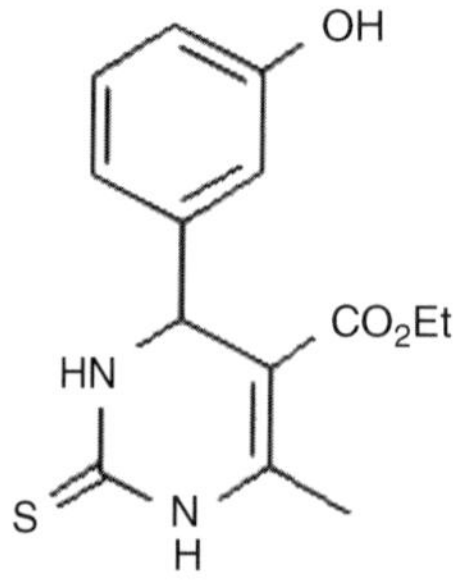

1 KSP IC_{50} = 14 µM
cell EC_{50} = 22 µM

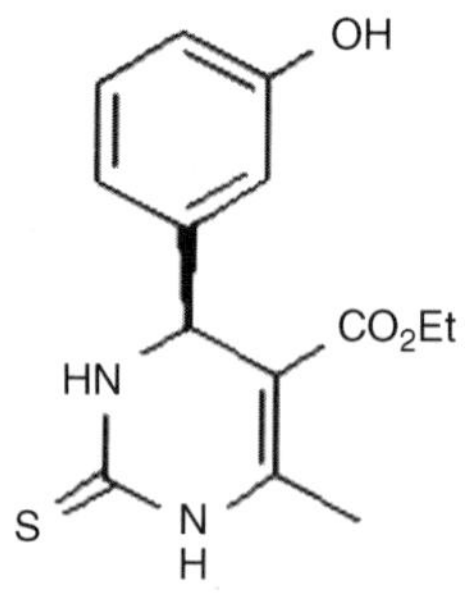

2 KSP IC_{50} = 7 µM
cell EC_{50} = 12 µM

Figure 4. Structures of monastrol (**1**) and (*S*)-monastrol (**2**) from Harvard University.

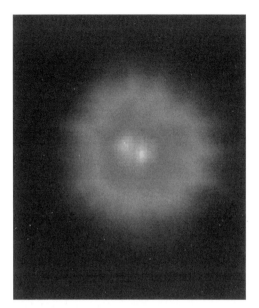

Figure 5. Micrograph of a monopolar spindle phenotype caused by KSP inhibition. Centrosomes are shown in orange, microtubules are shown in green, and chromosomes are shown in blue. (See color insert.)

it binds primarily to the ADP-bound form of KSP and prevents the MT-stimulated release of ADP. Interestingly, presteady-state kinetic data support the existence of a KSP-monastrol-nucleotide-MT complex near enough to the ATP hydrolysis conformation that the hydrolysis step becomes reversible [37]. To explain this, the authors of this work hypothesize that monastrol binds to both ATP- and ADP-bound forms of KSP, and that binding to the KSP-ATP-MT complex causes a stabilization of the hydrolysis transition state that allows reversal of the chemical step and an apparent slowing of ATP to ADP conversion. However, the KSP-ADP-monastrol complex has low intrinsic affinity for MTs and spends little time in a kinetically competent form. This proved to be an advantage for crystallization efforts of the native, nucleotide-bound form and, ultimately, the determination of an X-ray structure of the ADP-KSP-monastrol ternary complex (Fig 6) [38]. When compared with an earlier reported ADP-KSP structure [39], it soon became apparent not only where this molecule bound but also that the binding pocket is seen only in the presence of the compound (Figs 7 and 8). Hence, the binding site was termed an "induced fit" pocket, so-called because the inhibitor plays a significant role in shaping the binding site it occupies. Interestingly, despite the apparent flexibility of this binding pocket, monastrol exhibited a eudismic ratio of 15:1 in favor of the (S)-enantiomer (**2**).

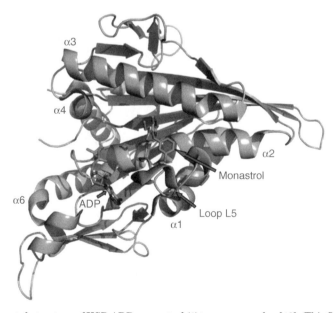

Figure 6. X-ray crystal structure of KSP-ADP-monastrol (**1**) ternary complex [40]. (This figure is available in full color at http://mrw.interscience.wiley.com/emrw/9780471266945/home.)

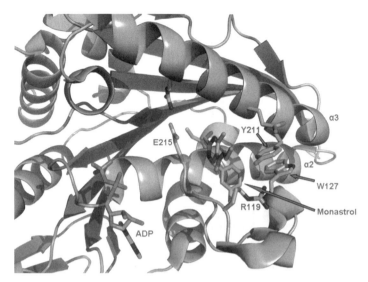

Figure 7. X-ray crystal structure of monastrol (1) bound to KSP at the "induced fit" pocket. The side chains most involved in creating the binding pocket are shown in detail. Included among those are E215 and Y211 from helix α2 and R119 and W127 from loop L5. (This figure is available in full color at http://mrw.interscience.wiley.com/emrw/9780471266945/home.)

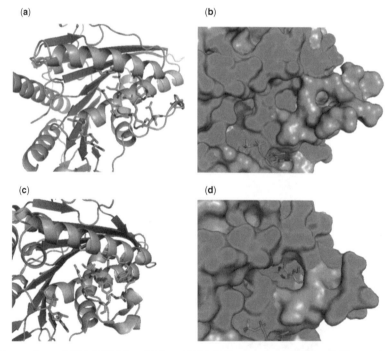

Figure 8. Comparison of KSP-ADP ((a) and (b)) and KSP-ADP-monastrol (1) ((c) and (d)) complexes. (a) In the inhibitor-free X-ray crystal structure, several side chains of helices α2 and α3 reside within the allosteric binding pocket, and loop L5 is unstructured. (b) The resulting protein surface (colored by secondary structure) does not reveal the existence of the monastrol binding pocket. (c) When bound to monastrol, the side chains of helices α2 and α3, particularly Y211, move to accommodate the inhibitor while loop L5 folds inward to allow its backbone and side chains, especially W127, to interact with the monastrol's phenol group. (d) The protein surface view clearly shows a relatively deep and well-formed binding pocket for monastrol. (See color insert.)

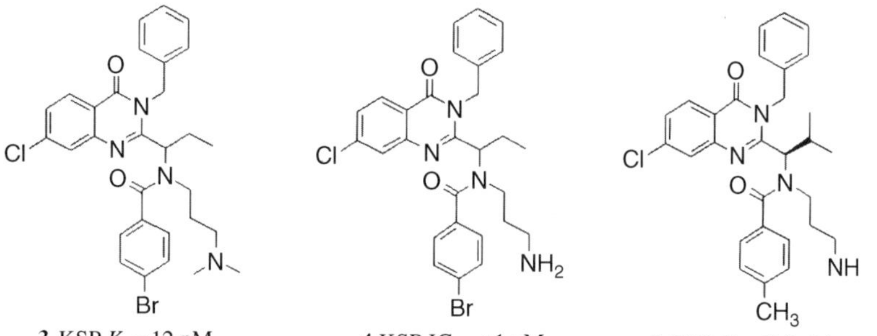

3 KSP K_i = 12 nM
SKOV3 IC$_{50}$ = 126 nM

4 KSP IC$_{50}$ < 1 nM
SKOV3 IC$_{50}$ = 17 nM

5 KSP K_i = 0.6 nM
SKOV3 IC$_{50}$ = 1 nM

Figure 9. Structures of quinazolinone-based inhibitors including CK0106023 (**3**) and ispinesib CK0238273/SB-715992 (**5**) from GSK and Cytokinetics.

As shown in Figs 6 and 7, monastrol binds to a site distal from both the nucleotide and the MT binding surfaces [40]. This induced fit pocket exists in the wedge between helices α2 and α3 and is separated from solvent mainly by side chains in loop L5. The α2-α3 wedge is one of the regions of the protein that undergoes conformational changes in going from the inhibitor-free structure to the inhibitor-bound structure. A conclusion that has been drawn from this is that monastrol locks KSP in an intermediate state in the procession cycle and keeps it free in solution owing to the complex's low affinity for MTs.

A key element in the tractability of KSP as a drug target is the fact that monastrol, besides binding to a specific site on KSP, was also shown to bind specifically to KSP among kinesins. The kinesin superfamily contains over forty members with many showing a significant degree of conservation at the ATP and MT binding regions [41]. The position where monastrol binds, however, has little sequence homology with other members of the superfamily, particularly with respect to the size and composition of loop L5. Thus, an opportunity presented itself for drug discovery efforts in the context of a highly selective binding site on a protein that is required for progression through mitosis and has no other apparent function.

4.2. Quinazolinone-Based Inhibitors

Not long after the initial report of monastrol, a second structure was divulged by a group at Cytokinetics. Called CK0106023 (**3**), it was an optimized structure derived from hits found in a biochemical high-throughput screen using MT-stimulated ATP turnover as the readout (Fig 9) [42]. This molecule was described as having a similar mode of inhibition to monastrol, including the finding that it was not competitive with respect to either ATP or MTs. In addition, it was shown that the (S)-enantiomer was >1000-fold more potent than its distomer, a much greater eudismic ratio than the modest 15-fold preference for the (S)-enantiomer of monastrol. Furthermore, it was reported to be >1000-fold selective over other mitotic kinesins and nonmitotic kinesins, against which it showed no measurable activity.

CK0106023 was shown to be significantly more potent (IC$_{50}$ = 12 nM) than monastrol (IC$_{50}$ = 14 µM) and was shown to inhibit the growth of a human ovarian carcinoma cell line (SKOV3 IC$_{50}$ = 126 nM). Perhaps more importantly, CK0106023 was shown to inhibit other tumor cell lines that included MDR-overexpressing cell lines NCI/ADR, HCT-15, and A2780ADR, all three of which are refractory to treatment with other antimitotic agents such as *paclitaxel*. An elevated mitotic index with figures consistent with failure to form a bipolar spindle was noted, confirming specific on-target activity. Following i.v. administration in mice bearing SKOV3 ovarian tumor xenografts, this compound was shown to reduce tumor growth by 71% overall and gave one partial regression at 12.5 mg/kg, half its maximum tolerated dose (MTD). Perhaps more importantly, it reduced tumor growth to a greater extent than *paclitaxel* at its MTD of 20 mg/kg. Monopolar spindle mitotic figures were noted in excised tumor tissue from treated animals, in agreement with tissue culture results, confirming an on-target effect.

The chemical optimization of this quinazolinone series has not yet been published, but aspects of the evolution toward the first clinical candidate have been divulged in the form of oral and poster presentations. For example, it was noted that the terminal dimethylamino group of CK0106023 underwent rapid phase I metabolism to give the terminal primary amine (**4**). Assay of this metabolite revealed that it was not only metabolically more stable but also 10-fold more potent than its dimethyl parent. Continued exploration led to CK0238273/SB-715992 (**5**, ispinesib), which was found to have a KSP $IC_{50} < 1$ nM and an SKOV3 $IC_{50} = 1$ nM [43]. Key elements in this effort involved replacement of the bromo group from the *para*-position of the benzamide group and elaboration of the ethyl side chain to an isopropyl group, each of which further increased potency.

4.2.1. Ispinesib CK0238273/SB-715992 (**5**), USAN name ispinesib, was the first KSP inhibitor to enter clinical trials. Ispinesib was confirmed to be an allosteric inhibitor that binds at the motor domain and was reported to be >70,000-fold selective for KSP against other members of the kinesin family [44]. In tissue culture, this compound was found to have high levels of potency (IC_{50} and $GI_{50} < 30$ nM) against a wide range of murine and human tumor cell lines encompassing both solid and hematologic tissue-derived malignancies. However, the spectrum of activity appeared to be distinct from that of the taxane antimitotics. In preclinical tumor model studies, activity against a broad spectrum of human xenograft and murine syngeneic tumor models was reported [45]. Interestingly, some tumor cell lines that appeared to be sensitive in tissue culture were found to be refractory *in vivo* [46]. Among them were mammary carcinoma MX-1 and colon carcinoma HT-29 tumors, which were found to arrest in mitosis in response to ispinesib, as evidenced by the pharmacodynamic marker histone H3 phosphorylation, in excised tissue but nevertheless recovered and continued to grow [47]. Further analysis in tissue culture revealed that IC_{90}, but not IC_{50} data, were correlated with *in vivo* results, a result that served to guide further studies with this agent.

Additional preclinical tumor xenograft studies were conducted to identify beneficial combinations of ispinesib with other, approved antitumor agents. Working with the MX-1 human breast carcinoma cell line in preparation for breast cancer-targeted clinical trials, combinations of ispinesib dosed at 30 mg/m^2 with docetaxel at 30 mg/m^2 and ispinesib dosed at 15 mg/m^2 with capecitabine at 1500 mg/m^2 exhibited improved efficacy relative to either drug dosed as a single agent [48,49]. In addition, the murine P388 lymphocytic leukemia cell line was found to be more sensitive to the combination of ispinesib and cisplatin than to either agent alone [50]. Ispinesib was dosed as a single agent and in combination with trastuzumab, lapatinib, doxorubicin, and capecitabine in preclinical models of breast cancer, including estrogen receptor (ER) positive, human epidermal growth factor receptor 2 (HER-2) positive and triple negative (ER, progesterone receptor, and HER-2 negative) tumor lines [51]. Combinations of ispinesib with all four standard of care agents were found to be efficacious against the six tumors studied, producing better efficacy in combination than with either compound dosed as a single agent.

In preclinical studies with ispinesib it was noted that efficacy improved as this agent was administered on a pulsatile basis, with doses every fourth day (q4d) providing the best results [46]. Based on this observation, Phase I clinical trials employed dosing schedules with 1 h i.v. infusions of ispinesib administered daily 3 times every 3 weeks, once weekly 3 times every 4 weeks, and once every 3 weeks (Table 1). In one Phase I trial, ispinesib was dosed on a once every 21-day (q21) schedule in 45 patients with a variety of solid tumors [52]. Dose-limiting toxicities (DLTs) of prolonged neutropenia and febrile neutropenia were observed at a dose of 21 mg/m^2 and the MTD was determined to be 18 mg/m^2. Overall, the drug was well tolerated on this schedule, and disease stabilization, ranging from 7 to 43 weeks, was observed in 15 of 45 patients, with evi-

Table 1. Results of Phase I Trials with Ispinesib as Single Agent

Disease (No. of Patients)	Description	DLT	MTD	Best Reported
Advanced solid tumors (45)	i.v. administration 1 h infusion once every 21 days (q21d)	Neutropenia	18 mg/m^2	SD for 7–43 weeks in 15 patients
Advanced solid tumors (30)	i.v. administration 1 h infusion days 1, 8, and 15 every 28 days	Neutropenia	7 mg/m^2	SD for 16 to >32 weeks in 8 patients
Advanced solid tumors (27)	i.v. administration 1 h infusion days 1, 2, and 3 every 21 days	Neutropenia	6 mg/m^2	SD for 4 and 5 cycles in 2 patients
Pediatric solid tumors (24)	i.v. administration 1 h infusion days 1, 8, and 15 every 28 days	Hyperbilirubinemia, Increased hepatic transaminases	9 mg/m^2	SD for 3–7 cycles in 3 patients
Leukemia (36)	i.v. administration 1 h infusion days 1, 2, and 3 every 21 days	Hyperbilirubinemia, Increased hepatic transaminases, mucositis	10 mg/m^2	Rapid clearing of blast counts in 3 patients

SD: stable disease.

dence of minimal response (<50% tumor mass decrease) in 3 patients. The pharmacokinetics (PK) of ispinesib at the MTD showed a mean Cl of 98 mL/min, V_{ss} of 256 L, and median $t_{1/2}$ of 34 h. Based on these observations, 18 mg/m^2 on a q21d schedule was selected for Phase II clinical trials.

Several more dose dense Phase I trials were also conducted in adult and pediatric patients with solid tumors and adult patients with acute leukemia (Table 1). In the first trial, ispinesib was dosed i.v. in 30 patients with solid tumors on a 28-day cycle using once-weekly doses for three consecutive weeks, followed by one week of rest [53]. Dose-limiting toxicity was Grade (Gr) 3 neutropenia at 8 mg/m^2 and MTD was 7 mg/m^2. Disease stabilization was observed in 8 out of 30 patients, ranging from 16 to over 32 weeks. Similar to the q21-day schedule, PK at the MTD showed a median Cl of 94 mL/min and a median $t_{1/2}$ of 33 h. In the second trial in adult solid tumor patients [54], ispinesib was dosed i.v. in 27 patients on a days 1, 2, and 3 every 21-day schedule. An MTD of 6 mg/m^2 was established with DLT of neutropenia reported at 8 mg/m^2. Stable disease (SD) was reported as best outcome for two patients, lasting 4 and 5 cycles, respectively. PK was linear and comparable with that observed at 18 mg/m^2 on the q21d schedule. The monopolar spindle phenotype observed in tissue culture experiments and preclinical models was also observed in tumor biopsy samples from patients in Phase I trials.

A Phase I trial in pediatric cancer patients was conducted also. Patients in the age group 1–21 years with refractory or recurrent solid tumors, including CNS tumors and lymphoma [55], were treated with ispinesib as a 1 h i.v. infusion weekly 3 times every 28 days. Ispinesib dose was escalated from 5 to 12 mg/m^2 until DLT was encountered. Twenty-four patients were enrolled, with a median age of 10 years; the most common tumor types enrolled were central nervous system (CNS) tumors ($n = 10$) and sarcomas ($n = 5$). Dose-limiting toxicities included neutropenia, hyperbilirubinemia, and increased hepatic transaminases. MTD was determined to be 9 mg/m^2. No objective responses were observed but SD was reported in three patients (ependymoblastoma, seven courses; alveolar soft part sarcoma, five courses; and anaplastic astrocytoma, three courses). Substantial interpatient variability in PK prevented drawing conclusions.

A Phase I trial was also conducted in adult patients with relapsed/refractory acute myeloid (AML) or acute lymphoid leukemia (ALL) or advanced myelodysplastic syndrome (MDS) [56]. Thirty-six patients were treated with ispinesib dosed as a 1 h i.v. infusion daily for three consecutive days every 21 days; dosing started at 1.5 mg/m^2/day and were escalated to 13.3 mg/m^2/day. Dose-limiting toxicities included increased hepatic transaminases, hyperbilirubinemia, and mucositis. The MTD was determined to be 10 mg/m^2/day given on three consecutive days. Three patients with very high leukemic blast counts at baseline had rapid clearing of their blasts during treatment; two patients had late bone marrow responses with less than 5% blasts; however, none of the patients had normalization of their blood counts.

Phase Ib clinical trials were conducted with ispinesib in combination with docetaxel, capecitabine and carboplatin (Table 2). For the combination with docetaxel, DLT was prolonged Gr 4 neutropenia [57,58]. The optimally tolerated regimen (OTR) for the combination was 10 mg/m^2 of ispinesib and 60 mg/m^2 docetaxel, each administered once every 21 days. Ispinesib and docetaxel plasma concentrations were consistent with those previously reported for each agent as monotherapy. Of 24 patients in the trial, 13 had a best response of SD that lasted 2.25–7.5 months. Two of 14 prostate cancer patients had a prostate-specific antigen (PSA) partial response (PR).

Results from the Phase I combination trial of ispinesib and capecitabine indicated that ispinesib could be administered at 18 mg/m^2 every 21 days with capecitabine at 2000 mg/m^2 daily for 14 of 21 days [59]. In contrast to the combination with docetaxel, the MTD of ispinesib as a single agent was tolerated in combination with full dose capecitabine. Ispinesib exposure was not altered when given in combination with capecitabine. A single DLT of prolonged Gr 4 neutropenia was reported. Eight patients (3 breast, and 1 each of tongue, colorectal, bladder, thyroid, and salivary gland) exhibited SD (2–6.75 months).

For the carboplatin-ispinesib combination, the OTR was 18 mg/m^2 for ispinesib and a target AUC value of 6 mg*min/mL for carboplatin dosed together on a q21-day schedule. There was no evidence of drug–drug interactions [60], and DLTs included prolonged Gr 4 neutropenia, Gr 3 febrile neutropenia, and Gr 4 thrombocytopenia. Best responses included 1 PR for a patient with breast cancer and SD for 13 out of 28 patients (duration 3–9 months).

Table 2. Results of Phase Ib Trials with Ispinesib in Combination with Standard of Care Agents

Disease (No. of Patients)	Description	DLT	OTR	Best Reported Outcome
Advanced solid tumors (24)	Ispinesib Docetaxel Once every 21 days (q21d)	Neutropenia	10 mg/m^2 ispinesib 60 mg/m^2 docetaxel	SD for >18 weeks in 7 patients
Advanced solid tumors (16)	Ispinesib Once every 21 days (q21d) Capecitabine daily 14 of 21 days	Neutropenia	18 mg/m^2 ispinesib 2000 mg/m^2 capecitabine	SD for 2–6.75 months in 8 patients
Advanced solid tumors (28)	Ispinesib Carboplatin Once every 21 days (q21d)	Neutropenia Thrombocytopenia Febrile neutropenia	6 mg/m^2 carboplatin 18 mg/m^2 ispinesib	PR in 1 patient SD for 3–9 months in 13 patients

SD: stable disease; PR: partial response.

A total of 9 Phase II clinical trials have been conducted with ispinesib dosed at 18 mg/m^2 on a q21d schedule as a single agent [61]. Among the cancers studied were locally advanced or metastatic breast cancer, platinum-refractory and platinum-sensitive nonsmall cell lung cancer, platinum/taxane refractory and sensitive relapsed ovarian cancer, metastatic colorectal cancer, squamous cell carcinoma of the head and neck, hormone-refractory prostate cancer, hepatocellular cancer, renal cell carcinoma, and melanoma. Results for these studies are shown in Table 3. The best responses have been PR in the breast and ovarian cancer settings. The most promising results were observed in women with locally advanced or metastatic breast cancer, all of whom were previously treated with an anthracycline and a taxane. In Stage I of this trial, PR, confirmed by RECIST [62], was observed in 3 of 33 evaluable patients [63]. Responses ranged from 46% to 68% tumor size reduction that endured from 7.1 weeks to 13.4 weeks. In Stage II of this study, an additional 12 patients were enrolled and a fourth PR was reported, for an overall response rate of 9% [64]. In the ovarian cancer study, a 70% decrease in tumor size that persisted for 42 weeks was observed in one patient [65]. The other studies showed best outcomes of SD, except for the hormone refractory prostate cancer trial, where no PRs or SD were observed [66–73]. Ispinesib plasma concentrations were consistent with those observed in Phase I trials and adverse events were manageable and predictable [52,63].

Analysis of toxicokinetic data from a Phase I trial of ispinesib dosed every 21 days revealed that dose-limiting Gr 4 neutropenia reached a nadir at 1 week postdose and completely reversed within 14 days without apparent cumulative toxicity upon continued dosing [74]. Based upon on this finding and the promising data from the Phase II breast cancer trial, Cytokinetics advanced ispinesib into a Phase I/II clinical trial studying patients with locally advanced or metastatic

Table 3. Results of Phase II trials with Ispinesib as Single Agent

Disease (No. of Patients)	Trial Results	Adverse Events
Locally advanced or metastatic breast cancer (59)	PR lasting 7.1–30 weeks in 4 of 45 evaluable patients	Neutropenia
Platinum-refractory and -sensitive nonsmall cell lung cancer (47)	SD in 2–6 cycles in 10 patients from platinum sensitive arm and 2–4 cycles in 6 patients from platinum refractory arm	Neutropenia
Platinum/taxane refractory and sensitive relapsed ovarian cancer (22)	PR of 70% decrease in tumor size lasting 42 weeks in 1 patient	Neutropenia
Metastatic colorectal cancer (31)	Progression-free survival: 7 mg/m^2 weekly three times q28d: 49 days 18 mg/m^2 q21d: 37 days	Neutropenia
Squamous cell carcinoma of the head and neck (20)	SD for 3–5 cycles in 5 patients	Neutropenia
Metastatic hepatocellular (15)	SD for 2.6–9.5 months in 7 patients	Neutropenia Increased hepatic transaminases
Metastatic or recurrent malignant melanoma (17)	SD for median 2.8 months in 6 patients	Neutropenia
Hormone refractory prostate cancer (21)	No PR or SD	Neutropenia Anemia Fatigue
Renal cell carcinoma (20)	SD in 6 patients for 2 to >12 months	Anemia Elevated creatinine

SD: stable disease; PR: partial response.

breast cancer previously untreated with chemotherapy for recurrence or metastatic disease, with ispinesib dosed on a new schedule that consists of 1 h i.v. infusions on days 1 and 15 of a 28-day cycle. Interim results for the ongoing Phase I portion of the trial reported on 16 patients; 6 patients receiving 12 mg/m^2 reported no DLT and 2 of 7 patients receiving 14 mg/m^2 had DLTs of Gr 3 increases in hepatic transaminases after the second dose [75]. Further evaluation at the 14 mg/m^2 level is planned to determine if the DLT events are dose limiting. Of the 15 patients evaluable for efficacy, the best responses in this study were three patients with >30% reduction in target lesions, one at the 10 mg/m^2 dose level, and the two at the 12 mg/m^2 dose level. One of these patients had a confirmed PR by RECIST with duration of 24+ weeks. Ten patients had SD, 4 with duration of ≥4 months. The most common Gr 3/4 event was neutropenia.

Overall experience with ispinesib in more than 500 cancer patients indicates that it is well tolerated. The primary toxicity is neutropenia, often Gr 3 or 4, but infrequently associated with fever or sepsis. Anemia and thrombocytopenia are infrequent and rarely severe. Mild gastrointestinal toxicity (nausea, vomiting and diarrhea) is common, occurring in approximately 40% of patients but almost always Gr 1-2. Mild increases in hepatic transaminases and alkaline phosphatase, usually Gr 1, have been reported in up to 15% of patients. Neurotoxicity and alopecia have been reported infrequently.

4.3. KSP Inhibitors Structurally Related to Ispinesib

Many pharmaceutical companies have initiated research efforts using ispinesib as a chemical starting point, and a myriad of analogs have been reported, primarily within the patent literature (see also Knight and Parrish for a recent review of KSP inhibitors) [76]. In most cases, the differing structural element has been the incorporation of alternates for the quinazolinone core, including both bicyclic and moncyclic ring systems. The most prominent contributors have been GlaxoSmithKline (GSK) and Cytokinetics [77–84], Merck & Co. [85–89], Chiron (Novartis) [90–92], Bristol-Myers Squibb [93,94], AstraZeneca [95], and ArQule [96]. In addition to the many replacements for the quinazolinone ring of ispinesib, modification of other regions of the molecule has also been the focus of research efforts of several groups, including Chiron (Novartis) [97,98], GSK and Cytokinetics [99–102], and Merck & Co. [103].

4.3.1. CK929866/SB-743921 A second KSP inhibitor, CK929866/SB-743921 (**6**) (Fig 10), was reported by Cytokinetics and GSK as a backup to ispinesib. In biochemical and cellular assays this compound was more potent than ispinesib [104]. The high level of KSP selectivity was also retained along with activity against a wide range of tumor cell lines and the monopolar spindle mitotic arrest phenotype. The key factors that distinguished

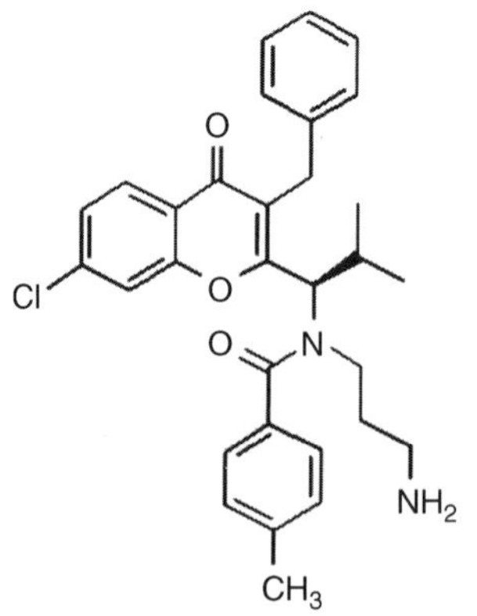

CK929866/SB-743921 (**6**)
KSP K_i = 0.1 nM
cell IC$_{50}$ = 0.2 nM

Figure 10. Structure of CK929866/SB-743921 from GSK and Cytokinetics.

CK929866/SB-743921 from ispinesib were a shorter half life across several species and efficacy at lower doses in most models tested [105]. In preclinical studies, CK929866/SB-743921 was a potent inhibitor of 7 non-Hodgkin lymphoma cell lines ($IC_{50} \leq 10$ nM) and was efficacious in a xenograft model with the one line tested (Ly1), exhibiting a 100% response rate, 60% of which were complete regressions [106]. Based on these results, CK929866/SB-743921 was progressed into clinical trials against both Hodgkin and non-Hodgkin lymphomas.

Utilizing the most frequently studied dosing schedule for ispinesib, a Phase I trial was conducted with CK929866/SB-743921 administered as a 1 h i.v. infusion q21d in 20 patients with advanced solid tumors (Table 4) [107]. The MTD on this schedule was 4 mg/m^2, and the most frequent DLT was prolonged neutropenia. One PR was observed in a patient with cholangiocarcinoma dosed at the MTD and SD lasting $\geq$4 cycles was observed in six patients [108]. At this dose, the median Cl was 21 mL/min, and $t_{1/2}$ was 28 h. Based on the kinetics of neutropenia, CK929866/SB-743921, like ispinesib, was further assessed in a Phase I/II clinical trial exploring a more dose dense every 2-week schedule. This study is currently enrolling patients with Hodgkin and non-Hodgkin lymphoma with CK929866/SB-743921. Since the predominant dose-limiting toxicity has been neutropenia, MTD will be determined with and without prophylactic granulopoietic factor support (i.e., G-CSF) [106]. For the 39 patients treated without prophylactic G-CSF, DLT was neutropenia and MTD was 6 mg/m^2. As of this writing, continued dose escalation with prophylactic G-CSF has reached 9 mg/m^2 with 1 of 2 patients having prolonged neutropenia. Of patients with Hodgkin lymphoma dosed at or above 6 mg/m^2, 2 of 14 had PRs of 53% and 65% reduction in dominant lesions. There were no reports of drug-related neurotoxicity or alopecia greater than Grade 1.

4.3.2. AZD4877 AstraZeneca has published several patents which cover heterocyclic fused pyrimidinones of the general structure **7** (Fig 11). The filings claim predominantly isothiazolo- and isoxazolo-pyrimidinones, exemplified by **8**. Although the structure and preclinical data of their lead KSP inhibitor have not been published, AstraZeneca initiated four Phase I clinical trials between 2006 and 2008 in the United States, Canada, and Japan to determine the safety, tolerability, and pharmacokinetics of AZD4877 in patients with aggressive and/or late-stage cancers [113]. The initial results from two of these trials were presented at the 2008 ASCO Annual Meeting [114,115]. In one dose escalation study, patients were treated with AZD4877 for three successive weekly doses, followed by one week of rest. Grade 3 or 4 neutropenia was noted as the DLT at doses $\geq$36 mg, but the effects were rapidly reversible [115]. No other significant dose or mechanism related Gr 3 or 4 toxicities occurred and neuropathy was not noted.

In another trial, AZD4877 was administered to 18 patients twice weekly based on preclinical data which showed increased efficacy with greater drug exposure at equivalent doses under this dosing regimen (Table 5) [114]. The dose-limiting toxicity in this study was Gr 4 neutropenia lasting longer than 4 days in 2/2 patients at the 15 mg dose. Patients dosed at 2, 4, 7, and 11 mg exhibited no evidence of dose-limiting toxicity.

Table 4. Results of Phase I Trials with CK929866/SB-743921

Disease (No. of Patients)	Trial Results	Adverse Events
Advanced solid tumors (20)	SD in 6 patients >4 cycles	Neutropenia
	PR in 1 patient with cholangiocarcinoma at cycle 10	Nausea
		Fatigue
Non-Hodgkin or Hodgkin lymphoma (39)	PR in 2 patients with Hodgkin lymphoma, 53% and 65% reduction in dominant lesions	Neutropenia

SD: stable disease; PR: partial response.

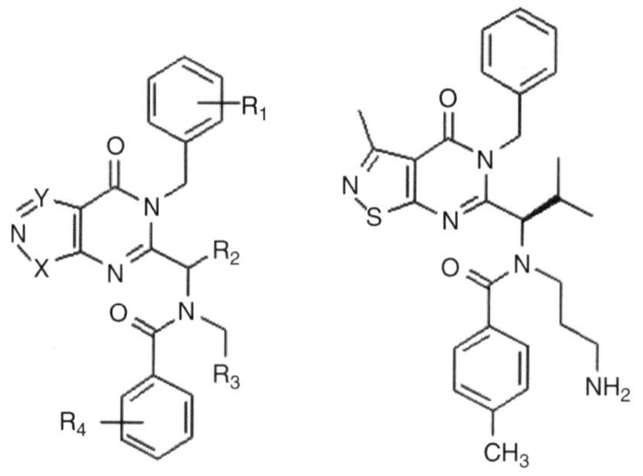

Figure 11. Structure of AstraZeneca's KSP inhibitor template (**7**) and a representative isothiazolopyrimidinone example (**8**).

Table 5. Results of Phase I Trials with AZD4877 as Single Agent

Disease (No. of Patients)	Description	DLT	MTD	Best Reported Outcome
Advanced cancer (29)	i.v. administration 1 h infusion days 1, 8, and 15 every 28 days	Neutropenia	30 mg	None reported
Advanced cancer (18)	i.v. administration 1 h infusion days 1,4, 8, and 11 every 21 days	Neutropenia	11 mg	SD for 2–7 cycles in 4 patients

[a] SD: stable disease.

Radiographically stable disease was observed in four patients ranging from two to seven treatment cycles with the most sustained effect in bladder cancer. All other patients had progressive disease.

In both trials, AZD4877 showed a linear pharmacokinetic profile with a half-life of 15–27 h across the dose groups. The observed neutropenia correlated with exposure and the MTDs were determined to be 30 mg on the weekly schedule and 11 mg on the twice-weekly regimen. Consistent with the mechanism of action, monopolar mitotic spindles (monoasters) were observed in peripheral blood mononuclear cells (PBMCs) following treatment. A Phase I/II study to measure tolerability and pharmacokinetics of a weekly 3 times schedule and a Phase II study to determine the antitumor efficacy of AZD4877 in patients with bladder cancer was initiated [113], however, AZD4877 has since been discontinued as a potential cancer therapy to treat solid tumors [116].

4.4. Dihydropyrazoles and Related Compounds

4.4.1. MK-0731 Over the last few years, Merck & Co. has published a series of papers detailing the discovery and optimization of novel dihydropyrazole and dihydropyrrole-based inhibitors of KSP, culminating in the discovery of MK-0731 (Fig 12) that has been studied in the clinic [117–125]. The discovery of MK-0731 began with the identification of two moderately active 3,5-diaryl-4,5-dihydropyrazoles (**10** and **11**) following a high-throughput screen of their in-house compound collection (ATPase $IC_{50} = 3.6$ and 6.9 µM, respectively) [117]. Initial chemistry efforts revealed that a combination of

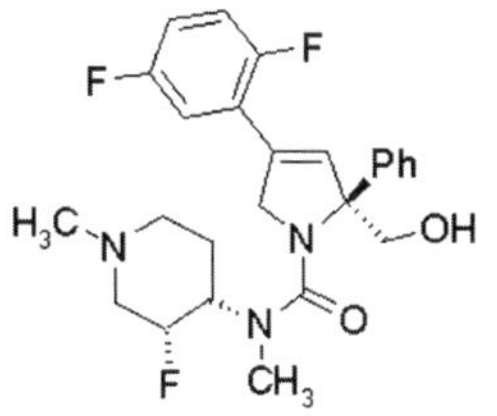

MK-0731 (9)

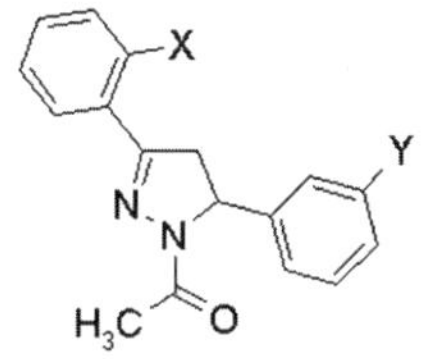

10 X = Cl, Y = H; KSP IC_{50} = 3600 nM
11 X = H, Y = OH; KSP IC_{50} = 6900 nM
12 X = Cl, Y = OH; KSP IC_{50} = 450 nM

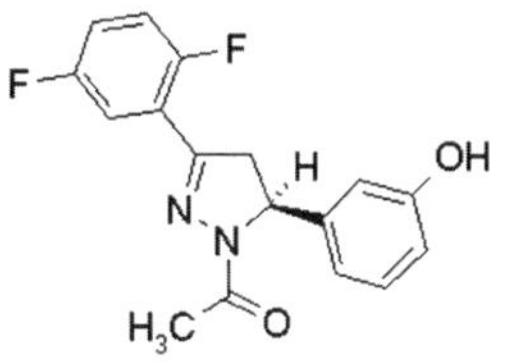

13 KSP IC_{50} = 26 nM
A2780 EC_{50} = 15 nM

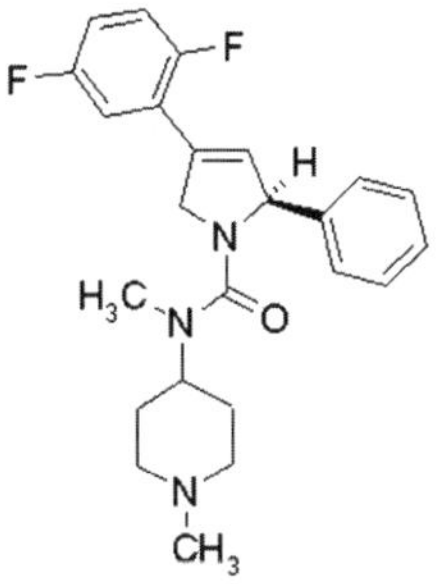

14 KSP IC_{50} = 4 nM
in vivo EC_{90} = 400 nM
hERG IC_{50} = 1200 nM

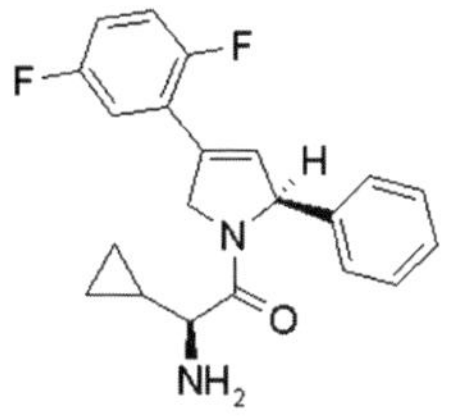

15 KSP IC_{50} = 2 nM
A2780 EC_{50} = 9 nM
hERG IC_{50} = 3500 nM
aq. sol. > 10 mg/mL

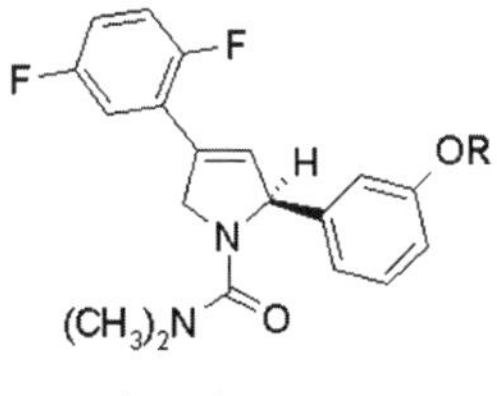

16 (R = H)
KSP IC_{50} = 7 nM
A2780 EC_{50} = 22 nM
hERG IC_{50} = 33,000 nM
aq. sol. < 1 mg/mL
17 (R = PO_3^{2-})
KSP IC_{50} > 1000 nM
aq. sol. > 20 mg/mL

Figure 12. Structures of dihydropyrazole and dihydropyrrole KSP inhibitors from Merck & Co.

aryl group substituents (chloro and hydroxy) to give **12** led to a 10-fold increase in potency and provided a more viable starting point for further elaboration. Further exploration around both aryl groups, as well as the *N*1-acyl moiety, resulted in the identification of dihydropyrazole **13**, which showed good activity in biochemical and cellular assays (KSP $IC_{50} = 26$ nM; A2780 $EC_{50} = 15$ nM). Similar to CK-0106023 (see above), the enzyme showed a preference for the (*S*)-enantiomer (**13**) over its (*R*)-antipode (KSP $IC_{50} = 4100$ nM). Importantly, dihydropyrazole **13** also exhibited caspase-3 activation, a well-known marker of apoptosis, and the monopolar spindle phenotype characteristic of KSP inhibition.

Modifications to the dihydropyrazole core were also examined, including a dihydropyrrole scaffold in which the hybridization of the dihydropyrrole carbons bearing the aryl groups was maintained [118,119]. Investigation of this series of inhibitors furnished dihydropyrroles **14** and **15**, which exhibited good biochemical and cellular activity, as well as a robust pharmacodynamic (PD) effect [125]. In an A2780 human tumor xenograft assay in mice, researchers at Merck quantified the PD effect of various compounds with *in vivo* EC_{90} data, as measured by the minimum plasma concentration required to induce the maximum extent of mitotic arrest. The basic amine on the *N*1-acyl group of **14** imparted good aqueous solubility, important for intravenous formulation, although inhibition of the human *ether-a-go-go* related gene (hERG) channel was also introduced (hERG $IC_{50} = 1.2\,\mu$M). The researchers found that the hERG activity could be attenuated and the KSP potency maintained by a combination of neutral *N*1-acyl groups and reintroduction of the phenol moiety (e.g., **16**) [119]. The phenol was then utilized in a phosphate prodrug strategy to provide **17**, itself inactive against KSP, but shown to be both highly soluble (>20 mg/mL) and rapidly converted to the parent molecule in rat and dog pharmacokinetic studies ($t_{1/2} <$ 0.1 h).

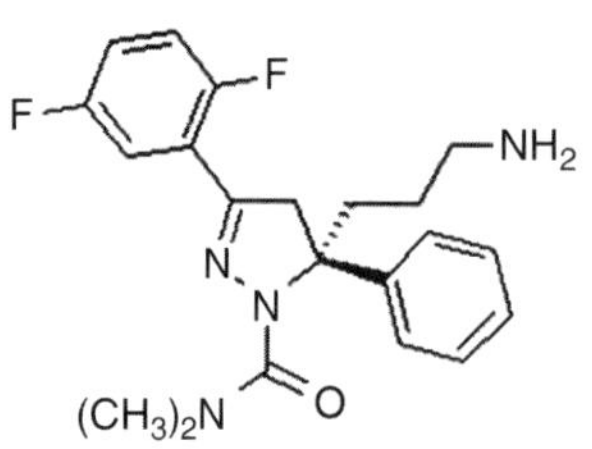

18 KSP IC$_{50}$ = 1.9 nM
A2780 EC$_{50}$ = 5.2 nM
hERG IC$_{50}$ = 19,000 nM
aq. sol. > 12 mg/mL
MDR ratio = 490

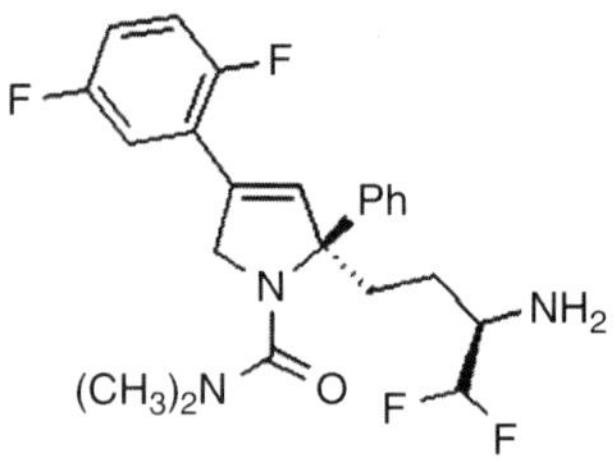

19 KSP IC$_{50}$ = 1.9 nM
A2780 EC$_{50}$ = 23 nM
hERG IC$_{50}$ = 16,000 nM
aq. sol. > 10 mg/mL
MDR ratio = 5

Figure 13. Structures of KSP inhibitors from Merck.

Analogs from both the dihydropyrazole and the dihydropyrrole series were found to not be competitive with ATP or MT, consistent with an allosteric mode of binding. This was confirmed with cocrystal structures of dihydropyrazole **12** and dihydropyrrole **15** with KSP-ADP, each of which bound within the induced fit loop L5 binding region [118,120]. Use of this X-ray crystallographic information, in combination with molecular modeling, led to the design of inhibitors that incorporated polar functionality at the pyrazole C5-position, which could potentially fill unoccupied space and interact with the enzyme backbone. Introduction of a C5 aminopropyl group provided dihydropyrazole **18** (Fig 13), which showed improved potency and pharmacokinetics compared to previous dihydropyrazole derivatives and maintained good aqueous solubility. However, these improvements were accompanied by attenuated activity in multidrug resistant (MDR) cell lines due to efflux by P-glycoprotein (Pgp), represented here by the ratio of MDR overexpressing (+) and MDR normal (−) cellular activity. This was overcome by modulating the basicity of the terminal amino group through β-fluorination, giving rise to dihydropyrrole **19**, which exhibited much improved activity in Pgp-overexpressing cell lines and an MDR ratio of 5 (i.e., five times less active in the MDR cell line) [121]. However, these modifications resulted in a concomitant loss in activity in the *in vivo* PD assay (EC$_{90}$ = 1425 nM).

As discussed above, although compounds such as **13** and **19** demonstrated that it was possible to obtain water soluble dihydropyrrole-based KSP inhibitors with good activity in MDR cell lines, **13** exhibited undesirable hERG channel affinity and **19** showed relatively poor *in vivo* activity. Researchers at Merck sought a new derivative to address these issues. Based on previous SAR studies, a hydroxymethyl moiety was introduced into **13**, which resulted in a compound (**20**) (Fig 14) that retained potent KSP inhibition (KSP IC$_{50}$ = 6 nM) and *in vivo* PD activity (EC$_{90}$ = 125 nM), coupled with attenuated hERG affinity (hERG IC$_{50}$ = 14.6 µM). However, incorporation of the hydroxymethyl also led to an MDR ratio of 21. Since it had already been demonstrated in previous studies that modulation of amine basicity could lead to an improved P-gp efflux profile, efforts were focused on reducing the pK_a to a value between 6.5 and 8 by introduction of a β-fluorine atom. Introduction of fluorine at the C-3 position of the piperidine ring gave rise to a set of four enatiomerically and diastereomerically pure derivatives, which ultimately led to the identification of MK-0731 [125]. Interestingly, the β-fluorine of MK-0731 was predicted to sit in an axial conformation and have a more subtle effect on the pK_a than its equatorial counterpart. Indeed, introduction of the fluorine modulated the pK_a from 8.8 (**20**; *des*-fluoro) to 7.6 (**MK-0731**; axial fluoro) and 6.6 (**21**; equatorial fluoro). Although both fluoro deri-

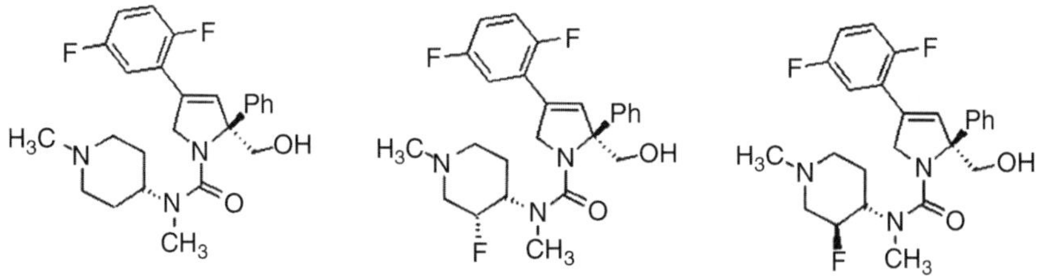

20
(des-fluoro)

KSP IC$_{50}$ = 6 nM
in vivo EC$_{90}$ = 125 nM
hERG IC$_{50}$ = 14,600 nM
MDR ratio = 21

MK-0731 (9)
(axial fluoro)

KSP IC$_{50}$ = 2.2 nM
A2780 IC$_{50}$ = 5.3 nM
in vivo EC$_{90}$ = 94 nM
hERG IC$_{50}$ = 20,500 nM
MDR ratio = 4.5

21
(equatorial fluoro)

KSP IC$_{50}$ = 12 nM
A2780 IC$_{50}$ = 17 nM
hERG IC$_{50}$ = 36,900 nM
MDR ratio = 2.4

Figure 14. Structures of KSP inhibitors from Merck.

vatives showed the desired improvement in MDR ratio, MK-0731 was more potent in biochemical and cellular assays and was selected for further progression.

As with other derivatives from the series, MK-0731 was determined not to be competitive with ATP or MT and was highly selective for KSP over other structurally related kinesins (>20,000-fold). The expected binding mode (allosteric loop L5 binding pocket) and the axial orientation of the fluorine atom were confirmed by cocrystallization of MK-0731 with KSP-ADP (Figs 15 and 16) [126]. Key features of the binding of MK-0731 include the difluorophenyl moiety situated in a large, hydrophobic region of the binding site, as well as

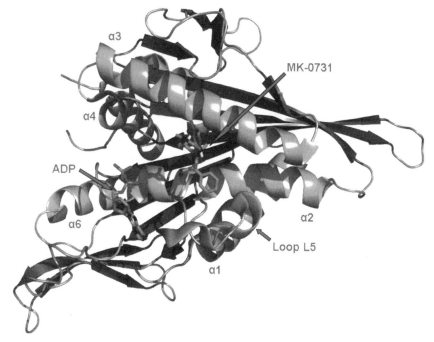

Figure 15. X-ray crystal structure of Merck inhibitor MK-0731 bound to KSP-ADP [126]. (This figure is available in full color at http://mrw.interscience.wiley.com/emrw/9780471266945/home.)

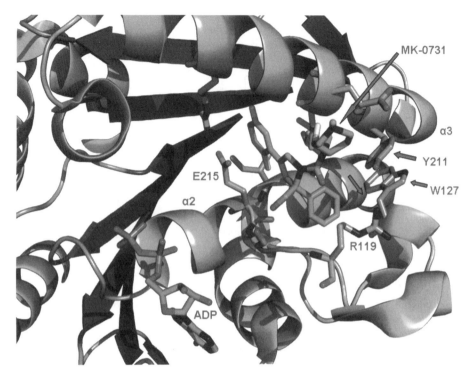

Figure 16. Closeup of interactions between Merck inhibitor MK-0731 and KSP-ADP. Included are the key side chains of helices α2 and α3 and loop L5 that interact with MK-0731 to form the induced fit pocket. As with monastrol, side chains Y211, W127, and R119 and the backbone of loop L5 shift relative to the unbound KSP-ADP structure. (See color insert.)

hydrogen bonds formed between the hydroxymethyl group and the backbone carbonyl oxygen of the G117 residue.

In xenograft studies in mice, MK-0731 showed good efficacy in both taxane-sensitive (A2780 or KB-3-1) and taxane-resistant (PTX10 or KB-v) tumor cell lines. MK-0731 also exhibited an acceptable developability profile, characterized by excellent aqueous solubility (>50 mg/mL at pH 4.3) and a moderate-to-high plasma clearance with a half-life between 1 and 2 h in three preclinical species (Table 6).

Table 6. Preclinical PK Parameters of MK-0731

PK Parameter	Rat	Dog	Monkey
Clp (mL/min/kg)	67	15	23
V_{ss} (L/kg)	3.0	1.6	2.3
$t_{1/2}$ (h)	1	2	1

In 2005, Merck & Co. initiated a Phase I clinical trial with MK-0731 to evaluate its safety, tolerability, and PK profile in patients with recurrent or nonresponsive solid tumors for which there is no existing standard therapy. MK-0731 was administered as a 24 h infusion at doses of 6–48 mg/m^2 once every 21 days to 35 patients. Prolonged Grade 4 neutropenia was observed in four patients treated with $\geq$24 mg/m^2 leading to an expansion of cohorts at lower doses [127,128]. Treatment with MK-0731 at the maximum tolerated dose of 17 mg/m^2 was well tolerated with a consistent dose-limiting toxicity of myelosupression (Table 7). However, MK-0731 exhibited nonlinear PK with drug concentrations decreasing mono- or biexponentially and terminal half-lives between 4 and 22 h, indicating that steady state concentrations may not be achieved at the end of the 24 h infusion. Stable disease was noted in 16 patients in the study, but enrollment

Table 7. Results of Phase I Trials with MK-0731 as Single Agent

Disease (No. of Patients)	Description	DLT	MTD	Best Reported Outcome
Advanced cancer (35)	i.v. administration 24 h infusion once every 21 days (q21d)	Myelosuppression	17 mg/m^2	SD for 4–10 cycles in 16 patients

SD: stable disease.

in the second part of the trial was closed and MK-0731 is no longer listed in Merck's pipeline [129].

4.4.2. ARRY-520 KSP inhibitor templates structurally related to the dihydropyrazoles and dihydropyrroles have also appeared in the patent literature, including a series of 2,3-dihydro-[1,3,4]oxadiazoles and 2,3-dihydro-[1,3,4]thiadiazoles (e.g., **22** and **23**) (Fig 17) by Array Biopharma [130,131]. Array has presented preclinical data on one of its most advanced inhibitors, ARRY-520 (**23**), and has progressed it into clinic trials [132–136].

ARRY-520 is a highly potent inhibitor of KSP (KSP IC$_{50}$ = 6 nM) with selectivity (>1000-fold) over other kinesins and a good developability profile, including good aqueous solubility (>4 mg/mL), low CYP inhibition (>25 uM for five isozymes), and good i.v. PK [132]. Although an X-ray cocrystal structure of ARRY-520 with KSP-ADP has not been published, ARRY-520 displays kinetics (uncompetitive with ATP, noncompetitive with MT) consistent with binding in the allosteric loop L5 region of the enzyme.

ARRY-520 exhibited good antiproliferative activity in a panel of human and rat tumor cell lines (cell EC$_{50}$ values = 0.4–14 nM), including a variety of leukemias and solid tumors, and caused an arrest in the G2/M phase of the cell cycle both by phosphohistone H3 (pHH3) accumulation and by FACS analysis [134]. Cells treated with ARRY-520 also displayed formation of monopolar spindles, the characteristic phenotype of KSP inhibition. In preclinical animal models, ARRY-520 reportedly exhibited efficacy comparable to ispinesib [132]. Tumor regressions with some complete responses and cures were observed in several subcutaneous mouse xenograft models, including HT-29 (colon), A2780 (ovarian), and K-562 (chronic myelogenous leukemia). In addition, following a single dose of ARRY-520 at its MTD, a sustained PD effect was observed for 4 days as measured by elevated levels of pHH3 and the presence of monopolar spindles.

In 2007, Array Biopharma progressed ARRY-520 into an open label, dose escalating Phase I clinical trial in patients with advanced cancer [137]. This initial study was designed to determine the maximum tolerated dose and the preliminary PK and safety profiles of ARRY-520. The starting dose was 2.5 mg/m^2, administered as a single i.v. infusion on day 1 of each 21-day treatment cycle and although this trial is now closed, no clinical data have been

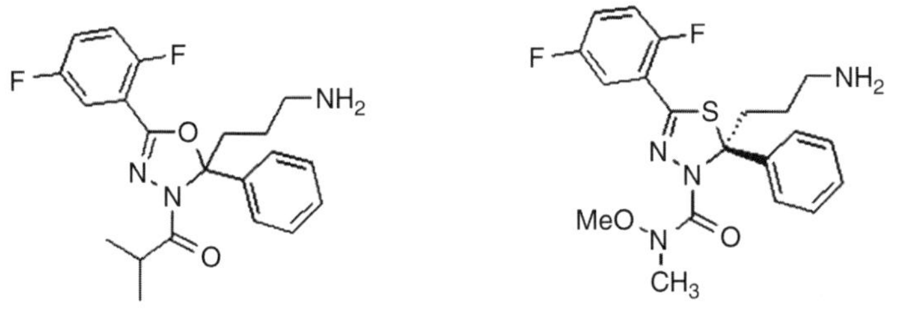

Figure 17. Structures of KSP inhibitors from Array Biopharma.

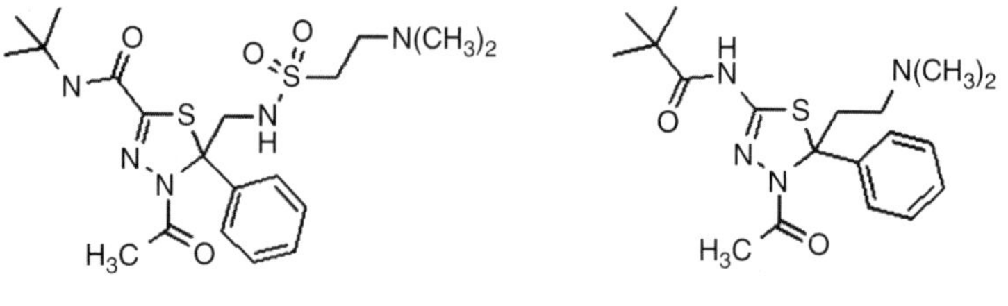

Figure 18. Structures of KSP inhibitors from Kyowa Hakko Kogyo.

reported. More recently, a Phase Ib/II study was initiated in patients with acute leukemia to assess the safety, tolerability, and preliminary efficacy of ARRY-520. In this study, the drug is administered i.v., although on a potentially more aggressive treatment plan, up to three times weekly based on clinical observations. No data from this study have been reported, but completion of the trial is expected in the first half of 2009. In addition, Array Biopharma plans to begin a Phase II clinical trial in 2009 to treat multiple myeloma [138].

4.4.3. LY2523355 A series of 2,3-dihydro-[1,3,4] thiazoles has also been claimed by Kyowa Hakko Kogyo (e.g., **24** and **25**) (Fig 18) [139,140]. A KSP inhibitor (LY2523355) of unknown structure has been licensed from Kyowa exclusively to Eli Lilly for clinical development. Lilly has since initiated two Phase I clinical trials to study LY2523355 in patients with either acute leukemia or advanced (Stage 3 or 4) cancer [141,142]. Both trials are currently open, although no additional information has been reported.

4.5. Additional "Induced Fit" Pocket Inhibitors

4.5.1. EMD-534085 Merck Serono has also progressed its KSP inhibitor, EMD-534085, into clinical trials. Although the structure of EMD-534085 has not yet been disclosed, several patent applications based around a tetrahydroquinoline core have been published recently by researchers at Merck Serono [143–145]. The most extensive of these is based on the hexahydropyranoquinoline (HHPQ) scaffold (**26**), including compound **27** (Fig 19) [143]. A 192 g synthesis of **27** has appeared in a manufacturing process patent [146], as well as in a patent describing its pharmaceutical formulation [147].

In 2006, Merck KGaA presented some of the preclinical data from the HHPQ series at the 97th AACR Annual Meeting [148]. Following a high-throughput screen, derivatives from this series were identified as potent, selective inhibitors of KSP and shown by X-ray crystallography and Biacore® binding studies to bind in the allosteric loop L5 pocket. Inhibition of cellular proliferation in a variety of tumor cell lines (e.g., Colo205, HCT116, and

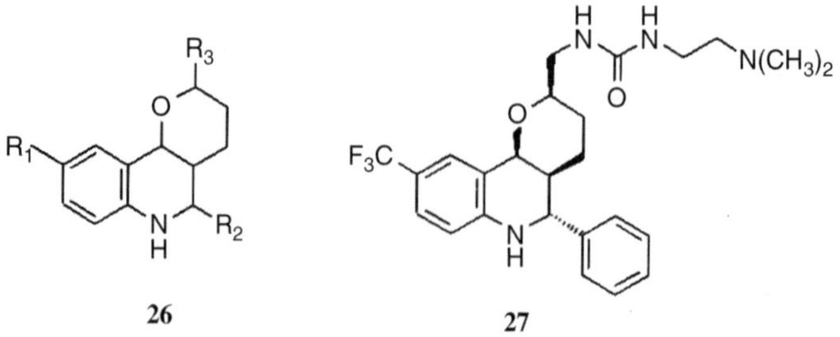

Figure 19. Tetrahydroquinoline KSP inhibitors by Merck Serono.

SW707 colon cancer lines and MDA-MB468 breast carcinoma cells) was noted with observed mitotic arrest and subsequent apoptosis. Following either intraperitoneal or oral dosing in mice bearing human tumor xenografts, HHPQ analogs exhibited efficacy and a pharmacodynamic response, as measured by pHH3 response in tumor biopsies. As expected from the absence of KSP from postmitotic neurons, these compounds were superior to taxane-based agents in a rat model for peripheral neuropathy.

By October 2007, Merck Serono had initiated a Phase I clinical trial to evaluate the effects of EMD-534085 in patients with solid tumors and hematological malignancies [149]. The trial is currently open, although no additional information has been reported.

4.6. ATP-Competitive Inhibitors

As described above, the majority of reported KSP inhibitors bind in an "induced fit" binding pocket formed by loop L5 and the α2 and α3 helices of the motor domain. Consistent with this allosteric mode of binding, these compounds exhibit ATP-uncompetitive and MT-noncompetitive mechanisms of action. Traditionally, the initial small-molecule inhibitors of ATPases (e.g., kinases) compete with ATP for access to the nucleotide binding site of the enzyme. Sections 4.6.1 and 4.6.2 describe efforts to discover ATP-competitive agents of KSP.

4.6.1. ATP-Competitive Inhibitors from GlaxoSmithKline and Cytokinetics

Researchers at GlaxoSmithKline and Cytokinetics identified a colorectal tumor cell line (HCT116) resistant to ispinesib and other compounds known to bind in the loop L5 allosteric pocket [150]. DNA sequencing of KSP from these cells revealed a D130V mutation in the loop L5 region of the enzyme. This was in agreement with other studies in which site-directed mutants of loop L5 were shown to be refractory to monastrol and S-trityl-L-cysteine [151]. Although the cell line was generated in vitro and its relevance in the clinic has yet to be determined, these studies highlight the potential need to discover KSP inhibitors with distinct binding modes. Scientists at GSK and Cytokinetics subsequently screened a collection of cell active KSP inhibitors against the isolated KSP D130V mutant enzyme to potentially identify compounds with alternate mechanisms to those already found. They discovered the biphenylsulfonamide derivative **28**, which showed equipotent activity against both WT and mutant KSP (Fig 20) [152]. In contrast to ispinesib, steady state enzymology studies showed that **28** was ATP-competitive and MT-uncompetitive, indicating mutually exclusive binding with regard to ATP. Although **28** exhibited only modest antiproliferative activity in the ovarian cancer cell line SKOV3 (IC$_{50}$ = 6 µM), it did induce monopolar spindle formation and provided a promising starting point for further elaboration.

A comprehensive lead optimization effort revealed the series had a relatively flat SAR profile. In general, only simple modifications to the sulfonamide moiety resulted in improved activity, whereas changes to the biphenyl core and trifluoromethyl group afforded analogs with similar or attenuated potency. Replacement of the sulfonamide moiety led to the identification of sulfamide **29** and lactam **30** [152,153], which showed improved KSP inhibitory potency compared to **28** (Fig 21). Importantly, both inhibitors displayed ATP-competitive, MT-uncompetitive behavior and, in contrast to quinazolinone **31**, an analog structurally related to ispinesib, exhibited po-

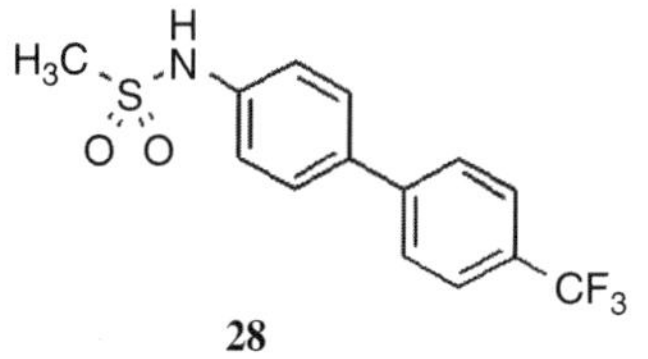

WTKSP K_i = 120 nM
D130V KSP K_i = 110 nM
SKOV3 IC$_{50}$ = 6130 nM

28

Figure 20. Structure of the lead ATP-competitive inhibitor from GlaxoSmithKline and Cytokinetics.

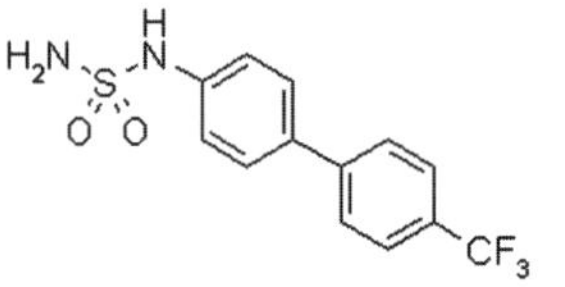

29
KSP (WT) IC$_{50}$ = 6.2 nM
KSP (D130V) IC$_{50}$ = 7 nM
HCT116 (WT) IC$_{50}$ = 403 nM
HCT116 (D130V) IC$_{50}$ = 5.4 nM

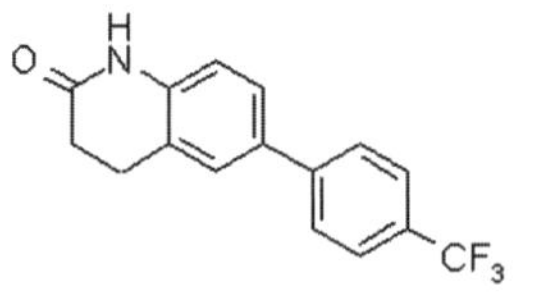

30
KSP (WT) IC$_{50}$ = 1.7 nM
KSP (D130V) IC$_{50}$ = 2 nM
HCT116 (WT) IC$_{50}$ = 36 nM
HCT116 (D130V) IC$_{50}$ = 0.5 nM

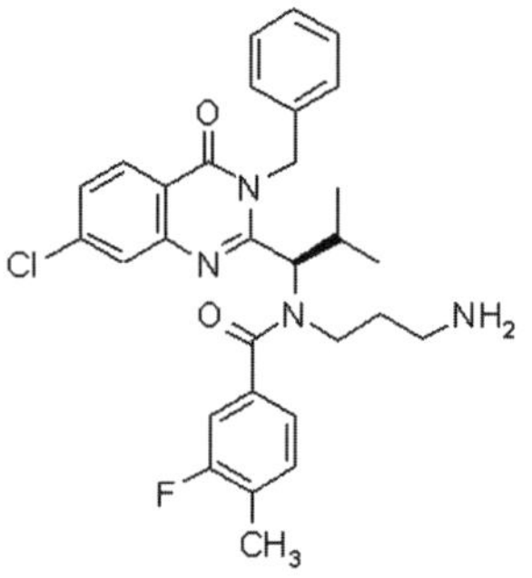

31
KSP (WT) IC$_{50}$ < 1 nM
KSP (D130V) IC$_{50}$ = 350 nM
HCT116 (WT) IC$_{50}$ = 2 nM
HCT116 (D130V) IC$_{50}$ = 1372 nM

Figure 21. Structures of ATP-competitive inhibitors and ispinesib analog from GSK and Cytokinetics.

tent inhibition of the mutant form of KSP (KSP D130V) as well. Consistent with their biochemical activity profiles, **29** and **30** showed antiproliferative activity in both WT (HCT116) and mutant (HCT116 D130V) cancer cell lines (although the disproportionately lower potency of **29** against the wild type cell line is not well explained), whereas **31** suffered a nearly 100-fold decrease in activity in the mutant line. Curiously, the ATP-competitive inhibitors showed ~10-fold better potency in HCT116 D130V cells compared to WT cells, a fact which the authors rationalized by the nearly 20-fold higher K_m for ATP of the mutant enzyme. Since **29** and **30** are competitive with respect to ATP, the lower affinity of KSP for ATP in the mutant enzyme would result in the observed relative increase in activity.

Based on its combination of potency and pharmacokinetic profile, sulfamide **29** was evaluated for *in vivo* efficacy in a HCT116 D130V human tumor xenograft study in mice. Following intraperitoneal dosing on a once-every-4-day schedule, **29** showed dose-dependent inhibition of tumor growth, including complete regressions at the highest dose tested (125 mg/kg). In comparison, and as expected based on its *in vitro* inhibition profile, quinazolinone **31** exhibited only a modest delay in tumor growth under the same dosing schedule at its MTD. Both compounds were also evaluated in a WT KSP colorectal tumor cell line, Colo205, historically one of the most sensitive to KSP inhibitors. Significant antitumor activity was not observed with **29** in this study, presumably due to its attenuated potency in cell lines that carry WT KSP (Colo205 IC$_{50}$ = 181 nM). In contrast, and again as expected, quinazolinone **31** was much more effective against these tumors with complete and partial regressions noted. Although these *in vivo* studies demonstrate the potential for ATP-competitive inhibitors to overcome possible resistance to loop L5 KSP inhibitors, substantially more potent compounds may be required to inhibit tumors bearing WT KSP in an environment containing high intracellular ATP concentrations.

Researchers at GSK noted that cocrystallization of the biphenyl inhibitors with KSP was unsuccessful, presumably due to the presence of large amounts of microtubules in the crystallization studies, an additive that was necessary due to the MT-uncompetitive kinetics. In lieu of these results, they utilized inhibitor-resistant cells, site-directed mutagenesis, and photoaffinity labeling to determine the inhibitor binding locus [153]. Wild-type (HCT116) and mutant (HCT116 D130V) cells were used to select for variants that were resistant to lactam **30** and upon subsequent RNA isolation and gene sequencing, four single amino acid

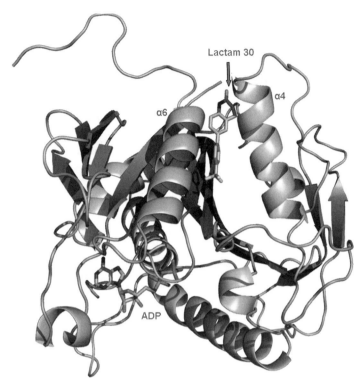

Figure 22. Model of lactam **30** (cyan) bound to KSP at the interface of helices α4 and α6. The nucleotide binding site is illustrated here with the structure of ADP (gray) [153]. (See color insert.)

point mutations were identified (I19M, A356T, I299F, and an undefined amino acid change at I332) along the interface of the α4 and α6 helices. Interestingly, a fifth point mutation was noted in the mutant cell line, namely that D130V reverted back to the wild-type residue (V130D). Two of the four constructs, A356T and I299F, were prepared by site-directed mutagenesis and **30** was found to be 39- and 64-fold less potent against these mutants, respectively. Concurrently, photoaffinity labeling studies of KSP with a biphenyl diazirine analog led to a single adduct within the KSP motor domain at L295, which resides along the α4 interface with α6. The authors concluded, based on cumulative data from all experiments, that the binding site for these biphenyl inhibitors was outside the nucleotide pocket and between the α4 and the α6 helices (Figs 22 and 23) [153]. They suggested that these inhibitors block a conformational transition that allows ATP binding and hydrolysis through interaction with the enzyme at the α4 and α6 helical interface.

4.6.2. ATP-Competitive Inhibitors from Merck & Co In 2008, Merck & Co. also reported the discovery of ATP-competitive inhibitors of KSP [154]. Following a high-throughput screen of some of their compound libraries, researchers at Merck discovered a novel thiazole-based inhibitor template (Fig 24), exemplified by the modestly active **32** (KSP IC$_{50}$ 11 μM). A subsequent lead optimization effort led to the identification of inhibitors with improved potency, including thiazole **33** (KSP IC$_{50}$ = 540 nM). Thiazole **33** was extensively characterized and, like the GSK biphenyl inhibitors, shown to behave in an ATP-competitive and MT-uncompetitive manner under steady-state conditions with K_i values of 115 and 170 nM, respectively. Importantly, thia-

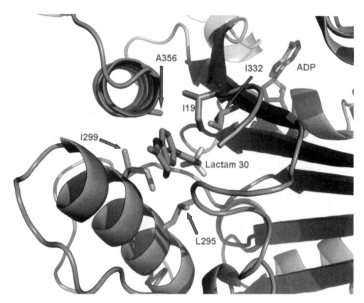

Figure 23. Closeup of modeled lactam **30** interactions with KSP showing putative interations with residues A356, I19, I299, I332, and L295. (This figure is available in full color at http://mrw.interscience.wiley.com/emrw/9780471266945/home.)

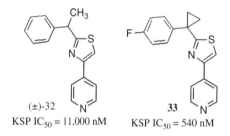

Figure 24. Structures of ATP-competitive compounds from Merck & Co.

zole **33** also induced mitotic arrest in A2780 cells and caused the formation of monopolar spindles, the characteristic KSP inhibitor phenotype. Despite the fact that the nucleotide binding site is highly conserved among kinesins, **33** exhibited good selectivity over the panel of 10 kinesins tested. As expected, based on the differences between kinesin and protein kinase binding sites, **33** was also much less active against a panel of 36 protein kinases. The initial cocrystallization of **33** with KSP was unsuccessful, but the authors proposed that the thiazole inhibitors may bind in the KSP nucleotide binding site, a hypothesis that was rationalized by potentially favorable π-stacking and hydrophobic interactions. More definitive experiments to support this hypothesis have not yet been reported.

5. CONCLUSION

Cancer cells, like all proliferating cells, require mitosis for survival and propagation. Disruption of the mitotic cycle can lead to apoptosis and cell death and has become a validated method for treating cancer patients. The taxanes and vinca alkaloids, while effective, result in undesired side effects associated with microtubule poisons, including peripheral neuopathy. Since KSP, a mitotic kinesin, has no direct effect on tubulin polymerization, many of these deleterious effects can be circumvented, making KSP an attractive therapeutic target.

Since the discovery of monastrol in 1999, scientists from industry and academic laboratories have discovered novel inhibitors of KSP, some of which are currently under evaluation in Phase I and Phase II clinical trials for treatment of solid tumors and hematologic malignancies. Treatment of patients with ispinesib, the first KSP inhibitor to enter the clinic, led to disease stabilization in several cancer types and objective responses in breast cancer patients. As anticipated, tumor biop-

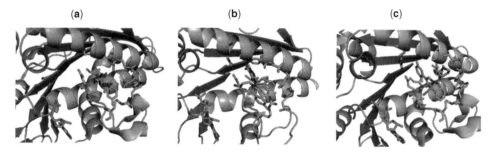

Figure 25. Comparison of ATP-uncompetitive inhibitors in the α2-α3-L5 allosteric binding site. The interactions between inhibitors monastrol (a), BMS inhibitor N-(3-aminopropyl)-N-((3-benzyl-5-chloro-4-oxo-3,4-dihydropyrrolo[2,1-f][1,2,4]triazin-2-yl)(cyclopropyl)methyl)-4-methylbenzamide [155,156] (b), and Merck & Co. inhibitor MK-0731 (c) and KSP induce nearly identical shifts in key KSP side chains and loop L5. Despite having distinct pharmacophores, all cause biochemically similar inhibition of KSP via their conserved interactions with the "induced fit" binding pocket. (This figure is available in full color at http://mrw.interscience.wiley.com/emrw/9780471266945/home.)

sies confirmed the monopolar spindle phenotype and treatment was not limited by neuropathy or other neuerotoxicities. Other small-molecule inhibitors have also progressed into clinical trials, including MK-0731, AZD4877 [155,156], ARRY-520, LY2523355, and EMD-534085, although only limited data have been reported.

A majority of the KSP inhibitors reported to date have been shown to bind in an "induced fit" allosteric binding pocket formed by loop L5 and the α2 and α3 helices of the motor domain (Fig 25). These compounds exhibit ATP-uncompetitive and MT-noncompetitive steady state kinetics. More recently, researchers have discovered inhibitors that bind to KSP in a second, distinct allosteric binding pocket that blocks productive ATP binding and hydrolysis. Although these novel ATP-competitive compounds have not yet progressed into the clinic, the discovery of KSP inhibitors with a novel mechanism of action expands the opportunities for targeting KSP for anticancer therapy, either as a treatment for potential resistance generated by the ATP-uncompetitive loop L5 binders or as a monotherapy.

6. ACKNOWLEDGMENTS

We wish to thank Neysa Nevins for help providing the KSP structural representations and Stephen Schauer for providing the mitotic cell micrograph.

REFERENCES

1. Kim AJ, Endow SA. J Cell Sci 2000;113: 3681–3682.
2. Schliwa M, editor. Molecular Motors. Weinheim, Germany: Wiley-VCH; 2003.
3. Bergnes G, Brejc K, Belmont L. Curr Top Med Chem 2005;5:127–145.
4. Wood KW, Cornwell WD, Jackson JR. Curr Opin Pharmacol 2001;1:370–377.
5. Jordan MA, Wilson L. Nat Rev Cancer 2004;4: 253–265.
6. Gunderson GG, Cook TA. Curr Opin Cell Biol 1999;11:81–94.
7. Quasthoff S, Hartung HP. J Neurol 2002;249: 9–17.
8. Rowinsky EK, Eisenhauer EA, Chaudhry V, Arbuck SG, Donehower RC. Semin Oncol 1993;20:1–15.
9. Brady RC, Schibler MJ, Cabral F. Methods Enzymol 1986;134:217–225.
10. Andersen JS, Wilkinson CJ, Mayor T, Mortensen P, Nigg EA, Mann M. Nature 2003;426: 570–574.
11. Skop AR, Liu H, Yates J, Meyer BJ III, Heald R. Science 2004;305:61–66.
12. Hirokawa N. Science 1998;279:519–526.
13. Inoué S, Salmon ED. Mol Biol Cell 1995;6: 1619–1640.
14. Endow SA. Eur J Biochem 1999;262:12–18.
15. Desai A, Verma S, Mitchison TJ, Walczak CE. Cell 1999;96:69–78.
16. Hunter AW, Wordeman L. J Cell Sci 2000;113: 4379–4389.
17. Sawin KE, LeGuellec K, Philippe M, Mitchison TJ. Nature 1992;359:540–543.

18. Whitehead CM, Rattner JB. J Cell Sci 1998;111:2551–2561.
19. Blangy A, Lane HA, d'Herin P, Harper M, Kress M, Nigg EA. Cell 1995;83:1159–1169.
20. Gaglio T, Saredi A, Bingham JB, Hasbani MJ, Gill SR, Schroer TA, Compton DA. J Cell Biol 1996;135:399–414.
21. Mak J, Freedman R, Beraud C. Utilization of gene expression profiles to identify mitotic kinesins. Presented at the 2002 AACR Annual Meeting, San Francisco, CA, Apr 2002; Poster 5375.
22. Miki H, Setou M, Kaneshiro K, Hirokawa N. Proc Natl Acad Sci 2001;98:7004–7011.
23. Kashina A, Baskin R, Cole D, Wedaman K, Saxton W, Scholey JM. Nature 1996;379:270–272.
24. Turner J, Anderson R, Guo J, Beraud C, Sakowicz R, Fletterick R. RCSB Protein Data Bank. Available at http://www.rcsb.org/pdb. 1I6. Released 2001 July 18.
25. Vale RD, Milligan RA. Science 2000;288:88–95.
26. Hirose K, Akimaru E, Akiba T, Endow SA, Amos LA. Molecular Cell 2006;23:913–923.
27. Kikkawa M. Trends Cell Biol 2008;18:128–135.
28. Marx A, Muller J, Mandelkow EM, Hoenenger A, Mandelkow E. J Muscle Res Cell Motil 2006;27:127–137.
29. Kaseda K, Crevel I, Hirose K, Cross RA. EMBO Rep 2008;9:761–765.
30. Amos LA, Hirose K. J Cell Sci 2009;120:3919–3927.
31. Valentine MT, Gilbert SP. Curr Opin Cell Biol 2007;19:75–81.
32. Korneev MJ, Lakämper S, Schmidt CF. Eur Biophys J 2007;36:675–681.
33. Krzyslak TC, Grabe M, Gilbert SP. J Biol Chem 2008;283:2078–2087.
34. Kapitein LC, Kwok BH, Weinger JS, Schmidt CF, Kapoor TM, Peterman EJG. J Cell Biol 2008;182:421–428.
35. Kapitein LC, Peterman EJ, Kwok BH, Kim JH, Kapoor TM, Schmidt CF. Nature 2005;435:114–118.
36. Mayer TU, Kapoor TM, Haggarty SJ, King RW, Schreiber SL, Mitchison TJ. Science 1999;286:971–974.
37. Cochran JC, Gatial JE III, Kapoor TM, Gilbert SP. J Biol Chem 2005;280:12658–12667.
38. Yan Y, Sardana V, Xu B, Homnick C, Halczenko W, Buser CA, Schaber M, Hartman GD, Huber HE, Kuo LC. J Mol Biol 2004;335:547–554.
39. Turner J, Anderson R, Guo J, Beraud C, Fletterick R, Sakowicz R. J Biol Chem 2001;276:25496–25502.
40. Maliga Z, Mitchison TJ. RCSB Protein Data Bank. Available at http://www.rcsb.org/pdb. 1X88. Released 2005 Nov 15.
41. Grant BJ, McCammon JA, Caves LS, Cross RA. J Mol Biol 2007;368:1231–1248.
42. Sakowicz R, Finer JT, Beraud C, Crompton A, Lewis E, Fritsch A, Lee Y, Mak J, Moody R, Turincio R, Chabala JC, Gonzales P, Roth S, Weitman S, Wood KW. Cancer Res 2004;64:3276–3280.
43. Bergnes G, Ha E, Yao B, Smith WW, Tochimoto T, Lewis ER, Lee YY, Moody R, Turincio RA, Finer JT, Wood KW, Sakowicz R, Crompton AM, Chabala JC, Morgans DJ Jr, Sigal NH, Sabry JH. Mitotic kinesin-targeted antitumor agents: discovery, lead optimization and antitumor activity of a series of novel quinazolinones as inhibitors of kinesin spindle protein (KSP). Presented at the 2002 AACR Annual Meeting, San Francisco, CA, Apr 2002; Poster 3648.
44. Lad L, Luo L, Carson JD, Wood KW, Hartman JJ, Copeland RA, Sakowicz R. Biochemistry 2008;47:3576–3585.
45. Gonzales P, Bienek A, Boehme M, Rivali C, Roth S, Wynne H, Piazza G, Weitman S, Wood K, Crompton A, Moody R, Schauer S, Feng B, Bergnes G, Sigal N, Morgans D. Breadth of anti-tumor activity of CK0238273 (SB-715992), a novel inhibitor of the mitotic kinesin KSP. Presented at the 2002 AACR Annual Meeting, San Francisco, CA, Apr 2002; Poster 1337.
46. Johnson RK, McCabe FL, Cauder E, Innlow L, Whitacre M, Winkler JD, Bergnes G, Feng B, Smith WW, Morgans D, Wood K, Jackson JR. SB-715992, a potent and selective inhibitor of KSP mitotic kinesin, demonstrates broad-spectrum activity in advanced murine tumors and human tumor xenografts. Presented at the 2002 AACR Annual Meeting, San Francisco, CA, Apr 2002; Poster 1335.
47. Jackson JR, Gilmartin AG, Williams T, McCabe FL, Caulder E, Inlow L, Whitacre M, Mattern M, Bergnes G, Feng B, Smith WW, Morgans D, Ward K, Smith B, Wood KW, Moody RS, Belmont L, Schauer S, Johnson RK, Winkler JD. A pharmacodynamic marker of mitosis demonstrates the anti-mitotic activity of SB-715992, an inhibitor of the mitotic kine-

sin KSP. Presented at the American Association for Cancer Research (AACR) Annual Meeting, Apr 2002.
48. Rodon J, Till E, Patnaik A, Takimoto C, Kathman S, Williams D, Vasist L, Bowen C, Hodge J, Dar M, Tolcher A. Phase I study of lspinesib in combination with capecitabine in patients with advanced solid tumors. Presented at the 18th Annual EORTC-NCI-AACR Symposium on Molecular Targets and Cancer Therapeutics, Prague, Czech Republic, Nov 2006.
49. Blagden S, Seebaran G, Molife R, Payne M, Reid A, Protheroe A, Kathman S, Williams D, Bowen C, Hodge J, Dar M, de Bono J, Middleton M. Phase I study of ispinesib (SB-715992) in combination with docetaxel in patients with advanced solid tumors. Presented at the 17th AACR-NIH-EORTC International Conference on Molecular Targets and Cancer Therapeutics, Philadelphia, PA, 2005 Nov 17; Poster C80.
50. Sutton D, Diamond M, Onori J, Zhang SY, Giardiniere M, Faucette L, Belmont L, Wood KW, Jackson JR, Huang P. Cisplatin enhances the activity of ispinesib, a novel KSP inhibitor, against murine P388 lymphocytic leukemia. Presented at the AACR-NCI-EORTC, Philadelphia, PA, Nov 2005.
51. Purcell JW, Reddy M, Davis J, Martin S, Samayoa K, Vo H, Thomsen K, Bean P, Wood KW, Cases S. Ispinesib (SB-715992) a kinesin spindle protein (KSP) inhibitor has single agent activity and enhances the efficacy of standard-of-care Therapies in pre-clinical models of breast cancer. Presented at the 31st Annual San Antonio Breast Cancer Symposium (SABCS), San Antonio, TX, Dec 2008.
52. Chu Q, Holen KD, Rowinsky EK, Wilding G, Volkman JL, Orr JB, Williams DD, Hodge JP, Kerfoot CA, Sabry J. A phase I study of novel kinesin spindle protein (KSP) inhibitor, SB-715992 administered intravenously once every 21 days. Presented at the 40th ASCO Annual Meeting, New Orleans, LA, June 2004; Poster 2078.
53. Burris H III, LoRusso P, Jones S, Guthrie T, Orr JB, Williams DD, Hodge JP, Bush M, Sabry J. Phase I trial of novel kinesin spindle protein (KSP) inhibitor SB-715992 IV days 1, 8, 15 q 28 days. Presented at the 40th ASCO Annual Meeting, New Orleans, LA, June 2004; Poster 2004.
54. Heath EI, Alousi JP, Eder M, Valdivieso LS, Vasist L, Appleman P, Bhargava AD, Colevas AD, LoRusso PM, Shapiro GA. Phase I dose escalation trial of ispinesib (SB-715992) administered days 1-3 of a 21-day cycle in patients with advanced solid tumors. Presented at the American Society of Clinical Oncology (ASCO), Atlanta, GA, June 2006.
55. Souid A, Dubowy RL, Greenwald Triplett D, Ingle AM, Sun J, Blaney SM, Adamson PC. Pediatric phase I trial and pharmacokinetic (PK) study of ispinesib (SB715992): a children's oncology group phase I consortium study. Presented at the American Society of Clinical Oncology, Chicago, IL, June 2008. Abstract 10014.
56. Arfons LM, Kirschbaum MH, Synold TW, Lazarus HM, Kindwall-Keller TL, Laughlin MJ, Conlan MG, Seroogy J, Doyle L, Cooper BW. A phase I study of the kinesin spindle protein inhibitor ispinesib (SB-715992) in relapsed/refractory acute leukemia. Presented at the 50th American Society of Hematology Annual Meeting, San Francisco, CA, Dec 2008. Abstract 4308.
57. Blagden SP, Molife RR, Seebaran A, Payne M, Reid AHM, Protheroe AS, Vasist LS, Williams DD, Bowen C, Kathman SJ, Hodge JP, Dar MM, de Bono JS, Middleton MR. Br J Cancer 2008;98:894–899.
58. Kathman SJ, Williams DH, Hodge JP, Dar M. Cancer Chemother Pharmacol 2009;63: 469–476.
59. Calvo E, Chu Q, Till E, Rowinsky E, Patnaik A, Takimoto C, Kathman S, Williams D, Bowen C, Hodge J, Dar M, Tolcher A., Phase I study of ispinesib (SB-715992) in combination with capecitabine in patients with advanced solid tumors. Presented at the 17th AACR-NIH-EORTC International Conference on Molecular Targets and Cancer Therapeutics, Philadelphia, PA, 2005 Nov 17; Poster C85.
60. Jones S, Plummer E, Burris H, Razak A, Meluch A, Bowen C, Williams D, Vasist L, Hodge J, Dar M, Calvert A. Phase I study of ispinesib in combination with carboplatin in patients with advanced solid tumors. Presented at the 42nd ASCO Annual Meeting, Atlanta, GA, June 2006; Poster 2027.
61. www.cytokinetics.com.
62. Therasse P, Arbuck SG, Eisenhauer EA, Wanders J, Kaplan RS, Rubinstein L, Verweij J, Van Glabbeke M, van Oosterom AT, Christian MC, Gwyther SG. J Natl Cancer Inst 2000;92:205–216.
63. Miller K, Ng C, Ang P, Brufsky AM, Lees SC, Dees EC, Piccart M, Verrill M, Wardley A, Loftiss J, Yeoh S, Hodge J, Williams D, Dar

M, Ho PTC. Phase II, open label study of SB-715992 (ispinesib) in subjects with advanced or metastatic breast cancer. Presented at the 28th Annual San Antonio Breast Symposium, San Antonio, TX, Dec 2005; Poster 1089.

64. http://www.cytokinetics.com/press_releases/release/pr_1183058254.

65. Shahin M, Braly P, Rose P, Malpass T, Bailey H, Alvarez RD, Hodge J, Bowen C, Buller R. A phase II, open-label study of ispinesib (SB-715992) in patients with platinum/taxane refractory or resistant relapsed ovarian cancer. Presented at the American Society of Clinical Oncology, Chicago, IL, June 2007.

66. El-Khoueiry AB, Iqbal S, Singh DA, D'Andre S, Ramanathan RK, Shibata S, Yang DY, Lenz HJ, Synold T, Gandara DR. A randomized phase II non-comparative study of ispinesib given weekly or every three weeks in metastatic colorectal cancer. A California Cancer Consortium Study (CCC-P). Presented at the American Society of Clinical Oncology (ASCO), Atlanta, GA, June 2006.

67. Tang PA, Siu LL, Chen EX, Hotte SJ, Chia S, Schwarz JK, Pond GR, Johnson C, Colevas AD, Synold TW, Vasist LS, Winquist E. Invest New Drugs 2008;26:257–264.

68. http://www.cytokinetics.com/press_releases/release/pr_1143752038.

69. Knox JJ, Gill S, Synold TW, Biagi JJ, Major P, Feld R, Cripps C, Wainman N, Eisenhauer E, Seymour L. Invest New Drugs 2008;26: 265–272.

70. Lee CW, Bélanger K, Rao SC, Petrella TM, Tozer RG, Wood L, Savage KJ, Eisenhauer EA, Synold TW, Wainman N, Seymour L. Invest New Drugs 2008;26:249–255.

71. Beekman KW, Dunn R, Colevas D, Davis N, Clark J, Agamah E, Thomas S, Nichols K, Redman B, Stadler W. University of Chicago Consortium phase II study of ispinesib (SB-715992) in patients (pts) with advanced renal cell carcinoma (RCC). Presented at the American Society of Clinical Oncology, Chicago, IL, June 2007.

72. Beer TM, Goldman B, Synold TW, Ryan CW, Vasist L, Van Veldhuizen PJ Jr, Dakhil SR, Lara PN Jr, Drelichman A, Hussain MHA, Crawford ED. Clin Genitourin Cancer 2008;6:103–109.

73. Lee RT, Beekman KE, Hussain M, Davis NB, Clark JI, Thomas SP, Nichols KF, Stadler WM. Clin Genitourin Cancer 2008;6:21–24.

74. Williams DD, Kathman SJ, Chu QSC, Holen KD, Rowinsky EK, Wilding G, Mudd PN Jr, Herendeed JM, Orr JB, Pandite LN, A phase I study (KSP10001) of SB-715992, a novel kinesin protein spindle (KSP) inhibitor: pharmacokinetic (PK)/pharmacodynamic (PD) modeling of absolute neutrophil counts (ANC). Presented at the EORTC-NCI-AACR, Sept 2004.

75. Gómez H, Castaneda C, Philco M, Pimentel P, Falcón S, Escandón R, Saikali K, Conlan M, Seroogy J, Wolff A. A phase I-II trial of Ispinesib, a kinesin spindle protein inhibitor, dosed every two weeks in patients with locally advanced or metastatic breast cancer previously untreated with chemotherapy for metastatic disease or recurrence. Presented at the 31st Annual San Antonio Breast Cancer Symposium (SABCS), San Antonio, TX, Dec 2008.

76. Knight SD, Parrish CA. Curr Top Med Chem 2008;8:888–904.

77. Morgans DJC, Knight SD, Newlander KA, Dhanak D, Zhou H-J, Adams ND. WO2003094839. A2. 2002 May 9.

78. McDonald A, Lu P-P, Bergnes G, Morgans DJ Jr, Dhanak D, Knight SD. WO2003106426. A1. 2002 June 14.

79. McDonald A, Morgans DJ Jr, Bergnes G, Dhanak D, Knight SD. WO2004006865. A2. 2002 July 17.

80. McDonald A, Zhou H-J. WO2004024086. A2. 2002 Sept 13.

81. Bergnes G, Qian X, Morgans DJ Jr, Knight SD, Dhanak D., WO2004032840. A2. 2002 Oct 11.

82. Zhou H-J, McDonald A, Bergnes G, Morgans DJ Jr, Chabala JC, Knight SD, Dhanak D. WO2004064741. A2. 2003 Jan 17.

83. McDonald A, Bergnes G, Feng B, Morgans DJ Jr, Knight SD, Newlander KA, Dhanak D, Brook CS. WO2003088903. 2003 Oct 30.

84. Bergnes G. WO2005046588. 2005 May 26.

85. Fraley ME, Garbaccio RM. WO2003049678. A2. 2001 Dec 6.

86. Fraley ME, Hartman GD, Hoffman WF. WO2003049679. 2001; Dec 6.

87. Coleman PJ, Fraley ME, Hoffman WF. WO2004039774. A2. 2002 May 23.

88. Fraley ME, Hartman GD, Hoffman WF. WO2003050064. A2. 2003 June 19.

89. Fraley ME, Hartman GD. WO2003050122. A2. 2003 June 19.

90. Wang W, Constantine R, Lagniton L.US patent 2,005,228,002. 2005 Apr 6.

91. Boyce RS, Guo H, Mendenhall KG, Walter AO, Wang W, Xia Y.US patent 2,006,084,687. 2006 Apr 20.

92. Barsanti PA, Xia Y, Wang W, Mendenhall KG, Lagniton LM, Ramurthy S, Phillips MC, Subramanian S, Boyce R, Brammeier NM, Constantine R, Duhl D, Walter AO, Abrams TJ, Renhowe PA. US patent. 20,070,037,853. 2006 Aug 9.
93. Lombardo LJ, Bhide RS, Kim KS, Lu S. WO2003099286. A1. 2002 May 21.
94. Kim KS.US patent 2,005,176,717. 2005 Feb 1.
95. Aquila B, Block MH, Davies A, Ezhuthachan J, Filla S, Luke RW, Pontz T, Theoclitou M-E, Zheng X. WO2004078758. 2003 Mar 7.
96. Liu J, Ali SM, Ashwell MA, Ye P, Guan Y, Ng S, Palma R, Yohannes D. WO2009002808. 2008 Dec 31.
97. Wang W, Lagniton LM, Constantine RN, Burger MT. WO2004111058. 2003 May 30.
98. Wang W, Constantine RN, Lagniton LM, Pecchi S, Burger MT, Desai MC. WO2004113335. 2003 June 20.
99. Bergnes G, Smith WW, Yao B, Morgans DJ Jr, McDonald A. WO2004009036. A2. 2002 July 23.
100. Bergnes G, Morgans DJ Jr. WO2004034972. A2. 2002 Sept 30.
101. Feng B, Bergnes G, Morgans DJC, Dhanak D, Knight SD, Darcy MG. WO2003097053. A1. 2002 May 9.
102. Bergnes G, Dhanak D, Knight SD, Lu P-P, Morgans DJ Jr, Newlander KA. WO2004055008. A1, 2002 Dec 13.
103. Fraley ME, Hoffman WF. WO2003039460. A2. 2001 Nov 7.
104. Belmont L, Sutton D, Wood KW. Sequence dependent anti-tumor activity of *bortezomib* and the KSP inhibitor, SB-743921, in a solid tumor xenograft model. Presented at the 18th Annual EORTC-NCI-AACR Symposium on Molecular Targets and Cancer Therapeutics, Prague, Czech Republic, Nov 2006.
105. Jackson JR, Gilmartin A, Dhanak D, Knight S, Luo L, Sutton D, Caulder E, Diamond M, Giardiniere M, Zhang S, Huang P. A second generation KSP inhibitor, SB-743921, is a highly potent and effective therapeutic in preclinical models of cancer. Presented at the AACR Molecular Diagnostics in Cancer Therapeutics Conference, Chicago, IL, Sept 2006.
106. Gerecitano J, O'Connor OA, Van Deventer H, Hainsworth J, Leonard JP, Goy A, Afanasyev B, Chen MM, Saikali K, Conlan MG, Escandón R, Wolff A. A phase I/II trial of the kinesin spindle protein (KSP) inhibitor SB-743921 administered on days 1 and 15 every 28 days without and with prophylactic G-CSF in non-Hodgkin or Hodgkin lymphoma. Presented at the 50th American Society of Hematology Annual Meeting, San Francisco, CA, Dec 2008. Abstract 1563.
107. Holen KD, Belani CP, Wilding G, Ramalingam S, Heideman JL, Ramanathan RK, Bowen CJ, Williams DD, Hodge JP. Phase I study to determine tolerability and pharmacokinetics (PK) of SB-743921, a novel kinesin spindle protein (KSP) inhibitor. Presented at the 42nd ASCO Annual Meeting, Atlanta, GA, June 2006. Abstract 2000.
108. http://www.cytokinetics.com/press_releases/release/pr_1149513207.
109. Aquila B, Block MH, Davies A, Ezhuthachan J, Pontz T, Russell DJ, Theoclitou M-E, Zheng X. WO2006018627. 2006 Feb 23.
110. Aquila B, Block MH, Davies A, Ezhuthachan J, Pontz T, Russell DJ, Theoclitou M-E, Zheng X. WO2006018628. 2006 Feb 23.
111. Block MH, Davies A, Russell DJ, Theoclitou M-E. WO206008523. 2006 Jan 26.
112. Brown JL, Huszar D, McCoon PE. WO2008122798. 2008 Oct 16.
113. www.clinicaltrials.gov.
114. Stephenson JJ, Lewis N, Martin JC, Ho A, Li J, Wu K, Pace L, Eder JP, Schwartz GK. Phase I multicenter study to assess the safety, tolerability, and pharmacokinetics of AZD4877 administered twice weekly in adult patients with advanced solid malignancies. Presented at the American Society of Clinical Oncology, Chicago, IL, June 2008.
115. Infante JR, Spratlin JL, Kurzrock R, Eckhardt SG, Burris HA III, Puchalski TA, Li J, Wu K, Ochs J, Herbst RS. Clinical, pharmacokinetic (PK), pharmacodynamic findings in a phase I trial of weekly (wkly) intravenous AZD4877 in patients with refractory solid tumors. Presented at the American Society of Clinical Oncology, Chicago, IL, June 2008.
116. http://www.astrazeneca-annualreports.com/2008/directors_report/resources_skills/dev_pipeline.html.
117. Cox CD, Breslin MJ, Mariano BJ, Coleman PJ, Buser CA, Walsh ES, Hamilton K, Huber HE, Kohl NE, Torrent M, Yan Y, Kuo LC, Hartman GD. Bioorg Med Chem Lett 2005;15:2041–2045.
118. Fraley ME, Garbaccio RM, Arrington KL, Hoffman WF, Tasber ES, Coleman PJ, Buser CA, Walsh ES, Hamilton K, Fernandes C, Schaber MD, Lobell RB, Tao W, South VJ, Yan Y, Kuo C, Prueksaritanont T, Shu C, Torrent M,

Heimbrook DC, Kohl NE, Huber HE, Hartman GD. Bioorg Med Chem Lett 2006;16: 1775–1779.

119. Garbaccio RM, Fraley ME, Tasber ES, Olson CM, Hoffman WF, Arrington KL, Torrent M, Buser CA, Walsh ES, Hamilton K, Schaber MD, Fernandes C, Lobell RB, Tao W, South VJ, Yan Y, Kuo LC, Prueksaritanont T, Slaughter DE, Shu C, Heimbrook DC, Kohl NE, Huber HE, Hartman GD. Bioorg Med Chem Lett 2006;16:1780–1783.

120. Cox CD, Torrent M, Breslin MJ, Mariano BJ, Whitman DB, Coleman PJ, Buser CA, Walsh ES, Hamilton K, Schaber MD, Lobell RB, Tao W, South VJ, Kohl NE, Yan Y, Kuo LC, Prueksaritanont T, Slaughter DE, Li C, Mahan E, Lu B, Hartman GD. Bioorg Med Chem Lett 2006;16:3175–3179.

121. Cox CD, Breslin MJ, Whitman DB, Coleman PJ, Garbaccio RM, Fraley ME, Zrada MM, Buser CA, Walsh ES, Hamilton K, Lobell RB, Tao W, Abrams MT, South VJ, Huber HE, Kohl NE, Hartman GD. Bioorg Med Chem Lett 2007;17:2697–2702.

122. Coleman PJ, Schreier JD, Cox CD, Fraley ME, Garbaccio RM, Buser CA, Walsh ES, Hamilton K, Lobell B, Rickert K, Tao W, Diehl RE, South VJ, Davide JP, Kohl NE, Yan Y, Kuo L, Prueksaritanont T, Li C, Mahan EA, Fernandez-Metzler C, Salata JJ, Hartman GD. Bioorg Med Chem Lett 2007;17:5390–5395.

123. Garbaccio RM, Tasber ES, Neilson LA, Coleman PJ, Fraley ME, Olson C, Bergman J, Torrent M, Buser CA, Rickert K, Walsh ES, Hamilton K, Lobell RB, Tao W, South VJ, Diehl RE, Davide JP, Yan Y, Kuo LC, Li C, Prueksaritanont T, Fernandez-Metzler C, Mahan EA, Slaughter DE, Salata JJ, Kohl NE, Huber HE, Hartman GD. Bioorg Med Chem Lett 2007;17:5671–5676.

124. Roecker AJ, Coleman PJ, Mercer SP, Schreier JD, Buser CA, Walsh ES, Hamilton K, Lobell RB, Tao W, Diehl RE, South VJ, Davide JP, Kohl NE, Yan Y, Kuo LC, Li C, Fernandez-Metzler C, Mahan EA, Prueksaritanont T, Hartman GD. Bioorg Med Chem Lett 2007;17:5677–5682.

125. Cox CD, Coleman PJ, Breslin MJ, Whitman DB, Garbaccio RM, Fraley ME, Buser CA, Walsh ES, Hamilton K, Schaber MS, Lobell RB, Tao W, Davide JP, Diehl RE, Abrams MT, South VJ, Huber HE, Torrent M, Prueksaritanont T, Li C, Slaughter DE, Mahan E, Fernandez-Metzler C, Yan Y, Kuo LC, Kohl NE, Hartman GD. J Med Chem 2008;51: 4239–4252.

126. Yan Y. RCSB Protein Data Bank. Available at http://www.rcsb.org/pdb. 3CJO. Released 2008 July 1.

127. Stein MN, Rubin EH, Scott PD, Fernandez R, Agrawal NG, Hsu K, Walker A, Holen K, Wilding G. Phase I clinical and pharmacokinetic (PK) trial of the kinesin spindle protein (KSP) inhibitor MK-0731 in cancer patients. Presented at the 42nd ASCO Annual Meeting, Atlanta, GA, June 2006. Abstract 2001.

128. Stein MN, Tan A, Taber K, Fernandez R, Agrawal NG, Vandendries E, Hsu K, Walker A, Holen K, Wilding G. Phase I clinical and pharmacokinetic (PK) trial of the kinesin spindle protein (KSP) inhibitor MK-0731 in patients with solid tumors. Presented at the 43rd ASCO Annual Meeting, Chicago, IL, June 2007. Abstract 2548.

129. http://www.merck.com/finance/pipeline.swf.

130. Hans J, Wallace EM, Zhao Q, Lyssikatos JP, Aicher T, Laird E, Robinson J, Allen S. WO2006044825. 2006 Apr 27.

131. Hans J, Wallace EM, Zhao Q, Allen S, Laird E, Lyssikatos JP, Robinson JE, Corrette CP, Delisle RK. US patent 2,006,247,178. 2006 May 1.

132. Woessner R, Corrette C, Allen S, Hans J, Zhao Q, Aicher T, Lyssikatos J, Robinson J, Poch G, Hayter L, Cox A, Lee P, Winkler J, Koch K, Wallace E. ARRY-520, a KSP inhibitor with efficacy and pharmacodynamic activity in animal models of solid tumors. Presented at the 2007 AACR National Meeting, Los Angeles, CA, Apr 2007; Poster 1433.

133. Woessner R, Tunquist B, Chlipala E, Humphries M, Trawick D, Cox A, Lee P, Winkler J, Walker D. ARRY-520, a KSP inhibitor, with potent *in vitro* and *in vivo* efficacy and pharmacodynamic activity in models of multiple myeloma. Presented at the 20th Annual EORTC-NCI-AACR Symposium on Molecular Targets and Cancer Therapeutics, Geneva, Switzerland, Oct 2008.

134. Lemieux C, DeWolf W, Voegtli W, Delisle RK, Laird E, Wallace E, Woessner R, Corrette C, Allen S, Hans J, Zhao Q, Aicher T, Lyssikatos J, Robinson J, Koch K, Winkler J, Gross S. ARRY-520, a novel, highly selective KSP inhibitor with potent anti-proliferative activity. Presented at the 2007 AACR National Meeting, Los Angeles, CA, Apr 2007; Poster 5590.

135. Carter B, Mak DH, Schober WD, Woessner R, Gross S, Harris D, Estrov Z, Andreeff M. Inhibition of KSP by ARRY-520 induces cell cycle block and cell death via the mitochondrial pathway in AML cells. Presented at the 49th

American Society of Hematology Annual Meeting, Atlanta, GA, Dec 2007.
136. Woessner R, Cox A, Koch K, Lee P, Sumeet R, Tunquist B, Walker D, Winkler J. In vivo and pharmacodynamic profiling of the KSP inhibitor ARRY-520 supports potent activity in hematological cancers and drug-resistant tumors. Presented at the 100th Annual AACR Meeting, Denver, CO, Apr 2009. Abstract 4703.
137. http://clinicaltrials.gov/ct2/show/NCT00462358?intr=%22ARRY-520%22&rank=1.
138. www.arraybiopharma.com.
139. Murakata C, Amishiro N, Ino Y, Yamamoto J, Atsumi T, Nakai R, Nakano T. WO2005035512. 2004 Oct 8.
140. Murakata C, Kato K, Yamamoto J, Nakai R, Okamoto S, Ino Y, Kitamura Y, Saitoh T, Katsuhira T. WO2006101102. A1. 2006 Sept 28.
141. http://www.nmcca.org/patients/clinicaltrials.htm.
142. http://investor.lilly.com/downloads/elli27858.pdf.
143. Schiemann K, Anzali S, Drosdat H, Emde U, Finsinger D, Gleitz J, Hock B, Reubold H, Zenke F. WO2005063735. A1. 2005 July 14.
144. Schiemann K, Bruge D, Buchstaller H, Emde U, Finsinger D, Amendt C, Zenke F. DE102005027168. A1, 2006 Dec 14.
145. Schiemann K, Bruge D, Buchstaller H, Finsinger D, Staehle W, Amendt C, Emde U, Zenke F. WO2006002726. A1. 2006 Jan 12.
146. Schiemann K, Emde U, Schlueter T, Saal C, Maiwald M. WO2007147480. A2. 2007 Dec 27.
147. Bauer H, Rasenack N, Ammer K, Kemmerer B, Heil K. WO2008138459. 2008 Nov 20.
148. Finsinger D, Zenke F, Amendt C, Schiemann K, Emde U, Knöchel T, Bomke J, Gleitz J, Wilm C, Meyring M, Osswald M, Funk J. Discovery, synthesis and characterization of hexahydropyranoquinolines as a novel class of selective inhibitors of the mitotic kinesin Eg5. Presented at the 97th annual meeting of the American Association for Cancer Research (AACR), Washington, DC, Apr 2006; 5713.

149. Merck Serono Fall 2007 Presentation Roadshow, www.merck.de/servlet/PB/menu/1101280/index.html.
150. Jackson JR, Auger KR, Gilmartin A, Eng WK, Luo L, Concha N, Parrish C, Sutton D, Diamond M, Giardiniere M, Zhang S-Y, Huang P, Wood KW, Belmont L, Lee Y, Bergnes G, Anderson R, Brejc K, Sakowicz R. A resistance mechanism for the KSP Inhibitor ispinesib implicates point mutations in the compound binding site. Presented at the 17th AACR-NCI-EORTC International Conference on Molecular Targets and Cancer Therapeutics, Philadelphia, PA, Nov 2005 Poster C207.
151. Brier S, Lemaire D, DeBonis S, Forest E, Kozielski F. J Mol Biol 2006;360:360–376.
152. Parrish CA, Adams ND, Auger KR, Burgess JL, Carson JD, Chaudhari AM, Copeland RA, Diamond MA, Donatelli CA, Duffy KJ, Faucette LF, Finer JT, Huffman WF, Hugger ED, Jackson JR, Knight SD, Luo L, Moore ML, Newlander KA, Ridgers LH, Sakowicz R, Shaw AN, Sung CM, Sutton D, Wood KW, Zhang S, Zimmerman MN, Dhanak D. J Med Chem 2007;50:4939–4952.
153. Luo L, Parrish CA, Nevins N, McNulty DE, Chaudhari AM, Carson JD, Sudakin V, Shaw AN, Lehr R, Zhao H, Sweitzer S, Lad L, Wood KW, Sakowicz R, Annan RS, Huang PS, Jackson JR, Dhanak D, Copeland RA, Auger KR. Nat Chem Bio 2007;3:722–726.
154. Rickert KW, Schaber M, Torrent M, Neilson LA, Tasber ES, Garbaccio R, Coleman PJ, Harvey D, Zhang Y, Yang Y, Marshall G, Lee L, Walsh ES, Hamilton K, Buser CA. Arch Biochem Biophys 2008;469:220–231.
155. Kim KS, Lu S, Cornelius LA, Lombardo LJ, Borzilleri RM, Schroeder GM, Sheng C, Rovnyak G, Crews D, Schmidt RJ, Williams DK, Bhide RS, Traeger SC, McDonnell PA, Mueller L, Sheriff S, Newitt JA, Pudzianowski AT, Yang Z, Wild R, Lee FY, Batorsky R, Ryder JS, Ortega-Nanos M, Shen H, Gottardis M, Roussell DL. Bioorg Med Chem Lett 2006;16:3937–3942.
156. Sheriff S. RCSB Protein Data Bank. Available at http://www.rcsb.org/pdb. 2GM1. Released 2006 June 27.

CNS CANCERS

ROBERTSON GRAEME
Siena Biotech SpA,
Siena, Italy

1. INTRODUCTION

There are 12 main categories and more than 100 subcategories of primary brain tumors with overlapping and differing pathobiological characteristics. CNS oncology is further compounded by the presence of secondary, recurrent, and metastatic tumors that contribute to the range of pathobiologies. Tumor heterogeneity and drug delivery across the blood–brain barrier (BBB) further compound the difficulties in efficiently tackling CNS tumors.

Brain tumors are generally named according to either their cellular origins (e.g., gliomas) or the area in which they present (e.g., pituitary adenoma or meningioma) [1]. Glial tumors are referred to as gliomas (often subdivided into astrocytomas and oligodendrogliomas), whereas those in which neuronal cells predominate are called neuroblastomas or primitive neuroectodermal tumors [2].

Brain tumors are usually classified using the WHO grading [3] based on cell origin and cellular function, from the least aggressive (benign) to the most aggressive (malignant). Genetic alterations, pathological features, and associated signaling pathways are also now being used to describe tumor populations [4]. Histopathological classification of CNS tumors, together with cytogenetic and molecular alterations is thus providing a better understanding of the molecular and clinical pathogenesis to help refine tumor and patient classification [5]. Malignant CNS tumors display glial [6] or neuronal phenotypes or a mixture of cell types and are among the most aggressive and lethal of cancer types. There are also large differences between primary, secondary (including metastatic), and recurrent CNS tumors and most CNS tumors contain a mixture of cell types [7].

Modeling CNS tumors characteristics in both *in vitro* and *in vivo* settings to allow the progression of compounds with the right physicochemical properties for penetrating the blood–brain barrier are key focus areas in the generation of molecular target-based therapies.

Malignant CNS tumors are characterized by being highly heterogeneous, aggressive and frequently invasive, leading to poor overall patient survival times [8]. Primary CNS tumors, including gliomas and medulloblastomas, are among the most chemoresistant and malignant tumors [9] accounting for approximately 1.3% of all cancers and 2.2% of all cancer-related deaths [10]. Current therapeutic approaches are focused on molecular target-based approaches (small molecule and vaccine) to compliment the current management regimens via combinations of surgery, radiotherapy, and cytotoxic agents. For example, in patients with glioblastomas, the introduction of concomitant chemotherapy using the cytotoxic agent, temozolomide, has resulted in a prolongation of survival times [11]. In addition, the identification of molecular markers, such as methylation status of MGMT in glioblastomas, or the translocation of 1p and 19q in oligodendrogliomas, have allowed differentiation of patients better placed to respond to treatment. The heterogeneity and plasticity of CNS tumors together with their ability to evade treatment, however, remains a significant challenge [12]. Median survival times remain poor prompting investigation of additional molecular target-based therapies [13] and in particular combination molecular-target therapies. Combination therapy approaches will, it is hoped, provide better coverage of tumor heterogeneity and resistance and thus provide improved treatments in the short-to-medium term. However, more basic research is required to better understand the evolutionary (developmental) underpinnings of tumor initiation to fully tackle tumor suppression. Progress in therapy and several of the unmet needs in current approaches to the treatment of glioblastoma have recently been reviewed [14,15]. Developing knowledge on possible molecular targets among the network of dysregulated signaling pathways is driving the development of small-molecule inhibitors of key cellular pathways manifest in CNS tumorigenesis [16].

Malignant pediatric CNS tumors are the most common solid tumors and the leading cancer-related deaths in children [17]. The current therapies carry significant side effects and improvements in patient survival and quality of survival need to be elicited without undue impact on developmental processes [18,19] and impact on neurocognitive function. Treatment of pediatric CNS tumors especially in young children is particularly challenging due to the vulnerability of the developing brain to treatment-related toxicity or potential for impact on developmental signaling pathways. For example, clinical trials of in pediatric patients with refractory brain tumors of SU5416 a novel small molecule tyrosine kinase inhibitor of VEGFR 1 and 2 (Fig. 1) were stopped for side-effects which included drug-related toxicities of grade III liver enzyme abnormalities, arthralgia, and hallucinations [20].

The most common malignant pediatric CNS tumor subtypes include ependymomas, high-grade gliomas, medulloblastoma, and supratentorial primitive neuroectodermal tumors. High-grade tumors of the nervous system are among the most common and most chemoresistant neoplasms of childhood and adolescence [21] and remain difficult to cure despite recent advances in imaging, neurosurgery, and radiation. Current treatment modalities have demonstrated only modest survival benefit [22].

Brain tumors also present challenging medical problems in adults where neurotoxicity [23] is frequently a major dose-limiting side effect. Seizures, peritumoral edema, venous thromboembolism, fatigue, and cognitive dysfunction can complicate the treatment of patients with either primary or metastatic brain tumors [24]. Neurotoxicity is also a problem in the treatment of many peripheral tumors [25] leading to the concept termed "chemobrain" [26] as one of the consequences of chemotherapy in general. Neuropsychological testing has confirmed various aspects of cognitive impairment [27]. Cognitive impairment is not the only manifestation of neurotoxicity induced by chemotherapy; alterations in sensory and motor function are also observed. Epilepsy is common in patients with brain tumors and often an epileptic seizure is the presenting symptom although late seizures may also occur [28]. Epilepsy is most common in developmental tumors, slow-growing tumors (low-grade gliomas), hemorrhagic tumors, and multiple metastases [29], and is an important morbidity factor.

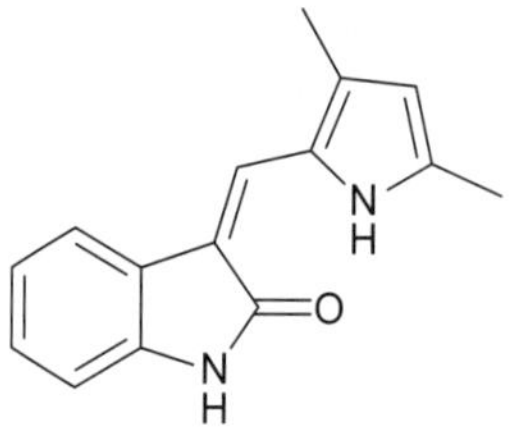

Overall, there are few if any effective therapies for many tumors of the nervous system, partly in consequence of their location and the role of the brain microenvironment, partly due to the characteristics of the cells that initiate CNS tumor development [30–33]. However, recent advances in understanding of the pathobiology and tumor heterogeneity, and the development of mouse models including transgenic models [34,35] are providing insights into underlying pathways and mechanisms [36] that should lead to the development of more effective interventions.

Faced not only with cell heterogeneity but also with the plasticity of stem cells to adapt to selection criteria (treatment), the current cancer model systems do not fully recapitulate the complexity of the human setting, perhaps, partially explaining the lack of power of these models in predicting clinical outcomes [37].

2. BRAIN TUMOR INITIATION: BRAIN TUMOR STEM CELLS

2.1. Tumor Initiation

The propensity of malignant CNS tumors for self-renewal proliferation, migration/invasion, and their resistance to both chemo- and radiotherapy [38] are thought to be driven by a small subpopulation of tumor cells, termed

Figure 1. SU5416.

cancer stem cells (CSCs) with stem or progenitor cell properties ("cancer stem cells") [39].

Cytotoxic therapies used in current therapies to the debulk tumor cells fail to obliterate the relatively quiescent and resistant CSC compartment, thereby allowing these cells to survive and drive tumor recurrence [40]. Cancer stem cells have a distinct capability to self-renew, and give rise to progeny of multiple neuroepithelial lineages.

Evidence for the existence of cancer stem or progenitor cells was first obtained in hematological malignancies and, more recently, for solid tumors including not only brain cancers but also breast, colon, and pancreatic cancers. Stem cells have been detected in glioblastoma, medulloblastoma, and ependymoma. Brain tumor stem cells are typically located in the hippocampus, the subventricular zone and to a lesser extent in the cerebellum [41]. For an excellent review of stem cells, cells of origin, and cancer stem cells in the generation of CNS tumors, see Ref. [2]. Tumor stem-like cells have been shown to initiate and propagate neoplasms with high efficiency. When they are removed from the bulk tumor mass, regrowth is rarely observed; thus, implicating cancer stem/progenitor cells as key therapeutic entry points.

Many if not most CNS cancers display a hierarchy of differentiation states within the tumor cell population. The signaling networks driving tumor formation and malignancy commonly overlap with developmental processes driven by normal stem cells function, so therapies will need to differentiate between normal and aberrant signaling of stem cells. Cancer stem cells can be derived from normal stem cells and maintain the plasticity and adaptability of these cell types. Alterations in the cellular and genetic mechanisms that control adult neurogenesis might thus contribute to brain tumorigenesis [42] in particular glioblastoma.

Brain tumors are highly infiltrative with a stem-like phenotype and can demonstrate angiogenesis-independent tumor growth. Data relating to the uncoupling of invasion and angiogenesis suggest two independent mechanisms can drive tumor progression underlining the need for developing therapies that specifically target the stem-like cell pools in brain tumors [43].

Gliomas are thought to originate from tumor stem or progenitor cells [44] where mutations and disruptions in the intracellular (typically developmental) pathways of glial progenitors provoke pathogenesis and provide mechanisms for tumor initiation and progression [45] (Fig. 2). Parallel to the role that normal stem cells play in organogenesis, cancer stem cells are thought to be crucial for tumorigenesis [46]. Indeed the marker phenotypes, morphologies, and migratory properties of cells in gliomas strongly resemble glial progenitors [47]. Comparisons have been established between the phenotypic plasticity of glial progenitors and the responses to growth factors in promoting proliferation and migration of normal and glioma cells.

A small population of CNS tumor stem cells, which form neurospheres and possess the capacity for self-renewal, has been recently identified in adult and pediatric brain tumors. They differentiate into phenotypically diverse populations, including neuronal, astrocytic, and oligodendroglial cells *in vitro* and recapitulate original tumors *in vivo*. Furthermore, the relationship and difference of cell proliferation between tumor stem cells and normal stem cells have been reviewed. Malignant brain tumor stem cells appear to have a higher proliferative capacity than stem/progenitor cells in benign brain tumors [32].

Brain tumors such as medulloblastomas or glioblastomas are believed to derive from aberrant transformation of neural stem cells (NSCs) into tumor stem or stem-like cells. CNS tumors show often exhibit areas of divergent differentiation and a full explanation of the role of tumor stem cells is still emerging. The concept of brain–tumor stem cells could provide new insights for future therapies, particularly if the main capacity for self-renewal of tumor cells and tumor growth is driven by this subset of cancer cells [48].

In vitro functional assays have shown that gliomas contain a mixture of glial progenitor cells and their progeny [49]. Pilocytic astrocytomas contain slowly dividing oligodendrocyte-type 2 astrocyte/oligodendrocyte precursor cells in keeping with their benign behavior, whereas both glioblastomas and oligodendrogliomas contain neoplastic glial restricted pre-

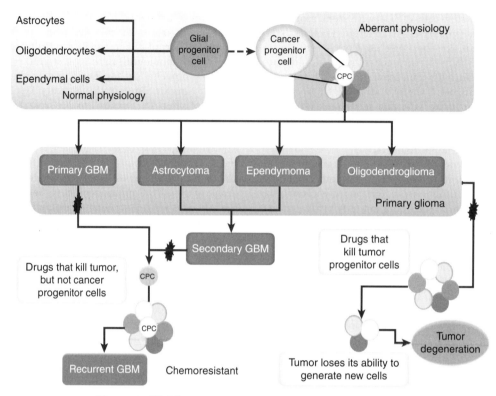

Figure 2. Glial Progenitors – Normal vs. Aberrant Physiology.

cursor cells reflecting their malignant nature. Normal glial/neuronal cells are not the sole source of brain tumor stem cells [42]. They can also arise due to conversion of more differentiated progeny such as astroglia, oligodendroglia that undergo *trans*-differentiation processes [50], greatly expanding the potential target of oncogenic mutations [51].

Targeting these cells is therefore an important focus in the development of CNS tumor therapies both for controlling tumor initiation and potentially for combating resistance to conventional therapies [52] not only by sensitizing tumors to current therapies but also in deriving more efficient molecular target-based therapies. A contributing factor in the resistance to current glioblastoma therapies may depend on alterations in apoptosis-related proteins such as overexpression of antiapoptotic factors or silencing of key death effectors rending them strongly resistant to death receptor ligands and inflammatory cytokines. Brain tumor stem cells may exploit similar mechanisms to escape the effect of cytotoxic drugs. Indeed, apoptosis mechanisms offer numerous targets for molecular-targeted approaches, including p53, Bax, Bcl-2, caspases, growth factor receptors, PI-3-kinase, Akt, and apoptosis inhibitors. Alternatively, as reviewed below the cell cycle pathway is also an area of focus. In both apoptosis and cell cycle pathways, a range of approaches is under investigation including antisense approaches, vaccine approaches, monoclonal antibodies, and small-molecule modulators, or inhibitors [53].

Tumor stem-like cells have also been isolated from benign tumors; for example, sphere-forming, self-renewable, and multipotent stem-like cells have been isolated from pituitary adenomas [54]. These cells show increased levels of stem cell-associated gene products, antiapoptotic proteins, and pituitary progenitor cell markers. Similar to CSCs isolated from glioblastomas, they are more resistant to chemotherapeutics and are pituitary tumor-initiating cells when transplantated into SCID mice. Cancer stem-like cells might have important functions in chemoresistance, for example, glioma stem cells are more aggressive in recurrent tumors that display a more malignant progression than primary tumors [55]. In melanoma, one of the most aggressive forms of cancer with a continuously growing incidence rate [56], chemotherapy has limited efficacy in part due to the existence of cancer stem-like cells.

Evidence that glioblastomas, which comprise a highly heterogeneous tumor population, are composed of tumor cells and small portion a highly tumorigenic stem cells with a low proliferation rate have been reviewed [57]. Glioma cancer stem cells are phenotypically similar to the normal stem cells, express genes characteristic of neural stem cells, and posses the self-renewal potential (see Section 2).

Animal model studies are focusing on orthotopic models where highly infiltrative brain tumors with stem-like phenotypes have been established which better mimic the human and clinical setting. For example, Bjerkvig and his group have developed orthotopic models of human glioblastomas in immunodeficient rats. Highly infiltrative glioblastoma tumors with a stem-like phenotype have been established, which show reduced chemosensitivity and are more chemoresistant than angiogenic tumors derived from the same patients [58].

2.2. CNS Tumors and Their Microenvironment

There is considerable evidence that the tissue microenvironment plays an important role in tumor growth and progression. CNS tumors contain a complex interplay between the neoplastic compartment and neighboring vasculature, inflammatory cells, immune cells, growth factors, and stromal elements [59]. Distortion of the microenvironment by tumor cells can promote growth, recruit nonmalignant cells that provide physiological resources, and facilitate invasion (including the spread of metastases to the brain). Histologically similar tumors formed within different brain regions are molecularly distinct with distinct populations of site-restricted progenitor cells implying interaction between the cell of origin, the tumor microenvironment [60,61], and specific cancer-causing genetic changes [62]. Thus, glioblastoma stem cells seem to be dependent on cues from aberrant vascular niches that mimic the normal neural stem cell niche [46].

Disruption of the tumor microenvironment via targets/pathways in the vasculature, inflammatory, or immune cells may therefore provide alternative or additional therapeutic targets to combat tumor progression. Several approaches are being pursued to target glioblastoma microenvironments including aberrant TGF-β signaling, aberrant vasculogenesis, tumor invasion mechanisms, and pathways associated with hypoxia [61].

There is a significant difference in the therapeutic responses of glioblastoma and medulloblastoma. Medulloblastomas respond to radiation of their microenvironment, and evade of cell death by the nestin-expressing cells in the perivascular stem cell niche presumable driven by aberrant p53 and/or phosphoinositide 3-kinase (PI3K) signaling. It remains to be determined whether this mechanism of evasion applies to the more complex stem cell niche and tumor bulk of gliomas [63].

Experimental studies including *in vivo* models have demonstrated that formation of brain metastases is facilitated by vascular, rather than neuronal, processes. This vessel interaction was adhesive in nature implicating the vascular basement membrane (VBM) as the active substrate for tumor cell growth in the brain. Accordingly, VBM promoted adhesion and invasion of malignant cells and was sufficient for tumor growth prior to any evidence of angiogenesis. Blockade or loss of the β1-integrin subunit in tumor cells prevented adhesion to VBM and attenuated metastasis establishment and growth *in vivo*. These findings may help inform the design of effective molecular therapies for patients with fatal CNS malignancies [64].

2.3. CNS Tumors and Signaling Pathways

Rapid proliferation and high motility are hallmarks of both normal development and tumorigenic processes implying that development pathways are probable targets for oncogenic transformations. If aberrant control of developmental pathways is responsible for propagation and/or maintenance of cancer stem cells, it is conceivable that exogenous pathway regulators may have novel therapeutic roles for brain cancer [65].

Among the developmental signaling networks prevalent in CNS cancer progenitor cells are epidermal growth factor (EGF)/EGFR [66], stem cell factor (SCF)/KIT, Hedgehog, Notch [67], and/or Wnt/beta-catenin [68,69] combined with the increased DNA repair mechanisms and ABC transporter-mediated drug efflux [38]. Mutations or changes in expression levels, in both spontaneous and experimental brain tumor cell lines, often target stem cell genes or genes lying within their functional pathways. The main examples being the Hedgehog and Wnt pathways which are currently areas of high scientific and drug discovery endeavor. Approaches to the discovery of new therapeutic strategies [70] based on many of these pathways are summarized later.

Several stem cell genes are often overexpressed in brain tumors, even if they are not mutated, suggesting a role in the generation of cancer stem cells from more differentiated precursors, or for cancer stem cell maintenance [51].

New *in vitro* and *in vivo* models for interrogating brain tumor cells in stem cell conditions have been developed that provide important new opportunities for elucidating the key pathways responsible for driving the proliferation of these cells [71].

The developmental signaling pathways Hedgehog and Wnt play key roles in both development and in various tumor types including CNS tumors both play critical roles in stem cell properties [72]. Brain tumor initiating cells exhibit many stem-like properties, including self-renewal, extended proliferation, and multipotency, and are both phenotypically and genetically similar to normal neural stem cells. Aberrant expression of developmental pathways, such as Wnt, Hedgehog, Notch, transforming growth factor-β (TGF-β) [73,74], and bone morphogenetic protein (BMP), have been demonstrated in brain tumors, and modulation of aberrant signaling within these pathways is being investigated to target brain tumor stem-like cells. For example, changes in the regulation of axin1 and β-catenin components of Wnt signaling pathway principally, downregulation of axin expression and upregulation of β-catenin have been shown to be critical components of aberrant Wnt signaling in neuroepithelial brain tumors [75]. Therapeutic strategies to inhibit CNS tumors via modulators of these pathways will need to minimize neural stem-related side effects and may require detailed understanding of the disregulation of multiple pathopathways in the tumorigenic setting [65].

The Hedgehog pathway is involved in both embryonic development and adult tissue homeostasis. Hedgehog signaling plays a central role in the proliferative control and differentiation of both embryonic stem cells and adult stem cells. Thus, alterations in the Hedgehog pathway that provoke aberrant signaling play a role in the development and propagation of several tumor types notably medulloblastoma [76].

The canonical Wnt cascade has emerged as a critical regulator of stem cells often causally associated with tumor progression. Alterations in the tightly regulated self-renewal mediated by Wnt signaling in stem and progenitor cells promote tumor development and malignant proliferation. Insights into the role of Wnt signaling in both stem cell and cancer cell maintenance and growth in the intestinal, epidermal, and hematopoietic systems may serve to distinguish between normal and aberrant function of this developmental pathway [68]. Aberrant Wnt signaling has been implicated in a range of CNS tumors including PNET [77], medulloblastoma [78], and malignant gliomas including glioblastoma [79]. See Wnt signaling section for more details.

3. BRAIN TUMOR CLASSIFICATION

Brain tumors are divided into primary, secondary (progression from lower grade tumors or metastatic tumors), and recurrent tumors, which can have significantly different tumor profiles and phenotypes [80]. Collectively, CNS tumors comprise of neuroepithelial tumors, tumors of cranial and paraspinal nerves, tumors of the meninges, lymphomas, hematopoietic neoplasms, germ cell tumors, tumors of the sellar region, and metastasis to the brain. They are generally categorized using the WHO classification system [3], derived from criteria based on histological characteristics and genetic typing, including increased cellular density, mitosis, vascular proliferation, and necrosis.

Gliomas are the most frequent CNS tumors and are divided into astrocytic, oligodendrocytic, and mixed gliomas. To better characterize gliomas and potentially provide a data source for molecular targeted-therapies a joint collaboration between the National Cancer Institute and National Institute of Neurological Disorders and Stroke has created a bioinformatics database "The Repository for Molecular Brain Neoplasia Data (REMBRANDT)" [81], accessible via http://caintegrator-info.nci.nih.gov/rembrandt that integrates genomic and clinical data from clinical trials of patients with gliomas. REMBRANDT also supports the Glioma Molecular Diagnostic Initiative's (GMDI) that aims to provide pathologic classification schemas for gliomas.

Current therapies and chemotherapy options for each subtype are summarized in the relevant classifications and then detailed under pertinent targets and/or signaling pathways.

4. PRIMARY BRAIN TUMORS

Primary brain tumors are the most common pediatric neoplasms and although rare in adults they have a disproportionate impact because of their poor prognosis [82]. Primary tumors arise in many areas of the brain and central nervous system. Their distribution based the WHO histological classification is shown in Table 1.

A particular problem for treatment of CNS tumors is the evolution of the tumorigenic phenotype during tumor progression coupled with the ability of brain tumors to evade treatment [82]. For example, treatment of glioblastoma xenografts with the anti-EGFR antibody, panitumumab, only partially inhibited tumor growth as xenografts rapidly reverted to the HGF/c-Met signaling pathway [12]. Co-

Table 1. Brain and CNS Tumors by WHO Histological Classification

Tumor Type	Cases	Percentage
Meningoma	34,291	34,6
Glioblastoma	19,707	19,9
Pituitary tumors	12,038	12,2
Tumors of cranial and paraspinal nerves	8,631	8,7
Unclassified and other tumors	6,347	6,4
Anaplastic astrocytoma	4,615	4,7
Other neuroepithelial tumors	3,740	3,8
Lymphoma	2,485	2,5
Pilocytic astrocytoma	1,685	1,7
Oligodendroglioma	1,454	1,5
Ependymoma/anaplastic ependymoma	1,888	1,9
Embryonal/primitive/ medulloblastoma	1,023	1,0
Anaplastic oligodendroglioma	674	0,7
Germ cell tumors	412	0,4
Total	98,990	

Data from Central Brain Tumor Registry of the United States Statistical Report 2004–2005.

treatment with panitumumab and the HGF antibody, AMG 102, prevented this escape leading to significant tumor inhibition through an apoptotic mechanism, consistent with the induction of oncogenic shock. These results illustrate that glioblastoma cells can achieve drug resistance by rapidly changing the targets/pathways driving their oncogenic phenotype.

4.1. Neuroepithelial Tumors

Malignant (neuroepithelial tumors) gliomas comprise a spectrum of different tumor subtypes, including anaplastic astrocytoma, anaplastic oligodendroglioma, and glioblastoma. They share some basic features such as preferential location in cerebral hemispheres, diffuse infiltration of brain tissue, and aggressive tumor growth. The diffuse infiltration (invasive) nature of gliomas is regarded as one of the main reasons for poor therapeutic success. Widespread infiltration makes complete surgical removal of gliomas impossible. Invasion of glioma cells requires interaction with the extracellular matrix and with surrounding cells of the healthy brain tissue. Vascular proliferates and tissue necroses are characteristic features of malignant gliomas, in particular glioblastoma and are most likely the consequence of a rapidly increasing hypoxic tumor mass [83].

4.2. Astrocytic Tumors (Gliomas) (Grades I–IV)

Gliomas (astrocytomas) are the most common CNS neoplasms (in both adults and children) accounting for more than 60% of all primary brain tumors [84]. Gliomas include tambours that are composed predominantly of astrocytes (astrocytomas), oligodendrocytes (oligodendrogliomas), ependymal cells (ependymomas) and mixtures of various glial cells. Although most CNS tumors are of glial origin, it is unclear whether tumor cells result from the transformation of immature precursor cells or from the dedifferentiation of mature glial cells (see Section 2).

The main subtypes of gliomas are described using the WHO grading system [85] that classifies gliomas into four grades based on the degree of malignancy, as determined by histopathological criteria. The main histological subtype of grade I gliomas are pilocytic astrocytomas, diffuse astrocytomas, oligodendrogliomas and oligoastrocytomas (low-grade (II) or high-grade (III and IV) tumors). Glioblastomas correspond to grade IV astrocytomas. Glioma subtypes exhibit distinct histological and molecular profiles [80,86].

Grades II–IV gliomas are malignant and diffusely infiltrate throughout the brain. Malignant gliomas (WHO grades II–IV) form approximately 80% of primary malignant CNS tumors.

4.3. WHO Grade I (e.g., Pilocytic Astrocytomas)

Low-grade gliomas are rare primary brain tumors although pilocytic astrocytoma (PA) is the most common pediatric brain tumor [87]. The most common variants are pilocytic and low-grade astrocytomas, oligodendrogliomas, and mixed oligoastrocytomas located in the cerebral hemispheres [88]. Grade I gliomas are slow growing, along with a better prognosis, than other CNS primary neoplasms typically occurring in young patients [89]. While pilocytic astrocytoma represents a benign and potentially surgically curable neoplasm that rarely undergoes malignant transformation, diffuse astrocytoma is a surgically incurable low-grade malignancy, prone to further malignant progression and eventual fatality.

4.4. WHO Grade II Low-Grade Astrocytomas

Advances in immunohistochemistry and pathology classification schemes have led to the recognition of diverse pathologies among low-grade gliomas potentially requiring different treatments ranging from complete surgical resection, repeat resection, and treatment with adjuvant therapies [90]. Balancing effective use of radiotherapy and chemotherapy in the management of low-grade tumors against side effects remains controversial. Some advances have, however, been made in chemotherapy for low-grade gliomas, such as low-grade astrocytomas and oligodendrogliomas [91].

Mutations in key molecular pathways, such as the p53-MDM2-p21 and p16-p15-CDK4-CDK6-RB pathways, are associated with astrocytoma development and progression. Two-thirds of low-grade astrocytomas, for example, have p53 mutations [92].

4.5. WHO Grade III (e.g., Anaplastic Astrocytoma)

Malignant astrocytomas are aggressive neoplasms and patients with anaplastic astrocytomas have short median survival despite maximal multimodality therapy that involves surgical resection, radiotherapy, and temozolomide [93]. Anaplastic astrocytomas [94] are highly aggressive tumors that display a high-level of molecular and genetic heterogeneity. Anaplastic astrocytomas constitute about 10% of all gliomas are histologically distinct from other malignant gliomas and characterized by an abundance of pleomorphic astrocytes with evidence of mitosis [95]. Predictive and prognostic factors are lacking because of their unclear pathophysiology and variable clinical outcome [96].

Maximal resective surgery is traditionally complemented by radiation therapy. Chemotherapy is typically used on patients with tumor recurrence, when their functional status is congruent with further treatment. New target-based approaches based on differentiation agents, antiangiogenic targets, matrix-metalloproteinase inhibitors, and signal transduction inhibitors are being investigated in clinical trials in an effort to compliment current therapies.

4.6. WHO Grade IV (e.g., Primary Glioblastoma and Secondary Glioblastoma)

Among astrocytic gliomas, glioblastoma multiforme, or WHO Grade IV astrocytoma, is the most common, most malignant, and most aggressive human brain tumor. Glioblastoma multiforme poses a unique challenge to therapy due to aggressive proliferation, high cell motility and resistance to apoptosis [6] leading to a propensity for invasion and proliferation. Astrocytic gliomas infiltrate throughout the brain and are largely resistant to radiation and chemotherapy. Patient prognosis remains very poor and although advances have been made with the alkylating agents carmustine and temozolomide (TMZ) [97]. Glioblastoma is also characterized by diverse causative genotypes and a high level of heterogeneity.

Malignant gliomas such as glioblastoma multiforme (GBM) present some of the greatest challenges in the treatment of cancer patients. Even with aggressive surgical resections using state-of-the-art preoperative and intraoperative neuroimaging, along with recent advances in radiotherapy and chemotherapy, the prognosis for GBM patients remains dismal: median survival after diagnosis is about 14 months. Standard treatment includes resection, followed by concurrent chemotherapy and radiotherapy.

Glioblastomas are divided into primary and secondary subtypes. The majority of cases (>90%) are primary glioblastomas that develop *de novo*, without clinical or histological evidence of less malignant precursor lesions. Secondary glioblastomas develop progressively from low-grade astrocytomas or anaplastic astrocytomas [98]. Primary and secondary glioblastoma, although sharing many similar histological features, constitute distinct disease subtypes, affecting patients of different age and developing through different genetic pathways. Primary and secondary glioblastomas also differ significantly in their pattern of promoter methylation and in expression profiles at both the RNA and the protein levels, together with variations in observed mutations [99,100].

Primary glioblastoma, in contrast with secondary glioblastoma, is characterized [101] by the amplification of epidermal growth factor receptor gene (EGFR) that is often associated with deletion of the CDKN2Ap16/Arf gene encoding two tumor suppressors, p16 an inhibitor of CDK4 and p14ARF a negative regulator of MDM2, and absence of p53 mutation. Indeed, primary glioblastomas with simultaneous EGFR and p53 alterations are significantly associated with worse survival [102]. Secondary glioblastoma feature mutually exclusive aberrations in p16 and p14ARF together with mutations in the p53 tumor suppressor gene [103]. The impact of the differences in pathopharmacology between primary and secondary glioblastoma and the underlying genomic and proteomic alterations have significant implications; particularly, for the development of novel, targeted therapies [104].

Pediatric glioblastoma, frequently situated in the brain stem, form a third pathogenetically distinct group from their adult counterparts [105]. Pediatric gliomas are clinically, histologically, and molecularly very heterogeneous and although histologically indistinguishable from their adult counterparts display different pathobiology.

Recently, concomitant and adjuvant chemoradiotherapy with temozolomide (Fig. 3) has become the standard treatment for newly diagnosed (primary) glioblastoma as well as in recurrent or progressive malignant gliomas [106]. Preclinical investigations suggest synergism or additivity with radiotherapy in glioma cell lines.

Temozolomide (Fig. 3) [97] is capable of depleting cancer stem cells in glioblastomas in methylguanine methyltransferase (MGMT) negative cells but not MGMT positive

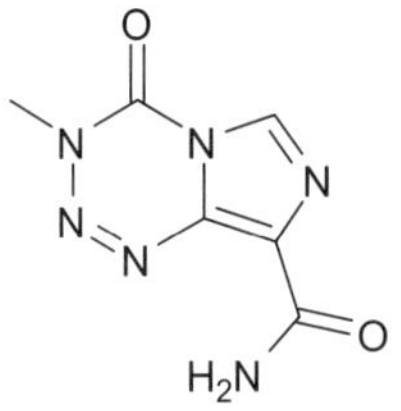

Figure 3. Temozolomide.

cells [107] and epigenetic inactivation MGMT seems to be the strongest predictive marker for outcome in patients treated with temozolomide-based chemotherapy. Patients whose tumors do not have MGMT promoter methylation are less likely to benefit from the addition of temozolomide chemotherapy and require alternative treatment strategies [108]. However, several independent DNA repair mechanisms can restore the integrity of alkylated DNA bases, and thus contribute to drug resistance and subsequent therapy failure.

Recent work suggests that glioblastomas develop as cellular and functional hierarchies through small subpopulations of stem cell-like cancer cells that are responsible for tumor initiation and maintenance. Such cells also appear to possess enhanced DNA repair capacity compared to other cells within the tumors. It remains to be determined what role cancer stem-like cells represent as a target population for therapy against glioma resistance toward current alkylating agent-based chemotherapies, and as therapeutic entry points that may lead to first-line therapies [58]. To date, glioblastoma has proved resistant to targeted chemotherapeutic approaches via single targets since glioblastomas display diffuse infiltration within the brain and tend to recur despite extensive debulking via surgery followed by radio- and chemo-therapy [58]. For example, small-molecule kinase inhibitors are common agents in the treatment of cancers, but monotherapy with selective kinase inhibitors has so far demonstrated limited efficacy in nondifferentiated GBM patient populations.

A better understanding of relevant molecular mechanisms and signaling pathways has allowed for rational targeting of specific pathways of repair, signaling, invasion, and angiogenesis. Tyrosine kinase inhibitors, such as vatalanib (PTK787) and vandetinib (ZD6474), the integrin inhibitor cilengitide (Fig. 4), monoclonal antibodies such as bevacizumab and cetuximab, are among other agents showing effect in newly diagnosed glioblastoma [108]. However, pan-active (multi-targeted) kinase inhibitors or combination of single-targeted kinase inhibitors with one another or with traditional cytotoxics [109] or other targeted small molecules will probably be necessary to better overcome resistance.

Targeted therapies directed against pathways that are upregulated in high-grade gliomas have so far shown limited clinical activity as single agents in clinical studies. The exceptions are trials with agents inhibiting the vascular endothelial growth factor (VEGF) signaling system such as bevacizumab and AZD2171, cediranib (Fig. 10) an oral tyrosine kinase inhibitor of VEGFR [110]. These trials showed high response rates (which might be due to vessel normalization similar to the effects of steroid treatment) and promising 6-month progression-free survival rates in glioblastoma multiforme. MRI techniques have shown that AZD2171 normalized tumor vessels in recurrent glioblastoma patients in a prolonged but reversible manner. Basic FGF, SDF1α, and viable circulating endothelial cells (CECs) increased when tumors escaped treatment, and circulating progenitor cells (CPCs) increased when tumors progressed after drug interruption suggesting that the timing of combination therapy may be critical for optimizing activity against this tumor [110].

Combination therapies together with multimodal treatments may increase patient survival [111] as the effectiveness of multidisciplinary approaches improves. For the many molecular-targeted agents, a critical review of their pathological role in glioblastoma is required, especially if combination regimens are investigated [112]. Recently, studies focused on identification of aberrant genetic events and signaling pathways, together with tumor stem cell identification and characterization, modulation of tumor immunological responses, and combination therapies, are helping build an understanding the underlying molecular biology [113]. Targeting nonoverlapping pathways, rather than a single agent

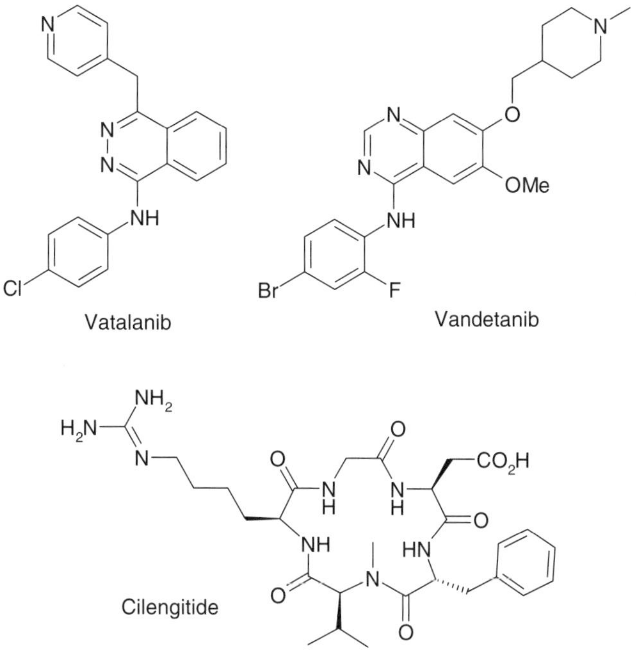

Figure 4. Compounds showing effect against newly diagnosed Glioblastoma.

approach, is more likely to be effective. Also the role of ensuring penetration of not only an intact blood–brain barrier but also a blood–tumor barrier should also be more prominent in compound design [114,115].

4.7. Ependymoma (Grades I–III)

Ependymomas represent a heterogeneous group of rare tumors of neuroectodermal origin ranging from myxopapillary ependymoma and subependymoma (grade I), to ependymoma (grade II) and anaplastic ependymoma (grade III) [116]. Ependymomas represent 6–9% of primary CNS neoplasms and account for 30% of primary CNS neoplasms in children younger than 3-year old. Resectioning and radiotherapy are the front-line treatments but are often associated with significant reductions in neurologic, endocrine, and cognitive function [117]. Cytokines and growth factors have been identified as predictors or correlates of radiation effects in a number of systems and it has been shown that proinflammatory cytokines and growth factors (such as EGF and VEGF) decline significantly following radiotherapy in localized ependymona although the level of decline varied significantly according to tumor location [118].

There is limited evidence that postoperative chemotherapy (platinum-, nitrosourea-, or temozolomide-based chemotherapy) improves patient outcome [119]. Knowledge relating to potential therapeutic targets is limited owing to the rarity of the disease [120]. In addition, ependymomas show a range of histological, molecular, and clinical variables [121]. Moreover, the selection of targets for small-molecule therapies may not be straightforward, for example, while EGFR amplified gliomas respond to gefitinib, EGFR expressing ependymoma remain resistant [122].

4.8. Oligodendroglial Tumors (Grade II to III)

Oligodendrogliomas are rare, diffusely infiltrating tumors comprising approximately 4–5% of brain tumors and 5–20% of gliomas. Oligodendrogliomas arise in the white matter of cerebral hemispheres, and have better sensitivity to treatment and prognosis than many other gliomas, although the best approach for newly diagnosed anaplastic oligodendroglial tumors is unclear [123]. Chemotherapy [124] with either vincristine [125] or temozolomide constitutes a standard for recurrent/progressive disease [126]. Current second-line chemotherapy results are, however, modest.

Identifying the cellular origins of high-grade gliomas can be difficult and oligodendrogliomas can share common features with anaplastic astrocytomas. However, while it may be difficult to distinguish these tumors using histological criteria, proteomic studies have been able to differentiate GBM from other glial tumors. Thus, selective biomarkers expressed in GBMs but not in oligodendrogliomas have been found [127]. Correlation among pathology, genetic, and epigenetic profiles and clinical outcomes in oligodendroglial tumors was recently demonstrated [128] and some options for molecular target-based therapies are emerging [129].

5. EMBRYONAL TUMORS

Central nervous system primitive neuroectodermal tumors (CNS PNETs) develop from primitive (undifferentiated) nerve cells to form high-grade, predominantly pediatric tumors. Medulloblastoma a primitive neuroectodermal tumor of the cerebellum is the most common type of pediatric malignant embryonal brain tumor (up to 30% of all solid brain tumors) [130]. Medulloblastoma-like neuroblastomas that are malignant embryonal pediatric tumors of the peripheral nervous system have deficiencies in its apoptotic processing. In addition, normal development and tumorigenesis share several common characteristics. Both processes involve altered proliferation, differentiation, migration, and apoptosis. A major challenge for targeted therapies especially those targeting developmental pathways will therefore be to avoid affecting normal progenitor cells. This is especially important for embryonal tumors that are particularly prevalent in pediatric patients [131].

Present multimodality treatment [19] of medulloblastoma includes surgery, radiotherapy, and chemotherapy (alkylating agents and platinum compounds, such as lomustine, cyclophosphamide, and cisplatin are the main treatments) [132–134]. Clinical prognosis is poor and resistance is common with 5-year survival rates of 70–80% for standard-risk patients, and 55–76% for high-risk patients. Radiation induced side effects can be serve with the risk of side effects inversely correlated with age [135]. Patients treated for medulloblastoma often develop cognitive and endocrine deficits with an increased risk of secondary tumors later in life. Thus, treatment of medulloblastoma remains restricted by the toxicity of conventional treatments and by the infiltrative nature of medulloblastomas. Current research is focused on the signaling pathways regulating medulloblastoma tumor formation [130] such as the Hedgehog pathway. Medulloblastoma patients are currently categorized by combined clinical and histological criteria. Recent advances in the molecular biology, however, have shown that distinct subtypes of medulloblastoma exist [136] and that therefore classification of medulloblastoma on histology and clinical criteria alone may be insufficient. The various distinct subtypes of medulloblastoma likely arise from different precursor cell populations. These precursor cell populations that form the cerebellum are susceptible to mutations in their developmental signaling pathways [137]. Risk stratification based on a combination of histopathological evaluation may therefore require additional molecular biology analysis. Of the medulloblastoma variants recognized by histological classification, patients with desmoplastic medulloblastoma and medulloblastoma with extensive nodularity show better survival rates than those with classical medulloblastoma.

Genomic and proteomic studies are revealing a more detailed classification based on the key signaling networks driving medulloblastoma progression [138]. In particular, aberrant signaling in the Hedgehog and Wnt developmental pathways and insulin-like growth factor (IGF) pathways has been implicated in the pathogenesis of medulloblastoma.

Recent animal model observations also show that medulloblastoma can result from defects in DNA repair pathways that lead to genomic instability in neural progenitor cells [138]. Studies of mRNA expression profiles have characterized five medulloblastoma subtypes with distinct genetic profiles, pathway signatures and clinicopathological features [139]. These subtypes are differentiated by Wnt or Hedgehog signaling abnormalities, or expression of neuronal differentiation or photoreceptor genes. Clinicopathological features significantly different between the five subtypes included metastatic disease and age at diagnosis and histology. Interestingly, patients with Wnt-activated tumors have a more favorable prognosis. The elements that contribute to Wnt-driven tumor progression in medulloblastoma and other CNS PNETs largely do not, however, appear to be caused by mutations in β-catenin or APC [77].

Genetic alterations leading to constitutive activation of signaling pathways such as Wnt, Hedgehog, and Notch have been implicated in the transformation of both multipotent cerebellar stem cells and lineage-restricted progenitors into medulloblastoma [140]. Modulation of the Hedgehog and Notch pathways suppresses medulloblastoma growth both *in vitro* and *in vivo* and may thus prove effective in targeting the cancer stem cell subpopulation driving tumor initiation and long-term propagation [139].

Suppression of oncogenic Wnt-mediated signaling holds promise as an anticancer therapeutic strategy. A novel class of small-molecule Wnt response inhibitors (IWR-1 and 2, inhibitors of Wnt response) (Fig. 5) that antagonize Wnt signaling by stabilizing the Axin destruction complex have recently been identified via luciferase-based reporter assays [142].

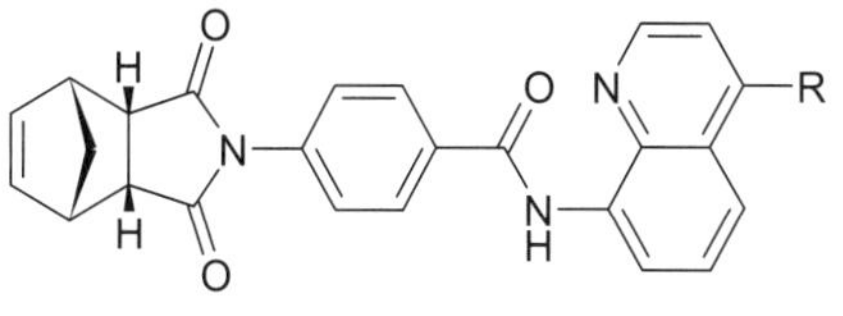

Figure 5. Inhibitors of Wnt-signaling response.

6. OTHER NEUROEPITHELIAL TUMORS

Various other rare tumors of neuroepithelial origin exist, including pineal parenchymal tumors (PPTs). PPTs are rare neuroepithelial tumors that arise from pineocytes or their precursors [143]. PPTs are subdivided into well-differentiated pineocytoma, poorly differentiated pineoblastoma, and PPT with intermediate differentiation (PPTID). Since pineal parenchymal tumors are rare only limited data regarding clinical outcomes and therapy options are available [144].

7. GERM CELL TUMORS

Intracranial germ cell tumors (GCTs), especially pineal tumors have unique growth sites, with characteristic subtypes and distinct histology [145]. Although long-term results from radiation therapy [146] are good, the potential for late effects makes the treatment controversial [147]. Localized germinomas are treated by conventional treatment with surgery and radiation therapy failed to survive longer than 3 years. Chemotherapy is usually reserved for disseminated germinomas. While chemotherapy alone has so far failed to inhibit tumor progression with a high rate of recurrence, combination trials with chemotherapy together with reduced dose and volume radiotherapy have demonstrate good event-free survival rates [148].

Germinomas generally have a good treatment prognosis following low-dose radiotherapy in combination with chemotherapy. Mature teratomas are curable by surgical resection alone but other germ cell tumors require combination therapies that include surgery, craniospinal radiation, and intensive chemotherapy (typically cisplatin) [149].

8. PITUITARY TUMORS

Pituitary tumors are the most common intrasellar tumors. The majority of pituitary tumors are andenomas [150] divided into clinically functioning [151] or nonfunctioning adenomas [152]. Although rarely fatal, they require treatment often via multiple surgical sectioning and/or radiation therapy with associated side effects. Pituitary adenomas are typically slow growing and histologically benign that do not exhibit aggressive behavior but instead grow via expansion generally via nonangiogenic mechanisms [153] approximately a third of pituitary adenomas are, however, invasive. Although pituitary adenomas generally occur sporadically, approximately 5% are familial [154]. Rare metastatic carcinomas are also known [155,156], particularly from small cell lung carcinomas.

Prolactin (PRL) secreting adenomas [157] are the most common form of pituitary tumor and cause amenorrhea, galactorrhea, and reproductive dysfunctions in females and hypogonadism, decreased libido, and impotence in males. Growth hormone (GH) secreting tumors [158] cause gigantism in children and acromegaly in adults. Additional pituitary adenomas include adrenocorticotropic hormone (ACTH) secreting adenomas [159] that produce Cushing's disease (hyperadrenalism) [160], thyrotropin secreting (TSH) adenomas [161], and follicle stimulating hormone (FSH)-secreting adenomas.

The treatment of a pituitary adenoma depends on the tumor type. Prolactinomas are treated with dopamine agonists such as cabergoline or bromocriptine, and other lesions are usually treated by transsphenoidal surgery, although chemotherapy options exist. Replacement of deficient hormones, or lowering of hormone hypersecretion is necessary for optimal functioning and some patients require multimodal therapy [162]. Radiotherapy may be used for residual or recurrent cases [163]. Pituitary adenomas can become clinically destructive, invade adjacent structures, and recur after treatment. Complete surgical resection is thus difficult without unacceptable neurological deficits and invasive adenomas require multimodal treatments [164].

Pituitary tumors offer numerous potential therapeutic targets that are typically different to those in other neoplasms [165]. Pituitary tumors express both somatostatin and dopamine receptors, agonists for these targets have seen widespread application against pituitary adenomas [166–168]. Somatostatin and dopamine are both negative controls of hormonal secretion in the anterior pituitary. Recently, there has been significant progress in determining pituitary pathophysiology, and the pathogenetic factors implicated in pituitary tumorigenesis. These data have been used to derive novel targeted therapies [169] including new somatostatin analogs, such as SOM230 (Fig. 6) [170].

Medical treatment with somatostatin analogs is a cornerstone of GH- and TSH-secreting tumors [171], while treatment with dopamine agonists is a cornerstone of prolactin-secreting tumors [172]. Dopamine agonists have also demonstrated some efficacy in patients with PRL- [173], GH-, and TSH-secreting adeno-

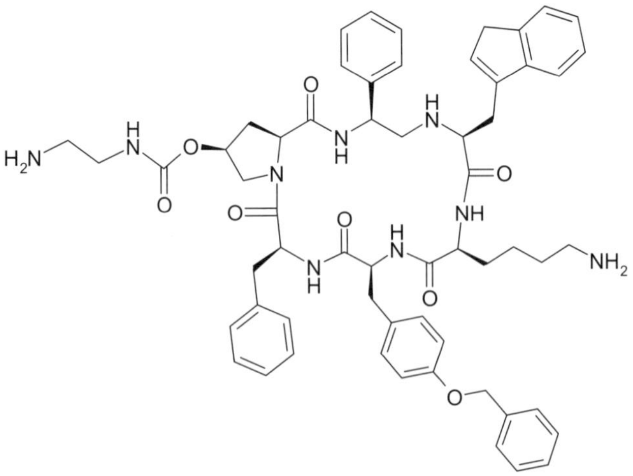

Figure 6. SOM230.

mas. The effectiveness of dopaminergic and somatostatinergic agonists in treatment of pituitary adenomas and the growing knowledge of mechanisms controlling pituitary secretions as well as cellular proliferation should permit better treatments, perhaps avoiding the use of invasive surgical procedures and/or radiotherapy [174].

Alternative approaches may be possible via nuclear receptors, including the estrogen receptor [175], PPAR [176], and the retinoic acid receptor [177], which are abundantly expressed in pituitary tumors. Modulators of nuclear receptors assisted by pharmacogenomic profiling may lead to improved treatments for pituitary tumors [178].

9. TUMORS OF THE COVERINGS OF THE BRAIN

9.1. Meningioma

Meningiomas are slow growing lesions and are the most common intracranial primary neoplasm in adults. They comprise about 25–35% of brain tumors in adults approximately 90% are benign, 2% are malignant while the remainder are atypical [179]. Overall incidence is increasing perhaps indicating impact of environmental risk factors or more sensitive diagnostic procedures such as radiologic imaging [180]. Although there are currently no effective chemotherapies, most meningiomas can effectively be treated surgically [181,182]. Atypical or malignant meningiomas and surgically inaccessible meningiomas may not be completely removed, for which a range of chemotherapies are considered, including hydoxyurea or antiprogesterone treatments [183]. Improved understanding of the underlying tumor genetics and molecular biology is also assisting the development of new therapeutic agents such as angiogenesis inhibitors, modulators of fundamental cell signaling pathways, and somatostatin analogs [182]. The histological characterization of meningiomas is an important predictor of tumor behavior and is frequently used to determine preferred therapies. The relationship between meningioma histological characteristics and behavior and hence prognosis is not always clear and additional prognostic markers that require immunohistochemical, cytogenetic, or molecular techniques are under investigation to improve tumor classification [184].

Although advances in surgery, radiation therapy and stereotactic radiosurgery have significantly improved the treatment of meningiomas, there remains an important subset of patients who remain refractory to conventional therapy. Treatment with chemotherapeutic agents such as hydroxyurea and alpha-interferon has provided minimal benefit. The role of emerging targeted therapies for meningiomas for recurrent or progressive meningiomas is therefore important [182,185,186]. Clinical trials are underway with a number of such agents, including imatinib, sunitib, and the somatostatin analog SOM230 (which is also under investigation for pituitary tumors, including Cushing's disease) [187]. In addition, the topoisomerase I inhibitor, Irinotecan (CPT-11) (Fig. 7) has demonstrated growth-inhibitory effects in meningiomas both *in vitro* and *in vivo*, being more effective against malignant than against primary meningioma [188].

10. TUMORS OF CRANIAL AND SPINAL NERVES

In the peripheral nervous system neurofibroma and schwannoma are the two most common glial tumors.

Neurofibromatosis (NF) is a common autosomal dominant genetic disorder, classified

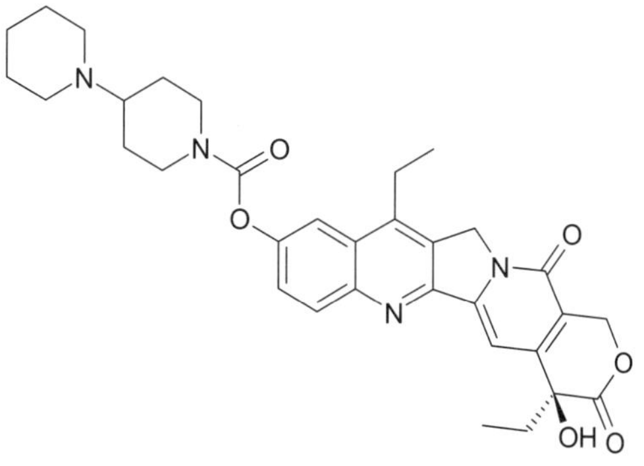

Figure 7. Irinotecan.

into two genetically distinct subtypes characterized by multiple cutaneous lesions and tumors of the peripheral and central nervous system. Neurofibromatosis type 1 (NF1), also referred to as Recklinghausen's disease presents with a variety of characteristic abnormalities of the skin and the peripheral nervous system. Neurofibromatosis type 2 (NF2), previously termed central neurofibromatosis, is much more rare [189]. Often, the first clinical signs of NF2 become apparent due to a sudden loss of hearing following the development of bi- or unilateral vestibular schwannomas. In addition, NF2 patients may suffer from further nervous tissue tumors such as meningiomas or gliomas [190].

Malignant peripheral nerve sheath tumors (MPNSTs) can develop in patients with underlying NF1, and usually arise as a result of malignant transformation of a preexisting plexiform neurofibroma. Microarray immunohistochemical studies of immortalized cell lines from primary, metastatic, and recurrent malignant peripheral nerve sheath tumors with underlying neurofibromatosis-1 [189] has identified multiple clinical, pathologic, and molecular markers useful for prognostic assessment of MPNSTs. Of particular interest was the association of Nm23-H1 expression with disease progression [191] expanding the family of "p53-mdm2-like" protein–protein interactions associated with oncology applications. As with many cancer types, the PI3K/Akt/mammalian target of rapamycin (mTOR) signaling pathway is a potential target for the treatment of malignant peripheral nerve sheath tumors. Combined treatment with PI3K/AKT and mTOR inhibition has been postulated to enhance mTOR inhibitors by also blocking induced prosurvival, protumorigenic signaling such as enhanced pAKT expression [192]. When Rapamycin (mTOR inhibitor), LY294002 (dual PI3K/mTOR inhibitor), and PI-103 (potent PI3K/AKT/mTOR inhibitor) (Fig. 8) were tested independently and together dual targeting of AKT and mTOR was shown to be significantly better that either modality alone.

10.1. Schwannoma

Meningiomas and schwannomas are the two most common mostly benign extra-axial intracranial tumors in adults. The mainstay therapies for these lesions [193] are microsurgery and radiotherapy although both Lapatinib and Gleevec (Fig. 9) have been in clinical trials for schwannoma.

Schwannomatosis is a rare tumor syndrome characterized by the presence of multiple schwannomas without the stigmata of

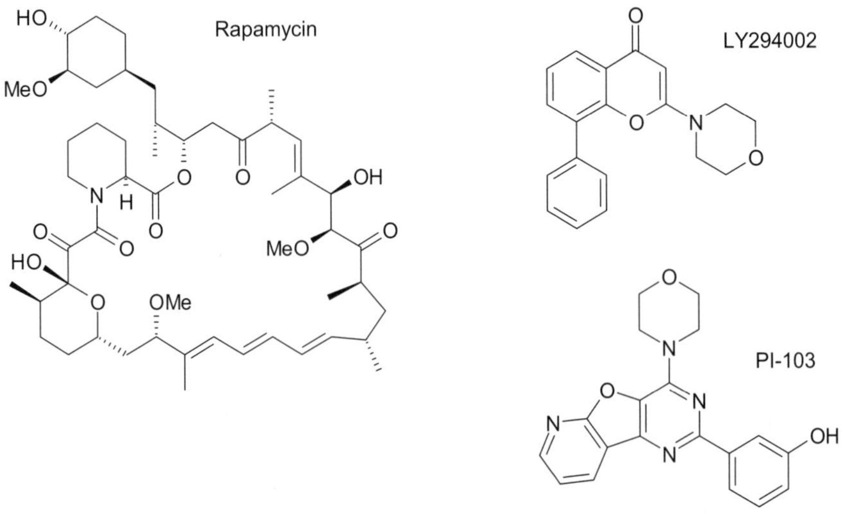

Figure 8. AKT and mTOR inhibitors used in Neurofibromatosis.

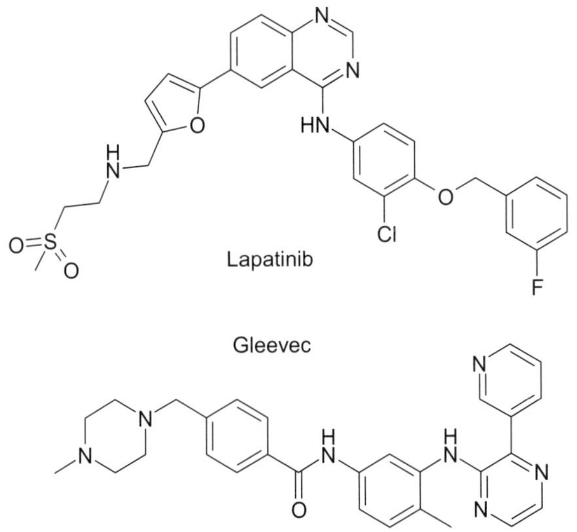

Figure 9. Compounds used against Schwannoma in the clinic.

neurofibromatosis type 1 or 2 for a review of clinical management of the disease see [194].

11. PRIMARY CNS LYMPHOMA

Primary CNS lymphomas (PCNSL) are rare (3–4% of primary CNS neoplasms) and aggressive B-cell tumors characterized clinically by the absence of systemic disease and distributed within the leptomeninges, spinal cord, and intraocular compartments. PCNSLs display unique structural, biological and histological characteristics [195] and are typically associated with a worse prognosis than other localized extranodal lymphomas with similar histological characteristics or systemic lymphomas. PCNSLs share overlapping features with systemic lymphoma, but have distinct gene and protein expression profiles [196].

Incidence of PCNSL has increased and occurs in both immunocompromized and immunocompetent patients. However, as PCNSL is relatively rare, the identification of molecular prognostic biomarkers and the definition of a standard therapeutic strategy have been challenging [197,198].

Primary CNS lymphoma in immunocompetent patients is associated with unique diagnostic, prognostic, and therapeutic issues and the management of this malignancy is different from other forms of extranodal non-Hodgkin's lymphoma. Primary CNS lymphoma may involve the neural tissue of the brain, cerebrospinal fluid, or eyes [199]. Recent findings suggest primary CNS lymphoma has both CNS-specific and systemic components with limited interchange. The more malignant behavior of tumor cells in the CNS suggests either influence of the CNS microenvironmental or a less malignant phenotype of the peripheral variant [200,201].

Despite recent therapeutic advances, these malignancies remain one of the worst prognoses among non-Hodgkin's lymphomas (NHL). Radiotherapy (RT) is the standard treatment; however, relapse usually occurred within a few months after RT, with a median survival of 14 months and a 5-year survival of approximately 15–24%. Although the introduction of systemic chemotherapy (high-dose methotrexate) has consistently improved survival, the prognosis of PCNSL is still poor [202].

Resection provides no therapeutic benefit whole-brain radiation therapy alone is insufficient for durable tumor control and is associated with a high risk of neurotoxicity (typically significant cognitive, motor, and autonomic dysfunction). Chemotherapy (particularly high-dose methotrexate) [201–204] and whole-brain radiation therapy together improve tumor response rates and survival compared with whole-brain radiation therapy alone but, this combined treatment carries a risk of delayed neurotoxicity [205]. Methotrexate-based multiagent chemotherapy without whole-brain radiation therapy is associated with similar tumor response rates and survival compared with regimens that include whole-brain radiation therapy, but show a lower risk of neurotoxicity [206,207] or treatment-related morbidity and mortality [197]. Unfortunately, drugs used in treating systemic non-Hodgkin's lymphoma have mostly proven ineffective against primary CNS lymphoma due to difficulties in crossing the blood–brain barrier [207]. Immunotherapy including, for example, treatment with intrathecal rituximab is an alternative approach that may have promise in refractory or relapsed disease [197].

12. METASTATIC BRAIN TUMORS

Although not a specific disease or tumor type metastases to the brain are an important part of CNS oncology, and occur in 15–20% of cancers. Lung [208], breast [209], renal, colorectal [210] carcinomas, and melanoma [211] are the most common primary tumors that metastasise to the brain, largely to the cerebrum or brain stem. Treatment options for these tumor types should therefore preferably also consider metastases to the brain and potentially the need to delivery therapeutic agents across the blood–brain barrier [212]. Although the majority of CNS metastases occur at an advanced stage of the peripheral disease, brain metastases are often the cause of presenting symptoms. CNS metastases are a major cause of morbidity and mortality affecting survival, neurocognition, speech, coordination, behavior, and quality of life.

Therapeutic approaches to treating CNS metastases include whole-brain radiation therapy, surgery, stereotactic radiosurgery, radiation sensitizers, chemotherapies [213,214], and systemic chemotherapy [215,216]. However, the key aim of most current treatments, which are often palliative, is preservation the neurological status of the patient [217]. Quality of life and preservation of cognitive functions are therefore key goals for new therapies [218,219] toward brain metastases. With increasing survival periods from primary peripheral cancers and advances, tumor detection via less invasive techniques such as CT and MRI an increase in metastatic lesions has been observed.

New multimodal treatment strategies are required to more fully tackle CNS metastases [220]. Potential targets include growth factor receptors and other protein tyrosine kinases, internal signal transduction pathways, ras activation, and matrix metalloprotease activity [215].

13. RECURRENT CNS TUMORS

Management of malignant gliomas remains poor largely, since glioma cells actively migrate to other brain regions, making them diffuse and elusive targets for effective surgical or radiotherapy management. Invasive malignant glioma cells show a decrease in their proliferation rates and a relative resistance to apoptosis and this may contribute to their resistance to conventional proapoptotic chemotherapy and radiotherapy. Resistance to apoptosis results from changes at the genomic, transcriptional, and posttranscriptional level of proteins, such as protein kinases and their transcriptional factor effectors. Recurrent tumors unlike primary tumors typically show resistance to apoptosis. Inhibiting the migration of malignant cells should restore a level of sensitivity to proapoptotic drugs. Recent series of studies have supported the concept that malignancy of gliomas involves crosstalks between cancer cells, their microenvironment, the vasculature and stem cells [112] that are potential targets for building comprehensive therapies.

Intrinsic chemoresistance to alkylating agents and or acquired resistance to cytotoxic and molecular target-based agents remain a major cause of treatment failure in malignant brain tumors. Alkylating agents, such as temozolomide (usually in combination with radiotherapy), have made significant advances in the treatment for brain tumors such as primary glioblastoma via induction of apoptosis. Recurrence usually occurs, however, and new targeted therapies and antiangiogenic treatments are required [221].

There are diverse mechanisms of chemoresistance relevant to malignant glioma and novel pharmacologic approaches are needed to overcome resistance to current treatments (alkylating agents) and to eventually replace them [222]. Hypoxia, for example, is implicated in many aspects of tumor growth, development, and angiogenesis. Invasion, apoptosis, chemoresistance, resistance to antiangiogenic therapy, and radiation resistance may all have a hypoxic component. The extent of the influence of hypoxia in these processes thus makes hypoxia-regulated proteins attractive therapeutic targets for gliomas and meningiomas [223]. Glioma stem cells differentially respond to hypoxia with distinct hypoxia-inducible factor (HIF) induction patterns. Targeting HIFs self-renewal, proliferation, and survival *in vitro*, attenuates the tumor initiation potential of glioma stem cells *in vivo* [224]. Hypoxia is readily recognizable in glioblastoma, as indicated by focal or extensive necrosis and vascular proliferation. The hypoxia-driven tumor response is associated with inflammation and highlights the importance of the tumor–host interaction or microenvironment interaction [225].

14. BRAIN TUMOR INVASIVENESS

The efficacy of treating malignant gliomas with adjuvant therapies remains largely unsuccessful not only due to the tumor heterogeneity but also due to the ability for tumor cells to evade treatment. The propensity of malignant gliomas to invade into adjacent and distant brain tissue remains a major challenge in the treatment and management of gliomas. Invasive tumor cells thus escape surgical removal and lethal radiation exposure or chemotherapy. The aggressive nature and invasiveness of malignant gliomas makes them highly destructive neurologically and combating tumor invasion is therefore a key area.

Invasive tumor cells escape surgical removal and, because of their reduced proliferation rate and increased resistance to apoptosis, are relatively resistant to radiation therapy and chemotherapy. Effective therapies against invasion are lacking and future approaches via compounds modulating the signaling pathways that mediate glioma invasion, proliferation, and apoptosis will be needed to tackle invasion in order to capitalize on successes with primary tumor treatments [226]. Improved understanding of biochemical and molecular criteria of glioma invasiveness may also provide clues to novel targets [227]. Invasiveness is not restricted to malignant gliomas, over a third of adenomas are invasive, again making complete surgical resection without unacceptable neurological deficits difficult [164,224].

Preclinical animal models that include aspects of tumor invasion such as orthotopic glioblastoma multiforme models using patient derived cell lines [228] should offer advantages for investigating the pathobiology of intracranial tumor growth and for monitoring systemic and intracranial responses to antitumor agents including invasion and angiogenesis [229].

A range of targets/mechanisms are involved in stimulating and controlling migration and invasiveness such as integrins [230], PPAR [231], endothelial growth factors [232], and cytokines [233].

Integrins mediate cell and extracellular matrix interactions, facilitating extracellular matrix dependent organization of the cytoskeleton and the activation of intracellular signaling that is required for the regulation of cell adhesion and migration. Since integrins facilitate extracellular matrix dependent organization of the cytoskeleton and activation of intracellular signaling in cell adhesion and migration processes, they are also thought to play a significant role in glioma invasion [230].

In addition to paracrine effects on endothelial cells, autocrine VEGF signaling may regulate invasiveness of glioblastoma cells [232] by normalizing abnormal tumor vasculature, and enhancing response to radiation and chemotherapy. Resistance develops leading to a highly infiltrative and invasive phenotype suggesting the need for combination with other modalities. For example, combined treatment with interleukin-6 (IL6) and VEGF inhibitors is synergistic and reduces global activity of major pathways of cell survival [234].

Advances in understanding the molecular determinants of glioma invasion and migration are highlighting new therapeutic strategies for inhibiting invasion in glioblastoma [235,236]. In particular, evidence for a role of the developmental signaling networks such as Hedgehog and Wnt as modulators of cell invasion. For example, a novel gain-of-function Gli1 splice variant that promotes migration and invasiveness of glioblastoma cells has recently been identified [237].

The inhibition of tumor angiogenesis could also be an efficient therapeutic strategy for the treatment of malignant gliomas although combined approaches to confront both invasion and angiogenesis are most likely required [238]. Gliomas induce prominent neovascularization, and microvascular proliferation is a malignancy-grading criterion. However, glioma cells can also invade the brain diffusely over long distances without necessarily requiring angiogenesis [239].

15. ANGIOGENESIS IN BRAIN TUMORS

Antiangiogenic approaches have the potential to contribute significantly in the treatment of glioblastoma tumors. Angiogenesis is a common feature of many tumor cells, and glioblastomas are among the most angiogenic. Despite multimodality treatment with surgery, radiation therapy, and chemotherapy, most patients with malignant glioma have a poor prognosis. Preclinical data indicate that angiogenesis is essential for the proliferation and survival of malignant glioma cells, which suggests that inhibition of angiogenesis, might be an effective therapeutic strategy.

The angiogenesis cascade is primarily initiated by hypoxia, leading to the formation of hyperpermeable tumor blood vessels via activation of the HIF pathway. Alternatively, it can be triggered by genetic factors. Glioblastomas exhibit extremely high levels of neovascularization, which may contribute to their extremely aggressive behavior [240]. Moreover, data derived from animal models suggest that angiogenesis inhibition may promote an invasive phenotype. This may represent an important mechanism of resistance to antiangiogenic therapies [238,241].

Clinical trials with antiangiogenic drugs have shown activity in recurrent malignant gliomas. Further investigations are necessary to determine whether these drugs have a role in first-line therapy. Potential toxicities associated with angiogenesis inhibition (such as hypertension, and GI, hematological, and cerebral toxicity) will need to be addressed [242].

Because of the lack of effective treatments for gliomas and their high vascularity there are numerous antiangiogenic approaches under investigation for glioma treatments. Current targeted approaches to inhibit angiogenesis include combination therapy with monoclonal antibodies (especially anti-VEGF), or receptor tyrosine kinase inhibitors (especially VEGFR), administered in conjunction with cytotoxic chemotherapy [243,244] antisense oligonucleotides or gene therapy. Clinical trials have or are being carried out with several angiogenesis inhibitors, including thalidomide, CC-5103 (a thalidomide analog), and the VEGFR tyrosine kinase inhibitor vatalanib (PTK787, ZK222584) (Fig. 10). Following treatment with antiangiogenic agents a decrease in vascular permeability and perfusion can be detected. For example, treatment with the anti-VEGF hmAb bevacizumab combined with cytotoxic chemotherapy showed a prolongation of progression-free survival, and a lower need for corticosteroids [245]. Similar results have been shown with small-molecule inhibitors of VEGFR tyrosine kinase [246], such as cediranib (AZD2171) (Fig. 10) [247].

Reliable parameters for understanding and measuring the effectiveness of antiangiogenic approaches are limited. In particular, reliable biomarkers that can predict which patients will benefit from treatment and that accurately indicate response and progression during therapy are still lacking. In addition, most patients treated with antiangiogenic drugs eventually progress, and there are no biomarkers that identify escape pathways so targeting resistance is difficult. A number of potential systemic, circulating, tissue, and imaging mechanism-based biomarkers have, however, emerged from recently completed Phase I–III studies, including VEGF polymorphisms, MRI-measured Ktrans, or circulating angiogenic proteins. Some have the potential to be pharmacodynamic (e.g., in-

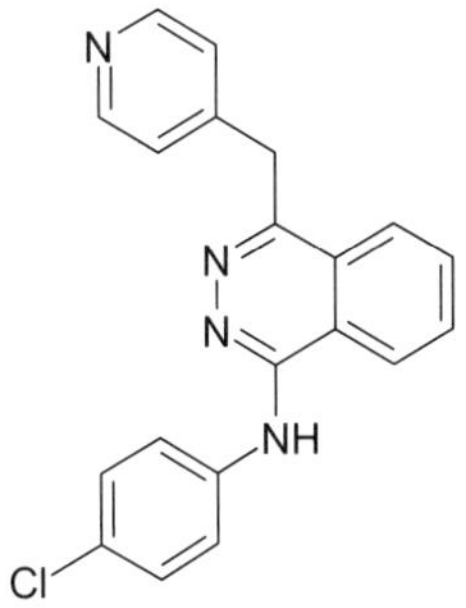

Figure 10. VEGFR inhibitors used in angiogenesis.

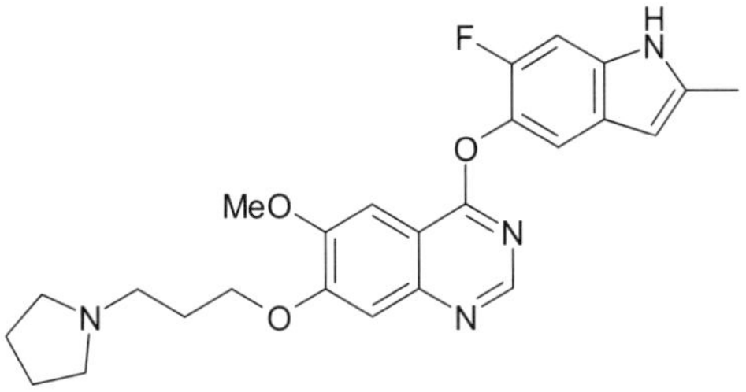

crease in circulating VEGF, or placental growth factor) while others have potential for predicting clinical benefit or identifying the escape pathways (e.g., SDF-1α or interleukin-6) [248].

Clinical trials of antiangiogenic VEGF agents in both recurrent and newly diagnosed malignant gliomas are ongoing in efforts to determine the best intervention opportunities [244]. Sunitinib is a multitargeted tyrosine kinase inhibitor with both antiangiogenic and antitumor activities due to inhibition of various receptor tyrosine kinases, including those important for angiogenesis (VEGFR and PDGFR). In preclinical animal models of glioblastoma, sunitinib (Fig. 11) potently inhibited angiogenesis and prolonged of survival [249].

Although antiangiogenic therapies appear promising, the duration of response with available regimens is modest with tumors able to overcome these therapies and to reestablish angiogenesis. As yet, there are therefore no effective treatments for patients via antiangiogenic therapies [245].

One consequence of antiangiogenic therapies is that vessel normalization may result in restoration of the BBB with consequences for the efficacy of chemotherapeutic agents [250]. For example, in an orthotopic mouse model of glioblastoma treatment with vandetanib (an angiogenesis inhibitor), temozolomide (a DNA alkylating agent), or a combination of these agents showed that vandetanib (Fig. 12) selectively inhibited angiogenic growth aspects of glioma and restored the BBB although it did not notably affect diffuse infiltrative growth and survival.

In combination studies, vandetanib antagonized the effects of temozolomide presumably by restoration of the BBB and obstruction of the delivery of temozolomide to tumor cells [251]. Thus, given the positive role of antiangiogenic compounds in neuro-oncology and the need for combination therapies a fuller consideration should be given to blood–brain barrier penetration in the design and development of compounds for brain tumor therapies.

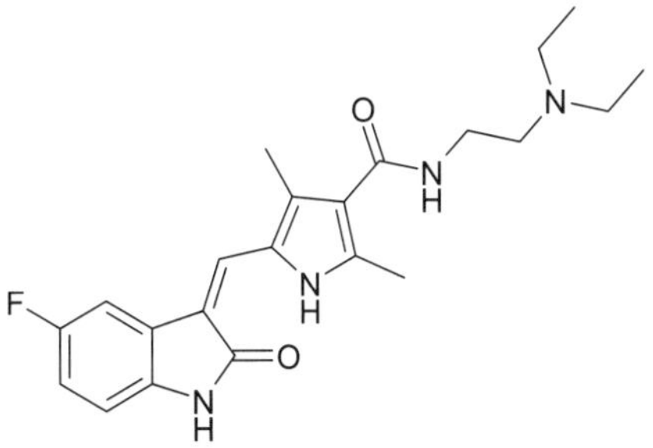

Figure 11. Sunitinib.

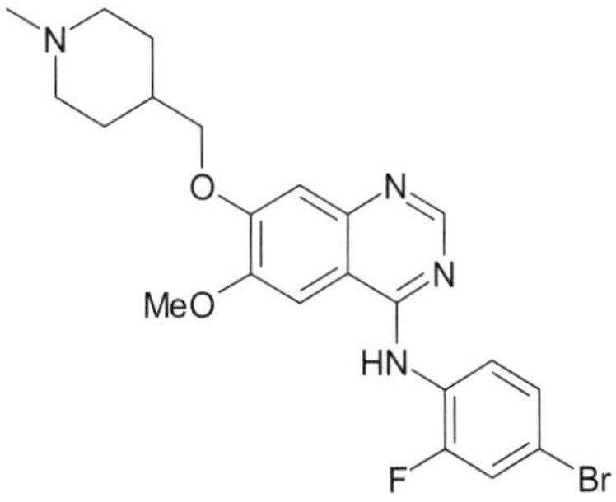

Figure 12. Vandetanib.

16. BRAIN TUMOR THERAPIES

The prognosis of patients with glioblastoma, anaplastic astrocytoma, and anaplastic oligodendroglioma remains poor despite standard treatment with radiotherapy and temozolomide. Molecularly targeted therapies hold the promise of providing new anticancer treatments that are more effective and less toxic than traditional cytotoxic chemotherapy. However, the development of targeted therapy for gliomas has been particularly challenging. The oncogenic process in these tumors is driven by several signaling pathways that are differentially activated or silenced with both parallel and converging complex interactions. Therefore, it has been difficult to identify prevalent targets that act as key promoters of oncogenesis and that can be successfully addressed by novel agents.

High-grade primary brain tumors remain refractory to conventional and multimodal approaches, including radiotherapy and cytotoxic chemotherapy. Molecular neuro-oncology has, however, begun to clarify the transformed phenotype of these malignant tumors and identify oncogenic pathways that might be amenable to small-molecule and antibody "targeted" therapy [252]. Preliminary efficacy results to date on single agent targeted therapy trials for malignant gliomas (including glioblastomas and anaplastic forms of astrocytomas, oligodendrogliomas, and oligoastrocytomas) have largely been disappointing. These targeted therapies included EGFR tyrosine kinase inhibitors (gefitinib and erlotinib), PDGFR inhibitors (imatinib), mTOR inhibitors (temsirolimus and everolimus), and VEGFR, protein kinase C β and other angiogenesis pathway inhibitors (vatalanib, bevacizumab, and Enzastaurin). These trials have also highlighted advances in translational research, including the interpretation of preclinical data, and specific issues in glioma trial design [253]. Current clinical trial protocols are focused on inhibiting multiple targets simultaneously either through utilization of less specific, multitargeting drugs, or better via combination of single-targeted drugs, with the aim of better confronting tumor resistance [254]. For example, combination strategies targeting angiogenesis through inhibition of the VEGFR pathway (e.g., bevacizumab combined with irinotecan) have demonstrated promising activity. Importantly, there is therefore considerable attention being given to the use of combination therapies rather than reliance on single agent chemotherapy approaches in an attempt to confront tumor heterogeneity their aggressive nature and their propensity for resistance [253].

16.1. Cytotoxic Agents

Cytotoxic agents together with radiotherapy remain the first line chemotherapy for many if not most CNS neoplasms [255].

16.1.1. Temozolomide The cytotoxic agent temozolomide has become a standard therapy for GBM and the role of temozolomide chemotherapy in the management of GBM has recently been reviewed [256]. The cytotoxic activity of temozolomide, and other alkylating agents, is believed to manifest largely by the formation of $O(6)$-methylguanine DNA adducts. Since $O(6)$-methylguanine methyltransferase promoter (MGMT) methylation status is a determinant of patient response stratification, MGMT methylation status may be used as a predictive factor in selecting patients most likely to benefit from such treatment. One of the barriers to successful treatment of malignant glioma is developed resistance to alkylating agents such as temozolomide. The primary mechanism of resistance to temozolomide is competition between the rate of DNA alkylation and the rate of MGMT protein synthesis. Several studies have shown that prolonged exposure to temozolomide can deplete MGMT activity in blood cells, potentially increasing antitumor activity. However, to date, there are limited data demonstrating the depletion of MGMT activity in tumor tissue exposed to temozolomide. Cumulative dosing of temozolomide and prolonged exposure are currently being investigated with the goal of improving antitumor activity and overcoming resistance. The regimen that provides the best balance between enhanced antitumor activity and acceptable hematologic toxicity, however, has yet to be determined [257].

16.2. Growth Factors, Their Receptors, and Downstream Cellular Signaling Pathways

Growth factor signaling pathways are often upregulated in brain tumors and contribute to oncogenesis through autocrine and paracrine mechanisms. Upregulation of growth factor receptor pathways can promote disregulation of a range of signaling pathways including; Ras/Raf, MEK, PI3K, Akt, and mTOR. A range of molecular-target agents have been investigated against growth factor signaling and Ras pathways including tyrosine kinase inhibitors (e.g., imatinib and erlotinib) and farnesyltransferase inhibitors (e.g., tipifarnib). Molecular therapeutic agents targeted to Raf, PI3K, and mTOR signaling include sorafenib, LY-294002, and temsirolimus, respectively [253].

Although often effective at reducing tumor mass, conventional treatment approaches such as surgical debulking, radiotherapy, and cytotoxic chemotherapies have failed to demonstrate full benefit to patients and malignant gliomas represent one of the most aggressive forms of brain cancer. Recent insights into the cellular and molecular biology of gliomas together with a better understanding of the deregulated molecular pathways of gliomas and their transformed phenotype(s) are starting to indicate targets amenable to molecular-targeted and antibody therapies. These include tyrosine kinase receptors and associated signal transduction pathways and insights into their role in tumor initiation, tumor maintenance, and vascular proliferation. These insights are providing an emerging picture of the targetable molecular phenotype of glioblastoma and have led to the development of molecules targeting key nodes in glioblastoma transduction pathways. Special attention has been focused on cellular and extracellular signaling pathways (EGFR, VEGF, RAS, PI3-K, and integrins) that have been implicated in malignant glioma pathobiology and key targets therein [258].

Loss of the tumor-suppressor phosphatase and tensin homolog (PTEN) and activation of the receptor tyrosine kinases (RTKs) EGFR, c-Met, PDGFR, and VEGFR are among the most common molecular dysfunctions associated with glioma malignancy. Knowledge of the molecular and functional interactions between PTEN and RTK pathways in glioma malignancy is helping refine combined RTK-targeted therapies, for example, with therapies aimed at counteracting the effects of PTEN loss, such as mTOR inhibition to design more effective therapies [259].

Multitargeted kinase inhibitors, novel monoclonal antibodies, and new vaccines have been developed. These new therapies for malignant gliomas include modulators of growth factors and their receptors (e.g., EGFR, VEGFR, and PDGFR), as well as the intracellular effector molecules that are downstream of these growth factors (e.g., Ras/Raf/mitogen-activated protein (MAP) kinase, phosphatidylinositol 3-kinase/AKT/mTOR, and protein kinase C). Novel small-molecule inhibitors have been shown to modify the activity of these kinase receptors and signaling pathways. Thus far, however, only a limited number of kinases have been investigated and drug activity has been comprehensively evaluated only in a limited number of different malignancies. Highlighting that one of the limiting factors for novel drug design and development is the incomplete knowledge of growth factor pathways and their functions in malignant glioma [260].

The concept of "signaling pathways" does not fully capture the complexity of intracellular signaling where crosstalk is common, for example, it is well established that protein kinase C (PKC) isozymes are involved in the proliferation of glioma cells linking many effectors from AKT, Wnt to inositol phosphate signaling (see, for example, Ref. [261]).

However, reports differ on which PKC isozymes are responsible for glioma proliferation. As a means to further elucidate this, the inhibition of PKC-α, PKC-β, and PKCμ with PD 406976 and their role in regulation of cell cycle, cell proliferation during glioma growth and development have been investigated [262].

16.3. Protein Kinase C

Activation of protein kinase C β (PKC β) has been repeatedly implicated in tumor-induced angiogenesis and the PKC β-selective inhibitor, Enzastaurin (LY317615) (Fig. 13) was

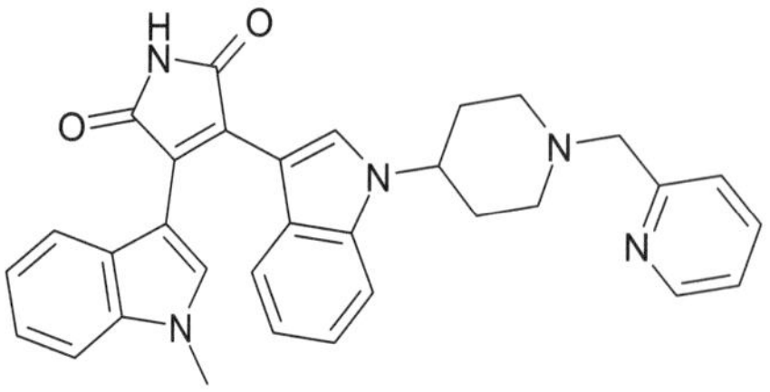

Figure 13. Enzastaurin.

initially shown to suppress angiogenesis in clinical trials. PKC isoforms have also been implicated in tumor cell proliferation, apoptosis, and tumor invasiveness [263]. *In vitro* in human tumor cells Enzastaurin also induces apoptosis and suppresses proliferation. In animal models, it significantly suppresses the growth of both glioblastoma and colon carcinoma xenografts. Since Enzastaurin treatment also suppresses GSK3β phosphorylation, this effect may serve as a pharmacodynamic marker for the range Enzastaurin activities, that is, direct suppression of proliferation and induction of cell death and its indirect effect in suppressing angiogenesis [264]. The use of GSK3β as a biomarker was supported by recent high-dose clinical trials where thrombocytopenia and prolonged QTc were dose-limiting toxicities [265].

The PKC inhibitor PD 406976 also showed growth inhibitory effects on a range of glioma cell lines (SVG, U-138MG, and U-373MG), but interestingly this was not as a result of apoptosis [262].

Strategies are generally lacking, however, for directing kinase inhibitor or pathway-specific therapies to individual patients most likely to benefit [266]. The response to targeted agents is determined not only by the presence of the key mutant kinases but also by other critical changes in the molecular circuitry of cancer cells, such as loss of key tumor suppressor proteins, the selection for kinase-resistant mutants, and the deregulation of feedback loops. Understanding these signaling networks, and studying them in patients, will be critical for developing rational combination therapies to suppress resistance for malignant glioma patients.

The efficacy of other novel targeted inhibitors such as deacetylase inhibitors and heat shock protein (HSP90) inhibitors, as well as new combination therapies have also been investigated [267]. Despite such multimodal therapeutic approaches, malignant gliomas have retained their dismal prognosis. However, molecular classification of primary glioma and elucidation of synergistic effects may eventually allow targeting of distinct glioma subsets and a more rational approach to combination therapies [268].

16.4. EGRF

Clinical studies have shown that HER-2/Neu is overexpressed in up to one-third of cancer patients. Additionally, in several cancer types, overexpression of a number of EGFRs and their ligands HB-EGF and amphiregulin have highlighted the role of the EGFR pathway in tumor pathobiology. For a general review of targeting the EGFR pathway for a range of cancer therapies see Ref. [269]. EGFR overexpression and mutant EGFRvIII expression occur in approximately 50% of glioblastoma patients [270], but monotherapy with anti-EGFR agents such as imatinib did not provide meaningful improvements in patient survival.

Anti-EGFR agents studied to date include kinase ATP-binding site inhibitors, monoclonal antibodies, and anti-EGFR vaccines. The EGFR inhibitors gefitinib and erlotinib have been evaluated in several clinical trials for recurrent high-grade gliomas with contrasting results. Initial studies failed to show significant effect against high-grade gliomas. Moreover, no clear molecular or clinical predictors have been identified [66]. As the EGFR signaling pathway is exceptionally complex, newer approaches targeting multiple points in the pathway are being developed to improve treatment efficacy [271]. Thus, insights gained to date from targeting not only EGFR but also PI3K/Akt/mTOR signaling in patients have led to a reconceptualization of some of the challenges and directions for targeted treatment. In particular, they have highlighted the adaptiveness or plasticity of tumor cells in evading targeted therapies [272]. Thus, although preclinical research has implicated EGFR pathway inhibition in

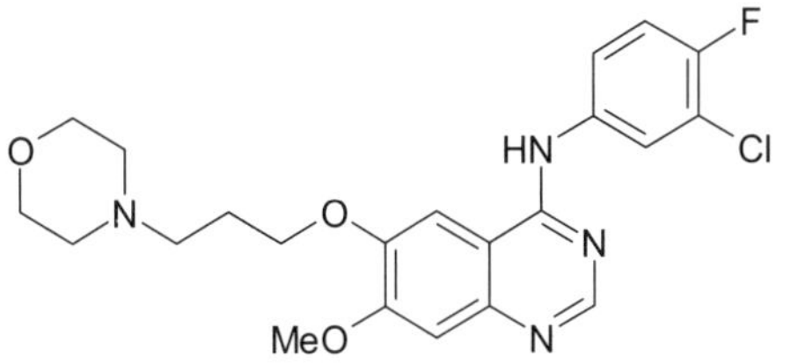

Figure 14. Geftinib (Iressa).

glioma invasion, proliferation, and angiogenesis, response in the clinical setting to EGFR tyrosine kinase inhibitors such as gefitinib (Iressa, ZD1839) (Fig. 14) has been disappointing.

One potential explanation may come from the range of mutations seen in gliomas or their adaptiveness. Investigation of responses to gefitinib on various parameters in malignant gliomas showed a reduction in cell invasion in EGFR amplified tumors and that PTEN loss of expression seems to be a determinant of resistance. Interestingly, however, inhibition of angiogenesis by gefitinib seems independent on the tumor EGFR genetic status [273].

Since cross-activation of the c-Met receptor tyrosine kinase by EGFRvIII has been observed the use of Met kinase inhibitors in combination with either an EFGR kinase inhibitor of cisplatin enhanced the cytotoxicity of EGFRvIII-expressing cells compared with treatment or with either compound alone. These results suggest that the clinical use of c-Met kinase inhibitors, such as SU112754 (Fig. 15) in combination with either EGFR inhibitors or standard chemotherapeutics might represent a previously undescribed therapeutic approach to overcome the observed chemoresistance in patients with GBMs expressing EGFRvIII [272].

16.5. HGF/MET

With the ability of kinases to act as nodes between signaling pathways the inhibition of aberrant neoplasm tyrosine kinase activity is an attractive strategy for the treatment of malignancy if specificity can be achieved and resistance overcome. Among these kinase pathways and nodes the hepatocyte growth factor (HGF)/mesenchymal-epithelial transition factor (Met) tyrosine kinase pathway inhibition could have significant anticancer potential and several Met inhibitors are nearing clinical trials [274]. For example, EGFRvIII and c-Met pathway inhibitors synergize against PTEN-null/EGFRvIII+ glioblastoma xenografts despite the lack of response to single agent approaches with either the neutralizing anti-HGF monoclonal antibody (L2G7) or erlotinib (Fig. 16) alone. Combination therapy with these agents produced a synergistic response against both subcutaneous and orthotopic U87-EGFRvIII xenografts.

The response to combination treatment in U87-EGFRvIII xenografts occurred in the absence of Akt and MAPK inhibition. These findings show that EGFR and c-Met pathway inhibitors synergize against PTEN-null/ EGFRvIII+ glioblastoma xenografts [275].

In glioblastoma c-Met and its ligand, hepatocyte growth factor (HGF), are frequently overexpressed while the tumor suppressor PTEN is often mutated. Since PTEN can interact with c-Met-dependent signaling, and PTEN loss amplifies c-Met-induced malignancy combining anti-HGF/c-Met approaches with PTEN restoration or mTOR inhibition may also provide synergistic benefits [276].

c-Met activation has also been implicated in medulloblastoma inducing tissue factor expression and activity [277]. Small-molecule Met kinase inhibitors such as XL-184 (Exe-

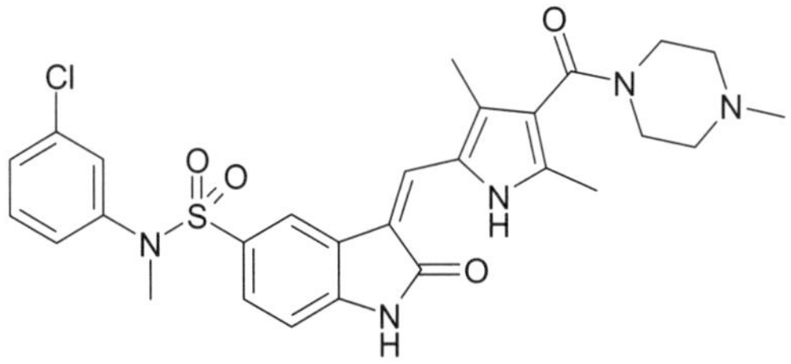

Figure 15. SU11274.

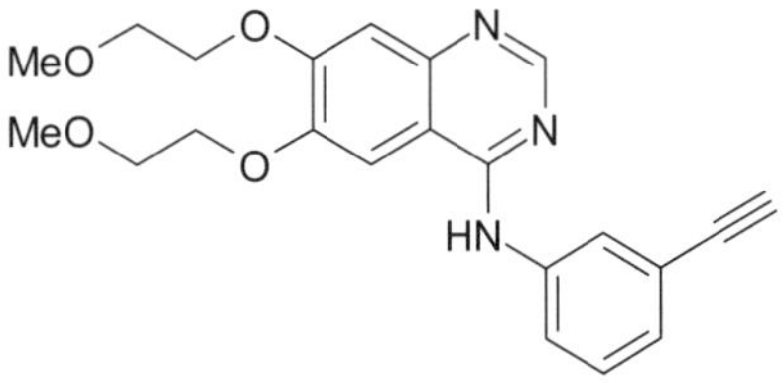

Figure 16. Erlotinib.

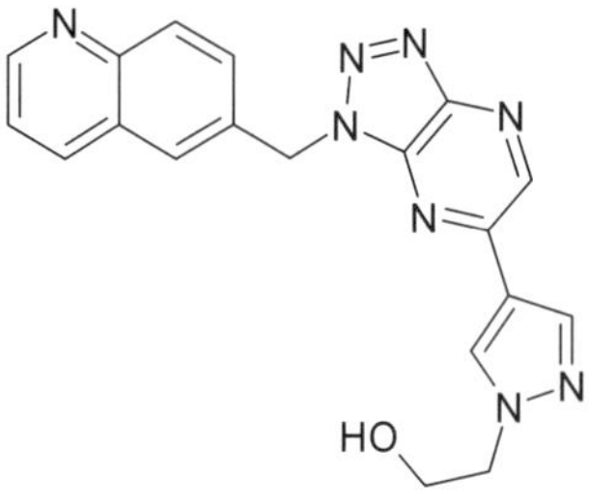

Figure 17. PF-04217903.

lexis, pan-tyrosine kinase inhibitor) [278] and PF-04217903 (Pfizer, c-Met selective) (Fig. 17) [279] are also under investigation in clinical trials against glioma and other neoplasms. The c-Met antibody CE-355621 has been shown to be active in a U87 MG mouse xenograft model [280].

16.6. VEGF Inhibitors

Rapidly dividing glioma cells maintain adequate oxygen and nutrient delivery through angiogenesis via co-opting existing host blood vessels or promoting the formation of new vessels. Angiogenesis is considered to be a regulating factor of vascular development and growth for malignant gliomas, including glioblastoma and anaplastic astrocytomas. Hypoxia induced activation of the HIF pathway leads to the production of VEGF and basic fibroblast growth factor. Vascular endothelial growth factor-A (VEGF-A) is perhaps the best known proangiogenic factor in glioblastoma and plays a key role in angiogenesis of several tumor types. VEGF-A is upregulated in glioblastoma in which it is produced by tumor cells as well as stromal and adjacent inflammatory cells. VEGF/VEGFR-2 is the predominant angiogenic signaling pathway in malignant gliomas and anti-VEGFR therapies form an important role in antiangiogenic molecularly targeted therapies for CNS tumors promoting rapid but reversible decrease in vascular permeability [281]. Anti-VEGFRs monoclonal antibodies decrease vascular permeability and perfusion, whereas kinase inhibitors interfere with cell communication, receptor signaling, and tumor growth [282]. Several completed, ongoing, and planned clinical trials are evaluating or have evaluated anti-VEGF strategies for malignant glioma patients [283].

Recent clinical trials combining bevacizumab (Avastin), an anti-VEGF antibody, with chemotherapy reported encouraging response rates. It should be kept in mind though that tumors can develop escape mechanisms. In particular invasive cells, which migrate away from the highly vascularized tumor core, are not targeted by antiangiogenic therapies. The success of antiangiogenic therapy will therefore rely on a combination strategy including chemotherapy and drugs that target invasive glioma cells [284].

Most brain tumors oversecrete VEGF, which leads to an abnormally permeable tumor vasculature. Agents that block the VEGF pathway are able to decrease vascular permeability and, thus, cerebral edema, by restoring the abnormal tumor vasculature to a more normal state. Decreasing cerebral edema minimizes the adverse effects of corticosteroids and could improve clinical outcomes [285].

Results indicate that glioma cells enhance endothelial progenitor cell (EPC) angiogenesis via VEGFR-2, not VEGFR-1, and that this is mediated by the MMP-9, Akt and ERK signal pathways [286].

16.7. VEGFR Kinase Inhibitors

Overactivity of VEGF and other effectors leads to neoplastic angiogenesis, several antiangiogenesis approaches therefore target VEGF and VEGFR including mAbs to VEGF (e.g., bevacizumab) and VEGF receptor tyrosine kinase inhibitors (e.g., vatalanib (Fig. 18) and sunitinib) [252]. The effect of ZD6474, a potent inhibitor of VEGF-receptor-2, against the neovascularization of malignant glioma was evaluated in combination with either radiotherapy or chemotherapy with temozolomide. ZD6474 in combination with these two standard modalities markedly decreased the

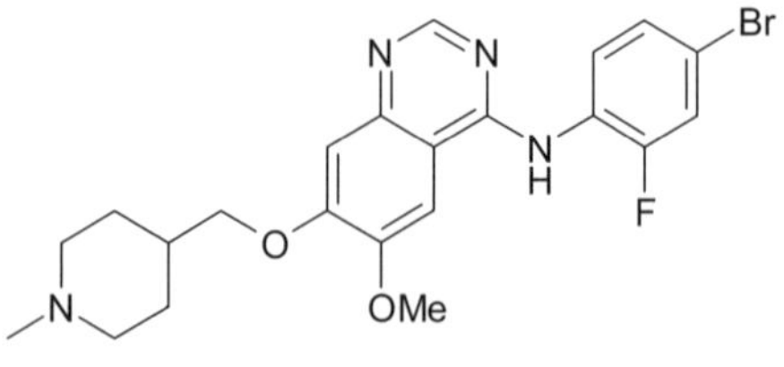

Figure 18. Vandetanib/ZD-6474.

growth of intracerebral experimental glioma [287].

16.8. PDGF

Platelet-derived growth factor (PDGF) promotes gliomagenesis through autocrine and paracrine loops, via expression of platelet-derived growth factor receptor alpha (PDGFRα) receptor on glioma cells and PDGFRβ in proliferating endothelial cells. In oligodendroglial tumors, however, although PDGFRα is expressed, expression levels were not useful in predicting tumor grade [288]. Immunohistochemically PDGFRα is a marker for the oligodendrocyte component of glial tumors such as oligoastrocytomas and oligodendrogliomas. Recent studies have shown that chronic PDGF signaling in glial progenitors can lead to the formation of oligodendrogliomas in mice, whereas chronic combined Ras and Akt signaling leads to astrocytomas. PDGF-mediated pathways are therefore targets of interest for therapeutic strategies.

16.9. PDGF Receptor

SU11657, toceranib (Fig. 19), an inhibitor of class III/V receptor tyrosine kinases, including PDGFR VEGFR, KIT, and FLT3 reduces cell proliferation and clonogenic survival of atypical and benign meningioma cells. SU11657 also increased radiosensitivity of human meningioma cells in clonogenic survival and cell number/proliferation assays.

The anticlonogenic and antiproliferative effects and the radiosensitization effects of SU11657 were more pronounced in atypical meningioma cells compared with benign meningioma cells [289].

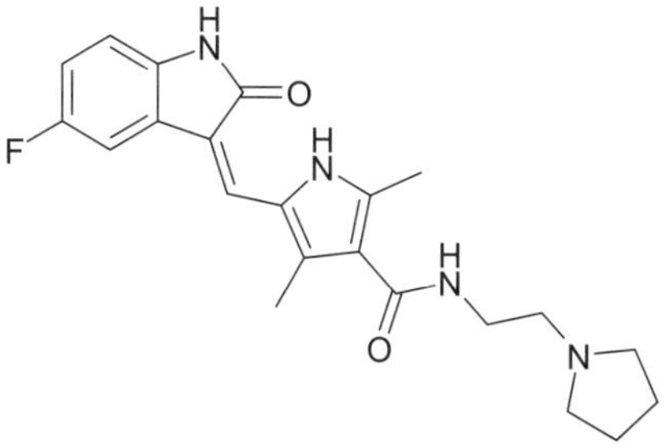

Figure 19. SU11657.

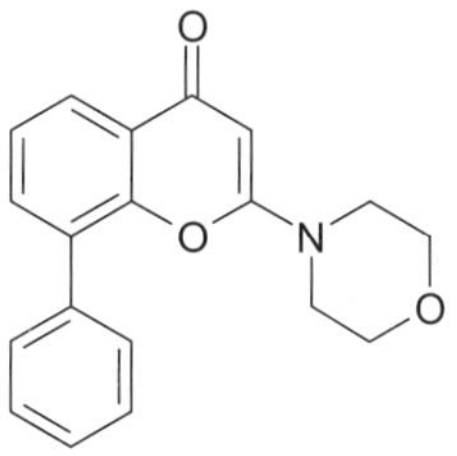

Figure 20. LY-294002.

16.10. PI3K/AKT Signaling

The PI3K/Akt pathway is often upregulated in brain tumors due to excessive stimulation by growth factor receptors and Ras. Loss of function of the tumor suppressor gene PTEN also frequently contributes to upregulation of PI3K/Akt. Several compounds, such as wortmannin and LY-294002 (Fig. 20), can target PI3K and inhibit activity of this pathway [290].

16.11. Phosphoinositide 3-kinase (PI3-kinase) Inhibitors

Gliomas frequently exhibit abnormalities in PI3K signaling. The molecular mechanism of action of the isoform-selective class I PI3K and mTOR inhibitor PI-103 has been investigated in human glioma cells [291]. In contrast to PI-103, LY294002, and PI-387 (GDC-0941) (Fig. 21) induced apoptosis, indicative of likely off-target effects. PI-103 interacted synergistically or additively with cytotoxic agents used in the treatment of glioma, namely, vincristine, 1,3-bis(2-chloroethyl)-1-nitrosourea (BCNU), and temozolomide. Compared to individual treatments, the combination of PI-103 with temozolomide significantly improved the response of U87MG glioma xenografts [292].

Optimization of this series lead to the resulting clinical development candidate PI-397 (GDC-0941) with improved antitumor efficacy over PI-103, following i.p. dosing in U87MG glioblastoma mice xenografts [293]. A useful review of PI3-K inhibitors is provided within the Pirmed paper on the identification of GDC-0941 [294].

Akt pathway activation is prevalent in CNS tumors including GBM, frequently via numerous upstream alterations including

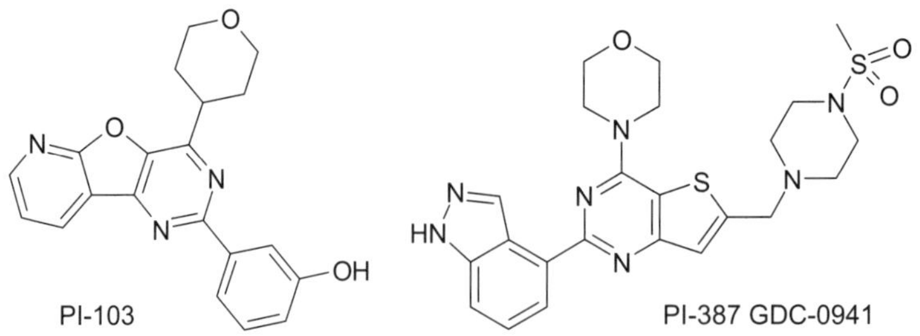

Figure 21. PI-3K inhibitors that induce apoptosis.

genomic amplification of epidermal growth factor receptor, PTEN deletion, or PIK3CA mutations. Inhibition of Akt inhibits growth of glioblastoma and glioblastoma stem-like cells. PI3-kinase/Akt small-molecule inhibitors, such as A-443654 (Fig. 22) tested in an isogenic cell culture system with an activated Akt pathway (secondary to a PIK3CA mutation) showed selective inhibition of cells with the mutant phenotype. A-443654 was also effective against a panel of glioblastoma cell lines promoting cell death via apoptosis. In an orthotopic rat model of glioblastoma multiforme A-443654 extended survival compared with controls. No resistance to glioblastoma multiforme cells grown in stem cell conditions was observed, however [295].

AKT signaling in glioblastoma can also be inhibited by targeting integrin-linked kinase (ILK) as demonstrated with the ILK inhibitor QLT0267 which demonstrated cell growth-inhibition at high concentrations, and reduced cellular invasion and angiogenesis at much lower concentrations in both *in vitro* invasion assays and against VEGF secretion [296,297]. Integrin-linked kinase has been shown to be key in the regulation of cell migration, as well as proliferation and apoptosis [298].

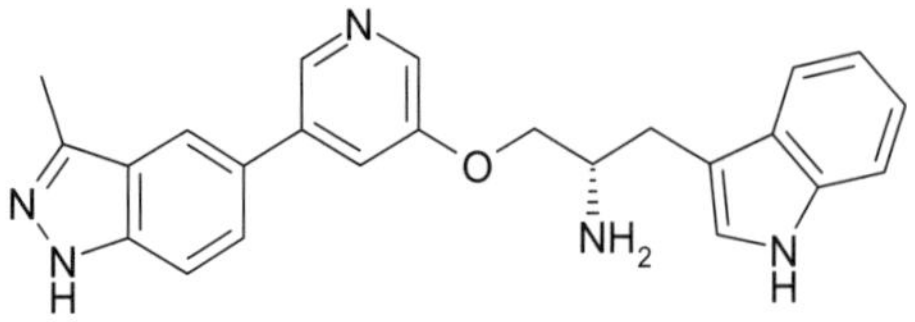

16.12. mTOR Signaling

The mammalian target of rapamycin (mTOR) is an important regulator of cell growth and metabolism and is often upregulated by Akt. Rapamycin itself has anticancer activity in PTEN-deficient glioblastoma, combination with PI3K pathway inhibitors could be particularly beneficial [299]. The rapamycin analog CCI-779 (Fig. 23) has been progressed to a range of clinical trials for glioblastoma and is currently in several Phase II studies [300,301].

NVP-BEZ235 (Fig. 24), a novel dual PI3K/mTOR inhibitor potently suppressed glioma cell proliferation by specifically inhibiting the activity of target proteins in the PI3K/Akt/mTOR signaling pathway. NVP-BEZ235 treatment led to G(1) cell cycle arrest and induced autophagy. Furthermore, expression of VEGF in glioma cells was significantly decreased, suggesting that NVP-BEZ235 may also exert an antiangiogenic effect [301].

However, the role of mTOR in tumor progression may be more complicated, since antagonism of the mTOR selectively mediates metabolic effects of epidermal growth factor receptor inhibition and protects human malignant glioma cells from hypoxia-induced cell death [302]. Antagonism of mTOR has been suggested as a strategy to augment the efficacy of EGFR inhibition by interfering with deregulated signaling cascades downstream of Akt. Antagonism of mTOR utilizing rapamycin or a small hairpin RNA-mediated gene silencing has been compared to EGFR inhibition alone or combined inhibition of EGFR and mTOR in human malignant glioma cells. In contrast to EGFR inhibition, mTOR antagonism neither induced cell death nor enhanced

Figure 22. A-443654.

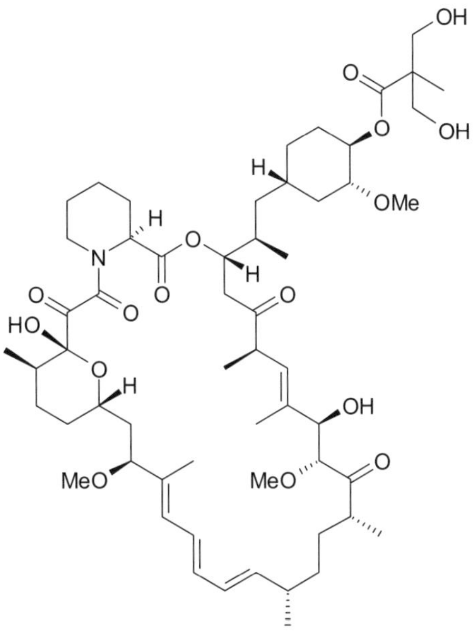

Figure 23. CCI-779 (Temsirolimus).

apoptosis induced by CD95 ligand or chemotherapeutic drugs. However, mTOR inhibition mimicked the hypoxia-protective effects of EGFR inhibition and dependent on the tumor microenvironment, mTOR inhibition may adversely affect outcome by protecting the hypoxic tumor cell fraction [303].

EGFR-driven gliomas differ in their PTEN status, and erlotinib blocked proliferation only in PTEN(wt) cells expressing EGFR, showing little effect as a monotherapy in PTEN(mt) glioma. However, cotreatment with the dual inhibitor of PI3Kα and mTOR (PI-103) [293,294] greatly augmented the antiproliferative efficacy of erlotinib in this setting. Combining PI-103 and erlotinib was superior either to monotherapy or to the therapy combining erlotinib with either rapamycin (an inhibitor of mTOR) or PIK-90 (an inhibitor of PI3Kα) (Fig. 25). These experiments show that a dual inhibitor of PI3Kα and mTOR augments the activity of EGFR blockade, offering a mechanistic rationale for targeting EGFR, PI3Kα, and mTOR in the treatment of EGFR-driven, PTEN-mutant glioma [304].

Monotherapies have proven largely ineffective for the treatment of glioblastoma, due to the ability for cancer cells to evade therapy and develop resistance. A range of tumorigenic pathways are known to be active in glioblastoma including, not only the developmental signaling pathways but also the Ras/Raf/mitogen-activated protein kinase and PI-3-kinase/AKT/mTOR pathways. Combination of the Raf inhibitor LBT613 and the TOR

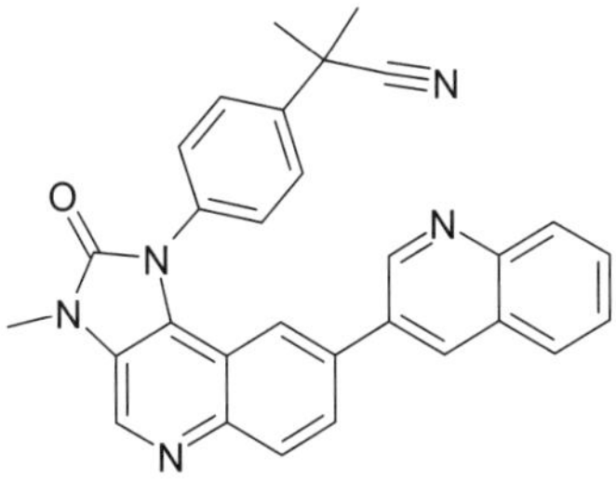

Figure 24. NVP-BEZ235.

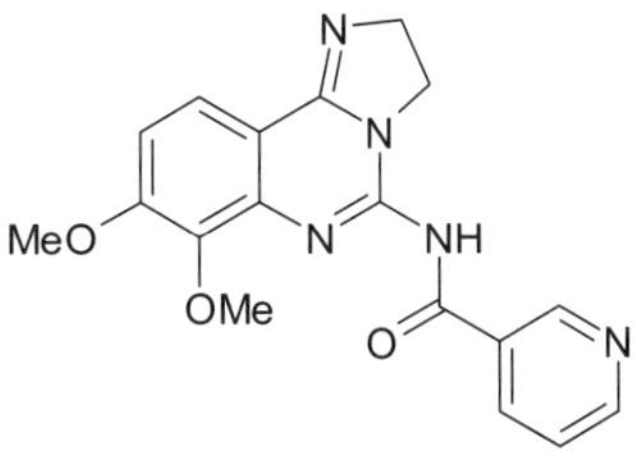

Figure 25. PIK-90.

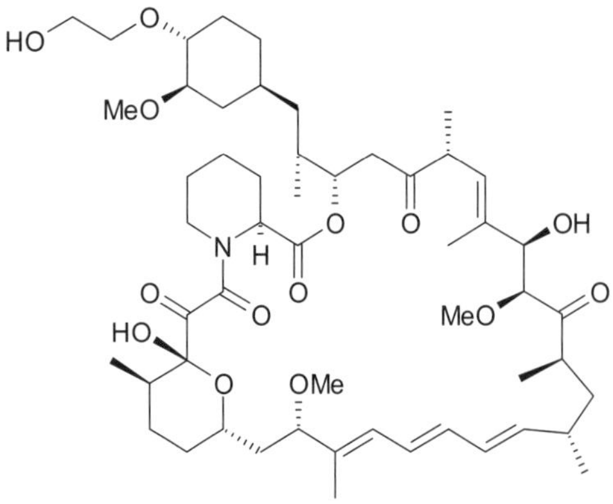

Figure 26. RAD001(Afinitor, Everolimus).

Figure 27. Zamestra.

inhibitor RAD001 (Afinitor, everolimus) (Fig. 26) inhibited the invasion of human glioma cells through matrigel to a greater degree than treatment with either drug alone. These data suggest that the combination of LBT613 and RAD001 reduces glioma cell proliferation and invasion and support examination of the combination of Raf and TOR inhibitors for the treatment of glioblastoma [305].

16.13. RAS/RAF/MEK/ERK Signaling

Growth factor signaling pathways are often upregulated in brain tumors and may contribute to oncogenesis through autocrine and paracrine mechanisms. Excessive growth factor receptor stimulation can also lead to overactivity of the Ras signaling pathway, which is frequently aberrant in brain tumors. Receptor tyrosine kinase inhibitors, antireceptor monoclonal antibodies and antisense oligonucleotides are targeted approaches under investigation as methods to regulate aberrant growth factor signaling pathways in brain tumors. Several receptor tyrosine kinase inhibitors, including imatinib mesylate (Gleevec), gefitinib (Iressa), and erlotinib (Tarceva), have entered clinical trials for high-grade glioma patients. Farnesyl transferase inhibitors, such as tipifarnib (Zarnestra) (Fig. 27), which impair processing of proRas and inhibit the Ras signaling pathway, have also entered clinical trials for patients with malignant gliomas [306].

Recent advances in compounds targeting the Ras/Raf/MEK/ERK pathway have been reviewed [307]. The first MEK inhibitor (CI-1040) lacked efficacy in clinical trials, its low toxicity prompted many companies to investigate the development of novel compounds with enhanced target potency, such as AZD-8330, RO-4987655, and SB590885 (Fig. 28).

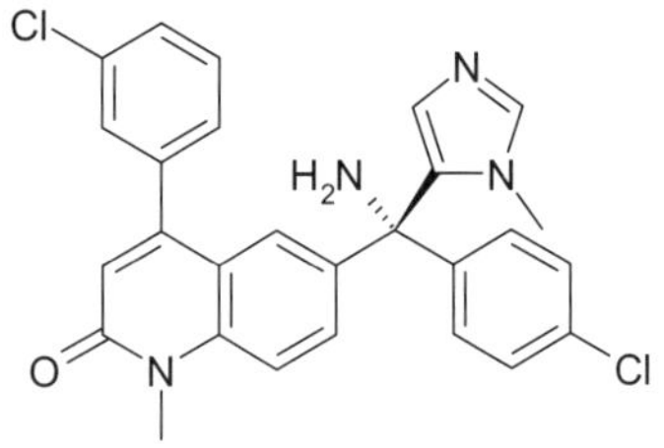

16.14. Rho/ROCK and MAPK Signaling Pathways

Inhibition of Rho A kinase (ROCK), a direct downstream effector of Rho A, by the ROCK inhibitor Y27632, or a dominant negative construct of ROCK, induced apoptosis in glioma cells. Indicating that the RhoA/ROCK pathway is involved negatively in the regulation of glioma cell death pathway [308].

The interactions between migrating glioma cells and myelinated fiber tracts are poorly understood. C6 glioma cells have been shown to migrate along myelinated chicken retinal axons expressing small GTPases of the Rho family and serine/threonine Rho-associated kinases (ROCKs). The ROCK1 isoform is also highly expressed in high-grade gliomas and glioma cells migrated faster *in vitro* along myelinated axons than on laminin-1. This migration can be specifically and reversibly blocked by Y27632 suggesting that the mechanisms underlying the migration of glioma cells on myelinated axons differ from those underlying the migration on extracellular matrix molecules such as laminin-1 [309].

In vitro studies using Y27632, and U0126 (Fig. 29), an upstream MAPK kinase inhibitor, alone or in combination in the human glioblastoma cell line LN-18, have shown that PDGF and fibronectin-induced cell proliferation were suppressed by pretreatment with either compound, but with the greatest reduction achieved by a combination of the two inhibitors. This suggests that Rho/ROCK signaling

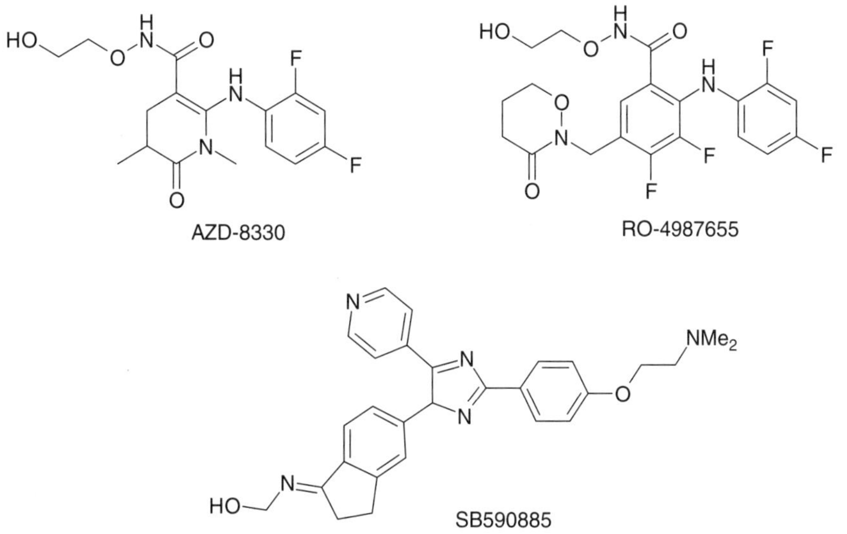

Figure 28. Ras/Raf/MEK/ERK pathway inhibitors.

is involved in glioblastoma migration and proliferation, and that this pathway may be linked to ERK signaling [310].

16.15. Hypoxia Inducible Factor Pathway

Glioblastoma is characterized intratumoral ischemic necrosis (i.e., pseudo-palisading necrosis) and activation of the HIF-1α pathway with consequent peritumoral microvascular proliferation and infiltrative behavior. Hypoxia plays an important role in driving tumor growth and in particular the angiogenic processes. HIF-1α expression levels are elevated in patients with transitional meningioma and glioblastoma, suggesting that although hypoxia is one of the most important and powerful stimuli for HIF-1α elevation and consequently angiogenesis. Other mechanisms may play roles in HIF-1α stimulation in benign brain tumors such as transitional meningoma [311]. Hypoxia-inducible factor (HIF-2α) and multiple HIF-regulated genes are preferentially expressed in glioma stem cells relative to nonstem tumor cells and normal neural progenitors. HIF-2α expression correlates with poor glioma patient survival. Targeting HIFs inhibits self-renewal, proliferation, and survival *in vitro*, and attenuates the tumor initiation potential of glioma stem cells *in vivo* [312]. HIF is a critical regulatory factor in the tumor microenvironment because of its central role in promoting proangiogenic and invasive properties. While HIF activation strongly promotes angiogenesis, the emerging vasculature is often abnormal, leading to a vicious cycle that causes further hypoxia and HIF upregulation [313]. HIF-1α induced by hypoxic stress is an essential event in the activation of glioma cell motility through alteration of invasion-related molecules [314]. A U251-derived orthotopic model replicated these feature and recapitulates the most sali-

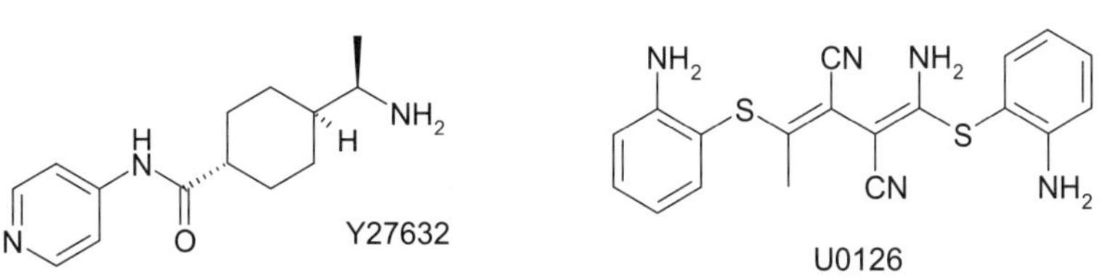

Figure 29. Y27632 (ROCK inhibitor) and U0126 (MAPK inhibitor).

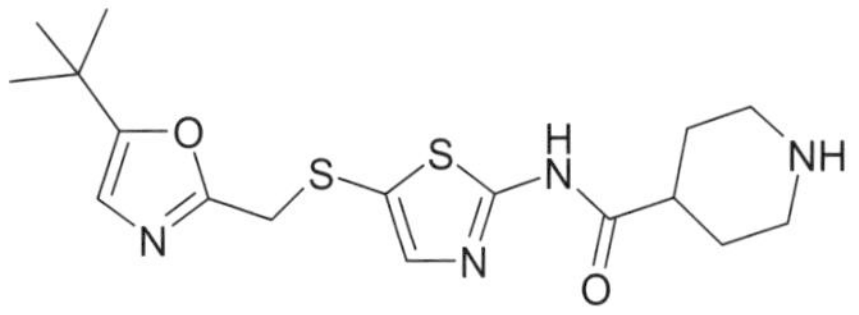

Figure 30. SNS-032, BMS-387032.

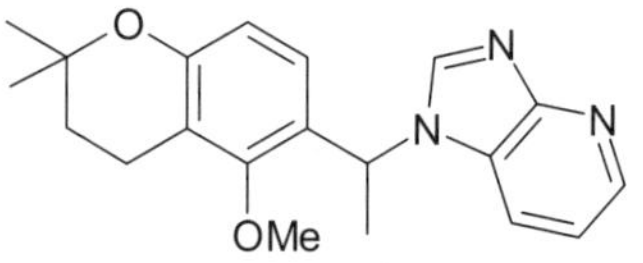

Figure 32. 103D5R.

ent pathobiological features reported for human glioblastoma [315].

Links of HIF-1α expression and its transregulating factors to gliomagenesis were revealed by studies on the CDK antagonist SNS-032 (Fig. 30) that prevented hypoxia-induced U87MG cell invasion by blocking the expression of HIF-1α. SNS-032 also blocked HIF-1α mediated transcription of COX-2, MMP-2, VEGF, and uPAR expression in U87MG cells in response to hypoxia, and blocked HIF-1α expression by a proteasome independent pathway. The effects were similar to those observed with HIF-1α siRNA which prevented cellular invasion by blocking HIF-1α expression and its downstream effectors [316].

Additional insights into the role of hypoxia in glioma pathogenesis have been provided by investigation of tumor responses to chemotherapeutic agents in a D54MG-derived glioma model with an inducible HIF-1 pathway. Treatment with the angiogenesis inhibitor ABT-869 (Fig. 31) showed that concomitant HIF-1 inhibition did not produce a synergistic benefit compared with ABT-869 treatment alone. The cytotoxic drug temozolomide, when used in combination with HIF-1α knockdown, exhibited an additive and likely synergistic therapeutic effect compared with the monotherapy of either treatment alone [317].

Functional inhibition of HIF-1 signaling has been investigated via screening using an alkaline phosphatase reporter gene under the control of hypoxia-responsive elements in human glioma cells LN229-HRE-AP stably expressing a hypoxia-responsive alkaline phosphatase reporter gene. In this system, a series of 2,2-dimethylbenzopyrans represented by 103D5R (Fig. 32) where found to strongly reduce HIF-1α protein synthesis, whereas HIF-1α mRNA levels and HIF-1α degradation were not affected [318].

103D5R inhibited the phosphorylation of Akt, Erk1/2, and stress-activated protein kinase/c-jun-NH$_2$-kinase, without changing the total levels of these proteins. Further studies on the mechanism of action of 103D5R will likely provide new insights into its validity/applicability for the pharmacologic targeting of HIF-1α for therapeutic purposes [319].

Functional screening using a luciferase-readout cell-based high-throughput screen in U251 human glioma identified several small-molecule inhibitors of the HIF-1 pathway [319]; among which NSC-609699 (Topotecan) (Fig. 33) selectively inhibited accumulation of HIF-1α and consequently the appearance of DNA binding activity, which correlates with inhibition of VEGF mRNA expression.

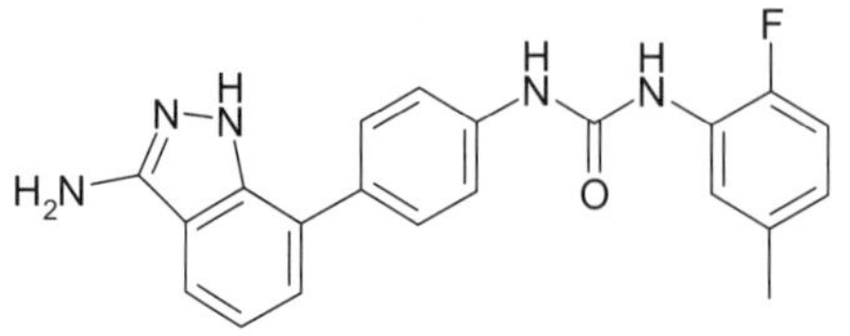

Figure 31. ABT-869, Linifanib.

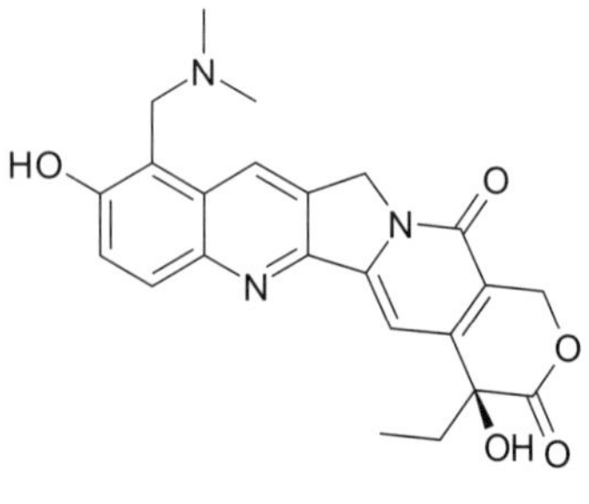

Figure 33. Topotecan.

Whether a direct inhibition of Topo-I activity is essential for inhibition of HIF-1α protein accumulation and VEGF expression was not investigated.

Glioblastomas contain cellular hierarchies with self-renewing cancer stem cells that can propagate tumors in secondary transplant assays. The potential significance of cancer stem cells in cancer biology was also demonstrated by studies showing their contribution to therapeutic resistance, angiogenesis and tumor dispersal. Importantly, it has been reported that physiologic oxygen levels differentially induce hypoxia inducible factor-2α (HIF2α) levels in cancer stem cells [320]. Thus, while HIF1α promotes the proliferation and survival of cancer cells it is also activated in normal neural progenitors (potentially suggesting a restricted therapeutic index), HIF2α is essential in only in cancer stem cells and is not expressed by normal neural progenitors. HIF2α may therefore have a specific role in promoting glioma tumorigenesis. Moreover, HIF2α also augmented the tumorigenic potential of the nonstem population. The plasticity of the nonstem glioma population and the stem-like phenotype emphasizes the importance of developing therapeutic strategies targeting the microenvironmental influence on the tumor in addition to cancer stem cells [320].

Several mechanisms have also been linked hypoxia mechanisms including the sphingosine kinase 1/sphingosine 1-phosphate (SphK1/S1P) pathway, which elicits various cellular processes including cell proliferation, cell survival, or angiogenesis. The SphK1/S1P pathway is a modulator of HIF-1α activity under hypoxic conditions [321] and could therefore represent an important target for intervention in cancer. Finally, an αvβ3/αvβ5 integrin-dependent loop of hypoxia autoregulation has also been elucidated in glioma. Targeting this, hypoxia loop may be crucial to optimizing radiotherapy efficiency [322].

16.16. Integrins

In malignant gliomas, integrin receptors seem to play a key role in cell adhesion and thus tumor invasive growth and angiogenesis. For example, expression studies revealed that, αvβ3 integrin is expressed not only in endothelial cells but also, to a large extent, in glial tumor cells [323]. Immunohistochemical analysis of adhesion molecules and matrix metalloproteinases (MMPs) in malignant CNS lymphomas suggest that adhesion molecules and MMPs are also essential for malignant lymphoma cell invasion from the vasculature into the brain parenchyma and that they may be the key determinants for malignant lymphoma cells to behave as primary CNS malignant lymphomas (PCNSML) or intravascular lymphomas (IVL) cells. Moreover, integrins have been shown to mediate adhesion of medulloblastoma cells to tenascin and activate pathways associated with survival and proliferation [324]. In addition, expressions of α9 and β1 integrins in combination with extracellular tenascin are necessary for medulloblastoma adhesion to the leptomeninges which could be blocked by targeted antibodies [325].

JSM6427 (Fig. 34) an inhibitor of α5β1 integrin modulates hypoxia-induced neovascularization, attenuates glioma growth and decreases the density of microglia at the tumor border in an experimental mouse model. Treatment with JSM6427 showed a significant reduction in tumor size as compared to control attenuating both microglial migration and proliferation [326].

The selective α5β1 antagonist, SJ749 (Fig. 35), inhibits glioblastoma angiogenesis (in both A172 and U87 cells) reducing both adhesion and migration of in a dose-dependent manner. Interestingly, the effects of SJ749 were cell line dependent (again highlighting the variability and heterogeneity of glioma cells). Under adherent conditions, administration of SJ749 reduced proliferation of A172 cells and led to the formation of neurospheres. U87 cell morphology was unaffected and proliferation was only slightly affected. In nonadherent conditions, the effect of SJ749 on tumor cell growth characteristics depends on the level of α5β1 expression [327].

The mode of action of the RGD mimetic, cilengitide (EMD 121974) (Fig. 36), a cyclic peptide antagonist of integrins αvβ3 and αvβ5 is thought to be mainly antiangiogenic, but may include direct effects on tumor cells, notably attachment, migration, invasion, and

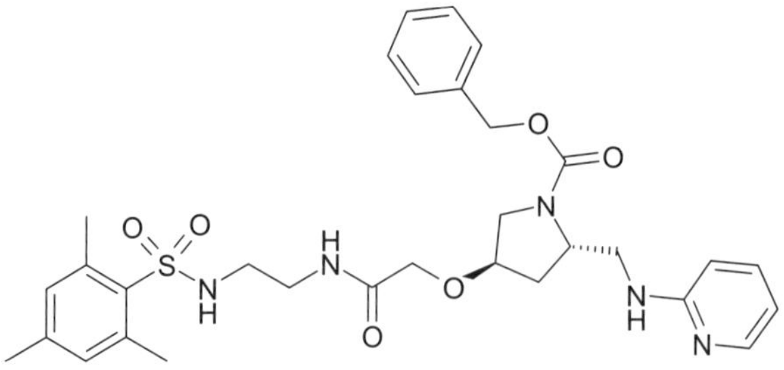

Figure 34. JSM6427.

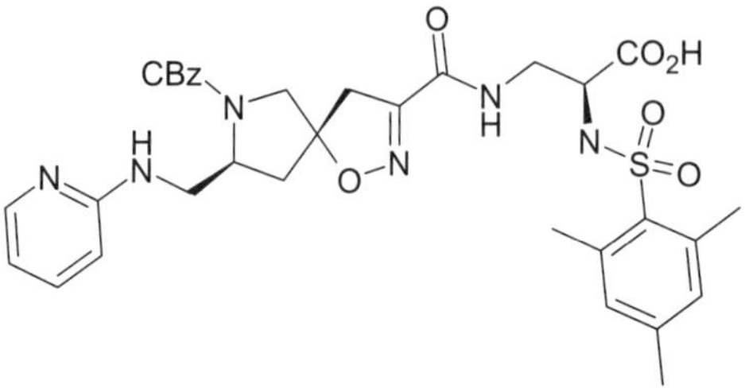

Figure 35. SJ749.

viability. Data suggest that the beneficial clinical effects derived from cilengitide *in vivo* may arise from altered perfusion that promotes temozolomide delivery to glioma cells.

In clinical trials for recurrent GBM, single-agent cilengitide showed antitumor benefits and minimal toxicity. Among newly diagnosed GBM patients, Phase II studies incorporating cilengitide into a standard radiotherapy/temozolomide regimen have shown encouraging activity with no increased toxicity [328]. The beneficial clinical effects derived from cilengitide *in vivo* may arise from altered perfusion that promotes temozolomide delivery to glioma cells [329]. Reports on such use of cilengitide in the adjuvant treatment of glioblastoma suggest that stratification by tumor MGMT expression status best defines the preferred (temozolomide) patient subpopulation. Cilengitide modulates attachment and viability of human glioma cells, but not sensitivity to irradiation or temozolomide *in vitro* [330]. For a review of integrins and other cell adhesion molecules and their role in brain invasiveness, see Ref. [237].

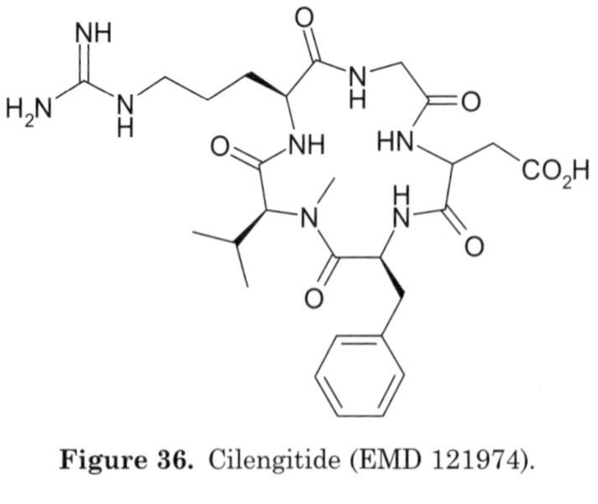

16.17. p53 Signaling Pathway

The gene TP53, encoding transcription factor p53, is mutated or deleted in half of human cancers, demonstrating the crucial role of p53 in tumor suppression and the control, in particular of apoptosis mechanisms [331]. Importantly, p53 inactivation in cancers can also result from the amplification/overexpression of its specific inhibitors MDM2 and MDM4 (MDMX). The presence of wild-type p53 in those tumors with MDM2 or MDM4 overexpression stimulates the search for new therapeutic agents to selectively reactivate it [332]. Therapeutic strategies to normalize p53 signaling in cells with mutant p53 include pharmacologic rescue of mutant protein, gene therapy approaches, small-molecule agonists of downstream inhibitory genes, antisense approaches, and oncolytic viruses. Other strategies include activation of normal p53 activity, inhibition of MDM2-mediated degradation of p53 and blockade of p53 nuclear export [331].

Figure 36. Cilengitide (EMD 121974).

Primary GBM, in contrast with secondary GBM, has been associated with the presence of EGFR amplification and absence of p53 mutations. Although most of the tumors showed a mutually exclusive pattern, concurrent alterations of EGFR and p53 were also detected. Survival analysis showed that CDK4 amplification with a worse clinical outcome, whereas MDM2, CDK6, PTEN, and p21 were not associated with patient survival [101].

16.18. MDM2 and MDMX

MDM2 is overexpressed in most cancers and is well evidenced in many peripheral tumor types (such as colorectal cancer) and retinoblastoma. Glioblastomas and astrocytomas show positive staining for MDM2 protein in up to 75% of tumor samples, and approximately 10% have MDM2 gene amplification [333]. Although rare, MDM2 overexpression is associated with short survival in adults with medulloblastoma [334] and high-grade ependymomas show overexpression of p53 and MDM2 but it is not clear that they are involved in tumor progression [335]. Amplification and overexpression of MDM4 is associated with a small fraction of human malignant gliomas that escape p53-dependent growth control [336]. A robust case for MDM2 inhibition in gliomagenesis is not as yet proven since although MDM2 amplification is related to poor prognosis [337] suitable ligands to investigate the role of MDM2 inhibition in CNS tumors are lacking (i.e., those that would penetrate the BBB). MDM2 and MDMX inhibitors therefore are included here since their possible potential as anti-CNS tumor remains to be explored. The p53-MDM2 and p53-MDMX interactions potentially represent two of the more tractable protein–protein interaction targets [338] although the impact of structural data is key to design and optimization of compounds [339].

The pioneering small-molecule MDM2 antagonist, Nutlin-3 (Fig. 37), has been used to probe downstream p53 signaling in many tumorigenic cell lines expressing wild-type p53. Cancer cells with MDM2 gene amplification are the most sensitive to Nutlin-3 both *in vitro* and *in vivo*. Nutlin-3 also showed good efficacy against tumors with normal MDM2 expression, suggesting that many of the patients with wild-type p53 tumors may benefit from antagonists of the p53-MDM2 interaction [340]. Nutlin-3 and related compounds have been advanced by Roche to clinical trials for hematological neoplasm and a range of solid tumors.

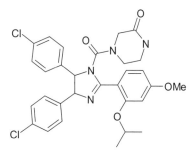

Figure 37. Nutlin-3 – First MDM2 inihibitor.

In addition, stabilization of p53 in cells lacking MDM4 with Nutlin-3 was sufficient to induce a cell death response [341].

Block of the inhibition of wild-type p53 by MDM2 has also been achieved with the compound MI-63 (Fig. 38) a potent nonpeptide small molecule. MI-63 as compared with Nutlin-3 was found to be more potent in the inhibition of cell proliferation/viability of rhabdomyosarcoma cells expressing wild-type p53 protein. Further, synergy was observed when MI-63 was used in combination with doxorubicin [342].

The analog MI-219 provoked temporal activation of p53 by a specific inhibition of MDM2 moreover MI-219 was selectively toxic to tumors and lead to complete inhibition of tumor growth in cancer cell lines with wild-type p53, such as osteosarcoma SJSA-1 cells [343]. Using *in silico* screening and small library synthesis, a series of 2-alkyl-3-aryl-3-

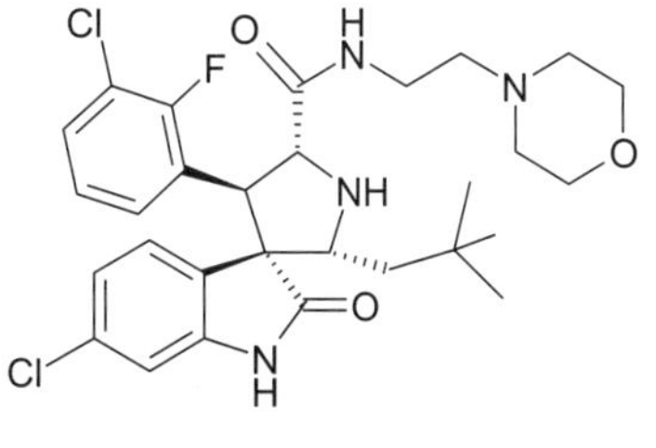

Figure 38. MI-63.

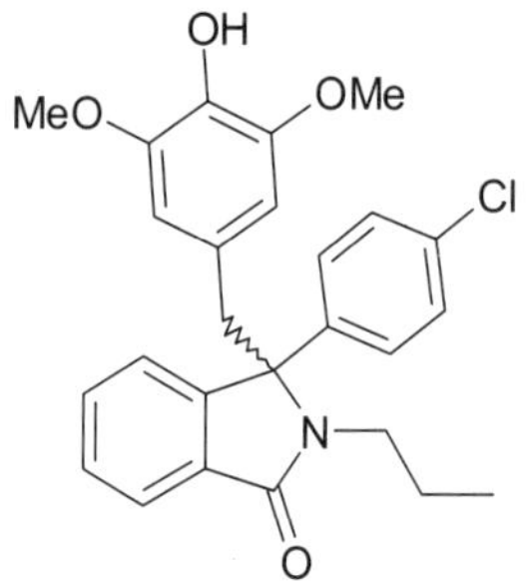

Figure 39. Alkyl-aryl-alkoxyisoindolinone MDM2 inhibitor.

alkoxyisoindolinones (Fig. 39) has been identified as inhibitors of the MDM2-p53 interaction. Two of the most potent compounds induced p53-dependent gene transcription, in a dose-dependent manner, in the MDM2 amplified SJSA human sarcoma cell line [344].

NSC 333003 (Fig. 40) has been identified from the NCI Diversity Set as an inhibitor of the MDM2-p53 protein–protein interaction by *in silico* docking (virtual screening). *E*- and *Z*-stereoisomers of NSC 333003 show near identical docking poses with the benzene ring and pyridyl ring loosely interchangeable suggesting that intramolecular hydrogen bonding does not occur [345].

Peptide inhibitors of MDM2 and MDMX have also been investigated in detail although probably not suitable for CNS tumors they have an important role in peripheral applications and in guiding the design of small-molecule inhibitors. For example, screening of a duodecimal peptide phage library against site-specifically biotinylated p53-binding domains of human MDM2 and MDMX identified several peptide inhibitors of the p53-MDM2/MDMX interactions with low nanomolar affinities. X-ray data are helping provide the structural basis for the design of both high-affinity peptide and small-molecule inhibitors of the p53 interactions with MDM2 and MDMX [346]. Current models of p53 regulation by MDM2 and MDMX together with rationales for the design of MDMX-specific therapeutics based on structural and biological functions have been reviewed [347].

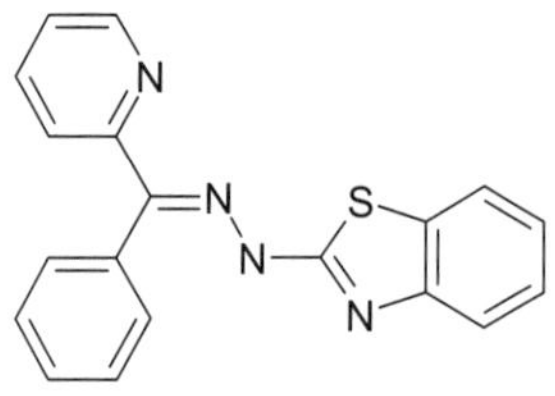

Finally another compound, ABT-737 (Fig. 41) targeting another α-helix binding groove interaction, that of Bcl-2 has also been investigated. Glioblastomas are characterized by high expression levels of antiapoptotic Bcl-2 family proteins. ABT-737 treatment released the proapoptotic Bax protein from its binding partner Bcl-2 and potently induced apoptotic cell death in glioblastoma cells *in vitro* and *in vivo*. The local administration of ABT-737 prolonged survival in an intracranial glioma xenograft model. Moreover, ABT-737 potentiated the cytotoxicity of the chemotherapeutic drugs, vincristine and etoposide, and of the death ligand TRAIL. ABT-737 was less effective at inhibition of proliferation and induction of apoptosis in glioma stem cells than in nonstem cell-like glioma cells [348].

16.19. Phosphatase and Tensin Homolog (PTEN)

PTEN is a tumor suppressor that antagonizes the PI3K/Akt/mTOR pathway by functioning as a lipid phosphatase. PTEN thus antagonizes PI3K activity and regulates numerous malignant cellular processes such as angiogenesis, motility, invasiveness, survival, and proliferation. In addition to lipid phosphatase activity, PTEN also dephosphorylates protein substrates and interacts with other key regulatory molecules. PTEN is frequently mutated in a range of cancers, and compromised PTEN function may contribute to gliomagenesis through disrupted regulation of proliferation, migration, invasion, angiogenesis, stem cell self-renewal, and regulation of other tumor suppressor pathways such as p53 [349]. Loss of PTEN function and subsequent activation of receptor tyrosine kinases such as EGFR, c-Met [275], PDGFR, and VEGFR are frequent molecular dysfunctions associated with glioma malignancy. PTEN can to counteract PI3K activation, as well as kinase activation of the MAPK pathway and kinase-

Figure 40. NSC-333003.

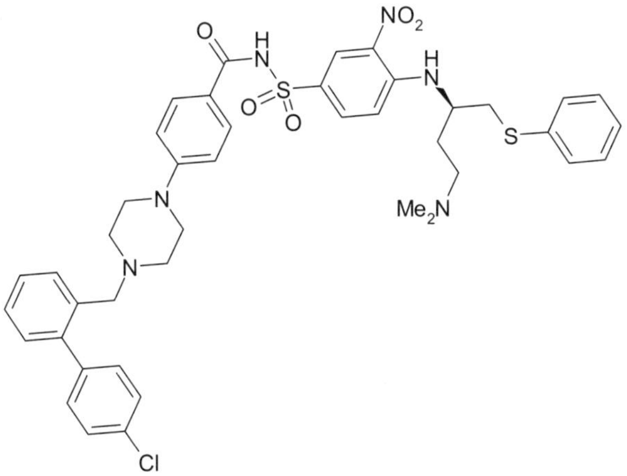

Figure 41. ABT-737.

dependent gene-expression regulation [258]. Deregulation of PI3K signaling pathways resulting from genetic alterations in the PTEN tumor suppressor gene at the level of loss of heterozygosity (LOH), mutation and methylation have been identified in at least 60% of glioblastoma. Loss of PTEN function by mutation or LOH correlates with poor survival in both anaplastic astrocytoma and glioblastoma [350].

Clinical findings in high-grade glioma suggest that PTEN gene alterations are associated with poor prognosis and may influence response to specific therapies. Thus, targeting PTEN and PTEN signaling may offer opportunities for combating not only cancer but also neurocognitive disorders [349,351].

PTEN also plays a role in tumors associated with aberrant developmental signaling pathways, for example, Sonic Hedgehog (Shh) and patched (Ptch1) expression levels are significantly higher in PTEN-expressing tumors than in PTEN-deficient tumors. Hyperactive Shh-Gli signaling in PTEN-coexpressing human GBMs is associated with reduced survival time [352].

16.20. Hedgehog Signaling Pathway

Hedgehog signaling plays an important role in embryonic development and adult tissue homeostasis and is one of several developmental pathways that are involved in deregulated signaling in stem cells resulting in tumorigenesis. Aberrant activation of the Hedgehog (Hh) pathway has been associated with numerous malignancies including Gorlin syndrome and basal cell carcinoma itself, medulloblastoma, prostate, pancreatic, and breast cancers [353]. *In vivo* evidence suggests the antagonism of aberrant Hedgehog signaling may provide a route to mechanism-based anticancer therapies [290]. Recent developments in targeting cell-surface proteins and intracellular targets from the hedgehog pathway have been reviewed [354]. Hedgehog signaling is triggered via a series of transmembrane receptors Patched (Ptch1) and Smoothened (Smo) of which the 7-TM Smo is the most investigated. Indeed, nearly all known Hedgehog pathway antagonists to date target the Smoothened receptor. Targeting tumors and especially medulloblastoma via inhibitors of the Sonic Hedgehog pathway is thus usually pursued via antagonists of the Smoothened receptor although some agents target downstream intervention points [355].

Medulloblastoma is the most common malignant pediatric brain tumor and multiple signaling pathways have been associated with medulloblastoma formation and growth. These include the developmental pathways Hedgehog, Notch, and Wnt as well as the receptor tyrosine kinase c-Met, erbB2, IGF-R, and TrkC and the oncoprotein Myc [140]. The Sonic Hedgehog signaling pathway seems to play an important role in the pathology of

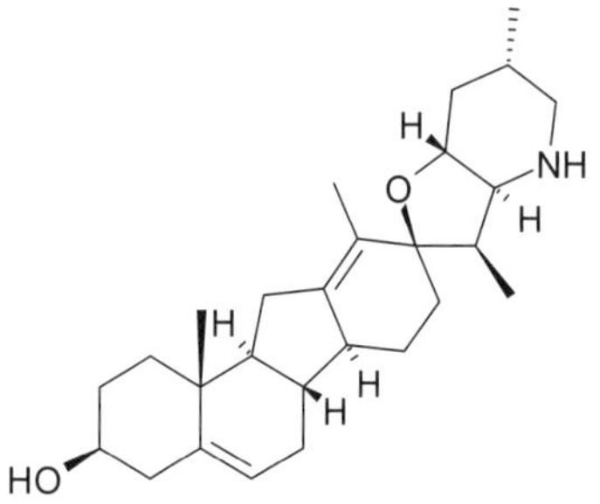

Figure 42. Cyclopamine.

this disease in both sporadic and some familial medulloblastomas. Small-molecule inhibitors of this pathway can regress tumors in transgenic mouse models of medulloblastoma [356].

The Hedgehog pathway was initially studied in preclinical research using cyclopamine a plant Veratrum alkaloid (Fig. 42) [357]. Cyclopamine showed a strong inhibition of systemic pancreatic metastases in spontaneously metastatic orthotopic xenograft models. Cyclopamine, however, has suboptimal pharmacokinetics for clinical studies.

A series of orally bioavailable cyclopamine analogs such as, IPI-269609, have been tested using *in vitro* and *in vivo* model systems. Single-agent IPI-269609 inhibited systemic metastases in orthotopic pancreatic xenografts [358].

IPI-269609 has a greater acid stability and better aqueous solubility compared to cyclopamine that has been optimized via A-ring modifications to improved potency and/or solubility. IPI-926 (Fig. 43) a novel semisynthetic cyclopamine analog was developed with substantially improved pharmaceutical properties and potency and improved pharmacokinetic profile. Complete tumor regression was observed in a Hedgehog-dependent medulloblastoma allograft model following oral administration IPI-926 [359].

Many Smo antagonists were identified via phenotypic screening in cell-based assays including a series of orally bioavailable 1-amino-4-benzylphthalazines (Fig. 44). Key compounds from this series displayed good pharmacokinetic profiles and also afforded tumor regression in a genetic mouse model of medulloblastoma [360].

Ortho-biphenyl carboxamides (Fig. 45) originally prepared as inhibitors of microsomal triglyceride transfer protein (MTP) have also been identified as novel inhibitors of the Hedgehog signaling pathway via antagonism of the Smoothened receptor [361].

Recently, the small-molecule SANT-2 was identified as a potent antagonist of Hh-signaling pathway via interference with the expression of the Hedgehog target gene Gli1. The derivative TC-132 (Fig. 46) that was slightly more potent than the parent compound SANT-2. However, although Gli expression levels indicated a 16-fold higher inhibiting activity than observed for cyclopamine, none of the tested compounds were able to induce the cyclopamine-specific phenotype [362].

Biochemical characterization of allosteric interactions of agonists and antagonists for Smo combined with functional data is providing a better understanding of small-molecule interactions with Smo and their influence on the Hedgehog pathway. Binding of two radioligands, [^{3}H]SAG-1.3 (agonist) and [^{3}H]cyclopamine (antagonist), was characterized using the human Smo receptor expressed in

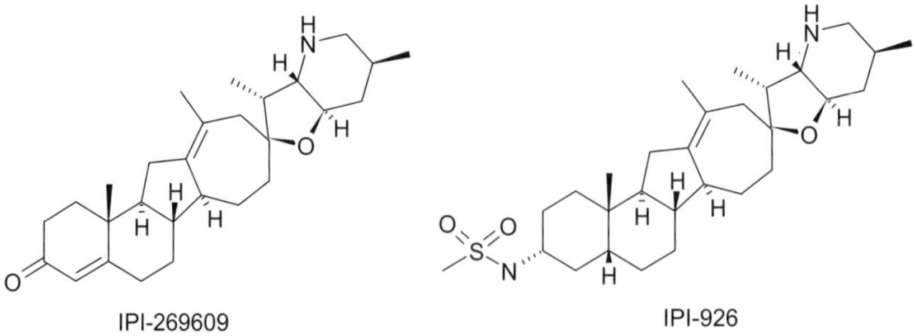

Figure 43. Cyclopamine derivatives - IPI-269609 and IPI-926.

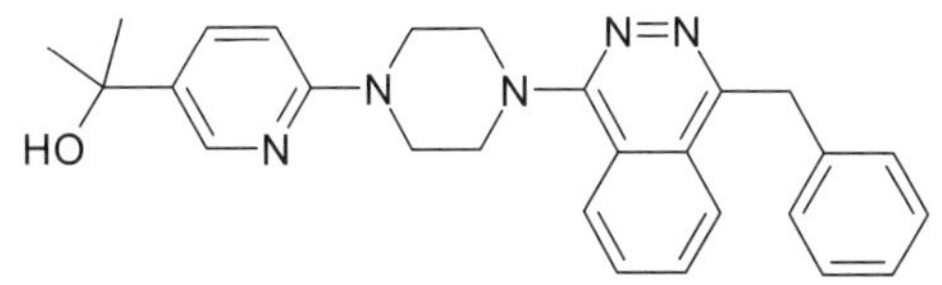

Figure 44. Amino-benzylphthalazine smoothened antagonists.

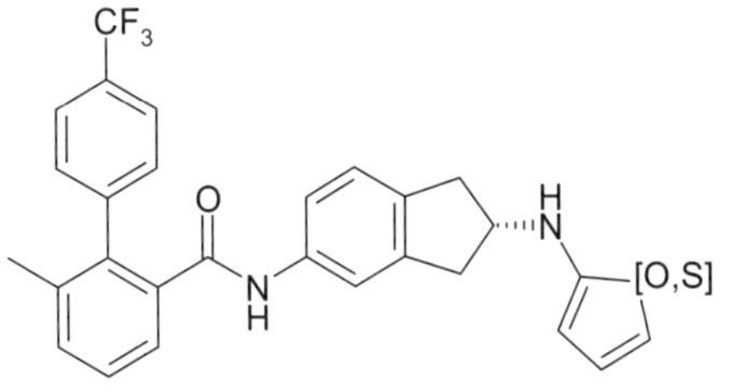

Figure 45. Ortho-biphenyl carboxamides smoothened antagonists.

HEK293F membranes. Full displacement of [^{3}H]cyclopamine was observed for all Smo ligands examined. SANT-1 (Fig. 47), an antagonist, did not, however, fully inhibit the binding of the agonist [^{3}H]SAG-1.3. Detailed radioligand binding analysis revealed that the antagonists, SANT-1 and SANT-2, modulated the Smo receptor via allosteric modulation [363].

Using Gli1-dependent GFP transgenic zebrafish and *in vitro* biochemical assays, two potent Smo inhibitors, SANT74 and 75 (Smoothened antagonist 74 and 75) (Fig. 48) have been identified by screening a small-molecule library based around the scaffold of Smo agonist SAG. Although structurally related to SAG, SANT74 and 75 exert opposite effects to SAG on Smo activity by regulating protein conformation.

This study represents the first demonstration of conformational regulation of Smo by small-molecule analogs, and could permit temporal control of Smo activity potentially useful for studying Hedgehog pathobiology [364].

Several companies are progressing Smo antagonist through clinical trials including Curis-Genentech whose antagonist GDC-0499 (Fig. 49) is currently in Phase II studies for a range of tumor indications including

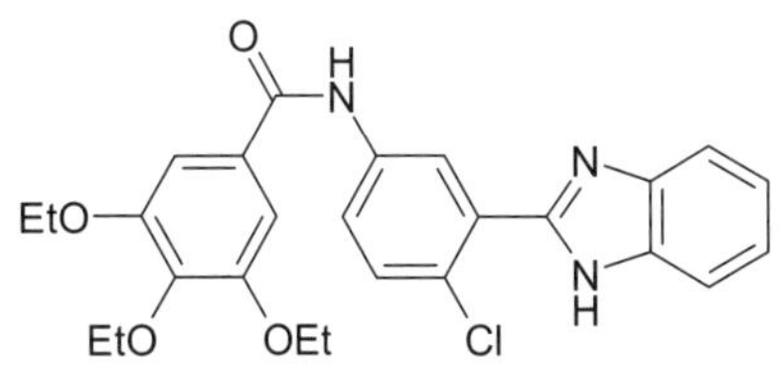

SANT-2

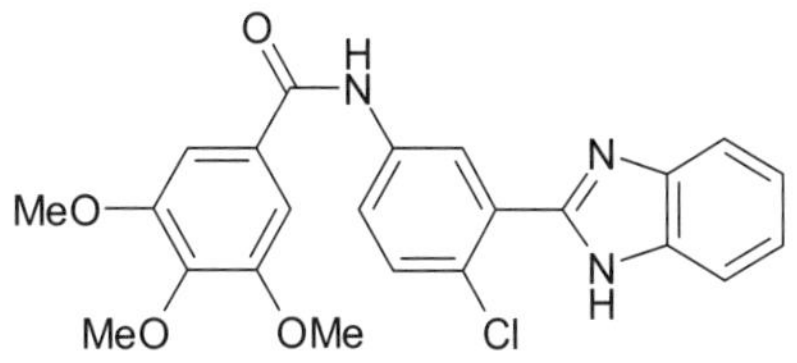

TC132

Figure 46. Antagonists of hedgehog signaling.

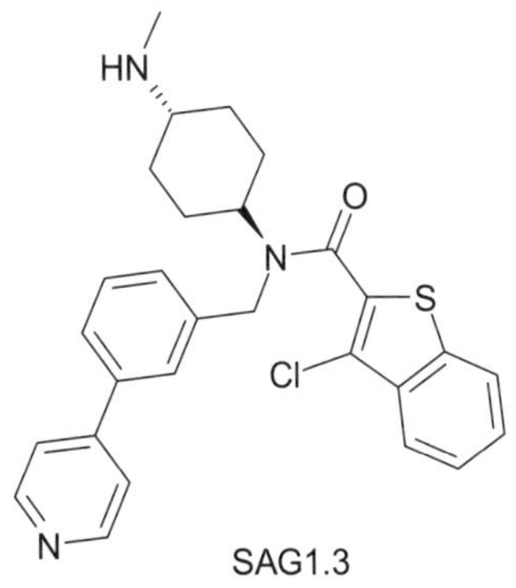

SAG1.3

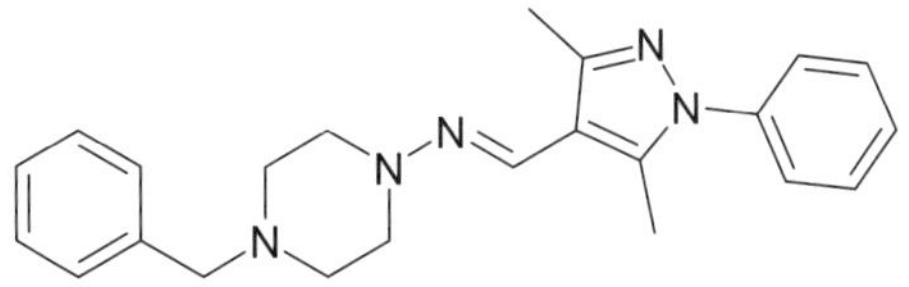

SANT-1

Figure 47. SAG1.3 Smo agonist and SANT-1 an allosteric Smo antagonist.

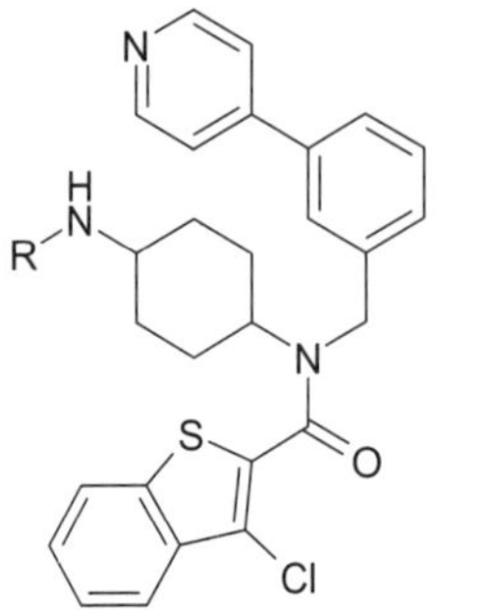

SANT74   R = CH₂CH=CH₂
SANT75   R = CH₂CH₂CH₃

Figure 48. Smoothened antagonists (SANT) 74 and 75.

Figure 50. LDE-225.

basal cell carcinoma, cancer, colorectal tumor, and medulloblastoma.

GDC-0449 has shown promising antitumor responses in early clinical studies of cancers driven by mutations in this pathway. As with most molecular-targeted approach tumor resistance occurs in some patients. Investigations into the mechanism of resistance in a medulloblastoma patient identified a mutation at a conserved aspartic acid residue of the Smo receptor. Interestingly, although the Asp-mutation had no effect on Smo signaling, the same amino acid mutation was also observed in a GDC-0449-resistant mouse model of medulloblastoma. Both *in vitro* and *in vivo* therefore this mutation disrupted the ability of GDC-0449 to bind SMO and suppress this pathway [365].

Novartis is progressing the Smo antagonist LDE-225 (Fig. 50) in Phase I and Phase II (for skin cancer) clinical trials.

Several other companies have compounds in various stages of preclinical research.

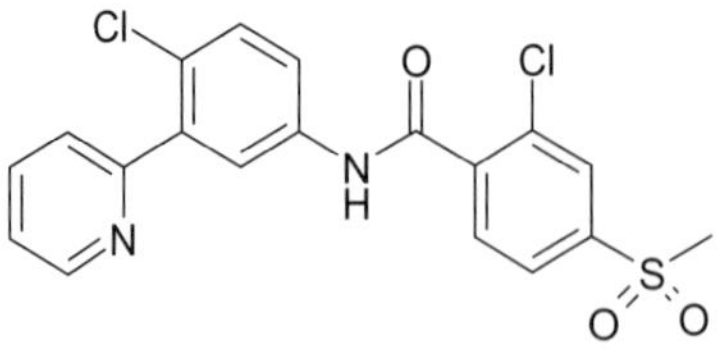

Figure 49. GDC-0449.

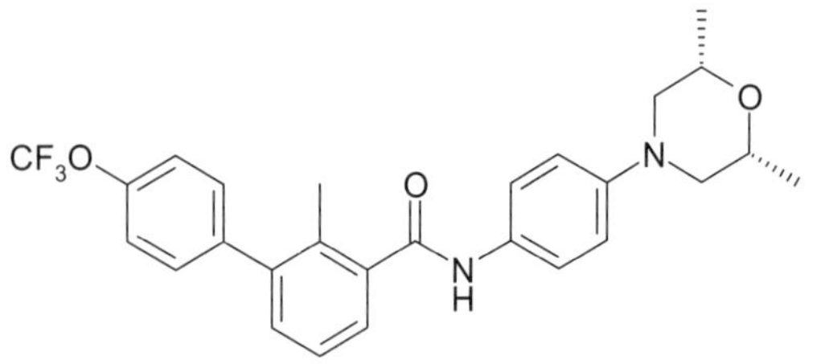

Although small molecules that suppress downstream effectors could comprehensively suppress Hedgehog pathway-dependent tumors there are limited reports of such compounds in the literature. Recently, four Hedgehog pathway antagonists were identified including two (HPI-2 and HPI-3) (Fig. 51) that can inhibit Hedgehog target gene expression induced by overexpression of the Gli transcription factors. Each inhibitor has a unique mechanism of action, and the phenotypes modulated include Gli processing, Gli activation, and primary cilia formation [366].

As with many single-agent targeted therapies combination with appropriate additional agents can provide advantages in targeting distinct cellular signaling mechanisms that potentially improve efficacy and reduce toxicity. Histone deacetylases (HDACs) can suppress cell growth via epigenetic modulation. Aberrant Hedgehog signaling promotes the initiation and progression of a range of neoplasia thus the HDAC and Hedgehog pathway potentially act cooperatively to regulate cellular proliferation. A combination of the HDAC inhibitor SAHA and the Smoothened antagonist SANT-1 supra-additively suppressed cellular proliferation and colony formation in pancreatic tumors via enhanced induction of apoptotic cell death, cell cycle arrest in G(0)/G(1) phase, and ductal epithelial differentiation [367].

Hedgehog signaling activates the Gli transcription factor family members. Since tyrosine kinase signaling potentiates Gli activity through PI3K-AKT-mediated GSK3 inactivation or RAS-STIL1-mediated SUFU inactivation, and GPCR signaling to Gs also repress Gli activity through adenylate cyclase-mediated PKA activation suggesting the com-

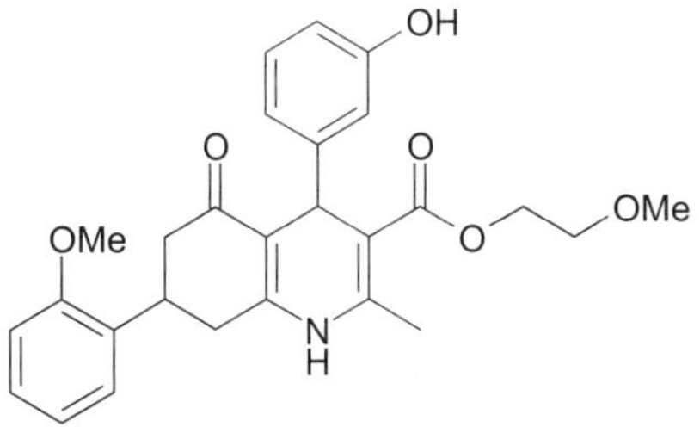

Figure 51. Hedgehog pathway antagonists.

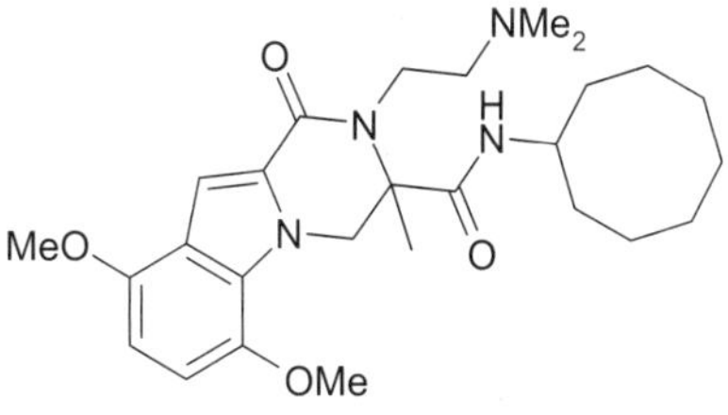

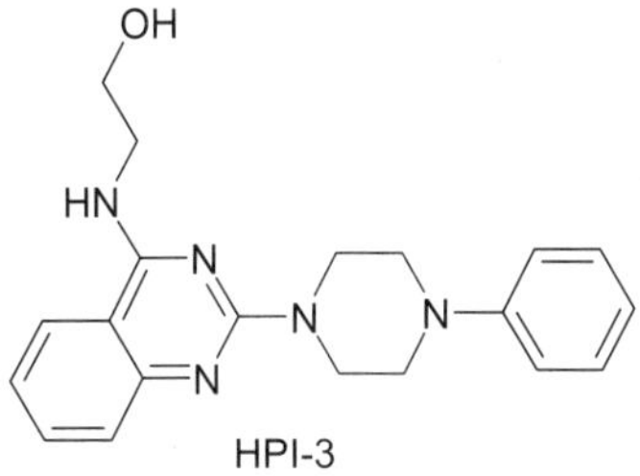

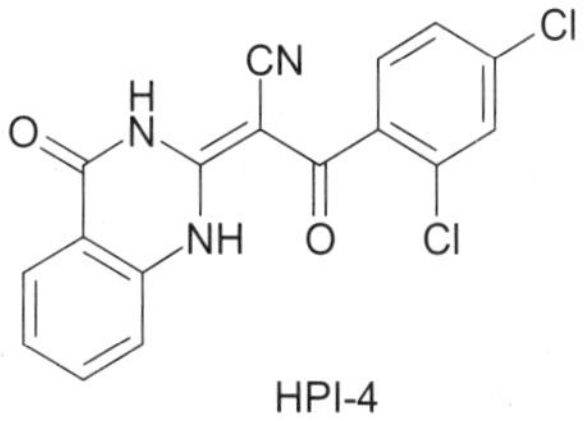

bination of Smo antagonists with Gli inhibitors, selective tyrosine kinase inhibitors, or GPCR modulators, and/or irradiation for more effective cancer therapies [368].

16.21. Wnt Signaling Pathway

Aberrant Wnt signaling has been implicated in a range of CNS tumors including PNET [77], medulloblastoma [78], and malignant gliomas [79]. The balance between WNT-FGF-Notch and BMP-Hedgehog signaling networks is important for the maintenance of homeostasis among stem and progenitor cells. Disruption of the stem cell signaling network results in pathological conditions, such as congenital diseases and cancer [369]. Importantly, Wnt is cross-linked to several other oncogenic pathways including the Hedgehog [370], and TGF-β/BMP [371] pathways. The nature of such signaling crosstalk is overwhelmingly complex and highly context-dependent moreover the complexity of signaling within a single pathway/network is also highly context dependent. For example, Wnt5a operates as either a tumor suppressor or a tumor stimulator, depending on the tumor type. In GBM, the overexpression of Wnt5a increases proliferation. Downregulation of Wnt5a expression as the result of RNA interference reduces proliferation *in vitro* and reduced tumorigenicity *in vivo* [372]. Wnt7a is involved in both neuronal differentiation and tumor suppression. Wnt7A overexpression decreased Wnt signaling in GBM and colon cancer cells. Its role, however, depends on which Frizzled (Fzd) receptor it binds to thus; Wnt7a binding to Fzd5 activates canonical Wnt signaling and increases cellular proliferation while Wnt7a signaling mediated by Fzd10 induces a noncanonical c-Jun kinase-responsive pathway [373].

Classical GPCR signaling is mediated by heterotrimeric guanine nucleotide-binding proteins (G proteins), GPCR kinases, and β-arrestins. Recently, it has become increasingly apparent that classical and atypical 7TM receptors such as the Frizzled receptors in Wnt signaling networks share both these factors as well as scaffolding proteins. This sharing of signaling components by agonists that bind classical and those that bind atypical 7TMRs establishes the possibility of extensive crosstalk between these receptor classes [374].

16.22. GSK3β Inhibitors

Cancers are driven by a population of cells with the stem cell properties of self-renewal and unlimited growth. As a subpopulation

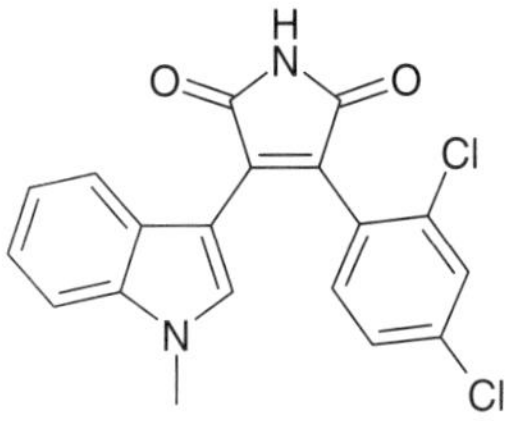

Figure 52. SB216673.

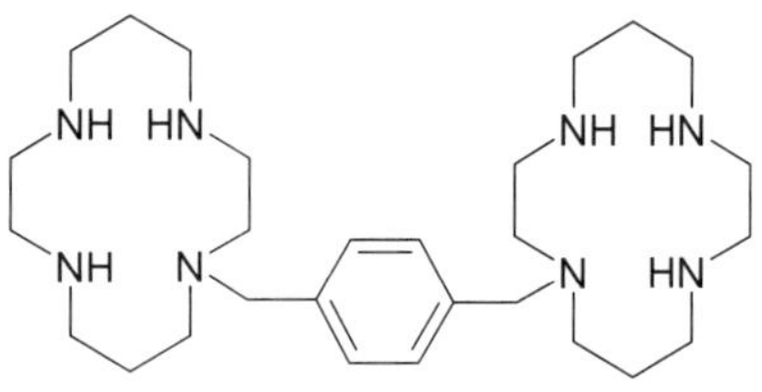

Figure 53. AMD3100.

within the tumor mass, these cells are believed to constitute a tumor cell reservoir. Aberrant Wnt signaling is one of the pathways contributing to the deregulation of stem cells renewal. Within the Wnt pathway the kinase glycogen synthase kinase 3β (GSK3 β) is one of the most studied intervention points. Interference with GSK3 β activity by siRNA, the specific inhibitor SB216763 (Fig. 52), or lithium chloride induced tumor cell differentiation. In addition, tumor cell apoptosis was enhanced, the formation of neurospheres was impaired, and clonogenicity reduced in a dose-dependent manner [375].

16.23. Chemokine Receptors

16.23.1. CXCR4 CXCR4 is highly expressed in a variety of brain tumors and activation of CXCR4 by its ligand CXCL12 (SDF-1) promotes survival, proliferation, and migration of both medulloblastoma- and glioblastoma–derived tumor cells [376]. CXCR4 is overexpressed in primary glioblastoma progenitor cells versus corresponding differentiated tumor cells and following administration of CXCL12, stimulates a specific and significant proliferative response in progenitors but not differentiated tumor cells, suggesting an important role for the CXCR4 signaling mechanism in glioblastoma CSC proliferation [377]. CXCR4 signaling is also important for the homing of neoplastic cells from the primary site to the target and plays a prominent role in peripheral metastasis to the brain. Brain metastases occur in about 25% of patients who die of cancer. The most common sources of brain metastases in adults are lung [378], breast [379–381], kidney, melanoma [378], and less frequently colorectal cancer [382].

Systemic administration of CXCR4 antagonists such as AMD3100 (Fig. 53) inhibits growth of intracranial glioblastoma and medulloblastoma xenografts by increasing apoptosis and decreasing the proliferation of tumor cells [383]. Moreover, combination treatment with BCNU followed by AMD3100 results in synergistic antitumor efficacy both *in vitro* and in an orthotopic GBM model. Treatment with subtherapeutic doses of BCNU in combination with AMD3100 resulted in tumor regression reflecting both increased apoptosis and decreased proliferation following combination therapy. Inhibition of CXCR4 thus synergizes with conventional cytotoxic therapies in a clinically relevant combinatorial strategy [376].

CXCR4 expression levels correlate with the degree of malignancy of human gliomas and may contribute to their rapid growth [384]. Expression of SDF-1 and its two receptors CXCR4 and CXCR7 has also been shown in a cohort of brain metastases [385]. The alternative SDF-1 receptor, CXCR7, has been shown to be expressed by both tumor and endothelial cells (both within the tumor and in the adjacent brain tissue) [385]. CXCR7 binds chemokines CXCL11 (I-TAC) and CXCL12 (SDF-1) but does not act as a classical chemoattractant receptor. It is unclear whether tumor progression is driven by chemokine homodimers or heterodimers. Using CCX771, a novel small molecule with high affinity and selectivity for CXCR7, it was shown that CXCR7 plays an essential role in the CXCL12/CXCR4-mediated transendothelial migration of tumor cells. Importantly, although CXCL11 is unable to stimulate directly the migration of these cells, it acts as a potent antagonist of their CXCL12-induced transendothelial migration (TEM). Furthermore, even though TEM is driven by CXCR4, the CXCR7 ligand CCX771 is substantially more potent at inhibiting it than the CXCR4 antagonist AMD3100 [386], suggesting a role for heterodimers [387]. The full role of CXCR7 and any potential for cooperativity between CXCL12 receptor subtypes is largely being determined in non-CNS tumor types and other diseases.

However, the multiple homeostatic functions of CXCR4 may preclude global inhibition as a therapeutic strategy. CXCR4 signaling may differ in normal and transformed cells. CXCR4 mediates unique signals in cancer cells as a consequence of abnormal counter-regulation resulting in novel biological responses. Targeting abnormal CXCR4 signaling might provide an antitumor effect without disturbing normal CXCR4 functions [388].

For a review of chemokine receptors as targets for cancer therapy, see Ref. [389].

16.24. PDE4

PDE4A is widely expressed in a range CNS tumors and promotes tumor growth in medulloblastoma, glioblastoma, oligodendroglioma, ependymoma, and meningioma. Overexpression of PDE4A1 results in increased growth rates in both medulloblastoma and glioblastoma cells. Treatment with the PDE4 inhibitor Rolipram (Fig. 54) in combination with first-line glioma therapy (temozolomide and conformal radiation therapy) enhanced the survival of mice in an orthotopic glioblastoma model [390]. Bioluminescence imaging studies again in an orthotopic model showed that whereas temozolomide and radiation therapy arrested intracranial tumor growth, the addition of Rolipram to this regimen resulted in tumor regression [391].

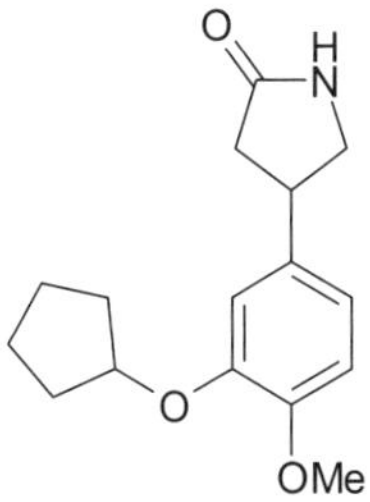

16.25. Nuclear Receptors (Estrogen Receptor, PPAR, RXR)

For a review on the use of nuclear receptor, modulators in approaches to the treatment of pituitary tumors including estrogen receptor, PPAR, and adrenergic receptor modulators, see Ref. [178].

16.26. Estrogen Receptor

Medulloblastoma arises from cerebellar granule cell-like precursors. *In vivo* and *in vitro* experiments using selective agonists revealed that the growth-stimulating actions of estradiol were mediated by estrogen receptor-β (ER-β) and not ER-α dependent mechanisms. Inhibition of ER-mediated signaling with the ER antagonist Faslodex (ICI182,780) (Fig. 55) blocked estrogen-mediated effects in both cell culture and xenograft models of human medulloblastoma [392,393].

16.27. Peroxisome Proliferator-Activated Receptors (PPAR)

Peroxisome proliferator-activated receptor γ (PPAR-γ) has been identified in transformed neural cells and PPAR-γ agonists decrease cell proliferation, stimulate apoptosis and induce morphological changes and expression of markers typical of a more differentiated phenotype in glioblastoma and astrocytoma cell lines. Experimental data from *in vitro* and *in vivo* models indicate that PPAR ligand activation regulates cell differentiation and induces cell growth arrest and apoptosis in a variety of cancer types. PPAR-γ agonists may interfere with glioblastoma growth and

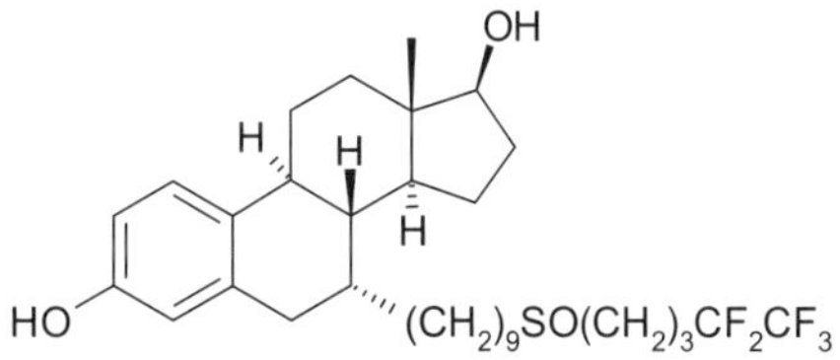

Figure 54. Rolipram.

Figure 55. Faslodex, Fulvestrant.

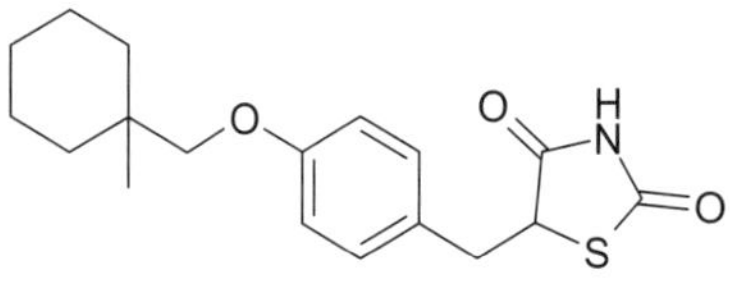

Figure 56. Ciglitazone.

malignancy and might be taken into account as novel antitumor drugs [394].

PPARs bind to DNA as heterodimers with retinoid X receptors (RXRs), therefore suggesting a potential for a combined effect of PPAR and RXR ligands. Indeed retinoid 6-OH-11-O-hydroxyphenantrene (IIF) potentiates the antiproliferative and anti-invasive effect of the PPAR-γ ligand ciglitazone (CGZ) (Fig. 56) in glioblastoma U87MG and melanoma G361 cells. The addition of IIF to ciglitazone resulted in a synergistic decrease of metalloproteinase expression (MMP2 and MMP9) together enhanced PPAR-γ expression. Combined treatment also markedly reduced proliferation and induced apoptosis. These *in vitro* findings were substantiated in a murine glioma model, where oral administration of PGZ and IIF resulted in significantly reduced tumor volume and proliferation [395,396].

Rosiglitazone (RGZ) (Fig. 57) exerts a suppressive effect on the cell viability in nonfunctioning pituitary adenomas. However, the lack of correlation between PPAR-γ expression and antitumoral effect of rostiglitazone suggests that the mechanism of action of this compound is independent of PPAR-γ expression levels [176].

PPAR-γ agonists, including thiazolidinediones (TDZs such as ciglitazone) and nonthiazolidinediones, block the motility and invasiveness of glioma cells and other highly migratory tumor entities [397]. The mechanism(s) by which PPAR-γ activators mediate their antimigratory and anti-invasive properties, however, remain elusive. Moreover, both troglitazone and a PPAR-γ inactive derivative Δ2-troglitazone (Δ2-TRO) potently inhibit glioma progression *in vivo* in a PPAR-γ independent manner [398].

Considering the promigratory properties of GBM, the transforming growth factor-β (TGF-β) signaling pathway has become a major therapeutic target. Analyses of resected glioma tissues revealed an intriguing correlation between tumor grade and the expression of TGF-β (1–3) as well as their receptors I and II. The effects of PPAR-γ agonists on glioma proliferation, migration, and brain invasion have been investigated using troglitazone (Fig. 58).

Troglitazone was shown to be a potent inhibitor of both glioma cell migration and brain invasion, but in a PPAR-γ-independent manner. The antimigratory property of troglitazone was in concordance with the transcriptional repression of TGF-β (1–3) and their receptors I and II and associated with reduced TGF-β release. Due to its capacity to counteract TGF-β release and glioma cell motility and invasiveness already at low micromolar doses, troglitazone represents a promising drug for adjuvant therapy of glioma and other highly migratory tumor entities [399].

Natural as well as synthetic PPAR agonists exhibit profound antineoplastic as well as redifferentiation effects in CNS tumors. The molecular understanding of the underlying mechanisms is still emerging, with partially controversial findings reported by a number of studies dealing with the influence of PPARs on treatment of tumor cells *in vitro*. Studies examining the effects of these drugs *in vivo* are just beginning to emerge. However, the ago-

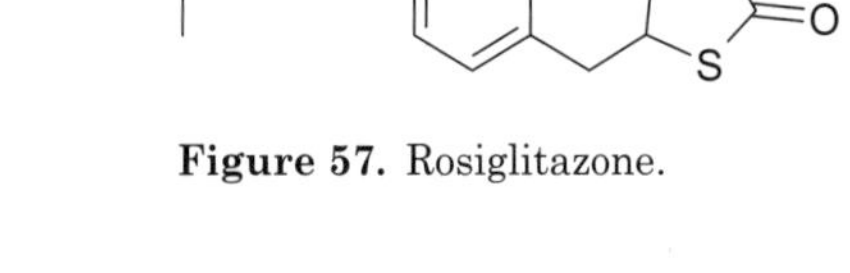

Figure 57. Rosiglitazone.

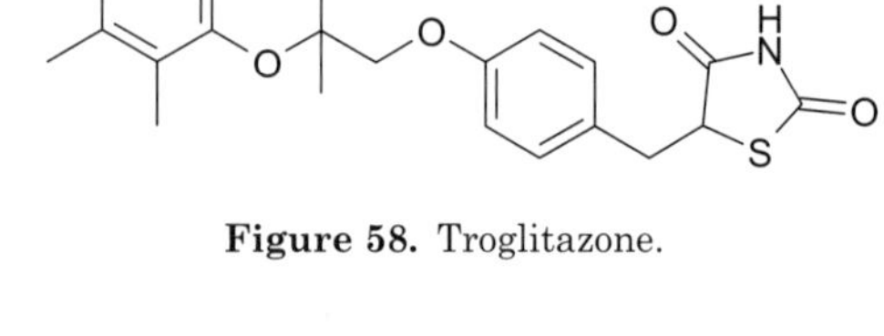

Figure 58. Troglitazone.

nists of PPARs, in particular the thiazolidinediones, seem to be promising candidates for new approaches in human CNS tumor therapy [400].

PPAR-γ agonists inhibited the proliferation and expansion of glioma and gliosphere cells in a dose-dependent manner. They also induced cell cycle arrest and apoptosis in association with the inhibition of EGF/bFGF signaling through Tyk2-Stat3 pathway and expression of PPAR-γ in gliosphere cells [401].

16.28. RXR

Although retinoids strongly inhibit proliferation and migration in primary cultures of human glioblastomas multiforme, RAR and RXR receptors have been little explored as potential targets for CNS tumors [402].

16.29. Transforming Growth Factor-β

Transforming growth factor (TGF)-βs and their family members, including BMPs, Nodal and activins [403], are implicated in the maintenance and differentiation of ES cells, somatic stem cells, and cancer stem cells including gliomas [73]. Transforming growth factor β (TGF-β) signaling plays an important role in the development of neural progenitor cells and in glioblastoma development from neural stem cells leading to cell growth, differentiation, and morphogenesis. These roles are, however, reversible and during pathological conditions the growth-inhibitory effect of TGF-β is replaced with its tumor promoting ability [404].

Glioblastomas are a rich source of immunosuppressants of which, transforming growth factor (TGF)-β is the most prominent. TGF-β not only interferes with multiple steps of afferent and efferent immune responses but also stimulates migration, invasion, and angiogenesis.

Several approaches have been adopted for antagonism of TGF-β including antisense strategies, inhibition of pro-TGF-β processing, scavenging TGF-β by decorin, or blocking TGF-β activity by TGF-β receptor 1 (TGF-β R1) kinase inhibitors such as SB-431542 (Fig. 59) [405]. SB-431542 is a small-molecule ATP-mimetic that inhibits several members of

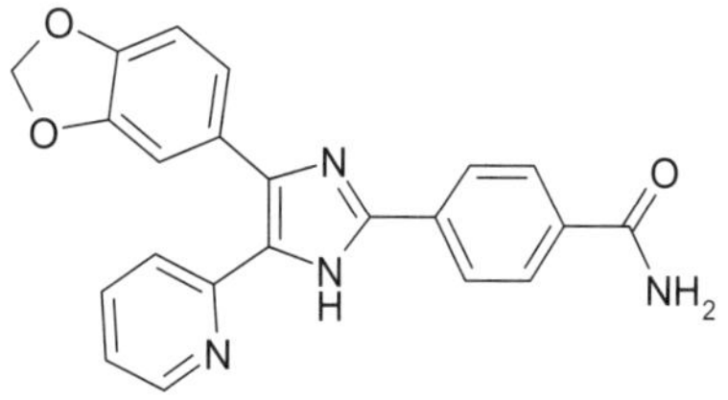

Figure 59. SB-431542.

the activin receptor-like kinase (ALK) family—specifically ALK5 (TGF-β type I receptor, TGF-βRI), ALK4 (activin type I receptor), and ALK7 (nodal type I receptor).

TGF-β-antagonistic treatment strategies could be particularly effective in conjunction with novel approaches of cellular immunotherapy and vaccination [406].

16.30. Bone Morphogenetic Proteins)

Like TGF-β, BMP show a dual role and have a proproliferative effect on early embryonic neural stem cells, and a prodifferentiative effect on postnatal neural stem cells.

The BMP-BMPR signaling system, plays a significant role in the activity of normal brain stem cells, but is also a key inhibitory regulator of GBM stem-like cells. BMPs, among which BMP4 elicits the strongest effect, activate their cognate receptors (BMPRs) and trigger a significant reduction in the stem-like precursors of glioblastomas (GBMs). Inhibition of the BMP receptors leads to a reduction in proliferation and increased expression of markers of neural differentiation [407,408]. However, BMPs have differing effects on different BTSC lines, either promoting or inhibiting an astrocytic-like differentiation. This latter effect is the result of epigenetic silencing of the BMP receptor 1B (BMPR1B) again highlighting the heterogeneity of brain tumor cells [409,410].

16.31. Heat Shock Protein Inhibitors

Studies into the role for HSP90α in CNS tumorigenesis, highlight the adaptive (or evolutionary) nature of tumor progenitor cells and their ability to evade treatment and migrate from hostile compartments. There are no re-

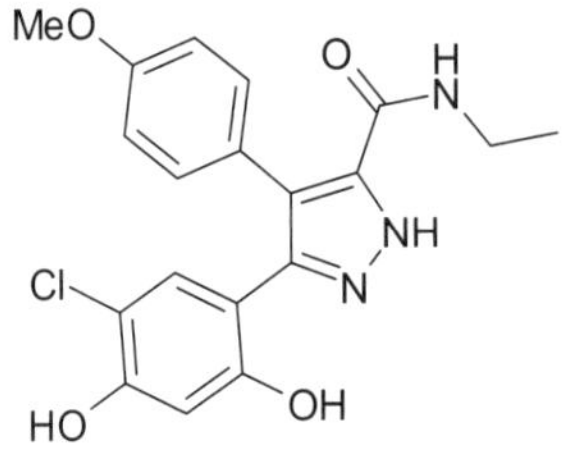

Figure 60. HSP90 inhibitors – VER-49009 and NVP-AUY922.

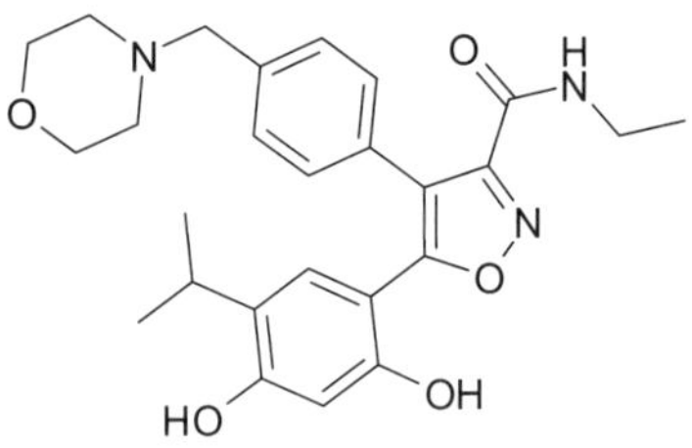

ports of HSP90α expression in normal human brain tissue; however, it is implicated in glioma tumorigenesis, providing a rationale for targeting HSP90α as a tumor target that is potentially not involved in normal brain function [411]. A role for HSP90 in CNS tumors was established with 17-allylamino-17-demethoxygeldanamycin (17-AAG, tanespimycin) that inhibits cell growth of human glioma cell lines and the growth of intracranial tumors. However, unlike many other compounds 17-AAG does not show synergistic effects with temozolomide in glioma models [412]. Synergistic effects have been shown, however, between 17-AAG and the PI3K kinase inhibitor LY294002 in promoting induced cell death, inhibiting proliferative activity and colony forming ability of human glioma cells [413]. These differences, perhaps, highlight not only the impact of combination therapies but also the need for a more fundamental revision to approaches to cancer treatment in understanding a priori which combinations may be most effective.

Tumor resistance may, however, remain an issue since acquired resistance to HSP90 inhibitors in glioblastoma has been demonstrate for 17-AAG but not, however, with the structurally unrelated HSP90 inhibitors, such as radicicol, the purine BIIB021, and the resorcinylic pyrazole/isoxazole amide compounds VER-49009, VER-50589, and NVP-AUY922 (Fig. 60) [414]. In some cases, this may be related to NAD(P)H/quinone oxidoreductase 1 (NQO1) expression/activity low NQO1 activity is correlated to resistance to 17-AAG various tumor types.

Whether such resistance can be overcome with novel HSP90 inhibitors is to be established. NVP-AUY922 is also effective under hypoxic conditions and depletes HIF-1α in glioblastoma xenografts, illustrating the potential of HSP90 inhibitors to also target angiogenic phenotypes [415]. For a review of the medicinal chemistry of HSP90 inhibitors and their investigation in various tumor settings including glioblastoma, see Ref. [416].

16.32. Other Small Molecules

Atrasentan (Fig. 61) an oral selective endothelin-A receptor antagonist inhibits cell proliferation and interfere with angiogenesis during glioma growth [417].

16.33. Therapeutic Vaccines for CNS Tumors

Despite the many overall advances in knowledge of cancer pathobiology and progress in therapeutic strategies, most CNS malignancies are still difficult to treat clinically. The

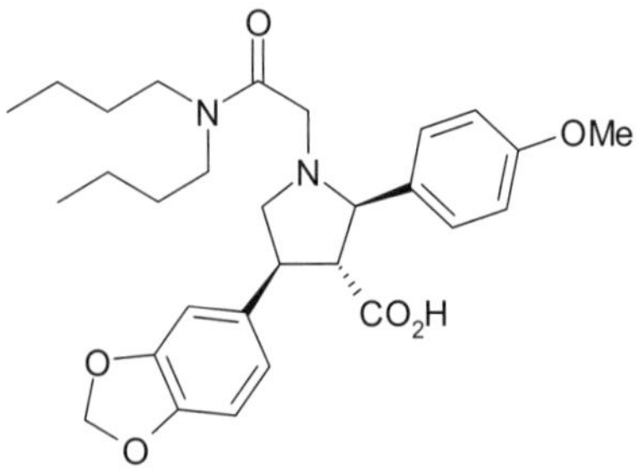

Figure 61. Altrasentan.

current combinations of aggressive surgical resection, radiation, and chemotherapy do not significantly improve long-term patient survival and the prognosis for patients with malignant primary brain tumors remains very poor. This can at various levels be attributed to the infiltrative nature of the disease, the CNS microenvironment, the ability of tumors to evade treatment, and resistance of tumors to chemotherapy. Moreover, the nonspecific nature of conventional therapy also plays a role and small-molecule cytotoxic treatments are often limited by damage to surrounding normal brain (in particular) and systemic tissues. There is thus an urgent need for the development of therapeutic strategies that more precisely target tumor cells. Among the therapeutic strategies under investigation immunotherapy could provide options for effective and focused therapy with the possibility of minimizing many of the side effects of current approaches [418]. Malignant gliomas have not, in the past, generally been considered an appropriate target for immunotherapy in part since they produce an immunosuppressive environment through multiple mechanisms. Setting up immunotherapy for an inherent immunosuppressive tumor located in the brain that is traditionally considered to be an immune-privileged environment requires detailed integration of both tumor immunology and neuro-oncology [419]. The major potential advantage of immunotherapy is the specific effect on tumor cells while minimizing side effects. Both autologous dendritic cells (DCs)-based immunotherapy and peptide-based approaches are being pursued. DCs are the most potent antigen-presenting cells (APCs). DCs-based immunotherapy may improve tumor response by increasing the survival rate and time. Malignant gliomas have heterogeneity of tissue associated antigens (TAAs). To find universal common antigens through different kinds of tumor culture may be the essential issue for tumor vaccine development in the future [420].

Reviews on the preclinical and clinical progress of in particular peptide-based and dendritic cell-based vaccination therapies for CNS malignancies, highlight progress to date and some of the challenges for developing effective vaccination strategies, such as abnormal immune molecules on glioma cells and abnormal lymphocyte populations within a glioma [421–423]. The success of vaccine-based approaches is thus dependent on understanding the mechanisms of immune regulation within the CNS, as well as in countering the broad defects in host cell-mediated immunity that malignant gliomas are known to elicit [418]. Both dendritic cell- and peptide-based immunotherapy strategies appear promising as an approach induce an antitumor immune response and increase patient survival without major side effects [424]. Given the heterogeneity of gliomas, it might also be necessary to evaluate the molecular genetic abnormalities in individual patient tumors and design novel immunotherapeutic strategies based on the pharmacogenomic findings to achieve effective therapy [425]. Overall, DC vaccines approaches to glioma are well tolerated with few documented side effects and median overall survival times are prolonged. However, several factors have limited the efficacy of DC vaccines [426]. Dendritic cell vaccines for cancer stem cells have also been investigated as an approach to glioblastoma. In particular DC vaccination against neurospheres restrained the growth of a highly infiltrating and aggressive model of glioma [427].

Although the brain has been long considered an immunologically privileged organ, there is increased awareness of and appreciation for the complex interplay between the nervous system and the immune system in the setting of many disease states, including CNS neoplasms [428]. Conventional therapies for glioblastoma multiforme fail to target tumor cells exclusively and their efficacy is ultimately limited by nonspecific toxicity or tumor resistance. Immunologic targeting of tumor-specific gene mutations may allow more precise eradication of neoplastic cells. For example, EGFRvIII is a consistent and tumor-specific mutation widely expressed in glioblastomas clinical trials on DC-based vaccines targeting the EGFRvIII antigen have shown that the EGFRvIII mutation is a safe and immunogenic tumor-specific target for immunotherapy [429]. The recurrence of EGFRvIII-negative tumors, however, highlights the need for targeting a broader repertoire of tumor-specific antigens [430].

17. OBSTACLES TO EFFICIENT THERAPIES

Molecular-targeted therapies tend to be orientated toward (based on) tumor phenotypes (such as angiogenesis or proliferation) rather than on tumor type (GBM, medulloblastoma, meningioma, etc.). Many of the current approaches are not specific to brain tumors and thus difficulty in crossing the blood–brain barrier is a recurrent issue. The major problems, however, are

- tumor heterogeneity (e.g., primary versus secondary GBM and the differences in their transcriptional profiles),
- patient heterogeneity (e.g., Glioma patients MGMT status and resistance to temozolomide), and
- lack of understanding at a molecular level of the genes driving tumor progression and hence a need for combination therapies.

For many compounds, therapeutic efficacy is limited not only by tumor evasion or other aspects of tumorigenesis but also because of difficulties crossing the BBB and obtaining therapeutic doses in the tumor environment. This is being confronted in two ways by designing better compound characteristics and from the onset targeting CNS tumors and secondly by approaches to chemotherapy delivery across the BBB.

18. BLOOD–BRAIN BARRIER (DESIGNING BETTER COMPOUNDS)

A brief introduction to some of the concepts of compound design is included here as they will play an increasing important role in the development of sustainable CNS tumor therapies. Many other articles have been published on compound design and drug-like properties such as [431]. For reasons of space and focus within this review, the reader is encouraged to investigate further this well-documented area more directly. Emphasis here is on the unique characteristics of delivering therapeutic entities into the brain.

A wide variety of models have been developed to predict blood–brain barrier penetration, most have focused on predicting drug total concentrations and expressing this as a simple brain:blood (or plasma) ratio. Various *in vitro* and *in silico* strategies have thus been developed to model drug action as a function of BBB properties, with models of various levels of predictivity, reliability, and throughput [432]. Focusing on blood:brain ratios is a somewhat flawed approach since it fails to address the need to understand the relationship between accesses of free drug to the requisite site of action [433]. Blood–brain barrier permeability is therefore only one of the important determinants in achieving efficacious CNS drug concentrations and it should not be considered in isolation. Optimal CNS penetration requires the correct balance of permeability (not only BBB penetration), a low potential for active efflux and the appropriate physicochemical properties that allow for drug partitioning and distribution into brain tissue [434]. Establishing adequate drug coverage in the target cells is determined by both BBB permeability and brain tissue binding with low efflux compounds being more effective. Compound design focusing on reducing efflux and facilitating passive permeability facilitates rapid brain penetration. Drug transporter analysis tends to focus on the importance of P-glycoprotein (P-gp) in limiting BBB penetration although other transporters are clearly involved (see Ref. [435]). Interactions mediated by P-gp can be the result of either inhibition or induction of P-gp [436] and usually compound design focuses on avoiding P-gp effects. However, drug transporters, such as P-glycoprotein (P-gp; ABCB1/MDR1) and ABCG2 (breast cancer resistance protein/mitoxantrone resistance protein), may be overexpressed in cancer cells, thus reducing intracellular drug concentrations, and may allow the evolution of point mutations that confer stronger drug resistance. Thus, while focus in drug design for neurodegeneration focuses on avoiding efflux from the brain, design criteria for neuro-oncology applications may need to consider P-gp as a molecular target in multidrug-resistant cancers [437]. The impact of the BBB in chemotherapies for CNS tumors and the approaches to enhance the drug exposure of brain tumors have been reviewed with a focus on the role of the BBB in primary malignant gliomas, its impact on brain metastases should also be noted [438].

Refinement of the concepts of drug-likeness in the design of compounds that target CNS diseases to include aspects of BBB permeability should increase the odds of compounds residing within CNS-accessible chemical space. A number of published examples where CNS activity and/or penetration characteristics have been a factor in library design have been reviewed [439] and introduce the concept of structure–brain exposure relationships [440]. Recent developments in prediction models including, cytochrome P450 metabolism, BBB permeability, CNS activity, and blockade of the hERG-potassium channel have also been reviewed [441].

Delivery of small-molecule drugs into the CNS is not the only problem faced in the treatment of CNS diseases. The difficulties for CNS therapies, especially those with biological macromolecules, such as proteins and nucleic acid constructs introduce further complications. The solutions for which may be common between chronic administration for neurodegenerative disease and CNS cancer therapies or may need to be focused based on the specific issues of the modality and the tumor type under investigation [442].

Relationships between physicochemical drug properties and toxicity are also critical in the design of molecular target modulators for CNS tumor control. From an analysis of a data set consisting of animal *in vivo* toleration studies on 245 preclinical Pfizer compounds; an increased likelihood of toxic events was found for less polar, more lipophilic compounds. This trend held across a wide range of types of toxicity and across a broad swath of chemical space [443].

19. TARGETED DELIVERY

While designing brain penetrant compounds is one approach it may not be applicable for all therapeutic modalities and targeted delivery across the BBB is a viable alternative. One approach is to develop chemically modified derivatives or chemically modified nanoparticulate vectors of drugs, capable of crossing the BBB [444]. Alternatively, more invasive methods of delivery, such as intra-arterial infusion with or without BBB disruption or direct intracerebral administration, with clysis or drug-impregnated wafers are also being actively investigated [445]. Chemical modification approaches include the "standard" approaches of glycosylation and pegylation. Drug-delivery vehicles, such as nanoparticles and liposomes, in particular nanoparticles are heavily investigated. Nanoparticles offer particular advantages since while systemically administered small-molecule chemotherapeutics maintain peak blood concentrations for only minutes, and generally do not accumulate to therapeutic concentrations within individual brain tumor cells. Spherical nanoparticles can maintain peak blood concentrations for several hours and could be used to deliver drugs to accumulate to therapeutic concentrations within individual brain tumor cells. The Gd-G5-doxorubicin dendrimer, an imageable nanoparticle bearing chemotherapy has already been demonstrated to be effective in an orthotopic RG-2 rodent model of malignant glioma [446].

Implantation of polymeric wafers to deliver a chemotherapeutic drug is a common strategy against CNS tumors. Modeling of the drug transport criteria of several known chemotherapies has analyzed the distribution of the drug in the CNS tumor with respect to physicochemical properties [447].

For a general review of delivery methods and vehicles including the above and targeting vectors that include natural ligands or the conjugation of a therapeutic peptide or protein with a targeting molecule that can induce transcytosis across blood–brain barrier endothelial cells, see Ref. [448].

As alternative to delivery across the BBB, barrier disruption and convection-enhanced delivery have emerged as potential delivery techniques for the treatment of malignant brain tumors. Clinical trials using these methods have been completed, but with mixed results. For a review of the clinically available methods and clinical trial results to date, see Ref. [449].

A variety of modalities have been investigated to deliver drugs to diffusely distributed CNS tumor cells. These include systemic, diffusive, and convection-enhanced delivery (CED) methods. Systemic delivery is limited by large molecules typically >500 Da and thus therapeutic concentrations are difficult to attain. Localized delivery approaches relying on diffusion and CED have therefore been used to circumvent the BBB. The strengths and weaknesses of these strategies and details on the use of predictive modeling in optimizing these delivery modalities for clinical application have been reviewed [450].

Antiangiogenic therapies may be a particular case since many of the targets of antiangiogenic compounds are also involved in BBB permeability. Several of the limitations of antiangiogenic drugs for CNS tumors including novel biomaterials and nanotechnological approaches have been reviewed [451].

20. MODELING BRAIN CANCERS

The role of cancer stem cells has generated significant interest in cell lines that better duplicate the pathobiology of CNS tumors particularly gliomas. The importance of cell lines in mimicking tumor cell populations and phenotypes is one aspect critical to the generation and use of effective models for CNS cancers. For example, renewable neurosphere formation is a defining characteristic of certain brain tumor initiating cells [32]. The relationship between neurosphere formation in cultures, tumorigenic capacity, and clinical outcomes shows that both renewable neurosphere formation and tumorigenic capacity are significantly associated with clinical outcome measures in both glioma and primary glioblastoma [452]. The growth and migration of glioblastoma has also been modeled in a three-dimensional mathematical model [453].

Of key importance is the use of animal models to in some way allow the connection between preclinical studies and predictive outcomes in the clinic. The importance of such translational use of animal models in CNS cancers and in particular orthotopic models (crossing BBB, tumor micro environment, metastasis, etc.) is of particular importance in pediatric CNS tumors [454]. For pediatric tumors, the challenge of tumor heterogeneity is compounded by the fact that tumors arise from progenitor cells in a developmental context so targeting abnormal signaling selectively is an important objective. This is a particular challenge, as neural progenitor cells show considerable temporal and spatial heterogeneity during development. Mouse models of pediatric CNS tumors are facilitating an understanding not only of the pathobiology of CNS pediatric tumors but also of the normal developmental processes to better focus pediatric cancer care [455].

Animal models need to be predictive and reliable tools that recapitulate the most salient pathobiological features reported for human GBM including tumor heterogeneity. Studies using a U251 orthotopic xenograft model in nude mice recapitulated the salient pathobiological features described for human GBM, while U87MG orthotopic xenografts proved to be very dissimilar from human GBM, showing expansile growth, occasional necrotic foci without pseudo-palisades, intratumoral lacunar pattern of angiogenesis [456]. Mouse glioma models based on human tumor genetic abnormalities and the resulting gliomas can show similar histology to endpoints measured in the clinical setting. This is particularly true of orthotopic models where several preclinical endpoints may be observable [457]. Accurate recapitulation of genetic alterations driving cell cycle regulation and signaling abnormalities in primary brain tumors in mouse models allows for *in vivo* modeling of brain tumors with similar histopathology, etiology, and biology [458]. MRI is a particular endpoint that could fulfill this goal. For example, MRI investigations into orthotopic brain tumor xenografts in immunocompromised mice from both patient tumor cells implanted immediately after surgery or from cultured human tumor lines showed significant correlations in key metabolites across species. Differences were observed, however, including lower lipids in animal models [459].

Recapitulating both the underlying genetics and the characteristic tumor-stroma microenvironment of brain cancers in genetically engineered mouse models (GEMMs) may offer distinct advantages over cell culture and xenograft systems in the preclinical testing of promising therapies. The characteristics of recently developed GEMMs for both glioma and medulloblastoma, and their potential use in preclinical research have been reviewed [34]. Genetically engineered mouse models of Nf1 have also been used for preclinical drug evaluation. In particular, an Nf1 optic glioma model showed potential to preclinically benchmark novel therapies that have a high likelihood of success in human clinical trials [460].

Metastases to the CNS are a particular case for preclinical models that permit investigation of the dual properties of control of experimental tumors and penetration of the BBB to be simultaneously explored in one system. For example, metastatic spread of melanoma to CNS is associated with dismal prognosis. Rodent models utilizing both murine and human melanoma cell lines have highlighted the complex biology of cerebral metastasis, involving apparent disease progression through the selection of subclones at each stage, eventually leading to disease in the brain [461].

Spontaneous CNS melanoma metastases can be generated from primary human tumors in mice. The tumors spontaneously metastasize and retain their phenotype but were found to have increased ability to proliferate in brain-conditioned medium and displayed enhanced adhesion to lung and brain endothelial cells [462]. This model has been used to investigate metronomic-based chemotherapeutic approaches [463].

Imaging and histopathology play an important role but the screening cascades used pre- and postanimal models should also move closer to representing the human disease setting.

21. IMAGING BRAIN TUMORS

Imaging of brain tumors in both the preclinical and the clinical setting allows noninvasive analysis of tumor characteristics in humans and animal models via a common modality and should allow the development of more relevant, that is, predictive CNS tumor animal models, for example, see Ref. [464]. The major advances have been in magnetic resonance imaging (MRI) and bioluminescence imaging (BI) that offer different but complimentary endpoints. Particularly, since advances in neurosurgical techniques, radiotherapy protocols and in particular the development of molecular-target drug treatments have generated an increasing need for reproducible, noninvasive, quantitative imaging biomarkers of which physiological MRI, BI, and PET molecular imaging will be particularly important [465].

Rodent tumor models are essential for developing translational therapeutic strategies in neuro-oncology. MRI offers tumor localization, volumetric measurement, and the potential for advanced physiologic imaging but is less well suited to high-throughput studies and has limited capacity to assess early tumor growth. Bioluminescence imaging is better suited for early identification of tumors and monitoring response of therapeutic intervention against tumor growth. The use of bioluminescence and MRI together to more fully characterize tumor growth, for example, in a mouse model of glioblastoma provides a multimodality strategy for selecting cohorts of animals with similar tumor growth patterns to improve the accuracy *in vivo* measurements of growth and treatment response in preclinical studies [466].

Noninvasive bioluminescence imaging has successfully been used to image in a wide range of CNS tumor types. For example, a rat model that closely resembles human glioblastoma in biological behavior and that utilizes bioluminescence imaging has been developed. Histological analysis of the rodent brains at varying stages postimplantation, allowed for statistically significant correlation between tumor size and luminescence readouts [467]. A brainstem-tumor model in rats using a bioluminescence imaging readout for monitoring tumor growth and response to therapy in athymic rats has also been reported [468]. Animal models of angiogenesis should allow analysis of the abnormal vascular structure to best monitor the effects of compound administration. An orthotopic glioma model of nude mice with a fluorescence readout has allowed demarcation of primary tumor margins and visualization of local invasion at the single-cell level. Abnormal vascular structure and glioma cells can be visualized concurrently by fluorescence microscopy allowing the study of physiologically relevant patterns of brain tumor invasion and angiogenesis *in vivo* [469]. Noninvasive bioluminescence imaging has successfully been used to image intracranial meningiomas in mice. The tumors grew in a fashion similar to that of aggressive meningiomas in humans, and exhibited the microscopic, immunohistochemical, and ultrastructural features characteristic of meningiomas. This animal model overcomes the main obstacle in studying intracranial meningiomas by enabling sequential noninvasive tumor measurement in a cost-effective manner [470]. Although both transgenic and xenograft mouse models of vestibular schwannomas have been previously reported none of these models replicate the intracranial location of these tumors. Intracranial vestibular schwannoma xenografts can, however, be successfully followed with bioluminescent imaging [471]. The value of bioluminescence imaging is also evidenced in a glioma model used for DC immunotherapy [472].

A murine model of radiation necrosis has been developed to facilitate investigation of imaging biomarkers that distinguish between radiation necrosis and tumor recurrence [473].

Finally, functional impairment is a common side effect of CNS tumor treatments [23]. Accurate modeling of neurological function is therefore of importance. A neurological scoring system for a rat brainstem glioma model has been developed that enables quantification of drug and tumor-related morbidity as a factor for functional performance during therapy [474].

22. CONCLUSIONS

CNS tumors represent a diverse collection of tumor types driven by a range of signaling inputs that often generate highly aggressive and at the same time heterogeneous tumor types. The effect of these tumors is particularly dramatic and impacts on children as well as adults. CNS tumors present within the nervous system there are particular challenges in the development of effective molecular-targeted therapies in dealing not only with the aggressive nature of these tumors and their ability to evade treatment but particularly in maintaining normal neuronal and brain functions.

Research to find effective treatments for CNS tumors is being facilitated by two key inputs a much greater understanding of tumor pathobiology and the potential role of developmental signaling pathways and tumor stem-like cells, and by the development of animal models that more adequately represent endpoints measurable in the clinic. Together, these advances should assist in the selection of better targets and the optimization of better compounds to progress into clinical trials.

REFERENCES

1. Collins VP. Brain tumours: classification and genes. J Neurol Neurosurg Psychiatr 2004;75 (Suppl 2): ii2–ii11.
2. Emmenegger BA, Wechsler-Reya RJ. Stem cells and the origin and propagation of brain tumors. J Child Neurol 2008;23(10): 1172–1178.
3. Louis DN, Ohgaki H, Wiestler OD, Cavenee WK, editors. World Health Organization Classification of Tumours of the Central Nervous System. Lyon: IARC; 2007.
4. Kanu OO, Hughes B, Di C, Lin N, Fu J, Bigner DD, Yan H, Adamson C. Glioblastoma multiforme oncogenomics and signaling pathways. Clin Med Oncol 2009;3:39–52.
5. Ware ML, Berger MS, Binder DK. Molecular biology of glioma tumorigenesis. Histol Histopathol 2003;18(1): 207–216.
6. Lino M, Merlo A. Translating biology into clinic: the case of glioblastoma. Curr Opin Cell Biol 2009;21(2): 311–316.
7. Schiff D, Principles and practice of neuro-oncology. New York: McGraw-Hill Professional; 2005.
8. Brem SS, Bierman PJ, Black P, Brem H, Chamberlain MC, Chiocca EA, DeAngelis LM, Fenstermaker RA, Friedman A, Gilbert MR, Glass J, Grossman SA, Heimberger AB, Junck L, Linette GP, Loeffler JJ, Maor MH, Moots P, Mrugala M, Nabors LB, Newton HB, Olivi A, Portnow J, Prados M, Raizer JJ, Shrieve DC,

Sills AK Jr. Central nervous system cancers. J Natl Compr Canc Netw 2008;6(5): 456–504.
9. Buckner JC, Brown PD, O'Neill BP, Meyer FB, Wetmore CJ, Uhm JH. Central nervous system tumors. Mayo Clin Proc 2007;82(10): 1271–1286.
10. Schor NF. Pharmacotherapy for adults with tumors of the central nervous system. Pharmacol Ther 2009;121(3): 253–264.
11. Hottinger AF, Stupp R. Therapeutic strategies for the management of gliomas. Rev Neurol 2008;164(6–7): 523–530.
12. Pillay V, Allaf L, Wilding AL, Donoghue JF, Court NW, Greenall SA, Scott AM, Johns TG. The plasticity of oncogene addiction: implications for targeted therapies directed to receptor tyrosine kinases. Neoplasia 2009;11(5): 448–458.
13. Wong ML, Kaye AH, Hovens CM. Targeting malignant glioma survival signalling to improve clinical outcomes. J Clin Neurosci 2007;14(4): 301–308.
14. Adamson C, Kanu OO, Mehta AI, Di C, Lin N, Mattox AK, Bigner DD. Glioblastoma multiforme: a review of where we have been and where we are going. Expert Opin Investig Drugs 2009;18(8): 1061–1083.
15. Mut M, Schiff D. Unmet needs in the treatment of glioblastoma. Expert Rev Anticancer Ther 2009;9(5): 545–551.
16. Fine HA. Promising new therapies for malignant gliomas. Cancer J 2007;13(6): 349–354.
17. Herrington B, Kieran MW. Small molecule inhibitors in children with malignant gliomas. Pediatr Blood Cancer 2009;53(3): 312–317.
18. Mueller S, Chang S. Pediatric brain tumors: current treatment strategies and future therapeutic approaches. Neurotherapeutics 2009;6 (3):570–586.
19. Lafay-Cousin L, Strother D. Current treatment approaches for infants with malignant central nervous system tumors. Oncologist 2009;14(4): 433–444.
20. Kieran MW, Supko JG, Wallace D, Fruscio R, Poussaint TY, Phillips P, Pollack I, Packer R, Boyett JM, Blaney S, Banerjee A, Geyer R, Friedman H, Goldman S, Kun LE. Macdonald T; Pediatric Brain Tumor Consortium. Phase I study of SU5416, a small molecule inhibitor of the vascular endothelial growth factor receptor (VEGFR) in refractory pediatric central nervous system tumors. Pediatr Blood Cancer 2009;52(2): 169–176.
21. Schor NF. New approaches to pharmacotherapy of tumors of the nervous system during childhood and adolescence. Pharmacol Ther 2009;122(1): 44–55.
22. Herrington B, Kieran MW. Small molecule inhibitors in children with malignant gliomas. Pediatr Blood Cancer 2009;53(3): 312–317.
23. Kannarkat G, Lasher EE, Schiff D. Neurologic complications of chemotherapy agents. Curr Opin Neurol 2007;20(6): 719–725.
24. Drappatz J, Schiff D, Kesari S, Norden AD, Wen PY. Medical management of brain tumor patients. Neurol Clin 2007;25(4): 1035–1071.
25. Taillibert S, Voillery D, Bernard-Marty C. Chemobrain: is systemic chemotherapy neurotoxic? Curr Opin Oncol 2007;19(6): 623–627.
26. Weiss B. Chemobrain: a translational challenge for neurotoxicology. Neurotoxicology 2008;29(5): 891–898.
27. Nelson CJ, Nandy N, Roth AJ. Chemotherapy and cognitive deficits: mechanisms, findings, and potential interventions. Palliat Support Care 2007;5(3): 273–280.
28. Dupont S. Epilepsy and brain tumours. Rev Neurol 2008;164(6–7): 517–522.
29. Rajneesh KF, Binder DK. Tumor-associated epilepsy. Neurosurg Focus 2009;27(2): E4.
30. Zhu Y, Parada LF. The molecular and genetic basis of neurological tumours. Nat Rev Cancer 2002;2(8): 616–626.
31. Hide T, Takezaki T, Nakamura H, Kuratsu J, Kondo T. Brain tumor stem cells as research and treatment targets. Brain Tumor Pathol 2008;25(2): 67–72.
32. Yao Y, Tang X, Li S, M2ao Y, Zhou L. Brain tumor stem cells: view from cell proliferation. Surg Neurol 2009;71(3): 274–279.
33. Piccirillo SG, Vescovi AL. Brain tumour stem cells: possibilities of new therapeutic strategies. Expert Opin Biol Ther 2007;7(8): 1129–1135.
34. Huse JT, Holland EC. Genetically engineered mouse models of brain cancer and the promise of preclinical testing. Brain Pathol 2009;19(1): 132–143.
35. Fults DW. Modeling medulloblastoma with genetically engineered mice. Neurosurg Focus 2005;19(5): E7.
36. Johannessen TC, Wang J, Skaftnesmo KO, Sakariassen PØ Enger PØ Petersen K, Øyan AM, Kalland KH, Bjerkvig R, Tysnes BB. Highly infiltrative brain tumours show reduced chemosensitivity associated with a stem cell-like phenotype. Neuropathol Appl Neurobiol 2009;35(4): 380–393.

37. Rich JN, Eyler CE. Cancer stem cells in brain tumor biology. Cold Spring Harb Symp Quant Biol 2008;73:411–420.
38. Ischenko I, Seeliger H, Schaffer M, Jauch KW, Bruns CJ. Cancer stem cells: how can we target them? Curr Med Chem 2008;15(30): 3171–3184.
39. Louis DN. Molecular pathology of malignant gliomas. Annu Rev Pathol 2006;1:97–117.
40. Das S, Srikanth M, Kessler JA. Cancer stem cells and glioma. Nat Clin Pract Neurol 2008; 4(8): 427–435.
41. Quiñones-Hinojosa A, Chaichana K. The human subventricular zone: a source of new cells and a potential source of brain tumors. Exp Neurol 2007;205(2): 313–324.
42. Vescovi AL, Galli R, Reynolds BA. Brain tumour stem cells. Nat Rev Cancer 2006;6(6): 425–436.
43. Sakariassen PØ Prestegarden L, Wang J, Skaftnesmo KO, Mahesparan R, Molthoff C, Sminia P, Sundlisaeter E, Misra A, Tysnes BB, Chekenya M, Peters H, Lende G, Kalland KH, Øyan AM, Petersen K, Jonassen I, van der Kogel A, Feuerstein BG, Terzis AJ, Bjerkvig R, Enger PØ. Angiogenesis-independent tumor growth mediated by stem-like cancer cells. Proc Natl Acad Sci USA 2006;103(44): 16466–16471.
44. Zaidi HA, Kosztowski T, DiMeco F, Quiñones-Hinojosa A. Origins and clinical implications of the brain tumor stem cell hypothesis. J Neurooncol 2009;93(1): 49–60.
45. Assanah M, Lochhead R, Ogden A, Bruce J, Goldman J, Canoll P. Glial progenitors in adult white matter are driven to form malignant gliomas by platelet-derived growth factor-expressing retroviruses. J Neurosci 2006;26 (25): 6781–6790.
46. Gilbertson RJ, Rich JN. Making a tumour's bed: glioblastoma stem cells and the vascular niche. Nat Rev Can 2007;7(10): 733–736.
47. Canoll P, Goldman JE. The interface between glial progenitors and gliomas. Acta Neuropathol 2008;116(5): 465–477.
48. Nern C, Sommerlad D, Acker T, Plate KH. Brain tumor stem cells. Recent Results Cancer Res 2009;171:241–259.
49. Colin C, Baeza N, Tong S, Bouvier C, Quilichini B, Durbec P, Figarella-Branger D. *In vitro* identification and functional characterization of glial precursor cells in human gliomas. Neuropathol Appl Neurobiol 2006;32(2): 189–202.
50. Tysnes BB, Bjerkvig R. Cancer initiation and progression: involvement of stem cells and the microenvironment. Biochim Biophys Acta 2007;1775(2): 283–297.
51. Nicolis SK. Cancer stem cells and "stemness" genes in neuro-oncology. Neurobiol Dis 2007;25(2): 217–229.
52. Eramo A, Ricci-Vitiani L, Zeuner A, Pallini R, Lotti F, Sette G, Pilozzi E, Larocca LM, Peschle C, De Maria R. Chemotherapy resistance of glioblastoma stem cells. Cell Death Differ 2006;13(7): 1238–1241.
53. Newton HB. Molecular neuro-oncology and the development of targeted therapeutic strategies for brain tumors. Part 5. Apoptosis and cell cycle. Expert Rev Anticancer Ther 2005; 5(2): 355–378.
54. Xu Q, Yuan X, Tunici P, Liu G, Fan X, Xu M, Hu J, Hwang JY, Farkas DL, Black KL, Yu JS. Isolation of tumour stem-like cells from benign tumours. Br J Cancer 2009;101(2): 303–311.
55. Huang Q, Zhang QB, Dong J, Wu YY, Shen YT, Zhao YD, Zhu YD, Diao Y, Wang AD, Lan Q. Glioma stem cells are more aggressive in recurrent tumors with malignant progression than in the primary tumor, and both can be maintained long-term *in vitro*. BMC Cancer 2008;8:304.
56. Rass K, Hassel JC. Chemotherapeutics, chemoresistance and the management of melanoma. G Ital Dermatol Venereol 2009;144(1): 61–78.
57. Altaner C. Glioblastoma and stem cells. Neoplasma 2008;55(5): 369–374.
58. Johannessen TC, Bjerkvig R, Tysnes BB. DNA repair and cancer stem-like cells: potential partners in glioma drug resistance? Cancer Treat Rev 2008;34(6): 558–567.
59. Hoelzinger DB, Demuth T. Berens ME: Autocrine factors that sustain glioma invasion and paracrine biology in the brain microenvironment. J Natl Cancer Inst 2007;99(21): 1583–1593.
60. Rubin JB. Only in congenial soil: the microenvironment in brain tumorigenesis. Brain Pathol 2009;19(1): 144–149.
61. Barcellos-Hoff MH, Newcomb EW, Zagzag D, Narayana A. Therapeutic targets in malignant glioblastoma microenvironment. Semin Radiat Oncol 2009;19(3): 163–170.
62. Gilbertson RJ, Gutmann DH. Tumorigenesis in the brain: location, location, location. Cancer Res 2007;67(12): 5579–5582.
63. Hambardzumyan D, Becher OJ, Holland EC. Cancer stem cells and survival pathways. Cell Cycle 2008;7(10): 1371–1378.

64. Carbonell WS, Ansorge O, Sibson N, Muschel R. The vascular basement membrane as "soil" in brain metastasis. PLoS One 2009;4(6): e5857.
65. Clark PA, Treisman DM, Ebben J, Kuo JS. Developmental signaling pathways in brain tumor-derived stem-like cells. Dev Dyn 2007;236(12): 3297–3308.
66. Brandes AA, Franceschi E, Tosoni A, Hegi ME, Stupp R. Epidermal growth factor receptor inhibitors in neuro-oncology: hopes and disappointments. Clin Cancer Res 2008;14(4): 957–960.
67. Pierfelice TJ, Schreck KC, Eberhart CG, Gaiano N. Notch, neural stem cells, and brain tumors. Cold Spring Harb Symp Quant Biol 2008;73:367–375.
68. Reya T, Clevers H. Wnt signalling in stem cells and cancer. Nature 2005;434(7035): 843–850.
69. Yu JM, Jun ES, Jung JS, Suh SY, Han JY, Kim JY, Kim KW, Jung JS. Role of Wnt5a in the proliferation of human glioblastoma cells. Cancer Lett 2007;257(2): 172–181.
70. Hide T, Takezaki T, Nakamura H, Kuratsu J, Kondo T. Brain tumor stem cells as research and treatment targets. Brain Tumor Pathol 2008;25(2): 67–72.
71. Dirks PB. Brain tumor stem cells: bringing order to the chaos of brain cancer. J Clin Oncol 2008;26(17): 2916–2924.
72. Sareddy GR, Panigrahi M, Challa S, Mahadevan A, Babu PP. Activation of Wnt/beta-catenin/Tcf signaling pathway in human astrocytomas. Neurochem Int 2009;55(5): 307–317.
73. Watabe T, Miyazono K. Roles of TGF-beta family signaling in stem cell renewal and differentiation. Cell Res 2009;19(1): 103–115.
74. Wick W, Naumann U, Weller M. Transforming growth factor-beta: a molecular target for the future therapy of glioblastoma. Curr Pharm Des 2006;12(3): 341–349.
75. Nikuševa Martić T, Pećina-Šlaus N, Kušec V, Kokotović T, Mušinović H, Tomas D, Zeljko M. Changes of AXIN-1 and beta-catenin in neuroepithelial brain tumors. Pathol Oncol Res 2010;16(1):75–79.
76. Medina V, Calvo MB, Díaz-Prado S, Espada J. Hedgehog signalling as a target in cancer stem cells. Clin Transl Oncol 2009;11(4): 199–207.
77. Rogers HA, Miller S, Lowe J, Brundler MA, Coyle B, Grundy RG. An investigation of WNT pathway activation and association with survival in central nervous system primitive neuroectodermal tumours (CNS PNET). Br J Cancer 2009;100(8): 1292–1302.
78. Fattet S, Haberler C, Legoix P, Varlet P, Lellouch-Tubiana A, Lair S, Manie E, Raquin MA, Bours D, Carpentier S, Barillot E, Grill J, Doz F, Puget S, Janoueix-Lerosey I, Delattre O. Beta-catenin status in paediatric medulloblastomas: correlation of immunohistochemical expression with mutational status, genetic profiles, and clinical characteristics. J Pathol 2009;218(1): 86–94.
79. Pu P, Zhang Z, Kang C, Jiang R, Jia Z, Wang G, Jiang H. Downregulation of Wnt2 and beta-catenin by siRNA suppresses malignant glioma cell growth. Cancer Gene Ther 2009;16(4): 351–361.
80. Figarella-Branger D, Colin C, Coulibaly B, Quilichini B, Maues De Paula A, Fernandez C, Bouvier C. Histological and molecular classification of gliomas. Rev Neurol 2008;164 (6–7): 505–515.
81. Madhavan S, Zenklusen JC, Kotliarov Y, Sahni H, Fine HA, Buetow K. Rembrandt: helping personalized medicine become a reality through integrative translational research. Mol Cancer Res 2009;7(2): 157–167.
82. Lampson LA. Targeted therapy for neuro-oncology: reviewing the menu. Drug Discov Today 2009;14(3–4): 185–191.
83. Preusser M, Haberler C, Hainfellner JA. Malignant glioma: neuropathology and neurobiology. Wien Med Wochenschr 2006;156(11–12): 332–337.
84. Riemenschneider MJ, Reifenberger G. Astrocytic tumors. Recent Results Cancer Res 2009;171:3–24.
85. Louis DN, Ohgaki H, Wiestler OD, Cavenee WK, Burger PC, Jouvet A, Scheithauer BW, Kleihues P. The 2007 WHO classification of tumours of the central nervous system. Acta Neuropathol 2007;114(2): 97–109.
86. Janzer RC. Neuropathology and molecular pathology of glioma. Rev Med Suisse 2009;5 (211): 1501–1504.
87. Tibbetts KM, Emnett RJ, Gao F, Perry A, Gutmann DH, Leonard JR. Histopathologic predictors of pilocytic astrocytoma event-free survival. Acta Neuropathol 2009;117(6): 657–665.
88. Stieber VW. Low-grade gliomas. Curr Treat Options Oncol 2001;2(6): 495–506.
89. Perry A. Pathology of low-grade gliomas: an update of emerging concepts. Neuro Oncol 2003;5(3): 168–178.
90. Bristol RE. Low-grade glial tumors: are they all the same? Semin Pediatr Neurol 2009; 16(1): 23–26.

91. Paleologos NA. Chemotherapy for low-grade gliomas. Expert Rev Neurother 2005;5(Suppl 6): S21–S24.
92. Chawengchao B, Petmitr S, Ponglikitmongkol M, Chanyavanich V, Sangruji T, Theerapuncharoen V. Detection of a novel point mutation in the p53 gene in grade II astrocytomas by PCR-SSCP analysis with additional Klenow treatment. Anticancer Res 2001;21(4A): 2739–2743.
93. Sathornsumetee S, Rich JN, Reardon DA. Diagnosis and treatment of high-grade astrocytoma. Neurol Clin 2007;25(4): 1111–1139, x.
94. Stupp R, Reni M, Gatta G, Mazza E, Vecht C. Anaplastic astrocytoma in adults. Crit Rev Oncol Hematol 2007;63(1): 72–80.
95. See SJ, Gilbert MR. Anaplastic astrocytoma: diagnosis, prognosis, and management. Semin Oncol 2004;31(5): 618–634.
96. Compostella A, Tosoni A, Blatt V, Franceschi E, Brandes AA. Prognostic factors for anaplastic astrocytomas. J Neurooncol 2007;81(3): 295–303.
97. Villano JL, Seery TE, Bressler LR. Temozolomide in malignant gliomas: current use and future targets. Cancer Chemother Pharmacol 2009;64(4): 647–655.
98. Ohgaki H, Kleihues P. Genetic pathways to primary and secondary glioblastoma. Am J Pathol 2007;170(5): 1445–1453.
99. Rao SK, Edwards J, Joshi AD, Siu IM, Riggins GJ. A survey of glioblastoma genomic amplifications and deletions. J Neurooncol 2010;96(2):169–179.
100. Tso CL, Freije WA, Day A, Chen Z, Merriman B, Perlina A, Lee Y, Dia EQ, Yoshimoto K, Mischel PS, Liau LM, Cloughesy TF, Nelson SF. Distinct transcription profiles of primary and secondary glioblastoma subgroups. Cancer Res 2006;66(1): 159–167.
101. Gu J, Liu Y, Kyritsis AP, Bondy ML. Molecular epidemiology of primary brain tumors. Neurotherapeutics 2009;6(3): 427–435.
102. Ruano Y, Ribalta T, de Lope AR, Campos-Martín Y, Fiaño C, Pérez-Magán E, Hernández-Moneo JL, Mollejo M, Meléndez B. Worse outcome in primary glioblastoma multiforme with concurrent epidermal growth factor receptor and p53 alteration. Am J Clin Pathol 2009;131(2): 257–263.
103. Hegi ME, Murat A, Lambiv WL, Stupp R. Brain tumors: molecular biology and targeted therapies. Ann Oncol 2006;17(Suppl 10): x191–x200.
104. Kumar HR, Zhong X, Sandoval JA, Hickey RJ, Malkas LH. Applications of emerging molecular technologies in glioblastoma multiforme. Expert Rev Neurother 2008;8(10): 1497–1506.
105. Pfister S, Witt O. Pediatric gliomas. Recent Results Cancer Res 2009;171:67–81.
106. Tilleul P, Brignone M, Hassani Y, Taillandier L, Taillibert S, Cartalat-Carel S, Borget I, Chinot O. Prescription guidebook for temozolomide usage in brain tumors. Bull Cancer 2009;96(5): 579–589.
107. Beier D, Röhrl S, Pillai DR, Schwarz S, Kunz-Schughart LA, Leukel P, Proescholdt M, Brawanski A, Bogdahn U, Trampe-Kieslich A, Giebel B, Wischhusen J, Reifenberger G, Hau P, Beier CP. Temozolomide preferentially depletes cancer stem cells in glioblastoma. Cancer Res 2008;68(14): 5706–5715.
108. Stupp R, Hegi ME, Gilbert MR, Chakravarti A. Chemoradiotherapy in malignant glioma: standard of care and future directions. J Clin Oncol 2007;25(26): 4127–4136.
109. Sathornsumetee S, Reardon DA. Targeting multiple kinases in glioblastoma multiforme. Expert Opin Investig Drugs 2009;18(3): 277–292.
110. Batchelor TT, Sorensen AG, di Tomaso E, Zhang WT, Duda DG, Cohen KS, Kozak KR, Cahill DP, Chen PJ, Zhu M, Ancukiewicz M, Mrugala MM, Plotkin S, Drappatz J, Louis DN, Ivy P, Scadden DT, Benner T, Loeffler JS, Wen PY, Jain RK. AZD2171, a pan-VEGF receptor tyrosine kinase inhibitor, normalizes tumor vasculature and alleviates edema in glioblastoma patients. Cancer Cell 2007; 11(1): 83–95.
111. Lefranc F, Rynkowski M, DeWitte O, Kiss R. Present and potential future adjuvant issues in high-grade astrocytic glioma treatment. Adv Tech Stand Neurosurg 2009;34:3–35.
112. Brandsma D, van den Bent MJ. Molecular targeted therapies and chemotherapy in malignant gliomas. Curr Opin Oncol 2007;19(6): 598–605.
113. Kanu OO, Mehta A, Di C, Lin N, Bortoff K, Bigner DD, Yan H, Adamson DC. Glioblastoma multiforme: a review of therapeutic targets. Expert Opin Ther Targets 2009;13(6): 701–718.
114. Muldoon LL, Soussain C, Jahnke K, Johanson C, Siegal T, Smith QR, Hall WA, Hynynen K, Senter PD, Peereboom DM, Neuwelt EA. Chemotherapy delivery issues in central nervous system malignancy: a reality check. J Clin Oncol 2007;25(16): 2295–2305.
115. Declèves X, Amiel A, Delattre JY, Scherrmann JM. Role of ABC transporters in the chemore-

sistance of human gliomas. Curr Cancer Drug Targets 2006;6(5): 433–445.
116. Reni M, Gatta G, Mazza E, Vecht C. Ependymoma. Crit Rev Oncol Hematol 2007;63(1): 81–89.
117. Merchant TE, Fouladi M. Ependymoma: new therapeutic approaches including radiation and chemotherapy. J Neurooncol 2005;75(3): 287–299.
118. Merchant TE, Li C, Xiong X, Gaber MW. Cytokine and growth factor responses after radiotherapy for localized ependymoma. Int J Radiat Oncol Biol Phys 2009;74(1): 159–167.
119. Malkin MG. Chemotherapy of ependymoma. In: Newton HB, editor. Hand-Book of Brain Tumor Chemotherapy. London: Academic Press, Elsevier; 2006. p 426–431.
120. Rudà R, Gilbert M, Soffietti R. Ependymomas of the adult: molecular biology and treatment. Curr Opin Neurol 2008;21(6): 754–761.
121. Hasselblatt M. Ependymal tumors. Recent Results Cancer Res 2009;171:51–66.
122. Geoerger B, Gaspar N, Opolon P, Morizet J, Devanz P, Lecluse Y, Valent A, Lacroix L, Grill J, Vassal G. EGFR tyrosine kinase inhibition radiosensitizes and induces apoptosis in malignant glioma and childhood ependymoma xenografts. Int J Cancer 2008;123(1): 209–216.
123. Bromberg JE, van den Bent MJ. Oligodendrogliomas: molecular biology and treatment. Oncologist 2009;14(2): 155–163.
124. Paleologos NA, Fahey C. Chemotherapy of oligodendrogliomas. In: Newton HB, editor. Hand-Book of Brain Tumor Chemotherapy. London: Academic Press, Elsevier; 2006. p 371–381.
125. van den Bent MJ. Anaplastic oligodendroglioma and oligoastrocytoma. Neurol Clin 2007; 25(4): 1089–1109, ix–x.
126. Van den Bent MJ, Reni M, Gatta G, Vecht C. Oligodendroglioma. Crit Rev Oncol Hematol 2008;66(3): 262–272.
127. Bouamrani A, Ternier J, Ratel D,et al. Direct-tisse SELDI-TOF mass spectrometry analysis: a new application for clinical proteomics. Clin Chem 2006;52(11): 2103–2106.
128. Kuo LT, Kuo KT, Lee MJ, Wei CC, Scaravilli F, Tsai JC, Tseng HM, Kuo MF, Tu YK. Correlation among pathology, genetic and epigenetic profiles, and clinical outcome in oligodendroglial tumors. Int J Cancer 2009;124(12): 2872–2879.
129. Ney DE, Lassman AB. Molecular profiling of oligodendrogliomas: impact on prognosis, treatment, and future directions. Curr Oncol Rep 2009;11(1): 62–67.
130. Rossi A, Caracciolo V, Russo G, Reiss K, Giordano A. Medulloblastoma: from molecular pathology to therapy. Clin Cancer Res 2008;14(4): 971–976.
131. Johnsen JI, Kogner P, Albihn A, Henriksson MA. Embryonal neural tumours and cell death. Apoptosis 2009;14(4): 424–438.
132. Newton HB. Chemotherapy of medulloblastoma. In: Newton HB, editor. Hand-Book of Brain Tumor Chemotherapy. London: Academic Press, Elsevier; 2006. p 407–425.
133. Gottardo NG, Gajjar A. Chemotherapy for malignant brain tumors of childhood. J Child Neurol 2008;23(10): 1149–1159.
134. Grill J, Bhangoo R. Recent development in chemotherapy of paediatric brain tumours. Curr Opin Oncol 2007;19(6): 612–615.
135. Wlodarski PK, Jozwiak J. Therapeutic targets for medulloblastoma. Expert Opin Ther Targets 2008;12(4): 449–461.
136. Gilbertson RJ, Ellison DW. The origins of medulloblastoma subtypes. Annu Rev Pathol 2008;3:341–365.
137. Gulino A, Arcella A, Giangaspero F. Pathological and molecular heterogeneity of medulloblastoma. Curr Opin Oncol 2008;20(6): 668–675.
138. Saran A. Medulloblastoma: role of developmental pathways, DNA repair signaling, and other players. Curr Mol Med 2009;9(9): 1046–1057.
139. Kool M, Koster J, Bunt J, Hasselt NE, Lakeman A, van Sluis P, Troost D, Meeteren NS, Caron HN, Cloos J, Mrsić A, Ylstra B, Grajkowska W, Hartmann W, Pietsch T, Ellison D, Clifford SC, Versteeg R. Integrated genomics identifies five medulloblastoma subtypes with distinct genetic profiles, pathway signatures and clinicopathological features. PLoS One 2008;3(8): e3088.
140. Guessous F, Li Y, Abounader R. Signaling pathways in medulloblastoma. J Cell Physiol 2008;217(3): 577–583.
141. Fan X, Eberhart CG. Medulloblastoma stem cells. J Clin Oncol 2008;26(17): 2821–2827.
142. Lu J, Ma Z, Hsieh JC, Fan CW, Chen B, Longgood JC, Williams NS, Amatruda JF, Lum L, Chen C. Structure–activity relationship studies of small-molecule inhibitors of Wnt response. Bioorg Med Chem Lett 2009;19(14): 3825–3827.

143. Sato K, Kubota T. Pathology of pineal parenchymal tumors. Prog Neurol Surg 2009;23:12–25.
144. Schild SE, Scheithauer BW, Schomberg PJ, Hook CC, Kelly PJ, Frick L, Robinow JS, Buskirk SJ. Pineal parenchymal tumors. Clinical, pathologic, and therapeutic aspects. Cancer 1993;72(3): 870–880.
145. Kyritsis AP. Management of primary intracranial germ cell tumors. J Neurooncol 2010;96(2):143–149.
146. Shibamoto Y. Management of central nervous system germinoma: proposal for a modern strategy. Prog Neurol Surg 2009;23:119–129.
147. Aoyama H. Radiation therapy for intracranial germ cell tumors. Prog Neurol Surg 2009;23:96–105.
148. Matsutani M. Pineal germ cell tumors. Prog Neurol Surg 2009;23:76–85.
149. Sawamura Y. Strategy of combined treatment of germ cell tumors. Prog Neurol Surg 2009;23:86–95.
150. Osamura RY, Kajiya H, Takei M, Egashira N, Tobita M, Takekoshi S, Teramoto A. Pathology of the human pituitary adenomas. Histochem Cell Biol 2008;130(3): 495–507.
151. Popović-Brkić V. Advances in understanding pituitary adenomas. Horm Res 2009;71(Suppl 2): 75–77.
152. Korbonits M, Carlsen E. Recent clinical and pathophysiological advances in non-functioning pituitary adenomas. Horm Res 2009;71 (Suppl 2): 123–130.
153. Cohen AB, Lessell S. Angiogenesis and pituitary tumors. Semin Ophthalmol 2009;24(3): 185–189.
154. Tichomirowa MA, Daly AF, Beckers A. Familial pituitary adenomas. J Intern Med 2009; 266(1): 5–18.
155. Sidibé EH. Pituitary carcinoma. Anatomic and clinical features of cases reported in literature. Neurochirurgie 2007;53(4): 284–288.
156. Kars M, Roelfsema F, Romijn JA, Pereira AM. Malignant prolactinoma: case report and review of the literature. Eur J Endocrinol 2006;155(4): 523–534.
157. Molitch ME. Prolactin-secreting tumors: what's new? Expert Rev Anticancer Ther 2006;6(Suppl 9): S29–S35.
158. Lania A, Spada A. G-protein and signalling in pituitary tumours. Horm Res 2009;71(Suppl 2): 95–100.
159. Libé R, Bertherat J. Molecular genetics of adrenocortical tumours, from familial to sporadic diseases. Eur J Endocrinol 2005;153(4): 477–487.
160. Alexandraki KI, Grossman AB. Pituitary-targeted medical therapy of Cushing's disease. Expert Opin Investig Drugs 2008;17(5): 669–677.
161. Losa M, Fortunato M, Molteni L, Peretti E, Mortini P. Thyrotropin-secreting pituitary adenomas: biological and molecular features, diagnosis and therapy. Minerva Endocrinol 2008;33(4): 329–340.
162. Vance ML. Pituitary adenoma: a clinician's perspective. Endocr Pract 2008;14(6): 757–763.
163. Mathioudakis N, Salvatori R. Pituitary tumors. Curr Treat Options Neurol 2009;11(4): 287–296.
164. Hornyak M, Couldwell WT. Multimodality treatment for invasive pituitary adenomas. Postgrad Med 2009;121(2): 168–176.
165. Sonabend AM, Musleh W, Lesniak MS. Oncogenesis and mutagenesis of pituitary tumors. Expert Rev Anticancer Ther 2006;6(Suppl 9): S3–S14.
166. Ferone D, Gatto F, Arvigo M, Resmini E, Boschetti M, Teti C, Esposito D, Minuto F. The clinical-molecular interface of somatostatin, dopamine and their receptors in pituitary pathophysiology. J Mol Endocrinol 2009;42 (5): 361–370.
167. Saveanu A, Jaquet P. Somatostatin–dopamine ligands in the treatment of pituitary adenomas. Rev Endocr Metab Disord 2009;10(2): 83–90.
168. de Bruin C, Feelders RA, Lamberts SW, Hofland LJ. Somatostatin and dopamine receptors as targets for medical treatment of Cushing's syndrome. Rev Endocr Metab Disord 2009; 10(2): 91–102.
169. Asa SL, Ezzat S. The pathogenesis of pituitary tumors. Annu Rev Pathol 2009;4:97–126.
170. Fedele M, De Martino I, Pivonello R, Ciarmiello A, Del Basso De Caro ML, Visone R, Palmieri D, Pierantoni GM, Arra C, Schmid HA, Hofland L, Lombardi G, Colao A, Fusco A. SOM230, a new somatostatin analogue, is highly effective in the therapy of growth hormone/prolactin-secreting pituitary adenomas. Clin Cancer Res 2007;13(9): 2738–2744.
171. Melen-Mucha G, Lawnicka H, Kierszniewska-Stepien D, Komorowski J, Stepien H. The place of somatostatin analogs in the diagnosis and treatment of the neuoroendocrine glands tu-

mors. Recent Pat Anticancer Drug Discov 2006;1(2): 237–254.
172. Colao A, Filippella M, Pivonello R, Di Somma C, Faggiano A, Lombardi G. Combined therapy of somatostatin analogues and dopamine agonists in the treatment of pituitary tumours. Eur J Endocrinol 2007;156(Suppl 1): S57–S63.
173. Lamberts SW, Hofland LJ. Future treatment strategies of aggressive pituitary tumors. Pituitary 2009;12(3): 261–264.
174. Drutel A, Caron P, Archambeaud F. New medical treatments in pituitary adenomas. Ann Endocrinol 2008;69(Suppl 1): S16–S28.
175. Burdman JA, Pauni M, Heredia Sereno GM, Bordón AE. Estrogen receptors in human pituitary tumors. Horm Metab Res 2008;40(8): 524–527.
176. Winczyk K, Kunert-Radek J, Gruszka A, Radek M, èawnicka H, Pawlikowski M. Effects of rosiglitazone-peroxisome proliferators-activated receptor gamma (PPARgamma) agonist on cell viability of human pituitary adenomas *in vitro*. Neuro Endocrinol Lett 2009;30(1): 107–110.
177. Labeur M, Paez-Pereda M, Arzt E, Stalla GK. Potential of retinoic acid derivatives for the treatment of corticotroph pituitary adenomas. Rev Endocr Metab Disord 2009;10(2): 103–109.
178. Heaney AP. Targeting pituitary tumors. Horm Res 2007;68(Suppl 5): 132–136.
179. Rockhill J, Mrugala M, Chamberlain MC. Intracranial meningiomas: an overview of diagnosis and treatment. Neurosurg Focus 2007; 23(4): E1.
180. Campbell BA, Jhamb A, Maguire JA, Toyota B, Ma R. Meningiomas in 2009: controversies and future challenges. Am J Clin Oncol 2009;32(1): 73–85.
181. Marosi C, Hassler M, Roessler K, Reni M, Sant M, Mazza E, Vecht C. Meningioma. Crit Rev Oncol Hematol 2008;67(2): 153–171.
182. Norden AD, Drappatz J, Wen PY. Advances in meningioma therapy. Curr Neurol Neurosci Rep 2009;9(3): 231–240.
183. Sioka C, Kyritsis AP. Chemotherapy, hormonal therapy, and immunotherapy for recurrent meningiomas. J Neurooncol 2009;92(1): 1–6.
184. Commins DL, Atkinson RD, Burnett ME. Review of meningioma histopathology. Neurosurg Focus 2007;23(4): E3.
185. Newton HB, Chemotherapy of meningiomas. Newton HB, Hand-Book of Brain Tumor Chemotherapy. London: Academic Press, Elsevier; 2006. p 463–474.
186. Wen PY, Drappatz J. Novel therapies for meningiomas. Expert Rev Neurother 2006;6(10): 1447–1464.
187. Boscaro M, Ludlam WH, Atkinson B, Glusman JE, Petersenn S, Reincke M, Snyder P, Tabarin A, Biller BM, Findling J, Melmed S, Darby CH, Hu K, Wang Y, Freda PU, Grossman AB, Frohman LA, Bertherat J. Treatment of pituitary-dependent Cushing's disease with the multireceptor ligand somatostatin analog pasireotide (SOM230): a multicenter, phase II trial. J Clin Endocrinol Metab 2009;94(1): 115–122.
188. Gupta V, Su YS, Samuelson CG, Liebes LF, Chamberlain MC, Hofman FM, Schönthal AH, Chen TC. Irinotecan: a potential new chemotherapeutic agent for atypical or malignant meningiomas. J Neurosurg 2007;106(3): 455–462.
189. Asthagiri AR, Parry DM, Butman JA, Kim HJ, Tsilou ET, Zhuang Z, Lonser RR. Neurofibromatosis type 2. Lancet 2009;373(9679): 1974–1986.
190. Gerber PA, Antal AS, Neumann NJ, Homey B, Matuschek C, Peiper M, Budach W, Bölke E. Neurofibromatosis. Eur J Med Res 2009;14(3): 102–105.
191. Zou C, Smith KD, Liu J, Lahat G, Myers S, Wang WL, Zhang W, McCutcheon IE, Slopis JM, Lazar AJ, Pollock RE, Lev D. Clinical, pathological, and molecular variables predictive of malignant peripheral nerve sheath tumor outcome. Ann Surg 2009;249(6): 1014–1022.
192. Fang Y, Elahi A, Denley RC, Rao PH, Brennan MF, Jhanwar SC. Molecular characterization of permanent cell lines from primary, metastatic and recurrent malignant peripheral nerve sheath tumors (MPNST) with underlying neurofibromatosis-1. Anticancer Res 2009;29(4): 1255–1262.
193. Zou CY, Smith KD, Zhu QS, Liu J, McCutcheon IE, Slopis JM, Meric-Bernstam F, Peng Z, Bornmann WG, Mills GB, Lazar AJ, Pollock RE, Lev D. Dual targeting of AKT and mammalian target of rapamycin: a potential therapeutic approach for malignant peripheral nerve sheath tumor. Mol Cancer Ther 2009; 8(5): 1157–1168.
194. Asthagiri AR, Helm GA, Sheehan JP. Current concepts in management of meningiomas and schwannomas. Neurol Clin 2007;25(4): 1209–1230, xi.

195. Huang JH, Simon SL, Nagpal S, Nelson PT, Zager EL. Management of patients with schwannomatosis: report of six cases and review of the literature. Surg Neurol 2004;62(4): 353–361.
196. Hochberg FH, Baehring JM, Hochberg EP. Primary CNS lymphoma. Nat Clin Pract Neurol 2007;3(1): 24–35.
197. Algazi AP, Kadoch C, Rubenstein JL. Biology and treatment of primary central nervous system lymphoma. Neurotherapeutics 2009;6(3): 587–597.
198. Kadoch C, Treseler P, Rubenstein JL. Molecular pathogenesis of primary central nervous system lymphoma. Neurosurg Focus 2006; 21(5): E1.
199. DeAngelis LM. Primary central nervous system lymphoma: coming or going? Blood 2009; 113(19): 4483–4484.
200. McCann KJ, Ashton-Key M, Smith K, Stevenson FK, Ottensmeier CH. Primary central nervous system lymphoma: tumor-related clones exist in the blood and bone marrow with evidence for separate development. Blood 113 (19): 2009; 4677–4680.
201. DeAngelis LM, Iwamoto FM. An update on therapy of primary central nervous system lymphoma. Hematology Am Soc Hematol Educ Program 2006;1311–316.
202. Ferreri AJ, Reni M. Primary central nervous system lymphoma. Crit Rev Oncol Hematol 2007;63(3): 257–268.
203. Ekenel M, DeAngelis LM. Treatment of primary central nervous system lymphoma. Curr Neurol Neurosci Rep 2007;7(3): 191–199.
204. El Kamar FG, Abrey LE, Chemotherapy for primary central nervous system lymphoma. Newton HB, editor. Hand-Book of Brain Tumor Chemotherapy Academic Press, Elsevier; London 2006. p 395–406.
205. Ekenel M, Iwamoto FM, Ben-Porat LS, Panageas KS, Yahalom J, DeAngelis LM, Abrey LE. Primary central nervous system lymphoma: the role of consolidation treatment after a complete response to high-dose methotrexate-based chemotherapy. Cancer 2008;113(5): 1025–1031.
206. Gerstner E, Batchelor T. Primary CNS lymphoma. Expert Rev Anticancer Ther 2007;7(5): 689–700.
207. Morris PG, Abrey LE. Therapeutic challenges in primary CNS lymphoma. Lancet Neurol 2009;8(6): 581–592.
208. Schwer AL, Gaspar LE. Update in the treatment of brain metastases from lung cancer. Clin Lung Cancer 2006;8(3): 180–186.
209. Wadasadawala T, Gupta S, Bagul V, Patil N. Brain metastases from breast cancer: management approach. J Cancer Res Ther 2007;3(3): 157–165.
210. Mongan JP, Fadul CE, Cole BF, Zaki BI, Suriawinata AA, Ripple GH, Tosteson TD, Pipas JM. Brain metastases from colorectal cancer: risk factors, incidence, and the possible role of chemokines. Clin Colorectal Cancer 2009;8(2): 100–105.
211. McWilliams RR, Rao RD, Buckner JC, Link MJ, Markovic S, Brown PD. Melanoma-induced brain metastases. Expert Rev Anticancer Ther 2008;8(5): 743–755.
212. Shibui S. Treatment of metastatic brain tumours. Int J Clin Oncol 2009;14(4): 273–274.
213. Suh JH, Chao ST, Vogelbaum MA. Management of brain metastases. Curr Neurol Neurosci Rep 2009;9(3): 223–230.
214. Shibamoto Y, Sugie C, Iwata H. Radiotherapy for metastatic brain tumors. Int J Clin Oncol 2009;14(4): 281–288.
215. Newton HB. Chemotherapy for the treatment of metastatic brain tumors. Expert Rev Anticancer Ther 2002;2(5): 495–506.
216. Walbert T, Gilbert MR. The role of chemotherapy in the treatment of patients with brain metastases from solid tumors. Int J Clin Oncol 2009;14(4): 299–306.
217. Guillamo JS, Emery E, Busson A, Lechapt-Zalcman E, Constans JM, Defer GL. Current management of brain metastases. Rev Neurol 2008;164(6–7): 560–568.
218. Aragon-Ching JB, Zujewski JA. CNS metastasis: an old problem in a new guise. Clin Cancer Res 2007;13(6): 1644–1647.
219. Soffietti R, Rudà R, Trevisan E. Brain metastases: current management and new developments. Curr Opin Oncol 2008;20(6): 676–684.
220. Kosmas C, Tsakonas G, Mylonakis N. Treatment strategies in CNS metastases. Expert Opin Pharmacother 2008;9(12): 2087–2098.
221. Franceschi E, Tosoni A, Bartolini S, Mazzocchi V, Fioravanti A, Brandes AA. Treatment options for recurrent glioblastoma: pitfalls and future trends. Expert Rev Anticancer Ther 2009;9(5): 613–619.
222. Sarkaria JN, Kitange GJ, James CD, Plummer R, Calvert H, Weller M, Wick W. Mechanisms of chemoresistance to alkylating agents in malignant glioma. Clin Cancer Res 2008; 14(10): 2900–2908.
223. Jensen RL. Brain tumor hypoxia: tumorigenesis, angiogenesis, imaging, pseudoprogres-

sion, and as a therapeutic target. J Neurooncol 2009;92(3): 317–335.

224. Li Z, Bao S, Wu Q, Wang H, Eyler C, Sathornsumetee S, Shi Q, Cao Y, Lathia J, McLendon RE, Hjelmeland AB, Rich JN, Cancer Cell 2009;15(6): 501–513.

225. Murat A, Migliavacca E, Hussain SF, Heimberger AB, Desbaillets I, Hamou MF, Rüegg C, Stupp R, Delorenzi M, Hegi ME. Modulation of angiogenic and inflammatory response in glioblastoma by hypoxia. PLoS One 2009;4(6): e5947.

226. Salhia B, Tran NL, Symons M, Winkles JA, Rutka JT, Berens ME. Molecular pathways triggering glioma cell invasion. Expert Rev Mol Diagn 2006;6(4): 613–626.

227. Nakada M, Nakada S, Demuth T, Tran NL, Hoelzinger DB, Berens ME. Molecular targets of glioma invasion. Cell Mol Life Sci 2007;64 (4): 458–478.

228. Bakker A, Caricasole A, Gaviraghi G, Pollio G, Robertson G, Terstappen GC, Salerno M, Tunici P. How to achieve confidence in drug discovery and development: managing risk (from a reductionist to a holistic approach). ChemMedChem 2009;4(6): 923–933.

229. Xie Q, Thompson R, Hardy K, DeCamp L, Berghuis B, Sigler R, Knudsen B, Cottingham S, Zhao P, Dykema K, Cao B, Resau J, Hay R, Vande Woude G.F.. A highly invasive human glioblastoma pre-clinical model for testing therapeutics. J Transl Med 2008;6:77.

230. D'Abaco GM, Kaye AH. Integrins: molecular determinants of glioma invasion. J Clin Neurosci 2007;14(11): 1041–1048.

231. Grommes C, Landreth GE, Sastre M, Beck M, Feinstein DL, Jacobs AH, Schlegel U, Heneka MT. Inhibition of in vivo glioma growth and invasion by peroxisome proliferator-activated receptor gamma agonist treatment. Mol Pharmacol. 2006;70(5):1524–1533.

232. Lucio-Eterovic AK, Piao Y, de Groot JF. Mediators of glioblastoma resistance and invasion during antivascular endothelial growth factor therapy. Clin Cancer Res 2009;15(14): 4589–4599.

233. Wang H, Lathia JD, Wu Q, Wang J, Li Z, Heddleston JM, Eyler CE, Elderbroom J, Gallagher J, Schuschu J, Macswords J, Cao Y, McLendon RE, Wang XF, Hjelmeland AB, Rich JN. Targeting interleukin 6 signaling suppresses glioma stem cell survival and tumor growth. Stem Cells 2009;27(10): 2393–2404.

234. Saidi A, Hagedorn M, Allain N, Verpelli C, Sala C, Bello L, Bikfalvi A, Javerzat S. Combined targeting of interleukin-6 and vascular endothelial growth factor potently inhibits glioma growth and invasiveness. Int J Cancer 2009;125(5): 1054–1064.

235. Drappatz J, Norden AD, Wen PY. Therapeutic strategies for inhibiting invasion in glioblastoma. Expert Rev Neurother 2009;9(4): 519–534.

236. Newton HB. Molecular neuro-oncology and the development of targeted therapeutic strategies for brain tumors. Part 3. Brain tumor invasiveness. Expert Rev Anticancer Ther 42004; (5): 803–821.

237. Lo HW, Zhu H, Cao X, Aldrich A, Ali-Osman F. A novel splice variant of GLI1 that promotes glioblastoma cell migration and invasion. Cancer Res 69:2009; 6790–6798.

238. Chi A, Norden AD, Wen PY. Inhibition of angiogenesis and invasion in malignant gliomas. Expert Rev Anticancer Ther 2007;7(11): 1537–1560.

239. Lamszus K, Kunkel P, Westphal M. Invasion as limitation to anti-angiogenic glioma therapy. Acta Neurochir Suppl 2003;88:169–177.

240. Anderson JC, McFarland BC, Gladson CL. New molecular targets in angiogenic vessels of glioblastoma tumours. Expert Rev Mol Med 2008;10:e23.

241. Martin V, Liu D, Gomez-Manzano C. Encountering and advancing through antiangiogenesis therapy for gliomas. Curr Pharm Des 2009;15(4): 353–364.

242. Dietrich J, Norden AD, Wen PY. Emerging antiangiogenic treatments for gliomas: efficacy and safety issues. Curr Opin Neurol 2008;21(6): 736–744.

243. Chamberlain MC. Antiangiogenic blockage: a new treatment for glioblastoma. Expert Opin Biol Ther 2008;8(10): 1449–1453.

244. Reardon DA, Desjardins A, Rich JN, Vredenburgh JJ. The emerging role of anti-angiogenic therapy for malignant glioma. Curr Treat Options Oncol 2008;9(1): 1–22.

245. Norden AD, Drappatz J, Wen PY. Novel antiangiogenic therapies for malignant gliomas. Lancet Neurol 2008;7(12): 1152–1160.

246. Wong ET, Brem S. Antiangiogenesis treatment for glioblastoma multiforme: challenges and opportunities. J Natl Compr Canc Netw 6 (5): 2008; 515–522.

247. Dietrich J, Wang D, Batchelor TT. Cediranib: profile of a novel anti-angiogenic agent in patients with glioblastoma. Expert Opin Investig Drugs 2009;18(10): 1549–1557.

248. Jain RK, di Tomaso E, Duda DG, Loeffler JS, Sorensen AG, Batchelor TT. Biomarkers of

248. ... reponse and resistance to antiangiogenic therapy. Nat Rev Neurosci 2007;8(8): 610–622.
249. de Boüard S, Herlin P, Christensen JG, Lemoisson E, Gauduchon P, Raymond E, Guillamo JS. Antiangiogenic and anti-invasive effects of sunitinib on experimental human glioblastoma. Neuro Oncol 2007;9(4): 412–423.
250. Verhoeff JJ, van Tellingen O, Claes A, Stalpers LJ, van Linde ME, Richel DJ, Leenders WP, van Furth WR. Concerns about anti-angiogenic treatment in patients with glioblastoma multiforme. BMC Cancer 2009;9(1): 444.
251. Claes A, Wesseling P, Jeuken J, Maass C, Heerschap A, Leenders WP. Antiangiogenic compounds interfere with chemotherapy of brain tumors due to vessel normalization. Mol Cancer Ther 2008;7(1): 71–78.
252. Newton HB. Small-molecule and antibody approaches to molecular chemotherapy of primary brain tumors. Curr Opin Investig Drugs 2007;8(12): 1009–1021.
253. Omuro AM, Faivre S, Raymond E. Lessons learned in the development of targeted therapy for malignant gliomas. Mol Cancer Ther 2007;6(7): 1909–1919.
254. Omuro AM. Exploring multi-targeting strategies for the treatment of gliomas. Curr Opin Investig Drugs 2008;9(12): 1287–1295.
255. Nieder C, Mehta MP, Jalali R. Combined radio- and chemotherapy of brain tumours in adult patients. Clin Oncol (R Coll Radiol) 2009;21(7): 515–524.
256. Dehdashti AR, Hegi ME, Regli L, Pica A, Stupp R. New trends in the medical management of glioblastoma multiforme: the role of temozolomide chemotherapy. Neurosurg Focus 2006; 20(4): E6.
257. Wick W, Platten M, Weller M. New (alternative) temozolomide regimens for the treatment of glioma. Neuro Oncol 2009;11(1): 69–79.
258. Jagannathan J, Prevedello DM, Dumont AS, Laws ER. Cellular signaling molecules as therapeutic targets in glioblastoma multiforme. Neurosurg Focus 2006;20(4): E8.
259. Abounader R. Interactions between PTEN and receptor tyrosine kinase pathways and their implications for glioma therapy. Expert Rev Anticancer Ther 2009;9(2): 235–245.
260. Ren H, Yang BF, Rainov NG. Receptor tyrosine kinases as therapeutic targets in malignant glioma. Rev Recent Clin Trials 2007;2(2): 87–101.
261. Vogt PK, Hart JR. Akt demoted in glioblastoma. Sci Signal 2009;2(67): pe26.
262. Russell C, Acevedo-Duncan M. Effects of the PKC inhibitor PD 406976 on cell cycle progression, proliferation, PKC isozymes and apoptosis in glioma and SVG-transformed glial cells. Cell Prolif 2005;38(2): 87–106.
263. Teicher BA. Protein kinase C as a therapeutic target. Clin Cancer Res 2006;12(18): 5336–5345.
264. Graff JR, McNulty AM, Hanna KR, Konicek BW, Lynch RL, Bailey SN, Banks C, Capen A, Goode R, Lewis JE, Sams L, Huss KL, Campbell RM, Iversen PW, Neubauer BL, Brown TJ, Musib L, Geeganage S, Thornton D. The protein kinase Cbeta-selective inhibitor, Enzastaurin (LY317615.HCl), suppresses signaling through the AKT pathway, induces apoptosis, and suppresses growth of human colon cancer and glioblastoma xenografts. Cancer Res 2005;65(16): 7462–7469.
265. Kreisl TN, Kim L, Moore K, Duic P, Kotliarova S, Walling J, Musib L, Thornton D, Albert PS, Fine HA. A phase I trial of Enzastaurin in patients with recurrent gliomas. Clin Cancer Res 2009;15(10): 3617–3623.
266. Huang TT, Sarkaria SM, Cloughesy TF, Mischel PS. Targeted therapy for malignant glioma patients: lessons learned and the road ahead. Neurotherapeutics 2009;6(3): 500–512.
267. Mercer RW, Tyler MA, Ulasov IV, Lesniak MS. Targeted therapies for malignant glioma: progress and potential. BioDrugs 2009;23(1): 25–35.
268. Kapoor GS, O'Rourke DM. Receptor tyrosine kinase signaling in gliomagenesis: pathobiology and therapeutic approaches. Cancer Biol Ther 2003;2(4): 330–342.
269. Johnston JB, Navaratnam S, Pitz MW, Maniate JM, Wiechec E, Baust H, Gingerich J, Skliris GP, Murphy LC, Los M. Targeting the EGFR pathway for cancer therapy. Curr Med Chem 2006;13(29): 3483–3492.
270. Huang PH, Xu AM, White FM. Oncogenic EGFR signaling networks in glioma. Sci Signal 2009;2(87): re6.
271. Voelzke WR, Petty WJ, Lesser GJ. Targeting the epidermal growth factor receptor in high-grade astrocytomas. Curr Treat Options Oncol 2008;9(1): 23–31.
272. Huang PH, Mukasa A, Bonavia R, Flynn RA, Brewer ZE, Cavenee WK, Furnari FB, White FM. Quantitative analysis of EGFRvIII cellular signalling networks reveals a combinatorial therapeutic strategy for glioblastoma. Proc Natl Acad Sci USA 2007;104(31): 12867–12872.

273. Guillamo JS, de Boüard S, Valable S, Marteau L, Leuraud P, Marie Y, Poupon MF, Parienti JJ, Raymond E, Peschanski M. Molecular mechanisms underlying effects of epidermal growth factor receptor inhibition on invasion, proliferation, and angiogenesis in experimental glioma. Clin Cancer Res 2009;15(11): 3697–3704.

274. Naran S, Zhang X, Hughes SJ. Inhibition of HGF/MET as therapy for malignancy. Expert Opin Ther Targets 2009;13(5): 569–581.

275. Lal B, Goodwin CR, Sang Y, Foss CA, Cornet K, Muzamil S, Pomper MG, Kim J, Laterra J. EGFRvIII and c-Met pathway inhibitors synergize against PTEN-null/EGFRvIII+ glioblastoma xenografts. Mol Cancer Ther 2009; 8(7): 1751–1760.

276. Li Y, Guessous F, DiPierro C, Zhang Y, Mudrick T, Fuller L, Johnson E, Marcinkiewicz L, Engelhardt M, Kefas B, Schiff D, Kim J, Abounader R. Interactions between PTEN and the c-Met pathway in glioblastoma and implications for therapy. Mol Cancer Ther 2009;8(2): 376–385.

277. Provençal M, Labbé D, Veitch R, Boivin D, Rivard GE, Sartelet H, Robitaille Y, Gingras D, Béliveau R. c-Met activation in medulloblastoma induces tissue factor expression and activity: effects on cell migration. Carcinogenesis 30(7): 2009; 1089–1096.

278. Bannen LC, Chan DSM, Dalrymple LE, Jammalamadaka VKRG, Leahy JW, Mac MB, Mann G, Mann, LW, Nuss JM, Parks JJ, Wang Y, Xu W,c-Met modulators and method of use. WO2006014325.

279. Timofeevski SL, McTigue MA, Ryan K, Cui J, Zou HY, Zhu JX, Chau F, Alton G, Karlicek S, Christensen JG, Murray BW. Enzymatic characterization of c-Met receptor tyrosine kinase oncogenic mutants and kinetic studies with aminopyridine and triazolopyrazine inhibitors. Biochemistry 2009;48(23): 5339–5349.

280. Tseng JR, Kang KW, Dandekar M, Yaghoubi S, Lee JH, Christensen JG, Muir S, Vincent PW, Michaud NR, Gambhir SS. Preclinical efficacy of the c-Met inhibitor CE-355621 in a U87 MG mouse xenograft model evaluated by 18F-FDG small-animal PET. J Nucl Med 2008;49(1): 129–134.

281. de Groot JF, Yung WK. Bevacizumab and irinotecan in the treatment of recurrent malignant gliomas. Cancer J 2008;14(5): 279–285.

282. Argyriou AA, Giannopoulou E, Kalofonos HP. Angiogenesis and anti-angiogenic molecularly targeted therapies in malignant gliomas. Oncology 2009;77(1): 1–11.

283. Reardon DA, Wen PY, Desjardins A, Batchelor TT. Vredenburgh JJ Glioblastoma multiforme: an emerging paradigm of anti-VEGF therapy. Expert Opin Biol Ther 2008;8(4): 541–553.

284. Miletic H, Niclou SP, Johansson M, Bjerkvig R. Anti-VEGF therapies for malignant glioma: treatment effects and escape mechanisms. Expert Opin Ther Targets 2009;13(4): 455–468.

285. Gerstner ER, Duda DG, di Tomaso E, Ryg PA, Loeffler JS, Sorensen AG, Ivy P, Jain RK, Batchelor TT. VEGF inhibitors in the treatment of cerebral edema in patients with brain cancer. Nat Rev Clin Oncol 2009;6(4): 229–236.

286. Zhang J, Zhao P, Fu Z, Chen X, Liu N, Lu A, Li R, Shi L, Pu P, Kang C, You Y. Glioma cells enhance endothelial progenitor cell angiogenesis via VEGFR-2, not VEGFR-1. Oncol Rep 2008;20(6): 1457–1463.

287. Sandström M, Johansson M, Bergström P, Bergenheim AT, Henriksson R. Effects of the VEGFR inhibitor ZD6474 in combination with radiotherapy and temozolomide in an orthotopic glioma model. J Neurooncol 2008;88(1): 1–9.

288. Majumdar K, Radotra BD, Vasishta RK, Pathak A. Platelet-derived growth factor expression correlates with tumor grade and proliferative activity in human oligodendrogliomas. Surg Neurol 2009;72(1): 54–60.

289. Milker-Zabel S, Zabel-du Bois A, Ranai G, Trinh T, Unterberg A, Debus J, Lipson KE, Abdollahi A, Huber PE. SU11657 enhances radiosensitivity of human meningioma cells. Int J Radiat Oncol Biol Phys 2008;70(4): 1213–1218.

290. Newton HB. Molecular neuro-oncology and development of targeted therapeutic strategies for brain tumors. Part 2. PI3K/Akt/PTEN, mTOR, SHH/PTCH and angiogenesis. Expert Rev Anticancer Ther 2004;4(1): 105–128.

291. Powis G, Ihle NT, Yung WK. Inhibiting PI-3-K for glioma therapy. Cell Cycle 2009;8(3): 335.

292. Guillard S, Clarke PA, Te Poele R, Mohri Z, Bjerke L, Valenti M, Raynaud F, Eccles SA, Workman P. Molecular pharmacology of phosphatidylinositol 3-kinase inhibition in human glioma. Cell Cycle 2009;8(3): 443–453.

293. Raynaud FI, Eccles SA, Patel S, Alix S, Box G, Chuckowree I, Folkes A, Gowan S, De Haven Brandon A, Di Stefano F, Hayes A, Henley AT, Lensun L, Pergl-Wilson G, Robson A, Saghir N, Zhyvoloup A, McDonald E, Sheldrake P, Shuttleworth S, Valenti M, Wan NC, Clarke PA,

Workman P. Biological properties of potent inhibitors of class I phosphatidylinositide 3-kinases: from PI-103 through PI-540, PI-620 to the oral agent GDC-0941. Mol Cancer Ther 2009;8(7): 1725–1738.

294. Folkes AJ, Ahmadi K, Alderton WK, Alix S, Baker SJ, Box G, Chuckowree IS, Clarke PA, Depledge P, Eccles SA, Friedman LS, Hayes A, Hancox TC, Kugendradas A, Lensun L, Moore P, Olivero AG, Pang J, Patel S, Pergl-Wilson GH, Raynaud FI, Robson A, Saghir N, Salphati L, Sohal S, Ultsch MH, Valenti M, Wallweber HJ, Wan NC, Wiesmann C, Workman P, Zhyvoloup A, Zvelebil MJ, Shuttleworth SJ. The identification of 2-(1H-indazol-4-yl)-6-(4-methanesulfonyl-piperazin-1-ylmethyl)-4-morpholin-4-yl-thieno[3,2-d]pyrimidine (GDC-0941) as a potent, selective, orally bioavailable inhibitor of class I PI3 kinase for the treatment of cancer. J Med Chem 2008;51(18): 5522–5532.

295. Gallia GL, Tyler BM, Hann CL, Siu IM, Giranda VL, Vescovi AL, Brem H, Riggins GJ. Inhibition of Akt inhibits growth of glioblastoma and glioblastoma stem-like cells. Mol Cancer Ther 2009;8(2): 386–393.

296. Koul D, Shen R, Bergh S, Lu Y, de Groot JF, Liu TJ, Mills GB, Yung WK. Targeting integrin-linked kinase inhibits Akt signaling pathways and decreases tumor progression of human glioblastoma. Mol Cancer Ther 2005;4(11): 1681–1688.

297. Edwards LA, Woo J, Huxham LA, Verreault M, Dragowska WH, Chiu G, Rajput A, Kyle AH, Kalra J, Yapp D, Yan H, Minchinton AI, Huntsman D, Daynard T, Waterhouse DN, Thiessen B, Dedhar S, Bally MB. Suppression of VEGF secretion and changes in glioblastoma multiforme microenvironment by inhibition of integrin-linked kinase (ILK). Mol Cancer Ther 2008;7(1): 59–70.

298. D'Abaco GM, Kaye AH. Integrin-linked kinase: a potential therapeutic target for the treatment of glioma. J Clin Neurosci 2008;15(10): 1079–1084.

299. Cloughesy TF, Yoshimoto K, Nghiemphu P, Brown K, Dang J, Zhu S, Hsueh T, Chen Y, Wang W, Youngkin D, Liau L, Martin N, Becker D, Bergsneider M, Lai A, Green R, Oglesby T, Koleto M, Trent J, Horvath S, Mischel PS, Mellinghoff IK, Sawyers CL. Antitumor activity of rapamycin in a Phase I trial for patients with recurrent PTEN-deficient glioblastoma. PLoS Med 2008;5(1): e8.

300. Kuhn JG, Chang SM, Wen PY, Cloughesy TF, Greenberg H, Schiff D, Conrad C, Fink KL, Robins HI, Mehta M, DeAngelis L, Raizer J, Hess K, Lamborn KR, Dancey J, Prados MD. North American Brain Tumor Consortium and the National Cancer Institute. Pharmacokinetic and tumor distribution characteristics of temsirolimus in patients with recurrent malignant glioma. Clin Cancer Res 2007;13(24): 7401–7406.

301. Chang SM, Wen P, Cloughesy T, Greenberg H, Schiff D, Conrad C, Fink K, Robins HI, De Angelis L, Raizer J, Hess K, Aldape K, Lamborn KR, Kuhn J, Dancey J, Prados MD. North American Brain Tumor Consortium and the National Cancer Institute. Phase II study of CCI-779 in patients with recurrent glioblastoma multiforme. Invest New Drugs 2005;23(4): 357–361.

302. Liu TJ, Koul D, LaFortune T, Tiao N, Shen RJ, Maira SM, Garcia-Echevrria C, Yung WK. NVP-BEZ235, a novel dual phosphatidylinositol 3-kinase/mammalian target of rapamycin inhibitor, elicits multifaceted antitumor activities in human gliomas. Mol Cancer Ther 2009;8(8): 2204–2210.

303. Ronellenfitsch MW, Brucker DP, Burger MC, Wolking S, Tritschler F, Rieger J, Wick W, Weller M, Steinbach JP. Antagonism of the mammalian target of rapamycin selectively mediates metabolic effects of epidermal growth factor receptor inhibition and protects human malignant glioma cells from hypoxia-induced cell death. Brain 2009;132(Pt 6): 1509–1522.

304. Fan QW, Cheng CK, Nicolaides TP, Hackett CS, Knight ZA, Shokat KM, Weiss WA. A dual phosphoinositide-3-kinase alpha/mTOR inhibitor cooperates with blockade of epidermal growth factor receptor in PTEN-mutant glioma. Cancer Res 2007;67(17): 7960–7965.

305. Hjelmeland AB, Lattimore KP, Fee BE, Shi Q, Wickman S, Keir ST, Hjelmeland MD, Batt D, Bigner DD, Friedman HS, Rich JN. The combination of novel low molecular weight inhibitors of RAF (LBT613) and target of rapamycin (RAD001) decreases glioma proliferation and invasion. Mol Cancer Ther 2007;6(9): 2257–2449.

306. Newton HB. Molecular neuro-oncology and development of targeted therapeutic strategies for brain tumors. Part 1. Growth factor and Ras signaling pathways. Expert Rev Anticancer Ther 2003;3(5): 595–614.

307. Wong KK. Recent developments in anti-cancer agents targeting the Ras/Raf/MEK/ERK pathway. Recent Pat Anticancer Drug Discov 2009;4(1): 28–35.

308. Rattan R, Giri S, Singh AK, Singh I. Rho/ROCK pathway as a target of tumor therapy. J Neurosci Res 2006;83(2): 243–255.
309. Oellers P, Schröer U, Senner V, Paulus W, Thanos S. ROCKs are expressed in brain tumors and are required for glioma-cell migration on myelinated axons. Glia 2009;57(5): 499–509.
310. Zohrabian VM, Forzani B, Chau Z, Murali R, Jhanwar-Uniyal M. Rho/ROCK and MAPK signaling pathways are involved in glioblastoma cell migration and proliferation. Anticancer Res 2009;29(1): 119–123.
311. Kaynar MY, Sanus GZ, Hnimoglu H, Kacira T, Kemerdere R, Atukeren P, Gumustas K, Canbaz B, Tanriverdi T. Expression of hypoxia inducible factor-1alpha in tumors of patients with glioblastoma multiforme and transitional meningioma. J Clin Neurosci 2008;15(9): 1036–1042.
312. Li Z, Bao S, Wu Q, Wang H, Eyler C, Sathornsumetee S, Shi Q, Cao Y, Lathia J, McLendon RE, Hjelmeland AB, Rich JN. Hypoxia-inducible factors regulate tumorigenic capacity of glioma stem cells. Cancer Cell 2009;15(6): 501–513.
313. Kaur B, Khwaja FW, Severson EA, Matheny SL, Brat DJ, Van Meir EG. Hypoxia and the hypoxia-inducible-factor pathway in glioma growth and angiogenesis. Neuro Oncol 2005;7(2): 134–153.
314. Fujiwara S, Nakagawa K, Harada H, Nagato S, Furukawa K, Teraoka M, Seno T, Oka K, Iwata S, Ohnishi T. Silencing hypoxia-inducible factor-1alpha inhibits cell migration and invasion under hypoxic environment in malignant gliomas. Int J Oncol 2007;30(4): 793–802.
315. Radaelli E, Ceruti R, Patton V, Russo M, Degrassi A, Croci V, Caprera F, Stortini G, Scanziani E, Pesenti E, Alzani R. Immunohistopathological and neuroimaging characterization of murine orthotopic xenograft models of glioblastoma multiforme recapitulating the most salient features of human disease. Histol Histopathol 2009;24(7): 879–891.
316. Ali MA, Reis A, Ding LH, Story MD, Habib AA, Chattopadhyay A, Saha D. SNS-032 prevents hypoxia-mediated glioblastoma cell invasion by inhibiting hypoxia inducible factor-1alpha expression. Int J Oncol 2009;34(4): 1051–1060.
317. Li L, Lin X, Shoemaker AR, Albert DH, Fesik SW, Shen Y. Hypoxia-inducible factor-1 inhibition in combination with temozolomide treatment exhibits robust antitumor efficacy in vivo. Clin Cancer Res 2006;12(15): 4747–4754.
318. Tan C, de Noronha RG, Roecker AJ, Pyrzynska B, Khwaja F, Zhang Z, Zhang H, Teng Q, Nicholson AC, Giannakakou P, Zhou W, Olson JJ, Pereira MM, Nicolaou KC, Van Meir EG. Identification of a novel small-molecule inhibitor of the hypoxia-inducible factor 1 pathway. Cancer Res 2005;65(2): 605–612.
319. Rapisarda A, Uranchimeg B, Scudiero DA, Selby M, Sausville EA, Shoemaker RH, Melillo G. Identification of small molecule inhibitors of hypoxia-inducible factor 1 transcriptional activation pathway. Cancer Res 2002;62(15): 4316–4324.
320. Heddleston JM, Li Z, McLendon RE, Hjelmeland AB, Rich JN. The hypoxic microenvironment maintains glioblastoma stem cells and promotes reprogramming towards a cancer stem cell phenotype. Cell Cycle 2009;8(20): 3274–3284.
321. Ader I, Malavaud B, Cuvillier O. When the sphingosine kinase 1/sphingosine 1-phosphate pathway meets hypoxia signaling: new targets for cancer therapy. Cancer Res 2009;69(9): 3723–3726.
322. Skuli N, Monferran S, Delmas C, Favre G, Bonnet J, Toulas C, Cohen-Jonathan ME. Alphavbeta3/alphavbeta5 integrins-FAK-RhoB: a novel pathway for hypoxia regulation in glioblastoma. Cancer Res 2009;69(8): 3308–3316.
323. Schnell O, Krebs B, Wagner E, Romagna A, Beer AJ, Grau SJ, Thon N, Goetz C, Kretzschmar HA, Tonn JC, Goldbrunner RH. Expression of integrin alphavbeta3 in gliomas correlates with tumor grade and is not restricted to tumor vasculature. Brain Pathol 2008;18(3): 378–386.
324. Kinoshita M, Izumoto S, Hashimoto N, Kishima H, Kagawa N, Hashiba T, Chiba Y, Yoshimine T. Immunohistochemical analysis of adhesion molecules and matrix metalloproteinases in malignant CNS lymphomas: a study comparing primary CNS malignant and CNS intravascular lymphomas. Brain Tumor Pathol 2008;25(2): 73–78.
325. Fiorilli P, Partridge D, Staniszewska I, Wang JY, Grabacka M, So K, Marcinkiewicz C, Reiss K, Khalili K, Croul SE. Integrins mediate adhesion of medulloblastoma cells to tenascin and activate pathways associated with survival and proliferation. Lab Invest 2008;88(11): 1143–1156.
326. Färber K, Synowitz M, Zahn G, Vossmeyer D, Stragies R van Rooijen N Kettenmann H An alpha5beta1 integrin inhibitor attenuates glioma growth. Mol Cell Neurosci 2008;39(4): 579–585.

327. Maglott A, Bartik P, Cosgun S, Klotz P, Rondé P, Fuhrmann G, Takeda K, Martin S, Dontenwill M. The small alpha5beta1 integrin antagonist, SJ749, reduces proliferation and clonogenicity of human astrocytoma cells. Cancer Res 2006;66(12): 6002–6007.

328. Maurer GD, Tritschler I, Adams B, Tabatabai G, Wick W, Stupp R, Weller M. Cilengitide modulates attachment and viability of human glioma cells, but not sensitivity to irradiation or temozolomide in vitro. Neuro Oncol 2009; 11(6): 747–756.

329. Reardon DA, Nabors LB, Stupp R, Mikkelsen T. Cilengitide: an integrin-targeting arginine-glycine-aspartic acid peptide with promising activity for glioblastoma multiforme. Expert Opin Investig Drugs 2008;17(8): 1225–1235.

330. Chamberlain MC. Cilengitide: does it really represent a new targeted therapy for recurrent glioblastoma? J Clin Oncol 2009;27(11): 1921.

331. Newton HB. Molecular neuro-oncology and the development of targeted therapeutic strategies for brain tumors. Part 4. p53 signaling pathway. Expert Rev Anticancer Ther 2005; 5(1): 177–191.

332. Toledo F, Wahl GM. MDM2 and MDM4: p53 regulators as targets in anticancer therapy. Int J Biochem Cell Biol 2007;39(7–8): 1476–1482.

333. Rayburn ER, Ezell SJ, Zhang R. Recent advances in validating MDM2 as a cancer target. Anticancer Agents Med Chem 2009;9(8): 882–903.

334. Giordana MT, Duó D, Gasverde S, Trevisan E, Boghi A, Morra I, Pradotto L, Mauro A, Chió A. MDM2 overexpression is associated with short survival in adults with medulloblastoma. Neuro Oncol 2002;4(2): 115–122.

335. Sharma MC, Ghara N, Jain D, Sarkar C, Singh M, Mehta VS. A study of proliferative markers and tumor suppressor gene proteins in different grades of ependymomas. Neuropathology 2009;29(2): 148–155.

336. Riemenschneider MJ, Büschges R, Wolter M, Reifenberger J, Boström J, Kraus JA, Schlegel U, Reifenberger G. Amplification and overexpression of the MDM4 (MDMX) gene from 1q32 in a subset of malignant gliomas without TP53 mutation or MDM2 amplification. Cancer Res 1999;59(24): 6091–6096.

337. Khatri RG, Navaratne K, Weil RJ. The role of a single nucleotide polymorphism of MDM2 in glioblastoma multiforme. J Neurosurg 2008;109(5): 842–848.

338. Shangary S, Wang S. Small-molecule inhibitors of the MDM2-p53 protein–protein interaction to reactivate p53 function: a novel approach for cancer therapy. Annu Rev Pharmacol Toxicol 2009;49:223–241.

339. Murray JK, Gellman SH. Targeting protein–protein interactions: lessons from p53/MDM2. Biopolymers 2007;88(5): 657–686.

340. Tovar C, Rosinski J, Filipovic Z, Higgins B, Kolinsky K, Hilton H, Zhao X, Vu BT, Qing W, Packman K, Myklebost O, Heimbrook DC, Vassilev LT. Small-molecule MDM2 antagonists reveal aberrant p53 signaling in cancer: implications for therapy. Proc Natl Acad Sci USA 2006;103(6): 1888–1893.

341. Barboza JA, Iwakuma T, Terzian T, El-Naggar AK, Lozano G. Mdm2 and Mdm4 loss regulates distinct p53 activities. Mol Cancer Res 2008; 6(6): 947–954.

342. Canner JA, Sobo M, Ball S, Hutzen B, DeAngelis S, Willis W, Studebaker AW, Ding K, Wang S, Yang D, Lin J. MI-63: a novel small-molecule inhibitor targets MDM2 and induces apoptosis in embryonal and alveolar rhabdomyosarcoma cells with wild-type p53. Br J Cancer 2009;101(5): 774–781.

343. Shangary S, Qin D, McEachern D, Liu M, Miller RS, Qiu S, Nikolovska-Coleska Z, Ding K, Wang G, Chen J, Bernard D, Zhang J, Lu Y, Gu Q, Shah RB, Pienta KJ, Ling X, Kang S, Guo M, Sun Y, Yang D, Wang S. Temporal activation of p53 by a specific MDM2 inhibitor is selectively toxic to tumors and leads to complete tumor growth inhibition. Proc Natl Acad Sci USA 2008;105(10): 3933–3938.

344. Hardcastle IR, Ahmed SU, Atkins H, Farnie G, Golding BT, Griffin RJ, Guyenne S, Hutton C, Källblad P, Kemp SJ, Kitching MS, Newell DR, Norbedo S, Northen JS, Reid RJ, Saravanan K, Willems HM, Lunec J. Small-molecule inhibitors of the MDM2-p53 protein–protein interaction based on an isoindolinone scaffold. J Med Chem 2006;49(21): 6209–6221.

345. Lawrence HR, Li Z, Yip ML, Sung SS, Lawrence NJ, McLaughlin ML, McManus GJ, Zaworotko MJ, Sebti SM, Chen J, Guida WC. Identification of a disruptor of the MDM2-p53 protein–protein interaction facilitated by high-throughput in silico docking. Bioorg Med Chem Lett 2009;19(14): 3756–3759.

346. Pazgier M, Liu M, Zou G, Yuan W, Li C, Li C, Li J, Monbo J, Zella D, Tarasov SG, Lu W. Structural basis for high-affinity peptide inhibition of p53 interactions with MDM2 and MDMX. Proc Natl Acad Sci USA 2009;106(12): 4665–4670.

347. Wade M, Wahl GM, Targeting Mdm2 and Mdmx in cancer therapy: better living through medicinal chemistry? Mol Cancer Res 2009; 7(1): 1–11.
348. Tagscherer KE, Fassl A, Campos B, Farhadi M, Kraemer A, Böck BC, Macher-Goeppinger S, Radlwimmer B, Wiestler OD, Herold-Mende C, Roth W. Apoptosis-based treatment of glioblastomas with ABT-737, a novel small molecule inhibitor of Bcl-2 family proteins. Oncogene 2008;27(52): 6646–6656.
349. Endersby R, Baker SJ. PTEN signaling in brain: neuropathology and tumorigenesis. Oncogene 2008;27(41): 5416–5430.
350. Koul D. PTEN signaling pathways in glioblastoma. Cancer Biol Ther 2008;7(9): 1321–1325.
351. Haas-Kogan D, Stokoe D. PTEN in brain tumors. Expert Rev Neurother 2008;8(4): 599–610.
352. Xu Q, Yuan X, Liu G, Black KL, Yu JS. Hedgehog signaling regulates brain tumor-initiating cell proliferation and portends shorter survival for patients with PTEN-coexpressing glioblastomas. Stem Cells 2008;26(12): 3018–3026.
353. Xie J. Hedgehog signaling pathway: development of antagonists for cancer therapy. Curr Oncol Rep 2008;10(2): 107–113.
354. Kiselyov AS. Targeting the hedgehog signaling pathway with small molecules. Anticancer Agents Med Chem 2006;6(5): 445–449.
355. Mahindroo N, Punchihewa C, Fujii N. Hedgehog-Gli signaling pathway inhibitors as anticancer agents. J Med Chem 2009;52(13): 3829–3845.
356. Romer J, Curran T. Targeting medulloblastoma: small-molecule inhibitors of the Sonic Hedgehog pathway as potential cancer therapeutics. Cancer Res 2005;65(12): 4975–4978.
357. Chen JK, Taipale J, Cooper MK, Beachy PA. Inhibition of Hedgehog signaling by direct binding of cyclopamine to Smoothened. Genes Dev 2002;16(21): 2743–2748.
358. Feldmann G, Fendrich V, McGovern K, Bedja D, Bisht S, Alvarez H, Koorstra JB, Habbe N, Karikari C, Mullendore M, Gabrielson KL, Sharma R, Matsui W, Maitra A. An orally bioavailable small-molecule inhibitor of Hedgehog signaling inhibits tumor initiation and metastasis in pancreatic cancer. Mol Cancer Ther 2008;7(9): 2725–2735.
359. Tremblay MR, Lescarbeau A, Grogan MJ, Tan E, Lin G, Austad BC, Yu LC, Behnke ML, Nair SJ, Hagel M, White K, Conley J, Manna JD, Alvarez-Diez TM, Hoyt J, Woodward CN, Sydor JR, Pink M, MacDougall J, Campbell MJ, Cushing J, Ferguson J, Curtis MS, McGovern K, Read MA, Palombella VJ, Adams J, Castro AC. Discovery of a potent and orally active hedgehog pathway antagonist (IPI-926). J Med Chem 2009;52(14): 4400–4418.
360. Miller-Moslin K, Peukert S, Jain RK, McEwan MA, Karki R, Llamas L, Yusuff N, He F, Li Y, Sun Y, Dai M, Perez L, Michael W, Sheng T, Lei H, Zhang R, Williams J, Bourret A, Ramamurthy A, Yuan J, Guo R, Matsumoto M, Vattay A, Maniara W, Amaral A, Dorsch M, Kelleher JF 3rd. 1-Amino-4-benzylphthalazines as orally bioavailable smoothened antagonists with antitumor activity. J Med Chem 2009;52(13): 3954–3968.
361. Peukert S, Jain RK, Geisser A, Sun Y, Zhang R, Bourret A, Carlson A, Dasilva J, Ramamurthy A, Kelleher JF. Identification and structure–activity relationships of ortho-biphenyl carboxamides as potent Smoothened antagonists inhibiting the Hedgehog signaling pathway. Bioorg Med Chem Lett 2009;19(2): 328–331.
362. Büttner A, Seifert K, Cottin T, Sarli V, Tzagkaroulaki L, Scholz S, Giannis A. Synthesis and biological evaluation of SANT-2 and analogues as inhibitors of the Hedgehog signaling pathway. Bioorg Med Chem 2009;17(14): 4943–4954.
363. Rominger CM, Bee WL, Copeland RA, Davenport EA, Gilmartin A, Gontarek R, Hornberger KR, Kallal LA, Lai Z, Lawrie K, Lu Q, McMillan L, Truong M, Tummino PJ, Turunen B, Will M, Zuercher WJ, Rominger DH. Evidence for allosteric interactions of antagonist binding to the smoothened receptor. J Pharmacol Exp Ther 2009;329(3): 995–1005.
364. Yang H, Xiang J, Wang N, Zhao Y, Hyman J, Li S, Jiang J, Chen JK, Yang Z, Lin S. Converse conformational control of smoothened activity by structurally related small molecules. J Biol Chem 2009;284(31): 20876–20884.
365. Yauch RL, Dijkgraaf GJ, Alicke B, Januario T, Ahn CP, Holcomb T, Pujara K, Stinson J, Callahan CA, Tang T, Bazan JF, Kan Z, Seshagiri S, Hann CL, Gould SE, Low JA, Rudin CM, de Sauvage FJ. Smoothened mutation confers resistance to a Hedgehog pathway inhibitor in medulloblastoma. Science 2009;326 (5952): 572–574.
366. Hyman JM, Firestone AJ, Heine VM, Zhao Y, Ocasio CA, Han K, Sun M, Rack PG, Sinha S, Wu JJ, Solow-Cordero DE, Jiang J, Rowitch DH, Chen JK. Small-molecule inhibitors re-

veal multiple strategies for Hedgehog pathway blockade. Proc Natl Acad Sci USA 2009;106 (33): 14132–14137.
367. Chun SG, Zhou W, Yee NS. Combined targeting of histone deacetylases and Hedgehog signaling enhances cytoxicity in pancreatic cancer. Cancer Biol Ther 2009;8(14): 1328–1339.
368. Katoh Y, Katoh M. Hedgehog target genes: mechanisms of carcinogenesis induced by aberrant Hedgehog signaling activation. Curr Mol Med 2009;9(7): 873–886.
369. Katoh M. Networking of WNT, FGF, Notch, BMP, and Hedgehog signaling pathways during carcinogenesis. Stem Cell Rev 2007;3(1): 30–38.
370. Taylor MD, Zhang X, Liu L, Hui CC, Mainprize TG, Scherer SW, Wainwright B, Hogg D, Rutka JT. Failure of a medulloblastoma-derived mutant of SUFU to suppress WNT signaling. Oncogene 2004;23(26): 4577–4583.
371. Guo X, Wang XF. Signaling cross-talk between TGF-beta/BMP and other pathways. Cell Res 2009;19(1): 71–88.
372. Yu JM, Jun ES, Jung JS, Suh SY, Han JY, Kim JY, Kim KW, Jung JS. Role of Wnt5a in the proliferation of human glioblastoma cells. Cancer Lett 2007;257(2): 172–181.
373. Carmon KS, Loose DS. Secreted frizzled-related protein 4 regulates two Wnt7a signaling pathways and inhibits proliferation in endometrial cancer cells. Mol Cancer Res 2008;6 (6): 1017–1028.
374. Force T, Woulfe K, Koch WJ, Kerkelä R. Molecular scaffolds regulate bidirectional crosstalk between Wnt and classical seven-transmembrane-domain receptor signaling pathways. Sci STKE 2007;2007(397): pe41.
375. Korur S, Huber RM, Sivasankaran B, Petrich M, Morin P Jr, Hemmings BA, Merlo A, Lino MM. GSK3beta regulates differentiation and growth arrest in glioblastoma. PLoS One 2009;4(10): e7443.
376. Redjal N, Chan JA, Segal RA, Kung AL. CXCR4 inhibition synergizes with cytotoxic chemotherapy in gliomas. Clin Cancer Res 2006;12(22): 6765–6771.
377. Ehtesham M, Mapara KY, Stevenson CB, Thompson RC. CXCR4 mediates the proliferation of glioblastoma progenitor cells. Cancer Lett 2009;274(2): 305–312.
378. Murakami T, Cardones AR, Hwang ST. Chemokine receptors and melanoma metastasis. J Dermatol Sci 2004;36(2): 71–78.
379. Hinton CV, Avraham S, Avraham HK. Role of the CXCR4/CXCL12 signaling axis in breast cancer metastasis to the brain. Clin Exp Metastasis 2010;27(2):97–105.
380. Cheng X, Hung MC. Breast cancer brain metastases. Cancer Metastasis Rev 2007;26(3–4): 635–643.
381. Lee BC, Lee TH, Avraham S, Avraham HK. Involvement of the chemokine receptor CXCR4 and its ligand stromal cell-derived factor 1alpha in breast cancer cell migration through human brain microvascular endothelial cells. Mol Cancer Res 2004;2(6): 327–338.
382. Mongan JP, Fadul CE, Cole BF, Zaki BI, Suriawinata AA, Ripple GH, Tosteson TD, Pipas JM. Brain metastases from colorectal cancer: risk factors, incidence, and the possible role of chemokines. Clin Colorectal Cancer 2009;8(2): 100–105.
383. Rubin JB, Kung AL, Klein RS, Chan JA, Sun Y, Schmidt K, Kieran MW, Luster AD, Segal RA. A small-molecule antagonist of CXCR4 inhibits intracranial growth of primary brain tumors. Proc Natl Acad Sci USA 2003;100(23): 13513–13518.
384. Bian XW, Yang SX, Chen JH, Ping YF, Zhou XD, Wang QL, Jiang XF, Gong W, Xiao HL, Du LL, Chen ZQ, Zhao W, Shi JQ, Wang JM. Preferential expression of chemokine receptor CXCR4 by highly malignant human gliomas and its association with poor patient survival. Neurosurgery 61(3): 2007; 570–578.
385. Salmaggi A, Maderna E, Calatozzolo C, Gaviani P, Canazza A, Milanesi I, Silvani A, DiMeco F, Carbone A, Pollo B. CXCL12, CXCR4 and CXCR7 expression in brain metastases. Cancer Biol Ther 2009;8(17): 1608–1614.
386. Zabel BA, Wang Y, Lewén S, Berahovich RD, Penfold ME, Zhang P, Powers J, Summers BC, Miao Z, Zhao B, Jalili A, Janowska-Wieczorek A, Jaen JC, Schall TJ. Elucidation of CXCR7-mediated signaling events and inhibition of CXCR4-mediated tumor cell transendothelial migration by CXCR7 ligands. J Immunol 2009;183(5): 3204–3211.
387. Levoye A, Balabanian K, Baleux F, Bachelerie F, Lagane B. CXCR7 heterodimerizes with CXCR4 and regulates CXCL12-mediated G protein signaling. Blood 2009;113(24): 6085–6093.
388. Rubin JB. Chemokine signaling in cancer: one hump or two? Semin Cancer Biol 2009;19(2): 116–122.
389. Wu X, Lee VC, Chevalier E, Hwang ST. Chemokine receptors as targets for cancer therapy. Curr Pharm Des 2009;15(7): 742–757.
390. Goldhoff P, Warrington NM, Limbrick DD, Hope A, Woerner BM, Jackson E, Perry A,

Piwnica-Worms D, Rubin JB. Targeted inhibition of cyclic AMP phosphodiesterase-4 promotes brain tumor regression. Clin Cancer Res 2008;14(23): 7717–7725.

391. Schmidt AL, de Farias CB, Abujamra AL, Brunetto AL, Schwartsmann G, Roesler R. Phosphodiesterase-4 inhibition and brain tumor growth. Clin Cancer Res 2009;15(9): 3238.

392. Belcher SM, Ma X, Le HH. Blockade of estrogen receptor signaling inhibits growth and migration of medulloblastoma. Endocrinology 2009;150(3): 1112–1121.

393. Belcher SM. Blockade of estrogen receptor signaling to improve outlook for medulloblastoma sufferers. Future Oncol 2009;5(6): 751–754.

394. Benedetti E, Galzio R, Cinque B, Biordi L, D'Amico MA, D'Angelo B, Laurenti G, Ricci A, Festuccia C, Cifone MG, Lombardi D, Cimini A. Biomolecular characterization of human glioblastoma cells in primary cultures: differentiating and antiangiogenic effects of natural and synthetic PPARgamma agonists. J Cell Physiol 2008;217(1): 93–102.

395. Papi A, Tatenhorst L, Terwel D, Hermes M, Kummer MP, Orlandi M, Heneka MT. PPARgamma and RXRgamma ligands act synergistically as potent antineoplastic agents in vitro and in vivo glioma models. J Neurochem 2009;109(6): 1779–1790.

396. Papi A, Rocchi P, Ferreri AM, Guerra F, Orlandi M. Enhanced effects of PPARgamma ligands and RXR selective retinoids in combination to inhibit migration and invasiveness in cancer cells. Oncol Rep 2009;21(4): 1083–1089.

397. Strakova N, Ehrmann J, Dzubak P, Bouchal J, Kolar Z. The synthetic ligand of peroxisome proliferator-activated receptor-gamma ciglitazone affects human glioblastoma cell lines. J Pharmacol Exp Ther 2004;309(3): 1239–1247.

398. Seufert S, Coras R, Tränkle C, Zlotos DP, Blümcke I, Tatenhorst L, Heneka MT, Hahnen E. PPAR gamma activators: off-target against glioma cell migration and brain invasion. PPAR Res 2008;2008:513943.

399. Coras R, Hölsken A, Seufert S, Hauke J, Eyüpoglu IY, Reichel M, Tränkle C, Siebzehnrübl FA, Buslei R, Blümcke I, Hahnen E. The peroxisome proliferator-activated receptor-gamma agonist troglitazone inhibits transforming growth factor-beta-mediated glioma cell migration and brain invasion. Mol Cancer Ther 2007;6(6): 1745–1754.

400. Tatenhorst L, Hahnen E, Heneka MT. Peroxisome proliferator-activated receptors (PPARs) as potential inducers of antineoplastic effects in CNS tumors. PPAR Res 2008;2008:204514.

401. Chearwae W, Bright JJ. PPARgamma agonists inhibit growth and expansion of CD133+ brain tumour stem cells. Br J Cancer. 99(12): 2008; 2044–2053.

402. Bouterfa H, Picht T, Kess D, Herbold C, Noll E, Black PM, Roosen K, Tonn JC. Retinoids inhibit human glioma cell proliferation and migration in primary cell cultures but not in established cell lines. Neurosurgery 2000; 46(2): 419–430.

403. Tsuchida K, Nakatani M, Hitachi K, Uezumi A, Sunada Y, Ageta H, Inokuchi K. Activin signaling as an emerging target for therapeutic interventions. Cell Commun Signal 2009;7:15.

404. Golestaneh N, Mishra B. TGF-beta, neuronal stem cells and glioblastoma. Oncogene 2005; 24(37): 5722–5730.

405. Hjelmeland MD, Hjelmeland AB, Sathornsumetee S, Reese ED, Herbstreith MH, Laping NJ, Friedman HS, Bigner DD, Wang XF, Rich JN. SB-431542, a small molecule transforming growth factor-beta-receptor antagonist, inhibits human glioma cell line proliferation and motility. Mol Cancer Ther 2004;3(6): 737–745.

406. Wick W, Naumann U, Weller M. Transforming growth factor-beta: a molecular target for the future therapy of glioblastoma. Curr Pharm Des 2006;12(3): 341–349.

407. Piccirillo SG, Vescovi AL. Bone morphogenetic proteins regulate tumorigenicity in human glioblastoma stem cells. Ernst Schering Found Symp Proc 2006; (5): 59–81.

408. Piccirillo SG, Reynolds BA, Zanetti N, Lamorte G, Binda E, Broggi G, Brem H, Olivi A, Dimeco F, Vescovi AL. Bone morphogenetic proteins inhibit the tumorigenic potential of human brain tumour-initiating cells. Nature 2006; 444(7120): 761–765.

409. Nakano I, Saigusa K, Kornblum HI. BMPing off glioma stem cells. Cancer Cell 2008;13(1): 3–4.

410. Lee J, Son MJ, Woolard K, Donin NM, Li A, Cheng CH, Kotliarova S, Kotliarov Y, Walling J, Ahn S, Kim M, Totonchy M, Cusack T, Ene C, Ma H, Su Q, Zenklusen JC, Zhang W, Maric D, Fine HA. Epigenetic-mediated dysfunction of the bone morphogenetic protein pathway inhibits differentiation of glioblastoma-initiating cells. Cancer Cell 2008;13(1): 69–80.

411. Shervington A, Cruickshanks N, Lea R, Roberts G, Dawson T, Shervington L. Can the lack of HSP90alpha protein in brain normal tissue and cell lines, rationalise it as a possible

therapeutic target for gliomas? Cancer Invest 2008;26(9): 900–904.
412. Gaspar N, Sharp SY, Pacey S, Jones C, Walton M, Vassal G, Eccles S, Pearson A, Workman P. Acquired resistance to 17-allylamino-17-demethoxygeldanamycin (17-AAG, tanespimycin) in glioblastoma cells. Cancer Res 2009;69(5): 1966–1975.
413. Sauvageot CM, Weatherbee JL, Kesari S, Winters SE, Barnes J, Dellagatta J, Ramakrishna NR, Stiles CD, Kung AL, Kieran MW, Wen PY. Efficacy of the HSP90 inhibitor 17-AAG in human glioma cell lines and tumorigenic glioma stem cells. Neuro Oncol 2009;11(2): 109–121.
414. Premkumar DR, Arnold B, Jane EP, Pollack IF. Synergistic interaction between 17-AAG and phosphatidylinositol 3-kinase inhibition in human malignant glioma cells. Mol Carcinog 2006;45(1): 47–59.
415. Eccles SA, Massey A, Raynaud FI, Sharp SY, Box G, Valenti M, Patterson L, de Haven Brandon A, Gowan S, Boxall F, Aherne W, Rowlands M, Hayes A, Martins V, Urban F, Boxall K, Prodromou C, Pearl L, James K, Matthews TP, Cheung KM, Kalusa A, Jones K, McDonald E, Barril X, Brough PA, Cansfield JE, Dymock B, Drysdale MJ, Finch H, Howes R, Hubbard RE, Surgenor A, Webb P, Wood M, Wright L, Workman P. NVP-AUY922: a novel heat shock protein 90 inhibitor active against xenograft tumor growth, angiogenesis, and metastasis. Cancer Res 2008;68(8): 2850–2860.
416. Drysdale MJ, Brough PA. Medicinal chemistry of Hsp90 inhibitors. Curr Top Med Chem 2008;8(10): 859–868.
417. Phuphanich S, Carson KA, Grossman SA, Lesser G, Olson J, Mikkelsen T, Desideri S, Fisher JD. New Approaches to Brain Tumor Therapy (NABTT) CNS Consortium. Phase I. Safety study of escalating doses of atrasentan in adults with recurrent malignant glioma. Neuro Oncol 2008;10(4): 617–623.
418. Mitchell DA, Fecci PE, Sampson JH. Immunotherapy of malignant brain tumors. Immunol Rev 2008;222:70–100.
419. Van Gool S, Maes W, Ardon H, Verschuere T, Van Cauter S, De Vleeschouwer S. Dendritic cell therapy of high-grade gliomas. Brain Pathol 2009;19(4): 694–712.
420. Cho DY, Lin SZ, Yang WK, Hsu DM, Lee HC, Lee WY, Liu SP. Recent advances of dendritic cells (DCs)-based immunotherapy for malignant gliomas. Cell Transplant 2009;CT-2064: pii.
421. Ebben JD, Rocque BG, Kuo JS. Tumour vaccine approaches for CNS malignancies: progress to date. Drugs 2009;69(3): 241–249.
422. Luptrawan A, Liu G, Yu JS. Dendritic cell immunotherapy for malignant gliomas. Rev Recent Clin Trials 2008;3(1): 10–21.
423. Wheeler CJ, Black KL. DCVax-Brain and DC vaccines in the treatment of GBM. Expert Opin Investig Drugs 2009;18(4): 509–519.
424. Yamanaka R. Dendritic-cell- and peptide-based vaccination strategies for glioma. Neurosurg Rev 2009;32(3): 265–273.
425. Yamanaka R. Cell- and peptide-based immunotherapeutic approaches for glioma. Trends Mol Med 2008;14(5): 228–235.
426. Gustafson MP, Knutson KL, Dietz AB. Therapeutic vaccines for malignant brain tumors. Biologics 2008;2(4): 753–761.
427. Pellegatta S, Finocchiaro G. Dendritic cell vaccines for cancer stem cells. Methods Mol Biol 2009;568:233–247.
428. Das S, Raizer JJ, Muro K. Immunotherapeutic treatment strategies for primary brain tumors. Curr Treat Options Oncol 2008;9(1): 32–40.
429. Sampson JH, Archer GE, Mitchell DA, Heimberger AB, Herndon JE 2nd, Lally-Goss D, McGehee-Norman S, Paolino A, Reardon DA, Friedman AH, Friedman HS, Bigner DD. An epidermal growth factor receptor variant III-targeted vaccine is safe and immunogenic in patients with glioblastoma multiforme. Mol Cancer Ther 2009;8(10): 2773–2779.
430. Sampson JH, Archer GE, Mitchell DA, Heimberger AB, Bigner DD. Tumor-specific immunotherapy targeting the EGFRvIII mutation in patients with malignant glioma. Semin Immunol 2008;20(5): 267–275.
431. Di L, Kerns EH, Carter GT. Drug-like property concepts in pharmaceutical design. Curr Pharm Des 2009;15(19): 2184–2194.
432. Vastag M, Keseru GM. Current *in vitro* and *in silico* models of blood–brain barrier penetration: a practical view. Curr Opin Drug Discov Dev 2009;12(1): 115–124.
433. Abbott NJ, Dolman DE, Patabendige AK. Assays to predict drug permeation across the blood–brain barrier, and distribution to brain. Curr Drug Metab 2008;9(9): 901–910.
434. Jeffrey P, Summerfield S. Assessment of the blood–brain barrier in CNS drug discovery. Neurobiol Dis 2010;37(1): 33–37.
435. Baumert C, Hilgeroth A. Recent advances in the development of P-gp inhibitors. Anticancer Agents Med Chem 2009;9(4): 415–436.

436. Türk D, Szakács G. Relevance of multidrug resistance in the age of targeted therapy. Curr Opin Drug Discov Dev 2009;12(2): 246–252.
437. Liu X, Chen C, Smith BJ. Progress in brain penetration evaluation in drug discovery and development. J Pharmacol Exp Ther 2008; 325(2): 349–356.
438. de Vries NA, Beijnen JH, Boogerd W, van Tellingen O. Blood–brain barrier and chemotherapeutic treatment of brain tumors. Expert Rev Neurother 2006;6(8): 1199–1209.
439. Hitchcock SA. Blood–brain barrier permeability considerations for CNS-targeted compound library design. Curr Opin Chem Biol 2008; 12(3): 318–323.
440. Hitchcock SA, Pennington LD. Structure–brain exposure relationships. J Med Chem 2006;49(26): 7559–7583.
441. Hutter MC. In silico prediction of drug properties. Curr Med Chem 2009;16(2): 189–202.
442. Barchet TM, Amiji MM. Challenges and opportunities in CNS delivery of therapeutics for neurodegenerative diseases. Expert Opin Drug Del 2009;6(3): 211–225.
443. Hughes JD, Blagg J, Price DA, Bailey S, Decrescenzo GA, Devraj RV, Ellsworth E, Fobian YM, Gibbs ME, Gilles RW, Greene N, Huang E, Krieger-Burke T, Loesel J, Wager T, Whiteley L, Zhang Y. Physiochemical drug properties associated with in vivo toxicological outcomes. Bioorg Med Chem Lett 2008;18(17): 4872–4875.
444. Juillerat-Jeanneret L. The targeted delivery of cancer drugs across the blood–brain barrier: chemical modifications of drugs or drug-nanoparticles? Drug Discov Today 2008;13(23–24): 1099–1106.
445. Mathieu D, Fortin D. Chemotherapy and delivery in the treatment of primary brain tumors. Curr Clin Pharmacol 2007;2(3): 197–211.
446. Sarin H. Recent progress towards development of effective systemic chemotherapy for the treatment of malignant brain tumors. J Transl Med 2009;7:77.
447. Arifin DY, Lee KY, Wang CH. Chemotherapeutic drug transport to brain tumor. J Control Release 2009;137(3): 203–210.
448. Newton HB. Advances in strategies to improve drug delivery to brain tumors. Expert Rev Neurother 2006;6(10): 1495–1509.
449. Bidros DS, Vogelbaum MA. Novel drug delivery strategies in neuro-oncology. Neurotherapeutics 2009;6(3): 539–546.
450. Stukel JM, Caplan MR. Targeted drug delivery for treatment and imaging of glioblastoma multiforme. Expert Opin Drug Deliv 2009;6(7): 705–718.
451. Benny O, Pakneshan P. Novel technologies for antiangiogenic drug delivery in the brain. Cell Adh Migr 2009;3(2): 224–229.
452. Laks DR, Masterman-Smith M, Visnyei K, Angenieux B, Orozco NM, Foran I, Yong WH, Vinters HV, Liau LM, Lazareff JA, Mischel PS, Cloughesy TF, Horvath S, Kornblum HI. Neurosphere formation is an independent predictor of clinical outcome in malignant glioma. Stem Cells 2009;27(4): 980–987.
453. Eikenberry SE, Sankar T, Preul MC, Kostelich EJ, Thalhauser CJ, Kuang Y. Virtual glioblastoma: growth, migration and treatment in a three-dimensional mathematical model. Cell Prolif 2009;42(4): 511–528.
454. Mislow JM, Friedlander RM. Targeted oncogenes create a new realistic mouse model of glioblastoma multiforme. Neurosurgery 2009;64(5): N13.
455. Dyer MA. Mouse models of childhood cancer of the nervous system. J Clin Pathol 2004;57(6): 561–576.
456. Radaelli E, Ceruti R, Patton V, Russo M, Degrassi A, Croci V, Caprera F, Stortini G, Scanziani E, Pesenti E, Alzani R. Immunohistopathological and neuroimaging characterization of murine orthotopic xenograft models of glioblastoma multiforme recapitulating the most salient features of human disease. Histol Histopathol 2009;24(7): 879–891.
457. Hu X, Holland EC. Applications of mouse glioma models in preclinical trials. Mutat Res 2005;576(1–2): 54–65.
458. Fomchenko EI, Holland EC. Mouse models of brain tumors and their applications in preclinical trials. Clin Cancer Res 2006;12(18): 5288–5297.
459. Rosol M, Harutyunyan I, Xu J, Melendez E, Smbatyan G, Finlay JL, Krieger MD, Gonzalez-Gomez I, Reynolds CP, Nelson MD, Erdreich-Epstein A, Blüml S. Metabolism of orthotopic mouse brain tumor models. Mol Imaging 2009;8(4): 199–208.
460. Hegedus B, Banerjee D, Yeh TH, Rothermich S, Perry A, Rubin JB, Garbow JR, Gutmann DH. Preclinical cancer therapy in a mouse model of neurofibromatosis-1 optic glioma. Cancer Res 2008;68(5): 1520–1528.
461. Cranmer LD, Trevor KT, Bandlamuri S, Hersh EM. Rodent models of brain metastasis in melanoma. Melanoma Res 2005;15(5): 325–356.

462. Cruz-Munoz W, Man S, Xu P, Kerbel RS. Development of a preclinical model of spontaneous human melanoma central nervous system metastasis. Cancer Res 2008;68(12): 4500–4505.

463. Cruz-Munoz W, Man S, Kerbel RS. Effective treatment of advanced human melanoma metastasis in immunodeficient mice using combination metronomic chemotherapy regimens. Clin Cancer Res 2009;15(15): 4867–4874.

464. Newton HB, Ray-Chaudhury A, Cavaliere R. Brain tumor imaging and cancer management: the neuro-oncologists perspective. Top Magn Reson Imaging 2006;17(2): 127–136.

465. Waldman AD, Jackson A, Price SJ, Clark CA, Booth TC, Auer DP, Tofts PS, Collins DJ, Leach MO, Rees JH. National Cancer Research Institute Brain Tumour Imaging Subgroup. Quantitative imaging biomarkers in neuro-oncology. Nat Rev Clin Oncol 2009; 6(8): 445–454.

466. Jost SC, Collins L, Travers S, Piwnica-Worms D, Garbow JR. Measuring brain tumor growth: combined bioluminescence imaging-magnetic resonance imaging strategy. Mol Imaging 2009;8(5): 245–253.

467. Bryant MJ, Chuah TL, Luff J, Lavin MF, Walker DG. A novel rat model for glioblastoma multiforme using a bioluminescent F98 cell line. J Clin Neurosci 2008;15(5): 545–551.

468. Hashizume R, Ozawa T, Dinca EB, Banerjee A, Prados MD, James CD, Gupta N. A human brainstem glioma xenograft model enabled for bioluminescence imaging. J Neurooncol 2010;96(2):151–159.

469. Zhang X, Zheng X, Jiang F, Zhang ZG, Katakowski M, Chopp M. Dual-color fluorescence imaging in a nude mouse orthotopic glioma model. J Neurosci Methods 2009;181(2): 178–185.

470. Ragel BT, Elam IL, Gillespie DL, Flynn JR, Kelly DA, Mabey D, Feng H, Couldwell WT, Jensen RL. A novel model of intracranial meningioma in mice using luciferase-expressing meningioma cells. Laboratory investigation. J Neurosurg 2008;108(2): 304–310.

471. Neff BA, Voss SG, Allen C, Schroeder MA, Driscoll CL, Link MJ, Galanis E, Sarkaria JN. Bioluminescent imaging of intracranial vestibular schwannoma xenografts in NOD/SCID mice. Otol Neurotol 2009;30(1): 105–111.

472. Maes W, Deroose C, Reumers V, Krylyshkina O, Gijsbers R, Baekelandt V, Ceuppens J, Debyser Z, Van Gool SW. *In vivo* bioluminescence imaging in an experimental mouse model for dendritic cell based immunotherapy against malignant glioma. J Neurooncol 2009;91(2): 127–139.

473. Jost SC, Hope A, Kiehl E, Perry A, Travers S, Garbow JR. A novel murine model for localized radiation necrosis and its characterization using advanced magnetic resonance imaging. Int J Radiat Oncol Biol Phys 2009;75(2): 527–533.

474. Thomale UW, Tyler B, Renard V, Dorfman B, Chacko VP, Carson BS, Haberl EJ, Jallo GI. Neurological grading, survival, MR imaging, and histological evaluation in the rat brainstem glioma model. Childs Nerv Syst 2009; 25(4): 433–441.

KINASE INHIBITORS: APPROVED DRUGS AND CLINICAL CANDIDATES

Robert H. Bradbury
Cancer and Infection Research, AstraZeneca, Macclesfield, UK

1. INTRODUCTION

Regulatory and signaling networks that control fundamental cellular processes such as vascularization, growth, and proliferation are significantly enhanced in tumor cells, in response to factors that include genetic make up, age, and exposure to environmental carcinogens [1]. Over the past two decades, protein kinases involved in a wide range of cellular signaling pathways have proved a tractable class of drug target [2], and by the end of 2008 nine small-molecule kinase inhibitors had been approved by regulatory authorities for treatment of human hematological or solid tumors. Of the more than 150 kinase inhibitors currently undergoing clinical development, a majority are being developed for oncology indications [3].

Protein kinases fall into three broad classes, according to cellular location of the kinase domain and identity of the amino acid of the substrate that is phosphorylated (tyrosine or serine/threonine) [4]. Whereas receptor tyrosine kinases constitute the intracellular kinase domain of a transmembrane receptor that is activated by extracellular binding of a ligand, most commonly a growth factor, nonreceptor tyrosine kinases and serine–threonine kinases are located in the cytoplasm and are activated by more diverse pathways.

Structural relationships have been mapped for a significant proportion of the more than 500 protein kinases identified in the human genome [5,6], with the majority of protein kinase inhibitors that have progressed to clinical evaluation targeting the adenosine triphosphate (ATP) binding site. As reviewed in the following chapter and depicted in illustrative crystal structures (Chapter 6.9, Fig. 2a and d), all kinase enzymes consist of a bilobal cleft that binds ATP and inhibitors typically contain a donor–acceptor motif that forms key hydrogen bonds with amino acid residues located in the adenine binding pocket [7]. Since these residues constitute part of the hinge that connects the amino- and carboxy-terminal kinase domains, the donor–acceptor motif is often termed the hinge binder. Donor–acceptor motifs embedded in some approved kinase inhibitor drugs are depicted in Fig. 1.

Also located within the ATP binding site are polar ribose and phosphate binding pockets, together with a hydrophobic pocket that can commonly be accessed by substituents appended to the core donor–acceptor motif to attenuate inhibitor selectivity. In the case of inhibitors that bind to an inactive conformation of the kinase [3], sometimes termed prevention of activation (POA) inhibitors, an additional hydrophobic binding site is formed by movement of the activation loop to the so-called DFG-out position (named after the conserved amino acid motif located at the start of the activation loop). Compounds that bind to an inactive kinase conformation may show time-dependent inhibition, a property that may have implications for selectivity and duration of pharmacological effect in the clinic [8].

Sensitive or resistant mutants can arise from changes to amino acid residues within the ATP binding site, notably to a residue referred to as the gatekeeper at the entrance to the selectivity pocket that leads to a resistant form of the kinase. Modulation of the cellular potency, physical properties, and pharmacokinetic profile of kinase inhibitors may be achievable by attaching polar groups that can project beyond the ATP binding site to a surface region of the protein termed the solvent channel.

The kinase profiles and clinical utility of inhibitors granted approval for oncology indications by the FDA during the period 2001–2008 are listed in Table 1. A debate underlying the trade-off between antitumor efficacy and toxicity that is fundamental to cancer chemotherapy concerns the relative merits of selective and multitargeted kinase inhibitors [9–11]. Although selective kinase inhibitors may offer the clearest indication that effects seen both in the preclinical models and in the clinic are a consequence of activity versus the target kinase, inhibition of a single

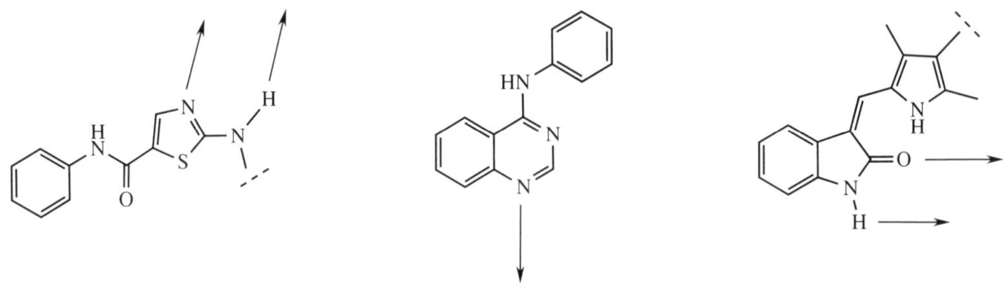

2-Aminothiazole-4-carboxamide, e.g, dasatinib

4-Anilinoquinazoline, e.g, gefitinib

3-Pyrrolidenyl indolin-2-one, e.g, sunitinib

Figure 1. Illustrative kinase inhibitor donor–acceptor motifs.

kinase may not be sufficient to achieve clinical efficacy due to inherent redundancy of signaling pathways or ability of tumors to acquire resistance [9]. Kinase selectivity profiles cited in this chapter are taken from journal publications or review articles and may not reflect the entirety of the human kinome [5,6] or that efficacious concentrations relative to the cellular potency of a specific kinase can be achieved in animal models or the clinic.

An analysis of clinical data for target-selective, pathway-selective, and extended spectrum inhibitors suggests drugs with activity against multiple kinases give greater clinical benefit [10]. However, side effects such as cardiotoxicity that reflect activity against other kinases are more likely for less selective agents [11]. Allosteric kinase inhibitors that target an alternative, non-ATP-competitive binding site have the potential to provide greater specificity and fewer off-target side effects, and a number of allosteric inhibitors have now progressed to early stage clinical trials [13].

In contrast with the discovery of historical intravenously administered cytotoxic drugs, the medicinal chemistry of kinase inhibitor drug programs has come more into line with other areas of therapy, with emphasis shifting toward orally bioavailable drugs with pharmacokinetics suitable for once or twice a day dosing commensurate with round-the-clock inhibition of drug target [14], and with a property profile that not only limits toxic effects on proliferating tissue such as bone marrow but also minimizes risks due to effects such as cardiac arrhythmia potential [15], cytochrome P450 liability [16], and variable absorption [17].

Table 1. FDA-Approved Oncology Kinase Inhibitors

Drug	FDA Approval[a]	Clinical Utility[b]	Kinase Inhibition Profile [11]
Imatinib (GLEEVEC™)	2001	CML, GIST	Bcr-Abl, PDGFRα/β, c-Kit
Gefitinib (IRESSA™)	2003[c]	NSCLC	EGFR
Erlotinib (TARCEVA™)	2004	NSCLC, pancreatic	EGFR
Sorafenib (NEXAVAR™)	2005	RCC, HCC	B/C-Raf, VEGFR2/3, PDGFRβ, c-Kit, Flt3
Dasatinib (SPRYCEL™)	2006	ALL, CML	Bcr-Abl, Src, PDGFRα/β, c-Kit
Sunitinib (SUTENT™)	2006	RCC, GIST	VEGFR1/2/3, PDGFRα/β, c-Kit, Flt3, CSF-1R
Lapatinib (TYKERB™)	2007	Breast	EGFR, erbB2
Nilotinib (TASIGNA™)	2007	CML	Bcr-Abl, PDGFRα/β, c-Kit
Temsirolimus (TORISEL™)	2007	RCC	mTOR[d]

[a] Year of first FDA approval [12].
[b] As approved by FDA [12].
[c] Use in the USA subsequently restricted to patients who continued to benefit from the drug.
[d] Binds to intracellular FKBP12 and complex inhibits mTOR.

Table 2. Comparison of Molecular Parameters of Kinase Inhibitor Oncology Clinical Candidates with Marketed Oral Drugs[a]

Drug	Clinical Kinase Inhibitor Set	Marketed Oral Drug Set
Number compounds in set	45	Approximately 400
Mean M Wt	457	338
Mean LogD$_{7.4}$ (calculated)	2.9	0.93
Mean number rotatable bonds	6.4	4.8

[a] Adapted from Ref. [18].

Comparison of the molecular parameters of a set of 45 orally bioavailable, kinase inhibitor oncology clinical candidates with a larger set of marketed oral drugs showed statistically significant differences in properties [18]. The kinase inhibitor set were on an average larger (>110 Da), more lipophilic (>1.5 log units), and more complex (approximately two more rotatable bonds) than marketed oral drugs (Table 2), differences that seem likely to reflect pursuit of increased potency and kinase selectivity through exploitation of lipophilic pockets in the ATP binding site region.

As mentioned earlier, a polar group attached to a kinase inhibitor template may improve cellular potency, physical properties, and pharmacokinetic profile, a structural change that also contributes to increased molecular weight. The association between lipophilicity and off-target toxicity proposed in recent publications [19,20] may be of less significance for kinase inhibitors in oncology compared with other areas of therapy, since recent year-on-year surveys of discontinued oncology clinical candidates suggest lack of efficacy rather than toxicity more frequently accounts for attrition [21–23].

After brief overviews of the design of Bcr-Abl and Flt3 inhibitors for treatment of hematological tumors, this chapter is structured around the effect of kinase inhibitor drugs on the "hallmarks of cancer," [24] the acquired cellular capabilities that collectively promote malignant growth of solid tumors—self-sufficiency in growth signals, insensitivity to antigrowth signals, evasion of apoptosis, sustained angiogenesis, tissue invasion, and metastasis. Although an over simplification due to the complexities of kinase function and inhibitor profile, this structure nevertheless provides a convenient framework for discussing compounds that have progressed to the clinic. The chapter continues by reviewing growth factor and antiangiogenic kinase inhibitors, approaches that have now been validated in large-scale clinical trials, and then proceeds to discuss inhibitors of kinases implicated in the cell cycle, survival signaling, and tissue invasion and metastasis, which are currently in earlier stage clinical trials.

The majority of these more recent kinase inhibitor drugs target solid tumors, thus expanding the range of treatable tumors beyond the hematological tumors targeted by Bcr-Abl inhibitors. Discussion in this chapter encompasses approximately 20 oncology kinase drug targets and 60 associated inhibitors with disclosed chemical structure that have been approved by regulatory authorities or are believed to be undergoing clinical trials at the time of completing this review in early 2009 [25,26]. In the interest of brevity, drug candidates understood to be no longer under development are largely excluded, as are most compounds with undisclosed chemical structure.

To illustrate themes in kinase inhibitor medicinal chemistry, also included in this chapter are synopses of evolution from lead to candidate drug distilled from original articles. Commencement of research on kinase inhibitors in the pharmaceutical industry in the 1990s coincided with introduction of high-throughput screening, and the resulting leads highlighted the hydrogen-bond donor–acceptor motif essential for interaction with the ATP site. While high-throughput screening has continued to play a significant role in uncovering kinase inhibitor leads, an increasing proportion of leads appears to have been derived from existing inhibitor temp-lates, or from alternative approaches such as directed libraries, fragment and virtual screening. With greater availability of crystal structures of inhibitor-bound complexes, structural information and homology modeling have been increasingly used as tools to aid prioritization of positions on the lead structure that may be explored to optimize po-tency, kinase selectivity, and overall property profile.

2. Bcr-Abl INHIBITORS FOR TREATMENT OF CHRONIC MYELOGENOUS LEUKEMIA

Expression of Bcr-Abl, a constitutively active form of Ableson (Abl) tyrosine kinase, is the hallmark of chronic myelogenous leukemia (CML) [27,28]. CML is associated with the presence of a genetic abnormality known as the Philadelphia chromosome. Fusion of a piece of chromosome 9 that contains a portion of the ABL gene with a piece of chromosome 22 that contains part of the BCR gene generates BCR-ABL, an oncogene that leads to the expression of Bcr-Abl, the constitutively active form of the Abl kinase that drives the uncontrolled growth of white blood cells. Although this genetic abnormality was first observed in CML, there is also a Philadelphia chromosome-positive form of acute lymphoblastic leukemia (ALL).

Since CML is characterized by a single oncogene, it had become clear by the early 1990s that the disease is a prime candidate for inhibition of kinase activity as targeted therapy. The Bcr-Abl inhibitor imatinib (GLEEVEC™), first approved by the FDA in 2001, emerged as the pioneering kinase inhibitor drug and dramatically improved the treatment of patients with early stage CML [29,30]. Imatinib is also an inhibitor of c-Kit, a kinase constitutionally active in gastrointestinal stromal tumors (GIST), and FDA approval for this indication followed in 2002 [31].

Although imatinib quickly became the first-line treatment for CML, it also soon became apparent that patients with late-stage disease often developed resistance to the drug due to point mutations in the catalytic domain of Abl kinase [32]. Subsequent research on Bcr-Abl inhibitors has focused on drugs to overcome resistance [27,28]. Despite the remarkable success of imatinib in treating hematological tumors, the efficacy of protein kinase inhibitors subsequently approved by the FDA has proved more limited in treatment of more complex solid tumors in which significant heterogeneity exists at the molecular and cellular levels [9,10].

Figure 2 summarizes the design of imatinib from an anilinopyrimidine lead identified in a screen for inhibitors of protein kinase C (PKC), a serine–threonine kinase [29,30]. During optimization of the lead structure, it was found that introduction of an amide group at the 3-position in the aniline ring gave improved activity against tyrosine kinases such as Bcr-Abl, and that substitution of a methyl group at the 6-position led to selectivity versus PKC. Finally, addition of a polar N-methyl piperazine moiety gave improved water solubility and bioavailability commensurate with oral human dosing.

Analysis of an X-ray crystal structure of the imatinib-Abl complex (Chapter 6.9, Fig. 2b) confirmed that imatinib binds to the inactive form of the kinase domain, with the anilinopyrimidine moiety occupying the ATP binding site and the N-methyl piperazine moiety lying along a surface exposed pocket [33]. Further analysis suggested that the potency and selectivity of imatinib could be improved by incorporating alternative groups to the N-methyl piperazine moiety. Although detailed SAR has not been disclosed, this approach led to nilotinib (TASIGNA™) (Fig. 2) [34–37], a second generation inhibitor approved by the FDA in 2007 for treatment of imatinib-resistant CML and active against all Bcr-Abl mutants except the T315I mutant that arises from changing the hydrogen-bonding Thr-315 gatekeeper residue to a lipophilic Ile residue.

An alternative approach to counter imatinib resistance is to use inhibitors that bind Bcr-Abl with less stringent conformational requirements than imatinib, since resistant mutants include those that impair the ability of the kinase domain of Abl to undergo the conformational changes necessary for imatinib binding, or that favor the active conformation to which imatinib is unable to bind. While specific differences between the inactive conformations of Abl and Src kinases account for the lack of binding of imatinib to Src-family kinases, the active conformations of Abl and Src kinases are more similar and consequently many inhibitors that bind to the active conformation of Src also inhibit Abl and some

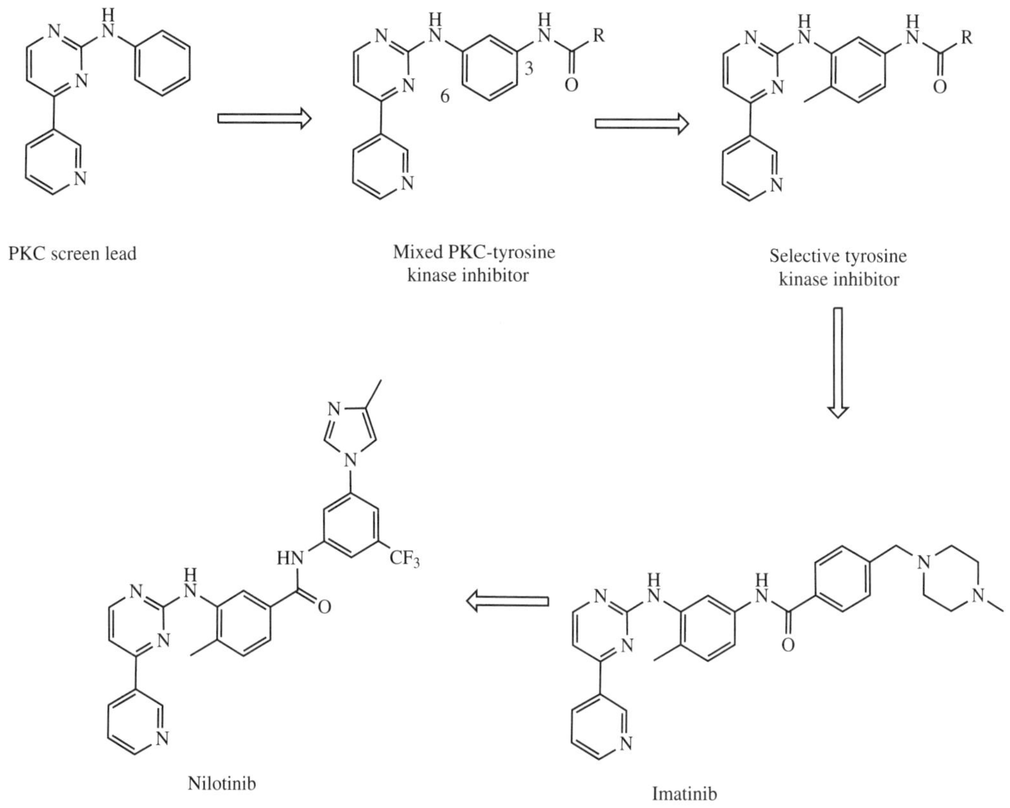

Figure 2. Design of Bcr-Abl inhibitors imatinib [29,30] and nilotinib [36].

imatinib-resistant mutants [38], an observation exploited in the design of dasatinib.

The chemical start point that led to dasatinib (SPRYCEL™) (Fig. 3) originated in a weakly active aminothiazole carboxamide lead identified during screening for inhibitors of Lck, a Src-family kinase implicated in acute and chronic T-cell-mediated autoimmune and inflammatory disorders [39,40]. Optimization of the aryl carboxamide moiety for Lck inhibitory activity revealed a narrow structural preference for a 2,6-disubstituted aryl ring. Modification of the aminothiazole moiety again showed a narrow SAR pattern from which improved Lck inhibitors emerged. Substitution of the amino group with heterocycles such as 2,6-dimethylpyrimidine gave a further improvement in potency, and profiling in a kinase selectivity panel demonstrated potent inhibition of Src-family and Bcr-Abl kinases, along with antiproliferative activity versus CML tumor cell lines.

Further work focused on antiproliferative activity and mouse oral exposure screening to select compounds for evaluation in human tumor xenograft efficacy models, with substitution of a hydroxyethyl piperazine moiety in the pyrimidine ring giving the best balance of cellular potency and pharmacokinetic properties. Like nilotinib, dasatinib inhibits clinically relevant imatinib-resistant Abl kinase domain mutants, with the exception of the T315I mutant [38,41].

Comparative data in Bcr-Abl wild type and imatinib-resistant cellular assays showed that dasatinib is approximately 20- and 375-fold more potent than nilotinib and imatinib, respectively [38]. Dasatinib was granted fast track approval by the FDA in 2006 for treatment of adults in all phases of CML with resistance or intolerance to prior therapy, and also for treatment of adults with Philadelphia chromosome-positive form of ALL. The oral exposure and high intrinsic potency of

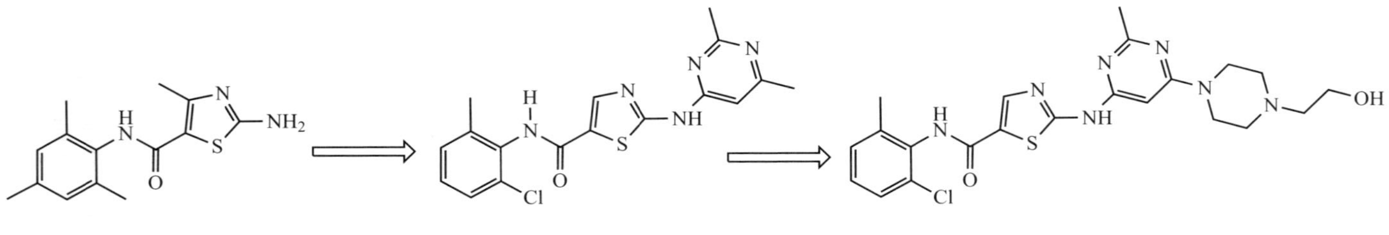

Lck screen lead

Bcr-Abl/Src-family inhibitor

Dasatinib

Figure 3. Design of Bcr-Abl inhibitor dasatinib [39,40].

Table 3. Flt3 Inhibitors in Clinical Trials

Drug	Highest Phase [25,26]	Kinase Inhibition Profile
Lestaurtinib	III	Flt3, TrkA/B/C, RET [44]
Midostaurin	III	Flt3, PKC, VEGFR1/2/3, PDGFRβ, FGFR1, c-Kit [49]
Tandutinib	II	Flt3, PDGFRβ, c-Kit [45]

dasatinib translates to efficacy at lower doses than imatinib and nilotinib in patients.

In contrast with the binding mode of imatinib and nilotinib, X-ray crystallography showed that dasatinib binds to the active form of Abl kinase (Chapter 7.12, Figs 2a and 6b) [42], with the aminothiazole carboxamide moiety occupying the ATP site and forming critical hydrogen bonds that include the Thr-315 gatekeeper interaction. Although it has been proposed from molecular modeling that the significantly increased binding of dasatinib reflects ability of dasatinib to recognize multiple states of Bcr-Abl [42], recently reported solution conformations of Abl complexed with imatinib, nilotinib, and dasatinib only show evidence of binding of dasatinib to the active form [43].

The conserved hydrogen-bonding interactions of imatinib, nilotinib, and dasatinib with the Thr-315 gatekeeper provide a powerful rationalization of the lack of activity of these drugs against CML tumors resistant due to the T315I mutation. While combination of Bcr-Abl inhibitors with other anticancer agents provides an option for treatment of tumors with this resistance mechanism, the pursuit of Bcr-Abl inhibitors with activity against the T315I mutant continues. In the interest of brevity, this overview of Bcr-Abl inhibitors is limited to approved drugs, but classes of inhibitor in early clinical trials include further inhibitors of Abl and Src-family kinases (Section 8.1), Aurora kinase inhibitors (Section 6.4) and non-ATP competitive inhibitors. The status of leading compounds has recently been reviewed [28].

3. Flt3 INHIBITORS FOR TREATMENT OF ACUTE MYELOGENOUS LEUKEMIA

Acute myelogenous leukemia (AML), the most common form of adult leukemia, is a hematological tumor that is also driven by genetic abnormalities, notably chromosomal translocations [44,45]. Constitutive activating mutations of FMS-like tyrosine kinase 3 represent the most frequent genetic events in AML, occurring in approximately 30% of cases, and also occurring in a majority of childhood acute leukemia [46]. These activating mutations regulate cell proliferation, differentiation and survival via downstream signaling mechanisms that include the MAPK and PI3K/PKB/mTOR pathways (Sections 4 and 7).

The staurosporine derivative lestaurtinib [43] and the quinazoline-based inhibitor tandutinib (Table 3, Fig. 4), originally characterized as a PDGFRβ inhibitor [45,47,48], are two agents that have progressed to late-stage clinical trials for treatment of AML. Lestaurtinib is also a potent inhibitor of tropomysin receptor kinases (TrkA/B/C), for which there is also evidence of a role in AML [45]. The extended spectrum inhibitor midostaurin is a further example of a Flt3 inhibitor in late-stage clinical trials, albeit in treatment of B-cell lymphoma where efficacy could more plausibly relate to inhibition of protein kinase C isoforms [49].

4. GROWTH FACTOR SIGNALING INHIBITORS

Growth factors regulate cellular proliferation, differentiation, apoptosis, migration, and invasion by binding to cognate receptors expressed on the surface of specific cells. These receptors contain an extracellular ligand binding domain, a single transmembrane domain, and an intracellular tyrosine kinase catalytic domain. Growth factor binding activates the receptor tyrosine kinase and initiates the mitogen-activated protein kinase (MAPK) signaling pathway that involves activation of G-protein Ras, Raf, MEK1 and 2, and ERK1 and 2. Phosphorylation of downstream proteins then determines overall cellular response

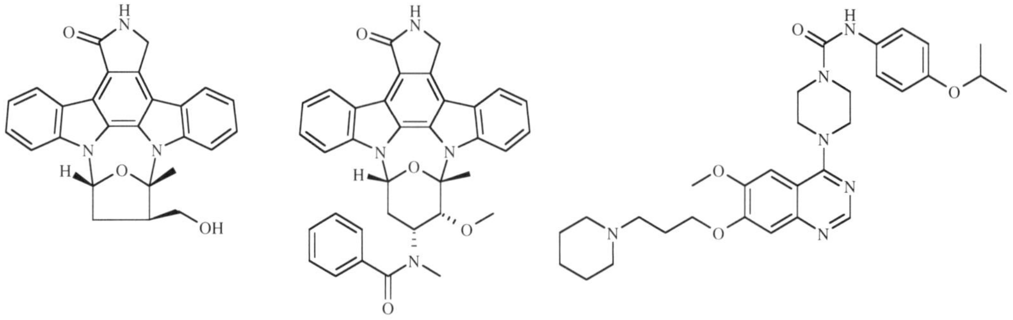

Lestaurtinib   Midostaurin   Tandutinib

Figure 4. Chemical structures of Flt3 inhibitors lestaurtinib, midostaurin, and tandutinib.

such as growth [4,50]. Other downstream signaling modules can also be activated, including the PI3K/PKB/mTOR pathway associated with propagation of survival signals (Section 7). This section covers approved drugs and clinical candidates that target the erbB family of growth factor receptor tyrosine kinases or B-Raf/MEK1 and 2 as downstream signaling targets.

4.1. Inhibitors of erbB-Family Kinases

There are more than 70 members of the tyrosine kinase family of growth factor receptors. Strongly implicated in human cancers are over expression and/or mutation of many of these receptors, including epidermal growth factor receptor (EGFR, erbB1), erbB2, VEGFR2, PDGFR, IGF-1R, c-MET, and RET [4,50]. The role of the EGFR and erbB2 signaling pathways is specific to cell growth and therapeutic validation was first provided by trastuzumab (HERCEPTIN™), a humanized monoclonal antibody to erbB2 approved by the FDA in 1998 that increases survival time in patients with metastatic breast tumors that over express erbB2 [51].

The EGFR antibody cetuximab (ERBITUX™) has subsequently been approved in Europe and the USA for treatment of metastatic colorectal cancer (CRC) and advanced squamous cell cancer of the head and neck [52]. To date, three reversible, ATP-dependent kinase inhibitor drugs targeting erbB-family pathways have been approved by the FDA, and a further three irreversible inhibitors are currently undergoing mid-late phase clinical trials (Table 4).

Over expression and/or activating mutations of EGFR have been detected in many human cancers, including nonsmall cell lung cancer (NSCLC), breast, ovary, colon, prostate, and head and neck, and are associated

Table 4. EGFR and erbB2 Inhibitors Approved or in Late Phase Clinical Trials

Drug	FDA Approval[a]	Clinical Utility[b]	Target Kinase
Gefitinib (IRESSA™)	2003[c]	NSCLC	EGFR
Erlotinib (TARCEVA™)	2004	NSCLC, pancreatic	EGFR
Lapatinib (TYKERB™)	2007	Breast	EGFR, erbB2
BIBW-2992 (TOVOK™)	III	NSCLC	EGFR, erbB2
PF-00299804[d]	II	NSCLC	EGFR, erbB2
Neratinib	II	Breast	EGFR, erbB2

[a] Year of first FDA approval [12] or most advanced clinical trial [25,26].
[b] As approved by FDA or most advanced clinical trial [25,26].
[c] Use in the USA subsequently restricted to patients who continued to benefit from the drug.
[d] Undisclosed chemical structure.

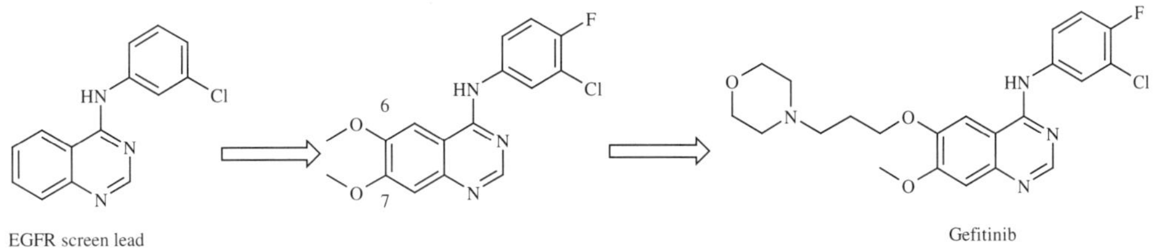

Figure 5. Design of EGFR inhibitor gefitinib [53,54].

with poor prognosis [50]. In the same era as research on Bcr-Abl inhibitors, it had become apparent by the early 1990s that inhibition of EGFR was an attractive drug target with potential for treatment of a range of solid tumors.

The design of gefitinib, the archetypal ATP-dependent EGFR-selective inhibitor, from a 4-anilinoquinazoline lead identified from directed screening is outlined in Fig. 5 [53,54]. SAR work showed a small lipophilic *meta* substituent in the aniline ring to be important for potent inhibition of EGFR kinase. Further substitution in the aniline ring and at the 6- and 7-positions of the quinazoline gave an inhibitor with optimized cellular potency, but relatively poor physical properties and exposure in rodents. Introduction of a basic side chain at the quinazoline 6-position addressed these issues and led to gefitinib [55,56]. As outlined later (Sections 5, 6.4 and 8.1), the 4-anilinoquinazoline template has proved a prolific start point for design of inhibitors of a range of kinases that includes VEGFR2, Aurora A/B, and Src.

A majority of the other erbB-family inhibitors that have received regulatory approval or progressed to clinical trials (Fig. 6) are based on a 4-anilinoquinazoline or closely related template. Medicinal chemistry relating to EGFR-selective inhibitor erlotinib has not been disclosed, but implicit in the chemical structure of erlotinib are a lipophilic *meta* substituent in the aniline ring and polar

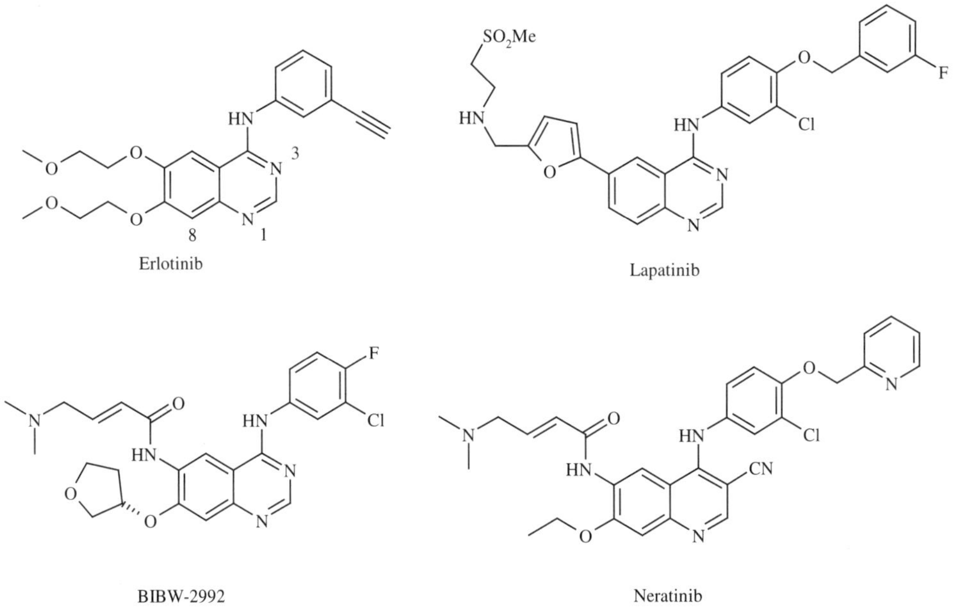

Figure 6. Chemical structures of erbB-family inhibitors erlotinib, lapatinib, BIBW-2992, and neratinib.

substituents in the quinazoline 6- and 7-positions. Crystal structures of erlotinib and gefitinib bound to the active form of the catalytic domain of EGFR kinase (Chapter 6.9, Fig. 4a) confirm that the quinazoline occupies the ATP site [57,58], with the N1 and C8 edge of the quinazoline orientated toward the hinge region, the quinazoline N3 participating in a water-mediated interaction with the hydroxyl of Thr-766, the *meta*-substituted aniline ring directed into the hydrophobic selectivity pocket and the quinazoline 6- and 7-substituents positioned at the solvent interface.

In the case of gefitinib, objective responses were observed in large Phase II clinical trials in 12–18% of patients with advanced refractory NSCLC [59,60]. However, a subsequent Phase III trial of gefitinib in patients with advanced refractory NSCLC failed to demonstrate a significant survival advantage over placebo in the overall population, although subgroup analysis showed a survival benefit for gefitinib in patients of Asian ethnicity and in patients who had never smoked [61]. In light of these Phase III data, use of gefitinib was restricted in the United States to patients who had already benefited from the drug.

In contrast, a Phase III clinical trial with erlotinib demonstrated an overall survival benefit relative to placebo in patients with pretreated NSCLC. The reason for the difference in outcome between these two studies with erlotinib and gefitinib is not clear, but may relate to differences in the study populations recruited [62,63]. More recently, a Phase III study demonstrated the noninferiority of gefitinib relative to docetaxel in terms of overall survival in patients with pretreated advanced NSCLC. Compared with docetaxel, gefitinib was also found to have a more favorable tolerability and quality of life profile [64].

Evidence has since emerged that various EGFR mutations confer increased sensitivity to inhibition by gefitinib and erlotinib [65–67], a premise supported by increased incidence of mutations in tumor samples taken from patients with a higher objective response rate, most notably samples from patients of Asian ethnicity [68]. Although an X-ray crystal structure of gefitinib in complex with a sensitive mutant showed minimal changes in binding mode compared with wild-type EGFR, direct binding measurements showed that gefitinib binds 20-fold more tightly to the mutant form [58]. Conversely, as seen with Bcr-Abl inhibitors, clinical use of EGFR inhibitors has led to emergence of resistant mutants, such as the T790M gatekeeper mutant [69].

Variation of the substitution pattern in the aniline ring modulates the selectivity profile of kinase inhibitors based on the 4-anilinoquinazoline template. In the case of erbB-family kinases, incorporation of an extended *para* substituent in the aniline ring gives dual EGFR-erbB2 inhibition, an observation central to the design of lapatinib [70–72]. The structure of EGFR co-crystallized with lapatinib (Chapter 7.12, Figs 2c and 4b) reveals significant differences from the erlotinib and gefitinib structures, in that lapatinib binds to the inactive form of EGFR with the extended aniline moiety occupying an enlarged selectivity pocket [73]. Similar features are seen in a recently reported crystal structure of lapatinib in complex with erbB4 [74], consistent with the observed slow off-rate of lapatinib reflecting the significant structural re-organization needed to disengage the deeply buried extended aniline from within the enlarged selectivity pocket. A detailed account has been published of the biochemical screening, kinetic profiling and preclinical studies that led to lapatinib [71].

The three EGFR-erbB2 inhibitors currently in mid-late phase clinical trials (Table 4, Fig. 6) are all irreversible inhibitors. In the case of BIBW-2992 [75–78], a pendant acrylamide Michael acceptor attached to the 6-position of the 4-anilinoquinazoline template forms a covalent bond with a nucleophilic cysteine residue adjacent to the ATP binding site (Cys-773 in EGFR, Cys-805 in erbB2). Although the chemical structure of PF-00299804 has not been disclosed, it may be related to canertinib, a prototype acrylamide-based inhibitor no longer being developed [79,80].

Neratinib is a related inhibitor based on a 3-cyano-4-anilinoquinoline template that incorporates an extended aniline moiety similar to that present in lapatinib [81,82]. An X-ray crystal structure of neratinib in complex with the T790M mutant shows the EGFR kinase adopts an inactive conformation [83], in which the cyano group interacts at the ATP binding site through displacement of the bridging water molecule to Thr-776, and as seen with lapatinib the extended aniline occupies an enlarged hydrophobic selectivity pocket. In addition to these noncovalent interactions, the expected covalent bond is observed between the cysteine residue and the crotonamide Michael acceptor.

Recent evidence suggests these irreversible inhibitors may have potential to overcome resistant mutants such as the EGFR T790M gatekeeper mutant [83–85]. It has been proposed that resistant mutants reflect increased sensitivity to ATP [83], and that when covalently bound, irreversible inhibitors such as neratinib are no longer in equilibrium with ATP. Of concern, however, remains the potential for mechanism-based toxicity with irreversible inhibitors, since at least 10 other kinases have a reactive cysteine residue at an equivalent position adjacent to the ATP binding site [83,86].

4.2. B-Raf Inhibitors

Raf kinases are serine–threonine kinases that exist as three isoforms, A-Raf, B-Raf, and C-Raf (Raf-1) [87–89]. The B-Raf gene is one of the most frequently mutated in human cancers, particularly in melanoma, whereas mutations in A-Raf and C-Raf are extremely rare. The most prominent mutation results in a single substitution of a valine residue in the activation loop of the kinase domain by a glutamic acid residue (V600E), leading to a constitutively active kinase. The crystal structure of the B-RafV600E mutant protein reveals that this mutation destabilizes the inactive conformation, favoring the activated state of the kinase.

This section outlines the design of the B-Raf inhibitor sorafenib (NEXAVAR™), a drug that originated from screening for inhibitors of the C-Raf isoform [87,88]. Further characterization showed sorafenib to be a potent multikinase inhibitor with a profile that includes wild type and mutant B-Raf, proangiogenic receptor tyrosine kinases (VEGFR2/3, PDGFRβ) and other kinases involved in tumorigenesis (c-Kit, Flt3).

Phases II and III clinical trials of sorafenib as a single agent or in combination with cytotoxic agents in patients with advanced melanoma gave cases of stable disease, but the Phase III primary endpoint of progression-free survival was not met [89]. No relationship was seen between mutant B-RafV600E status and disease stabilization. Sorafenib was however approved in 2005 for treatment of renal cell carcinoma (RCC), a notoriously difficult malignancy to combat, in which B-Raf mutations are not seen but where pathophysiology is strongly dependent on angiogenesis. Although it is likely that the efficacy of sorafenib versus renal cell carcinoma is related more to inhibition of multiple proangiogenic kinases (Section 5) than to effects on B-Raf, it is appropriate to include sorafenib here since the medicinal chemistry program that led to sorafenib targeted Raf inhibition [87,88].

The design of sorafenib is summarized in Fig. 7 [87,88], and also included is the chemical structure of the related B-Raf inhibitor RAF-265 (TKI-258) [90]. Starting from a weakly active diaryl urea lead identified from a C-Raf high-throughput screen, modification of the aryl groups gave a significant increase in potency. Introduction of a pyridin-4-yloxy substituent in one of the aryl rings provided a probe compound orally bioavailable in mice and active in xenograft models, and further optimization of the aryl moieties then led to sorafenib.

X-Ray crystal structures of sorafenib in complex with wild type and mutant B-Raf showed similarity to the interaction of imatinib with Abl referred to earlier (Section 2), in that the inhibitor stabilizes the inactive conformation of the kinase [91]. The structures indicate that the distal 4-pyridinyl substituent occupies the ATP site and that the urea forms two key hydrogen bonds with backbone aspartate and glutamate residues, thereby directing the lipophilic trifluoromethyl-substituted aromatic ring into the hydrophobic pocket.

RAF-265 is structurally related to sorafenib, with the trifluoromethyl imidazole group replacing the distal carboxamide substituent and the benzimidazole moiety serving as a urea isostere. The compound is a potent

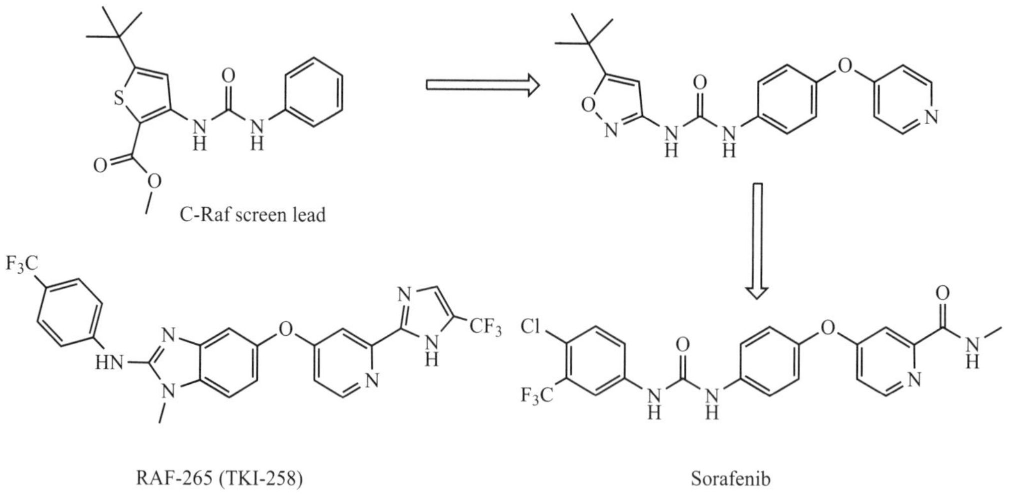

Figure 7. Design of B-Raf inhibitor sorafenib [88,89] and chemical structure of RAF-265.

inhibitor of B-Raf, VEGFR2, PDGFRβ, and c-Kit that has been reported to be more potent than sorafenib in animal models [90]. Clinical trials are ongoing to assess the safety and efficacy of RAF-265 in patients with melanoma and RCC. More potent and selective B-Raf inhibitors are reported to be undergoing preclinical evaluation [89], but the role of B-Raf mutations in development and progression of melanoma remains to be validated in the clinic.

4.3. MEK Inhibitors

The two principal substrates for Raf are MEK1 and MEK2, dual specificity kinases that can phosphorylate both serine/threonine and tyrosine residues [92]. MEK1 and MEK2 share 80% sequence identity in the catalytic domain and 100% identity within the ATP binding site. Although the two MEK isoforms play different roles in development, their role in oncogenesis has not been differentiated and inhibitors generally target both isoforms.

A number of MEK inhibitors have progressed to Phase I/II clinical trials (Fig. 8) [92], and all are allosteric, ATP uncompetitive inhibitors that bind to a pocket specific to MEK isoforms. The progenitor of all these inhibitors was a N-phenyl anthranilic acid lead (Fig. 8) identified via a high-throughput "kinase cascade" assay [93]. Optimization of the diphenylamine substituents and replacement of the carboxylic acid, initially with a hydroxamic acid and then with a hydroxamate ester, gave CI-1040. Lack of efficacy of CI-1040 in Phase II clinical trials due to low systemic exposure was attributed to metabolic instability and solubility limitations [93]. In combination with further refinement of a substituent on the diphenylamine moiety, polar hydroxamate esters gave >100-fold improvement in cellular potency and solubility and led to PD 0325901 [93].

X-Ray crystal structures of close analogues of PD 0325901 in complex with MEK1 and MEK2 bear little resemblance to bound structures of ATP-dependent inhibitors and confirm the diphenylamine moiety occupies an allosteric hydrophobic pocket adjacent to the ATP binding site [94]. Although formed by relatively simple conformational changes induced by binding of ATP, this pocket appears unique to MEK1 and MEK2 and accounts for the high degree of kinase selectivity seen with PD 0325901. While development of PD 0325901 has been discontinued, three other structurally related allosteric MEK inhibitors shown in Fig. 8 are currently being evaluated in Phase I/II clinical trials in patients with various tumors [95–99].

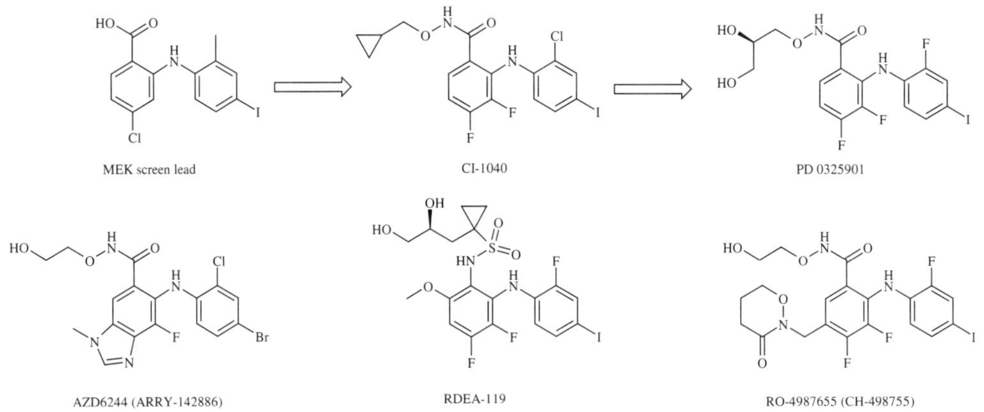

Figure 8. Design of MEK inhibitors CI-1040 and PD 0325901 [93], and chemical structures of AZD6244, RDEA-119, and RO4987655.

5. VEGFR2 INHIBITORS AS ANTIANGIOGENIC AGENTS

Growth of solid tumors depends on nutrients and oxygen received via angiogenesis (also termed neovascularization), the process by which new microcapilliaries are formed from existing vasculature. Angiogenesis also enables tumor cells to enter the circulation and spread to other organs (metastasis), and has been shown to be a key prognostic that correlates with tumor progression, disease severity, and capacity for metastasis [100,101].

In healthy adults, angiogenesis is a complex process controlled by a balance between endogenous proangiogenic and antiangiogenic factors. Tumor-induced angiogenesis is mediated by growth factors that include vascular endothelial growth factor (VEGF), platelet-derived growth factor (PDGF) and fibroblast growth factor (FGF). Binding of these ligands to cognate cell surface receptors activates the corresponding intracellular receptor tyrosine kinases and initiates the signaling cascade that leads to angiogenesis.

VEGF ligands bind to three receptors—VEGFR1 (also termed Flt1), VEGFR2 (KDR, Flk1), and VEGFR3 (Flt4). As understanding of these receptors has developed, VEGFR2 has become recognized as the receptor that plays the major role in triggering endothelial cell proliferation and tumor vascularization. Validation of VEGFR2 as an oncology drug target was initially provided by bevacizumab (AVASTIN™), a humanized monoclonal VEGF antibody approved by the FDA in 2004 for treatment of solid tumors [102].

Blockade of VEGFR2 signaling by small-molecule kinase inhibitors has since been shown to inhibit angiogenesis, tumor progression and metastasis in a number of preclinical and clinical studies. Approximately 25 structurally diverse VEGFR2 inhibitors are reported to be undergoing clinical development [103], and some of these agents also inhibit other kinases that commonly include PDGFRα/β and c-Kit. In the interest of brevity, this section is limited to small-molecule inhibitors that have been approved by the FDA or have progressed to Phase III clinical trials (Table 5, Figs 9–12).

First approved by the FDA in 2005 and initially characterized as a B-Raf inhibitor, sorafenib (Section 4.2) is a multitargeted inhibitor with clinical efficacy that is most likely due to inhibition of proangiogenic VEGFR and PDGFR kinases [87,88]. Also approved by the FDA in the following year was sunitinib (Fig. 9), a further multitargeted agent with proven clinical efficacy in treatment of RCC and GIST refractory or intolerant to imatinib [104,105].

Sunitinib originated from a weakly active VEGFR2 screen lead [106,107], a 3-pyrrolide-

Table 5. Tyrosine Kinase Inhibitors with Angiogenic Properties Approved or in Late-Phase Clinical Trials

Drug	FDA Approval[a]	Clinical Utility[b]	Kinase Inhibition Profile [103]
Sorafenib (NEXAVAR™)[c]	2005	RCC, HCC	VEGFR2/3, PDGFRβ, B/C-Raf, c-Kit, Flt3
Sunitinib (SUTENT™)	2006	RCC, GIST	VEGFR1/2/3, PDGFRα/β, c-Kit, Flt3, CSF-1R
Vandetanib (ZACTIMA™)	III	NSCLC, thyroid	VEGFR2/3, EGFR, RET
Cediranib (RECENTIN™)	III	CRC, Glioblastoma	VEGFR1/2/3, c-Kit
Pazopanib (ARMALA™)	III	RCC, breast,[d] sarcoma	VEGFR1/2/3, PDGFRα/β, c-Kit
Axitinib	III	RCC, thyroid	VEGFR1/2/3, PDGFRβ, c-Kit
Motesanib	III	NSCLC	VEGFR1/2/3, PDGFRβ, c-Kit
Vatalanib	III	CRC[e]	VEGFR1/2
Brivanib	III	CRC	VEGFR1/2, FGFR1

[a] Year of first FDA approval [12] or most advanced clinical trial [25,26].
[b] As approved by FDA or most advanced clinical trial [25,26].
[c] Initially characterized as B-Raf inhibitor, see Section 4.2.
[d] Combination with lapatinib.
[e] Phase III trial discontinued, Phase II trials reported to be continuing versus other tumors.

nyl indolin-2-one that exists exclusively as the geometrical isomer depicted in Fig. 9 due to intramolecular hydrogen bonding between the indolinone carbonyl and the pyrrole NH. Substitution of the pyrrole ring with methyl groups significantly improved enzymic and cellular potency and led to SU5416 (semaxinib), a VEGFR2 inhibitor that advanced to Phase III clinical trials but was later discontinued due to severe toxicity [104,107].

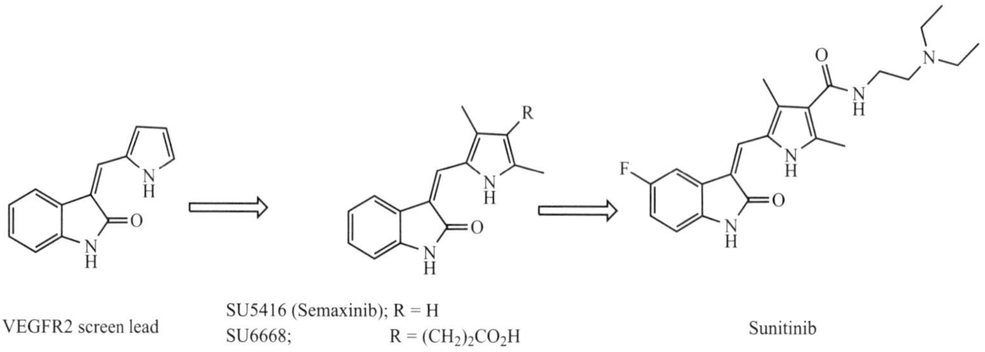

VEGFR2 screen lead   SU5416 (Semaxinib); R = H   Sunitinib
                      SU6668;   R = (CH$_2$)$_2$CO$_2$H

Figure 9. Design of VEGFR inhibitor sunitinib [106,107].

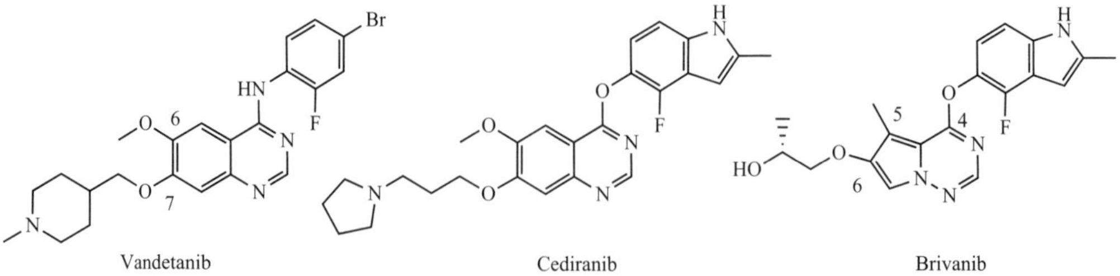

Vandetanib   Cediranib   Brivanib

Figure 10. Chemical structures of VEGFR inhibitors vandetanib, cediranib, and brivanib.

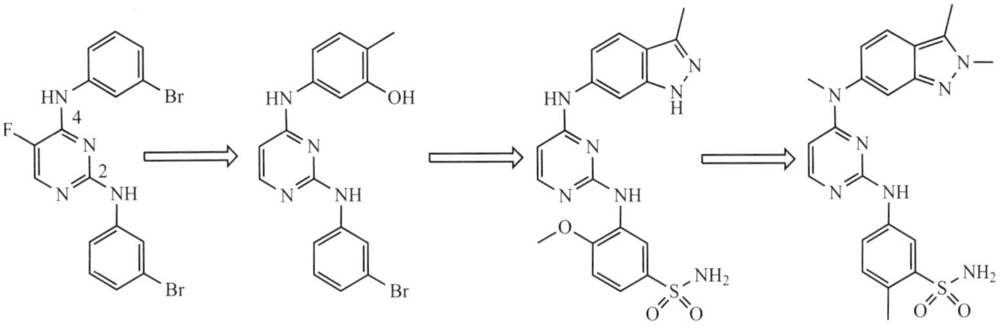

VEGFR2 screen lead

Pazopanib

Figure 11. Design of VEGFR inhibitor pazopanib [117].

Appending a carboxyethyl group to the pyrrole ring gave SU6668 [108], a PDGFRβ inhibitor that also progressed to clinical trials but was subsequently discontinued due to poor pharmacokinetics. Finally, introduction of a carboxamide substituent on the pyrrole ring and elaboration of the amide to include a basic group led to sunitinib [107], a soluble and orally bioavailable inhibitor of VEGFR and PDGFR kinases. Further characterization showed the kinase inhibitory profile of sunitinib to include c-Kit, Flt3, and CSF-1R, and clinical evaluation of sunitinib is ongoing in treatment of tumor types in which the role of these kinases is implicated [104,105].

The 4-anilinoquinazoline template that led to the EGFR kinase inhibitors gefitinib and erlotinib (Section 4.1) provided the start point for the VEGFR/EGFR/RET inhibitor vandetanib (Fig. 10). In an early example of use of robotic synthesis, variation of the substitution pattern in the aniline moiety established that small substituents at the 2- and 4-positions shifted the kinase inhibitory profile in favor of VEGFR2 [109]. From a wide range of substituents of differing lipophilic, electronic, and steric characters that could be accommodated at the 7-position of the quinazoline ring, a moiety containing an *N*-methyl piperidine was identified as providing a good balance of potency, solubility and pharmacokinetic properties. Fine-tuning of the substituent at the *para* position in the aniline ring led to vandetanib [110,111], an inhibitor of VEGF-dependent tumor angiogenesis and EGFR- and RET-dependent tumor cell proliferation that has progressed to Phase III clinical trials in patients with advanced NSCLC. Encouraging activity has also been reported in Phase II trials in patients with metastatic hereditary

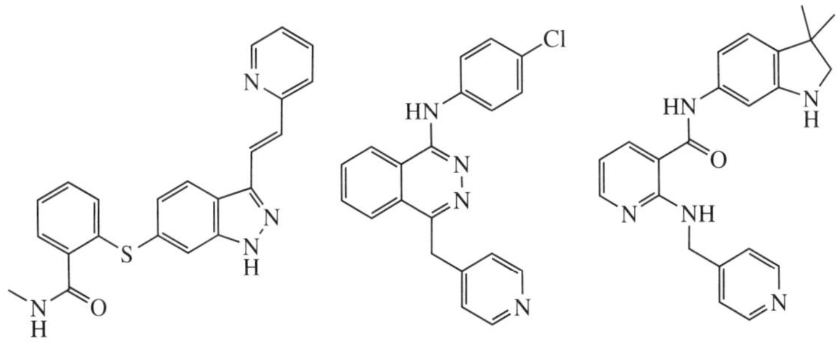

Axitinib Vatalani b Motesani b

Figure 12. Chemical structures of VEGFR inhibitors axitinib, vatalanib, and motesanib.

medullary thyroid cancer, a tumor type in which RET-dependent survival is strongly implicated [111].

Details of the medicinal chemistry evolution of cediranib (Fig. 10) [112,113] have not been disclosed, but based on the binding mode proposed for this class of inhibitor [110], in which the aniline moiety occupies the selectivity pocket and the quinazoline 7-substituent extends to the outside of the protein, improvements in VEGFR potency and changes in selectivity profile observed for cediranib may primarily reflect modification of the aniline. The high cellular potency of cediranib equates with efficacious doses in animal models that are significantly lower than reported for other VEGFR inhibitors [112], and cediranib has now progressed to Phase III clinical trials in patients with CRC and recurrent glioblastoma.

Also depicted in Fig. 10 is the chemical structure of brivanib, an inhibitor of VEGFR2 and FGFR1 that is undergoing Phase III clinical trials as the corresponding alaninate prodrug in patients with CRC. Brivanib originated from a medicinal chemistry program in which a pyrrolotriazine template was designed as a replacement for the quinazoline core present in EGFR and VEGFR2 inhibitors [114–116]. Incorporation of the indolyloxy substituent found in cediranib at the equivalent 4-position of the pyrrolotriazine, along with a methyl group at the 5-position, gave a potent VEGFR2 inhibitor that was optimized for aqueous solubility, pharmacokinetic properties and in vitro safety profile through introduction of polar 6-substituents. Compounds with neutral substituents were found to display reduced affinity for the hERG ion channel and weaker activity against a panel of CYP450 isoforms [115].

The optimized analog brivanib showed potent VEGFR2 and FGFR1 inhibition and in vivo efficacy in mouse xenograft models. However, oral administration to rats as a micronized suspension produced significantly lower systemic exposure than obtained from solution formulations and drug exposure did not increase proportionally with dose, a potential development issue attributed to low aqueous solubility [116]. To address this issue, amino ester derivatives were investigated and the alaninate ester was identified as a soluble, orally absorbed prodrug that is stable in human plasma but cleaved to brivanib in the liver.

Pazopanib, an inhibitor of VEGFR1/2/3, PDGFRα/β and c-Kit that originated from a moderately potent 2,4-bisanilinopyrimidine screen lead (Fig. 11) [117], has also progressed to Phase III clinical trials [118]. Potency of the bisanilinopyrimidine screen lead was significantly improved by replacement of the 4-anilino substituent with a 4-methyl-3-hydroxy aniline, a structural change based on overlay with a potent VEGFR2 inhibitor of the 4-anilinoquinazoline class. The resulting inhibitor had poor rodent oral bioavailability that was attributed to rapid glucuronidation or sulfation of the phenolic hydroxyl, an issue addressed by replacement of the 4-methyl-3-hydroxy aniline with a 6-amino-3-methylindazole moiety. Key changes that then led to pazopanib-included methylation of the 4-amino nitrogen to improve permeability and reduce metabolism, modification of the methyl indazole to obviate inhibition of CYP450 isoforms and fine-tuning of the 2-anilino substituent.

Axitinib [119], vatalanib [120], and motesanib [121] (Fig. 12) are three further VEGFR inhibitors that have progressed to late-stage clinical trials. While the medicinal chemistry evolution of axitinib and motesanib has not been disclosed, vatalanib constitutes a rare example of a screen lead that subsequently entered clinical development [120]. Phase III clinical trials of axitinib and motesanib are ongoing, and although Phase III clinical trials of vatalanib failed to show benefit in patients with CRC, Phase II trials are reported to be continuing versus other tumor types [25,26].

6. CELL CYCLE INHIBITORS

Aberrant regulation of the cell cycle is the hallmark of cancer that drives proliferation of tumor cells [122]. The two key phases of the cell cycle are the DNA synthesis (S) phase during when the genome is duplicated and the mitosis (M) phase, a complex process that segregates the two copies of the genome prior to cell division. Like other signaling pathways, regulation is achieved through modification of the phosphorylation status of the pathway components by protein kinases and phosphatases.

In contrast with traditional cytotoxic agents that prevent cell division by interfering with DNA synthesis or compromising the mitotic spindle, inhibitors of cell cycle protein kinases aim to target regulatory components with aberrant activity in transformed cells. This section briefly reviews inhibitors of four classes of cell cycle kinases—cyclin-dependent kinases, isoforms of which play fundamental roles in both key phases of the cell cycle, checkpoint kinases that regulate DNA replication and damage repair, and polo-like kinase 1 and Aurora kinases that are involved in spindle assembly and chromosome processing during mitosis.

As yet, the majority of these inhibitors have only progressed as far as Phase I/II clinical trials and efficacy in larger scale Phase III trials remains to be demonstrated. In contrast with the majority of kinase inhibitors described in this chapter, to minimize side effects episodic intravenous infusion is the preferred route of administration for many cell cycle inhibitors and the consequent need for high aqueous solubility has significantly influenced medicinal chemistry programs in this area.

6.1. Cyclin-Dependent Kinase Inhibitors

Cyclin-dependent kinases (CDKs) are serine–threonine kinases comprising a family of nine isoforms, of which CDK1, 2, 4, 6, and 9 have been implicated in regulation of various phases of the cell cycle through formation of complexes of a CDK catalytic subunit and a cyclin regulatory subunit. The role of CDKs in the cell cycle is more complex and diverse than initially perceived [122,123], and validation of CDKs as oncology drug targets has been complicated by the lack of sufficiently selective inhibitors. Reported X-ray crystal structures show a strong sequence homology between CDK isoforms [124], and it is therefore unsurprising that full selectivity for specific isoforms has not been achieved with the ATP-dependent inhibitors that have progressed to the clinic.

Table 6. Cyclin Dependent Kinase Inhibitors in Clinical Trials

Drug	Highest Phase [25,26]	CDK Isoform Targeted[a]	CDK Isoform Profile [122,123]
Alvocidib (flavopiridol)	III	Pan-CDK	Pan-CDK
Seliciclib	II	CDK2	CDK2/7/9
PD 0332991	II	CDK4	CDK4/6
BMS-387032 (SNS-032)	I	CDK2	CDK2/7/9
R547 (RO-4584820)	I	CDK1/2/4	CDK1/2/4
AT7519	I	CDK2	CDK1/2/4/5

[a] Isoform targeted in medicinal chemistry program.

Although medicinal chemistry programs that led to CDK inhibitors currently undergoing clinical trials often targeted a specific isoform, assays that subsequently became available revealed a variety of isoform profiles for these structurally diverse inhibitors (Table 6, Figs 13–17) [122,123]. Clinical efficacy against solid tumors has not been observed so far with the two most advanced agents, alvocidib (flavopiridol) and seliciclib, either as monotherapy or in combination with other agents [122,123]. Alvocidib has been studied in a wide range of indications using a variety of dosing schedules, but promising results have only emerged in treatment of chronic lymphocytic leukemia (CLL) [125,126]. These studies indicate that unusually high plasma concentrations of alvocidib may be required for efficacy due to high plasma protein binding [123]. It is likely that efficacy of alvocidib in CLL reflects effects on transcription and other targets rather than inhibition of CDKs [127,128].

Design of the CDK inhibitors listed in Table 6 involved a range of medicinal chemistry approaches. Whereas alvocidib and seliciclib originated from leads identified from screening of natural product extracts (rohitukine and olomoucine, respectively) (Fig. 13) [123,129], more recent clinical candidates have evolved from earlier less selective kinase templates and from high-throughput or fragment screening (Figs 14–17) [130–139].

BMS-387032 (SNS-032) emerged from a program that targeted CDK2 (Fig. 14)

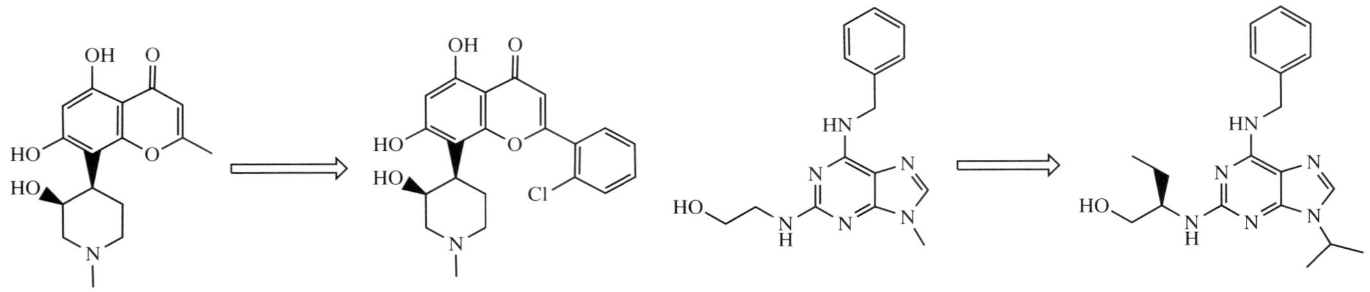

Rohitukine　Alvocidib (flavopiridol)　Olomoucine　Seliciclib

Figure 13. Design of CDK inhibitors alvocidib and seliciclib [123,129–132].

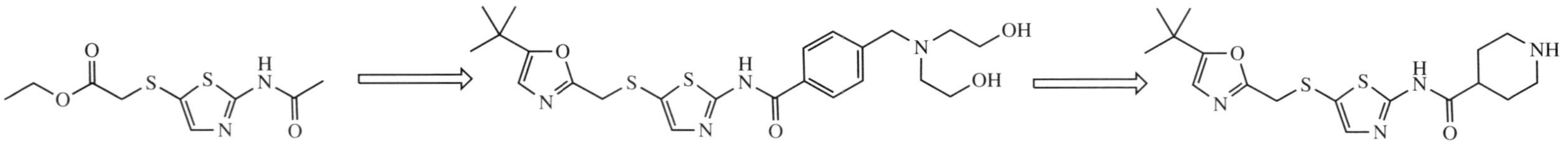

CDK2 screen lead　BMS-387032 (SNS-032)

Figure 14. Design of CDK2/7/9 inhibitor BMS-387032 [130–132].

CDK1/2/4 lead — CDK4 selective inhibitor — PD 0332991

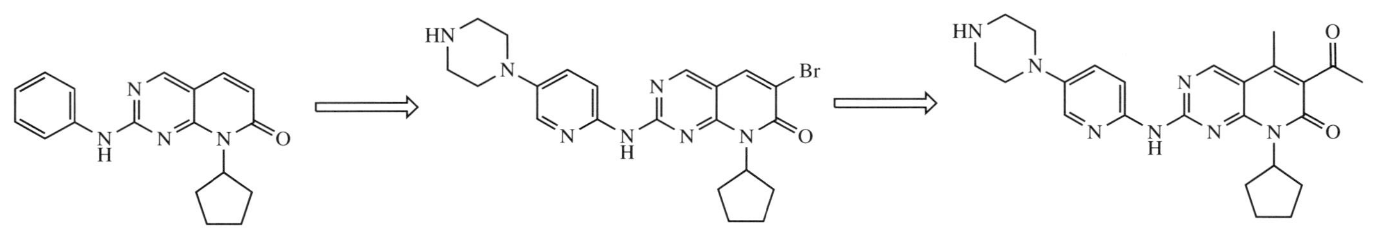

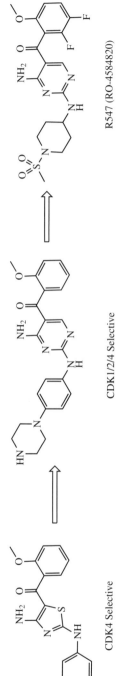

Figure 15. Design of CDK4/6 inhibitor PD 0332991 [133–135].

CDK4 Selective — CDK1/2/4 Selective — R547 (RO-4584820)

Figure 16. Design of CDK1/2/4 inhibitor R547 [137,138].

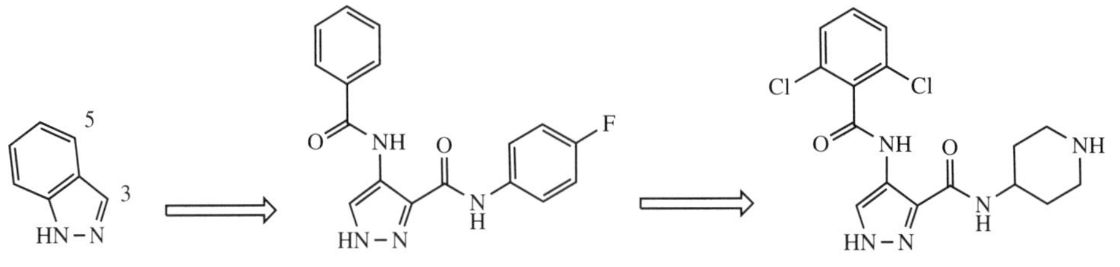

Figure 17. Design of CDK1/2/4/5 inhibitor AT7519 [139].

[130–132], thought at the time to be the key regulator of the DNA synthesis (S) phase of the cell cycle. More recent studies, however, suggest that the activity of BMS-387032 in cellular assays and preclinical models may reflect inhibition of other CDK isoforms [132]. High-throughput screening identified a 2-acylaminothiazole derivative as a moderately potent inhibitor that is selective for CDK2 relative to CDK1/4 isoforms. Replacement of the carboxylic ester substituent with an oxazole moiety to obviate plasma instability and elaboration of the acylamino group provided a probe compound, which showed a favorable mouse oral pharmacokinetic profile and demonstrated activity in murine antitumor models.

Key to optimizing physical and pharmacokinetic properties and antitumor efficacy was introduction of a basic end group designed to project toward the solvent-exposed region of CDK2, and examination of a range of truncated groups to lower molecular weight then led to BMS-387032. Formulation as a hemitartrate salt conferred aqueous solubility compatible with an intravenous dosing protocol in clinical studies. An X-ray crystal structure of BMS-387032 in complex with CDK2 confirms the aminothiazole template occupies the ATP site [131], with the oxazole substituent wrapping back toward the ribose pocket and the basic piperidine ring extending toward the exterior of the protein.

The design of PD 0332991 targeted the CDK4 isoform (Fig. 15) [133–135], based on developing evidence that this isoform plays a more significant role in mitosis than CDK2. Through insight from X-ray crystal structures of bound inhibitors, a lead with moderate potency against CDK1/2/4 was obtained by modification of an earlier arylamino pyridopyrimidone template that had given potent tyrosine kinase inhibitors [133]. Elaboration of the arylamino moiety and bromo substitution in the fused pyridone ring markedly improved potency and selectivity in favor of CDK4 [134]. Replacement of the bromo substituent with a range of small polar groups to improve physical properties and introduction of an adjacent methyl group to maintain CDK4 selectivity then led to PD 0332991 [134,135].

Further biochemical characterization showed PD 0332991 to be a potent inhibitor of CDK6, an isoform that may also have a key role in mitosis. Analysis of the crystal structure of PD 0332991 bound to human CDK6 suggests that the observed selectivity profile is due to relatively small conformational differences between CDK2 and CDK6 in the hinge region of the ATP binding site [136].

The medicinal chemistry program that led to R547 (RO-4584820) sought a potent inhibitor of CDK1/2/4 (Fig. 16) [137,138], on the premise that inhibition of multiple CDKs may be needed to sustain suppression of tumor growth in patients. Replacement of the diaminothiazole core of a series of CDK4-selective inhibitors [137] with a more hydrophilic 2,4-diamino-5-ketopyrimidine gave compounds with comparable activity versus CDK1/2/4 [138]. Truncation of the basic side chain and optimization of the substitution pattern in the aryl ring then culminated in R547, which displayed low nanomolar potency against CDK1/2/4. The X-ray crystal structure of R547 bound to CDK2 confirms that in a similar fashion to BMS-387032 and PD 0332991 the diaminopyrimidine motif fulfils key hydrogen-bonding interactions at the ATP binding site [138].

AT7519 originated from a fragment screening program (Fig. 17) [139], in which apo crystals of CDK2 were soaked with a low molecular weight screening set that included compounds identified from a virtual screen against the crystal structure of CDK2. Examination of the binding mode of a weakly active indazole fragment suggested introduction of substituents at the 3- and 5-positions of the indazole ring to pick up additional interactions at or near the ATP site, an idea best realized by simplification of the indazole core to a pyrazole that served to retain the key hydrogen-bonding interactions of the core. Modification of the pyrazole substituents to optimize potency and physical properties then led to AT7519, a compound with high aqueous solubility as the acetate or hydrochloride salt.

6.2. Checkpoint Kinase Inhibitors

As part of a mechanism that modulates arrest of the cell cycle and allows repair processes to be completed prior to progression of the cycle, checkpoint kinase 1 (Chk1) is a serine–threonine kinase that is activated in normal cells in response to DNA damage due to environmental factors such as UV light or errors in replication. In many tumor cells, checkpoint control is already partially disrupted, resulting in a greater reliance on the S- and G2-checkpoints regulated by Chk1. Inhibition of Chk1 should therefore abrogate cellular arrest and sensitize cells to the effect of DNA-damaging agents such as γ-irradiation and cytotoxic drugs, a

Table 7. Checkpoint Kinase Inhibitors in Clinical Trials

Drug	Highest Phase [25,26][a]	Kinase Inhibition Profile [140,141]
UCN-01	I[b]	Chk1, ALK, CDK1/2/4, PDK1, PI3K, PKC
SCH-900776	I	Chk1
PF-00477736	I	Chk1 > Chk2
AZD7762	I	Chk1/2

[a] In combination with DNA damaging agents.
[b] In Phase II as single agent.

hypothesis supported by *in vitro* studies involving combination of Chk1 inhibitors with established antimetabolite drugs [122,140]. Although the Chk2 isoform is also activated by DNA damage, the role of Chk2 in DNA repair is less well defined.

Uncertainties remain concerning the potential safety of Chk1 inhibitors, since Chk1 not only is activated in stressed cells but also plays an essential role in normal cell repair. Of agents undergoing clinical trials in combination with a cytotoxic drugs (Table 7, Figs 18 and 19), the hydroxystaurosporine derivative UCN-01 is the most studied [122,140]. Although UCN-01 inhibits numerous other kinases, the compound is relatively selective for Chk1 over Chk2 and potentiation of the effects of DNA-damaging agents is believed to reflect inhibition of Chk1. Even so, due to effects on other kinases the efficacy of this agent may not be predictive of clinical outcome with inhibitors that specifically target Chk1/2.

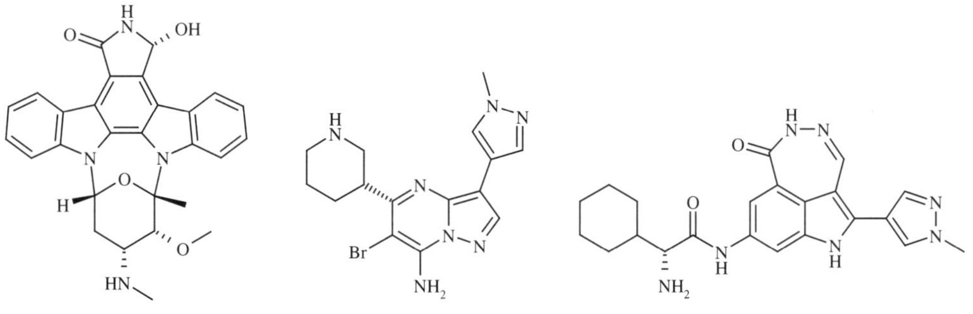

Figure 18. Chemical structures of Chk inhibitors UCN-101, SCH-900776, and PF-00477736.

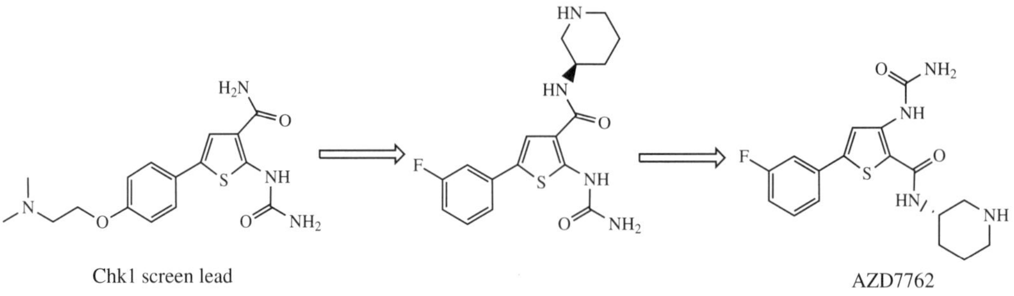

Chk1 screen lead → → AZD7762

Figure 19. Design of Chk1/2 inhibitor AZD7762 [143–145].

Selective Chk1 inhibitor SCH-900776 and Chk1/2 inhibitors PF-00477736 and AZD7762 (Figs 18 and 19) [141–143] are in early Phase I clinical trials. The evolution of AZD7762 has been outlined in part (Fig. 19) [143–145], starting from a Chk1 screen lead derived from a series of thiophene urea carboxamides that had been shown to inhibit IKK-2, a kinase implicated in autoimmune disease, and exploiting earlier knowledge that IKK-2 activity could be diminished by substitution of the carboxamide [145].

6.3. Polo-Like Kinase 1 Inhibitors

Polo-like kinases (PLKs) are a family of serine–threonine kinases with multiple functions in regulation of checkpoint processes and mitosis [122,146]. Of the four mammalian PLK isoforms, the role of PLK1 appears crucial for precise regulation of cell division, and there is clear evidence that elevated levels of PLK1 are associated with a broad range of human cancers [146]. An unusual aspect of the structure of PLK1 is the presence of two substrate binding sites, a kinase domain related to several members of the protein kinase superfamily and a "polo box" domain unique to PLK1 [146].

Of PLK1 inhibitors currently undergoing clinical trials (Table 8, Fig. 20), BI 2536 [147,148] is the most advanced compound and is being evaluated via intravenous infusion in Phase II trials as a treatment for several types of tumor [149]. Although medicinal chemistry evolution of the ATP-competitive inhibitors BI 2536 and more recent analog BI 6727 [150] has not been disclosed, the arylamino dihydropteridone core of these inhibitors strongly resembles the arylamino pyridopyrimidone template present in CDK4/6 inhibitor PD 0332991 (Section 6.1). An X-ray crystal structure of BI 2536 in complex with PLK1 suggests that the methoxy substituent in the aryl ring determines selectivity relative to non-polo-like kinases due to the presence in PLK1 of a specific small pocket in the hinge region [151].

Of the two other PLK1 inhibitors that have entered clinical trials [152,153], ON01910 is a non-ATP-competitive inhibitor that also inhibits a number of tyrosine kinases, and it is likely that this compound interacts with the kinase domain by occupying the substrate binding site rather than targeting the polo box domain [153].

Table 8. PLK1 Inhibitors in Clinical Trials

Drug	Highest Phase [25,26]	Kinase Inhibition Profile [122,146]
BI 2536	II	PLK1/2/3
BI 6727	I	Not disclosed
GSK461364	I	Not disclosed
ON01910	I	PLK1/2, Abl, CDK1, Flt-1, PDGFR, Src

6.4. Aurora Kinase Inhibitors

Aurora A, B, and C are a family of serine–threonine kinases that regulate distinct functions of the mitotic phase of the cell cycle [122,154,155]. Frequent over expression of Aurora A and/or Aurora B in tumors is indicative of a strong association with human malignancies. Whereas the Aurora A isoform regulates the processes of centrosome and spindle assembly, Aurora B controls chromosome alignment and cytokinesis.

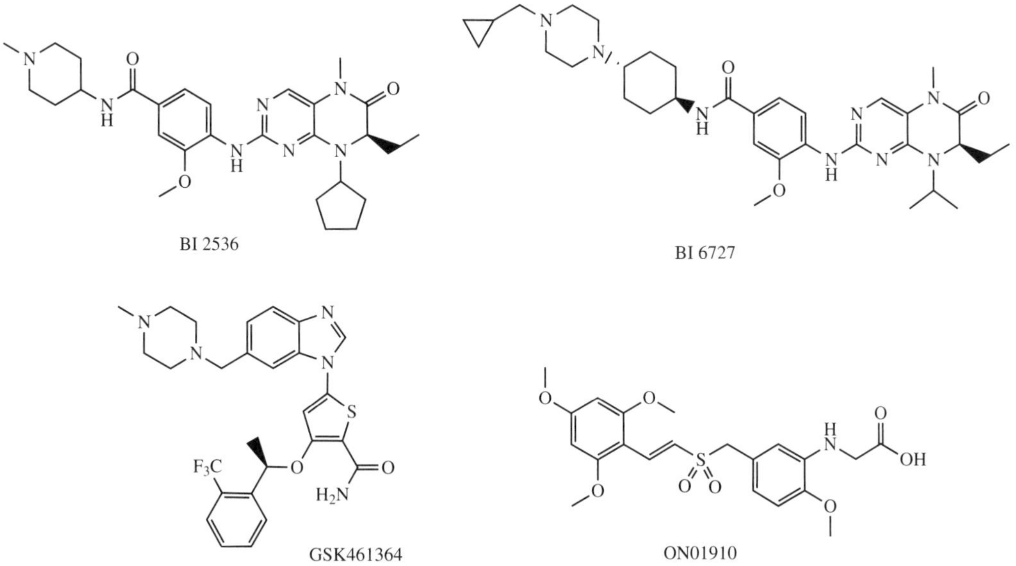

Figure 20. Chemical structures of PLK1 inhibitors BI 2536, BI 6727, GSK461364, and ON01910.

Structurally diverse Aurora kinase inhibitors representative of different Aurora isoform profiles are currently undergoing Phase I/II clinical trials (Table 9, Figs 21–24). As indicated in Table 9, a number of these compounds may be regarded as extended spectrum inhibitors since they also inhibit kinases other than Aurora isoforms. Small-molecule inhibitors that target the A and B isoforms in biochemical assays typically exhibit a phenotype consistent with Aurora B inhibition alone in cell-based systems [156], and it therefore remains unclear whether a particular isoform profile is preferred in oncology [157].

Of compounds listed in Table 9, medicinal chemistry has been disclosed relating to AZD1152, PHA-739358, and SNS-314 (Figs 21–23). Aurora B-selective inhibitor AZD1152 (Fig. 21) is based on the 4-anilinoquinazoline template present in EGFR and VEGFR inhibitors (Sections 4.1 and 5). Screening identified compounds with an extended benzamidoaniline motif as moderately potent Aurora A inhibitors [158,159]. X-Ray crystal structures of inhibitors bound to Aurora A confirmed binding at the ATP site and indicated scope to accommodate polar functionality at the 7-position of the quinazoline ring. Replacement of the aniline moiety with a range of more polar heterocycles that initially included aminopyrimidine then gave potent Aurora A/B inhibitors with significantly improved physical properties [159,160].

Extensive optimization of the extended aniline and the basic side chain, along with removal of the 6-methoxy substituent, pro-

Table 9. Aurora Kinase Inhibitors in Clinical Trials

Drug	Highest Phase [25,26]	Aurora A/B Isoform Profile	Other Kinases Inhibited
AZD1152	II	Aurora B	
PHA-739358	II	Aurora A/B	Abl, FGFR1, RET, TrkA [167]
AT9283	II	Aurora A/B	Abl, Flt3. JAK2 [169]
SNS-314	I	Aurora A/B	Flt3/4, TrkA/B [168]
MLN8054	I	Aurora A	
MLN8237	I	Aurora A	
CYC116	I	Aurora A/B	VEGFR2 [172]
PF-03814735	I	Aurora A/B	

Aurora-A screen lead

Potent Aurora-A/B inhibitor

AZD1152

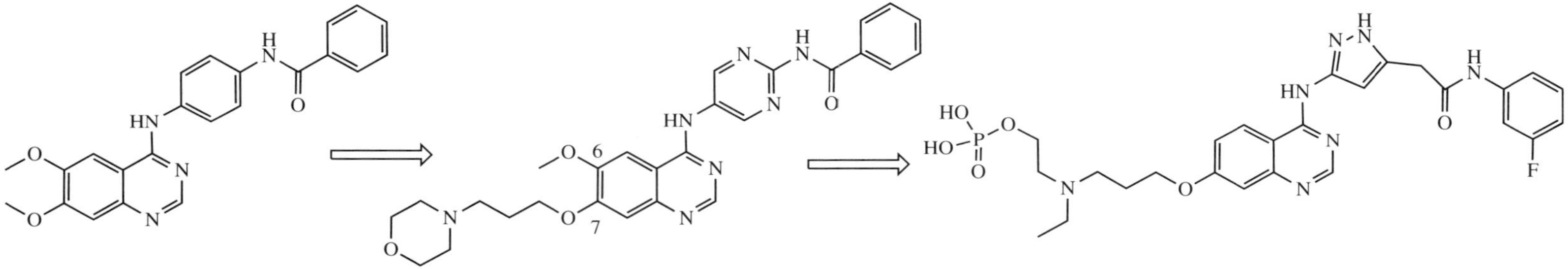

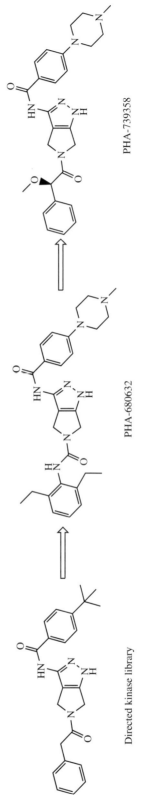

Figure 21. Design of Aurora B inhibitor AZD1152 [158–161].

Directed kinase library

PHA-680632

PHA-739358

Figure 22. Design of Aurora A/B inhibitor PHA-739358 [164,165].

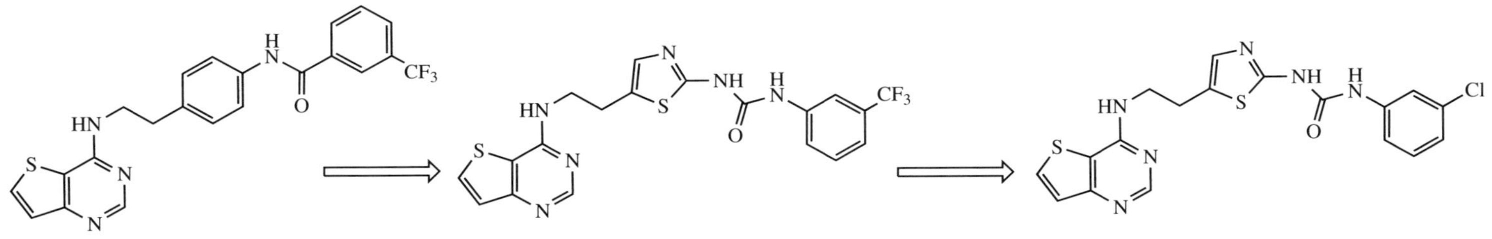

Aurora-A screen lead ⇒ Aurora-A/B inhibitor ⇒ SNS-314

Figure 23. Design of Aurora A/B inhibitor SNS-314 [168].

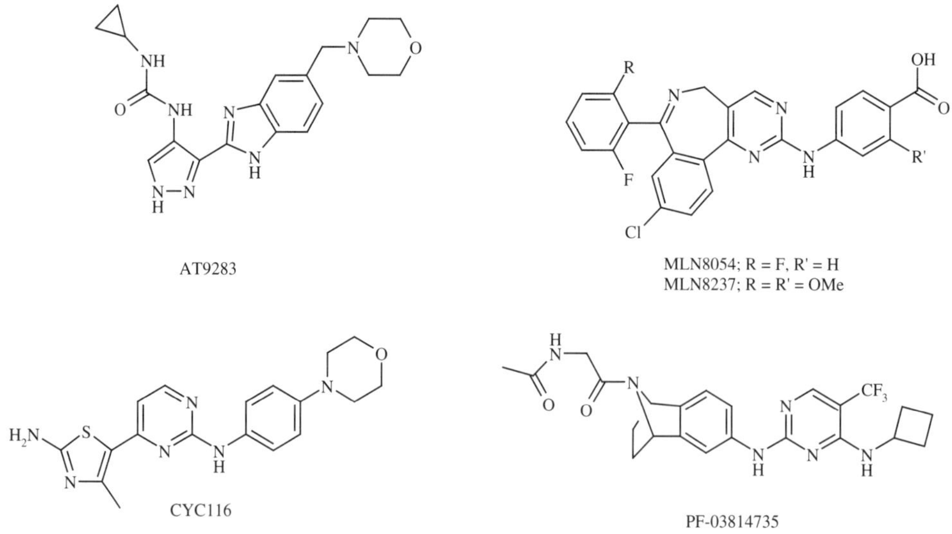

AT9283

MLN8054; R = F, R' = H
MLN8237; R = R' = OMe

CYC116

PF-03814735

Figure 24. Chemical structures of Aurora kinase inhibitors AT9283, MLN8054, MLN8237, CYC116, and PF-03814735.

vided Aurora B-selective inhibitors [161]. Since intravenous infusion was envisaged as the preferred route of administration, a pendant hydroxyl group was introduced into the basic side chain to enable derivatization as a phosphate ester prodrug that is rapidly cleaved *in vivo* by alkaline phosphatases [161]. Mechanistic studies confirmed that the mode of action of Aurora B-selective inhibitor AZD1152 reflects effects on cell division that are distinct from classical antimitotic drugs such as Taxol [162,163], and Phase I/II clinical trials have now commenced in patients with AML.

The Aurora A/B inhibitor PHA-739358 originated from a directed library based on a 3-aminopyrazole moiety previously utilized in CDK and p-38 kinase inhibitors, in which the hydrogen-bond donor–acceptor–donor motif is embedded within a bicyclic scaffold that provides a further handle for introducing structural diversity (Fig. 22) [164]. When screened against Aurora A, the kinase library gave a high hit rate characterized by a substituent such as *tert*-butyl at the *para* position of the benzamido group attached to the aminopyrazole. On the hypothesis that the *tert*-butyl substituent is directed to the outside of the kinase, a range of polar groups was introduced at the *para* position and served to increase both potency and aqueous solubility. Optimization of the substituent on the ring nitrogen gave prototype inhibitor PHA-680632 and further refinement, guided by insight from crystal structures of inhibitors bound to Aurora A, led to PHA-739358, a potent Aurora A/B inhibitor with water solubility compatible with intravenous infusion [165].

PHA-739358 has progressed to Phase II clinical trials in patients with metastatic hormone refractory prostate cancer. The compound is also a potent inhibitor of Abl, FGFR1, RET, and TrkA, all kinases that have been identified or proposed as viable targets for oncology drug design, and it has been suggested that the extended spectrum of this agent could offer promising therapeutic opportunities [165,166], albeit with the risk of decreased tolerance in patients. In addition, PHA-739358 is a potent inhibitor of several Bcr-Abl mutants, including the highly resistant T315I mutant highlighted earlier (Section 2), and Phase II clinical trials are under way in patients with relapsed CML. Cocrystallization of PHA-739358 with the T315I Abl kinase domain reveals that the inhibitor binds to the active form of the kinase, in a conformation that avoids the steric hindrance imposed on other Bcr-Abl inhibitors by the substitution of isoleucine for threonine that constitutes the T315I mutant [167].

The Aurora A/B inhibitor SNS-314 originated from an extended benzamide derivative identified from screening versus Aurora A (Fig. 23) [168]. Introduction of an aminothiazole moiety prompted a significant increase in potency and tuning the substituent on the terminal aryl ring then led to SNS-314. An X-ray crystal structure of SNS-314 in complex with Aurora A showed binding to the active form, with the thienopyrimidine moiety interacting with the hinge region and the extended urea moiety extending deep into the selectivity pocket.

Of other Aurora kinase inhibitors that have progressed to clinical trials (Fig. 24), AT9283 is an extended spectrum inhibitor that, like the CDK inhibitor AT7519 (Section 6.1), appears to have originated from a fragment screening approach [169]. MLN8054 and MLN8237 are Aurora A-selective inhibitors based on an unusual tricyclic template [170,171], and the clinical outcome from these agents will no doubt inform debate around the validity of selective Aurora A inhibition as an oncology target. Finally, CYC116 [172], an agent related to a series of CDK inhibitors [173], and PF-03814735 [174] are both Aurora A/B inhibitors that have recently entered clinical trials.

7. SURVIVAL SIGNALING INHIBITORS

Abnormal activation of the PI3K/PKB/mTOR signaling pathway, a key regulator of cellular growth, proliferation, survival, protein synthesis, and glucose metabolism is strongly implicated in a wide range of human tumor types [175,176]. A number of distinct or complementary events that can promote pathway activation include amplification of effects mediated by extracellular ligands such as insulin-like growth factor (IGF) and the associated receptor tyrosine kinase IGF-1R, amplification of phosphoinositide-3 kinase (PI3K) or activating mutation of the gene encoding for the PI3K p110α catalytic subunit, and over expression of downstream protein kinase B (PKB, also termed Akt) and the mammalian target of rapamycin (mTOR).

Known also as the survival or antiapoptotic pathway, the PI3K/PKB/mTOR pathway plays a key role in resistance of tumor cells to cytotoxic agents and more recent targeted drugs [175,176]. Blockade of this pathway could therefore not only inhibit proliferation and growth but also sensitize tumor cells to programed cell death (apoptosis). Although intervention in this signaling pathway has been validated in the clinic by the success of mTOR inhibitors derived from rapamycin, small-molecule candidate drugs that inhibit IGF-1R, PI3K, PKB, or mTOR have only entered clinical trials relatively recently.

7.1. Insulin-like Growth Factor-1 Receptor Kinase Inhibitors

Activation of IGF-1R triggers both the MAPK pathway that drives cell growth (Section 4) and the antiapoptotic PI3K/PKB/mTOR pathway [177–179]. A number of antibody-based therapeutics that specifically target the extracellular domain of IGF-1R have progressed to Phase II clinical trials [179], but due to high sequence homology with the closely related insulin receptor (InsR) a potential complication of developing small-molecule inhibitors of the kinase domain concerns the toxicological consequences of interfering with regulation of glucose homeostasis [177–179]. Medicinal chemistry has not been disclosed relating to NVP-AEW541 [180], OSI-906 [181], and BMS-754807 [182] (Fig. 25), the three small-molecule IGF-1R inhibitors with documented chemical structures that have progressed to Phase I clinical trials. Limited selectivity between IGF-1R and InsR has been seen in biochemical assays for these agents, but data from cellular assays and animal models suggest potential for separation between IGF-1R- and InsR-mediated effects with small-molecule kinase inhibitors [177,183].

7.2. Phosphoinositide-3 Kinase Inhibitors

Phosphoinositide-3 kinases are lipid kinases that catalyze phosphorylation of the 3-hydroxyl group of phosphoinositides [184,185]. The eight PI3Ks identified are divided into classes IA, IB, II, and III based on sequence homology and substrate preference, and are also part of a superfamily that includes Class IV protein kinases such as mTOR. The Class 1A subgroup (PI3Kα, PI3Kβ, PI3Kγ) primarily generates phosphoinositide-3,4,5-triphosphate,

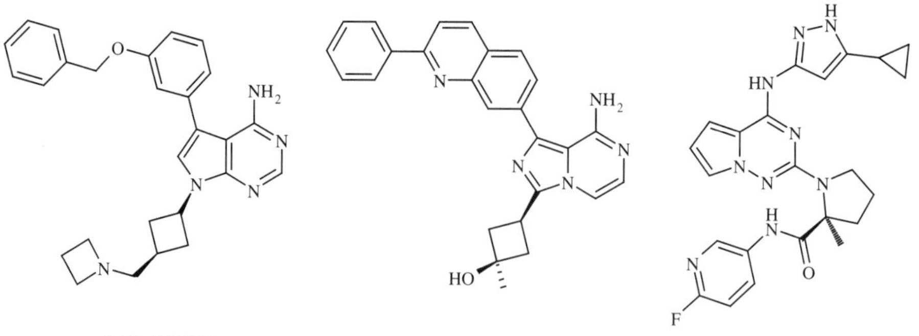

Figure 25. Chemical structures of IGF-1R inhibitors NVP-AEW541, OSI-906, and BMS-754807.

the second messenger that facilitates activation of downstream PKB.

Studies with PI3K inhibitors with a variety of isoform profiles have shown that inhibition of PI3Kα is essential to suppress growth in malignant cell lines, whereas the other isoforms have potential in disease areas that include inflammatory and autoimmune disease (PI3Kδ, PI3Kγ) and thrombosis (PI3Kβ) [184,185]. No PI3Kα-selective inhibitors have been reported, and the observation that other Class I isoforms can induce oncogenic transformation of cells may indicate that nonisoform-specific inhibitors could offer better therapeutic potential [184,185].

Four PI3K inhibitors with disclosed chemical structure have advanced to Phase I/II clinical trials (Figs 26-28). GDC-0941 is a specific inhibitor of Classes 1A and 1B PI3K isoforms that originated from a 6-hydroxy-4-morpholino-2-phenyl quinazoline lead identified through high-throughput screening against PI3K p110α (Fig. 26) [186,187]. Switching the hydroxyl group to the phenyl ring and changing to a thienopyrimidine template gave a significant increase in potency and subsequent work focused on improving physical and pharmacokinetic properties.

Inhibitor docking to a homology model of PI3K p110α based on the crystal structure of the p110γ isoform suggested that the oxygen atom of the morpoline ring forms a key hydrogen bond to the hinge and that a substituent at the 6-position of the thienopyrimidine ring would extend out of the ATP pocket into solvent, thus providing an option for improving physical properties. The resulting inhibitors still showed poor oral bioavailability that was attributed to glucuronidation of the phenolic hydroxyl, an issue addressed by replacement with an indazolyl moiety. Refinement of the polar thienopyrimidine 6-substituent then led to GDC-0941, for which a crystal structure in complex with PI3K p110γ confirmed the proposed binding mode.

The dual PI3K/mTOR kinase inhibitor NVP-BEZ235 evolved from a PDK1 imidazoquinoline screen lead (Fig. 27) [188,189]. Replacing the potentially metabolically labile cyanomethyl group with dimethylcyano-

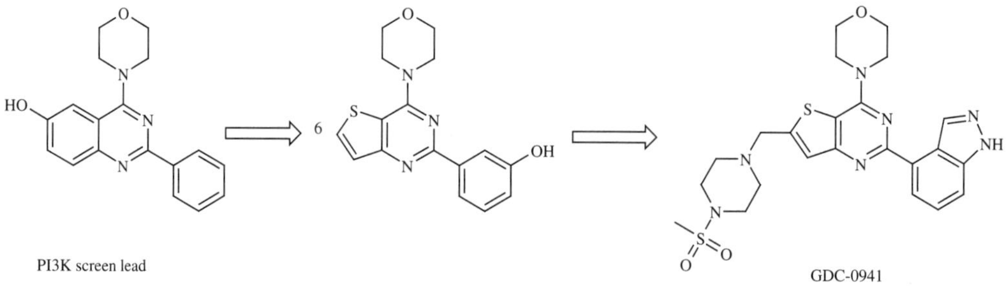

Figure 26. Design of PI3K inhibitor GDC-0941 [186,187].

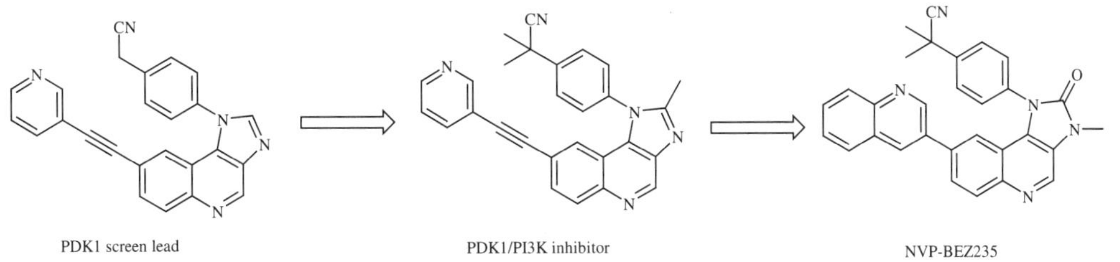

Figure 27. Design of PI3K/mTOR inhibitor NVP-BEZ235 [188,189].

methyl, and introducing a methyl substituent in the fused imidazole ring, reduced PDK1 inhibition and gave a compound equipotent against PI3K isoforms. A further change of the fused imidazole ring to *N*-methyl imidazolinone maintained PI3K inhibition but resulted in loss of PDK1 activity, an outcome rationalized from a PDK1 homology model as reflecting a repulsive electronic interaction between the imidazolinone carbonyl and the backbone of PDK1. Refinement of the quinoline substituent then led to NVP-BEZ235, subsequently characterized as a selective PI3K/mTOR kinase inhibitor [189].

As with IGF-1R inhibitors, effects on glucose homeostasis are of potential concern in clinical development of PI3Kα inhibitors, since under physiological conditions PI3Kα is a key downstream effector of the insulin receptor [184,185]. However, no effect of NVP-BEZ235 on insulin or glucose levels was reported at efficacious doses in a 13-week rat toxicity study [185], indicating that a therapeutic margin may be achievable relative to insulin-mediated effects.

Of the two other Class I PI3K inhibitors that have progressed to clinical trials (Fig. 28), PX-866 [184,185] is a stable derivative of the naturally occurring irreversible inhibitor wortmannin, while the thiazolidinedione GSK1059615 [190] is related to an earlier series of PI3Kγ-selective inhibitors, shown by X-ray crystallography to form a salt bridge between the thiazolidinedione moiety and residues at the ATP binding site of PI3Kγ [184].

7.3. Protein Kinase B Inhibitors

Protein kinase B (PKB, also termed Akt), a member of the AGC superfamily of kinases, is a serine–threonine kinase that is a central node of the complex cascade of signaling pathways, crosstalk mechanisms and feedback loops that regulate the activity of this kinase [175,176,191]. Three mammalian isoforms (PKBα, PKBβ, PKBγ) have been identified that are broadly expressed in a variety of malignancies, including breast, ovarian, prostate, pancreatic, and skin cancers, and are linked to an antiapoptotic effect in many cell types.

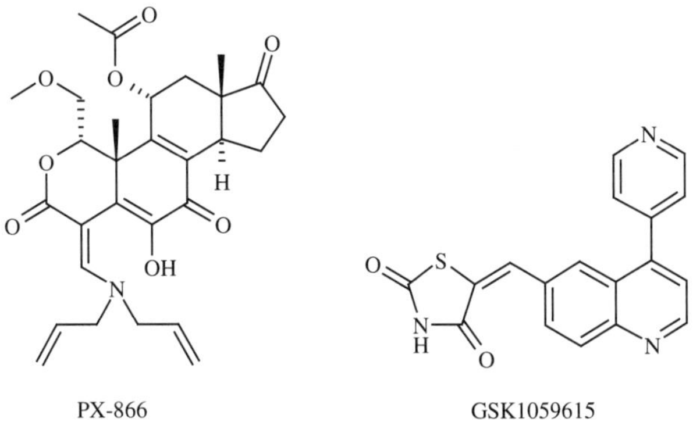

Figure 28. Chemical structures of PI3K inhibitors PX-866 and GSK1059615.

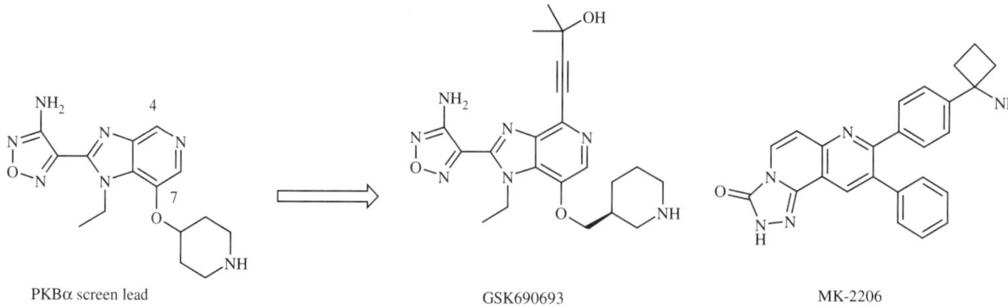

Figure 29. Design of ATP-competitive PKB inhibitor GSK690693 [192] and chemical structure of allosteric PKB inhibitor MK-2206.

Although a number of ATP-competitive inhibitors of PKB isoforms have been reported [191], only the PKBα/β/γ inhibitor GSK690693 (Fig. 29) has progressed to Phase I clinical trials [192]. Also depicted in Fig. 29 is the chemical structure of MK-2206, an allosteric inhibitor reported to bind to the PKB Pleckstrin homology domain, which has also recently entered clinical trials [193].

GSK690693 was designed starting from a PKBα screen lead based on an imidazopyridine template related in structure to ATP-competitive inhibitors of ROCK1 and MSK1 kinases, also members of the AGC superfamily [192]. Inspection of this lead docked into the ATP binding site of a model of PKBβ, generated from a crystal structure of a mutant PKB in complex with a peptide substrate, revealed that substitution at the 4-position of the imidazopyridine could grant access to a pocket in the back cleft of the kinase. Closer examination of the model suggested the opening into this pocket was narrow and that an alkyne spacer could be used to direct a range of substituents to explore interactions within the pocket, an idea that led to a methyl butynol substituent that substantially boosted PKB activity and increased selectivity relative to other AGC kinases.

Variation of the substituent at the 7-position of the imidazopyridine ring was shown to have significant effect on cellular potency, and GSK690693 was selected as representing the best combination of cellular potency and selectivity. Although GSK690693 showed poor oral bioavailability across species, excellent blood levels were achieved after intravenous administration and the compound entered clinical development as an intravenous agent.

A feature of the design of PKB inhibitor GSK690693 was observation of time-dependent inhibition of the kinase, a finding that necessitated use of assays involving a preincubation step and lower concentration of the kinase to obtain a more meaningful measure of potency [192]. Cocrystallization of GSK690693 with the kinase domain of PKBβ confirmed binding at the ATP site and suggested time-dependent inhibition may relate to displacement of a water molecule by the alkynol hydroxyl.

7.4. mTOR Inhibitors

Elucidation of the mode of action of rapamycin (Fig. 30), a naturally occurring macrolide antifungal agent first isolated in the 1970s, culminated in identification of a kinase dubbed mTOR that lies downstream of PKB in the survival signaling pathway [175,194,195]. Many years of research established that rapamycin binds to an intracellular protein termed FKBP12 to form a complex that acts as an allosteric inhibitor of mTOR. Poor aqueous solubility and chemical stability precluded use of rapamycin as an anticancer agent, an issue tackled by derivatization of the free hydroxyl substituent located in the cyclohexane ring pendant to the macrolide core.

Of more soluble, stable, and orally bioavailable analogs that have progressed to clinical trials (Fig. 30), temsirolimus (TORISEL™) [196,197] was approved by the FDA in 2007 for first-line treatment of poor prognosis patients with advanced RCC, while everoli-

Rapamycin;   R = H
Temsirolimus; R = (HOCH$_2$)$_2$C(CH$_3$)CO
Everolimus; R = HO(CH$_2$)$_2$
Deforolimus; R = (CH$_3$)$_2$PO

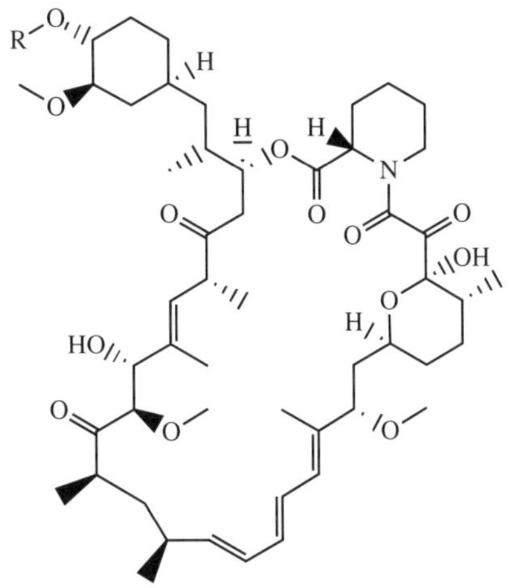

Figure 30. Chemical structures of mTOR inhibitors derived from rapamycin.

mus [195] and deforolimus [198] are undergoing Phase III trials. A majority of small-molecule inhibitors of mTOR are still at a preclinical stage [175], apart from the dual PI3K/mTOR inhibitor NVP BEZ235 that has recently entered development (Section 7.2) and selective mTOR inhibitor OSI-027 of undisclosed chemical structure [175]. Of the functionally distinct complexes of mTOR (TORC1 and TORC2), whereas only the catalytic domain of TORC1 is sensitive to rapamycin and analogs, small-molecule inhibitors may offer therapeutic advantage due to targeting of both domains [175].

8. INHIBITORS OF TUMOR INVASION AND METASTASIS

Tumor cell metastasis is a complex process that involves a number of phenotypic changes relating to cell-matrix and cell–cell adhesions, migration, and invasive capacity [199]. This chapter concludes with a brief overview of inhibitors of Src, focal adhesion kinase (FAK), and c-Met; the three kinases strongly implicated in tumor invasion and metastasis. Since each of these kinases has multiple cellular functions, inhibitors that have progressed to clinical trials may also have effects on a number of the other acquired tumor cell capabilities, making these kinases particularly attractive targets for oncology drug design.

8.1. Src Kinase Inhibitors

Src, the first identified proto-oncogene, is the most studied member of a family of nine non-receptor tyrosine kinases and is activated by signals initiated by binding of integrins and growth factors to the cell membrane [199–202]. Upregulation of Src kinase has been documented in many human tumor types, and since Src has a key role in bone resorption Src inhibitors may also provide an effective treatment for established bone metastasis. While Src is a key component of multiple signaling pathways that control cell growth, proliferation, survival, and invasion, the predominant consequences of increased Src activity in tumor cells are currently believed to be reduced cell adhesion, increased motility and promotion of an invasive phenotype [199–202].

Of Src inhibitors currently undergoing clinical trials (Table 10, Figs 31 and 32), bosutinib and saracatinib (AZD0530) are the most advanced [203–205]. Bosutinib is also a potent inhibitor of Bcr-Abl and is being evaluated in clinical trials involving patients with CML. Conversely, the Bcr-Abl inhibitor dasatinib, now approved by the FDA for treatment of CML (Section 2), is also a potent inhibitor of Src and ongoing clinical trials of dasatinib reflect tumor types in which Src inhibition may contribute to efficacy.

Bosutinib and saracatinib are further examples of clinical candidates that originated

Table 10. Src Inhibitors in Clinical Trials

Drug	Highest Phase [25,26]	Clinical Utility[a]	Kinase Inhibition Profile
Bosutinib	III	CML, breast	Src, Bcr-Abl, Fgr,[b] Lyn[b] [203]
Saracatinib (AZD0530)	II	Various solid tumors	Src, Bcr-Abl, Lck,[b] Yes[b] [205]
AP-24534[c]	I	CML	Src, Bcr-Abl[d]
KX2-391	I	Various solid tumors	Src, Bcr-Abl[d]

[a] Most advanced clinical trial [25,26].
[b] Src-family kinase.
[c] Chemical structure not disclosed, presumed related to AP-23464.
[d] Full profile not disclosed.

(Fig. 31) from 3-cyano-4-anilinoquinoline and 4-anilinoquinazoline templates first identified in design of ATP-competitive EGFR kinase inhibitors (Section 4.1) [203–206]. Variation of the anilino moiety identified substitution patterns that improved the interaction with the Src selectivity pocket, and cellular potency, physical and pharmacokinetic properties were then optimized by incorporating a basic group in the side chain at the 7-position of the quinoline or quinazoline ring. In the case of saracatinib, insight from docking of inhibitors to a homology model of Src provided the impetus to switch the quinazoline 6-substituent to the 5-position to probe access to the ribose pocket, a hypothesis subsequently confirmed by X-ray crystallography of an inhibitor-bound complex [205].

Two other Src inhibitors are have recently entered clinical trials (Fig. 32) [207–209]. Although not disclosed, the chemical structure of AP-24534 [207] is presumed to belong to the same chemical series as AP-23464 [210], a potent inhibitor of Src and Bcr-Abl derived from the same template as the CDK2 inhibitor seliciclib (Section 6.1) in which the hydroxyphenethyl and phosphine oxide substituents play a key role in determining kinase selectivity.

KX2-391 (also known as KX01) is a non-ATP competitive inhibitor of Src and Bcr-Abl [208,209], including resistant mutants such as T315I (Section 2), and is reported to interact with these kinases by occupying the substrate binding site [211]. It is the first substrate-targeted kinase inhibitor to enter clinical trials, and since the substrate binding site is less conserved than the ATP site in protein kinases an inhibitor such as KX2-0391 may be less likely to induce resistance in patients [211].

8.2. Focal Adhesion Kinase Inhibitors

Focal adhesions are found at the cell membrane where the cytoskeleton interacts with the proteins of the extracellular matrix [199,212]. Phosphorylation of FAK by Src in response to attachment to the extracellular matrix mediates formation of focal adhesions and promotes cellular motility and invasion. Over expression of FAK in a variety of human tumors such as melanoma, lymphoma, and multiple myeloma correlates with increased invasiveness and metastasis. FAK is a non-receptor tyrosine kinase and the only structurally related kinase is Pyk2, for which a role in tumorigenesis remains unclear. To date, two orally bioavailable, ATP-competitive FAK inhibitors, PF-562271 and PF-573228, closely related in chemical structure (Fig. 33) and reported to be selective for FAK and Pyk2, have entered clinical trials [213,214].

8.3. c-Met Kinase Inhibitors

c-Met is a receptor tyrosine kinase linked to several cell surface coreceptors that are activated by hepatocyte growth factor (HGF, also known as scatter factor). Signaling via c-Met has been shown to trigger a range of cellular responses, including angiogenesis, proliferation, invasion, and survival [215,216]. In particular, activation of c-Met can disrupt intracellular contacts and mobilize tumor cells away from their primary location toward the adjacent environment.

The distinct but complementary cellular activities of proliferation, motility and protection from apoptosis regulated by c-Met are collectively termed invasive growth, and over the past decade c-Met has thus emerged as an attractive oncology drug target. Aberrant reg-

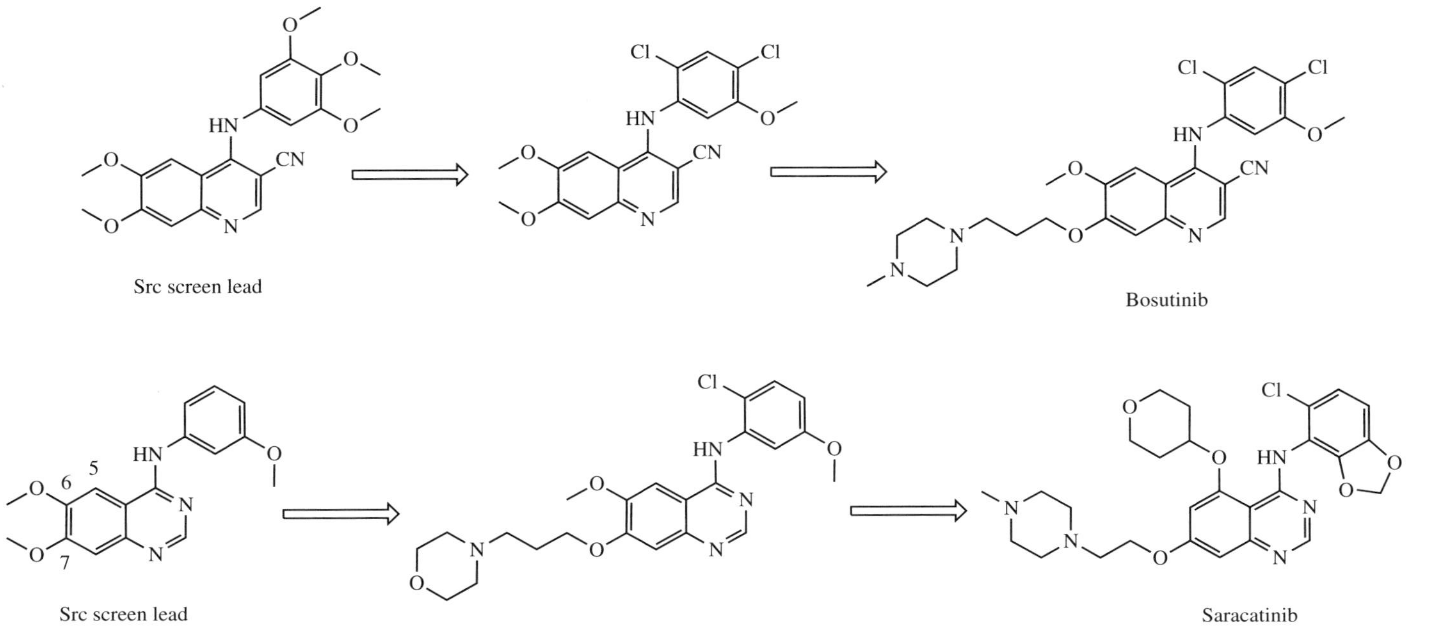

Figure 31. Design of Src inhibitors bosutinib [203,204] and saracatinib [205,206].

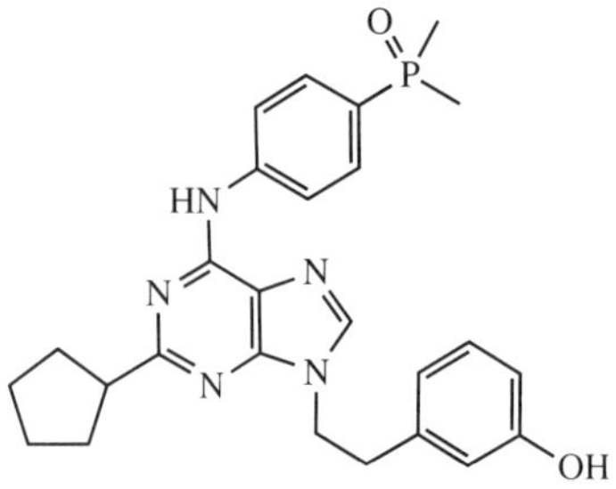

Figure 32. Chemical structures of Src inhibitors AP-23464 and KX2-391.

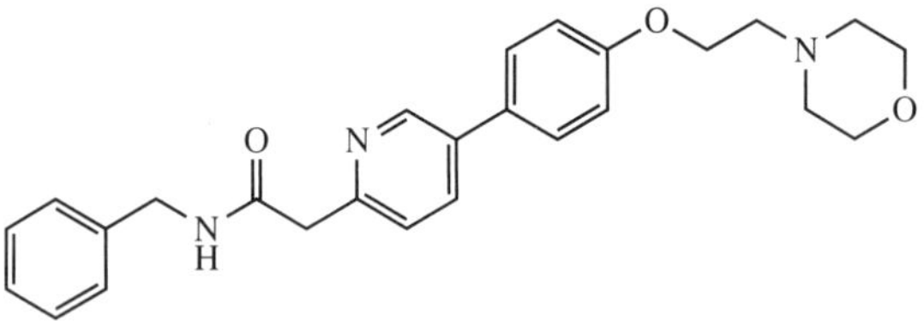

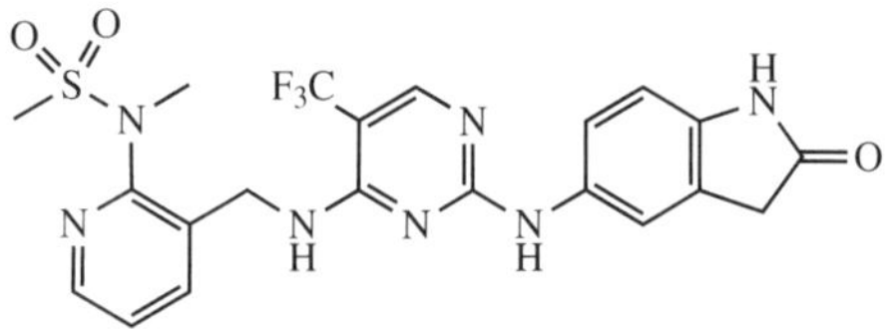

Figure 33. Chemical structures of FAK inhibitors PF-562271 and PF-573228.

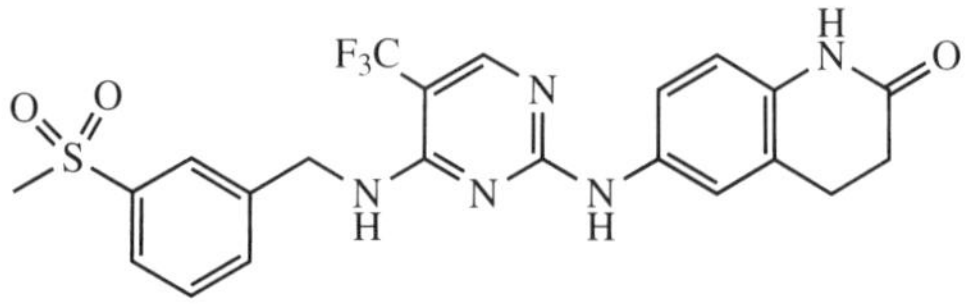

ulation of c-Met signaling has been implicated in a variety of human cancers, including prostate, gastric and a form of hereditary RCC [215,216]. It is only recently, however, that the first small-molecule inhibitors with disclosed chemical structures have entered clinical trials (Table 11, Figs 34 and 35).

The c-Met receptor tyrosine kinase subfamily is structurally distinct and other family members include the kinases RON, Axl, Mer, and Tyro3, all of which have a less well-understood role in tumorigenesis [215,216]. Of inhibitors that have progressed to the clinic, medicinal chemistry has only been disclosed relating to the design of BMS-777607, an inhibitor selective for c-Met and associated kinase family members that originated (Fig. 34) from a directed library based on a pyrrolotriazine template that serves as an ATP mimetic (see also Section 5) [217–220]. Changing the pyrrolotriazine to a 2-aminopyridine motif that can also bind to the hinge region of c-Met led to a significant improvement in potency and pharmacokinetic properties [217,218]. Based on insight from X-ray crystal structures, introduction of a pyridine 3-substituent provided access to the c-Met ribose pocket [220], while as reported for a related series of inhibitors based on a pyrrolopyridine template [219] the malonamide moiety could be incorporated in a conformationally constrained pyridinone ring to give BMS-777607 [220].

Other c-Met inhibitors listed in Table 11 and depicted in Fig. 35 have been disclosed in publications or at oncology conferences in the second half of 2008, and include compounds

Table 11. c-Met Inhibitors in Clinical Trials

Drug	Highest Phase [25,26]	Kinase Inhibition Profile [215,216]
BMS-777607	II	c-Met, RON,[a] Axl,[a] Mer,[a] Tyro3[a]
PF-02341066	I	c-Met, ALK
PF-04217903	I	c-Met
JNJ-38877605	I	c-Met
MGCD-265	I	c-Met, VEGFR1/2/3, Tie2, RON[a]
E-7050	I	c-Met, VEGFR2

[a] c-Met-family kinase.

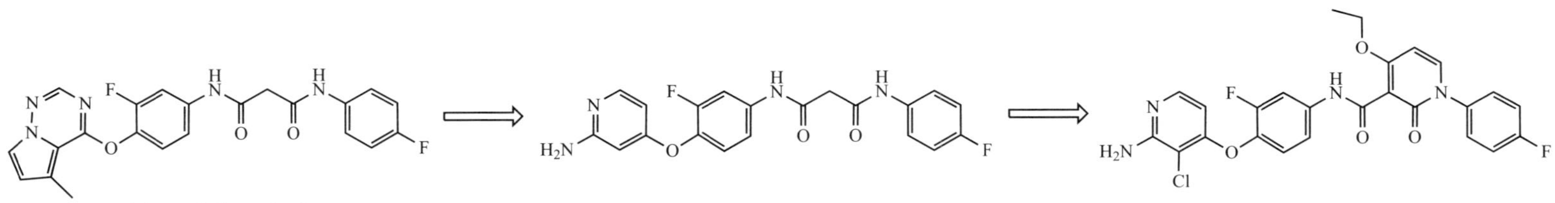

Figure 34. Design of c-Met inhibitor BMS-777607 [217–220].

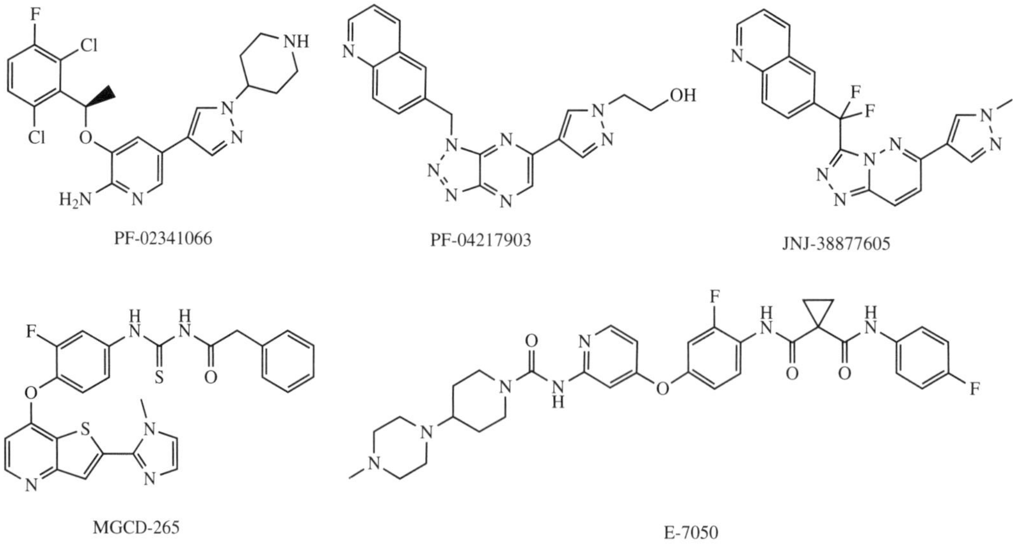

Figure 35. Chemical structures of c-Met inhibitors in Phase I clinical trials.

that are either relatively selective for c-Met [221–223] or display an extended spectrum of kinase activity [224,225]. As indication of the high level of commitment to c-Met as a therapeutic target, inhibitors of undisclosed structure that are also undergoing early stage clinical trials include the selective inhibitors SGX523 and ARQ197, and the extended spectrum inhibitors XL184, XL880, and MP470 [215,216].

9. CONCLUDING REMARKS

Medicinal chemistry during the past two decades has identified ATP-site directed inhibitors of protein kinases implicated in development of a wide range of human tumors. Commencement of kinase inhibitor research in the 1990s coincided with introduction of high-throughput screening, and the resulting leads highlighted the hydrogen-bond donor–acceptor motif essential for interaction with the ATP site. While high-throughput screening has continued to play a significant role in uncovering kinase inhibitor leads, an increasing proportion of leads has been derived from existing inhibitor templates, or from alternative approaches such as directed libraries, fragment, and virtual screening. With increasing availability of crystal structures of inhibitor-bound complexes, structural information and homology modeling have provided additional insight by prioritizing positions on the lead structure that can be varied to access ATP-site binding pockets and/or the solvent channel and thereby optimize potency, kinase selectivity, and overall property profile.

A debate underlying the trade-off between antitumor efficacy and toxicity that is fundamental to cancer chemotherapy concerns the relative merits of selective and multitargeted kinase inhibitors. Although drugs with activity against multiple kinases may give greater clinical benefit, side effects that reflect activity against other kinases may be more likely with less selective agents. Allosteric kinase inhibitors that target alternative, ATP uncompetitive binding sites have the potential to provide greater specificity and fewer off-target effects, and a number of allosteric inhibitors have now progressed to early stage clinical trials.

Bcr-Abl, growth factor and antiangiogenic kinase inhibitor drugs have all been validated in large-scale oncology clinical trials, and it is likely that drugs targeting the cell cycle, survival signaling and tumor invasion and metastasis will be approved by regulatory authorities in the coming years. Central to capitalizing on these advances will be more insightful understanding of tumor profiling, the resistance mechanisms, and the intricacies of the complex array of oncogenic kinase signaling cascades, to enable identification of optimal multitargeted inhibitors and combination therapies involving more selective agents.

ACKNOWLEDGMENTS

The author thanks Dr. Bryan Takasaki for access to a prototype version of a competitor intelligence database that greatly assisted with the tables of kinase inhibitors compiled for this chapter, and colleagues at AstraZeneca for critical review of the manuscript, particularly Drs. Andy Barker, Gillian Hill, Jason Kettle, Richard Luke, and Brian Tait.

ABBREVIATIONS

ALL	acute lymphoblastic leukemia
AML	acute myeloid leukemia
ATP	adenosine triphosphate
CDK	cyclin-dependent kinase
CHK	checkpoint kinase
CLL	chronic lymphocytic leukemia
CML	chronic myelogenous leukemia
CRC	colorectal cancer
CYP450	cytochrome P450
EGFR	epidermal growth factor receptor
FAK	focal adhesion kinase
FDA	Federal Drug Administration
FGFR	fibroblast growth factor receptor
Flt3	FMS-like tyrosine kinase 3
GIST	gastrointestinal stromal tumor
HCC	hepatocellular carcinoma
HGF	hepatocyte growth factor
IGF-1R	insulin-like growth factor-1 receptor
InsR	insulin receptor
MTD	maximum tolerated dose
mTOR	mammalian target of rapamycin
NSCLC	nonsmall cell lung cancer
PDGFR	platelet-derived growth factor receptor
PI3K	phosphoinositide-3 kinase
PKB	protein kinase B
PKC	protein kinase C
PLK	polo-like kinase
POA	prevention of activation
RCC	renal cell carcinoma
RTK	receptor tyrosine kinase
SAR	structure–activity relationships
Trk	tropomycin receptor kinase
VEGFR	vascular endothelial growth factor receptor

REFERENCES

1. Kufe DW, Pollock RE, Weichselbaum RR, Bast RC, Jr, Gansler TS, Holland JF, Frei E, III,editors. Cancer Medicine-6, Vol. 1. Hamilton: BC Decker; 2003.

2. Zhang J, Yang PL, Gray NS. Targeting cancer with small molecule kinase inhibitors. Nat Rev Cancer 2009;9:28–39.

3. Alton GR, Lunney EA. Targeting the unactivated conformations of protein kinases for small molecule drug design. Expert Opin Drug Discov 2008;3(6):595–605.

4. Blume-Jensen P, Hunter T. Oncogenic kinase signalling. Nature 2001;411:355–365.

5. Fabian A, Biggs H, Treiber DK, Atteridge E, Azimioara MD, Benedetti MG, Carter TA, Ciceri P, Edeen T, Floyd M, Ford JM, Galvin M, Gerlach JL, Grotzfeld RM, Herrgard S, Insko DE, Insko MA, Lai AG, Lelias J-M, Mehta SA, Milanov ZV, Velasco AM, Wodicka LM, Patel HK, Zarrinkar PP, Lockhart DJ. A small molecule-kinase interaction map for clinical kinase inhibitors. Nat Biotechnol 2005;23(3):329–336.

6. Karaman MW, Herrgard S, Treiber DK, Gallant P, Atteridge CE, Campbell BT, Chan KW, Ciceri P, Davis MI, Edeen PT, Faraoni R, Floyd M, Hunt JP, Lockhart DJ, Milanov ZV, Morrison MJ, Pallares G, Patel HK, Pritchard S, Wodicka LM, Zarrinkar PP. A quantitative analysis of kinase inhibitor selectivity. Nat Biotechnol 2008;26(1):127–132.

7. Liao JJ-L. Molecular recognition of protein kinase binding pockets for design of potent and selective kinase inhibitors. J Med Chem 2007;50(3):409–424.

8. Copeland RA, Pompliano DL, Meek TD. Drug-target residence time and its implica-

tions for lead optimization. Nat Rev Drug Discov 2006;5(9):730–739.
9. Kamb A, Wee S, Lengauer C. Why is cancer drug design so difficult? Nat Rev Drug Discov 2007;6:115–120.
10. LoRusso PM, Eder JP. Therapeutic potential of novel selective-spectrum kinase inhibitors in oncology. Expert Opin Investig Drugs 2008; 17(7):1013–1028.
11. Force T, Kerkela R. Cardiotoxicity of the new cancer therapeutics: mechanisms of, and approaches to, the problem. Drug Discov Today 2008;13(17/18):778–784.
12. Listing of FDA-approved oncology drugs: http://www.fda.gov/cder/cancer/approved.htm.
13. Bogoyevitch MA, Fairlie DP. A new paradigm for protein kinase inhibition: blocking phosphorylation without directly targeting ATP binding. Drug Discov Today 2007;12(15 & 16): 622–633.
14. Smith JK, Mamoon M, Duhe RJ. Emerging roles of targeted small molecule protein–tyrosine kinase inhibitors in cancer therapy. Oncol Res 2004;14(4–5):175–225.
15. Recanatini M, Poluzzi E, Masetti M, Cavalli A, de Ponti F. QT prolongation through hERG K+ channel blockade: current knowledge and strategies for the early prediction during drug development. Med Res Rev 2005;25 (2):133–166.
16. Wienkers LC, Heath TG. Predicting *in vivo* drug interactions from *in vitro* drug design data. Nat Rev Drug Discov 2005;4:825–833.
17. Li S, He H, Parthiban LJ, Yin H, Serajuddin ATM. IV-IVC considerations in the development of immediate-release oral dosage form. J Pharm Sci 2005;94(7):1396–1417.
18. Gill AL, Verdonk M, Boyle RG, Taylor R. A comparison of physicochemical property profiles of marketed oral drugs and orally bioavailable anti-cancer protein kinase inhibitors in clinical development. Curr Top Med Chem 2007;7(14):1408–1422.
19. Leeson PD, Springthorpe B. The influence of drug-like concepts on decision-making in medicinal chemistry. Nat Rev Drug Discov 2007;6(11):881–890.
20. Hughes JD, Blagg J, Price DA, Bailey S, DeCrescenzo GA, Devraj RV, Ellsworth E, Fobian YM, Gibbs ME, Gilles RW, Greene N, Huang E, Krieger-Burke T, Loesel J, Wager T, Whiteley L, Zhang Y. Physiochemical drug properties associated with *in vivo* toxicological outcomes. Bioorg Med Chem Lett 2008;18 (17):4872–4875.
21. Kelland L. Discontinued drugs in 2005: oncology drugs. Expert Opin Investig Drugs 2006; 15(11):1309–1318.
22. Williams R. Discontinued drugs in 2006: oncology drugs. Expert Opin Investig Drugs 2008; 17(3):269–283.
23. Williams R. Discontinued drugs in 2007: oncology drugs. Expert Opin Investig Drugs 2008; 17(12):1791–1816.
24. Hanahan D, Weinberg RA. The hallmarks of cancer. Cell 2000;100:57–70.
25. Open access database: http://clinicaltrials.gov.
26. Restricted access databases: http://integrity-prouscom; http://www.thomson-pharma.com.
27. Weisberg E, Manley PW, Cowan-Jacob SW, Hochhaus A, Griffin JD. Second generation inhibitors of BCR-ABL for the treatment of imatinib-resistant chronic myeloid leukemia. Nat Rev Cancer 2007;7(5):345–356.
28. Noronha G, Cao J, Chow CP, Dneprovskaia E, Fine RM, Hood J, Kang X, Klebansky B, Lohse D, Mak CC, McPherson A, Palanki MSS, Pathak VP, Renick J, Soll R, Zeng B. Inhibitors of ABL and the ABL-T315I mutation. Curr Top Med Chem 2008;8(10):905–921.
29. Capdeville R, Buchdunger E, Zimmerman J, Matter A. Glivec (STI1571, imatinib), a rationally developed, targeted anticancer drug. Nat Rev Drug Discov 2002;1:493–502.
30. Lydon NB, Druker BJ. Lessons learned from the development of imatinib. Leuk Res 2004;28 (Suppl 1):29–38.
31. Wiedmann MW, Caca K. Molecularly targeted therapy for gastrointestinal cancer. Curr Cancer Drug Targets 2005;5(3):171–193.
32. Gorre ME, Mohammed M, Ellwood K, Hsu N, Paquette R, Rao PN, Sawyers CL. Clinical resistance to STI-571 cancer therapy caused by BCR-ABL gene mutation or amplification. Science 2001;293:876–880.
33. Nagar B, Bornmann WG, Pellicena P, Schindler T, Veach DR, Miller WT, Clarkson B, Kuriyan J. Crystal structures of the kinase domain of c-Abl in complex with the small molecule inhibitors PD173955 and imatinib (STI-571). Cancer Res 2002;62(15):4236–4243.
34. Weisberg E, Manley PW, Breitenstein W, Brueggen J, Cowan-Jacob SW, Ray A, Huntly B, Fabbro D, Fendrich G, Hall-Meyers E, Kung AL, Mestan J, Daley GQ, Callahan L, Catley L, Cavazza C, Mohammed A, Neuberg D, Wright RD, Gilliland DG, Griffin JD. Char-

acterization of AMN107, a selective inhibitor of native and mutant Bcr-Abl. Cancer Cell 2005; 7(2):129–141.

35. Golemovic M, Verstovsek S, Giles F, Cortes J, Manshouri T, Manley PW, Mestan J, Dugan M, Alland L, Griffin JD, Arlinghaus RB, Sun T, Kantarjian H, Beran M. AMN107, a Novel aminopyrimidine inhibitor of Bcr-Abl, has in vitro activity against imatinib-resistant chronic myeloid leukemia. Clin Cancer Res 2005;11(13):4941–4947.

36. Weisberg E, Manley P, Mestan J, Cowan-Jacob S, Ray A, Griffin JD. AMN107 (nilotinib): a novel and selective inhibitor of BCR-ABL. Br J Cancer 2006;94(12):1765–1769.

37. Jabbour E, El Ahdab S, Cortes J, Kantarjian H. Nilotinib: a novel Bcr-Abl tyrosine kinase inhibitor for the treatment of leukemias. Expert Opin Investig Drugs 2008;17(7):1127–1136.

38. O'Hare T, Walters DK, Stoffregen EP, Jia T, Manley PW, Mestan J, Cowan-Jacob SW, Lee FY, Heinrich MC, Deininger MWN, Druker BJ. In vitro activity of Bcr-Abl inhibitors AMN107 and BMS-354825 against clinically relevant imatinib-resistant Abl kinase domain mutants. Cancer Res 2005;65 (11):4500–4505.

39. Lombardo LJ, Lee FY, Chen P, Norris D, Barrish JC, Behnia K, Castaneda S, Cornelius LAM, Das J, Doweyko AM, Fairchild C, Hunt JT, Inigo I, Johnston K, Kamath A, Kan D, Klei H, Marathe P, Pang S, Peterson R, Pitt S, Schieven GL, Schmidt RJ, Tokarski J, Wen M-L, Wityak J, Borzilleri RM. Design of N-(2-chloro-6-methyl-phenyl)-2-(6-(4-(2-hydroxyethyl)-piperazin-1-yl)-2-methylpyrimidin-4-ylamino)thiazole-5-carboxamide (BMS-354825), a dual Src/Abl kinase inhibitor with potent antitumor activity in preclinical assays. J Med Chem 2004;47(27):6658–6661.

40. Das J, Chen P, Norris D, Padmanabha R, Lin J, Moquin RV, Shen Z, Cook LS, Doweyko AM, Pitt Si, Pang S, Shen DR, Fang Q, de Fex HF, McIntyre KW, Shuster DJ, Gillooly KM, Behnia K, Schieven GL, Wityak J, Barrish JC. 2-Aminothiazole as a novel kinase inhibitor template. Structure–activity relationship studies toward the design of N-(2-chloro-6-methylphenyl)-2-[[6-[4-(2-hydroxyethyl)-1-piperazinyl)]-2-methyl-4-pyrimidinyl]amino)]-1,3-thiazole-5-carboxamide (dasatinib, BMS-354825) as a potent pan-Src kinase inhibitor. J Med Chem 2006;49(23):6819–6832.

41. McIntyre JA, Castaner J, Bayes M. Dasatinib: treatment of leukemia treatment of solid tumors Bcr-Abl and Src kinase inhibitor. Drugs Future 2006;31(4):291–303.

42. Tokarski JS, Newitt JA, Chang CYJ, Cheng JD, Wittekind M, Kiefer SE, Kish K, Lee FYF, Borzilleri R, Lombardo LJ, Xie D, Zhang Y, Klei HE. The structure of dasatinib (BMS-354825) bound to activated ABL kinase domain elucidates its inhibitory activity against imatinib-resistant ABL mutants. Cancer Res 2006;66(11):5790–5797.

43. Vajpai N, Strauss A, Fendrich G, Cowan-Jacob SW, Manley PW, Grzesiek S, Jahnke W. Solution conformations and dynamics of ABL kinase-inhibitor complexes determined by NMR substantiate the different binding modes of imatinib/nilotinib and dasatinib. J Biol Chem 2008;283(26):18292–18302.

44. Revill P, Serradell N, Bolos J, Rosa E. Lestaurtinib. Drugs Future 2007;32(3):215–222.

45. Cheng Y, Paz K. Tandutinib, an oral, small-molecule inhibitor of FLT3 for the treatment of AML and other cancer indications. IDrugs 2008;11(1):46–56.

46. Stubbs MC, Armstrong SA. FLT3 as a therapeutic target in childhood acute leukemia. Curr Drug Targets 2007;8(6):703–714.

47. Matsuno K, Ichimura M, Nakajima T, Tahara K, Fujiwara S, Kase H, Ushiki J, Giese NA, Pandey A, Scarborough RM, Lokker NA, Yu J-C, Irie J, Tsukuda E, Ide S, Oda S, Nomoto Y. Potent and selective inhibitors of platelet-derived growth factor receptor phosphorylation. 1. Synthesis, structure–activity relationship, and biological effects of a new class of quinazoline derivatives. J Med Chem 2002;45 (14):3057–3066.

48. Pandey A, Volkots DL, Seroogy JM, Rose JW, Yu J-C, Lambing JL, Hutchaleelaha A, Hollenbach SJ, Abe K, Giese NA, Scarborough RM. Identification of orally active, potent, and selective 4-piperazinylquinazolines as antagonists of the platelet-derived growth factor receptor tyrosine kinase family. J Med Chem 2002;45(17):3772–3793.

49. Podar K, Raab MS, Chauhan D, Anderson KC. The therapeutic role of targeting protein kinase C in solid and hematologic malignancies. Expert Opin Investig Drugs 2007;16 (10):1693–1707.

50. Choong NW, Ma PC, Salgia R. Therapeutic targeting of receptor tyrosine kinases in lung cancer. Expert Opin Ther Targets 2005;9: 533–559.

51. Yeon CH, Pegram MD. Anti-erbB-2 antibody trastuzumab in the treatment of HER2-ampli-

fied breast cancer. Invest New Drugs 2005;23: 391–409.
52. Langer CJ. Targeted therapy in head and neck cancer: state of the art 2007 and review of clinical applications. Cancer 2008;112 (12):2635–2645.
53. Ward WHJ, Cook PN, Slater AM, Davies DH, Holdgate GA, Green LR. Epidermal growth factor receptor tyrosine kinase. Investigation of catalytic mechanism, structure-based searching and design of a potent inhibitor. Biochem Pharmacol 1994;48(4):659–666.
54. Barker AJ, Gibson KH, Grundy W, Godfrey AA, Barlow JJ, Healy MP, Woodburn JR, Ashton SE, Curry BJ, Scarlett L, Henthorn L, Richards L. Studies leading to the identification of ZD1839 (Iressa): an orally active, selective epidermal growth factor receptor tyrosine kinase inhibitor targeted to the treatment of cancer. Bioorg Med Chem Lett 2001;11 (14):1911–1914.
55. Herbst RS, Fukuoka M, Baselga J. Timeline: gefitinib—a novel targeted approach to treating cancer. Nat Rev Cancer 2004;4(12):956–965.
56. Wakeling AE. Design and development of Iressa. In: Pinna LA, Cohen PTW, editors. Inhibitors of Protein Kinases and Protein Phosphatases, Vol. 167, Series: *Handbook of Experimental Pharmacology*. Berlin: Springer; 2005. p 443–450.
57. Stamos J, Sliwkowski MX, Eigenbrot C. Structure of the epidermal growth factor receptor kinase domain alone and in complex with a 4-anilinoquinazoline inhibitor. J Biol Chem 2002;277(48):46265–46272.
58. Yun C-H, Boggon TJ, Li Y, Woo MS, Greulich H, Meyerson M, Eck MJ. Structures of lung cancer-derived EGFR mutants and inhibitor complexes: mechanism of activation and insights into differential inhibitor sensitivity. Cancer Cell 2007;11(3):217–227.
59. Fukuoka M, Yano S, Giaccone G, Tamura T, Nakagawa K, Douillard J-Y, Nishiwaki Y, Vansteenkiste J, Kudoh S, Rischin D, Eek R, Horai T, Noda K, Takata I, Smit E, Averbuch S, Macleod A, Feyereislova A, Dong R-P, Baselga J. Multi-institutional randomized phase II trial of gefitinib for previously treated patients with advanced non-small-cell lung cancer. J Clin Oncol 2003;21(12):2237–2246.
60. Kris MG, Natale RB, Herbst RS, Lynch TJ, Jr, Prager D, Belani CP, Schiller JH, Kelly K, Spiridonidis H, Sandler A, Albain KS, Cella D, Wolf MK, Averbuch SD, Ochs JJ, Kay AC. Efficacy of gefitinib, an inhibitor of the epidermal growth factor receptor tyrosine kinase, in symptomatic patients with non-small cell lung cancer. A randomized trial. J Am Med Assoc 2003;290(16):2149–2158.
61. Thatcher N, Chang A, Parikh P, Rodrigues PJ, Ciuleanu T, von Pawel J, Thongprasert S, Tan EH, Pemberton K, Archer V, Carroll K. Gefitinib plus best supportive care in previously treated patients with refractory advanced non-small-cell lung cancer: results from a randomised, placebo-controlled, multicentre study (Iressa Survival Evaluation in Lung Cancer). Lancet 2005;366(9496):1527–1537.
62. Siegel-Lakhai WS, Beijnen JH, Schellens JHM. Current knowledge and future directions of the selective epidermal growth factor receptor inhibitors erlotinib (Tarceva) and gefitinib (Iressa). Oncologist 2005;10(8):579–589.
63. Blackhall F, Ranson M, Thatcher N. Where next for gefitinib in patients with lung cancer? Lancet Oncol 2006;7(6):499–507.
64. Kim ES, Hirsh V, Mok T, Socinski MA, Gervais R, Wu Y-L, Li L-Y, Watkins CL, Sellers MV, Lowe ES, Sun Y, Liao M-L, Osterlind K, Reck M, Armour AA, Shepherd FA, Lippman SM, Douillard J-Y. Gefitinib versus docetaxel in previously treated non-small-cell lung cancer (INTEREST): a randomised phase III trial. Lancet 2008;372(9652):1809–1818.
65. Giaccone G, Rodriguez JA. EGFR inhibitors: what have we learned from the treatment of lung cancer? Nat Clin Pract Oncol 2005;2: 554–561.
66. Hirsch FR, Varella-Garcia M, Bunn PA, Jr, Franklin WA, Dziadziuszko R, Thatcher N, Chang A, Parikh P, Pereira JR, Ciuleanu T, von Pawel J, Watkins C, Flannery A, Ellison G, Donald E, Knight L, Parums D, Botwood N, Holloway B. Molecular predictors of outcome with gefitinib in a phase III placebo-controlled study in advanced non-small-cell lung cancer. J Clin Oncol 2006;24(31):5034–5042.
67. Riely GJ, Politi KA, Miller VA, Pao W. Update on epidermal growth factor receptor mutations in non-small cell lung cancer. Clin Cancer Res 2006;12(24):7232–7241.
68. Thomas SK, Fossella FV, Liu D, Schaerer R, Tsao AS, Kies MS, Pisters KM, Blumenschein GR, Jr, Glisson BS, Lee JJ, Herbst RS, Zinner RG. Asian ethnicity as a predictor of response in patients with non-small-cell lung cancer treated with gefitinib on an expanded access program. Clin Lung Cancer 2006;7(5):326–331.
69. Costa DB, Nguyen K-SH, Cho BC, Sequist LV, Jackman DM, Riely GJ, Yeap BY, Halmos B,

Kim JH, Jaenne PA, Huberman MS, Pao W, Tenen DG, Kobayashi S. Effects of erlotinib in EGFR mutated non-small cell lung cancers with resistance to gefitinib. Clin Cancer Res 2008;14(21):7060–7067.

70. Petrov KG, Zhang Y-M, Carter M, Cockerill GS, Dickerson S, Gauthier CA, Guo Y, Mook RA, Rusnak DW, Walker AL, Wood ER, Lackey KE. Optimization and SAR for dual ErbB-1/ErbB-2 tyrosine kinase inhibition in the 6-furanylquinazoline series. Bioorg Med Chem Lett 2006;16(17):4686–4691.

71. Lackey KE. Lessons from the drug design of lapatinib, a dual ErbB1/2 tyrosine kinase inhibitor. Curr Top Med Chem 2006;6(5):435–460.

72. Moy B, Kirkpatrick P, Kar S, Goss P. Lapatinib. Nat Rev Drug Discov 2007;6:431–432.

73. Wood ER, Truesdale AT, McDonald OB, Yuan D, Hassell A, Dickerson SH, Ellis B, Pennisi C, Horne E, Lackey K, Alligood KJ, Rusnak DW, Gilmer TM, Shewchuk L. A unique structure for epidermal growth factor receptor bound to GW572016 (lapatinib): relationships among protein conformation, inhibitor off-rate, and receptor activity in tumor cells. Cancer Res 2004;64(18):6652–6659.

74. Qiu C, Tarrant MK, Choi SH, Sathyamurthy A, Bose R, Banjade S, Pal A, Bornmann WG, Lemmon MA, Cole PA, Leahy DJ. Mechanism of activation and inhibition of the HER4/ErbB4 kinase. Structure 2008;16(3):460–467.

75. Reid A, Vidal L, Shaw H, de Bono J. Dual inhibition of ErbB1 (EGFR/HER1) and ErbB2 (HER2/neu). Eur J Cancer 2007;43(3):481–489.

76. Eskens FALM, Mom CH, Planting AST, Gietema JA, Amelsberg A, Huisman H, van Doorn L, Burger H, Stopfer P, Verweij J, de Vries EGE. A phase I dose escalation study of BIBW 2992, an irreversible dual inhibitor of epidermal growth factor receptor 1 (EGFR) and 2 (HER2) tyrosine kinase in a 2-week on, 2-week off schedule in patients with advanced solid tumors. Br J Cancer 2008;98(1):80–85.

77. Li D, Ambrogio L, Shimamura T, Kubo S, Takahashi M, Chirieac LR, Padera RF, Shapiro GI, Baum A, Himmelsbach F, Rettig WJ, Meyerson M, Solca F, Greulich H, Wong K-K. BIBW2992, an irreversible EGFR/HER2 inhibitor highly effective in preclinical lung cancer models. Oncogene 2008;27(34):4702–4711.

78. Campas C, Castaner R, Bolos J. BIBW-2992. Dual EGFR/HER2 inhibitor, oncolytic. Drugs Future 2008;33(8):649–654.

79. Engelman JA, Zejnullahu K, Gale C-M, Lifshits E, Gonzales AJ, Shimamura T, Zhao F, Vincent PW, Naumov GN, Bradner JE, Althaus IW, Gandhi L, Shapiro GI, Nelson JM, Heymach JV, Meyerson M, Wong K-K, Jaenne PA. PF00299804, an irreversible pan-ERBB inhibitor, is effective in lung cancer models with EGFR and ERBB2 mutations that are resistant to gefitinib. Cancer Res 2007;67 (24):11924–11932.

80. Gonzales AJ, Hook KE, Althaus IW, Ellis PA, Trachet E, Delaney AM, Harvey PJ, Ellis TA, Amato DM, Nelson JM, Fry DW, Zhu T, Loi C-M, Fakhoury SA, Schlosser KM, Sexton KE, Winters RT, Reed JE, Bridges AJ, Lettiere DJ, Baker DA, Yang J, Lee HT, Tecle H, Vincent PW. Antitumor activity and pharmacokinetic properties of PF-00299804, a second-generation irreversible pan-erbB receptor tyrosine kinase inhibitor. Mol Cancer Ther 2008;7 (7):1880–1889.

81. Tsou H-R, Overbeek-Klumpers EG, Hallett WA, Reich MF, Floyd M, Brawner Johnson BD, Michalak RS, Nilakantan R, Discafani C, Golas J, Rabindran SK, Shen R, Shi X, Wang Y-F, Upeslacis J, Wissner A. Optimization of 6,7-disubstituted-4-(arylamino)quinoline-3-carbonitriles as orally active, irreversible inhibitors of human epidermal growth factor receptor-2 kinase activity. J Med Chem 2005;48(4):1107–1131.

82. Wissner A, Mansour TS. The development of HKI-272 and related compounds for the treatment of cancer. Arch Pharm 2008;341 (8):465–477.

83. Yun C-H, Mengwasser KE, Toms AV, Woo MS, Greulich H, Wong K-K, Meyerson M, Eck MJ. The T790M mutation in EGFR kinase causes drug resistance by increasing the affinity for ATP. Proc Nat Acad Sci USA 2008;105(6):2070–2075.

84. Minami Y, Shimamura T, Shah K, LaFramboise T, Glatt KA, Liniker E, Borgman CL, Haringsma HJ, Feng W, Weir BA, Lowell AM, Lee JC, Wolf J, Shapiro GI, Wong K-K, Meyerson M, Thomas RK. The major lung cancer-derived mutants of ERBB2 are oncogenic and are associated with sensitivity to the irreversible EGFR/ERBB2 inhibitor HKI-272. Oncogene 2007;26(34):5023–5027.

85. Godin-Heymann N, Ulkus L, Brannigan BW, McDermott U, Lamb J, Maheswaran S, Settleman J, Haber DA. The T790M "gatekeeper" mutation in EGFR mediates resistance to low concentrations of an irreversible EGFR inhibitor. Mol Cancer Ther 2008;7(4):874–879.

86. Hur W, Velentza A, Kim S, Flatauer L, Jiang X, Valente D, Mason DE, Suzuki M, Larson B,

Zhang Ji, Zagorska A, DiDonato M, Nagle A, Warmuth M, Balk SP, Peters EC, Gray NS. Clinical stage EGFR inhibitors irreversibly alkylate Bmx kinase. Bioorg Med Chem Lett 2008;18(22):5916–5919.

87. Smith RA, Dumas J, Adnane L, Wilhelm SM. Recent advances in the research and development of RAF kinase inhibitors. Curr Top Med Chem 2006;6(11):1071–1089.

88. Wilhelm S, Carter C, Lynch M, Lowinger T, Dumas J, Smith RA, Schwartz B, Simantov R, Kelley S. Design and development of sorafenib: a multikinase inhibitor for treating cancer. Nat Rev Drug Discov 2006;5(10):835–844.

89. Li N, Batt D, Warmuth M. B-Raf kinase inhibitors for cancer treatment. Curr Opin Investig Drugs 2007;8(6):452–456.

90. Stuart DD, Aardalen K, Venetsanakos E, Nagel T, Wallroth M, Batt DB, Ramurthy S, Poon D, Faure M, Lorenzana EG, Salangsang F, Dove J, Garrett EN, Aikawa M, Kaplan A, Amiri P, Renhowe P. RAF265 is a potent Raf kinase inhibitor with selective anti-proliferative activity in vitro and in vivo. Proc Am Assoc Cancer Res 2008;49: Abstract 4876.

91. Wan PT, Garnett MJ, Roe SM, Lee S, Niculescu-Duvaz D, Good VM, Jones CM, Marshall CJ, Springer CJ, Barford D, Marais R. Mechanism of activation of the RAF-ERK signaling pathway by oncogenic mutations of B-Raf. Cell 2004;116(6):855–867.

92. Wang J, Wilcoxen KM, Nomoto K, Wu S. Recent advances of MEK inhibitors and their clinical progress. Curr Top Med Chem 2007;7(14):1364–1378.

93. Barrett SD, Bridges AJ, Dudley DT, Saltiel AR, Fergus JH, Flamme CM, Delaney AM, Kaufman M, LePage S, Leopold WR, Przybranowski SA, Sebolt-Leopold J, Van Becelaere K, Doherty AM, Kennedy RM, Marston D, Howard WA, Smith Y, Warmus JS, Tecle H. The design of the benzhydroxamate MEK inhibitors CI-1040 and PD 0325901. Bioorg Med Chem Lett 2008;18(24):6501–6504.

94. Ohren JF, Chen H, Pavlovsky A, Whitehead C, Zhang E, Kuffa P, Yan C, McConnell P, Spessard C, Banotai C, Mueller WT, Delaney A, Omer C, Sebolt-Leopold J, Dudley DT, Leung IK, Flamme C, Warmus J, Kaufman M, Barrett S, Tecle H, Hasemann CA. Structures of human MAP kinase kinase 1 (MEK1) and MEK2 describe novel noncompetitive kinase inhibition. Nat Struct Mol Biol 2004;11(12):1192–1197.

95. Yeh TC, Marsh V, Bernat BA, Ballard J, Colwell H, Evans RJ, Parry J, Smith D, Brandhuber BJ, Gross S, Marlow A, Hurley B, Lyssikatos J, Lee PA, Winkler JD, Koch K, Wallace E. Biological characterization of ARRY-142886 (AZD6244), a potent, highly selective mitogen-activated protein kinase kinase 1/2 inhibitor. Clin Cancer Res 2007;13(5):1576–1583.

96. Davies BR, Logie A, McKay JS, Martin P, Steele S, Jenkins R, Cockerill M, Cartlidge S, Smith PD. AZD6244 (ARRY-142886), a potent inhibitor of mitogen-activated protein kinase/extracellular signal-regulated kinase 1/2 kinases: mechanism of action in vivo, pharmacokinetic/pharmacodynamic relationship, and potential for combination in preclinical models. Mol Cancer Ther 2007;6(8):2209–2219.

97. Adjei AA, Cohen RB, Franklin W, Morris C, Wilson D, Molina JR, Hanson LJ, Gore L, Chow L, Leong S, Maloney L, Gordon G, Simmons H, Marlow A, Litwiler K, Brown S, Poch G, Kane K, Haney J, Eckhardt SG. Phase I pharmacokinetic and pharmacodynamic study of the oral, small-molecule mitogen-activated protein kinase kinase 1/2 inhibitor AZD6244 (ARRY-142886) in patients with advanced cancers. J Clin Oncol 2008;26(13):2139–2146.

98. Hamatake R, Iverson C, Yeh L-T, Dadson C. RDEA119, a potent and highly specific MEK inhibitor is efficacious in mouse tumor xenograft studies. Proc Am Assoc Cancer Res 2008;49: Abstract 4878.

99. Lee L, Niu H, Rueger R, Igawa Y. A single-ascending dose study to assess the safety, tolerability, pharmacokinetics, and pharmacodynamics following oral administration of a MEK inhibitor, CH4987655 (RO4987655), in healthy volunteers: implications of clinical activity using a biological marker. Proc Am Assoc Cancer Res 2009;50: Abstract 3589.

100. Herbst RS. Therapeutic options to target angiogenesis in human malignancies. Expert Opin Emerg Drugs 2006;11(4):635–650.

101. Folkman J. Angiogenesis: an organizing principle for drug design? Nat Rev Drug Discov 2007;6(4):273–286.

102. Ferrara N, Hillan KJ, Gerber H-P, Novotny W. Design and development of bevacizumab, an anti-VEGF antibody for treating cancer. Nat Rev Drug Discov 2004;3:391–400.

103. Wedge SR, Jürgensmeier JM. VEGF receptor tyrosine kinase inhibitors for the treatment of cancer. In: Marme D, Fusenig N, editors. Tumour Angiogenesis: Basic Mechanisms and Cancer Therapy. Berlin: Springer; 2008; p 395–424.

104. Faivre S, Demetri G, Sargent W, Raymond E. Molecular basis for sunitinib efficacy and future clinical development. Nat Rev Drug Discov 2007;6(9):734–745.

105. Chow LQM, Eckhardt SG. Sunitinib: from rational design to clinical efficacy. J Clin Oncol 2007;25(7):884–896.

106. Sun L, Tran N, Tang F, App H, Hirth P, McMahon G, Tang C. Synthesis and biological evaluations of 3-substituted indolin-2-ones: a novel class of tyrosine kinase inhibitors that exhibit selectivity toward particular receptor tyrosine kinases. J Med Chem 1998;41(14):2588–2603.

107. Sun L, Liang C, Shirazian S, Zhou Y, Miller T, Cui J, Fukuda JY, Chu J-Y, Nematalla A, Wang X, Chen H, Sistla A, Luu TC, Tang F, Wei J, Tang C. Design of 5-[5-fluoro-2-oxo-1,2-dihydroindol-(3Z)-ylidenemethyl]-2,4-dimethyl-1H-pyrrole-3-carboxylic acid (2-diethylaminoethyl)amide, a novel tyrosine kinase inhibitor targeting vascular endothelial and platelet-derived growth factor receptor tyrosine kinase. J Med Chem 2003;46(7):1116–1119.

108. Sun L, Tran N, Liang C, Tang F, Rice A, Schreck R, Waltz K, Shawver LK, McMahon G, Tang C. Design, synthesis, and evaluations of substituted 3-[(3- or 4-carboxyethylpyrrol-2-yl)methylidenyl]indolin-2-ones as inhibitors of VEGF, FGF, and PDGF receptor tyrosine kinases. J Med Chem 1999;42(25):5120–5130.

109. Hennequin LF, Thomas AP, Johnstone C, Stokes ESE, Ple PA, Lohmann J-JM, Ogilvie DJ, Dukes M, Wedge SR, Curwen JO, Kendrew J, Lambert-van der Brempt C. Design and structure–activity relationship of a new class of potent VEGF receptor tyrosine kinase inhibitors. J Med Chem 1999;42(26):5369–5389.

110. Hennequin LF, Stokes ESE, Thomas AP, Johnstone C, Ple PA, Ogilvie DJ, Dukes M, Wedge SR, Kendrew J, Curwen JO. Novel 4-anilinoquinazolines with C-7 basic side chains: design and structure activity relationship of a series of potent, orally active, VEGF receptor tyrosine kinase inhibitors. J Med Chem 2002;45(6):1300–1312.

111. Herbst RS, Heymach JV, O'Reilly MS, Onn A, Ryan AJ. Vandetanib (ZD6474): an orally available receptor tyrosine kinase inhibitor that selectively targets pathways critical for tumor growth and angiogenesis. Expert Opin Investig Drugs 2007;16(2):239–249.

112. Wedge SR, Kendrew J, Hennequin LF, Valentine PJ, Barry ST, Brave SR, Smith NR, James NH, Dukes M, Curwen JO, Chester R, Jackson JA, Boffey SJ, Kilburn LL, Barnett S, Richmond GHP, Wadsworth PF, Walker M, Bigley AL, Taylor ST, Cooper L, Beck S, Juergensmeier JM, Ogilvie DJ. AZD2171: a highly potent, orally bioavailable, vascular endothelial growth factor receptor-2 tyrosine kinase inhibitor for the treatment of cancer. Cancer Res 2005;65(10):4389–4400.

113. Sorbera LA, Serradell N, Rosa E, Bolos J, Bayes M. Cediranib: VEGFR inhibitor antiangiogenic agent oncolytic. Drugs Future 2007;32(7):577–589.

114. Borzilleri RM, Cai Z-W, Ellis C, Fargnoli J, Fura A, Gerhardt T, Goyal B, Hunt JT, Mortillo S, Qian L, Tokarski J, Vyas V, Wautlet B, Zheng X, Bhide RS. Synthesis and SAR of 4-(3-hydroxyphenylamino)pyrrolo[2,1-f][1,2,4]triazine based VEGFR-2 kinase inhibitors. Bioorg Med Chem Lett 2005;15:1429–1433.

115. Bhide RS, Cai Z-W, Zhang Y-Z, Qian L, Wei D, Barbosa S, Lombardo LJ, Borzilleri RM, Zheng X, Wu LI, Barrish JC, Kim S-H, Leavitt K, Mathur A, Leith L, Chao S, Wautlet B, Mortillo S, Jayaseelan R, Sr, Kukral D, Hunt J, Kamath A, Fura A, Vyas V, Marathe P, D'Arienzo C, Derbin G, Fargnoli J. Design and preclinical studies of (R)-1-(4-(4-fluoro-2-methyl-1H indol-5-yloxy)-5-methylpyrrolo[2,1-f][1,2,4]triazin-6-yloxy)propanol (BMS-540215), an in vivo active potent VEGFR-2 inhibitor. J Med Chem 2006;49:2143–2146.

116. Cai Z-W, Zhang Y, Borzilleri RM, Qian L, Barbosa S, Wei D, Zheng X, Wu L, Fan J, Shi Z, Wautlet BS, Mortillo S, Jeyaseelan R, Kukral DW, Kamath A, Marathe P, D'Arienzo C, Derbin G, Barrish JC, Robl JA, Hunt JT, Lombardo LJ, Fargnoli J, Bhide RS. Design of brivanib alaninate (S)-(R)-1-(4-(4-fluoro-2-methyl-1H-indol-5-yloxy)-5-methylpyrrolo[2,1-f][1,2,4]triazin-6-yloxy)propan-2-yl)2-aminopropanoate), a novel prodrug of dual vascular endothelial growth factor receptor-2 and fibroblast growth factor receptor-1 kinase inhibitor (BMS-540215). J Med Chem 2008;51(6):1976–1980.

117. Harris PA, Boloor A, Cheung M, Kumar R, Crosby RM, Davis-Ward RG, Epperly AH, Hinkle KW, Hunter RN, III, Johnson JH, Knick VB, Laudeman CP, Luttrell DK, Mook RA, Nolte RT, Rudolph SK, Szewczyk JR, Truesdale AT, Veal JM, Wang L, Stafford JA. Design of 5-[[4-[(2,3-dimethyl-2H-indazol-6-yl)methylamino]-2-pyrimidinyl]amino]-2-methyl-benzenesulfonamide (pazopanib), a novel and potent vascular endothelial growth factor

receptor inhibitor. J Med Chem 2008;51(15):4632–4640.
118. Sonpavde G, Hutson TE, Sternberg CN. Pazopanib, a potent orally administered small-molecule multi-targeted tyrosine kinase inhibitor for renal cell carcinoma. Expert Opin Investig Drugs 2008;17(2):253–261.
119. Revill P, Mealy N, Bayes M, Bozzo J, Serradell N, Rosa E, Bolos J. Axitinib VEGFR/PDGFR tyrosine kinase inhibitor antiangiogenic agent. Drugs Future 2007;32(5):389–398.
120. Bold G, Altmann K-H, Frei J, Lang M, Manley PW, Traxler P, Wietfeld B, Brueggen J, Buchdunger E, Cozens R, Ferrari S, Furet P, Hofmann F, Martiny-Baron G, Mestan J, Roesel J, Sills M, Stover D, Acemoglu F, Boss E, Emmenegger R, Laesser L, Masso E, Roth R, Schlachter C, Vetterli W, Wyss D, Wood JM. New anilinophthalazines as potent and orally well absorbed inhibitors of the VEGF receptor tyrosine kinases useful as antagonists of tumor-driven angiogenesis. J Med Chem 2000;43(12):2310–2323.
121. Sorbera LA, Serradell N, Bolos J, Bozzo J, Bayes M. AMG-706 oncolytic antiangiogenic agent multikinase inhibitor. Drugs Future 2006;31(10):847–853.
122. Fischer PM. Cell cycle inhibitors in cancer: current status and future directions. In: Neidle Stephen, editor. Cancer Drug Design and Design. New York: Academic Press; 2008; p 253–283.
123. Misra RN. Clinical progress of selective cyclin-dependent kinase (CDK) inhibitors. Drugs Future 2006;31(1):43–52.
124. Schulze-Gahmen U, Kim S-H. Three-dimensional structures of cyclin-dependent kinases and their inhibitor complexes. In: Smith Paul J, Yue Eddy W, editors. Inhibitors of Cyclin-Dependent Kinases as Anti-Tumor Agents. Boca Raton, FL: CRC Press LLC; 2008; p 143–164.
125. Brown JR. Chronic lymphocytic leukemia: a niche for flavopiridol? Clin Cancer Res 2005;11(11):3971–3973.
126. Christian BA, Grever MR, Byrd JC, Lin TS. Flavopiridol in the treatment of chronic lymphocytic leukemia. Curr Opin Oncol 2007;19(6):573–578.
127. Blagosklonny MV. Flavopiridol, an inhibitor of transcription. Implications, problems and solutions. Cell Cycle 2004;3(12):1537–1542.
128. Getman CR, Bible KC. Molecular targets of flavopiridol, a promising new antineoplastic agent. Recent Res Dev Cancer 2004;6(Pt. 1):37–56.
129. Sharma PS, Sharma R, Tyagi R. Inhibitors of cyclin-dependent kinases: useful targets for cancer treatment. Curr Cancer Drug Targets 2008;8(1):53–75.
130. Kim KS, Kimball SD, Misra RN, Rawlins DB, Hunt JT, Xiao H-Y, Lu S, Qian L, Han W-C, Shan W, Mitt T, Cai Z-W, Poss MA, Zhu H, Sack JS, Tokarski JS, Chang CY, Pavletich N, Kamath A, Humphreys WG, Marathe P, Bursuker I, Kellar OKA, Roongta U, Batorsky R, Mulheron JG, Bol D, Fairchild CR, Lee FY, Webster KR. Design of aminothiazole inhibitors of cyclin-dependent kinase 2: synthesis, X-ray crystallographic analysis, and biological activities. J Med Chem 2002;45(18): 3905–3927.
131. Misra RN, Xiao H-Y, Kim KS, Lu S, Han W-C, Barbosa SA, Hunt JT, Rawlins DB, Shan W, Ahmed SZ, Qian L, Chen B-C, Zhao R, Bednarz MS, Kellar KA, Mulheron JG, Batorsky R, Roongta U, Kamath A, Marathe P, Ranadive SA, Sack JS, Tokarski JS, Pavletich NP, Lee Francis YF, Webster KR, Kimball SD. N-(Cycloalkylamino)acyl-2-aminothiazole inhibitors of cyclin-dependent kinase 2. N-[5-[[[5-(1,1-Dimethylethyl)-2-oxazolyl]methyl]thio]-2-thiazolyl]-4-piperidinecarboxamide (BMS-387032), a highly efficacious and selective antitumor agent. J Med Chem 2004;47(7):1719–1728.
132. Hunt JT. Design of BMS-387032, a potent cyclin-dependent kinase inhibitor in clinical development. In: Smith PJ, Yue EW, editors. Inhibitors of Cyclin-Dependent Kinases as Anti-Tumor Agents. Boca Raton, FL: CRC Press LLC; 2008; p. 251–264.
133. Barvian M, Boschelli DH, Cossrow J, Dobrusin E, Fattaey A, Fritsch A, Fry D, Harvey P, Keller P, Garrett M, La F, Leopold W, McNamara D, Quin M, Trumpp-Kallmeyer S, Toogood P, Wu Z, Zhang E. Pyrido[2,3-d]pyrimidin-7-one inhibitors of cyclin-dependent kinases. J Med Chem 2000;43(24):4606–4616.
134. VanderWel SN, Harvey PJ, McNamara DJ, Repine JT, Keller PR, Quin J III, Booth RJ, Elliott WL, Dobrusin EM, Fry DW, Toogood PL. Pyrido[2,3-d]pyrimidin-7-ones as specific inhibitors of cyclin-dependent kinase 4. J Med Chem 2005;48(7):2371–2387.
135. Toogood PL, Harvey PJ, Repine JT, Sheehan DJ, VanderWel SN, Zhou H, Keller PR, McNamara DJ, Sherry D, Zhu T, Brodfuehrer J, Choi C, Barvian MR, Fry W. Design of a potent and selective inhibitor of cyclin-depen-

dent kinase 4/6. J Med Chem 2005;48(7):2388–2406.

136. Lu H, Schulze-Gahmen U. Toward understanding the structural basis of cyclin-dependent kinase 6 specific inhibition. J Med Chem 2006;49(13):3826–3831.

137. Lovey A, Depinto W, Ding Q, Jiang N, Kim K, Yin X, Chu X-J, Bartkovitz D, Desai B, Smith M, Mullin J, McComas W, Graves B, Lukacs C, So S-S, Chen Y, Xiang Q. Design, synthesis, and biological evaluation of aminothiazoles as selective inhibitors of cyclin-dependent kinase 4 (CDK4). Abstracts of Papers, 229th ACS National Meeting, San Diego, CA, March 13–17, 2005. 2005.

138. Chu X-J, DePinto W, Bartkovitz D, So S-S, Vu BT, Packman K, Lukacs C, Ding Q, Jiang N, Wang K, Goelzer P, Yin X, Smith MA, Higgins BX, Chen Y, Xiang Q, Moliterni J, Kaplan G, Graves B, Lovey A, Fotouhi N. Design of [4-amino-2-(1-methanesulfonylpiperidin-4-ylamino)pyrimidin-5-yl](2,3-difluoro-6-methoxyphenyl)methanone (R547), a potent and selective cyclin-dependent kinase inhibitor with significant in vivo antitumor activity. J Med Chem 2006;49(22):6549–6560.

139. Wyatt PG, Woodhead AJ, Berdini V, Boulstridge JA, Carr MG, Cross DM, Davis DJ, Devine LA, Early TR, Feltell RE, Lewis EJ, McMenamin RL, Navarro EF, O'Brien MA, O'Reilly M, Reule M, Saxty G, Seavers LCA, Smith D-M, Squires MS, Trewartha G, Walker MT, Woolford AJ-A. Identification of N-(4-piperidinyl)-4-(2,6-dichlorobenzoylamino)-1H-pyrazole-3-carboxamide (AT7519), a novel cyclin dependent kinase inhibitor using fragment-based X-Ray crystallography and structure based drug design. J Med Chem 2008;51(16):4986–4999.

140. Ashwell S, Janetka JW, Zabludoff S. Keeping checkpoint kinases in line: new selective inhibitors in clinical trials. Expert Opin Investig Drugs 2008;17(9):1331–1340.

141. Parry D, Shanahan F, Davis N, Wiswell D, Seghezzi W, Pierce R, Hsieh Y, Paruch K, Guzi T. Targeting the replication checkpoint with a potent and selective CHK1 inhibitor. Proc Am Assoc Cancer Res 2009;50: Abstract 2490.

142. Blasina A, Hallin J, Chen E, Arango ME, Kraynov E, Register J, Grant S, Ninkovic S, Chen P, Nichols T, O'Connor P, Anderes K. Breaching the DNA damage checkpoint via PF-00477736, a novel small-molecule inhibitor of checkpoint kinase 1. Mol Cancer Ther 2008;7(8):2394–2404.

143. Zabludoff SD, Deng C, Grondine MR, Sheehy AM, Ashwell S, Caleb BL, Green S, Haye HR, Horn CL, Janetka JW, Liu D, Mouchet E, Ready S, Rosenthal JL, Queva C, Schwartz GK, Taylor KJ, Tse AN, Walker GE, White AM. AZD7762, a novel checkpoint kinase inhibitor, drives checkpoint abrogation and potentiates DNA-targeted therapies. Mol Cancer Ther 2008;7(9):2955–2966.

144. Janetka JW, Almeida L, Ashwell S, Brassil PJ, Daly K, Deng C, Gero T, Glynn RE, Horn CL, Ioannidis S, Lyne P, Newcombe NJ, Oza VB, Pass M, Springer SK, Su M, Toader D, Vasbinder MM, Yu D, Yu Y, Zabludoff SD. Design of a novel class of 2-ureido thiophene carboxamide checkpoint kinase inhibitors. Bioorg Med Chem Lett 2008;18(14):4242–4248.

145. Baxter A, Brough S, Cooper A, Floettmann E, Foster S, Harding C, Kettle J, McInally T, Martin C, Mobbs M, Needham M, Newham P, Paine S, St-Gallay S, Salter S, Unitt J, Xue Y. Hit-to-lead studies: the design of potent, orally active, thiophenecarboxamide IKK-2 inhibitors. Bioorg Med Chem Lett 2004;14(11):2817–2822.

146. Strebhardt K, Ullrich A. Targeting polo-like kinase 1 for cancer therapy. Nat Rev Cancer 2006;6(4):321–330.

147. Lenart P, Petronczki M, Steegmaier M, Di Fiore B, Lipp JJ, Hoffmann M, Rettig WJ, Kraut N, Peters J-M. The small-molecule inhibitor BI 2536 reveals novel insights into mitotic roles of polo-like kinase 1. Curr Biol 2007;17(4):304–315.

148. Steegmaier M, Hoffmann M, Baum A, Lenart P, Petronczki M, Krssak M, Guertler U, Garin-Chesa P, Lieb S, Quant J, Grauert M, Adolf GR, Kraut N, Peters J-M, Rettig WJ. BI 2536, a potent and selective inhibitor of polo-like kinase 1, inhibits tumor growth in vivo. Curr Biol 2007;17(4):316–322.

149. Von Pawel J, Reck M, Digel W, Kortsik C. Randomized phase II trial of two dosing schedules of BI 2536, a novel Plk-1 inhibitor, in patients with relapsed advanced or metastatic non-small-cell lung cancer (NSCLC). J Clin Oncol 2008;26 (15 Suppl): Abstract 8030.

150. Schöffski P, Awada A, Dumez H, Gil T. A phase I single dose escalation study of the novel polo-like kinase 1 inhibitor BI 6727 in patients with advanced solid tumours. Eur J Cancer Suppl 2008;6 (12): Abstract 36.

151. Kothe M, Kohls D, Low S, Coli R, Rennie GR, Feru F, Kuhn C, Ding Y-H. Selectivity-determining residues in Plk1. Chem Biol Drug Des 2007;70(6):540–546.

152. Blagden S, Olmos D, Sharma R, Barriuso J. A phase I first-in-human study of the polo-like kinase 1-selective inhibitor, GSK461364, in patients with advanced solid tumors. Eur J Cancer Suppl 2008;6 (12): Abstract 431.

153. Gumireddy K, Reddy MVR, Cosenza SC, Nathan RB, Baker SJ, Papathi N, Jiang J, Holland J, Reddy EP. ON01910, a non-ATP-competitive small molecule inhibitor of Plk1, is a potent anticancer agent. Cancer Cell 2005;7 (3):275–286.

154. Mahadevan D, Beeck S. Aurora kinase targeted therapeutics in oncology: past, present and future. Expert Opin Drug Discov 2007;2 (7):1011–1026.

155. Warner SL, Stephens BJ, Von Hoff DD. Tubulin-associated proteins: Aurora and polo-like kinases as therapeutic targets in cancer. Curr Oncol Rep 2008;10:122–129.

156. Yang H, Burke T, Dempsey J, Diaz B, Collins E, Toth J, Beckmann R, Ye X. Mitotic requirement for Aurora A kinase is bypassed in the absence of Aurora B kinase. FEBS Lett 2005;579(16):3385–3391.

157. Matthews N, Visintin C, Hartzoulakis B, Jarvis A, Selwood DL. Aurora A and B kinases as targets for cancer: will they be selective for tumors? Expert Rev Anticancer Ther 2006;6 (1):109–120.

158. Mortlock AA, Keen NJ, Jung FH, Heron NM, Foote KM, Wilkinson R, Green S. Progress in the development of selective inhibitors of Aurora kinases. Curr Top Med Chem 2005;5 (8):807–821.

159. Heron NM, Anderson M, Blowers DP, Breed J, Eden JM, Green S, Hill GB, Johnson T, Jung FH, McMiken HHJ, Mortlock AA, Pannifer AD, Pauptit RA, Pink J, Roberts NJ, Rowsell S. SAR and inhibitor complex structure determination of a novel class of potent and specific Aurora kinase inhibitors. Bioorg Med Chem Lett 2006;16(5):1320–1323.

160. Jung FH, Pasquet G, Lambert-van der Brempt C, Lohmann J-JM, Warin N, Renaud F, Germain H, De Savi C, Roberts N, Johnston T, Dousson C, Hill GB, Mortlock AA, Heron N, Wilkinson RW, Wedge SR, Heaton SP, Odedra R, Keen NJ, Green S, Brown E, Thompson K, Brightwell S. Design of novel and potent thiazoloquinazolines as selective Aurora A and B kinase inhibitors. J Med Chem 2006;49: 955–970.

161. Mortlock AA, Foote KM, Heron NM, Jung FH, Pasquet G, Lohmann J-JM, Warin N, Renaud F, De Savi C, Roberts NJ, Johnson T, Dousson CB, Hill GB, Perkins D, Hatter G, Wilkinson RW, Wedge SR, Heaton SP, Odedra R, Keen NJ, Crafter C, Brown E, Thompson Ka, Brightwell S, Khatri L, Brady MC, Kearney S, McKillop D, Rhead S, Parry T, Green S. Design, synthesis, and in vivo activity of a new class of pyrazolylamino quinazolines as selective inhibitors of Aurora B kinase. J Med Chem 2007;50(9):2213–2224.

162. Girdler F, Gascoigne KE, Eyers PA, Hartmuth S, Crafter C, Foote KM, Keen NJ, Taylor SS. Validating Aurora B as an anti-cancer drug target. J Cell Sci 2006;119(17):3664–3675.

163. Wilkinson RW, Odedra R, Heaton SP, Wedge SR, Keen NJ, Crafter C, Foster JR, Brady MC, Bigley A, Brown E, Byth KF, Barrass NC, Mundt KE, Foote KM, Heron NM, Jung FH, Mortlock AA, Boyle FT, Green S. AZD1152, a selective inhibitor of Aurora B kinase, inhibits human tumor xenograft growth by inducing apoptosis. Clin Cancer Res 2007;13(12): 3682–3688.

164. Fancelli D, Berta D, Daniela B, Simona C, Alexander C, Paolo C, Patrizia C, Cornel F, Barbara G, Patrizia G, Maria L, Mantegani S, Marsiglio A, Meroni M, Moll J, Pittala V, Roletto F, Severino D, Soncini C, Storici P, Tonani R, Varasi M, Vulpetti A, Vianello P. Potent and selective Aurora inhibitors identified by the expansion of a novel scaffold for protein kinase inhibition. J Med Chem 2005;48(8):3080–3084.

165. Fancelli D, Moll J, Varasi M, Bravo R, Artico R, Berta D, Bindi S, Cameron A, Candiani I, Cappella P, Carpinelli P, Croci W, Forte B, Giorgini ML, Klapwijk J, Marsiglio A, Pesenti E, Rocchetti M, Roletto F, Severino D, Soncini C, Storici P, Tonani R, Zugnoni P, Vianello P. 1,4,5,6-Tetrahydropyrrolo[3,4-c]pyrazoles: identification of a potent Aurora kinase inhibitor with a favorable antitumor kinase inhibition profile. J Med Chem 2006;49(24):7247–7251.

166. Modugno M, Casale E, Soncini C, Rosettani P, Colombo R, Lupi R, Rusconi L, Fancelli D, Carpinelli P, Cameron AD, Isacchi A, Moll J. Crystal structure of the T315I Abl mutant in complex with the Aurora kinases inhibitor PHA-739358. Cancer Res 2007;67(17):7987–7990.

167. Carpinelli P, Ceruti R, Giorgini ML, Cappella P, Gianellini L, Croci V, Degrassi A, Texido G, Rocchetti M, Vianello P, Rusconi L, Storici P, Zugnoni P, Arrigoni C, Soncini C, Alli C, Patton V, Marsiglio A, Ballinari D, Pesenti E, Fancelli D, Moll J. PHA-739358, a potent inhibitor of Aurora kinases with a selective target inhibi-

tion profile relevant to cancer. Mol Cancer Ther 2007;6(12 Pt 1):3158–3168.

168. Oslob JD, Romanowski MJ, Allen DA, Baskaran S, Bui M, Elling RA, Flanagan WM, Fung AD, Hanan EJ, Harris S, Heumann SA, Hoch U, Jacobs JW, Lam J, Lawrence CE, McDowell RS, Nannini MA, Shen W, Silverman JA, Sopko MM, Tangonan BT, Teague J, Yoburn JC, Yu CH, Zhong M, Zimmerman KM, O'Brien T, Lew W. Design of a potent and selective Aurora kinase inhibitor. Bioorg Med Chem Lett 2008;18(17):4880–4884.

169. Foran JM, Ravandi F, O'Brien SM, Borthakur G. Phase I and pharmacodynamic trial of AT9283, an Aurora kinase inhibitor, in patients with refractory leukemia. J Clin Oncol 2008;26 (Suppl 15): Abstract 2518.

170. Manfredi MG, Ecsedy JA, Meetze KA, Balani SK, Burenkova O, Chen W, Galvin KM, Hoar KM, Huck JJ, LeRoy PJ, Ray ET, Sells TB, Stringer B, Stroud SG, Vos TJ, Weatherhead GS, Wysong DR, Zhang M, Bolen JB, Claiborne CF. Antitumor activity of MLN8054, an orally active small-molecule inhibitor of Aurora A kinase. Proc Nat Acad Sci USA 2007;104(10):4106–4111.

171. Sells BT, Ecsedy AJ, Stroud S, Janowick D. MLN8237: an orally active small molecule inhibitor of Aurora A kinase in phase I clinical trials. Proc Am Assoc Cancer Res 2008;49: Abstract 237.

172. Hajduch M, Vydra D, Dzubak P, Dziechciarkova M, Stuart I, Zheleva D. *In vivo* mode of action of CYC116, a novel small molecule inhibitor of Aurora kinases and VEGFR2. Proc Am Assoc Cancer Res 2008;49: Abstract 5645.

173. McInnes C, Wang S, Anderson S, O'Boyle J, Jackson W, Kontopidis G, Meades C, Mezna M, Thomas M, Wood G, Lane DP, Fischer PM. Structural determinants of CDK4 inhibition and design of selective ATP competitive inhibitors. Chem Biol 2008;11(4):525–534.

174. Schöoffski P, Dumez H, Jones SF, Fowst C. Preliminary results of a phase I accelerated dose-escalation, pharmacokinetic and pharmacodynamic study of PF-03814735, an oral Aurora kinase A and B inhibitor, in patients with advanced solid tumors. Eur J Cancer Suppl 2008;6 (12): Abstract 282.

175. Yap TA, Garrett MD, Walton MI, Raynaud F, de Bono JS, Workman P. Targeting the PI3K-AKT-mTOR pathway: progress, pitfalls, and promises. Curr Opin Pharmacol 2008;8(4):393–412.

176. Garcia-Echeverria C, Sellers WR. Drug design approaches targeting the PI3K/Akt pathway in cancer. Oncogene 2008;27(41):5511–5526.

177. Garcia-Echeverria C. Medicinal chemistry approaches to target the kinase activity of IGF-1R. IDrugs 2006;9(6):415–419.

178. Hubbard RD, Wilsbacher JL. Advances towards the development of ATP-competitive small-molecule inhibitors of the insulin-like growth factor receptor (IGF-IR). ChemMedChem 2007;2(1):41–46.

179. Rodon J, De Santos V, Ferry RJ, Jr, Kurzrock R. Early drug development of inhibitors of the insulin-like growth factor-1 receptor pathway: lessons from the first clinical trials. Mol Cancer Ther 2008;7(9):2575–2588.

180. Scotlandi K, Manara MC, Nicoletti G, Lollini P-L, Lukas S, Benini S, Croci S, Perdichizzi S, Zambelli D, Serra M, Garcia-Echeverria C, Hofmann F, Picci P. Antitumor activity of the insulin-like growth factor-I receptor kinase inhibitor NVP-AEW541 in musculoskeletal tumors. Cancer Res 2005;65(9):3868–3876.

181. Mulvihill MJ, Ji Q-S, Rosenfeld-Franklin M, Buck E, Cooke A, Eyzaguirre A, Feng L, Foreman K, Landfair D, Mak G, O'Connor M, Pirritt C, Silva S, Turton R, Yao Y, Arnold L, Wild R, Pachter JA, Gibson NW. The design of OSI-906: a novel, potent, orally bioavailable imidazopyrazine-derived insulin-like growth factor-I receptor (IGF-1R) inhibitor with anti-tumor activity. Proc Am Assoc Cancer Res 2008;49: Abstract 4893.

182. Carboni JM, Wittman M, Yang Z, Fairchild C. BMS-754807, a small molecule inhibitor of IGF1R for clinical development. Proc Am Assoc Cancer Res 2009;50: Abstract 1742.

183. Wittman M, Carboni J, Attar R, Balasubramanian B, Balimane P, Brassil P, Beaulieu F, Chang C, Clarke W, Dell J, Eummer J, Frennesson D, Gottardis M, Greer A, Hansel S, Hurlburt W, Jacobson B, Krishnananthan S, Lee FY, Li A, Lin T-A, Liu P, Ouellet C, Sang X, Saulnier MG, Stoffan K, Sun Y, Velaparthi U, Wong H, Yang Z, Zimmermann K, Zoeckler M, Vyas D. Design of a 1*H*-benzoimidazol-2-yl-1*H*-pyridin-2-one (BMS-536924) inhibitor of insulin-like growth factor I receptor kinase with *in vivo* antitumor activity. J Med Chem 2005;48(18):5639–5643.

184. Verheijen JC, Zask A. Phosphatidylinositol 3-kinase (PI3K) inhibitors as anticancer drugs. Drugs Future 2007;32(6):537–547.

185. Maira S-M, Voliva C, Garcia-Echeverria C. Class IA phosphatidylinositol 3-kinase: from their biologic implication in human cancers to drug design. Expert Opin Ther Targets 2008;12(2):223–238.

186. Hayakawa M, Kaizawa H, Moritomo H, Koizumi T, Ohishi T, Okada M, Ohta M, Tsukamoto S, Parker P, Workman P, Waterfield M. Synthesis and biological evaluation of 4-morpholino-2-phenylquinazolines and related derivatives as novel PI3 kinase p110α inhibitors. Bioorg Med Chem 2006;14(20):6847–6858.

187. Folkes AJ, Ahmadi K, Alderton WK, Alix S, Baker SJ, Box G, Chuckowree IS, Clarke PA, Depledge P, Eccles SA, Friedman LS, Hayes A, Hancox TC, Kugendradas A, Lensun L, Moore P, Olivero AG, Pang J, Patel S, Pergl-Wilson GH, Raynaud FI, Robson A, Saghir N, Salphati L, Sohal S, Ultsch MH, Wallweber HJA, Wan NC, Wiesmann C, Workman P, Zhyvoloup A, Zvelebil MJ, Shuttleworth SJ. The identification of 2-(1H-indazol-4-yl)-6-(4-methanesulfonyl-piperazin-1-ylmethyl)-4-morpholin-4-yl-thieno[3,2-d]pyrimidine (GDC-0941) as a potent, selective, orally bioavailable inhibitor of class I PI3 kinase for the treatment of cancer. J Med Chem 2008;51(18):5522–5532.

188. Stauffer F, Maira S-M, Furet P, Garcia-Echeverria C. Imidazo[4,5-c]quinolines as inhibitors of the PI3K/PKB-pathway. Bioorg Med Chem Lett 2008;18(3):1027–1030.

189. Maira S-M, Stauffer F, Brueggen J, Furet P, Schnell C, Fritsch C, Brachmann S, Chene P, De Pover A, Schoemaker K, Fabbro D, Gabriel D, Simonen M, Murphy L, Finan P, Sellers W, Garcia-Echeverria C. Identification and characterization of NVP-BEZ235, a new orally available dual phosphatidylinositol 3-kinase/mammalian target of rapamycin inhibitor with potent in vivo antitumor activity. Mol Cancer Ther 2008;7(7):1851–1863.

190. Auger KR, Luo L, Knight S, Van Aller G. A novel inhibitor of phosphoinositide 3-kinase for the treatment of cancer. Eur J Cancer Suppl 2008;6 (12): Abstract 221.

191. Lindsley CW, Barnett SF, Layton ME, Bilodeau MT. The PI3K/Akt pathway: recent progress in the development of ATP-competitive and allosteric Akt kinase inhibitors. Curr Cancer Drug Targets 2008;8(1):7–18.

192. Heerding DA, Rhodes N, Leber JD, Clark TJ, Keenan RM, Lafrance LV, Li M, Safonov IG, Takata DT, Venslavsky JW, Yamashita DS, Choudhry AE, Copeland RA, Lai Z, Schaber MD, Tummino PJ, Strum SL, Wood ER, Duckett DR, Eberwein D, Knick VB, Lansing TJ, McConnell RT, Zhang SY, Minthorn EA, Concha NO, Warren GL, Kumar R. Identification of 4-(2-(4-Amino-1,2,5-oxadiazol-3-yl)-1-ethyl-7-{[(3S)-3-piperidinylmethyl]oxy}-1H-imidazo[4,5-c]pyridin-4-yl)-2-methyl-3-butyn-2-ol (GSK690693), a novel inhibitor of AKT kinase. J Med Chem 2008;51(18):5663–5679.

193. Trucksis M, Friedman E, Taylor A, Delgado L. A phase I single-rising dose study evaluating the safety, tolerability, pharmacokinetics and pharmacodynamics of an oral akt inhibitor in healthy male volunteers. Proc Am Assoc Cancer Res 2009;50: Abstract 3604.

194. Abraham RT, Eng CH. Mammalian target of rapamycin as a therapeutic target in oncology. Expert Opin Ther Targets 2008;12(2):209–222.

195. Fasolo A, Sessa C. mTOR inhibitors in the treatment of cancer. Expert Opin Investig Drugs 2008;17(11):1717–1734.

196. Rini B, Kar S, Kirkpatrick P. Temsirolimus. Nat Rev Drug Discov 2007;6(8):599–600.

197. Ma WW, Jimeno A. Temsirolimus. Drugs Today 2007;43(10):659–669.

198. Mita M, Sankhala K, Abdel-Karim I, Mita A, Giles F. Deforolimus (AP23573) a novel mTOR inhibitor in clinical development. Expert Opin Investig Drugs 2008;17(12):1947–1954.

199. Brunton VG, Frame MC. Src and focal adhesion kinase as therapeutic targets in cancer. Curr Opin Pharmacol 2008;8(4):427–432.

200. Johnson FM, Gallick GE. Src family nonreceptor tyrosine kinases as molecular targets for cancer therapy. Anticancer Agents Med Chem 2007;7(6):651–659.

201. Rucci N, Susa M, Teti A. Inhibition of protein kinase c-Src as a therapeutic approach for cancer and bone metastases. Anticancer Agents Med Chem 2008;8(3):342–349.

202. Benati D, Baldari CT. Src family kinases as potential therapeutic targets for malignancies and immunological disorders. Curr Med Chem 2008;15(12):1154–1165.

203. Boschelli DH, Boschelli F. Dual Src and Abl kinase inhibitor treatment of solid tumors treatment of CML and Ph+ ALL. Drugs Future 2007;32(6):481–490.

204. Boschelli DH. Exploitation of the 3-quinolinecarbonitrile template for Src tyrosine kinase inhibitors. Curr Top Med Chem 2008;8(10):922–934.

205. Hennequin LF, Allen J, Breed J, Curwen J, Fennell l, Green TP, Lambert-van der Brempt C, Morgentin R, Norman RA, Olivier A, Otterbein L, Ple PA, Warin N, Costello G. N-(5-Chloro-1,3-benzodioxol-4-yl)-7-[2-(4-methylpiperazin-1-yl)ethoxy]-5-(tetrahydro-2H-pyran-4-yloxy)quinazolin-4-amine, a novel, highly selective, orally available, dual-specific c-Src/Abl

kinase inhibitor. J Med Chem 2006;49 (22):6465–6488.

206. Ple PA, Green TP, Hennequin LF, Curwen J, Fennell M, Allen J, Lambert-van der Brempt C, Costello G. Design of a new class of anilinoquinazoline inhibitors with high affinity and specificity for the tyrosine kinase domain of c-Src. J Med Chem 2004;47(4):871–887.

207. Rivera VM, Xu Q, Wang F. Potent antitumor activity of AP24534, an orally active inhibitor of Bcr-Abl variants including T315I, in *in vitro* and *in vivo* models of chronic myeloid leukemia (CML). Blood 2008;110 (11): Abstract 1032.

208. Hangauer DG, Gelman I, Dyster L, Barnett A, Smolinski M, Hegab T, Gao L. KXO1: The first non-ATP competitive Src inhibitor for clinical development in oncology applications-1. Abstracts of Papers, 234th ACS National Meeting, Boston, MA, August 19–23, 2007.

209. Hangauer DG, Gelman I, Dyster L, Barnett A, Smolinski M, Hegab T, Gao L. KXO1: The first non-ATP competitive Src inhibitor for clinical development in oncology applications-2. Abstracts of Papers, 234th ACS National Meeting, Boston, MA, August 19–23, 2007.

210. Dalgarno D, Stehle T, Narula S, Schelling P, van Schravendijk MR, Adams S, Andrade L, Keats J, Ram M, Jin L, Grossman T, MacNeil I, Metcalf C III, Shakespeare W, Wang Y, Keenan T, Sundaramoorthi R, Bohacek R, Weigele M, Sawyer T. Structural basis of Src tyrosine kinase inhibition with a new class of potent and selective trisubstituted purine-based compounds. Chem Biol Drug Des (1): 672006; 46–57.

211. Ye G, Tiwari R, Parang K. Development of Src tyrosine kinase substrate binding site inhibitors. Curr Opin Investig Drugs 2008;9 (6):605–613.

212. Chatzizacharias NA, Kouraklis GP, Theocharis SE. Focal adhesion kinase: a promising target for anticancer therapy. Expert Opin Ther Targets 2007;11(10):1315–1328.

213. Roberts WG, Ung E, Whalen P, Cooper B, Hulford C, Autry C, Richter D, Emerson E, Lin J, Kath J, Coleman K, Yao L, Martinez-Alsina L, Lorenzen M, Berliner M, Luzzio M, Patel N, Schmitt E, LaGreca S, Jani J, Wessel M, Marr E, Griffor M, Vajdos F. Antitumor activity and pharmacology of a selective focal adhesion kinase inhibitor, PF-562271. Cancer Res 2008;68(6):1935–1944.

214. Slack-Davis JK, Martin KH, Tilghman RW, Iwanicki M, Ung EJ, Autry C, Luzzio MJ, Cooper B, Kath JC, Roberts WG, Parsons JT. Cellular characterization of a novel focal adhesion kinase inhibitor. J Biol Chem 2007;282 (20):14845–14852.

215. Liu Xi, Yao W, Newton RC, Scherle PA. Targeting the c-MET signaling pathway for cancer therapy. Expert Opin Investig Drugs 2008;17 (7):997–1011.

216. Comoglio PM, Giordano S, Trusolino L. Drug development of MET inhibitors: targeting oncogene addiction and expedience. Nat Rev Drug Discov 2008;7(6):504–516.

217. Schroeder GM, Chen X-T, Williams DK, Nirschl DS, Cai Z-W, Wei D, Tokarski JS, An Y, Sack J, Chen Z, Huynh T, Vaccaro W, Poss M, Wautlet B, Gullo-Brown J, Kellar K, Manne V, Hunt JT, Wong TW, Lombardo LJ, Fargnoli J, Borzilleri RM. Identification of pyrrolo[2,1-*f*][1,2,4]triazine-based inhibitors of Met kinase. Bioorg Med Chem Lett 2008;18(6):1945–1951.

218. Cai Z-W, Wei D, Schroeder GM, Cornelius LAM, Kim K, Chen X-T, Schmidt RJ, Williams DK, Tokarski JS, An Y, Sack JS, Manne V, Kamath A, Zhang Y, Marathe P, Hunt JT, Lombardo LJ, Fargnoli J, Borzilleri RM. Design of orally active pyrrolopyridine- and aminopyridine-based Met kinase inhibitors. Bioorg Med Chem Lett 2008;18(11):3224–3229.

219. Kim KS, Zhang L, Schmidt R, Cai Z-W, Wei D, Williams DK, Lombardo LJ, Trainor GL, Xie D, Zhang Y, An Y, Sack JS, Tokarski JS, Darienzo C, Kamath A, Marathe P, Zhang Y, Lippy J, Jeyaseelan R, Wautlet B, Henley B, Gullo-Brown J, Manne V, Hunt JT, Fargnoli J, Borzilleri RM. Design of pyrrolopyridine–pyridone based inhibitors of Met kinase: synthesis, X-ray crystallographic analysis, and biological activities. J Med Chem 2008;51(17):5330–5341.

220. Schroeder GM, An Y, Cai Z-W, Chen X-T, Clark C, Cornelius LAM, Dai J, Gullo-Brown J, Gupta A, Henley B, Hunt JT, Jeyaseelan R, Kamath A, Kim K, Lippy J, Lombardo LJ, Manne V, Oppenheimer S, Sack JS, Schmidt RJ, Shen G, Stefanski K, Tokarski JS, Trainor GL, Wautlet BS, Wei D, Williams DK, Zhang Y, Zhang Y, Fargnoli J, Borzilleri RM. Discovery of *N*-(4-(2-amino-3-chloropyridin-4-yloxy)-3-fluorophenyl)-4-ethoxy-1-(4-fluorophenyl)-2-oxo-1,2-dihydropyridine-3-carboxamide (BMS-777607), a selective and orally efficacious inhibitor of the Met kinase superfamily. J Med Chem 2009;52(5):1251–1254.

221. Zou HY, Li Q, Lee JH, Arango ME, McDonnell SR, Yamazaki S, Koudriakova TB, Alton G, Cui JJ, Kung P-P, Nambu MD, Los G, Bender SL, Mroczkowski B, Christensen JG. An orally

available small-molecule inhibitor of c-Met, PF-2341066, exhibits cytoreductive antitumor efficacy through antiproliferative and antiangiogenic mechanisms. Cancer Res 2007;67(9):4408–4417.
222. Zou YH, Li Q, Joseph L, Arango EM. PF-04217903, a novel selective c-Met kinase inhibitor with potent antitumor and anti-angiogenic properties *in vitro* and *in vivo*. Proc Am Assoc Cancer Res 2008;49: Abstract 4841.
223. Perera T, Lavrijssen T, Janssens B, Geerts T. JNJ-38877605: a selective Met kinase inhibitor inducing regression of Met-driven tumor models. Proc Am Assoc Cancer Res 2008;.49: Abstract 4837.
224. Beaulieu N, Beaulieu C, Dupont I, Nguyen H. Preclinical development of MGCD265, a potent orally active c-Met/VEGFR multi-target kinase inhibitor. Proc Am Assoc Cancer Res 2008;49: Abstract 4838.
225. Obaishi H, Nakagawa T, Yamaguchi A, Tohyama O. E7050: a novel small molecule inhibitor of the c-Met and VEGFR-2 tyrosine kinases. Proc Am Assoc Cancer Res 2008;49: Abstract 4846.

STRUCTURE-BASED DESIGN OF KINASE INHIBITORS: MOLECULAR RECOGNITION OF PROTEIN MULTIPLE CONFORMATIONS

JEFFREY JIE-LOU LIAO[1,2]
[1] University of Science and Technology of China, Hefei, Anhui, Department of Chemical Physics, China
[2] Duke University, Department of Chemistry, Durham, NC

1. INTRODUCTION

Protein kinases constitute one of the largest gene families in the human genome [1]. The kinase enzymes catalyze phosphorylation of serine, threonine, or tyrosine residue and regulate the majority of signal transduction pathways in cells. Deregulation of protein kinases is implicated in a number of diseases including cancer, diabetes, and inflammation [1]. Targeted inhibition of protein kinases has thereby become an attractive therapeutic strategy in the treatment of relevant diseases [2–4]. Tremendous efforts over the past decade have resulted in regulatory approval of several small-molecule anticancer protein kinase drugs so far (see Fig. 1 and Table 1), and several hundred additional inhibitors are presently undergoing clinical development.

Despite the remarkable achievements, design of selective small-molecule protein kinase inhibitors remains challenging because of the highly conserved nature of the kinase catalytic domain. Many kinase inhibitors have failed in preclinical or clinical development owing to intolerable toxicity relating to nonselectivity. Currently, a structure-based strategy for the discovery of selective protein kinase inhibitors involves targeting specific inactive conformations, as well as other specificity determinants, of protein kinases [3–6]. Systematic analysis of the crystal structures and sequences' variations of protein kinases can shed light on how the selectivity of kinase inhibitors can to some extent be achieved [7].

The occurrence of drug resistance represents another major obstacle in kinase-targeted anticancer therapy. As an inevitable consequence of the tumor response to drug treatment, acquired resistant mutations have been detected in the catalytic domain of c-Abl, c-Kit, and PDGFR in patients treated with imatinib [8–12], and EGFR treated with gefitinib or erlotinib [13–17]. These mutations preclude or weaken drug binding but retain the enzymatic function of the kinase. As shown in the c-Abl case, the acquired mutations resistant to imatinib diminish the specificity determinants and result in active or less specific inactive structural forms relative to the highly specific DFG-out conformation where the drug binds. The development of kinase inhibitors with a selectivity profile of broad spectrum seems necessary to thwart the drug-resistant mutation by targeting these less specific conformations [6]. Not surprisingly, second-line compounds thus developed bear a different resistant profile with narrow mutation spectrum compared to the frontline drug. The property of a drug that induces resistant mutations in the target is thus associated with its selectivity profile [18]. A drug discovery program aims to develop highly selective protein kinase inhibitors, but with the risk that acquired resistance may impair inhibitor binding by mutating specificity determinants or significant contact residues. As kinase enzymes adopt multiple conformations in the catalytic cycle, nevertheless, molecular recognition of these different conformations is of great help to optimize the selectivity and acquired resistant mutation profiles of kinase inhibitors [18].

This chapter first analyzes and classifies the binding pockets in the multiconformational catalytic cleft of protein kinases. Targeting these different conformations to develop protein kinase inhibitors with distinct but clinically useful profiles is then elucidated and exemplified. A structure-based rationale for the strategy to block the development of drug-resistant mutation is also provided.

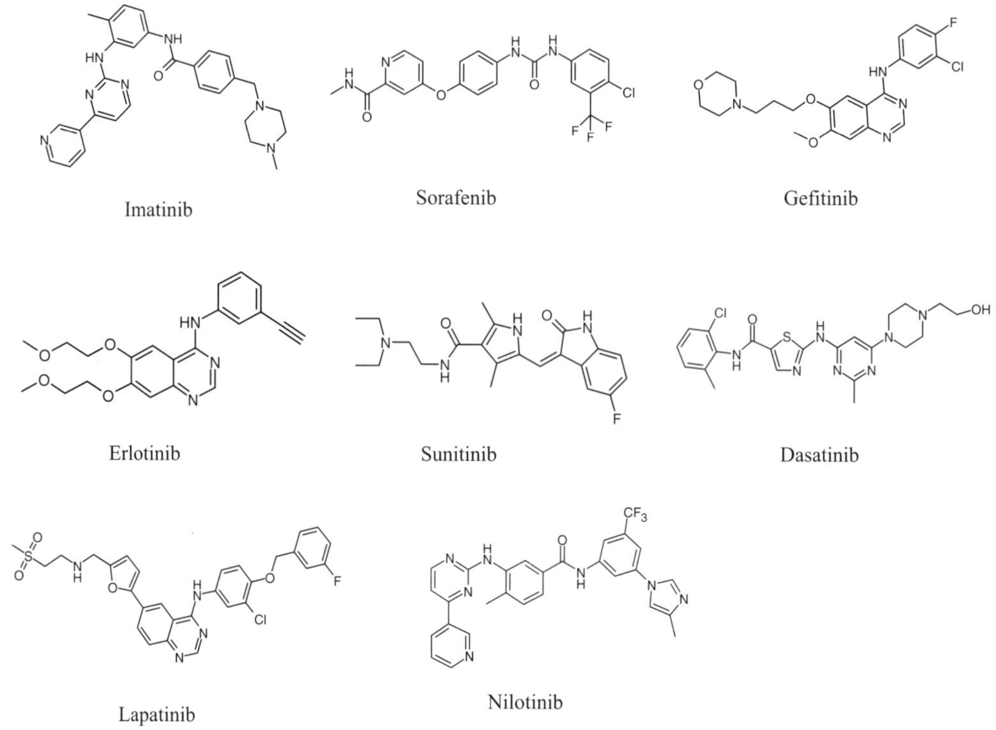

Figure 1. Structural formulas of small-molecule protein kinase drugs.

2. BINDING POCKETS IN THE MULTICONFORMATIONAL KINASE CATALYTIC DOMAIN

2.1. Binding Cleft in the Kinase Catalytic Domain

The kinase catalytic domain is a major focus of small-molecule inhibitor design for protein kinase targets. This domain consists of an N-terminal lobe (N-lobe) and a larger C-terminal lobe (C-lobe). The N- and C-lobes are connected with a linker region that contains a hinge (aqua in Fig. 2a and b). Here, the X-ray structure of c-Abl nonreceptor tyrosine kinase (PDB codes: 2GQG and 2HYY) [19] is used. A catalytic cleft formed between these two lobes is composed of two regions: the front cleft and the back cleft [7]. The front cleft contains predominantly the ATP binding site whereas the back cleft comprises elements important for regulation of kinase catalysis. An internal

Table 1. Small-Molecule Protein Kinase Drugs and Their Primary Targets

Generic Name	Primary Target	Disease	Year (Launched)
Imatinib	Abl, Kit, PDGFR	CML, GIST	2001/2002 CML/GIST
Gefitinib	EGFR	NSCLC	2003
Erlotinib	EGFR	NSCLC	2004
Sorafenib	Raf	Advanced RCC	2005
Sunitinib	Kit, VEGFR2, Flt3, PDGFR	GIST resistant to imatinib; advanced RCC	2006
Dasatinib	Abl, Src, PDGFRβ	CML resistant to imatinib	2006
Lapatinib	EGFR	NSCLC	2007
Nilotinib	Abl	CML resistant to imatinib	2007

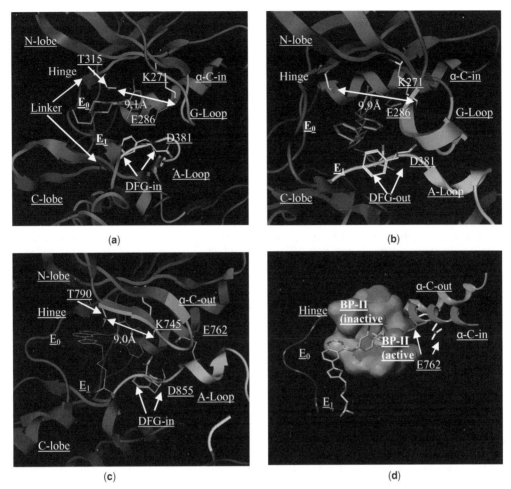

Figure 2. (a) Ribbon representation of the c-Abl catalytic domain in the active DFG-in and αC-in conformation (G-loop distorted) in complex with dasatinib (PDB code: 2GQG). (b) Ribbon representation of the c-Abl catalytic domain in the inactive conformation (DFG-out and αC-in) in complex with imatinib (PDB code: 2HYY). (c) Ribbon representation of the EGFR catalytic domain in the inactive conformation (DFG-in and αC-out) in complex with lapatinib (PDB code: 1XKK). (d) Superposition of the inactive (green; in complex with lapatinib) (PDB code: 1XKK) and active (red; PDB code: 1M17) clefts. The 3-fluorobenzyl group of lapatinib binds into the BP-II pocket (green), but clashes with E766 and M762 in the active conformation (red). (See color insert.)

gate between the front and the back cleft is formed by a gatekeeper residue (T315 in c-Abl) and a lysine (K271 in Abl). While ~25% protein kinases have a small gatekeeper residue such as threonine or alanine that makes the back cleft accessible for small-molecule binding [20,21], the majority of protein kinases possess a gatekeeper residue with a large side chain that blocks the entry of a small molecule into the back cleft through this gate. Interesting exceptions observed in crystallographic studies include Aura-A (gatekeeper: L210) [22], Tie-2 (I902) [23], and c-Met (L1157) [24], whose back cleft can be intruded by a small molecule through/under the internal gate.

A protein kinase that is flexible possesses multiple conformations. Conformational changes are functionally important for a protein kinase in the catalytic cycle. The multiconformations of a protein kinase can be characterized by variable structural forms of the

DFG motif and the αC helix [6], which are two key elements for the regulation of the kinase enzymatic activity [25]. While the DFG motif can be DFG-in, DFG-out, or DFG-out-like, the αC helix can adopt an αC-in or αC-out conformation. In a DFG-in conformation, the side chain of the DFG aspartate (Fig. 2a, D381 in c-Abl) is directed into the ATP binding site, and the aromatic ring of the phenylalanine (F382 in c-Abl) is positioned in the back cleft. In contrast, the phenylalanine aromatic ring of the DFG-out motif is positioned in the ATP binding site, and the aspartate side chain occupies the back cleft. Compared to the DFG-in conformation, the DFG-out backbone is significantly shifted to the ATP binding site, as demonstrated in the X-ray structures. Superposition of the Abl DFG-out and active crystal structures [19] shows that F382 in the DFG-out motif shifts to the ATP binding site by ~ 7.6 Å relative to its active counterpart, rendering the cleft basement accessible for compound binding. There exists a DFG-out-like conformation where the side chains of the phenylalanine and aspartate are positioned in a manner similar to a DFG-out one [6]. However, its main chain is positioned in an active-like fashion as the A-loop adopts an open conformation that mimics the activated form. As such, the C-lobe basement is inaccessible. The αC helix is highly mobile—it can adopt an αC-in conformation with its catalytic glutamate (E 286 in Abl in Fig. 2a and b) pointing into the ATP binding site. An αC-in conformation can occur in an active or inactive protein kinase as seen in Fig. 2a and b. When this helix moves outward, the catalytic glutamate is directed out of the catalytic cleft. This conformation is denoted αC-out (Fig. 2c). The αC-out conformation in the crystal structure of EGFR in complex with lapatinib [26] enlarges the hydrophobic pocket in the deep back cleft compared to the active αC-in conformation [27] (green, Fig. 2d).

The A-loop in a fully active state adopts an open conformation so that it is positioned away from the catalytic center (Fig. 2a), providing a platform for substrate binding. An active protein kinase adopts a DFG-in and αC-in conformation only. As such, the aspartate chelates Mg^{2+} and helps to orient the γ-phosphate for the phosphor transfer. The phenylalanine aromatic side chain in the active DFG-in motif is in contact with αC. This contact in many active kinases facilitates the formation of the Lys-Glu salt bridge for the kinase catalysis [25]. The conformational changes in the DFG motif and the αC helix can occur in a cooperative manner. For example, during the conformational switching between a DFG-out and a DFG-in c-Abl, an outward movement of αC is required to allow the bulky side chain of the DFG phenylalanine to flip freely without the steric hindrance imposed by the inward positioning of the αC glutamate. Combinations of different DFG and αC conformations have been observed, for example, in c-Abl [28] in crystallographic studies.

2.2. Binding Pockets in the Multiconformational Catalytic Cleft

The catalytic cleft contains distinct binding pockets. Detailed descriptions of these binding pockets are given elsewhere [7]. Two-dimensional mapping of the topological distributions of the binding pockets is shown in Fig. 3a and b. In this figure, A, R, and P represent the adenine, ribose, and phosphate pockets, respectively. The E_0 and E_1 regions in the front cleft, the BP-I pocket in the back cleft, and the DFG-in pocket (BP-II) are also shown in the figure. The BP-II pocket is also labeled DFG-in, DFG-out-like or DFG-out in Fig. 3a and b, respectively, according to the conformation of the DFG motif. The DFG-in pocket in the inactive kinase with the αC-out conformation is enlarged compared to that in the active conformation. However, the topological distribution of the binding pockets is similar. There exists a region (denoted K) in the deep front pocket, which is non-ATP contact in many cases. The K region is merged into the DFG-out pocket (Fig. 3b) with its bulky side chain of the phenylalanine flipping into the ATP binding site.

The kinase cleft conformation characterized by DFG and αC can significantly alter the topological distribution, accessibility, and size of the binding pockets, as seen in Fig. 3a and b [7]. For example, while the C-lobe basement in c-Abl with an DFG-in or DFG-out-like motif is inaccessible, the second gate formed

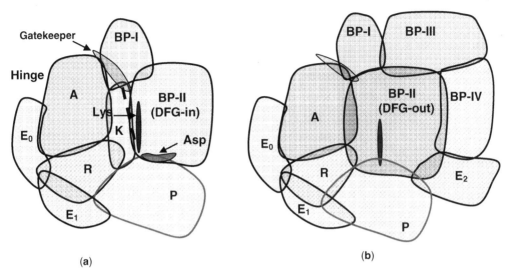

Figure 3. Topological distribution of the binding pockets in the catalytic cleft with (a) the DFG-in/DFG-out-like and αC-in/αC-out conformation and (b) the DFG-out and αC-in conformation. (This figure is available in full color at http://mrw.interscience.wiley.com/emrw/9780471266945/home.)

between M290 in αC and the DFG aspartate (D381) is open in the DFG-out c-Abl, allowing the binding pockets in the cleft basement accessible for small-molecule binding (Fig. 2b). As such, binding pockets, BP-III and BP-IV, emerge in the DFG-out conformation as shown in Fig. 3b.

There is an entrance region E_2 (Fig. 3b) at the αC side in the DFG-out conformation. An entrance located in the similar region has also been observed in the X-ray structures of mitogen-activated protein kinase kinase 1 (MEK1) and MEK2 (in the DFG-in conformation) complexed with their inhibitors, respectively [29]. This entrance allows a compound to bind into the back cleft in the enzyme highly specific in the kinase family (Fig. 20b in Ref. [7]).

Molecular recognition of the binding pockets in a multiconformational kinase target and their specificity plays an important role in the design of potent and selective kinase inhibitors. In a hydrophobic pocket whose dielectric constant is relatively small, hydrogen bonding interactions can significantly enhance inhibitor binding affinity [7]. Interestingly, a halogen bond can provide a contribute to binding affinity estimated to be about half to slightly more than that of an average hydrogen bond [30]. The importance of halogen bonding can be seen in the discovery of MEK inhibitors (see Fig. 20b in Ref. [7]) as well as other examples [31]. In general, it is relatively difficult to utilize ionic or polar interactions to enhance inhibitor binding affinity significantly in a hydrophilic pocket exposed to solvent because of the desolvation penalty. In contrast, a robust way to increase ligand binding affinity is to exploit hydrophobic interactions. The full burial of a methyl group in a hydrophobic pocket, for example, can contribute to the binding affinity by up to a factor of 10 at physiological temperature [32].

As the pharmacophoric feature of a binding pocket can significantly affect the protein–ligand interaction, the binding pockets in the catalytic cleft play different roles in enhancing the binding affinity of a kinase inhibitor. The adenine pocket or the DFG-out pocket in the deep cleft is often treated as a main scaffold H-bonded with a fragment of a small molecule. A structure-based method is thus applied in the design of kinase inhibitors with a core fragment to hit such a binding pocket, and then to expand with various chemical groups into different binding pockets [7]. This core-fragment expansion method using a focused fragment library provides an efficient way to discover potent kinase inhibitors compared to the high-throughput screening of a combinatorial library.

3. TARGETING PROTEIN MULTIPLE CONFORMATIONS FOR KINASE INHIBITOR DESIGN

Protein kinases exist in equilibrium between different conformational states in the catalytic cycle. Binding a kinase into one such conformation can trap the enzyme, either competing with ATP binding or shifting the equilibrium toward an inactive state, and thereby inhibiting the kinase activity. The size, availability, and topological distribution of the binding pockets in the multiconformational cleft are different. Multiple conformations of a protein kinase target provide an opportunity for design of kinase inhibitors with distinct but clinically useful profiles. While active kinase enzymes have highly homologous binding sites, for example, inactive protein kinases can adopt distinct structural forms. A potent protein kinase inhibitor fully recognizing binding pockets in an inactive kinase conformation usually has a relatively high degree of selectivity compared to an active kinase binder. On the other hand, a potent inhibitor of an active kinase target can be less vulnerable to acquired mutations resistant to a frontline drug as discussed in the next section. The structure-based strategy of targeting multiconformations of protein kinases is elucidated using EGFR, Kit, Abl, and c-Met, and their respective important small-molecule antagonists in the following sections.

EGFR is overexpressed in 40–80% of NSCLCs and is related to a poor prognosis [33]. Gefitinib [34], erlotinib [35], and lapatinib [36] are potent and selective EGFR inhibitors. Erlotinib binds in the active form of EGFR, whereas lapatinib targets the inactive enzyme. They both exploit the threonine gatekeeper (T790, Fig. 2c and d) to penetrate in the EGFR back cleft through passing this gate. The crystal structures for EGFR in complex with erlotinib (PDB code: 1M17) [27] and lapatinib (PDB code: 1XKK) [26] reveal the binding modes of the compounds shown in Fig. 4a and b. In both complexes, N1 of the quinazoline core accepts an H-bond from the amide nitrogen of M769 (numbering used in 1M17) in the hinge. The quinazoline C2 atom donates one weak H-bond to the carbonyl oxygen of the hinge residue E767. In EGFR complexed with erlotinib, the acetylene moiety at the C3 position of the phenyl ring is directed into the BP-I pocket, significantly improving the compound selectivity. Seen in Fig. 4a, the C6 and C7 substituents extend into the entrance regions. Compared with erlotinib, the key modification in lapatinib is at the 4' position of the aniline group: the large 4'-

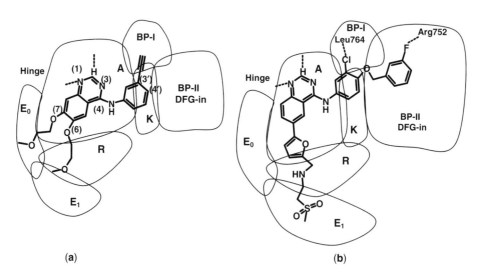

Figure 4. (a) Erlotinib binds into the active conformation (DFG-in and αC-in) of EGFR. (b) Lapatinib binds into the DFG-in and αC-out inactive conformation of EGFR. A dashed line represents a hydrogen bond or a halogen bond in this and the following figures.

(3-fluorobenzyl-oxy) substituent explores a bulky DFG-in pocket with the αC-out inactive conformation. As demonstrated in the EGFR–lapatinib complex, the 3'-chloro aniline group oriented into the BP-I pocket forms a halogen bond with the carbonyl oxygen of L764 in β_5 (1M17 numbering). The 3-fluorobenzyl group is extended into the DFG-in pocket with hydrogen bonded to the carbonyl oxygen of R752. The C6 substituent is directed into the E_1 region.

In contrast to most kinase family members, phosphorylation of EGFR is not required in its activation [37]. Whether this highly specific feature promotes the high degree of selectivity of erlotinib, geftinib, or lapatinib against EGFR over many other protein kinases remains elusive. Nevertheless, the fashion by which the enzyme is activated provides a possible rationale for the different selectivity profiles of erlotinib and lapatinib for EGFR over its isoform ErbB-2. Lapatinib is a potent dual inhibitor of inactive EGFR and ErbB-2, while erlotinib binds potently to active EGFR but not ErbB-2 although the kinase domains of EGFR and ErbB-2 are highly homologous (88% identical in sequence) [26,27]. Erlotinib is devoid of ErbB-2 inhibition, most likely due to the inability of ErbB-2 to adopt an active conformation seen in the apo and EGFR–erlotinib structures. While erlotinib binds in the active EGFR tightly, the drug would be packed loosely in the inactive form of the enzyme because the back cleft of the inactive conformation is enlarged owing to the outward rotation of αC discussed earlier.

Lapatinib has a very slow dissociation rate from EGFR compared with erlotinib and gefitinib [27]. Similarly, doramapimod (BIRB-796) also shows a slow dissociation from its targeted p38α, which plays a key role in producing proinflammatory cytokines such as tumor necrosis factor and interleukin-1 [38]. The calculated half-life for the dissociation of doramapimod from p38α is 23 h whereas the half-life is 0.1 s for the classic ATP-competitive compound SK&F86002 [39]. This slow dissociation can benefit *in vivo* efficacy of the inhibitors. The X-ray structure of doramapimod complexed with p38α shows that the compound binds into the DFG-out and αC-in conformation of the enzyme [39]. The slow kinetic profile of a compound is often linked to the conformational change from the bound state to the transition state with a relatively high energy barrier.

Kit, a type III receptor tyrosine kinase, is tightly regulated in normal cells. Binding the stem cell factor to Kit leads to receptor dimerization, followed by transphosphorylation in the Kit juxtamembrane region, and the activation of the kinase domain [40–42]. Several activating mutations in Kit are associated with the development and progression of GIST. Imatinib, originally developed for patients with CML, is a first-line drug used to treat GIST that targets the DFG-out and αC-in conformation of Kit as demonstrated in the X-ray structure (PDB code: 1T46) [40]. The mode of drug molecule binding is illustrated in Fig. 5a. Seen in this figure, the diaryl amide of imatinib forms two typical H-bonds between its amide nitrogen and oxygen atoms and the carboxylate side chain of E640 and the D810 backbone nitrogen within the DFG-out pocket. The 2-amino-pyrimidine fragment and the attached pyridine ring occupy the adenine pocket; the pyridine nitrogen accepts an H-bond from the main chain amide of the hinge residue C673, and the 2-amino nitrogen donates an H-bond to the side chain oxygen of the gatekeeper T670. Seen in Fig. 5a, the methyl group at the central phenyl ring inserts into hydrophobic pocket BP-I, whereas the methylpiperazinyl ring penetrates into the partially hydrophobic BP-IV pocket. Sunitinib, a FDA-approved drug for GIST treatment resistant to imatinib, preferentially binds to the front cleft of the active Kit tyrosine kinase. The mode of inhibitor binding into Kit from computer docking [6] is shown in Fig. 5b. The oxindole core binds into the adenine pocket with two hydrogen bonds between the oxindole N1 atom and the E671 carbonyl oxygen, and between the oxindole O_2 atom and the C673 amide nitrogen (Fig. 5b). The pyrrole ring is covered by L595 in β_1 and Y672 in the hinge, whereas the propionic acid substituent at the C3' position occupies the ribose pocket and interacts with N680 with a hydrogen bond in the E_1 region. In addition, the O_2 atom of the oxindole core forms an intramolecular H-bond with N1' on the pyrrole ring, helping to stabilize the planar conformation of the compound.

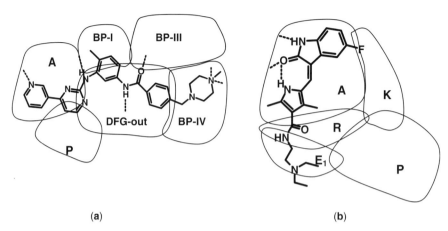

Figure 5. (a) Imatinib binds into the DFG-out and αC-in inactive conformation of Kit. (b) Sunitinib binds into the DFG-in and αC-in active conformation of Kit.

Imatinib and sunitinib display different selectivity profiles. Imatinib recognizes the DFG-out and αC-in conformation highly specific for a subset of protein kinases including Kit, Abl, and PDGFR over many other protein kinases. Sunitinib binds 73 of 113 distinct protein kinases demonstrated in an *in vitro* study [43], much more promiscuous than imatinib. Nevertheless, sunitinib is well tolerated in the clinic. Furthermore, the multitargeted nature of the drug is beneficial to its antitumor and antiangiogenic functions [44–47]. For example, sunitinib blocks the activity of multiple tyrosine kinases including VEGFR, activation of which results in endothelial cell proliferation and formation of new blood vessels. Inhibition of VEGFR by sunitinib plays an important role in its success for the treatment of advanced renal cell carcinoma (RCC) because the renal tumors are especially vascularized. Sunitinib inhibits most imatinib-resistant mutants and was FDA approved in 2006 for the treatment of GIST patients who are resistant to imatinib.

The c-Abl tyrosine kinase is located partially in the nucleus and its activity is tightly regulated. The fusion of the breakpoint cluster region (Bcr) gene to Abl facilitates the oligomerization of the kinase and its transautophosphorylation, resulting in the constitutive activation of the kinase often associated with CML [48]. Imatinib, the first FDA-approved small-molecule protein kinase drug, binds the DFG-out and αC-in conformation of Abl for the treatment of patients with CML. The drug molecule inhibits Abl in a similar mode by which it binds into Kit (see Fig. 5a). As shown in Fig. 2a and b, Abl encompasses multiple conformations where various inhibitors can bind. PD166326 [49] inhibits the protein into the DFG-out-like and αC-in inactive conformation where the A-loop is open [28]. The mode of inhibitor binding into Abl is shown in Fig. 6a. The amino-pyridopyrimidine core forms two canonical hydrogen bonds with M318 in the hinge. One chloro substituent of the dichlorophenyl group is positioned in BP-I, while another chloro group with the phenyl ring occupies the DFG pocket. The benzylhydroxyl substituent at the opposite end of the inhibitor extends into the E_0 region with the hydroxyl group H-bonded to the backbone of T319. Interestingly, PD166326 is nearly 100 times more potent than imatinib in inhibition of Bcr-Abl-dependent cell line growth, although it binds in Abl with a lower contact surface area and with fewer H-bonds [50]. This compound can inhibit Abl potently in both active and inactive states, whereas imatinib only binds in the highly specific DFG-out conformation. A fraction of the binding free energy of imatinib to Abl may be used to stabilize the A-loop conformation.

Dasatinib, a drug for adults with CML resistant to imatinib, can bind into multiple conformational states of Abl. The mode of drug binding into the active form of Abl [19,51] is

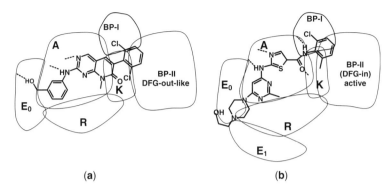

Figure 6. (a) PD166326 binds into DFG-out-like and αC-in inactive conformation of c-Abl. (b) Dasatinib binds into the active conformation of c-Abl.

illustrated in Fig. 6b. In this figure, dasatinib inserts the 2-chloro-6-methylbenzamide group into the BP-I and DFG pockets. The aminothiazole core binds in the adenine pocket with two hydrogen bonds linked to the hinge residue M318. One hydrogen bond is formed between the compound amide nitrogen and the hydroxyl oxygen of the gatekeeper residue T315. The substituted pyrimidine ring with the hydroxyethyl piperazine group extends into the E_0 region. An *in vitro* study shows that dasatinib inhibits 14 of 15 imatinib-resistant Abl mutants in the low nanomolar range [52].

PD166326 and dasatinib both are less selective than imatinib. For example, dasatinib is shown to inhibit about one-third of 148 protein kinases in the nanomolar range [43]. Interestingly, the kinase interaction map of dasatinib is somewhat complementary to that of staurospine, a highly nonselective compound targeting potently serine/threonine kinases. Despite its promiscuity, dasatinib is well tolerated in the clinic. The selectivity profile of a potent Abl inhibitor can be associated with its resistant mutation profile (see Fig. 2 in Ref. [18]). As discussed in the next section, inhibitors that possess distinct resistant-mutation profiles might offer potential therapies to thwart acquired mutations resistant to imatinib in the treatment of CML.

Many potent protein kinase inhibitors mimic ATP gearing to the hinge with two or three H-bonds. However, these ATP mimetics are often relatively promiscuous. A strategy to improve the inhibitor selectivity profile is to reduce the contribution of the hinge H-bonds to the binding affinity and compensate the free energy loss by developing H-bonding at another site, particularly in an inactive hydrophobic pocket. For example, the core of compound **1** (IC50 against c-Met is $\sim$150 nM) is H-bonded only to M1160 (medium strength) at the hinge of c-Met as shown in Fig. 7a. Another H-bond is formed between the hydroxyl group of the compound and the backbone oxygen of A1226 located at the inactive A-loop, which is highly specific, in the deep hydrophobic pocket. While the hinge H-bonding interaction is relatively weak, this O–H$\cdots$H type H-bond can contribute to the binding affinity by up to $\sim$5 kcal/mol. The benzyl group of the compound is extended into the phosphate pocket, forming a relatively loose π–π packing with F1089 and Y1230. The binding affinity of the compounds can be improved to low nanomolar by extending the aromatic group more into the π–π packing site with good selectivity profile. However, the resulting compounds suffer from metabolic instability. Unsurprisingly compound **1** was also observed to bind into another inactive c-Met conformation in a different mode with similar affinity. Guided by this binding mode, a series of low nanomolar c-Met compounds were discovered. Among them, compound **2**, whose binding mode is shown in Fig. 7b, has the best PK/potency profile [53].

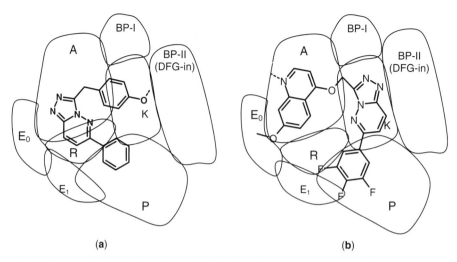

Figure 7. (a) Compound **1** binds into the autoinhibited conformation (inactive, DFG-in and αC-in) of c-Met. (b) Compound **2** binds into another inactive conformation (DFG-in and αC-in) of c-Met.

4. TARGETING MULTIPLE KINASE CONFORMATIONS FOR DESIGN OF SECOND-LINE KINASE INHIBITORS FOR THWARTING DRUG-RESISTANT MUTATION

4.1. Mechanism of Drug-Resistant Mutation

Although tremendous progress has been made in protein kinase drug discovery, drug-resistant mutation in treated patients has become a significant issue in kinase-targeted anticancer therapy. Imatinib-resistant mutation in Bcr-Abl is an extensively studied example. Twenty-five residue substitutions at 21 positions in the kinase domain have been observed in CML patients, precluding or weakening drug binding but retaining the enzymatic function of the kinase [54]. Seven of them are the drug–protein contact residues, among which T315I is a most frequently occurred mutation. This mutant alters the topological distribution of the accessible binding pockets, blocking the entry of the drug into the back cleft through the internal gate. In addition to the steric hindrance, this mutation in Abl also disrupts the hydrogen bonding interaction between the hydroxyl group of threonine and imatinib. Homologous mutations of the threonine gatekeeper also emerge in Kit (T670I), PDGFR (T674I) treated with imatinib, and EGFR (T790M) exposed to geftinib or erlotinib [17,55]. Other contact residue mutations include L248V and F317L in the adenosine pocket, F311T in BP-I, V298A and F359V in BP-II, and F382L in the DFG motif [54].

The majority of the above 25 mutations allosterically disrupt the contacts between the G-loop and the A-loop, destabilizing the DFG-out conformation. For example, the E255K mutation disrupts the interaction between E255 and K247. This mutation interferes with the optimal G-loop structure, thereby disrupting the contact between the G-loop and the A-loop. The L248V/R replacement most likely plays a dual role: while it reduces the target--drug contact, this substitution interferes with the interaction between L248 and F317 in the hinge, disturbing the G-loop. The Y253H mutation interferes with the aromatic–aromatic interaction between Y253 in the G-loop and F382 in the DFG motif, destabilizing the inactive conformation. Other resistant mutations in the G-loop including G250E and Q252H occur with a similar mechanism. Mutations in the A-loop such as H396P/R disrupt the closed autoinhibited A-loop and favor an open conformation as seen in the X-ray structure.

The allosteric drug-resistant mutations share a mechanism remarkably similar to that of the oncogenic activation of the autoinhibited B-Raf protein kinase [56]. Sorafenib traps B-Raf in the DFG-out conformation, which is stabilized by a number of interactions between the residues mostly situated in the G-loop and the A-loop. Important contact amino acids include V599 in the A-loop interacting with the aromatic ring of F467 in the G-loop, L596 with almost all the G-loop residues, and F594 in the DFG motif with Q580 and F582 in the catalytic loop [56]. The majority of over 30 activating oncogenic mutations in B-Raf emerge at these positions [57], disrupting the autoinhibited conformation. In particular, the V599E substitution as the hallmark of a conventional oncogene accounts for 90% of B-Raf mutations in human cancers. The replacement of V599 with a larger and charged glutamine relieves its hydrophobic contact with F467, activating B-Raf. Replacement of the G-loop glycines with bulkier residues such as G468V disturbs the interactions between the G- and A-loops, interfering with the DFG-out conformation. These findings could imply that similar allosteric drug-resistant mutations would occur in patients treated with sorafenib as evidence for further target validation. *In vitro* screening for imatinib-resistant mutations has identified a large number of mutants including all those observed in the clinic [58]. Interestingly, a number of mutations occur in the allosteric regions including the cap, the SH2–SH3 linker, and the SH3-kinase domain [58]. These mutants disrupt drug binding allosterically similar to those in the kinase domain discussed above.

4.2. Targeting Multiple Kinase Conformations for Thwarting Drug-Resistant Mutations

Molecular resistance in response to drug treatment is a major obstacle in kinase-targeted therapy. Strategies have emerged in the development of second-line protein kinase inhibitors to overcome the drug-resistant mutations. Many of the imatinib-resistant mutations in Abl discussed above confer a low or medium degree of resistance. Many of these mutations can be effectively suppressed by dose escalation of imatinib or by developing inhibitors analogous to imatinib with improved potency. Nilotinib, an optimized imatinib analog, is ~20 times more potent than imatinib against cell lines expressing Bcr-Abl [59–61]. The modification at the 3′ and 5′ positions of the aryl end of imatinib allows nilotinib to bind into the BP-III pocket, increasing the inhibitor binding affinity. The mode of inhibitor binding into Abl is described in Fig. 8a [6]. The cellular and biochemical experiments show that nilotinib inhibits the majority of the tested mutants except for the highly resistant mutations L248R, Y253H, E255K/V, and T315I [61,62]. Because nilotinib traps c-Abl into the DFG-out conformation as imatinib does, tumor cells treated with this compound would be expected to evolve the allosteric mutation mechanism of resistance as well.

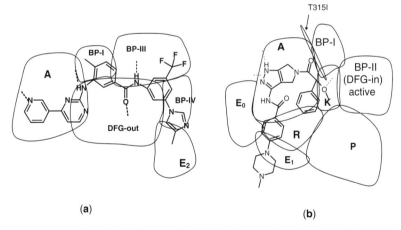

Figure 8. (a) Nilotinib binds in the DFG-out and αC-in inactive conformation of c-Abl. (b) PH-739358 binds into the active conformation of the T315I c-Abl mutant.

The allosteric resistant mutation in c-Abl leads to active or less specific inactive mutants. One approach to thwart the allosteric mutations is to develop inhibitors with the ability to tolerate the multiple conformations of Abl. Several such compounds including PD166326 [49] and dasatinib [51] were identified. PD166326 inhibits effectively many critical imatinib-resistant Abl mutants including Y253H, E255K, E255V, and H396P isolated from advanced CML patients [63–66]. Dasatinib is 325-fold more potent than imatinib against cells expressing wild-type Bcr-Abl and inhibits 14 of the 15 imatinib-resistant Abl mutants in the low nanomolar range [41,67]. Compared with imatinib, these second-line inhibitors are less selective because they bind into the aforementioned less specific conformations of Abl. In contrast with imatinib, these second-line compounds are potent dual Src/Abl inhibitors. Small-molecule compounds with dual activity against Abl and Src are especially attractive in the development of second-line Abl inhibitors because Src kinases are involved in Abl-activated leukemogenesis and have been implicated in imatinib resistance in some cases [68–70]. Dasatinib as well as PD166326 has a different resistant-mutation profile from imatinib. Overall, these inhibitors are less vulnerable to resistant mutations compared with imatinib. An *in vitro* screen shows that in the presence of 50 nM dasatinib, for example, 10 mutations emerge at six different amino acids, 8 of which are drug contact residues [71]. This profile is in contrast to that of imatinib where the majority of the mutants occur at the allosteric sites as discussed previously. Q252H and E255K in this 10-mutation set are allosteric, indicating that most likely dasatinib also binds into the DFG-out conformation. Furthermore, about half of the 10 mutants are not described in the imatinib case.

The majority of the mutants can be suppressed by dose escalation; however, the inhibitors at high concentration likely show undesirable off-target activity [71,72]. An appealing therapeutic approach to overcome this drawback is to use a combination of Abl inhibitors with distinct mutation profiles. An advantage of this strategy is that mutations resistant to one of the Bcr-Abl inhibitors may be susceptible to another component of the cocktail at low concentration. For example, when 50 nM dasatinib is combined with 5 μM imatinib, only one clone expressing T315I is recovered [71].

The T315I Abl mutant accounting for 10–15% detected mutations in the clinic confers complete resistance to the above and many other clinically available inhibitors. This mutation impedes the entry of an inhibitor into the back cleft through the internal gate. One way to override this gatekeeper resistant mutation is to develop small-molecule compounds to bind the enzyme into the front cleft. VX-680(MK-0457), an Aurora kinase inhibitor [73], showed the inhibitory activity against the T315I mutant in a Phase I clinical trial [74]. This compound inhibits potently the active form of Abl. The compound binds the active form of Abl with its cyclopropyl group into the back cleft through the internal gate, as demonstrated in the crystallographic study [75], while this cyclopropyl group is directed into the phosphate pocket in the front cleft because the T315I mutant renders the Abl back cleft inaccessible for small molecule [6]. The clinical trial of VX-680/MK-0457 was halted due to a potential heart safety issue in one patient. PHA-739358 [76], another potent Aurora kinase inhibitor, and SGX-930 [77] are two potent inhibitors against the c-Abl T315I mutant that are undergoing clinical development. Each of them binds into the front cleft of the active enzyme to avoid the steric hindrance the mutated gatekeeper residue. The binding mode of PHA-739358 is shown in Fig. 8b (PDB code: 2V7A) [78].

5. CONCLUDING REMARKS

Despite over 150 disease-associated potential kinases [1], validation of a druggable kinase target is not trivial. To date, only a limited number of protein kinases have been successfully targeted (see Table 1). The past decade has witnessed significant advances in molecular targeting of protein kinases. However, given only a limited number of protein kinases that have been successfully targeted so far, there is a high demand to disclose new druggable kinase targets. In particular, targeting kinase therapy remains challenging because the primary targets are often multifunctional and the drugs often lack selectivity as in the reported cardiotoxicity of imatinib [79] and sunitinib [80].

As targeting a specific inactive kinase conformation has become a major approach to develop highly selective inhibitors, binding a kinase target into its less specific conformational states plays an increasingly important role in the field. It seems likely that when the druggability of a kinase target is validated, its multiple conformations can be used to design inhibitors with distinct and useful clinical profiles as elucidated above. This is particularly the case where new agents are urgently needed to block the development of acquired resistance to a kinase drug in the clinic. Despite the remaining challenges, the cocktail of kinase drugs that target multiple kinase conformations currently represents the best option for anticancer therapy. This combination therapy can be extended to include agents with a mechanism of action different from targeting the kinase catalytic cleft.

Systematic analysis of the binding pockets in the multiconformational catalytic cleft offers insight into the design of small-molecule kinase inhibitors with distinct selectivity and resistant mutation profiles. This structure-based strategy can be of great help in the development of protein kinase inhibitors that serve as potential anti-kinase therapies as well as tools to exploit the druggable properties of kinase targets and their biological function.

ABBREVIATIONS

Abl	Abelson nonreceptor tyrosine kinase
Aurora	Aurora protein kinase
B-Raf	B-Raf kinase
c-Met	hepatic growth factor receptor tyrosine kinase
CML	chronic myelogenous leukemia
EGFR	epidermal growth factor receptor tyrosine kinase
Flt3	FMS-like tyrosine kinase 3
GIST	gastrointestinal stromal tumor
Kit	stem cell factor receptor tyrosine kinase
MEK	MAP kinase kinase
NSCLC	nonsmall cell lung cancer
PDGFR	platelet-derived growth factor receptor tyrosine kinase
p38α	p38α-MAP (mitogen-activated protein) kinase
RCC	renal cell carcinoma
Src	Src nonreceptor tyrosine kinase
Tie-2	Tie-2 receptor tyrosine kinase
VEGFR	vascular endothelial growth factor receptor

REFERENCES

1. Manning GD, Whyte B, Martinez R, et al. The protein kinase complement of the human genome. Science 2002;298:1912–1934.
2. Sebolt-Leopold JS, English JM. Mechanisms of drug inhibition of signaling molecules. Nature 2006;441:457–462.
3. Noble MEM, Endicott JA, Johnson LN. Protein kinase inhibitors: insights into drug design from structure. Science 2004;303:800–805.
4. Klebl BM, Muller G. Second-generation kinase inhibitors. Exp Opin Ther Targets 2005;9:975–993.
5. Liu Y, Gray NS. Rational design of inhibitors that bind to inactive kinase conformations. Nat Chem Biol 2006;2:358–364.
6. Liao JJ-L, Andrews RC. Targeting protein multiple conformations: a structure-based strategy for kinase drug design. Curr Top Med Chem 2007;7:1394–1407.
7. Liao JJ-L. Molecular recognition of protein kinase binding pockets for design of potent and selective kinase inhibitors. J Med Chem 2007;50:409–424.
8. Sawyers CL, Hochhaus A, Feldman E, et al. Imatinib induces hematologic and cytogenetic responses in patients with chronic myelogenous leukemia in myeloid blast crisis: results of a phase II study. Blood 2002;99:3530–3539.
9. Roche-Lestienne C, Soenen-Cornu V, Grardel-Duflos N, et al. Several types of mutations of the Abl gene can be found in chronic myeloid leukemia patients resistant to STI571, and they can pre-exist to the onset of treatment. Blood 2002;100:1014–1018.

10. Cools J, DeAngelo DJ, Gotlib J, et al. A tyrosine kinase created by fusion of the PDGFRα and FIP1L1 genes is a therapeutic target of imatinib in idiopathic hypereosinophilic syndrome. N Engl J Med 2003;348:1201–1214.
11. Tamborini E, Bonadiman L, Greco A, et al. A new mutation in the KIT ATP pocket causes acquired resistance to imatinib in a gastrointestinal stromal tumor patient. Gastroenterology 2004;127:294–299.
12. Deininger M, Buchdunger E, Druker BJ. The development of imatinib as a therapeutic agent for chronic myeloid leukaemia. Blood 2005;105:2640–2653.
13. Paez JG, Janne PA, Lee JC, et al. EGFR mutations in lung cancer: correlation with clinical response to gefitinib therapy. Science 2004;304:1497–1499 and 1500.
14. Pao W, Miller VA, Politi KA, et al. Acquired resistance of lung adenocarcinomas to gefitinib or erlotinib is associated with a second mutation in the EGFR kinase domain. PLoS Med 2005;2(e73): 225–235.
15. Kobayashi S, Boggon TJ, Dayaram T, et al. EGFR mutation and resistance of non-small-cell lung cancer to gefitinib. N Engl J Med 2005;352:786–792.
16. Greulich H, Chen TH, Feng W, et al. Oncogenic transformation by inhibitor-sensitive and resistant EGFR mutants. PLoS Med 2005;2:1167–1176.
17. Kosaka T, Yatabe Y, Endoh H, et al. Analysis of epidermal growth factor receptor gene mutation in patients with non-small cell lung cancer and acquired resistance to gefitinib. Clin Cancer Res 2006;12:5764–5769.
18. Liao JJL. Molecular targeting of protein kinases to optimize selectivity and resistance profiles of kinase inhibitors. Curr Top Med Chem 2007;7:1332–1335.
19. (a) Tokarski JS, Newitt J, Chang CYJ, et al. The structure of dasatinib (BMS-354825) bound to activated Abl kinase domain elucidates its inhibitory activity against imatinib-resistant Abl mutants. Cancer Res 2006;66:5790–5797. (b) Cowan-Jacob SW, Fendrich G, Floersheimer A, et al. Structural biology contributions to the discovery of drugs to treat chronic myelogenous leukaemia. Acta Crystallogr D 2007;63:80–93.
20. Liu Y, Shah K, Yang F, et al. A molecular gate which controls unnatural ATP analogue recognition by the tyrosine kinase v-Src. Bioorg Med Chem 1998;6:1219–1226.
21. Shewchuk L, Hassell A, Wisely B, et al. Binding mode of the 4-anilinoquinazoline class of protein kinase inhibitor: X-ray crystallographic studies of 4-anilinoquinazolines bound to cyclin-dependent kinase 2 and p38 kinase. J Med Chem 2000;43:133–138.
22. Heron NM, Anderson M, Blowers DP, et al. SAR and inhibitor complex structure determination of a novel class of potent and specific Aurora kinase inhibitors. Bioorg Med Chem Lett 2006;16:1320–1323.
23. Hodous BL, Geuns-Meyer SD, Hughes PE, et al. Evolution of a highly selective and potent 2-(pyridin-2-yl)-1,3,5-triazine Tie-2 kinase inhibitor. J Med Chem 2007;50:611–626.
24. Bellon SF, Kaplan-Lefko P, Yang Y, et al. c-Met inhibitors with novel binding mode show activity against several hereditary papillary renal cell carcinoma related mutations. J Biol Chem 2008;283:2675–2683.
25. Huse M, Kuriyan J. The conformational plasticity of protein kinases. Cell 2002;109:275–282.
26. Wood ER, Truesdale AT, McDonald OB, et al. A unique structure for epidermal growth factor receptor bound to GW572016 (Lapatinib): relationships among protein conformation, inhibitor off-rate, and receptor activity in tumor cells. Cancer Res 2004;64:6652–6659.
27. Stamos J, Sliwkowski MX, Eigenbrot C. Structure of the epidermal growth factor receptor kinase domain alone and in complex with a 4-anilinoquinazoline inhibitor. J Biol Chem 2002;277:46265–46272.
28. Levinson NM, Kuchment O, Shen K, et al. A Src-like inactive conformation in the abl tyrosine kinase domain. PLoS Biol 2006;4:0753–0767.
29. Ohren JF, Chen H, Pavlovsky A, et al. Structures of human map kinase kinase 1 (MEK1) and MEK2 describe novel noncompetitive kinase inhibition. Nat Struct Mol Biol 2004;11:1192–1197.
30. Auffinger P, Hays FA, Westhof E, et al. Halogen bonds in biological molecules. Proc Natl Acad Sci USA 2004;101:16789–16794.
31. Andrea VR, Ho PS. The role of halogen bonding in inhibitor recognition and binding by protein kinases. Curr Top Med Chem 2007;7:1336–1348.
32. Gohlke H, Klebe G. Approaches to the description and prediction of the binding affinity of small-molecule ligands to macromolecular receptors. Angew Chem Int Ed 2002;41: 2644–2676.
33. Arteaga CL. ErbB-targeted therapeutic approaches in human cancer. Exp Cell Res 2003;284:122–130.
34. Barker AJ, Gibson KH, Grundy W. Studies leading to the identification of ZD1839 (Iressa™): an orally active, selective epidermal

growth factor receptor tyrosine kinase inhibitor targeted to the treatment of cancer. Bioorg Med Chem Lett 2001;11:1911–1914.

35. Moyer JD, Barbacci EG, Iwata KK. Induction of apoptosis and cell cycle arrest by CP-358,774, an inhibitor of epidermal growth factor receptor tyrosine kinase. Cancer Res 1997;57:4838–4848.

36. Rusnak DW, Lackey K, Affleck K, et al. The effects of the novel, reversible epidermal growth factor receptor/ErbB-2 tyrosine kinase inhibitor, GW572016, on the growth of human normal and tumor-derived cell lines *in vitro* and *in vivo*. Mol Cancer Ther 2001;1:85–94.

37. Zhang X, Gureasko J, Shen K, et al. An allosteric mechanism for activation of the kinase domain of epidermal growth factor receptor. Cell 2006; 125:1137–1149.

38. Pearson G, Robinson F, Beers Gibson T, Xu BE, et al. Mitogen-activated protein (MAP) kinase pathways: regulation and physiological functions. Endocr Rev 2001;22:153–183.

39. Pargellis C, Tong L, Churchill L, Cirillo PF, et al. Inhibition of p38 MAP kinase by utilizing a novel allosteric binding site. Nat Struct Biol 2002;9:268–272.

40. Mol CD, Dougan DR, Schneider TR, et al. Structural basis for the autoinhibition and STI-571 inhibition of c-Kit tyrosine kinase. J Biol Chem 2004;279:31655–31663.

41. Zhang Z, Zhang R, Joachimiak A, Schlessinger J, et al. Crystal structure of human stem cell factor: implication for stem cell factor receptor dimerization and activation. Proc Natl Acad Sci USA 2000;97:7732–7737.

42. Heldin CH. Dimerization of cell surface receptors in signal transduction. Cell 1995;80: 213–223.

43. Fabian MA, Biggs WH, 3rd, Treiber DK, et al. A small molecule-kinase interaction map for clinical kinase inhibitors. Nat Biotechnol 2005;23:329–336.

44. Abrams TJ, Lee LB, Murray LJ, et al. SU11248 inhibits KIT and platelet-derived growth factor receptor β in preclinical models of human small cell lung cancer. Mol Cancer Ther 2003;2: 471–478.

45. Mendel DB, Laird AD, Xin X, et al. *In vivo* antitumor activity of SU11248, a novel tyrosine kinase inhibitor targeting vascular endothelial growth factor and platelet-derived growth factor receptors: determination of a pharmacokinetic/pharmacodynamic relationship. Clin Cancer Res 2003;9:327–337.

46. Eskens FA. Angiogenesis inhibitors in clinical development: where are we now and where are we going? Br J Cancer 2004;90:1–7.

47. Stadler WM. New targets, therapies, and toxicities: lessons to be learned. J Clin Oncol 2006;24:4–5.

48. Zhao X, Ghaffari S, Lodish H, Malashkevich VN, et al. Structure of the Bcr-Abl oncoprotein oligomerization domain. Nat Struct Biol 2002;9: 117–120.

49. Kraker AJ, Hartl BG, Barvian MR, et al. Biochemical and cellular effects of c-Src kinase-selective pyrido[2,3-*d*]pyrimidine tyrosine kinase inhibitors. Biochem Pharmacol 2000;60: 885–898.

50. Nagar B, Bornmann WG, Pellicena P, Schindler T, et al. Crystal structures of the kinase domain of c-Abl in complex with the small molecule inhibitors PD173955 and imatinib (Sti-571). Cancer Res 2002;62:4236–4243.

51. Lombardo LJ, Lee FY, Chen P, et al. Discovery of *N*-(2-chloro-6-methyl-phenyl)-2-(6-(4-(2-hydroxyethyl)-piperazin-1-yl)-2-ethylpyrimidin-4-ylamino) thiazole-5-carboxamide (BMS-354825), a dual Src/Abl kinase inhibitor with potent antitumor activity in preclinical assays. J Med Chem 2004;47:6658–6661.

52. Shah NP, Tran C, Lee FY, et al. Overriding imatinib resistance with a novel Abl kinase inhibitor. Science 2004;305:399–401.

53. Albrecht BK, Harmange JC, Bauer D, et al. Discovery and optimization of triazolopyridazines as potent and selective inhibitors of the c-Met kinase. J Med Chem 2008;22:2879–2882.

54. Nardi V, Azam M, Daley GQ. Mechanisms and implications of imatinib resistance mutations in Bcr-Abl. Curr Opin Hematol 2004;11:35–43.

55. Carter TA, Wodicka LM, Shah NP, et al. Inhibition of drug-resistant mutants of Abl, Kit, and EGF receptor kinases. Proc Natl Acad Sci USA 2005;102:11011–11016.

56. Wan PTC, Garnett MJ, Roe SM, et al. Cancer genome project mechanism of activation of the Raf-Erk signaling pathway by oncogenic mutations of B-Raf. Cell 2004;116:855–867.

57. Davies H, Bignell GR, Cox C, Stephens P, et al. Mutations of the BRAF gene in human cancer. Nature 2002;417:949–954.

58. Azam M, Latek RR, Daley GQ. Mechanisms of autoinhibition and STI-571/imatinib resistance revealed by mutagenesis of CR-Abl. Cell 2003;112:831–843.

59. Manley PW, Cowan-Jacob SW, Fendrich G, et al. Molecular interactions between the highly selective pan-Bcr-Abl inhibitor, AMN107, and the tyrosine kinase domain of Abl. Blood 2005;106:940a.

60. Golemovic M, Verstovsek S, Giles F, et al. AMN107, a novel aminopyrimidine inhibitor of Bcr-Abl, has *in vitro* activity against imatinib-resistant chronic myeloid leukemia. Clin Cancer Res 2005;11:4941–4947.
61. (a) Weisberg E, Manley P, Mestan J, et al. AMN107 (nilotinib): a novel and selective inhibitor of BCR-Abl. Br J Cancer 94:1765–1769. (b) Weisberg E. Manley P.W. Breitenstein W. et al. Characterization of AMN107, a selective inhibitor of native and mutant Bcr-Abl. Cancer Cell 2005;7:129–141.
62. O'Hare T, Walters DK, Stoffregen EP, et al. *In vitro* activity of Bcr-Abl inhibitors AMN107 and BMS-354825 against clinically relevant imatinib-resistant Abl kinase domain mutants. Cancer Res 2005;65:4500–4505.
63. La Rosée P, Corbin AS, Stoffregen EP, et al. Activity of the Bcr-Abl kinase inhibitor PD180970 against clinically relevant Bcr-Abl isoforms that cause resistance to imatinib mesylate (Gleevec, STI571). Cancer Res 2002;62: 7149–7153.
64. von Bubnoff N, Veach DR, Miller WT, et al. Inhibition of wild-type and mutant Bcr-Abl by pyrido-pyrimidine-type small molecule kinase inhibitors. Cancer Res 2003;63:6395–6404.
65. Huron DR, Gorre ME, Kraker AJ, et al. A novel pyridopyrimidine inhibitor of Abl kinase is a picomolar inhibitor of Bcr-Abl-driven K562 cells and is effective against STI571-resistant Bcr-Abl mutants. Clin Cancer Res 2003;9: 1267–1273.
66. von Bubnoff N, Veach DR, van der Kuip H, et al. A cell-based screen for resistance of Bcr-Abl-positive leukemia identifies the mutation pattern for PD166326, an alternative Abl kinase inhibitor. Blood 2005;105:1652–1659.
67. O'Hare T, Walters DK, Stoffregen EP, et al. Combined Abl inhibitor therapy for minimizing drug resistance in chronic myeloid leukemia: Src/Abl inhibitors are compatible with imatinib. Clin Cancer Res 2005;11:6987–6993.
68. Donato NJ, Wu JY, Stapley J, et al. Bcr-Abl independence and LYN kinase overexpression in chronic myelogenous leukemia cells selected for resistance to STI571. Blood 2003;101: 690–698.
69. Dai Y, Rahmani M, Corey SJ, et al. A Bcr/Abl-independent, Lyn-dependent form of imatinib mesylate (STI-571) resistance is associated with altered expression of Bcl-2. J Biol Chem 2004;279:34227–34239.
70. Martinelli G, Soverini S, Rosti G, Baccarani M. Dual tyrosine kinase inhibitors in chronic myeloid leukaemia. Leukemia 2005;19:1872–1879.
71. Burgess MR, Skaggs BJ, Shah NP, et al. Comparative analysis of two clinically active Bcr-Abl kinase inhibitors reveals the role of conformation-specific binding in resistance. Proc Natl Acad Sci USA 2005;102:3395–3400.
72. Azam M, Nardi V, Shakespeare WC, et al. Activity of dual Src-Abl inhibitors highlights the role of Bcr/Abl kinase dynamics in drug resistance. Proc Natl Acad Sci USA 2006;103: 9244–9249.
73. Harrington EA, Bebbington D, Moore J, et al. VX-680, a potent and selective small-molecule inhibitor of the Aurora kinases, suppresses tumor growth *in vivo*. Nat Med 2004;10: 262–267.
74. Tibes R, Giles F, McQueen T, et al. Translational *in vivo* and *in vitro* studies in patients (pts) with acute myeloid leukemia (AML), chronic myeloid leukemia (CML) and myeloproliferative disease (MPD) treated with MK-0457 (MK), a novel Aurora kinase, Flt3, JAK2, and Bcr-Abl inhibitor. Blood 2006;10: abstract 1362.
75. Young MA, Shah NP, Chao LH, et al. Structure of the kinase domain of an imatinib-resistant Abl mutant in complex with the Aurora kinase inhibitor VX-680. Cancer Res 2006; 66:1007–1014.
76. Gontarewicz A, Balabanov S, Keller G, et al. Simultaneous targeting of Aurora kinases and Bcr-Abl kinase by the small molecule inhibitor PHA-739358 is effective against imatinib-resistant Bcr-Abl mutations including T315I. Blood 2008;111(8): 4355–4364.
77. O'Hare T, Eide CA, Tyner JW, et al. SGX393 inhibits the CML mutant Bcr-Abl T315I and preempts *in vitro* resistance when combined with nilotinib or dasatinib. PNAS 2008;105: 5507–5512.
78. Modugno M, Casale E, Soncini C, et al. Crystal structure of the T315I Abl mutant in complex with the Aurora kinases inhibitor Pha-739358. Cancer Res 2007;67:7987–7990.
79. Kerkelä R, Grazette L, Yacobi R, et al. Cardiotoxicity of the cancer therapeutic agent imatinib mesylate. Nat Med 2006;12:908–916.
80. Chu TF, Rupnick MA, Kerkela R, et al. Cardiotoxicity associated with tyrosine kinase inhibitor sunitinib. Lancet 2007;370:2011–2019.

CANCER DRUG RESISTANCE: TARGETS AND THERAPIES

Barbara Zdrazil
Gerhard F. Ecker
Department of Medicinal Chemistry,
University of Vienna, Vienna, Austria

1. INTRODUCTION

Hundred years ago, Paul Ehrlich, the founder of chemotherapy, received the Nobel Price for Physiology or Medicine for his landmark immunological insights. Ehrlich postulated the existence of specific receptors (either associated with cells or distributed in the blood stream), which may be regarded as side chains that bind antigens ("side-chain theory of immunity," see timeline in Fig. 1) [1]. According to him, each type of receptors is attuned to one special group of drugs [2]. He declared "wir müssen zielen lernen, chemisch zielen lernen" ("we have to learn how to target chemically")—Ehrlich already suspected that the key for synthetic chemistry was to modify some starting material in various ways. After the discovery of the antisyphilitic activity of Salvarsan—an organic arsenic compound—in 1908 in Paul Ehrlich's laboratory, lead optimization led to the improved derivative Neosalvarsan in 1912. Biological activity of a lead compound for the first time was optimized through systematic modifications. This was the real beginning of chemotherapy [3,4].

It took some time—until the end of the World War II—that chemotherapy was introduced in clinical practice for cancer treatment. Gilman and coworkers treated a patient with non-Hodgkin lymphoma with nitrogen mustard, a chemical warfare agent that accidentally had caused lymphoid and myeloid suppression in humans during World War II. The therapy initially caused a dramatic antitumor effect, but by the time the third treatment was given, the tumor no longer responded to the chemotherapeutic treatment [4,5].

Since these early days of cancer chemotherapy, the increased knowledge of the cancer genome and the development of new drug discovery technologies, such as quantitative structure–activity relationships (QSAR), high-throughput screening (HTS), nuclear magnetic resonance (NMR), X-ray diffraction, and protein–ligand cocrystallography, have paved the way for targeted and multitargeted cancer therapeutics. Nevertheless, classical (unspecific cytotoxic) as well as targeted chemotherapy are often faced with one major obstacle that limits its success: drug resistance (tolerance).

2. OVERVIEW OF DRUG RESISTANCE MECHANISMS

By elucidation of the diverse resistance mechanisms to antineoplastic therapy, putative future drug targets can be studied. Some of the mechanisms may have the potential to be targeted, and to maintain or even improve selectivity of antitumor action [6]. Mechanisms of drug resistance that are associated with small molecules are illustrated in Fig. 2.

Those resistance mechanisms essentially include the following:

- Decreased intracellular concentration of the drug (e.g., ABC transporters)
- Alterations of the drug target (e.g., point mutations, overexpression of the target)
- Increased detoxification of the drug (e.g., glutathione conjugation)
- Decreased metabolic activation of the (pro)drug
- Decreased active uptake of the drug
- Activation of DNA repair systems
- Alterations in the cell-cycle checkpoint (e.g., p21)
- Changes in the ratio of pro- and anti-apoptotic proteins

Besides, we should distinguish between two basic types of antineoplastic drug tolerance mechanisms: *Intrinsic* resistance, which already exists at the time of the diagnosis prior to drug therapy, and *acquired (adaptive)* drug resistance, which appears later in the treatment. The latter is attributed to spontaneous genetic mutations and "negative" selection by cytotoxic chemotherapy—a phenomenon that

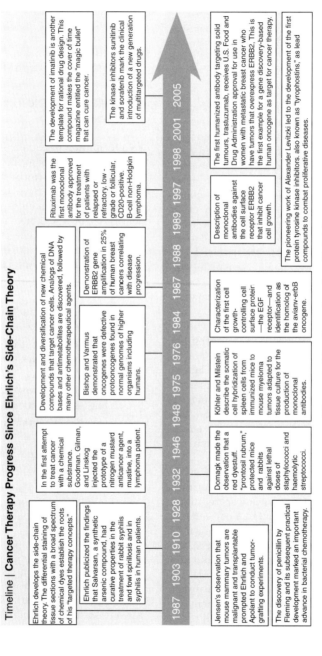

Figure 1. Timeline taken from Ref. [4].

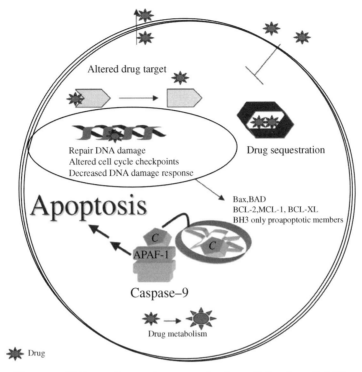

Figure 2. Different drug-resistance mechanisms. Taken from Ref. [7].

leads to selective deletion of the most responsive tumor cells but simultaneous survival of the least sensitive cells [6,8]. Since the tissues that are highly susceptible to chemotherapy include bone marrow and mucosal surfaces (e.g., of the gastrointestinal tract), their depletion actually limits treatment options [9]. Nevertheless, by trying to find a classification scheme for different drug resistance mechanisms, one should not lose sight of one attribute they all have in common: Resistance to cancer treatment should ultimately be regarded as a consequence to somatic mutations and genomic plasticity associated with cancer.

With respect to overcoming drug resistance, there are also two completely different cases. If the resistance observed is limited to the actual compound class used and, for example, is caused by a single point mutation of the target or of an enzyme involved in metabolic activation, then classical drug design approaches leading to distinctly modified compounds are the strategy of choice. On the other hand, there are mechanisms leading to multiple drug resistance. These include mainly the inability to enter apoptosis and the overexpression of drug efflux pumps, such as P-glycoprotein. For the latter, coadministration of respective inhibitors has been proposed and several compounds were tested in clinical trials. Interestingly, none of them has been marketed till now. Both principal approaches will be discussed in detail in this chapter, whereby we will restrict ourselves to more recent concepts.

3. TARGETED CHEMOTHERAPY AND RESISTANCE

Targeted chemotherapy includes the use of either monoclonal antibodies (MAbs) or small-molecule drugs, both of which interfere with tumor-specific (or tumor–associated) proteins to alter their signaling. The use of 'targeted small-molecule therapeutics' is a consequence of the manifold new insights into the molecular/somatic alterations that are present in tumors. It has initiated a second wave of anticancer drug development that is characterized by rational design of low molecular weight compounds that should specifically target crucial effectors involved in cell proliferation, invasion and metastasis, angiogenesis, and apoptosis [4].

During the early days of targeted cancer chemotherapy, the novel compounds were designed to target one single crucial oncoprotein in a highly specific fashion. Nowadays, cancer has been recognized as a multifactorial disease where there is multilevel cross-stimulation among the targets along several pathways of signal transduction that finally led to neoplasia. Thus, by blocking only one of these pathways, the other pathways involved in the manifestation of cancer (which are not blocked) could act as salvage mechanism for the cancer cell. Thus, a second generation of so-called "multitargeted" chemotherapeutics aims at the interference of a multitude of these pathways/oncoproteins that is expected to result in a broader antitumor effect.

3.1. Resistance to Tyrosine Kinase Inhibitors (TKIs)

More than 25 years ago, tyrosine kinases (TKs) have been found to be involved in tumor development and progression. Approximately 90 TKs are encoded by the human genome, and their overexpression or abnormal activation—caused by somatic mutation(s) of these genes—is generally accepted to be a characteristic feature of many cancers. Most of the known existing TKs (approximately 60) are receptor TKs (RTKs), consisting of an extracellular, a transmembrane, and an intracellular domain. The remaining ones (approximately 30) target nonreceptor TKs, which are located in the cytoplasm of the cell (e.g., the SRC-, and the ABL-family of nonreceptor TK).

Currently, TK inhibitors (TKIs) have become part of standard chemotherapy for specific tumors (in combination with conventional chemotherapy and radiotherapy) due to a number of good qualities: their capability to stabilize tumor progression and their minimal side effects. Classical RTKIs are small molecular weight molecules that bind intracellular and competitively to the ATP-binding catalytic site of RTKs. In contrast, antibody-based drugs compete with the endogenous ligand for binding to the extracellular domain [10–13].

However, after initial remission, most of the patients quickly develop drug resistance against TKIs. In principle, mechanisms that contribute to the progression of the disease may be primary (intrinsic)—such as in some cases of imatinib resistance (see below)—or acquired. In particular, they include secondary mutations and/or overexpression of the targeted kinase by gene amplification, increased drug efflux, altered drug metabolism, and activation of downstream salvage pathways [13,14]. Since in a high fraction of patients with resistance to TKI the acquired mutations of the TK-encoding genes is determined as the predominant hindrance for successful therapy, a second generation of TKI was designed to be active against these new mutations [12].

3.1.1. Small-Molecule TKIs

Resistance to Imatinib Mesylate (GLEEVEC, Novartis) The discovery of the Philadelphia chromosome in 1960—a chromosomal abnormality that results from a reciprocal translocation that juxtaposes the BCR gene (on chromosome 22) with the ABL gene (on chromosome 9)—was considered as a major milestone in the treatment of chronic myelogenous leukemia (CML) [15]. In patients with CML the generated fusion gene BCR-ABL encodes an oncoprotein (a cytoplasmatic TK) with deregulated (constitutively active) TK activity. Imatinib is a promiscuous (but selective) small-molecule inhibitor of the BCR-ABL kinase used in the therapy of CML, gastrointestinal stromal tumors (GISTs), and hypereosiophilic syndrome [13]. Its poly-specificity is oriented toward the inhibition of additional TKs, including the platelet-derived growth factor receptor (PDGFR) and the stem cell factor receptor (c-KIT) [12].

CML patients who are treated with imatinib in some cases retain the Philadelphia chromosome in most of their bone marrow cells, showing a form of intrinsic drug resistance [14]. Others first respond, but secondary acquire point-mutations in BCR-ABL (approximately 50% of acquired imatinib-resistant patients). These mutations frequently occur within amino acid sequences that encode important structural features of the TK,

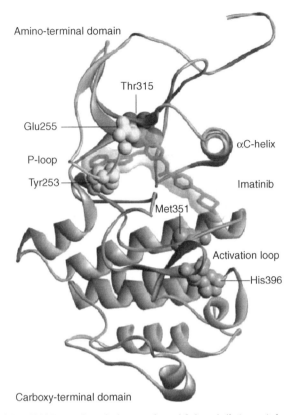

Figure 3. Structure of the Abl kinase domain in complex with imatinib (green): hotspots of frequent amino acid mutations are colored red (T315), yellow (p-loop, Y253, and E255), orange (activation loop, and H396), and pink (M351). Taken from Ref. [16]. (See color insert.)

such as the "gatekeeper" residue, the p-loop and the activation loop. Figure 3 illustrates the location of these mutational hotspots within the Abl kinase domain [16].

To a lesser extend gene amplification, overexpression at the mRNA or protein levels, drug efflux by ABC transporters, reduced drug-uptake by plasma sequestration, and activation of alternative pathways may play a role in the generation of imatinib resistance [17].

Thus, the development of so-called "second-generation" inhibitors (e.g., **dasatinib** [18], **nilotinib**, and **bosutinib** (Fig. 4)) yielded in effective inhibition of kinases, which harbor secondary mutations. Nilotinib binds—like imatinib—to the closed, inactive form of BCR-ABL but exhibits a 30-fold increased activity. In contrast, dasatinib binds to the catalytic domain in the biologically active conformation. As the mutations, which confer imatinib resistance, include those that lead to loss of the enzymes' ability to convert into the inactive form to allow imatinib binding, dasatinib is able to bind to the imatinib-resistant mutant enzyme. It has 300-fold increased activity (compared to imatinib) and a broad spectrum of activity against BCR-ABL, SRC kinases, c-KIT, and PDGFR [18].

Also for GIST, imatinib has become the standard therapy. Here, primarily activating mutations of the KIT and/or PDGF kinase domain(s) cause resistance and thus present limiting treatment obstacles. In such cases, a second-generation TKI approved for imatinib-refractory GIST, **sunitinib** is active against some of the KIT mutant cancer forms. Other forms of resistance that may affect (with a minor prevalence) the successful outcome of

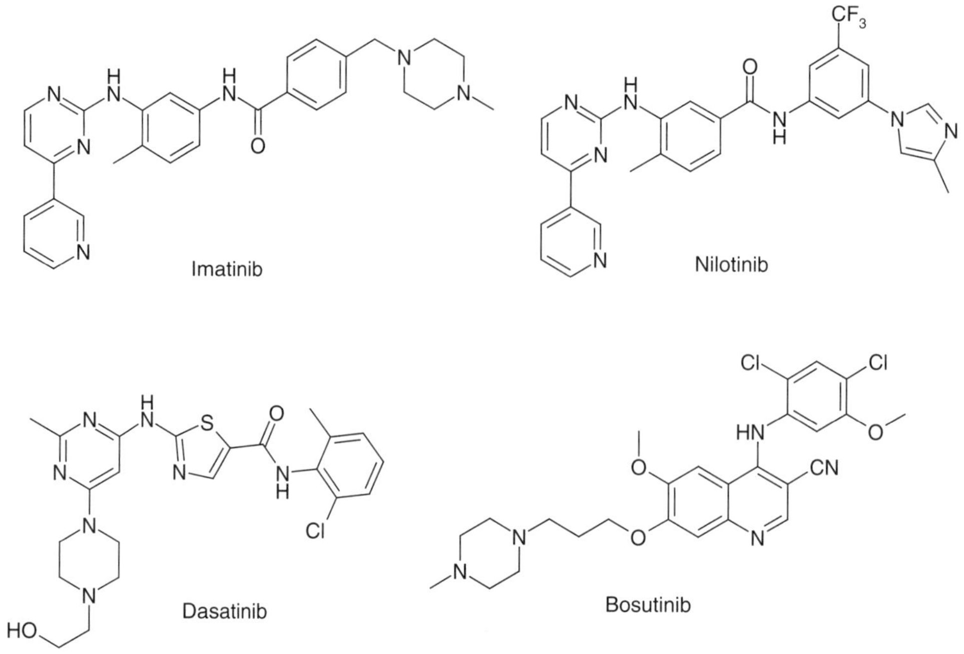

Figure 4. Chemical structures of selected tyrosine kinase inhibitors.

GIST treatment by imatinib are the same as being described for CML treatment.

However, the most frequent resistance mutation—the T315I mutation—is refractory against all clinically available TKIs. Recently, the Aurora kinase inhibitor **VX-680** (Merck) and the MAP kinase p38 inhibitor **BIRB-796** (Fig. 5) (Boehringer Ingelheim) have demonstrated inhibition of T315I mutants [19]. While BIRB-796 has been discontinued from development, more recent p38 inhibitors, such as VX-702 and SB-681323, are still under active development.

Alternatively, targeting the molecular chaperone heat shock protein 90 (HSP-90) or histone deacetylases (HDAC), or a combination of both could be effective in BCR-ABL mediated imatinib resistance. The geldanamycin derivative **17-allylamino-17-demethoxygeldanamycin** (17-AAG, National Cancer Institute), a HSP90 inhibitor, recently demonstrated good toxicity profile in phase I clinical trials [20,21]. **LAQ824** (Novartis), a cinnamyl hydroxamic acid analog inhibitor of HDAC, led to a decrease in BCR-ABL expression on the mRNA and protein level in patients with CML blast crisis. In addition, this HDAC inhibitor also diminished the association of HSP90 with BCR-ABL [22]. Moreover, in a recent study the combination of HDAC inhibitors with the Hsp90 antagonist 17-AAG shows synergistic effects and therefore may represent a novel strategy against Bcr-Abl$^+$ leukemias, including those resistant to imatinib [23].

Finally, there are efforts to target downstream signaling pathways of BCR-ABL (Ras/MAPK, Raf-1 and Mek, and the PI3K/Akt pathway). All these strategies and their promising drug candidates eventually alone—but most probably in combination with each other and with conventional chemotherapeutics—may postpone the apparently inalterable emergence of drug resistance and in that way optimize treatment outcome.

Resistance to EGFR Kinase Inhibitors Anti-EGFR drugs include small-molecule adenosine triphosphate-competitive inhibitors, as well as MAbs directed against the extracellular domain of the EGFR (see Section 3.1.2),

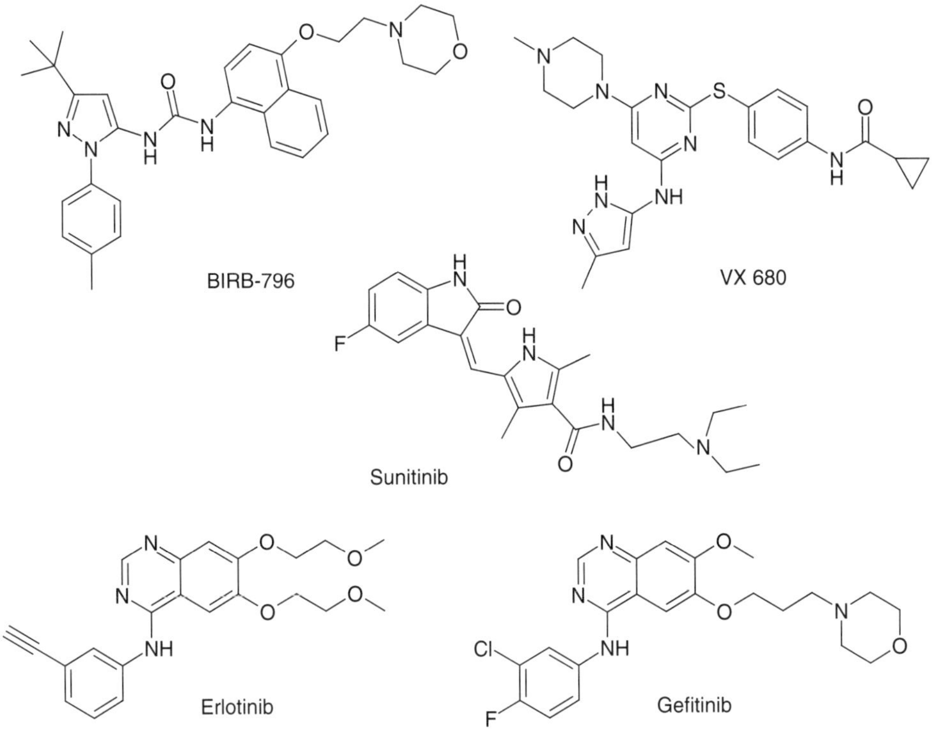

Figure 5. Chemical structures of kinase inhibitors.

which do not completely overlap regarding their mechanisms of action and their antitumor activity [12].

Small-molecule RTKI targeting EGFR (such as **erlotinib** and **gefitinib**) have proven to be effective in the therapy of nonsmall-cell lung cancer (NSCLC), and has resulted in cellular responses in patients with advanced pancreatic cancer [24], glioblastoma [25], colorectal carcinoma, head-and-neck cancer, and renal cell carcinoma [26]. Sensitivity to erlotinib or gefitinib can especially be expected in cases where activating mutations of the EGFR domain have driven the cancer development [27,28]. The most common mutation (ore than 40% of EGFR mutations in NSCLC), associated with responsiveness to erlotinib and gefitinib, is the L858R mutation [19].

However, an impressive initial response to the treatment with EGFR TKIs is often followed by resistance—in half of the cases, this happens due to a secondary mutation resulting in a threonine to methionine substitution at position 790 in the protein strand (T790M) [29]. In addition to frequent acquired T790M mutations in lung cancers, in a small number of cases, this mutation was detected prior to exposure to EGFR inhibitors (intrinsic) [30]. Interestingly, mutation of this conserved threonine residue located in the active site (also referred to as the "gatekeeper"), is also common in other structurally related kinases: BCR-ABL (T315I) and KIT (T670I) [19].

In an attempt to overcome T790M induced resistance, second-generation EGFR inhibitors (e.g., **HKI-272**, **EKB-569**; see Fig. 6) aim at irreversible blocking EGFR T790M signaling by binding to Cys773 [31]. Though, recently a novel secondary mutation (D761Y), as a response to anti-EGFR therapy, has been described [32].

Alternatively, to irreversible EGFR blockers there are a few strategies that may improve therapeutic efficacy by targeting other mechanisms of acquired resistance. Especially, downstream pathways of EGFR (phosphoinositol-3-kinase PI3K pathway) and amplification of the mesenchymal–epithelial

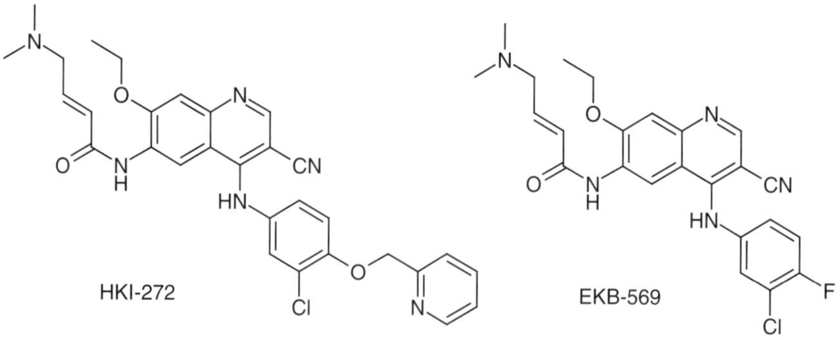

Figure 6. Chemical structures of selected EGRF inhibitors.

transition factor (MET) have been associated with secondary resistance to anti-EGFR therapy and may be interesting points of intersection [13].

Resistance to Angiogenesis Inhibitors Nowadays, the process of angiogenesis (neovascularization) during carcinogenesis is increasingly recognized as a rate-limiting secondary event. Thus, antiangiogentic therapy has been noted as an important milestone in cancer treatment and is becoming a component of a standard-of-care chemotherapy, especially for colorectal and renal cancers [33,34].

Above all, angiogenesis inhibitors which target the vascular endothelial growth factor (VEGF) proangiogenic signaling pathways demonstrated therapeutic efficacy. Most prominent representatives of these VEGF pathway inhibitors are **bevacizumab** (AVASTIN, Genentech/Roche), which is a ligand-trapping monoclonal antibody against the VEGF-RTK [see Section 3.1.2], **sorafenib** (NEXAVAR, Bayer/Onyx) and **sunitinib malate** (SUTENT, Pfizer)—two small-molecule multi-RTK inhibitors. Besides VEGFR inhibition sorafenib also interacts with the kinase activity of PDGFR, C-RAF and B-RAF, and c-KIT. In analogy, sunitinib targets additionally PDGFR, KIT, and fms-like TK3 [35]. Both RTK inhibitors are FDA approved for the treatment of advanced renal cell carcinoma (RCC). Sunitinib is also approved for patients with GIST who fail, or are intolerant of, therapy with imatinib mesylate [34].

However, though therapy with VEGF pathway inhibitors in many cases show demonstrable clinical benefit, sometimes progression of the tumor after initial response to the therapy is observed (adaptive resistance). Distinct (and partly interrelated) mechanisms that confer adaptive resistance may be basically attributed to the evasion of the antiangiogenic therapy. Hanahan et al. [34] essentially proposed the following mechanisms: revascularization by upregulation of alternative proangiogenic signaling factors (e.g., fibroblast growth factor (FGF)); decreased dependence on neovascularization by protection of the existing tumor vasculature (e.g., by increased pericyte coverage); and perivascular invasion. In addition, certain tumors also show intrinsic (preexisting) resistance to angiogenesis inhibitors, meaning that the pretherapeutic conditions do not allow any beneficial effect by the therapy.

As a strategy to circumvent such resistance mechanisms, the authors propose the combination of VEGF pathway inhibitors with anti-invasive and antimetastatic drugs, such as inhibitors of the proinvasive hepatocyte growth factor (HGF)—MET pathway, and drugs targeting the insulin-like growth factor 1 (IGF 1) receptor pathway. Secondly, simultaneous interference with parallel proangiogenic signaling pathways may be a fruitful strategy to prevent the emergence of resistance to antiangiogenic therapy [34].

3.1.2. Monoclonal Antibodies Targeting TK
The first MAb that was approved by the FDA for therapeutic use was **rituximab** (RITUXAN, Genentech) in 1997. This chimeric MAb binds to the pan-B-cell marker CD20 and is

used for the treatment of patients with relapsed or refractory low-grade or follicular, B-cell non-Hodgkin's lymphoma (NHL) [36]. Shortly later, in the year 1998, the humanized MAb **trastuzumab** (HERCEPTIN, Genentech) followed with an FDA approval for HER2 (ErbB2) overexpressing breast cancers. Successful clinical application of these two MAbs encouraged the assessment of further drug candidates into clinical trials. After the turn of the millennium, new MAbs were introduced in the clinics: **alemtuzumab** (FDA approval in 2001), which is directed against CD52 in patients with chronic lymphocytic leukemia, and **cetuximab** (FDA approval in 2004). The latter inhibits the EGFR tyrosine kinase (ErbB1) and is indicated in cases of metastatic colorectal cancer and head-and-neck tumors [37].

All this clinically approved MAbs directly attack the tumor cells by making use of different mechanisms of action: modulation of signaling pathways, antibody-dependent cellular cytotoxicity (ADCC), complement-dependent cytotoxicity (CDC), and immomodulation [37,38]. As a consequence, each mode of action may be responsible for one or multiple potential mechanism of resistance to antibody-based therapy. As examples, resistance mechanisms to rituximab and trastuzumab are pointed out below.

Rituximab leads to relevant clinical response to initial treatment in only approximately 50% of the patients. The manifold mechanisms that may contribute to rituximab resistance include Fc receptor polymorphism, CD20 modulation, and decreased ADCC. Consequently, potential mechanisms to overcome anti-CD20 resistance might be the use of engineered antibodies that facilitate Fc receptor binding, the use of cytokines (like interferon-α), radioimmunoconjugates and especially chemotherapy combinations [39]. In the case of trastuzumab, it is only approximately one-third of women under treatment that actually respond to monotherapy with this MAb [40]. Mechanisms of resistance related with failure of trastuzumab treatment are the activation of the PI3K/Akt pathway or via loss of PTEN function, a tumor surpressor and thus negative regulator of Akt. Here, the combination or sequential therapy with another chemotherapeutic drug (such as lapatinib) seem to be the most promising strategies [41].

In addition to the development of these MAbs, which all target membrane proteins in tumor cells, there has been effort to identify other targets in the microenvironment associated with the carcinogenic event. The approval of **bevacizumab** (AVASTIN, Genentech) in 2004 for the treatment of metastatic colorectal cancer (and later also for NSCLC and metastatic HER2-negative breast cancer) is an example for such a strategy. It is an antiangiogenic compound directed against the VEGFR. Resistance mechanisms that are associated with antiangiogenic therapy have been discussed in the Section 3.1.1 [34].

3.2. Endocrine Resistance in Breast Cancer

Approximately 70% of breast tumors express estrogen receptors (ERs), making them accessible for antiestrogens such as **tamoxifen** (Fig. 7). This nonsteroidal selective estrogen receptor modulator (SERM) blocks the ERs of tumor cells, but has an agonistic effect in other organs/tissues (liver, uterus, bone cells). In contrast, aromatase inhibitors (AI, estrogen synthase inhibitors) such as **exemestane**, **anastrozole**, and **letrozole** reduce the estrogen levels also in peripheral tissues [42]. The sequential use of exemestane demonstrated to be of advantage in the cases tamoxifen resistance. Such endocrine-insensitive states usually appear after 2–3 years of treatment with tamoxifen [43]. But not only partial estrogen receptor agonists but also "pure" antiestrogens (ER downregulators)—such as **fulvestrant**—lead to the acquisition of an endocrine-resistant state and an increase in their migratory and invasive capacity *in vitro* after chronic exposure [44].

Besides initiation and regulation of breast tumor growth by steroid hormones and ERs, also peptide growth factors and growth factor receptors (EGFR, HER2) are involved in this process. This is demonstrated by the fact that up to 25% of breast cancers are HER2-positiv and thus respond to trastuzumab treatment. It is generally believed that there exist tight interactions between these two signaling pathways. Especially cross-talk between ER,

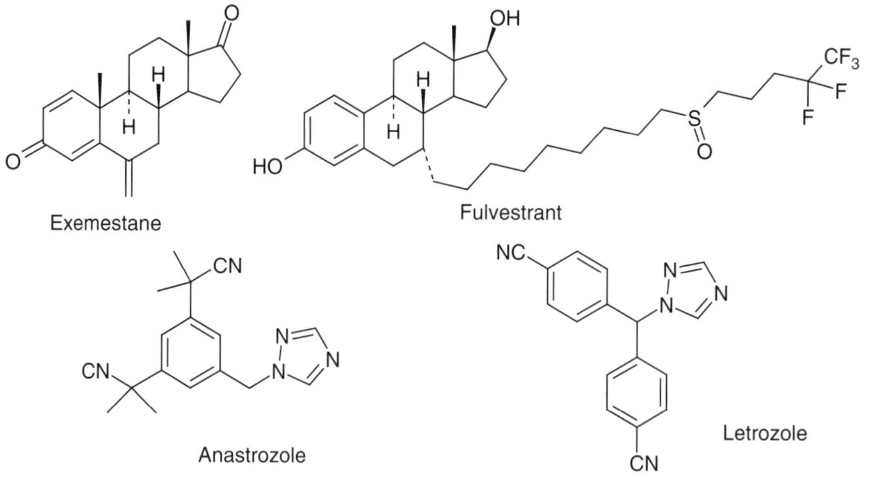

Figure 7. Chemical structures of selected ER inhibitors.

HER2, p38 and ERK at the time of tamoxifen resistance indicates that there are novel potential targets to overcome this resistance [45]. However, very recent investigation suggests no direct connection between ER and HER2 at a transcriptional level in the mediation of tamoxifen resistance. Instead, the authors postulate that PAX2 is a central key player in the ER-mediated repression of ErbB2. Accordingly, PAX2 mutations may be the main reason for increased ErbB2 expression and thus potentially determine response to tamoxifen. In addition, the linkage of the two breast cancer types (ER-positiv and HER2/ErbB2-positive) by the mechanism of repression of ErbB2 by ER-PAX2 suggests that evasion of this blocking could make an ER-positive tumor transform into an aggressive HER2-positive tumor [46].

To date, promising strategies to overcome resistance to endocrine therapy (besides the ones mentioned before) include the combination of trastuzumab with fulvestrant for treatment of cancers that express HER2 and ER, the simultaneous use of EGFR TKIs (like gefitinib) or other growth factor pathway inhibitors, targeting the PI3K/Akt pathway (which is often active in breast cancer), disruption of the ER function (e.g., farnesyltransferase inhibitors), and combination with angiogenesis inhibitors [42].

4. MULTIPLE DRUG RESISTANCE

Based on the knowledge about the multiple mechanisms that may confer multidrug resistance to cancer chemotherapy, a multitude of strategies has been evolved to deal with this problem. In this chapter, we will review only two of the most prominent approaches: targeting apoptosis pathways and reversal of resistance that is attributed to the function of ABC transporters. Others include for instance modulation of the methylation status of crucial proteins in tumorigenesis, or depletion of intracellular glutathione levels [47].

4.1. Targeting Proteins Involved in Apoptosis

Apoptosis—or programmed cell death—is an intrinsic mechanism of cells that leads to cell attrition. A balance between apoptosis and cell proliferation is necessary to maintain the tissue homeostasis. In many cancer patients, defects in the apoptosis pathways lead to the neoplastic transformation. Thus, many chemotherapeutic anticancer agents act by inducing apoptosis. However, genetic alterations in the apoptosis pathway may also result in apoptosis resistance [33,48].

Targeting this pathway to modulate drug resistance includes several strategies that take into account the potential different sites for therapeutic intervention. Death ligands (e.g., TRAIL) bind to death receptors on the cell surface that activate the caspase cascade and in that way initiate apoptosis [49]. Another approach is aimed at the downregulation of antiapoptotic proteins, such as Bcl-2 or Bcl-X_L. Here, the use of antisense oligonucleotides (ASOs) [50], small molecules, and RNA interference [51] are offering interesting possibilities.

Direct activation of caspases may be fulfilled by peptides that target the IAPs (inhibitors of apoptosis, e.g., survivin [52]). The conserved IAP protein family suppresses apoptosis by blocking caspases—proteins like Smac/DIABLO eliminate the inhibitory effect of IAPs and thus (re)activate apoptosis [53].

Finally, although controversial, there have been many efforts to restore wild-type p53 in tumor cells, a tumor surpressor gene that is often mutated in human cancer. Several small-molecule drugs have been demonstrated to restore a wild-type status to cells harboring mutant p53 and thus activate apoptosis [54].

4.2. Targeting Resistance Mediated by ABC Transporters

Regarding the different drug resistance mechanisms described in Chapter 2, the one that we most commonly are faced with in the laboratory is the enhanced extrusion of hydrophobic cytotoxic drugs by ABC transporters [55]. This ATP-binding cassette superfamily is the largest transporter gene family—consisting of 48 known human ABC genes, which can be divided into seven distinct subfamilies (designated A through G). Each functional unit of an ABC transporter consists of two cytoplasmatic, nucleotide binding domains (NBDs) and two transmembrane domains (TMDs) made up of α-helices. Driven by binding and hydrolysation of ATP at the NBDs, a wide variety of substrates are translocated across plasma membranes: sugars, amino acids, metal ions, peptides/proteins, hydrophobic compounds, and metabolites [56,57].

As these proteins are constitutively expressed in tissues that are important for absorption, metabolism and elimination (e.g., lung, GUT, liver, kidney), as well as in sanctuary site tissues (e.g., blood–brain barrier, placenta), their role as central key players in tissue defense has been increasingly recognized [58].

Besides active extrusion of xenobiotics out of normal cells, overexpression of ABC transporters in the cell membranes of multidrug-resistant tumor cells cause an active outward transport of chemotherapeutic agents. The emergence of multidrug resistance (MDR) is a major hindrance to the successful therapy of various forms of malignant and infectious diseases—not only in cancer treatment but also in other treatments. The concept of MDR describes simultaneous resistance (cross-resistance) of cancer cells toward a broad spectrum of structurally and also mechanistically unrelated cytotoxic drugs with different modes of action [59]. This implicates that after initial administration of a certain drug/drug combination, resistance may arise to agents to which the patient has not been exposed previously.

4.2.1. P-Glycoprotein (ABCB1)
In 1976, the first ABC transporter was discovered by Juliano and Ling [60] and called P-glycoprotein (P-gp, MDR1), because this cell surface glycoprotein was found in mutant cells displaying altered drug permeability (P stands for permeability). It was for the first time that one single protein was linked to a huge number of structurally diverse compounds to which it conferred resistance. However, not until 1986 the human *mdr1* gene was isolated and the evidence of its ability to confer alone the drug resistant phenotype was provided [61].

The broad spectrum of substances transported by P-gp contains a wide variety of natural product toxins such as vinca alkaloids, anthracyclines, epipodophyllotoxins, taxanes, and many more [55]. This phenomenon (often observed in ABC transporters) of a structurally unrelated ligand recognition pattern is called promiscuity or multispecificity. We prefer the latter term, as these transporters still show specificity toward distinct structural scaffolds and it has been demonstrated that predictive *in silico* models may be obtained [62].

Only 5 years after the discovery of ABCB1, verapamil—a calcium channel blocker—was found to block P-gp mediated transport and thus revert vincristin resistance in tumor cells [63]. After that an intensive research in the field of potential inhibitors of P-gp followed to invert drug resistance and reestablish sensitivity to standard therapeutic regimens. "First-generation" P-gp inhibitors include compounds that were already approved for other indications: two other classes of calcium channel blockers, benzothiazepines and 1,4-dihydropyridines, phenothiazines, quinine, tamoxifen, and cyclosporine A. Due to inherent cardiac or other toxicities of these drugs in the doses required for modulation of P-gp function [64], a "second generation" of inhibitors should fulfil the need of avoiding those limiting side effects. Examples are biricodar (VX-710), a derivative of the macrocyclic antibiotic FK-506, and the cyclosporine D analog Valspodar (PSC-833), which is able to block P-gp without having immunosuppressive effects or having an effect on Ca ion channels. However, also this class of agents has not reached the market due to interference with cytochrome P450 3A4 that often leads to limited drug clearance and therefore causes toxic plasma concentrations [65]. The "third generation" of MDR modulators was designed to satisfy the need of low pharmacokinetic interaction: laniquidar, tariquidar, zosuquidar, elacridar, and ONT-093 (Fig. 8). Although, the clinical studies of the majority of these compounds are not terminated or fully analyzed to date, it seems that also this generation of MDR modulators cannot meet the expectations. After more than 20 years and numerous clinical studies, there is no definitive proof that a PGP inhibitor effectively reverses drug resistance in humans. This is mainly due to the fact that the pharmacokinetic interactions observed with these agents have made it difficult to interpret efficacy.

Nevertheless, the search for more appropriate MDR modulators is going on, but now the focus is to target multiple ABC transporters (such as P-gp and ABCG2) at once. Though, one major drawback of such an approach may be an even greater potential of treatment limiting side effects.

Not only P-gp but also ABCG2 and MRP1 (multidrug resistance protein 1, ABCC1) may be determined as the most important proteins involved in the appearance of a MDR phenotype in cancer. Although, only modulators of these three ABC transporters have been evaluated in clinical trials, some emerging studies revealed the involvement of a lot more members of this superfamily: ABCA2, ABCB4 (MDR3), ABCB11 ("sister of P-gp"), ABCC2 (MRP2), ABCC3 (MRP3), ABCC6 (MRP6), and ABCC10 (MRP7) [55].

4.2.2. ABCC1 (Multidrug Resistance Protein 1) MRP1 was the second ABC transporter that was found to be involved in MDR [66]. It is highly expressed in stomach, lung, and brain and its physiological role as a high-affinity anion transporter makes it capable to translocate glutathione conjugates, such as LTC4, as well as glucuronate and sulphate conjugates [67,68]. With regard to its resistance profile, there is a substantial overlap with that of P-gp. It may be responsible for resistance to natural product drugs, such as vinca alkaloids, anthracyclines, and epipodophyllotoxines, and for mitoxantrone- and methotrexate-resistance. In contrast, high levels of resistance to taxanes or bisantrene have not been observed [69].

Like in the case of P-gp, a lot of effort has been made to find potential inhibitors of MRP1 to inverse MDR, but research is still in its infancy. Advances in the discovery of ABCC1 inhibitors have been reviewed recently [70].

4.2.3. ABCG2 (Breast Cancer Resistance Protein, MXR) The ABCG2 gene was first isolated from a breast cancer cell line and therefore called the breast cancer resistance protein (BCRP) gene. In human, it is highly expressed in the placenta, liver, intestine, kidney, in the lactating mammary gland, at the blood–brain barrier, and in hematopoietic stem cells [71]. In contrast to P-gp and the ABCC family, ABCG2 is a half-transporter forming homodimers to obtain functional units.

The range of substrate recognition of ABCG2 is almost as broad as that of P-gp. Moreover, many of the transported agents do simultaneously interact with P-gp [72]. However, results obtained from pharmacophore modeling and QSAR studies for the class of propafenones indicate that ABCG2 may be

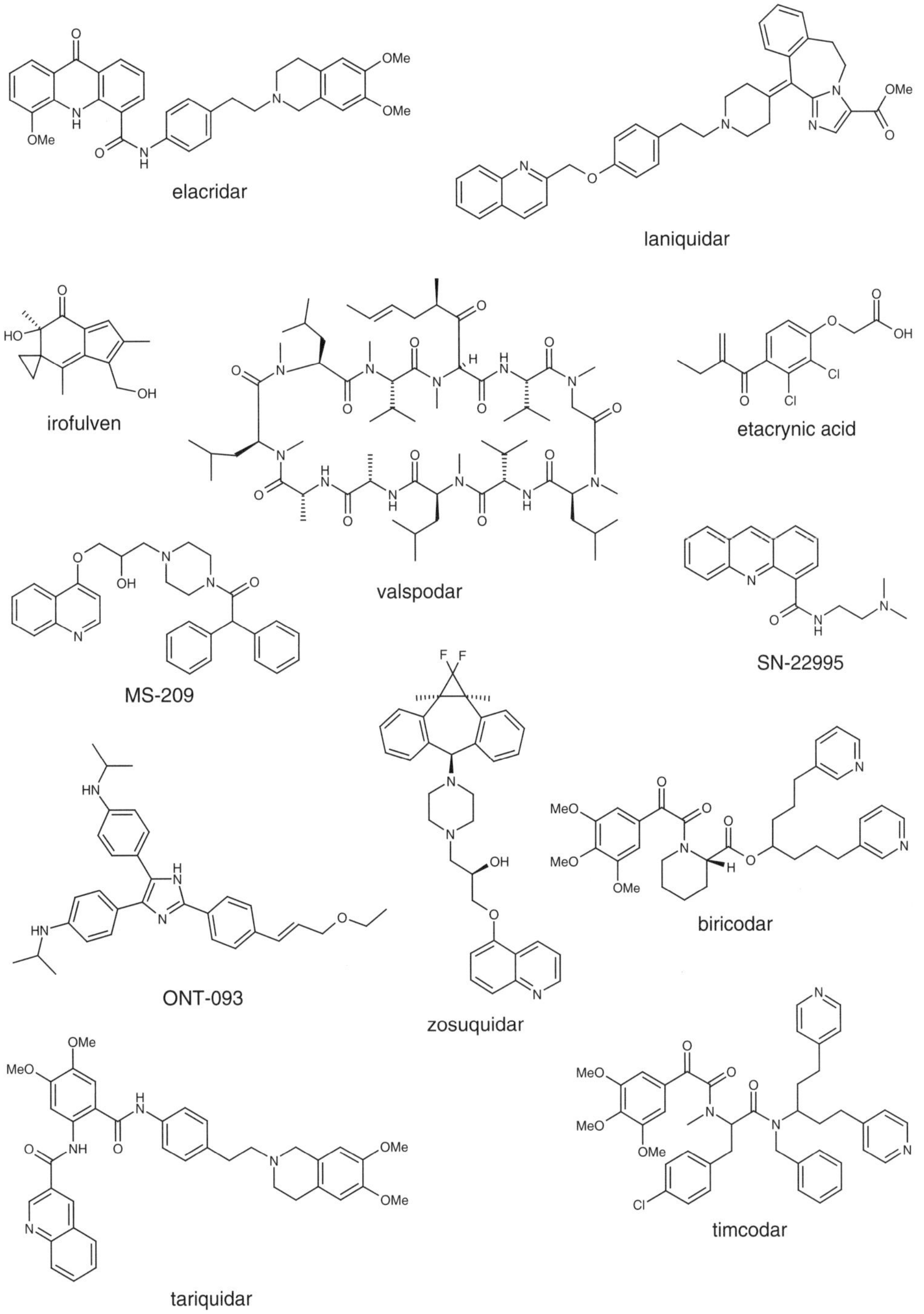

Figure 8. Chemical structures of selected P-gp inhibitors.

more tolerant to structural modification than ABCB1 [73].

4.2.4. SAR- and QSAR Studies on Inhibitors of ABC Transporters

In lead optimization programs, numerous QSAR studies on structurally homologous series of compounds have been performed. Especially, verapamil analogs, triazines, acridonecarboxamides, phenothiazines, thioxanthenes, flavones, dihydropyridines, propafenones, and cyclosporine derivatives have been extensively studied, and the results are summarized in several excellent reviews [74,75]. These studies pinpoint the importance of H-bond acceptors and their strength, the distance between aromatic moieties and H-bond acceptors as well as the influence of global physicochemical parameters, such as lipophilicity and molar refractivity. Systematic quantitative structure–activity relationship studies have been performed mainly on phenothiazines and propafenones [76]. The latter have been carried out using Hansch- and Free-Wilson analyses [77], hologram QSAR, CoMFA, and CoMSIA studies [78] as well as nonlinear methods [79] and similarity-based approaches [80]. Hansch-type correlation analyses normally lead to excellent correlations between lipophilicity and pIC_{50} values within structurally homologous series of compounds. However, this is not surprising as the interaction of ligands with P-gp is supposed to take place in the membrane bilayer. Thus, lipophilicity of the compounds triggers their concentration at the binding site rather than being a parameter important for ligand–protein interaction. However, Pajeva and Wiese demonstrated for both a series of phenothiazines and thioxanthenes [81] and for a subset of our propafenone-based library [82] that lipophilicity should also be regarded as a space directed property.

Although, all these QSAR studies give clear individual pictures and yield predictive models, the attempt to define distinct structural features necessary for high P-gp inhibitory activity leads to rather general features. Strong inhibitors are characterized by high lipophilicity (and/or molar refractivity) and possess at least two H-bond acceptors. Other features, such as H-bond donors, may act as additional interaction points. Furthermore, some steric constraints seem to apply in the vicinity of pharmacophoric structures.

This picture has been supported by various pharmacophore modeling studies, most comprehensivly studied by the group of Ekins [83,84]. They used several different training sets, such as inhibitors of digoxin transport, inhibitors of vinblastine binding, inhibitors of vinblastine accumulation, and inhibition of calcein accumulation. Not really surprising, all four models retrieved showed differences both in the number and type of features involved and in the spacial arrangement of these features. A consensus model, which correctly ranked all four data sets, consists of one H-bond acceptor, one aromatic feature, and two hydrophobic features. This further strengthens the hypothesis that toxins might bind to P-gp at different, but overlapping sites. This was also stressed out by Garrigues et al., who calculated the intramolecular distribution of polar and hydrophobic surfaces of a set of structurally diverse P-gp ligands and used the respective fields for superposition of the molecules. This led to the identification of two different, but partially overlapping binding pharmacophores [85].

We used a CATALYST model based on propafenone-type inhibitors for an *in silico* approach to identify new inhibitors of P-gp. The training set comprised 27 propafenone-type inhibitors of daunorubicin efflux and the model derived included one H-bond acceptor, two aromatic features, one hydrophobic area and one positively charged group. The model was validated with an additional 81 compounds from our in-house data set and subsequently used to screen the World Drug Index. After applying an additional shape filter, 32 structurally diverse hits were retrieved. Nine out of these 32 compounds have already been described as P-gp inhibitors [86]. Thus, it is rather likely that the other compounds selected also bind to P-gp.

4.2.5. Structural Aspects of ABC Transporters

ABC-transporters are membrane-spanning proteins, so for a long time there were no X-ray structures available for human transporters. Thus, the publication of the first bacterial homolog structure of a full-length ABC-transporter, the lipid A transporter MsbA from *Escherichia coli* representing an open

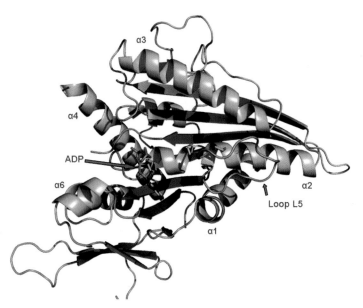

Figure 2 (Chapter 6). X-ray crystal structure of KSP-ADP complex in secondary structure rendering illustrating α-helical (blue), β-sheet (magenta), and unstructured (beige) regions. The bound molecule of ADP is shown (left of center) in stick form, colored by element [24].

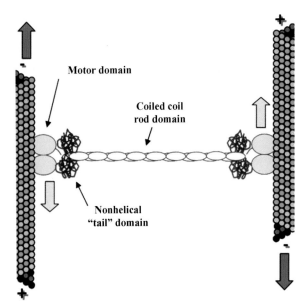

Figure 3 (Chapter 6). Schematic of tetrameric KSP bound to two oppositely polarized MTs. Four KSP motor domains (yellow) are paired at opposite ends of the tetramer and connected to each other via coiled coil and nonhelical (gray) domains. KSP motor domains process along MTs (green/blue) toward plus (+) ends (yellow arrows) and, as they do so, cause MTs to slide in opposite directions (teal arrows) forcing minus (−) ends to move apart.

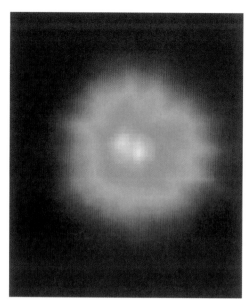

Figure 5 (Chapter 6). Micrograph of a monopolar spindle phenotype caused by KSP inhibition. Centrosomes are shown in orange, microtubules are shown in green, and chromosomes are shown in blue.

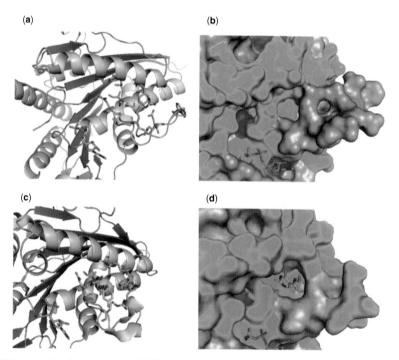

Figure 8 (Chapter 6). Comparison of KSP-ADP ((a) and (b)) and KSP-ADP-monastrol (1) ((c) and (d)) complexes. (a) In the inhibitor-free X-ray crystal structure, several side chains of helices $\alpha 2$ and $\alpha 3$ reside within the allosteric binding pocket, and loop L5 is unstructured. (b) The resulting protein surface (colored by secondary structure) does not reveal the existence of the monastrol binding pocket. (c) When bound to monastrol, the side chains of helices $\alpha 2$ and $\alpha 3$, particularly Y211, move to accommodate the inhibitor while loop L5 folds inward to allow its backbone and side chains, especially W127, to interact with the monastrol's phenol group. (d) The protein surface view clearly shows a relatively deep and well-formed binding pocket for monastrol.

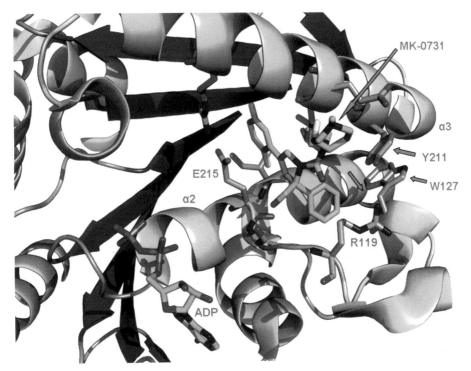

Figure 16 (Chapter 6). Closeup of interactions between Merck inhibitor MK-0731 and KSP-ADP. Included are the key side chains of helices $\alpha 2$ and $\alpha 3$ and loop L5 that interact with MK-0731 to form the induced fit pocket. As with monastrol, side chains Y211, W127, and R119 and the backbone of loop L5 shift relative to the unbound KSP-ADP structure.

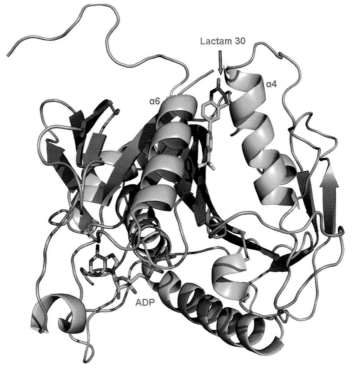

Figure 22 (Chapter 6). Model of lactam **30** (cyan) bound to KSP at the interface of helices $\alpha 4$ and $\alpha 6$. The nucleotide binding site is illustrated here with the structure of ADP (gray) [153].

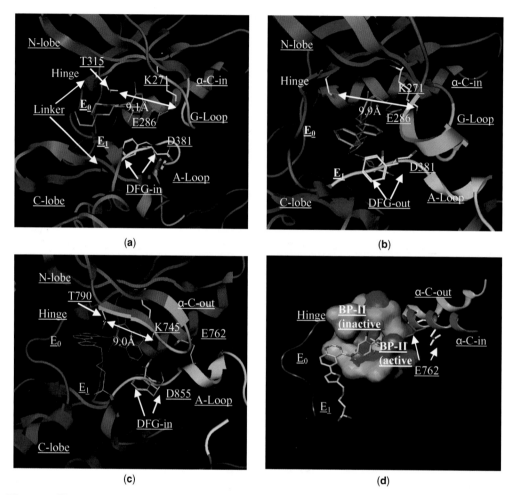

Figure 2 (Chapter 9). (a) Ribbon representation of the c-Abl catalytic domain in the active DFG-in and αC-in conformation (G-loop distorted) in complex with dasatinib (PDB code: 2GQG). (b) Ribbon representation of the c-Abl catalytic domain in the inactive conformation (DFG-out and αC-in) in complex with imatinib (PDB code: 2HYY). (c) Ribbon representation of the EGFR catalytic domain in the inactive conformation (DFG-in and αC-out) in complex with lapatinib (PDB code: 1XKK). (d) Superposition of the inactive (green; in complex with lapatinib) (PDB code: 1XKK) and active (red; PDB code: 1M17) clefts. The 3-fluorobenzyl group of lapatinib binds into the BP-II pocket (green), but clashes with E766 and M762 in the active conformation (red).

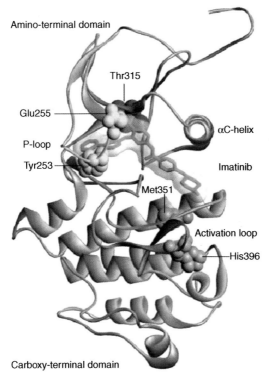

Figure 3 (Chapter 10). Structure of the Abl kinase domain in complex with imatinib (green): hotspots of frequent amino acid mutations are colored red (T315), yellow (p-loop, Y253, and E255), orange (activation loop, and H396), and pink (M351). Taken from Ref. [16].

Figure 9 (Chapter 10). Protein homology model of P-glycoprotein in the energized state. (Taken from Ref. [100].)

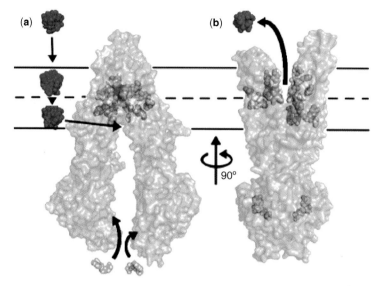

Figure 10 (Chapter 10). Proposed substrate translocation pathway for P-glycoprotein. Taken from Ref. [102].

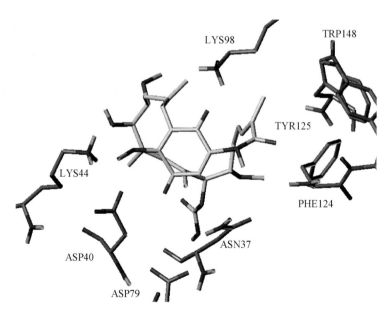

Figure 4 (Chapter 11). Cocrystal structure of GDA (yellow) bound to yeast Hsp90.

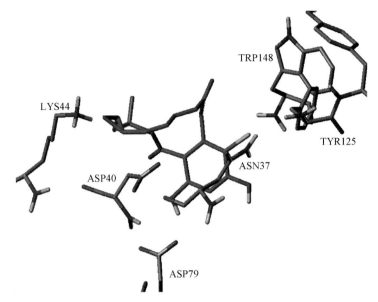

Figure 16 (Chapter 11). Cocrystal structure of RDC (magenta) bound to Hsp90.

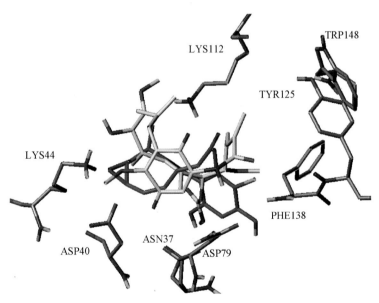

Figure 21 (Chapter 11). Superimposed cocrystal structures of radicicol (magenta) and geldanamcyin (yellow) bound to yeast Hsp90.

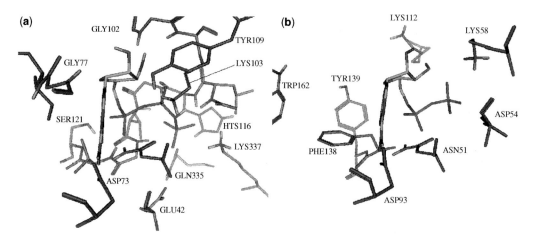

Figure 42 (Chapter 11). (a) Cocrystal structure of ADPNP bound to DNA Gyrase. (b) Cocrystal structure of the Hsp90 N-terminus bound to ADP.

conformation [87] immediately gave rise to protein homology modeling attempts. Also the subsequent appearance of MsbA from Vibrio cholerae showing the closed state [88] as well as the structure of MsbA from *Salmonella typhimurium*, which resembled the posthydrolytic conformation [89] seemed to provide versatile starting points for protein homology modeling of ABCB1 in various states of the catalytic cycle. However, a few years later the structures had to be withdrawn due to errors in data postprocessing [90], which also rendered the hitherto published homology models quite useless. Recently, both the corrected structures of MsbA [91] as well as two structures of a putative multidrug transporter from methicillin resistant *Staphylococcus aureus* (Sav1866) have been published [92,93]. Especially the structure of Sav1866, which exhibits a quite reasonable resolution (3.0 Å), has been immediately used for the generation of models of ABCB1 and ABCC5 [94–96]. The homology model from Globisch et al. was subsequently used for identification of putative binding sites for ligands. The authors were able to visualize multiple binding sites, which are mostly located in the transmembrane region of the model. However, Sav1866 has been crystallized in the posthydrolytic state, which might limit the usability in structure-based drug discovery approaches. We recently presented a Sav1866-based homology model that was refined in a data-driven structural modification approach to fit a vast collection of cross-linking data [97] (Fig. 9).

Very recently, Becker et al. published homology models of ABCB1 in different catalytic states by additionally utilizing the updated MsbA template structures [98]. They also performed docking of four different ligands into the nonenergized (nucleotide-free model) state and were able to show that all four ligands exhibit interactions with residues shown to be important. Although these first

Figure 9. Protein homology model of P-glycoprotein in the energized state. (Taken from Ref. [100].) (See color insert.)

attempts on structure based modeling of ABCB1-drug interactions seem to be quite promising, it has to be kept in mind that the templates currently show only a small sequence homology in the drug-binding regions (TMDs) and the only nonenergized structure of MsbA currently available does not resemble a full atom model but only represents the Cα-trace in a resolution >5 Å.

However, very recently the group of Geoffrey Chang published the first structure of P-glycoprotein (mouse P-gp) with a resolution of 3.8 Å [99]. The apo form shows an internal cavity of approximately 6000 Å^3 with a 30 Å separation of the two nucleotide-binding domains. The authors also solved two additional structures with the cyclic peptide inhibitors QZ59-RRR and QZ59-SSS bound to the protein. In both cases, the compounds are sandwiched between TM helices 6 and 12, which have been demonstrated various times as being important for drug binding. Combining now all the structure-based knowledge available, the substrate transport may be modeled as given in Fig. 10. This first structure of P-glycoprotein definitely will remarkably influence the whole field and we are convinced that this will form the basis for subsequent structure-based design studies.

5. CANCER STEM CELLS AND MDR

The important role of "stem cell-like" cancer cells in tumors of the hematopoietic system (like acute myeloid leukemia) has been suspected since the 1950. However, their implication in the emergence of solid tumors became evident only in 2003 [100]. According to the cancer stem cell (CSC) hypothesis, a small fraction of cancer cells—the so-called cancer stem cells and vast majority of the so-called "side population" (SP)—has the ability to self-renew and thus is believed to represent the pool of cells of origin that build up a tumor. However, by which factors this self-renewing capacity is determined and how cancer stem cells arise is not yet understood [59]. On principle, there are two mechanisms by which pluripotent stem cells in tumors might originate: either they could arise from organ stem cells that undergo malignant transformation or a more differentiated tumor cell acquires self-renewal capacity [101].

The term "side population" refers to the cells that are detected to one side of the majority of cells on a density plot during analysis by flow cytometry. SP cells do not accumulate the fluorescent dyes Hoechst 33342 and rhodamine 123, and thus can be separated from the vast mass of nontumorigenic cancer cells.

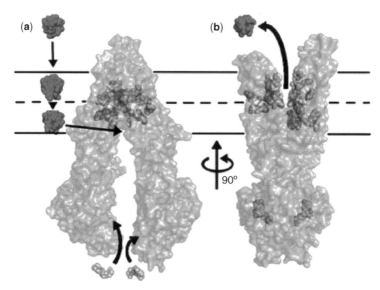

Figure 10. Proposed substrate translocation pathway for P-glycoprotein. Taken from Ref. [102]. (See color insert.)

Even though in the SP compartment there are stem cells and nonstem cells, strong evidence exists that predominantly the CSCs build up the SP. Besides, the SP phenotype—and also normal stem cells—are characterized by high levels of ATP-binding cassette transporters (mainly ABCB1 and ABCG2) that are expressed at the cell surfaces and thus lead to extrusion of the fluorescent dyes mentioned above [59]. It seems that none of these transporters is necessary for stem cell growth or maintenance, as knockout mice (who lack the corresponding genes) demonstrated to be viable, fertile, and have normal stem cell compartments [102]. However, through the expression of several ABC transporters, CSCs are able to confer resistance to drugs. Besides, they share other properties of normal stem cells: relative quiescence, active DNA-repair capacity, and resistance to apoptosis. All these characteristics are leading to a long lifespan of these cells and they provide evidence for the possibility of a so-called "cancer stem cell model of drug resistance." In this model, it is supposed that there exists a built-in population of drug-resistant CSCs, which (in contrast to the variably differentiated tumor cells) survive chemotherapy and repopulate the tumor. Alternatively, both compartments, CSCs and the vast mass of tumor cells, might show resistance to chemotherapy in some cases. This type of drug resistance corresponds to what we call "intrinsic resistance" [59].

Future considerations on drug resistance should take into account what we learned about CSCs within the last few years. Especially regarding the design of new drugs for its reversal/evasion, this new concept of how resistance might arise offers new therapeutic opportunities (see Chapter 6).

6. PERSPECTIVES FOR NEW STRATEGIES TO OVERCOME DRUG RESISTANCE

Nowadays, an ongoing effort in oncology concentrates on the development of novel drugs that target cancer cells in a new manner, and thus probably avoid the emergence of drug resistance. One task is to target key components of the cell-cycle machinery such as polo-like kinase 1 (PLK1) [103] and cyclin dependent kinases [104]. An alternative strategy focuses on blocking protein–protein interactions such as the regulation of p53 expression via MDM2/MDMX binding [105]. Regarding resistance mediated by P-glycoprotein, a new approach involves the interference with one of the regulatory steps in P-gp expression (ecteinascidin 743 under clinical trial) either than directly blocking the protein [106].

Additionally, it seems useful to reconsider drug application schemes. Especially combination therapies have proven to be useful in many cases. However, the question arises if the sequential application or the upfront administration of a combination of drugs (especially TKIs) under certain circumstances might be more beneficial for treatment outcome and retardation of the resistant phenotype [12].

A quantum leap still has to be surmounted in the fields of cancer diagnosis and the dedication of medical imaging technologies. New predictive biomarkers are urgently needed to classify patients correctly into responder/nonresponder to a special treatment (e.g., with trastuzumab) [107]. Furthermore, a genome wide expression profiling could give insights into the regulatory networks that are involved in the acquisition of drug resistance and thus lead to novel treatment strategies [13].

In the field of molecular imaging a step forward has to be made regarding the implementation of technologies such as nuclear magnetic resonance and optical imaging technologies. Thereby, not only the localization of the tumor but also the expression and activity of specific proteins and biological processes (such as apoptosis, and angiogenesis) can be visualized. Thus, information about the tumor behavior or therapy response is obtained [108].

Since cancer stem cells can be isolated from many tissues, purified and propagated, targeted therapies have the potential to be further improved. However, new cancer models are needed where future drugs can be directly tested for their ability to kill CSCs [59]. According to the stem cell hypothesis, new strategies should be focused on targeting these CSCs. Novel therapeutic opportunities include dual ABCB1/ABCG2 inhibitors (e.g., tariquidar), antibodies directed against ABC inhibitors, stem cell inhibitors like cyclopamine (targets the Hedgehog-Patched pathway), and immunotherapy [59].

ACKNOWLEDGMENTS

The authors gratefully acknowledge the financial support provided by the Austrian Science Fund, Grant No. F3502.

REFERENCES

1. Ehrlich P, Morgenroth J. Die Seitenkettentheorie der Immunität. Anleitung zu hygienischen Untersuchungen: nach den im Hygienischen Institut der königl. Ludwig-Maximilians-Universität zu München üblichen Methoden zusammengestellt 1902;3.Aufl:381–394.
2. Ehrlich P. Aus Theorie und Praxis der Chemotherapie. Folia Serologica 1911;7:697–714.
3. Satter H. Paul. Ehrlich Begründer der Chemotherapie. München: Leben–Werk–Vermächtnis; 1962.
4. Strebhardt K, Ullrich A. Paul Ehrlich's magic bullet concept: 100 years of progress. Nat Rev Cancer 2008;8(6):473–480.
5. Goodman LS, Wintrobe MM, Dameshek W, Goodman MJ, Gilman A, McLennan MT. Landmark article Sept. 21, 1946: Nitrogen mustard therapy. Use of methyl-bis(beta-chloroethyl) amine hydrochloride and tris(beta-chloroethyl)amine hydrochloride for Hodgkin's disease, lymphosarcoma, leukemia and certain allied and miscellaneous disorders. JAMA 1984;251(17):2255–2261.
6. Kroll DJ. Circumventing antineoplastic drug resistance: when tumor cells just say "no" to drugs. Am J Pharm Educ 1995;59(2):184–191.
7. Hazlehurst LA, Landowski TH, Dalton WS. Role of the tumor microenvironment in mediating de novo resistance to drugs and physiological mediators of cell death. Oncogene 2003;22(47):7396–7402.
8. Lu C, Shervington A. Chemoresistance in gliomas. Mol Cell Biochem 2008;312(1):71–80.
9. Chu E, DeVita VT. Principles of Medical oncology, In: DeVita VT, Hellman S, Rosenberg SA, editors. Cancer: Principles and Practice of Oncology. Lippincott: Philadelphia; 2001. p. 337–345.
10. Steeghs N, Nortier JWR, Gelderblom H. Small molecule tyrosine kinase inhibitors in the treatment of solid tumors: an update of recent developments. Ann Surg Oncol 2007;14 (2):942–953.
11. Krause DS, Van Etten RA. Tyrosine kinases as targets for cancer therapy. N Engl J Med 2005;353(2):172–187.
12. Baselga J. Targeting Tyrosine Kinases in Cancer: The Second Wave. Science 2006;312 (5777):1175–1178.
13. Engelman JA, Settleman J. Acquired resistance to tyrosine kinase inhibitors during cancer therapy. Curr Opin Genet Dev 2008;18 (1):73–79.
14. Shannon KM. Resistance in the land of molecular cancer therapeutics. Cancer Cell 2002;2 (2):99–102.
15. Sawyers CL. Chronic myeloid leukemia. N Engl J Med 1999;340(17):1330–1340.
16. Daub H, Specht K, Ullrich A. Strategies to overcome resistance to targeted protein kinase inhibitors. Nat Rev Drug Discov 2004;3 (12):1001–1010.
17. Burgess MR, Sawyers CL. Treating imatinib-resistant leukemia: the next generation targeted therapies. Sci World J 2006;6:918–930.
18. Kantarjian H, Jabbour E, Grimley J, Kirkpatrick P, Dasatinib. Nat Rev Drug Discov 2006;5 (9):717–718.
19. Carter TA, Wodicka LM, Shah NP, Velasco AM, Fabian MA, Treiber DK, Milanov ZV, Atteridge CE, Biggs WH, Edeen PT, Floyd M, Ford JM, Grotzfeld RM, Herrgard S, Insko DE, Mehta SA, Patel HK, Pao W, Sawyers CL, Varmus H, Zarrinkar PP, Lockhart DJ. Inhibition of drug-resistant mutants of ABL, KIT, and EGF receptor kinases. Proc Natl Acad Sci USA 2005;102(31):11011–11016.
20. Ramanathan RK, Egorin MJ, Eiseman JL, Ramalingam S, Friedland D, Agarwala SS, Ivy SP, Potter DM, Chatta G, Zuhowski EG, Stoller RG, Naret C, Guo J, Belani CP. Phase I and pharmacodynamic study of 17-(allylamino)-17-demethoxygeldanamycin in adult patients with refractory advanced cancers. Clin Cancer Res 2007;13(6):1769–1774.
21. Solit DB, Ivy SP, Kopil C, Sikorski R, Morris MJ, Slovin SF, Kelly WK, DeLaCruz A, Curley T, Heller G, Larson S, Schwartz L, Egorin MJ, Rosen N, Scher HI. Phase I trial of 17-allylamino-17-demethoxygeldanamycin in patients with advanced cancer. Clin Cancer Res 2007;13(6):1775–1782.
22. Nimmanapalli R, Fuino L, Bali P, Gasparetto M, Glozak M, Tao J, Moscinski L, Smith C, Wu J, Jove R, Atadja P, Bhalla K. Histone deace-

tylase inhibitor LAQ824 both lowers expression and promotes proteasomal degradation of Bcr-Abl and induces apoptosis of imatinib mesylate-sensitive or -refractory chronic myelogenous leukemia-blast crisis cells. Cancer Res 2003;63(16):5126–5135.

23. Rahmani M, Reese E, Dai Y, Bauer C, Kramer LB, Huang M, Jove R, Dent P, Grant S. Cotreatment with suberanoylanilide hydroxamic acid and 17-allylamino 17-demethoxygeldanamycin synergistically induces apoptosis in Bcr-Abl+ cells sensitive and resistant to STI571 (imatinib mesylate) in association with downregulation of Bcr-Abl, abrogation of signal transducer and activator of transcription 5 activity, and Bax conformational change. Mol Pharmacol 2005;67(4):1166–1176.

24. Moore M, Goldstein D, Hamm J, Figer A, Hecht J, Gallinger S, et al. Erlotinib plus gemcitabine compared to gemcitabine alone in patients with advanced pancreatic cancer. A phase III trial of the National Cancer Institute of Canada Clinical Trials Group [NCIC-CTG]. In: ASCO Annual Meeting 2005;23:2005.

25. Mellinghoff IK, Wang MY, Vivanco I, Haas-Kogan DA, Zhu S, Dia EQ, Lu KV, Yoshimoto K, Huang JHY, Chute DJ, Riggs BL, Horvath S, Liau LM, Cavenee WK, Rao PN, Beroukhim R, Peck TC, Lee JC, Sellers WR, Stokoe D, Prados M, Cloughesy TF, Sawyers CL, Mischel PS. Molecular determinants of the response of glioblastomas to EGFR kinase inhibitors. N Engl J Med 2005;353(19):2012–2024.

26. Baselga J, Arteaga CL. Critical update and emerging trends in epidermal growth factor receptor targeting in cancer. J Clin Oncol 2005;2(11):2445–2459.

27. Lynch TJ, Bell DW, Sordella R, Gurubhagavatula S, Okimoto RA, Brannigan BW, Harris PL, Haserlat SM, Supko JG, Haluska FG, Louis DN, Christiani DC, Settleman J, Haber DA. Activating mutations in the epidermal growth factor receptor underlying responsiveness of Non-small-cell lung cancer to gefitinib. N Engl J Med 2004;350(21):2129–2139.

28. Paez JG, Janne PA, Lee JC, Tracy S, Greulich H, Gabriel S, Herman P, Kaye FJ, Lindeman N, Boggon TJ, Naoki K, Sasaki H, Fujii Y, Eck MJ, Sellers WR, Johnson BE, Meyerson M. EGFR mutations in lung cancer: correlation with clinical response to gefitinib therapy. Science 2004;304(5676):1497–1500.

29. Pao W, Miller VA, Politi KA, Riely GJ, Somwar R, Zakowski MF, Kris MG, Varmus H. Acquired resistance of lung adenocarcinomas to gefitinib or erlotinib is associated with a second mutation in the EGFR kinase domain. PLoS Medicine 2005;2(3):e73.

30. Kosaka T, Yatabe Y, Endoh H, Kuwano H, Takahashi T, Mitsudomi T. Mutations of the epidermal growth factor receptor gene in lung cancer: biological and clinical implications. Cancer Res 2004;64(24):8919–8923.

31. Kwak EL, Sordella R, Bell DW, Godin-Heymann N, Okimoto RA, Brannigan BW, Harris PL, Driscoll DR, Fidias P, Lynch TJ, Rabindran SK, McGinnis JP, Wissner A, Sharma SV, Isselbacher KJ, Settleman J, Haber DA. Irreversible inhibitors of the EGF receptor may circumvent acquired resistance to gefitinib. Proc Natl Acad Sci USA 2005;102(21):7665–7670.

32. Balak MN, Gong Y, Riely GJ, Somwar R, Li AR, Zakowski MF, Chiang A, Yang G, Ouerfelli O, Kris MG, Ladanyi M, Miller VA, Pao W. Novel D761Y and common secondary T790M mutations in epidermal growth factor receptor–mutant lung adenocarcinomas with acquired resistance to kinase inhibitors. Clin Cancer Res 2006;12(21):6494–6501.

33. Hanahan D, Weinberg RA. The hallmarks of cancer. Cell 2000;100(1):57–70.

34. Bergers G, Hanahan D. Modes of resistance to anti-angiogenic therapy. Nat Rev Cancer 2008;8(8):592–603.

35. Gridelli C, Maione P, Del Gaizo F, Colantuoni G, Guerriero C, Ferrara C, Nicolella D, Comunale D, De Vita A, Rossi A. Sorafenib and sunitinib in the treatment of advanced non-small cell lung cancer. Oncologist 2007;12(2):191–200.

36. Smith MR, Rituximab (monoclonal anti-CD20 antibody): mechanisms of action and resistance. Oncogene 2003;22(47):7359–7368.

37. Schrama D, Reisfeld RA, Becker JC. Antibody targeted drugs as cancer therapeutics. Nat Rev Drug Discov 2006;5(2):147–159.

38. Carter P. Improving the efficacy of antibody-based cancer therapies. Nat Rev Cancer 2001;1(2):118–129.

39. Friedberg JW. Unique toxicities and resistance mechanisms associated with monoclonal antibody therapy. Hematology 2005;2005(1):329–334.

40. Vogel CL, Cobleigh MA, Tripathy D, Gutheil JC, Harris LN, Fehrenbacher L, Slamon DJ, Murphy M, Novotny WF, Burchmore M, Shak S, Stewart SJ, Press M. Efficacy and safety of trastuzumab as a single agent in first-line treatment of HER2-Overexpressing meta-

static breast cancer. J Clin Oncol 2002;20(3):719–726.
41. Berns K, Horlings HM, Hennessy BT, Madiredjo M, Hijmans EM, Beele K, Linn SC, Gonzalez-Angulo AM, Stemke-Hale K, Hauptmann M, Beijersbergen RL, Mills GB, van de Vijver MJ, Bernards R. A functional genetic approach identifies the PI3K pathway as a major determinant of trastuzumab resistance in breast cancer. Cancer Cell 2007;12(4):395–402.
42. Weinberg OK, Marquez-Garban DC, Pietras RJ. New approaches to reverse resistance to hormonal therapy in human breast cancer. Drug Resist Updat 2005;8(4):219–233.
43. Coombes RC, Hall E, Gibson LJ, Paridaens R, Jassem J, Delozier T, Jones SE, Alvarez I, Bertelli G, Ortmann O, Coate AS, Bajetta E, Dodwell D, Coleman RE, Fallowfield LJ, Mickiewicz E, Andersen J, Lonning PE, Cocconi G, Stewart A, Stuart N, Snowdon CF, Carpentieri M, Massimini G, Bliss JM, The Intergroup Exemestane Study. A randomized trial of exemestane after two to three years of tamoxifen therapy in postmenopausal women with primary breast cancer. N Engl J Med 2004;350(11):1081–1092.
44. Hiscox S, Jordan NJ, Jiang W, Harper M, McClelland R, Smith C, Nicholson RI. Chronic exposure to fulvestrant promotes overexpression of the c-Met receptor in breast cancer cells: implications for tumour-stroma interactions. Endocr Relat Cancer 2006;13(4):1085–1099.
45. Gutierrez MC, Detre S, Johnston S, Mohsin SK, Shou J, Allred DC, Schiff R, Osborne CK, Dowsett M. Molecular changes in tamoxifen-resistant breast cancer: relationship between estrogen receptor, HER-2, and p38 mitogen-activated protein kinase. J Clin Oncol 2005;23(11):2469–2476.
46. Hurtado A, Holmes KA, Geistlinger TR, Hutcheson IR, Nicholson RI, Brown M, Jiang J, Howat WJ, Ali S, Carroll JS. Regulation of ERBB2 by oestrogen receptor-PAX2 determines response to tamoxifen. Nature 2008;456(7222):663–666.
47. Fojo T, Bates S. Strategies for reversing drug resistance. Oncogene 2003;22(47):7512–7523.
48. Igney FH, Krammer PH. Death and anti-death: tumour resistance to apoptosis. Nat Rev Cancer 2002;2(4):277–288.
49. Keane MM, Ettenberg SA, Nau MM, Russell EK, Lipkowitz S. Chemotherapy augments TRAIL-induced apoptosis in breast cell lines. Cancer Res 1999;59(3):734–741.
50. Jansen B, Schlagbauer-Wadl H, Brown BD, Bryan RN, Van Elsas A, Muller M, Wolff K, Eichler H-G, Pehamberger H. bcl-2 Antisense therapy chemosensitizes human melanoma in SCID mice. Nat Med 1998;4(2):232–234.
51. Zamore PD. RNA interference: listening to the sound of silence. Nat Struct Mol Biol 2001;8(9):746–750.
52. Xing Z, Conway EM, Kang C, Winoto A. Essential role of survivin, an inhibitor of apoptosis protein, in T cell development, maturation, and homeostasis. J Exp Med 2004;199(1):69–80.
53. Chai J, Du C, Wu J-W, Kyin S, Wang X, Shi Y. Structural and biochemical basis of apoptotic activation by Smac/DIABLO. Nature 2000;406(6798):855–862.
54. Bullock AN, Fersht AR. Rescuing the function of mutant p53. Nat Rev Cancer 2001;1(1):68–76.
55. Szakacs G, Paterson JK, Ludwig JA, Booth-Genthe C, Gottesman MM. Targeting multidrug resistance in cancer. Nat Rev Drug Discov 2006;5(3):219–234.
56. Higgins CF. ABC transporters: from microorganisms to man. Annu Rev Cell Biol 1992;8(1):67–113.
57. Dean M, Hamon Y, Chimini G. The human ATP-binding cassette (ABC) transporter superfamily. J Lipid Res 2001;4(7):1007–1017.
58. Leslie EM, Deeley RG, Cole SPC. Multidrug resistance proteins: role of P-glycoprotein, MRP1, MRP2, and BCRP (ABCG2) in tissue defense. Toxicol Appl Pharmacol 2005;204(3):216–237.
59. Dean M, Fojo T, Bates S. Tumour stem cells and drug resistance. Nat Rev Cancer 2005;5(4):275–284.
60. Juliano RL, Ling V. A surface glycoprotein modulating drug permeability in Chinese hamster ovary cell mutants. Biochimi Biophys Acta 1976;455(1):152–162.
61. Roninson IB, Chin JE, Choi KG, Gros P, Housman DE, Fojo A, Shen DW, Gottesman MM, Pastan I. Isolation of human mdr DNA sequences amplified in multidrug-resistant KB carcinoma cells. Proc Natl Acad Sci USA 1986;83(12):4538–4542.
62. Ecker GF, Stockner T, Chiba P. Computational models for prediction of interactions with ABC-transporters. Drug Discov Today 2008;13(7–8):311–317.
63. Tsuruo T, Iida H, Tsukagoshi S, Sakurai Y. Overcoming of vincristine resistance in P388

leukemia *in vivo* and *in vitro* through enhanced cytotoxicity of vincristine and vinblastine by verapamil. Cancer Res 1981;41(5):1967–1972.

64. Raderer M, Scheithauer W. Clinical trials of agents that reverse multidrug resistance. A literature review. Cancer 1993;72(12):3553–3563.

65. Wandel CB, Kim R, Kajiji S, Guengerich FP, Wilkinson GR, Wood AJJ. P-glycoprotein and cytochrome P-450 3A inhibition: dissociation of inhibitory potencies. Cancer Res 1999;59(16):3944–3948.

66. Cole SP, Bhardwaj G, Gerlach JH, Mackie JE, Grant CE, Almquist KC, Stewart AJ, Kurz EU, Duncan AM, Deeley RG. Overexpression of a transporter gene in a multidrug-resistant human lung cancer cell line. Science 1992;258(5088):1650–1654.

67. Kruh G, Zeng H, Rea P, Liu G, Chen Z-S, Lee K, Belinsky M. MRP subfamily transporters and resistance to anticancer agents. J Bioenerg Biomembr 2001;33(6):493–501.

68. Deeley RG, Cole SPC. Substrate recognition and transport by multidrug resistance protein 1 (ABCC1). FEBS Lett 2006;580(4):1103–1111.

69. Schinkel AH, Jonker JW. Mammalian drug efflux transporters of the ATP binding cassette (ABC) family: an overview. Adv Drug Deliv Rev 2003;55(1):3–29.

70. Boumendjel A, Baubichon-Cortay H, Trompier D, Perrotton T, Di Pietro A. Anticancer multidrug resistance mediated by MRP1: recent advances in the discovery of reversal agents. Med Res Rev 2005;25(4):453–472.

71. van Herwaarden AE, Schinkel AH. The function of breast cancer resistance protein in epithelial barriers, stem cells and milk secretion of drugs and xenotoxins. Trends Pharmacol Sci 2006;27(1):10–16.

72. Bates S, Robey R, Miyake K, Rao K, Ross D, Litman T. The role of half-transporters in multidrug resistance. J Bioenerg Biomembr 2001;33(6):503–511.

73. Cramer J, Kopp S, Bates SE, Chiba P, Ecker GF. Multispecificity of drug transporters: Probing inhibitor selectivity for the human drug efflux transporters ABCB1 and ABCG2. Chem Med Chem 2007;2(12):1783–1788.

74. Raub TJ. P-glycoprotein recognition of substrates and circumvention through rational drug design. Mol Pharm 2006;3(1):3–25.

75. Pleban K, Ecker GF. Inhibitors of p-glycoprotein: lead identification and optimisation. Mini Rev Med Chem 2005;5(2):153–163.

76. Wiese M, Pajeva IK. Structure–activity relationships of multidrug resistance reversers. Curr Med Chem 2001;8(6):685–713.

77. Tmej C, Chiba P, Huber M, Richter E, Hitzler M, Schaper KJ, Ecker G. A combined Hansch/Free-Wilson approach as predictive tool in QSAR studies on propafenone-type modulators of multidrug resistance. Arch Pharm 1998;331(7–8):233–240.

78. Kaiser D, Smiesko M, Kopp S, Chiba P, Ecker GF. Interaction field based and hologram based QSAR analysis of propafenone-type modulators of multidrug resistance. Med Chem 2005;1(5):431–444.

79. Tmej C, Chiba P, Schaper KJ, Ecker G, Fleischhacker W. Artificial neural networks as versatile tools for prediction of MDR-modulatory activity. Adv Exp Med Biol 1999;457:95–105.

80. Klein C, Kaiser D, Kopp S, Chiba P, Ecker GF. Similarity based SAR (SIBAR) as tool for early ADME profiling. J Comput Aided Mol Des 2002;16(11):785–793.

81. Pajeva I, Wiese M. Molecular modeling of phenothiazines and related drugs as multidrug resistance modifiers: a comparative molecular field analysis study. J Med Chem 1998;41(11):1815–1826.

82. Pajeva IK, Wiese M. A comparatice molecular field analysis of propafenone-type modulators of cancer multidrug resistance. Quant Struct Act Relat 1998;17:301–312.

83. Ekins S, Kim RB, Leake BF, Dantzig AH, Schuetz EG, Lan LB, Yasuda K, Shepard RL, Winter MA, Schuetz JD, Wikel JH, Wrighton SA. Application of three-dimensional quantitative structure–activity relationships of P-glycoprotein inhibitors and substrates. Mol Pharmacol 2002;61(5):974–981.

84. Ekins S, Kim RB, Leake BF, Dantzig AH, Schuetz EG, Lan LB, Yasuda K, Shepard RL, Winter MA, Schuetz JD, Wikel JH, Wrighton SA. Three-dimensional quantitative structure–activity relationships of inhibitors of P-glycoprotein. Mol Pharmacol 2002;61(5):964–973.

85. Garrigues A, Loiseau N, Delaforge M, Ferte J, Garrigos M, Andre F, Orlowski S. Characterization of two pharmacophores on the multidrug transporter P-glycoprotein. Mol Pharmacol 2002;62(6):1288–1298.

86. Langer T, Eder M, Hoffmann RD, Chiba P, Ecker GF. Lead identification for modulators of multidrug resistance based on in silico screening with a pharmacophoric feature model. Arch Pharm 2004;337(6):317–327.

87. Chang G, Roth CB. Structure of MsbA from *E. coli*: a homolog of the multidrug resistance ATP binding cassette (ATP) transporters. Science 2001;293:1793–1800.
88. Chang G. Structure of MsbA form Vibrio cholerae: a multidrug resistance ABC transporter in a closed conformation. J Mol Biol 2003;330: 419–430.
89. Reyes CL, Chang G. Structure of the ABC transporter MsbA in complex with ADP•vanadate and lipopolysaccharide. Science 2005;308: 1028–1031.
90. Chang G, Retraction. Science 2006;314:1875b.
91. Ward A, Reyes CL, Yu J, Roth CB, Chang G. Flexibility in the ABC transporter MsbA: alternating access with a twist. Proc Natl Acad Sci USA 2007;104(48):19005–19010.
92. Dawson RJP, Locher KP. Structure of a bacterial multidrug ABC transporter. Nature 2006;443:180–185.
93. Dawson RJP, Locher KP. Structure of the multidrug ABC transporter Sav1866 from *Staphylococcus aureus* in complex with AMP-PNP. FEBS Lett 2007;581(5):935–938.
94. Globisch C, Pajeva IK, Wiese M. Identification of putative binding sites of P-glycoprotein based on its homology model. ChemMedChem 2008;3(2):280–295.
95. Ravna AW, Sylte I, Sager G. A molecular model of a putative substrate releasing conformation of multidrug resistance protein 5 (MRP5). Eur J Med Chem 2008;43(11):2557–2567.
96. Ravna AW, Sylte I, Sager G. Molecular model of the outward facing state of the human P-glycoprotein (ABCB1), and comparison to a model of the human MRP5 (ABCC5). Theor Biol Med Model 2007;4:33.
97. Stockner T, de Vries SJ, Bonvin AMJJ, Ecker GF, Chiba P. Data-driven homology modelling of P-glycoprotein in the ATP-bound state indicates flexibility of the transmembrane domains. FEBS J 2009;276(4):964–972.
98. Becker J-P, Depret G, van Bambeke F, Tulkens PM, Prevost M. Molecular models of human P-glycoprotein in two different catalyitc states. BMC Struct Biol 2009;9:3.
99. Aller SG, Yu J, Ward A, Weng Y, Chittaboina S, Zhuo R, Harrell PM, Trinh YT, Zhang Q, Urbatsch IL, Chang G. Structure of P-glycoprotein reveals a molecular basis for poly-specific drug binding. Science 2009;323(5922):1718–1722.
100. Schmidt C. Drug makers chase cancer stem cells. Nat Biotech 2008;26(4):366–367.
101. Clarke MF, Dick JE, Dirks PB, Eaves CJ, Jamieson CHM, Jones DL, Visvader J, Weissman IL, Wahl GM. Cancer stem cells: perspectives on current status and future directions: AACR Workshop on Cancer Stem Cells. Cancer Res 2006;66(19):9339–9344.
102. Schinkel AH, Smit JJM, van Tellingen O, Beijnen JH, Wagenaar E, van Deemter L, Mol CAAM, van der Valk MA, Robanus-Maandag EC, te Riele HPJ, Berns AJM, Borst P. Disruption of the mouse mdr1a P-glycoprotein gene leads to a deficiency in the blood–brain barrier and to increased sensitivity to drugs. Cell 1994;77(4):491–502.
103. Strebhardt K, Ullrich A. Targeting polo-like kinase 1 for cancer therapy. Nat Rev Cancer 2006;6(4):321–330.
104. de Carcer G, de Castro IP, Malumbres M. Targeting cell cycle kinases for cancer therapy. Curr Med Chem 2007;14(9):969–985.
105. Macchiarulo A, Giacche N, Carotti A, Baroni M, Cruciani G, Pellicciari R. Targeting the conformational transitions of MDM2 and MDMX: insights into dissimilarities and similarities of p53 recognition. J Chem Inf Model 2008;48(10):1999–2009.
106. Raguz S, Yague E. Resistance to chemotherapy: new treatments and novel insights into an old problem. Br J Cancer 2008;99(3):387–391.
107. Dalton WS, Friend SH. Cancer biomarkers: an invitation to the table. Science 2006;312 (5777):1165–1168.
108. Weissleder R. Molecular imaging in cancer. Science 2006;312(5777):1168–1171.

Hsp90 INHIBITORS

Brian S. J. Blagg[1]
M. Kyle Hadden[2]

[1] Department of Medicinal Chemistry, School of Pharmacy, University of Kansas, Lawrence, KS
[2] Department of Pharmaceutical Sciences, School of Pharmacy, University of Connecticut, Storrs, CT

As of 2008, more than 20 clinical trials had been launched toward the treatment of human cancers via Hsp90 inhibition. The 90 kDa heat shock proteins (Hsp90) are molecular chaperones required for the conformational maturation of a number of key signaling proteins that are often overexpressed and/or mutated in transformed cells. Consequently, small-molecule inhibitors of Hsp90 exert a combinatorial attack on cancer cells that may be comparable to the administration of multiple chemotherapeutic agents.

1. INTRODUCTION

Multiple signaling pathways are upregulated or constitutively activated in malignant cells leading to the six hallmarks of cancer: (1) self-sufficiency in growth signals, (2) insensitivity to antigrowth signals, (3) evasion of apoptosis, (4) limitless replicative potential, (5) sustained angiogenesis, and (6) tissue invasion/metastasis [1]. Several anticancer agents target individual enzymes/proteins involved in the above processes, but no clinically available antitumor agent is capable of simultaneously inhibiting all six hallmarks of cancer [2].

Since multiple pathways are dysfunctional in most cancers, and cancers accumulate new oncogenic mutations as they progress, a single drug is unlikely to be effective for most cancers. Consequently, there has been a rapid evolution toward the administration of combination therapies that simultaneously disrupt multiple therapeutic targets. An alternative to the use of combination therapy is the identification of a novel biological target that is substantially different in malignant cells than in normal cells and is responsible for mediating several commonly hijacked oncogenic pathways [3]. By inhibition of this biological target, one can disrupt multiple pathways while simultaneously achieving tumor cell specificity.

The 90 kDa heat shock proteins currently represent one of the most promising biological targets identified for the treatment of cancer [4–7]. As a molecular chaperone, Hsp90 is responsible for folding many of the proteins directly associated with malignant progression and hence, inhibition of the Hsp90 protein folding machinery results in a simultaneous attack on multiple oncogenic pathways [8]. Furthermore, studies have revealed that Hsp90 inhibitors accumulate in tumor cells more efficiently than in normal tissue, leading to differential selectivities of ~200-fold [9]. Currently, there are 27 clinical trials in progress and these studies have demonstrated that Hsp90 can be inhibited at doses that are well tolerated by patients [10,11].

Hsp90 is responsible for the conformational maturation of enzymes in all six hallmarks of cancer (Table 1). Because transformed cells rely on these pathways, several of these individual enzymes/proteins are therapeutically important anticancer targets [3]. Examples of proteins dependent upon Hsp90 for conformational maturation include Src-family kinases (oncogenic v-Src, Hck, and Lck) [12,13], Raf [14,15], receptor tyrosine kinases (ErbB1 and Her2) [16,17], mutant p53 (not normal p53) [18], telomerase [19–21], steroid hormone receptors [22–24], polo 1-kinase (PLK) [25,26], protein kinase B (AKT) [27,28], death domain kinase (RIP) [29,30], MET (kinase) [31], focal adhesion kinase (FAK) [32,33], as well as the aryl hydrocarbon receptor [34–36], elF2α kinases (HRI, PKR, GCN2, and PERK) [37,38], nitric oxide synthase [39–41], centrosomal proteins [25,42], and others. Consequently, Hsp90 inhibition provides an excellent target for the development of new cancer therapeutics because all six hallmarks of cancer can be simultaneously targeted by disruption of the Hsp90 protein folding machinery [43–48].

Table 1. Some Hsp90 Client Proteins Associated with the Six Hallmarks of Cancer [3]

Hallmark	Hsp90 Client Proteins
Self-sufficiency in growth signals	Raf-1, AKT, Her-2, MEK, Bcr-Abl, FLT-3, EGFR, IGF-1R, FGFR, KDR
Insensitivity to antigrowth signals	Wee 1, Myt 1, CDK4, CDK6, Plk
Evasion of apoptosis	RIP, AKT, mutant p53, c-MET, Apaf-1, Survivin
Limitless replicative potential	Telomerase (h-TERT)
Sustained angiogenesis	FAK, AKT, HIF-1α, VEGFR, FLT-3
Tissue invasion/metastasis	c-MET, MMP

1.1. The Hsp90-Mediated Protein Folding Process

The Hsp90-mediated protein folding process has not been fully resolved, but evidence suggests that a variety of cochaperones, immunophilins, and partner proteins are required for the conformational maturation of nascent polypeptides into biologically active three-dimensional structures (Fig. 1). The Hsp90 homodimeric species (**1.A**) interacts with a nascent client protein bound to Hsp70 to coordinate peptide transfer to Hsp90 [49,50]. The binary complex (**1.B**) then binds to cochaperones containing TPR (tetratricopeptide repeat) motifs, a highly conserved 34-amino acid sequence found in Hsp90 partner proteins), such as immunophilins with *cis/trans* peptidyl-prolyl-isomerase activity (FKBP51, FKBP52, or CyP A) [51–53], or protein phosphatase 5 bind to form a heteroprotein complex (**1.C**). This *activated* multiprotein complex binds ATP and "clamps" around the Hsp90 client protein (**1.D**) [54]. The cochaperone p23 is recruited to Hsp90 at this stage, and promotes ATP hydrolysis and stabilization of Hsp90's "clamped" high-affinity client-bound conformation (**1.E**). The ensemble of Hsp90 and its assistants promote folding of the bound client into its three-dimensional structure, and subsequently release the biologically active protein (**1.F**) [55].

When inhibitors of the Hsp90 protein folding process are present, Hsp90 is unable to "wrap" around the bound client protein. This results in an unstable protein complex (**1.G**) that is recognized by ubiquitin lygase, which ubiquitinylates the client protein for ultimate destruction by the proteasome [55–58]. Consequently, inhibitors of this process not only inhibit Hsp90 but also inhibit the conformational maturation of Hsp90-dependent client proteins, leading to their degradation and abolishment of their needed activity by growing tumors [59]. As a consequence of Hsp90 inhibition, these oncogenic proteins are

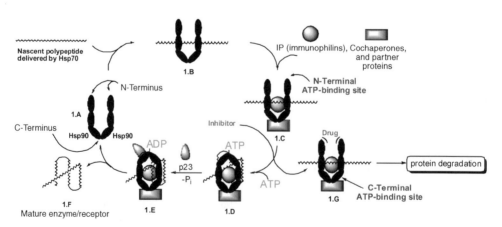

Figure 1. The Hsp90 protein folding process. (This figure is available in full color at http://mrw.interscience.wiley.com/emrw/9780471266945/home.)

uniformly degraded and may result in the administration of one drug for the treatment of cancer instead of cocktails enlisting multiple drugs to inhibit one or more of these individual targets.

1.2. Tumor Cell Selectivity

Previous studies have shown that Hsp90 inhibitors accumulate preferentially in tumor cells and exhibit high differential selectivity [3,9,60]. Recent studies have confirmed that Hsp90 in cancer cells (BT474, N87, SKOV3, and SkBr3; average $IC_{50} \sim 5$ nM) exhibit higher affinity for ligands than Hsp90 derived from normal cells (normal dermal fibroblasts, human renal epithelial cells, HMVEC, HUVEC, Hs578Bst, and PBMC; average $IC_{50} \sim 943$ nM) [61]. The enhanced affinity of Hsp90 for inhibitors was determined by immunoprecipitation of Hsp90 from both tumor and normal cells. Immunoprecipitated Hsp90 from tumor cells was found to reside in multiprotein complexes bound to client and partner proteins, whereas immunoprecipitated Hsp90 from normal cells was found primarily as the Hsp90 homodimeric species unbound to partner proteins. When the immunoprecipitated multiprotein complexes from tumors were incubated with ATP, a substantial increase in ATPase activity was observed compared to the homodimeric protein isolated from normal cells. Furthermore, the ATPase activity of tumor-derived Hsp90 multiprotein complexes was inhibited at lower concentrations of inhibitors than immunoprecipitated Hsp90 from normal cells. The enhanced affinity of tumorgenic Hsp90 for inhibitors is a result of the high population of Hsp90 heteroprotein complexes found in cancer cells, which are highly dependent upon the Hsp90 protein folding machinery for continual growth in hostile tumor microenvironments resulting from hypoxia, nutrient deprivation, acidosis, and mutated/overexpressed proteins [62]. In normal cells, Hsp90 can be found in its inactivated, homodimeric state, unbound to client and partner proteins, and this species manifests lower affinity for ligands than the highly abundant multiprotein complexes present in malignant cells [61].

1.3. Hsp90 Protein Structure and Function

In humans, Hsp90 exists as a homodimer, in which each monomeric unit contains three domains. These regions consist of a 25 kDa N-terminal ATP-binding domain, a 35 kDa middle domain, and a 12 kDa C-terminal dimerization domain that also includes a nucleotide binding site (Fig. 2). Since Hsp90 function depends upon the N-terminal domain to bind and hydrolyze ATP [63–65], small molecules that bind the N-terminus result an unproductive heteroprotein complex that is degraded by the ubiquitin-proteasome pathway [58,59,66–69]. The middle domain exhibits high affinity for cochaperones and client proteins [27,70–75] and is known to bind the γ-phosphate of ATP, when the nucleotide is bound to the N-terminus [74]. The structure of the Hsp90 C-terminus is characterized by the MEEVD terminal sequence, which binds cochaperones containing a TPR. Chadli and coworkers recently identified an additional location near the N-terminus to which proteins that contain a TPR can also bind [76].

As noted above, Hsp90 has been demonstrated to contain two nucleotide binding sites, one at each terminus. Although originally believed to be a client protein binding site, the Hsp90 N-terminal nucleotide binding site was later shown to bind ATP as determined by solution of the yeast Hsp90 cocrystal

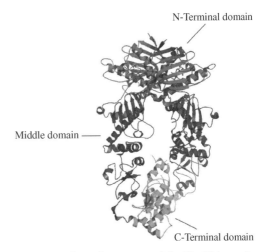

Figure 2. Crystal structure of the Hsp90 homodimer and the three domains. (This figure is available in full color at http://mrw.interscience.wiley.com/emrw/9780471266945/home.)

structure [77]. Unlike the C-terminal nucleotide binding site, the N-terminal ATP-binding site is responsible for providing the energy necessary for the protein folding process via hydrolysis of ATP [78]. In fact, only the N-terminus exhibits inherent ATPase activity, which can be measured independent of client or partner protein binding [79]. In contrast, the C-terminal nucleotide binding site appears to play an allosteric role, such that it promotes nucleotide displacement and client protein release from the N-terminal domain. However, no cocrystal structure of the Hsp90 C-terminus bound to a nucleotide or inhibitor has been solved, and its exact location remains undetermined. Nonetheless, both binding sites can be inhibited with small-molecule inhibitors that effectively block the Hsp90 protein folding process and can lead to potentially useful treatments for cancer.

2. INHIBITORS OF THE Hsp90 N-TERMINAL ATP-BINDING POCKET

2.1. Geldanamycin, an Hsp90 Inhibitor

Geldanamycin (GDA, Fig. 3) is a member of the ansamycin family of antibiotics originally isolated from the soil actinomycete *Streptomyces hygroscopis* in Kalamazoo, Michigan in 1970 [80]. The structural assignment of GDA as the first benzoquinone ansamycin was completed by Rinehart and colleagues [81]. Screened against various protozoa, bacteria, and fungi *in vitro*, GDA demonstrated modest growth inhibitory activity (MIC = 2 to >100 μg/mL). Interestingly, studies on its activity against two distinct cancer cell types identified GDA as a potent antitumor agent (<0.002 μg/mL).

Early exploration into the antitumor activity for which GDA exhibited both *in vitro* and *in vivo* was disappointing as the concentration required for activity was toxic in animal models. Interest in GDA's anticancer activity was renewed in the late 1980s when the related benzoquinone ansamycin herbimycin A was shown to inhibit the transforming activity of numerous tyrosine kinase oncogenes, including *src*, *ros*, *abl*, and *erb*B [82]. Herbimycin A also inhibited angiogenesis [83] and induced differentiation [84,85] in several *in vitro* systems, as a consequence of its ability to modulate the *src* family of tyrosine kinases [86,87]. Although GDA and herbimycin A did significantly decrease *v-src* kinase activity in certain cell lines [82,86,87], they were inactive against the purified protein [88], suggesting indirect inhibition of *v-src* kinase activity via an unknown mechanism.

Through affinity purification studies, Whitesell and Neckers determined that GDA binds reversibly to a 90 kDa protein, which was subsequently identified as Hsp90 [89]. Furthermore, it was demonstrated that GDA inhibits src-Hsp90 complex formation in both lysate and whole cells, suggesting the Hsp90 molecular chaperone is responsible for the conformational maturation of *v-src* and therefore its activity is dependent upon Hsp90. After identification of GDA as an Hsp90 inhibitor, it was utilized for the elucidation of

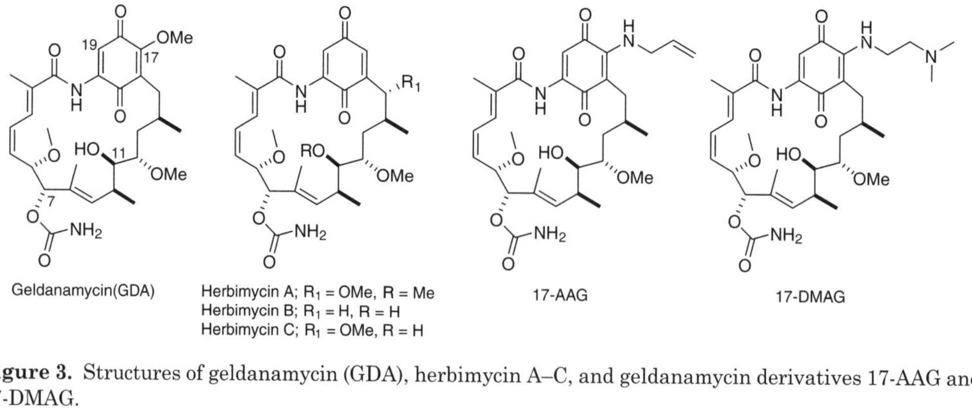

Geldanamycin(GDA)

Herbimycin A; R_1 = OMe, R = Me
Herbimycin B; R_1 = H, R = H
Herbimycin C; R_1 = OMe, R = H

17-AAG

17-DMAG

Figure 3. Structures of geldanamycin (GDA), herbimycin A–C, and geldanamycin derivatives 17-AAG and 17-DMAG.

more than 100 Hsp90-dependent client proteins [90]. The first cocrystal structure of GDA bound to Hsp90 was reported in 1997 and led to the hypothesis that GDA was acting as a peptide mimic of an Hsp90 client protein substrate, thus preventing the protein-complex formation [91]. This hypothesis was later determined incorrect as the cocrystal structure verified GDA binds to the ATP-binding site located in the N-terminus of Hsp90 [77]. Several key interactions were observed from the cocrystal structure of yeast Hsp90 bound to GDA (Fig. 4). First, the quinone moiety of GDA projects into the phosphate region of the binding pocket toward the outer surface of the protein and participates in five hydrogen-bonding interactions. This orientation suggested that the 17-methoxy group was not essential for binding and that modification at this position would likely have little or no effect on binding. In addition, GDA (similar to ATP) binds Hsp90 in a bent, cup-shaped conformation and contains a *cis*-amide bond, distinctive from its unbound structure in which it adopts a flat conformation and a *trans*-amide bond.

The cocrystal structure of a GDA derivative, 17-DMAG (see below) has led to increased insight into the conformation adopted by the ansamycins upon binding Hsp90 [92]. The 17-dimethylaminoethylamino functionality of 17-DMAG is oriented into the solvent, away from the pocket, supporting the previous findings that indicate C17-substitutions would have little effect on inhibitory activity. Targeted molecular dynamics stimulations and energetic analysis of GDA suggested that the key step for Hsp90 binding to GDA is isomerization of the amide bond from *trans* to *cis* in the bound conformation. This interconversion is believed to result in a 2.2–6.4 kcal/mol entropic penalty. Researchers have further suggested that analogs of GDA that contain a *cis*-amide or a predisposed bent, C-shaped conformation will exhibit >1000-fold increase in Hsp90 affinity.

While GDA demonstrates modest inhibitory activity against purified recombinant Hsp90 ($IC_{50} \sim 2.5\,\mu M$) in biochemical assays, its antiproliferative activity ($IC_{50} \sim 10$–50 nM) and its ability to induce degradation of Hsp90 client protein levels in whole cells is significantly greater. However, *in vivo* studies with GDA have proven that the redox-active quinone moiety produces hepatotoxicity unrelated to its Hsp90 inhibitory activity [93,94]. It

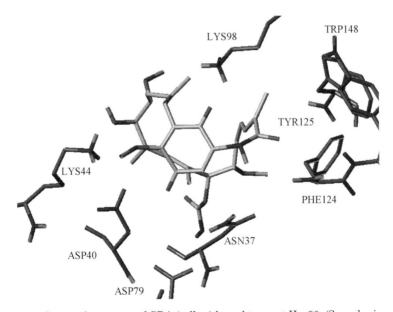

Figure 4. Cocrystal structure of GDA (yellow) bound to yeast Hsp90. (See color insert.)

has been reported that GDA is a substrate for flavin-dependent reductases that transform the quinone into a semiquinone, which then generates superoxide radicals upon exposure to oxygen. As such, analogs of GDA have focused on increasing its potency while reducing the redox-active behavior.

It is important to note that reported K_d values for GDA binding to the N-terminus of Hsp90 have covered a wide range (0.06–1.2 µM). These data most likely reflect the various methods utilized to measure GDA-Hsp90 binding. The K_d value for GDA and its analogs described in the following sections represent the value reported in that particular experiment. In addition, the method utilized to determine binding affinity is provided.

2.1.1. 17-Position Geldanamycin Derivatives

In an attempt to stabilize the quinone and reduce the redox potential of GDA, researchers at Pfizer prepared a series of GDA and dihydrogeldanamycin (DHGDA) derivatives in which electron-donating groups were introduced at the 17-position [95]. Because of the electrophilic nature of the quinone, the 17-position readily undergoes Michael addition/β-elimination of methanol in the presence of various amines to yield 17-amino-17-desmethoxy GDA analogs (e.g., **1**, Fig. 5). Condensation of GDA with a variety of diamines also affords derivatives at both the C17- and the C19-positions. A series of bi- and tri-cyclic derivatives of the quinone, including the guanidine derivatives (**2**), and geldanoxazinone (**3**) and geldanazine (**4**), were also prepared and evaluated. Depletion of the known Hsp90 client protein, Her2, in the Her2-overexpressing human breast cancer cell line SKBr3 was used to evaluate the in vitro efficacy of these analogs. Several of the most active analogs were then evaluated for in vivo activity in a mouse xenograft model. While the in vitro and in vivo data for these analogs did not follow a

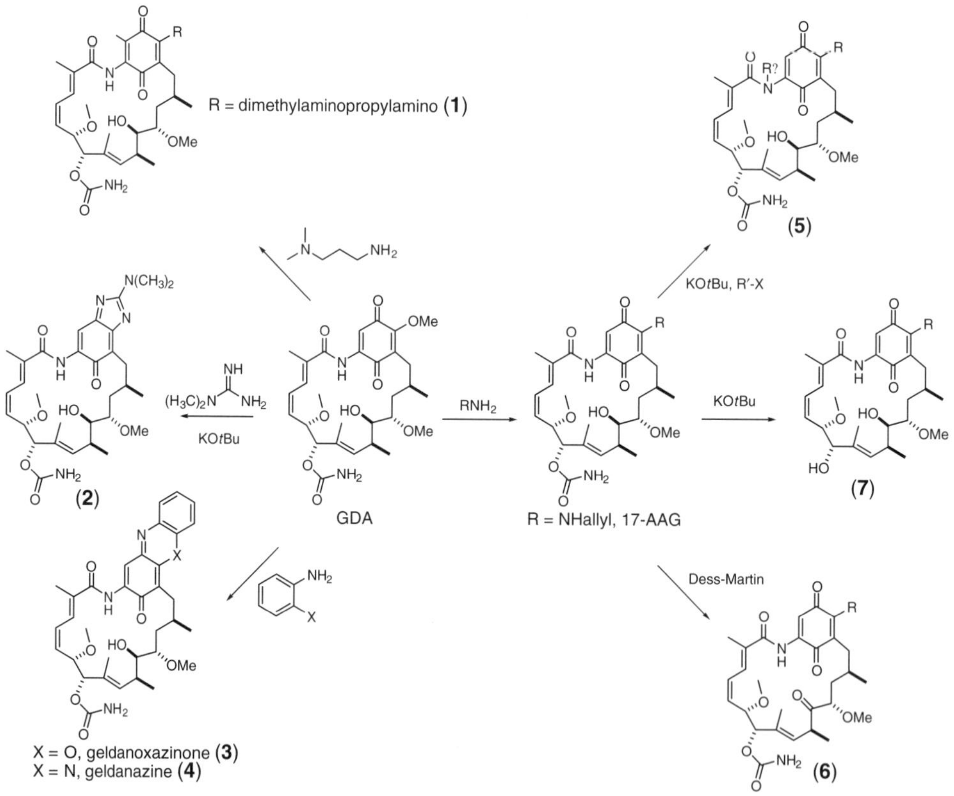

Figure 5. Early GDA derivatives.

strict corollary, it was identified that C17-substituents were particularly well tolerated and, in particular, that 17-allylamino-17-desmethoxygeldanamycin (17-AAG) was the best candidate for further evaluation.

Based on these results, researchers extended structure–activity relationship (SAR) studies to include GDA and 17-AAG analogs that incorporated modification at several other positions. A series of N22-substituted analogs was explored (**5**), however, HPLC analysis of the drug-treated cell culture medium showed that acylated variants readily hydrolyzed *in vitro* [96]. While the researchers noted these compounds may serve as viable prodrugs of GDA, they were unsuccessful at identifying more potent analogs, therefore further derivation at N22 was not pursued. Acyl derivatives of the 11-hydroxyl moiety generally resulted in decreased *in vitro* activity. Interestingly, oxidation to the corresponding ketone (**6**) resulted in analogs that retained inhibitory activity. Modification to the C7-carbamate afforded derivatives (**7**) with reduced *in vitro* activity, which is not surprising based on the cocrystal structure of Hsp90 bound to GDA. In fact, cleavage of the carbamate resulted in complete loss of inhibitory activity and simple *N*-ethyl derivatives produced greater than 100-fold loss in activity. Finally, opening of the ansa ring completely abolished Hsp90 inhibitory activity *in vitro*. Evaluation of the most potent GDA analogs *in vivo* demonstrated that GDA analogs with simple C17-modifications were most potent [95,96].

Continued exploration into the mechanism of action for 17-AAG confirmed that it too bound the N-terminal ATP-binding site of Hsp90 [97]. While 17-AAG bound with less affinity than GDA ($K_d = 7.2\,\mu M$ versus $0.17\,\mu M$, respectively, via competitive binding of purified Hsp90 with immobilized GDA), both compounds caused comparable antitumor effects against human cancer cell lines. In addition, 17-AAG demonstrated an improved toxicity profile *in vivo* compared to GDA, supporting its potential for clinical development [98]. 17-AAG (an investigational drug known as tanespimycin, KOS-953) entered clinical trials in 1999 and has been the subject of numerous Phase I and Phase II clinical trials for a variety of human cancers [99,100]. Despite its early introduction into the clinic, it continues to maintain some of the toxicity associated with GDA. In addition, 17-AAG is poorly water soluble and requires complex formulation strategies for delivery.

Collaborative efforts between Kosan Biosciences and the National Cancer Institute to develop improved analogs of GDA and 17-AAG led to 17-(dimethylaminoethylamino)-17-demethoxygeldanamycin (17-DMAG, see Fig. 3). The addition of the tertiary amine, which is readily ionized into the water-soluble ammonium ion, produced a promising new Hsp90 inhibitor. Preclinical evaluation of 17-DMAG demonstrated improved pharmacokinetic properties compared to 17-AAG, including good oral bioavailability, reduced plasma protein binding, and limited quantitative metabolism [101]. The *in vitro* and *in vivo* effects of 17-DMAG closely mimicked those observed for 17-AAG, verifying Hsp90 as the biological target [102,103]. 17-DMAG decreased levels of the Hsp90 client proteins phosphorylated AKT, cyclin D1, and cRaf-1 in separate human melanoma cancer cell lines [102]. It was also active in mice xenografts of pancreatic, melanoma, and lung cancers, with dose-dependent reduction of tumor volume for each model [103]. Promising early results for 17-DMAG (an investigational new drug known as Alvespimycin and KOS-1022) in Phase I clinical trials led to the announcement that phase II trials for the treatment of Her2-positive metastatic breast cancer were eminent. However, owing to an overall unfavorable toxicity profile, Kosan announced to halt the clinical development of 17-DMAG in March 2008 [100].

Recent attempts to convert GDA into a more drug-like compound with improved pharmacokinetic properties focused on the incorporation of amides, carbamates, ureas, and aryl moieties at the 17-position (Fig. 6) [104]. For 17-position amide analogs, aromatic derivatives were found to be more potent than the corresponding aliphatic analogs and decreasing the ring size from phenyl to furyl or thienyl resulted in a twofold decrease in inhibitory activity. Electron-donating substituents at the *ortho*- or *meta*-positions resulted in improved activity whereas electron-withdrawing moieties im-

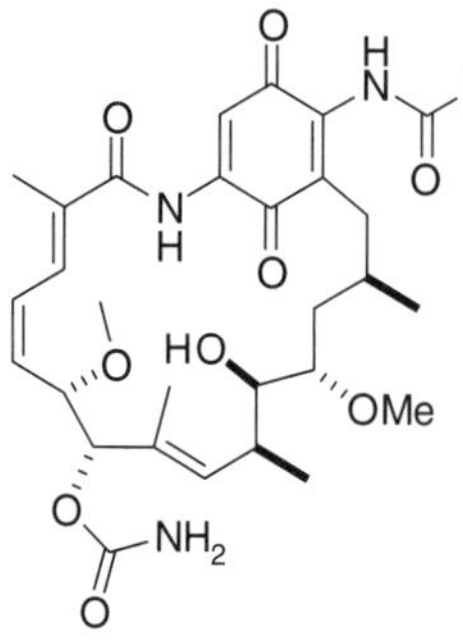

R = phenyl (**8**)
R = *o*-anisyl (**9**)
R = *m*-anisyl (**10**)

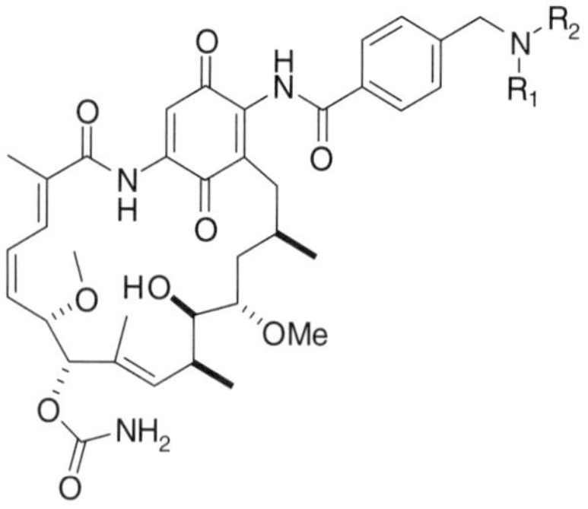

NR₁R₂ = benzylethylamino (**11**)
NR₁R₂ = 1-(4-(4-chlorophenyl))piperazinyl (**12**)

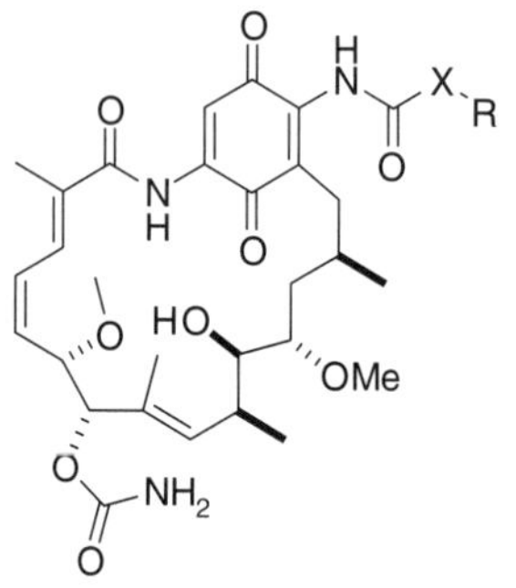

17-position carbamates and ureas
X = O or NH
R = alkyl

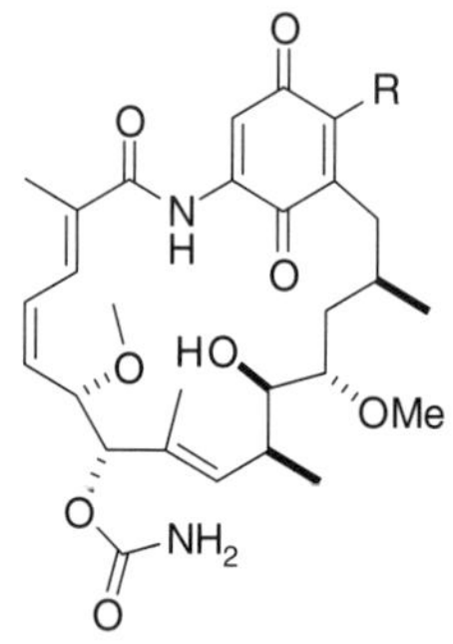

17-position aryl
R = phenyl or thienyl

Figure 6. 17-Amide, aryl, carbamate, and urea derivatives of GDA.

proved activity, when present in the *para*-position. However, the most active amide analogs prepared (**8**, **9**, and **10**) remained 13-fold less effective than 17-AAG at decreasing Her2 levels *in vitro* (IC$_{50}$ values 180–200 and 15 nM, respectively). Extension of the amide analogs by taking advantage of the improved pharmacokinetic properties demonstrated by 17-DMAG resulted in the incorporation of various ionizable amino groups in the *para*-position of the 17-aryl substituent. While improved water solubility was observed for this series of analogs, an overall improvement in potency was not. The most potent water-soluble analog, **11**, contained a benzylethylamino side chain and decreased Her2 levels with an IC$_{50}$ value of 140 nM. In addition, a comparison of cyclic amino side chains showed that 4-arylpiperazines were >15-fold more active than the corresponding piperidine analog. The inhibitory activity of the carbamate, urea, and aryl 17-GDA analogs was comparable to that identified for the amides (IC$_{50}$ range 50–500 nM). A survey of the most potent analogs in a variety of Hsp90 inhibitory assays identified the water-soluble derivatives, **11** and **12**, as lead compounds for further characterization *in vivo*. Chronic IP administration of both **12** (lung) and **11** (glioblastoma and lung) in mouse xenograft models resulted in significant reduction of tumor volume.

Recently, researchers at Infinity Pharmaceuticals identified and characterized the hydroquinone hydrochloride salt of 17-AAG, **IPI-504**, as a potent, water-soluble Hsp90 inhibitor [105,106] (Fig. 7). Protonation of the aniline nitrogen present in the 17-AAG hydroquinone sufficiently decreases electron den-

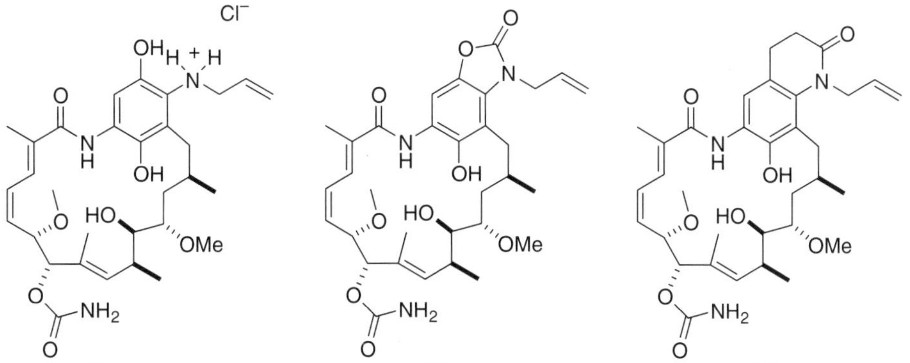

Figure 7. Hydroquinone derivatives of GDA.

sity within the aromatic ring, thereby reducing oxidative potential of the hydroquinone, and allowing for its stable isolation as the hydrochloride salt. Several GDA analogs containing 17-amino substituents were converted to the corresponding hydrochloride salts and tested for Hsp90 inhibition. In an attempt to prevent the quinone/hydroquinone equilibria normally associated with GDA and lock the analogs into the hydroquinone oxidation state, reduced GDA derivatives were prepared (**13** and **14**). These derivatives were significantly less active (~10-fold) in Hsp90 binding assays compared to the "unlocked" compounds, suggesting that the freely rotatable, redox-active component is necessary for optimal affinity and inhibition. Further *in vitro* and *in vivo* development identified IPI-504 as a more potent inhibitor of Hsp90 than 17-AAG [106]. IPI-504 reproducibly bound Hsp90 with approximately twofold higher affinity than 17-AAG, $EC_{50} = 63$ versus 119 nM, respectively. Cancer cell lines were found to be more susceptible to both IPI-504 and 17-AAG than their normal cell counterparts. Combination therapy of IPI-504 and bortezomib, a clinically useful proteasome inhibitor, resulted in synergistic cytotoxic effects, strongly supporting the use of Hsp90 inhibitors in combination studies. In addition, IV administration of IPI-504 significantly reduced tumor progression in an *in vivo* mouse xenograft model of multiple myeloma. Most importantly, both *in vitro* and *in vivo*, IPI-504 interconverts to 17-AAG through the establishment of an oxidative-reductive equilibrium. This finding demonstrates that the active form of the extensively studied 17-AAG and other GDA analogs is in fact the hydroquinone species. Because of its improved pharmacokinetic properties and its potent Hsp90 inhibition, IPI-504 has entered into several Phase I and Phase II clinical trials [99,100].

2.1.2. Other Semisynthetic GDA Derivatives

A structure-based approach for probing the effects of displacing key conserved water molecules identified via the GDA-Hsp90 cocrystal structure led to a series of studies involving 7-carbamate modifications [107]. These results determined the conserved water molecules and the 7-carbamate are important for optimal binding. Only one derivative, **15**, containing a 7-hydroxamate and the 17-dimethylaminoethylamino moiety of 17-DMAG retained weak Hsp90 inhibitory activity and exhibited a K_d of 18 μM (17-DMAG, $K_d = 0.5$ μM) (Fig. 8). The 7-hydroxamate substituent is not suspected to displace water molecules and likely accounts for its ability to bind Hsp90. All other analogs prepared with 7-carbamate substitutions predicted to displace the conserved waters, including pyrrole, imidazole, and cyclic carbamate derivatives, were devoid of activity.

A series of simplified GDA analogs that included an "unnatural" 18-membered analog was also prepared and evaluated for Hsp90 inhibitory activity [108]. Molecular modeling of this analog, **16**, showed that it adopted the bent conformation required for optimal Hsp90 binding and inhibition (see Fig. 4). Biological evaluation of these analogs showed only weak

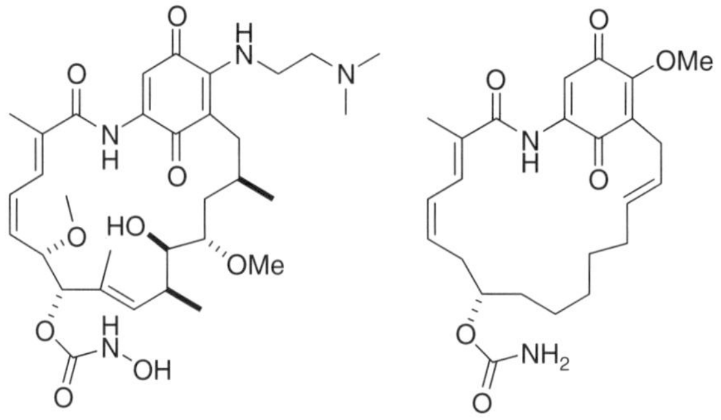

Figure 8. Synthetic derivatives of GDA.

Hsp90 inhibitory activity, demonstrating that removal of multiple functionalities is detrimental to inhibitory activity. Of interest, antiproliferative effects in human colon cancer cells with these compounds was observed at modest concentrations (25–40 µM), however, the researchers noted this may be independent of Hsp90 inhibition.

2.1.3. Bioengineered GDA Analogs Because the published total syntheses of GDA have proven unamenable to full-scale SAR determinations [109,110], several groups have taken a complementary approach to producing novel GDA analogs based on either the analysis of GDA fermentation products or through genetic engineering of the biosynthetic pathway from S. hygroscopicus strains [111,112]. The fermentation products isolated from recombinant strains of herbimycin-producing bacteria supplemented with GDA produced five novel GDA analogs that were tested for cytotoxic activities against SKBr3 human breast cancer cells (Fig. 9) [111]. The most active product afforded was 15-hydroxygeldanamycin (15-OHGDA), which exhibited an IC$_{50}$ value of 0.71 µM against SKBr3 cells. KOSN-1633, KOSN-1645, and KOSN-3163 were less active (IC$_{50}$ values 1.5–3.1 µM) and methyl geldanamycinate was completely inactive in vitro, further highlighting the importance of the flexible macrocycle.

Genetic engineering of the geldanamycin polyketide synthase (GDA-PKS) gene cluster from S. hygroscopicus was undertaken to explore modifications of GDA at positions other than the widely studied 17-position (Fig. 10) [112]. Substitutions of acyltransferase domains in six of the seven GDA-PKS modules led to isolation of several new desmethyl GDA derivatives. The analog that bound Hsp90 with the greatest affinity was KOSN1559, which manifested a K_d of 16 nM. Although KOSN1559 binding to Hsp90 was greatly improved compared to 17-AAG (I_d = 1300 nM), its in vitro cytotoxic activity against SKBr3 cells was significantly reduced (IC$_{50}$ = 860 nM). For the KOSN analogs tested, IC$_{50}$ values for in vitro cytotoxicity ranged from 470 to >5000 nM, compared to 17-AAG (IC$_{50}$ = 33 nM), suggesting that even slight changes in conformation and removal of a single methyl group can greatly effect Hsp90 binding. It is important to note that the phenyl ring of KOSN1559 does not possess the redox-active quinone moiety, providing an alternative and less toxic scaffold upon which new GDA derivatives can be pursued.

Inactivation of the carbamoyltransferase gene of the GDA biosynthetic pathway led to the production of two new 4,5-dihydrogeldanamycin derivatives, **17** and **18** (Table 2) [113,114]. In order to explore SAR for this series of compounds, numerous semisynthetic derivatives were prepared and screened for antiproliferative activity against human breast (SKBR3) and ovarian (SK-Ov3) cancers [114]. These analogs incorporated modifications to the 7-carbamate, 11-hydroxyl, and 17-methoxy functionalities. Not surprisingly, the most active compounds identified from this series mimicked 17-AAG and 17-DMAG (Table 2). For the 4,5-dihydrogeldanamycin analogs, 7,11-bis-carbamylation yielded inactive deri-

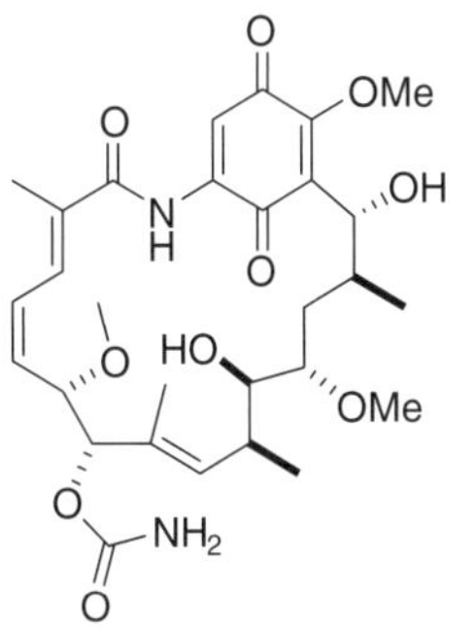

15-Hydroxygeldanamycin

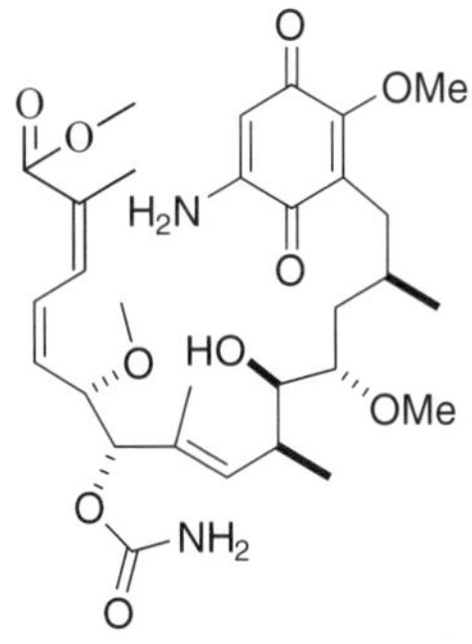

Methyl geldanamycinate

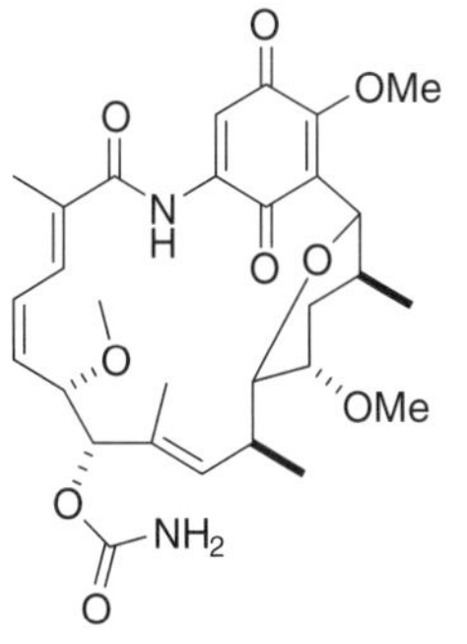

KOSN-1633

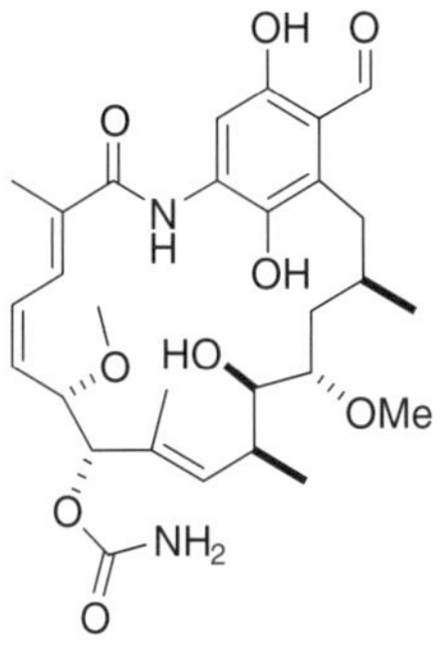

KOSN-1645

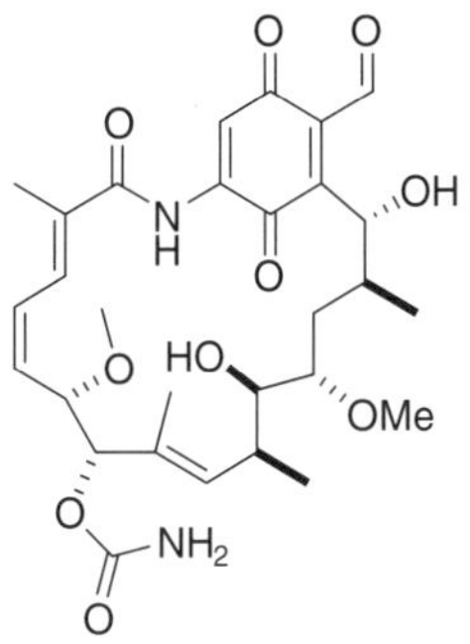

KOSN-1646

Figure 9. GDA analogs produced from fermentation.

vatives when the allylamino side chain of 17-AAG was introduced (**19**), however, the addition of tertiary amides at the 17-position restored cytotoxic activity (**20** and **21**). Replacement of the 11-carbamate with the alcohol also yielded derivatives that exhibited potent Hsp90 inhibitory activity (**22–24**). Surprisingly, the bis-carbamate of the diethylaminoethylamino 4,5-dihydro GDA derivative (**27**) was inactive against SKBr3 cells at the concentrations tested. However, cyclization of the 17-amine side chain of **27** to the corresponding pyrrolidine, **28**, restored cytotoxicity. Three analogs, **21**, **26**, and **28** were chosen for further analysis of Hsp90 inhibitory properties. Each compound induced degradation of Hsp90 client proteins in a time-dependent manner and **28** completely abolished Her2 levels after 8h of exposure.

Other groups have performed preliminary studies on the biosynthesis of the benzoquinone moiety of GDA [115,116]. In particular, mutational strains of GDA producing bacteria have been utilized as a method to prepare analogs that introduce additional functionalities to the quinone ring. While several novel GDA analogs have been prepared via this technique, their *in vitro* Hsp90 inhibitory activities have not yet been reported.

2.1.4. GDA Dimers and Conjugates The homodimeric nature of Hsp90 has led several research groups to study the inhibitory activity of GDA dimers [117,118]. The first reported series of dimers linked two GDA monomers through various carbon tethers [117]. GDA dimers that incorporated linkers $\leq$ seven carbons retained Hsp90 inhibitory activity comparable to GDA (IC_{50} values = 60–70 and 45 nM, respectively). Interestingly, one compound, **29**, demonstrated selective degradation of Her2 compared to Raf-1, which are both Hsp90 clients involved in oncogenic signaling pathways (Fig. 11). Additional studies demonstrated that **29** induced selective degradation of Her-family kinases and showed increased

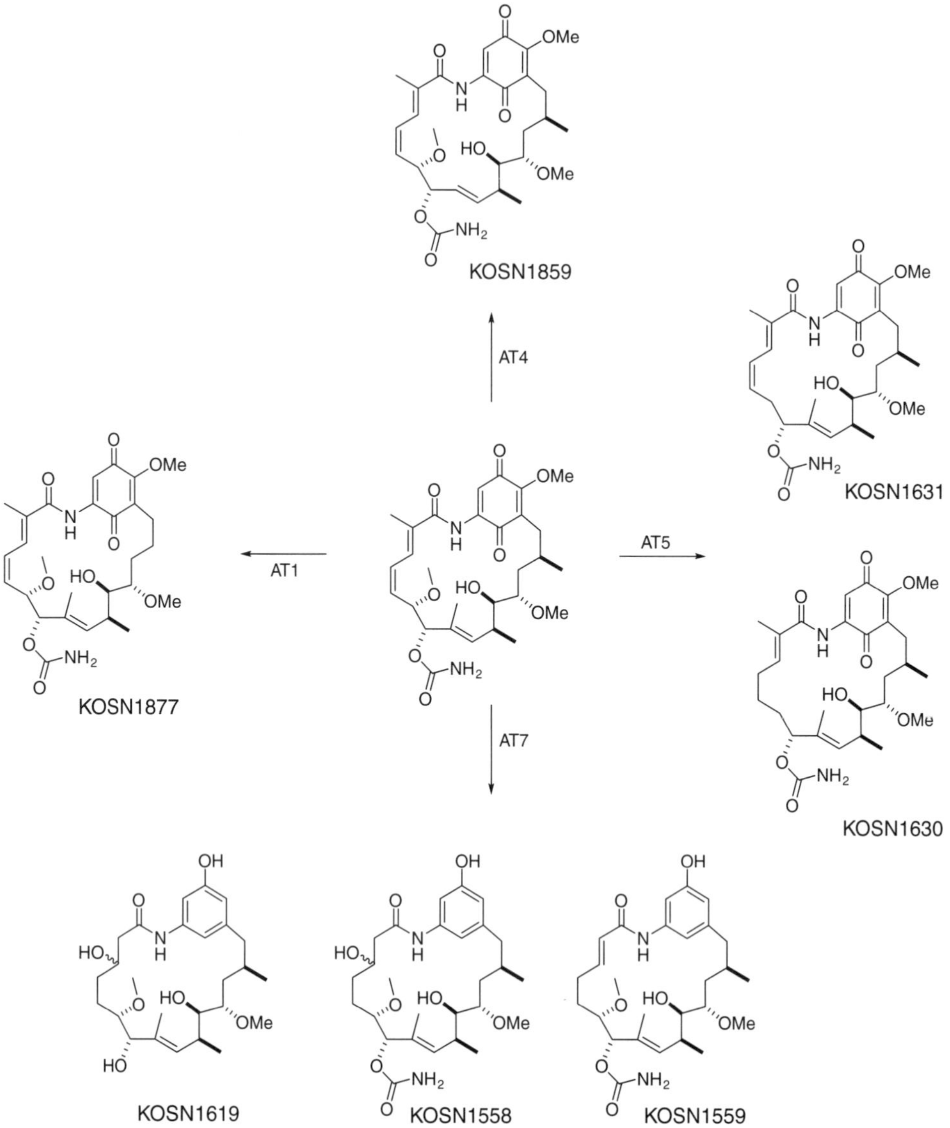

Figure 10. GDA derivatives prepared through bioengineering (Adapted from GDA28).

antiproliferative activity against tumor cell lines that overexpress Her2. Continued studies to optimize the linker of GDA dimers led to identification of several analogs that exhibited improved inhibitory activity compared to **29**, IC$_{50}$ values <20 nM [118]. Because **CF237** and **CF483** were able to induce degradation of a wide-range of Hsp90 client proteins, they were more thoroughly investigated. These dimers exhibited antiproliferative effects at low nanomolar concentrations and were more potent than 17-AAG upon 24 h of exposure. In addition, these dimers delayed recovery of the oncogenic signaling proteins to basal levels compared to 17-AAG. **CF237** reduced tumor growth in a mouse xenograft model of lung cancer, verifying these GDA dimers maintain potent *in vivo* anticancer activity.

Various GDA conjugates have also been prepared to target specific Hsp90 client proteins, specifically the steroid hormone receptors [119,120]. The most active linkers for

Table 2. Derivatives of GDA and 4,5-Dihydro GDA

7–24   **25–28**, 17–AAG, 17–DMAG

Compound	R	R_1	R_2	SKBr-3	SK-OV3
17	H	H	OMe	>10	>10
18	$CONH_2$	H	OMe	1.4	>10
19	$CONH_2$	$CONH_2$	-NH-CH₂-CH=CH₂	>10	10.5
20	$CONH_2$	$CONH_2$	-NH-CH₂CH₂-N(CH₃)	0.32	5.09
21	$CONH_2$	$CONH_2$	-NH-CH₂CH₂-N(Et)₂	0.01	1.14
22	$CONH_2$	H	-NH-CH₂-CH=CH₂	1.02	10.22
23	$CONH_2$	H	-NH-CH₂CH₂-N(CH₃)	0.03	1.54
24	$CONH_2$	H	-NH-CH₂CH₂-N(Et)₂	0.02	1.66
25	$CONH_2$	$CONH_2$	-NH-CH₂-CH=CH₂	0.05	6.97
26	$CONH_2$	$CONH_2$	-NH-CH₂CH₂-N(CH₃)	0.02	0.67
27	$CONH_2$	$CONH_2$	-NH-CH₂CH₂-N(Et)₂	>10	2.95
28	$CONH_2$	$CONH_2$	-NH-CH₂CH₂-pyrrolidine	0.51	2.07
17-AAG	$CONH_2$	H	-NH-CH₂-CH=CH₂	0.02	5.16
17-DMAG	$CONH_2$	H	-NH-CH₂CH₂-N(CH₃)	0.01	0.04

a series of GDA-estradiol hybrids contained the unsaturated *trans* or alkyne tethers. Saturation of the alkyl linker resulted in complete loss of Hsp90 inhibitory activity. Compound **30** reduced levels of both estrogen and nonestrogen (Her2) client proteins in MCF-7 cells (Fig. 12) [119]. Interestingly, **30** exhibited little or no effect on Raf levels,

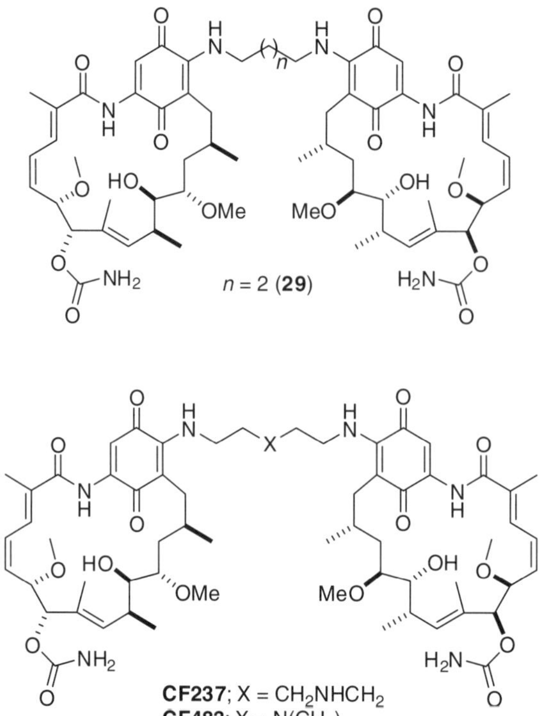

Figure 11. GDA dimers.

indicating that hybridization of these two moieties can result in selective client protein degradation. Based on these results, a similar series of GDA-testosterone hybrids was synthesized and evaluated [120]. Similar to the estradiol hybrids, the alkyne linker in **31** manifested the highest level of cytotoxicity against LNCaP cells, an androgen receptor-dependent human prostate cancer cell line ($IC_{50} = 100$ nM). Longer or shorter saturated and unsaturated alkyl chains were inactive ($IC_{50} > 1\,\mu$M) and established the five-carbon linker as optimal. While **31** demonstrated reduced cytotoxicity compared to GDA, its ability to induce degradation of the mutated androgen receptor suggests its potential use in advanced stages of hormone-independent prostate cancers.

The ability of GDA to modulate Hsp90 function also makes it an ideal ligand for the exploitation of low affinity ligands for other therapeutic targets. Chiosis and colleagues prepared conjugates of GDA and LY292223, a small-molecule inhibitor of phosphoinositol-3 kinase (PI3K), in hopes that the bifunctional ligand could selectively regulate PI3 kinase activity [121]. These hybrids, with linkers varying in composition, length, and site of attachment to LY292223 demonstrated improved efficacy and specificity toward the inhibition of PI3K and PI3K-related proteins. The most selective analog, **LY6-GM**, exhibited a twofold increase in selectivity for DNA-dependent protein kinase (DNA-PK) versus PI3K. However, the most potent heterodimeric ligand, **PI3K-2-GM** showed increased inhibition of both PI3K and DNA-PK.

The anti-Her2 monoclonal antibody Herceptin has shown clinical efficacy for carcinomas that overexpress Her2, an Hsp90-dependent client protein [122]. In an attempt to improve the anticancer activity of Herceptin, the GDA derivative 17-aminopropylamino-geldanamycin (17-APA-GA) was attached to lysine residues on the antibody through a stable linker [122,123]. The Herceptin-GDA (H-GDA) immunoconjugate, exhibited 10–200-fold increased antiproliferative activity against var-

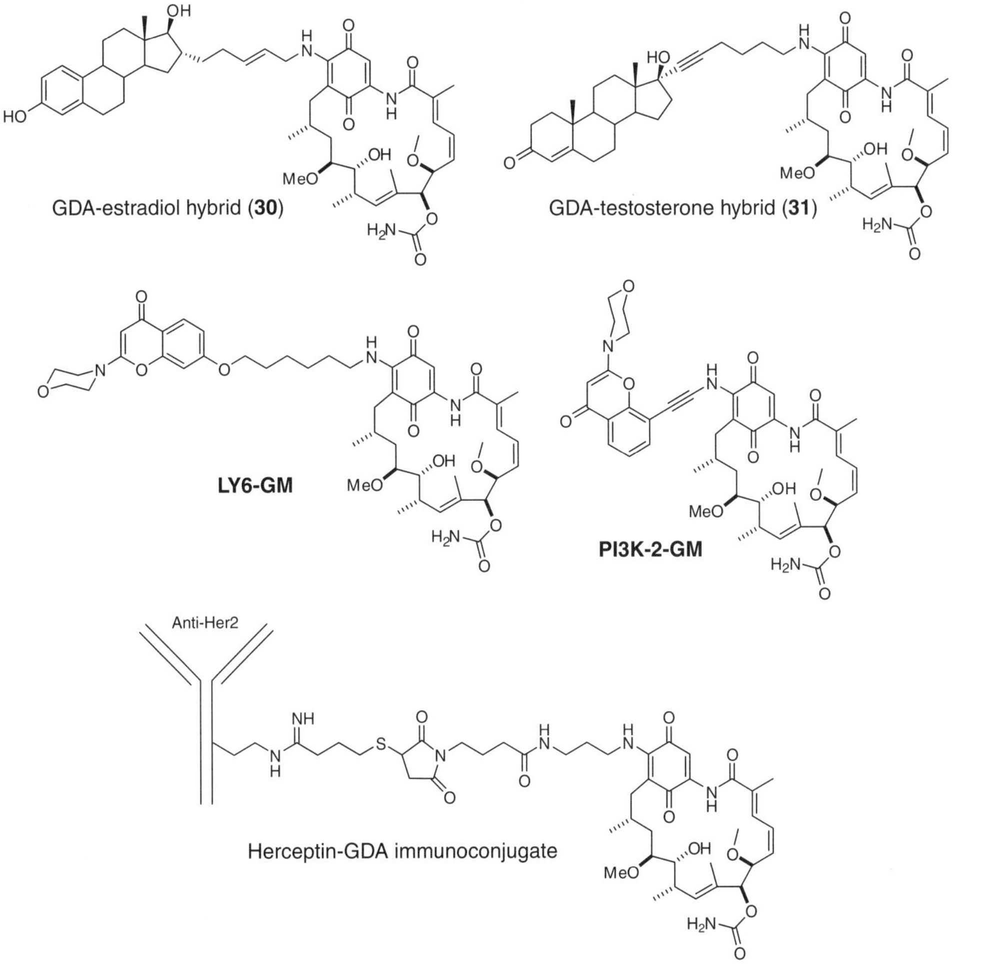

Figure 12. GDA conjugates.

ious cancer cell lines. In addition, H-GDA treatment prolonged survival of xenograft-bearing mice compared to Herceptin treatment alone through decreased tumor growth and induced tumor regression. These results suggest that immunoconjugates of this type can provide a complementary approach toward treatment of some cancers.

2.1.5. Related Benzoquinone Ansamycins
Several ansamycin antibiotics related to GDA have been reported and produce similar biological activities. These natural products include the macbecins [124], herbimycins [125], reblastatin [126], and the TAN compounds [127] shown in Fig. 13. To date, except herbimycin A (discussed above) and macbecin [128], there is limited experimental evidence to support these compounds inhibit Hsp90 *in vitro*.

Macbecin I was only recently characterized as a potent Hsp90 inhibitor [128]. It binds with higher affinity to the N-terminus of Hsp90 than GDA, ($K_d = 0.24$ and $1.2\,\mu M$, respectively, via isothermal titration calorimetry) and is more potent at inhibiting the Hsp90 ATPase activity with an IC_{50} value of $2\,\mu M$ compared to $7\,\mu M$ for GDA. Macbecin was evaluated for antiproliferative activity

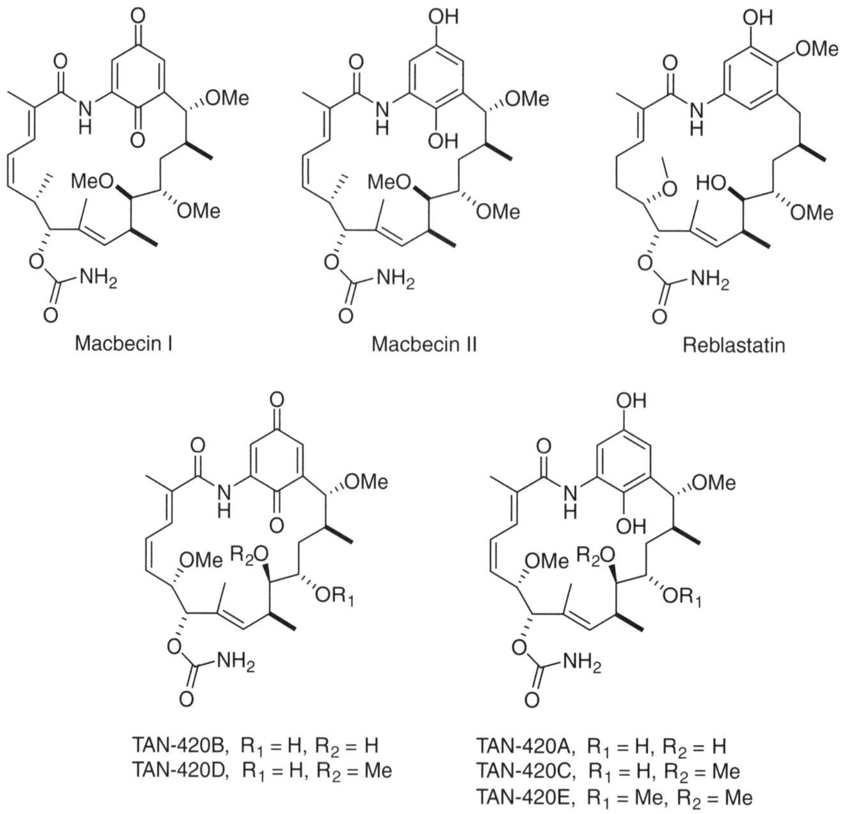

Figure 13. Related benzoquinone ansamycins.

against a panel of 38 human cancer cell lines and demonstrated an average IC_{50} of 0.4 µM. In western blot analyses, macbecin demonstrated significant degradation of Her2 and c-Raf at 1 µM with concomitant increases in Hsp70 levels, a hallmark of N-terminal inhibition. Macbecin also reduced tumor volume and doubling time in a mouse xenograft model of human prostate cancer, establishing *in vivo* anticancer activity. The cocrystal structure of macbecin in complex with the Hsp90 N-terminus shows that it maintains six of the key H-bond interactions responsible for GDA binding to Hsp90. Researchers hypothesize that the increase in binding affinity displayed by macbecin is the result of an extra hydrogen bond with Hsp90 compared to GDA, nine versus eight H-bonds, respectively. In addition, targeted molecular dynamics simulations predict that macbecin overcomes a lower energy barrier to adopt the bound, bent conformation, presumably through the lack of a C17-substituent.

After establishing macbecin as an Hsp90 inhibitor, researchers utilized a biosynthetic approach to generate analogs that were predicted to exhibit improved activity [129]. Genetic engineering of the biosynthetic machinery responsible for the production of macbecin resulted in a series of bacterial mutants that generated several new analogs, including a nonquinone compound (**32**) that demonstrated improved binding affinity for Hsp90 with a K_d value of 3 nM (Fig. 14). In addition, **32** showed enhanced water solubility and demonstrated comparable antiproliferative activity over a range of six human cancer cell lines (average IC_{50} = 20 nM). In contrast to the quinone-containing ansamycins, compound **32** exhibited *in vivo* antitumor efficacy at concentrations at which no toxicity was observed, suggesting its potential as a clinical candi-

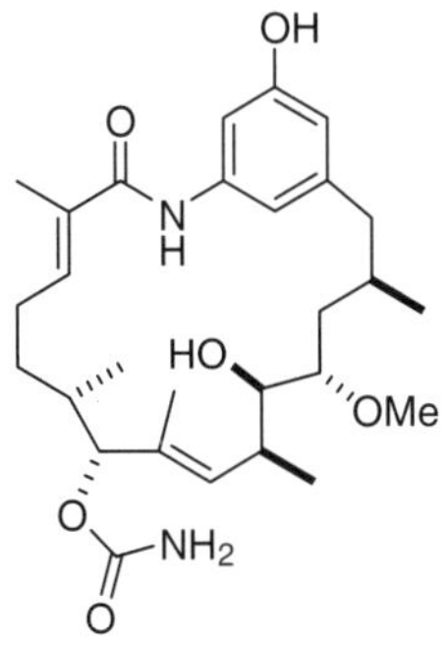

Figure 14. Genetically engineered nonquinone macbecin analog.

date. The cocrystal structure of **32** in complex with Hsp90 demonstrated the molecule adopted the same bent conformation as that exhibited by GDA, 17-AAG, and macbecin. The contributing factor that leads to improved binding is believed to result from saturation of the 4,5-olefin. Saturation in this region of the molecule results in a higher population of the bent conformer in solution, thus reducing the entropic barrier.

Screening of a *Streptomyces* strain isolated from Indonesian soil provided the first solid evidence in support of reblastatin as an Hsp90 inhibitor (see Fig. 13) [130]. Reblastatin was isolated along with several other structurally related ansamycins and exhibited potent Hsp90 inhibitory activity. Reblastatin bound with high affinity to the Hsp90 N-terminal ATP-binding site with a K_d of 2.3 nM (GDA, $K_d = 57$ nM, via DELFIA assay) and induced degradation of Hsp90 client proteins Her2 and Raf-1 at submicromolar concentrations (IC$_{50}$ 330 nM). In addition, reblastatin inhibited the growth of several human cancer cell lines with an average IC$_{50}$ value of ∼1 μM. Solution of the crystal structure of reblastatin revealed that in its native state, it exists in a cup-shaped conformation and the amide bond resides in the *cis*-conformation, suggesting that it may have an inherent advantage over GDA and related analogs.

2.1.6. Summary of GDA Analogs
The potent *in vitro* and *in vivo* activity of GDA and its analogs has led numerous groups to develop research projects around its optimization and development. However, the inherent toxicity associated with the benzoquinone moiety has hindered its clinical development. The identification of 17-position substituted analogs that demonstrate reduced redox cycling has produced a number of active compounds that exhibit improved toxicity profiles. Recent identification by two separate groups, that the related benzoquinone ansamycins that do not contain the quinone moiety and possess potent antitumor activity, has provided new opportunities for investigation. In addition, it appears as though future analogs must maintain the ability to adopt the bent conformation, which is necessary for high-affinity binding and optimal Hsp90 inhibitory activity. A summary of key GDA analogs and their pertinent Hsp90 inhibitory properties is provided in Table 3.

2.2. Radicicol

2.2.1. Radicicol, an Hsp90 Inhibitor
Radicicol (RDC, Fig. 15) is a 14-membered macrolide isolated from the culture broth of *Monosporium bonorden* in 1953 as an antifungal antibiotic [131]. Subsequent studies aimed at investigation of its antitumor activity confirmed that RDC possessed the ability to suppress transformation of *src*, *ras*, and *mos* oncogenes [132,133] and was therefore considered a specific tyrosine kinase inhibitor. Two independent groups later demonstrated the ability of RDC to inhibit Hsp90 and exert its antitumor activity via this mechanism [134,135]. In addition, RDC demonstrated competitive inhibition versus GDA for binding to the N-terminus of purified human Hsp90 and significantly decreased cellular levels of Hsp90-dependent oncogenic signaling proteins in human breast cancer cell lines [134]. These experiments confirmed that like geldanamycin, RDC exerts its anticancer activity through inhibition of the Hsp90 protein folding machinery.

The crystal structure of RDC bound to yeast Hsp90 confirmed that similar to GDA, it bound Hsp90 at the N-terminal ATP-binding pocket and provided insight into the key binding interactions of the resorcinolic ring (Fig. 16) [77]. The 3-phenol forms hydrogen-bonding interactions with Asp79 through a

Table 3. Biological Activities of GDA and Analogs[a]

Compound	Hsp90 Binding K_d (µM)	Antiproliferation IC_{50} (µM)	Client Protein Degradation IC_{50} (µM)	In Vivo Activity
GDA	0.2	0.1	0.1	Yes
17-AAG	0.034	0.123	0.2	Yes
17-DMAG	0.062	0.053	0.05	Yes
IPI-504	0.063	0.03	0.02	Yes
KOSN1559	0.016	0.86	ND	ND
28	ND	1	10	ND
29	ND	0.05	0.06	ND
CF237	ND	0.2	0.013	Yes
Herceptin-GDA	ND	0.6	0.2	Yes
Macbecin	0.24	0.4	5	Yes
32	0.003	0.02	ND	Yes
Reblastatin	0.0023	1	0.3	ND

[a] Values were gathered from various references and reflect general activities of these compounds.
ND: not disclosed.

conserved water molecule, mimicking key interactions of the natural adenine base. In addition, the 5-phenol forms a hydrogen bond with Leu34. The chloro-substituent partially fills a hydrophobic pocket adjacent to Phe124 that is important for optimal Hsp90 binding and inhibition. In fact, monocillin I, which lacks the chlorine atom present on RDC, exhibits substantially lower affinity for Hsp90 than RDC.

RDC exhibits higher affinity for full-length homodimeric Hsp90 than GDA ($K_d = 19$ and 1.2 µM, respectively, via isothermal titration calorimetry) and is a significantly more potent inhibitor in the purified recombinant Hsp90 ATPase assay [77]. RDC is not redox active, does not exhibit hepatotoxicity like GDA and its derivatives and has demonstrated potent cytotoxicity against retinoblastoma-negative cells known to be resistant to 17-AAG. However, two major issues have hindered the development of RDC as a clinically useful anticancer agent. First, RDC has yet to demonstrate differential selectivity toward tumorgenic Hsp90 heteroprotein complexes similar to GDA derivatives. Second, RDC lacks antitumor activity *in vivo* [136]. It has been hypothesized that the allylic epoxide and the $\alpha,\beta,\gamma,\delta$-unsaturated ketone undergo rapid metabolism *in vivo* to result in inactive metabolites that exhibit little or no affinity for Hsp90. To substantiate this hypothesis, it has been demonstrated that thiol-derived nucleophiles inactivate RDC through a similar mechanism [137].

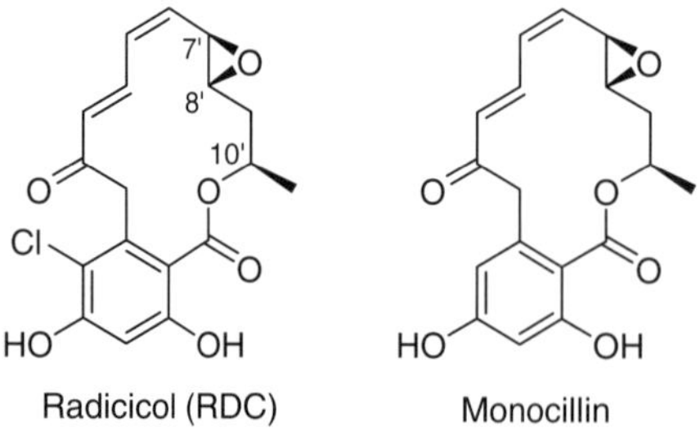

2.2.2. Analogs of RDC In an attempt to decrease the electrophilic nature of these functionalities, a series of 6-oxime and halohydrin derivatives of RDC was prepared and evaluated for anticancer activity [137–140]. The first of these derivatives reported was the potent *N*-oxime, KF25706, and its significantly less active analog KF29163 (Fig. 17) [140]. Both of these derivatives were synthesized and evaluated as a mixture of both the *E*- and the *Z*-isomers. While KF25706 demonstrated potent antiproliferative activity comparable to RDC against a wide variety of human cancer cell lines, KF29163 was significantly less active, indicating the epoxide is important for activity. KF25706 also decreased levels of several oncogenic Hsp90 client proteins *in vitro*, including p185[erbB2],

Figure 15. Structures of radicicol (RDC) and monocillin I.

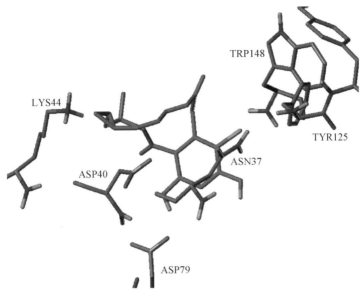

Figure 16. Cocrystal structure of RDC (magenta) bound to Hsp90. (See color insert.)

Raf-1, Cdk4, and p53. In addition, KF25706 depleted levels of Raf-1 and Cdk4 in an *in vivo* mouse xenograft model of human breast cancer, whereas both KF29163 and RDC did not. In an attempt to determine which oxime stereoisomer was responsible for Hsp90 inhibitory activity, the *E*- and *Z*-isomers of KF55823 were purified and evaluated [139], leading to identification of the *E*-oxime (KF58333) as the biologically active compound. KF58333 exhibited a 3–12-fold improvement in antiproliferative activity against human

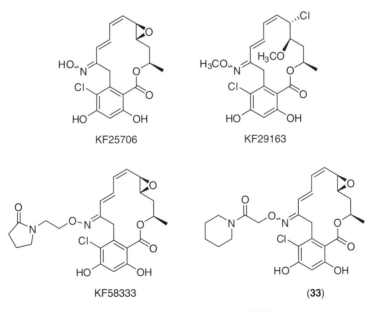

Figure 17. *N*-Oxime analogs of RDC.

breast cancer cell lines compared to the corresponding Z-isomer. In addition, it maintained potent inhibition of tumor growth in xenograft models, whereas the Z-isomer was inactive.

The increased stability and in vivo activity associated with KF25706 and KF58333 led Ikuina and coworkers to develop a series of O-carbamoylmethyloxime derivatives of RDC that focused on optimization of the oxime group and to probe SAR for this class of Hsp90 inhibitors [138]. Similar to previous studies, these compounds were prepared as a mixture of both E- and Z-isomers. Their antiproliferative activity was measured against normal cells and their transformed counterparts. This series of RDC oxime analogs was also studied for their ability to inhibit v-Src induced tyrosine autophosphorylation. In general, the oximes were 2–3-fold more active against transformed cell lines, providing a small therapeutic window for development of these compounds. The introduction of polar functional groups, including hydroxyl and carbamoyl groups, reduced antiproliferative activity for these molecules. Incorporation of an aromatic or heteroaromatic ring produced analogs that ranged from 2–11-fold less active than the parent compound, KF25706. One analog, **33** (1/6 E/Z ratio), displayed in vitro activity warranting further evaluation as the lead N-oxime RDC derivative. Compound **33** demonstrated better activity at Raf-1 depletion and inhibition of Erk2 phosphorylation, compared to KF25706. In an attempt to further clarify the activity of the E- and Z-isomers, **33** was synthesized in a 1/2:E/Z ratio and evaluated for Hsp90 inhibition. This compound showed improved antiproliferative activity and selectivity for transformed cell lines, indicating the E isomer is more potent. Finally, **33** (1/2:E/Z) demonstrated improved in vivo activity in mouse xenograft models compared to KF25706, providing a new lead compound for the N-oxime RDC analogs.

Although total syntheses of RDC have been reported [141], recent work by Danishefsky and coworkers led to succinct production of the natural product that ultimately afforded the first SAR between RDC and Hsp90 [142–145]. Evaluation of inhibitory activities for RDC analogs that contained epimers at the 7′-, 8′-, and 10′-positions demonstrated the natural configuration is optimal for Hsp90 inhibition [143]. Deviation from the absolute configuration of the natural product led to 40–94-fold decreases in Hsp90 inhibitory activity. In an attempt to replace the electrophilic epoxide with a more stable isostere, a cyclopropyl analog of RDC (cyclopro-paradicicol, c-RDC) was prepared (Fig. 18). The cocrystal structure of RDC bound to Hsp90 suggests the epoxide oxygen of RDC forms a hydrogen bond with Lys44, and may be critical for inhibitory activity. The antiproliferative and Her2 degradatory activities of c-RDC were found to be comparable to RDC in the estrogen receptor positive human breast cancer cell line MCF-7, suggesting that the conformation exerted on the macrocycle by the three-membered ring is more important than the hydrogen bond formed between the epoxide and the Hsp90. Not surprisingly, only the c-RDC analog containing the natural configuration at the 7′-, 8′-, and 10′-positions demonstrated in vitro inhibitory activity.

Further improvements toward the synthesis of RDC involved the "ynolide method" (Scheme 1), which provided several new c-RDC analogs for SAR studies [145]. As noted previously, the conformation of RDC appears

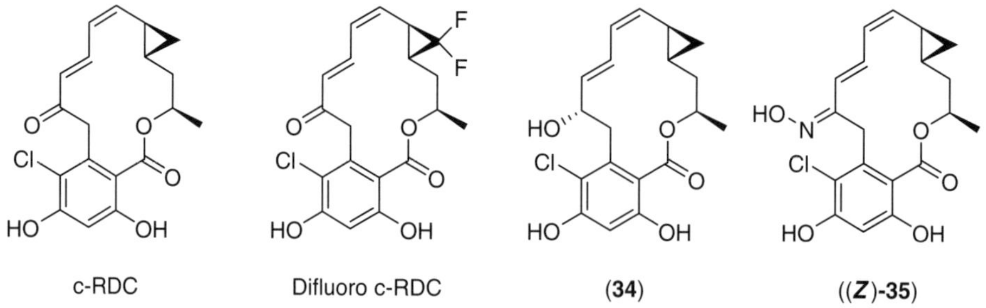

Figure 18. Cycloproparadicicol (c-RDC) and representative analogs.

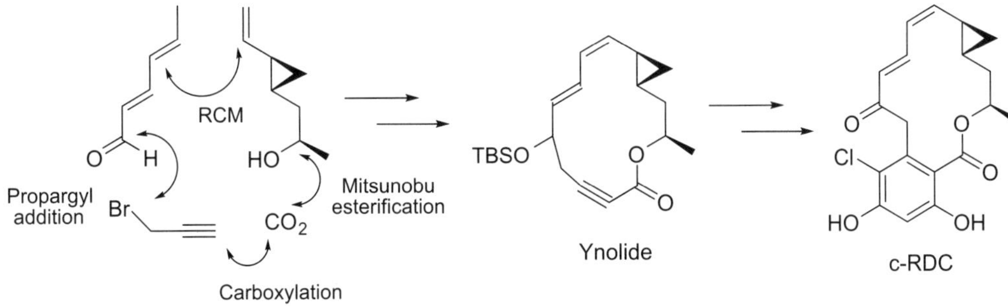

Scheme 1. Ynolide approach to cycloproparadicicol.

to be the most important feature for Hsp90 inhibition. This finding is further supported by the difluoro c-RDC derivative, which was proposed to reestablish hydrogen-bond interactions with Lys44 that are lost upon replacement of the epoxide with a cyclopropane ring. Surprisingly, the difluoro analog exhibited lower activity ($IC_{50} = 10\,\mu M$). Furthermore, reduction of the c-RDC ketone to the corresponding alcohol, **34**, also resulted in ~8-fold decrease in inhibitory activity. In contrast to the results previously observed for the 6-oxime derivatives of RDC, the Z-oxime [(**Z**)-**35**] of c-RDC was threefold more active than the E-isomer. The *in vivo* inhibitory activity of these analogs has not yet been reported.

Conformational analyses of RDC and analogs have provided important information about the bioactive conformation of this Hsp90 inhibitory class [146]. A molecular dynamics/minimization study of RDC and several related resorcinolic macrolides provided data to support these analogs adopt the bioactive cup-shaped conformation, resembling RDC. Results from these studies further supported previous findings suggesting that the key feature for Hsp90 inhibitory activity is the conformation of the epoxide and 10-methyl substituent. Three distinct conformations were identified for radicicol, and the lowest energy conformer represented the bioactive conformation (C-shape) as confirmed through solution of the cocrystal structure. Both inversion of configuration at the epoxide and conversion to the diol resulted in analogs in which the lowest energy conformer was no longer C-shaped. Inversion of stereochemistry at the 10-position also generated data consistent with experimentally observed inactive analogs. Cycloproparadicicol and the N-oxime, **33**, were susbequently evaluated for their ability to adopt the C-shaped bioactive conformation of RDC. Both of these analogs were found to exhibit a high propensity for the active conformation. Interestingly, the Z-isomer of **33** demonstrated a more favorable conformational bias toward the C-shape, consistent with previous findings for the c-RDC oximes described above. Finally, replacement of the epoxide with an olefin, as found in the pochinin class of compounds (see below) provided only modest disfavoring of the C-shape conformation.

Recent preparation of a series of synthetic macrolactones with simplified ring structures has provided additional insight into the conformational requirements for Hsp90 binding by RDC [147]. Removal of the C10-methyl group on analogs that maintain either the epoxide (**36**) or the olefin (**37**) resulted in ~5–10-fold reduction in Hsp90 inhibitory activity as measured by ATPase assays (Fig. 19). Conversion of the resorcinol phenols to the corresponding methoxy substituents completely abolished inhibitory activity, indicating the hydrogen-bond network maintained by these moieties is critical. In addition, 15- (**38**) and 16-membered (**39**) ring homologs were prepared and evaluated. Both ring-expanded derivatives maintained the ability to bind Hsp90, with **38** demonstrating only modest reduction in inhibitory activity compared to RDC ($IC_{50} = 0.76\,\mu M$ versus $0.2\,\mu M$, respectively).

2.2.3. Related 14-Membered Resorcinolic Macrolides
Several other compounds isolated and characterized as 14-membered resorcino-

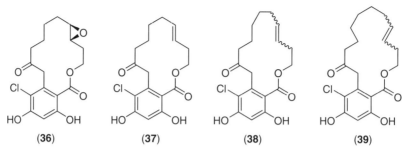

Figure 19. Simplified macrolactone RDC analogs.

lic macrolides have been evaluated for Hsp90 inhibitory activity (Fig. 20). The aigailomycins were recently isolated from the marine fungus *Aigialus parvus* BCC 5311 [148]. The most active of these, aigialomycin D, exhibited potent antimalarial and antitumor activity. Its structural similarity to radicicol led to several syntheses of the compound and determination of its ability to inhibit Hsp90 [145,149,150]. Although aigialomycin D demonstrated potent inhibition of CDK1, CDK5, and GSK3 kinases, it did not inhibit Hsp90.

Pochonin D was isolated from the fermentation broth of the fungi *Pochonia chlamydosporia* as one of a series of 14-membered macrolides including RDC and monocillin [151]. A molecular dynamics/minimization study of pochonin D within the Hsp90 N-terminal ATP-binding site demonstrated significant overlap with RDC and prompted Winssinger and colleagues to develop a polymer-supported synthesis of the pochonins [146]. Pochonin D exhibited high affinity for the N-terminal Hsp90 ATP-binding site and was only fourfold less active than RDC (K_d = 80 and 20 nM, respectively). Not surprisingly, deschloro pochonin D was inactive, further highlighting the importance of the chloro substituent on the resorcinol ring. Pochinin A has also been

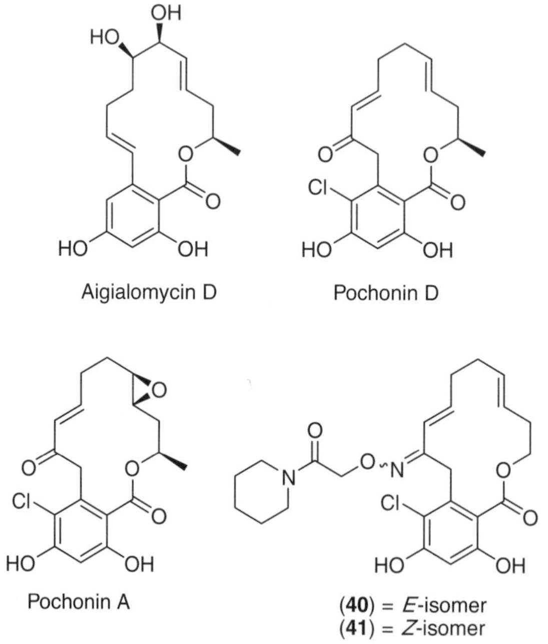

Figure 20. RDC-related 14-membered resorcinolic macrolides.

prepared and shown to be an Hsp90 inhibitor ($IC_{50} = 90$ nM) [152]. Conversion of the pochinin A epoxide to the corresponding diol abolished all inhibitory activity. The divergent synthesis of a pochinin library aimed at modification of the ketone resulted in preparation of N-oxime analogs that exhibits improved Hsp90 inhibitory activities [153]. The most potent of these analogs, **40**, incorporated the same N-oxime identified for the RDC analog, **33**. **40** exhibited significant increase in binding affinity for Hsp90 and Hsp90-dependent client protein degradation (Her2) compared to both RDC and pochinin D. It also demonstrated low nanomolar cytotoxicity in human breast cancer cell lines ($IC_{50} = 125$ and 320 nM in SKBr3 and HCC1954, respectively). In addition, treatment of BT-474 xenograft bearing mice with **40** resulted in dose-dependent inhibition of tumor growth. Consistent with the results for the RDC oximes, the *E*-isomer was more active than the corresponding *Z*-isomer, **41**. While **41** bound Hsp90 with only a threefold decrease in affinity, it demonstrated a 10–20-fold decrease in *in vitro* assays.

2.2.4. Summary of Structure–Activity Relationships for RDC and Analogs

The RDC series of compounds have provided insights into the key features necessary for Hsp90 inhibition. The most important feature for designing RDC analogs appears to be maintenance of the natural conformation of the macrocycle, including the absolute configuration at the 10′-position. The chlorine and free phenols of the resorcinol are also necessary for optimal activity. In addition, the liabilities associated with the electrophilic nature of RDC can be reduced through derivation of the $\alpha,\beta,\gamma,\delta$-unsaturated ketone, most notably to the corresponding N-oxime. However, a discrepancy exists as to which of the oxime configurations is responsible for *in vivo* activity. A summary of key RDC analogs and their Hsp90 inhibitory activities is summarized in Table 4.

2.3. Geldanamycin and Radicicol Chimeras

As noted previously, both GDA and RDC maintain inherent functionalities that make their continued development as Hsp90 inhibitors problematic. The superimposed cocrystal structures of GDA and RDC (Fig. 21) show the resorcinol ring of RDC to occupy the same binding region as the carbamate of geldanamycin. This is also the region in which the purine ring of ATP interacts with the binding pocket [77]. In contrast, the quinone of GDC resides in the same region as the catalytic residues necessary for ATP hydrolysis, adjacent to the protein–solution interface. It was proposed by Blagg and coworkers that compounds containing the quinone moiety of GDA and the resorcinol ring of RDC may exhibit improved binding interactions with the ATP-binding site based on their ability to take advantage of key hydrogen-bonding interactions at each location. In addition, these analogs would provide SAR for the natural products themselves, without their total syntheses. Radanamycin (Fig. 22), a macrocylic chimera of GDA and RDC that contains these two moieties linked through an amide

Table 4. Biological Activities of Pertinent RDC Analogs[a]

Compound	Hsp90 Binding K_d (μM)	Antiproliferation IC_{50} (μM)	Client Protein Degradation IC_{50} (μM)	*In Vivo* Activity
RDC	0.019	0.1	0.05	No
KF25706	0.01	0.05	0.2	Yes
KF58333	ND	0.005	0.02	Yes
33	ND	0.008	0.025	Yes
c-RDC	ND	0.054	0.2	ND
(Z)-35	ND	0.098	0.2	ND
38	0.13	10	Yes	ND
Pochonin D	0.08	10	3.5	ND
40	0.021	0.2	0.035	ND

[a] Values were gathered from various references and reflect general activities of these compounds.
ND: not disclosed.

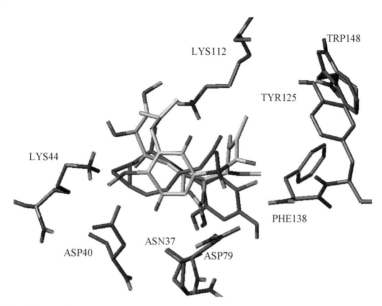

Figure 21. Superimposed cocrystal structures of radicicol (magenta) and geldanamcyin (yellow) bound to yeast Hsp90. (See color insert.)

bond derived from GDA and an ester bond derived from RDC was designed *in silico* to determine whether this class of inhibitors could complement the binding interactions of each natural product. In addition, the seco-derivatives, radamide and radester, were designed and prepared to determine the optimal tether length for connecting the quinone and resorcinol moieties.

The first chimeric compound synthesized and evaluated was the seco-amide derivative, radamide [154]. Radamide inhibited the inherent ATPase activity of Hsp90 at a level comparable to that of GDA (IC_{50} = 5.9 and 2.5 μM, respectively). Interestingly, the hydroquinone precursor to radamide was more active at Hsp90 inhibition than the corresponding quinone and was found to manifest an IC_{50} of 1.9 μM, indicating hydrogen-bond donors at this position provide analogs with improved inhibitory activities. This was further supported through preparation of a radamide analog that contained a trimethoxy-substituted ring, which yielded a compound with greatly reduced activity (IC_{50} > 50 μM). This trend continued in cellular assays as the radamide HQ was more potent at inducing degradation of the Hsp90 client protein, Her2, in MCF-7 human breast cancer cells, exhibiting an IC_{50} of ~40 μM, whereas the IC_{50} for radamide in this assay was ~70 μM. The synthetic route to prepare radamide provided a succinct method for the preparation of numerous analogs that could further elucidate SAR for the GDA quinone.

Following the preparation of radamide, the seco-ester derivative, radester, was prepared and evaluated for Hsp90 inhibitory activity [155]. The antiproliferative activity of radester and its hydroquinone in MCF-7 cells supported the activity previously identified for radamide, that is, the hydroquinone was approximately twofold more active than the quinone (IC_{50} = 14 and 7 μM, respectively). In addition, the ability of these two compounds to degrade Hsp90 client proteins (Raf and Her2) paralleled their cytotoxicities.

Upon identification of radamide and radester as Hsp90 inhibitors, the macrocyclic analog, radanamycin, was synthesized and evaluated [156,157]. Utilizing the synthetic route optimized for each of the seco-derivatives, along with the knowledge that the hydroquinone species is more active *in vitro*, radanamycin was prepared and evaluated [156]. Structural NMR studies provided evidence

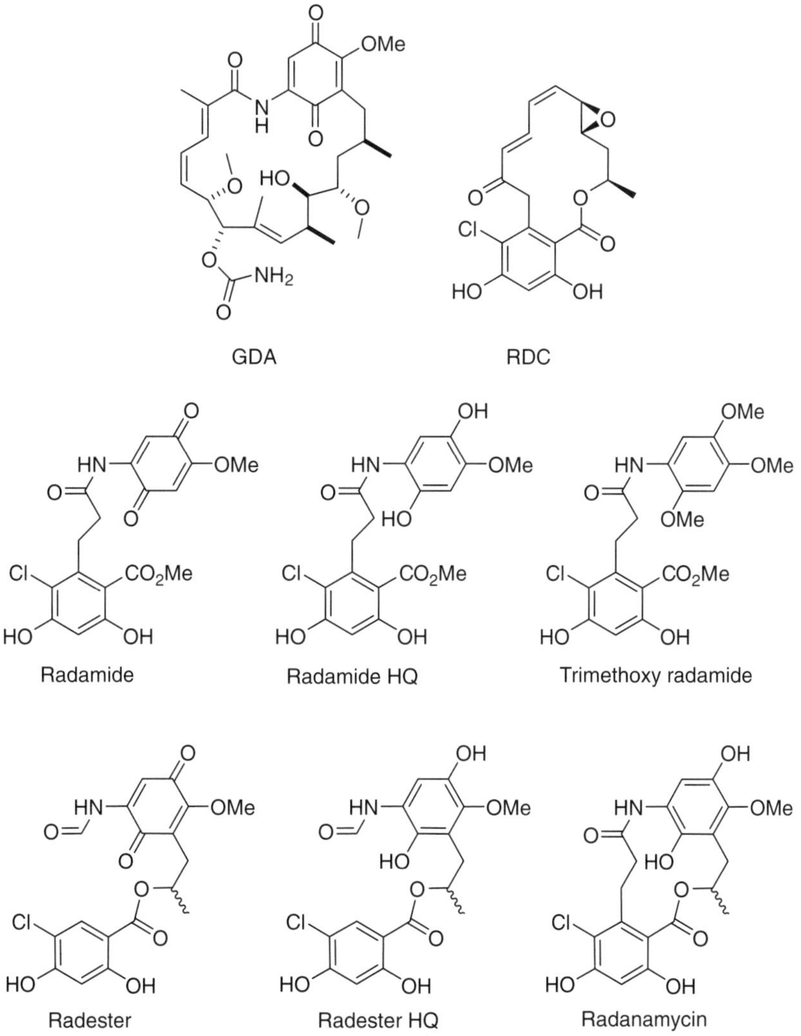

Figure 22. Chimeric inhibitors of Hsp90.

that radanamycin adopted the bent conformation observed for RDC and GDA upon binding Hsp90. In vitro analysis of radanamycin demonstrated potent antiproliferative activity against MCF-7 cells ($IC_{50} = 1.2\,\mu M$) and its ability to degrade the Hsp90 client proteins Her2 and Akt ($IC_{50} \sim 2\,\mu M$).

After establishing the chimeric molecules as Hsp90 inhibitors, expanded SAR studies were performed to identify key functionalities necessary for optimal binding and potency [158]. Radester analogs focused on quinone derivatives, analogs that contained varying tether lengths between the resorcinol and the quinone, optimization of the secondary methyl group, and the preparation of chloride replacements on the resorcinol ring (**42–45**). Consistent with results obtained for radamide, incorporation of a trimethoxy aromatic in lieu of the quinone produced decreased antiproliferative (IC_{50} values 12–100 μM) and Her2 degradatory activities (IC_{50} values 25–80 μM) compared to the quinone counterpart. No significant differences in Hsp90 inhibitory activity were observed upon alteration of tether length and consequently, subsequent analogs maintained the two-carbon linker found in radester. While slight differences in inhibitory activity could be observed for the quinone and hydroquinone analogs, the hydroquinones were generally more active. Attempts to probe the π-rich hydrophobic pocket occupied by the resorcinol chlorine atom demonstrated that increased size resulted in increased potency. Finally, the simple formamide on the quinone ring was more efficacious than larger, more hydrophobic substituents. Radester analog, **46**, was further evaluated for Hsp90 inhibitory activity via its ability to induce degradation of the client proteins, Her2 and Raf, while simultaneous inducing Hsp70 levels (Fig. 23). Its activity paralleled that observed in the Her2 ELISA assay and manifested IC_{50} values of 1–10 µM.

Radanamycin analogs were also prepared and evaluated to determine key SAR for this series of chimeras [158]. Generally, the macrocyclic analogs were more effective inhibitors than the seco-derivatives. The hydroquinones were more active than the corresponding quinones with at least a twofold increase in activity. Increasing tether length for the amide carbon chain derived from radamide resulted in significant increases in inhibitory activity. In contrast, the length of the ester linkage was less tolerable to modification. In addition, the secondary methyl group increased inhibitory activity compared to its desmethyl counterpart, similar to the radester series. Two key analogs, **47** and **48**, demonstrated concentration-dependent degradation of the Hsp90 client proteins, Her2 and Akt, with concomitant induction of Hsp90 levels. The structures and inhibitory activities of key radester and radanamycin analogs are provided in Fig. 23 and Table 4.

More recently, three series of radamide analogs have been prepared and evaluated for their Hsp90 inhibitory activities [159,160]. The first series focused on (1) replacing the quinone with less redox-active moieties to probe SAR at each position of the aromatic ring and (2) optimizing tether length between the quinone and the resorcinol rings [159]. As noted for radester and radanamycin analogs, SAR results for the radamide analogs identified the four-carbon linker as optimal for inhibitory activity. In addition, no difference in activity was observed *in vitro* for the

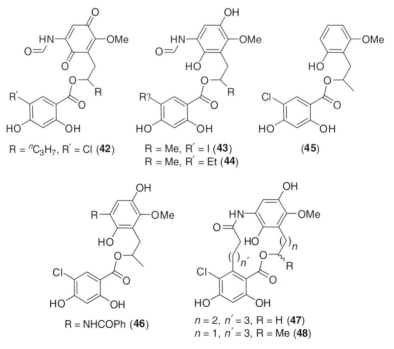

Figure 23. Important radester and radanamycin analogs (CHI5).

hydroquinone and quinone. Furthermore, no quinone mimic from this series displayed improved activity compared to the parent quinone. The most potent saturated chain analog, **49**, demonstrated low micromolar activity in both antiproliferation ($IC_{50} = 11.9–13.0\,\mu M$) and Her2 degradation ($IC_{50} = 13.0\,\mu M$) assays. Administration of compound **49** to MCF-7 breast cancer cells also resulted in the induced degradation of Hsp90 client proteins Her2, Akt, and Raf in a concentration-dependent manner.

The second series of radamide analogs developed were conformationally biased derivatives designed to adopt a bent conformation for improved Hsp90 binding [159, 160]. Molecular modeling studies have suggested that upon binding to Hsp90, GDA is twisted into a bent conformation by isomerization of the amide bond (*trans* → *cis*), which results in an entropic penalty between 2.2 and 6.4 kcal/mol [92]. Further studies have suggested that analogs of GDA that contain a predisposed *cis*-amide bond should exhibit >1000-fold increased affinity for Hsp90. Because RDC also binds Hsp90 in a bent conformation it has been hypothesized that molecules adopting this conformation will bind Hsp90 with little entropic penalty and greater affinity, similar to RDC binding to Hsp90. For this reason, conformationally biased radamide analogs were designed to contain an α,β-unsaturated amide bond in either a *cis* or a *trans* orientation and also exhibit a bent conformation [159]. Surprisingly, the conformationally biased quinone analog **50** (Fig. 24), was less active than **49** in both antiproliferation and Her2 degradation assays ($IC_{50} = 25–39$ and $35\,\mu M$, respectively), suggesting increased rigidity prevents the quinone moiety from forming the necessary hydrogen-bond interactions. As a general trend, the *cis*-oriented analogs exhibited greater antiproliferative activity compared to the corresponding saturated or *trans* compounds. Interestingly, the analog that exhibited the most cellular growth inhibition against MCF-7 cells, **51** ($IC_{50} = 7.0\,\mu M$) with a *para*-nitrile appendage, was completely inactive against SKBr3 cells, suggesting the potential for an alternative target for this class of analogs. Both **47** and **50** induced degradation of Hsp90 client proteins *in vitro*,

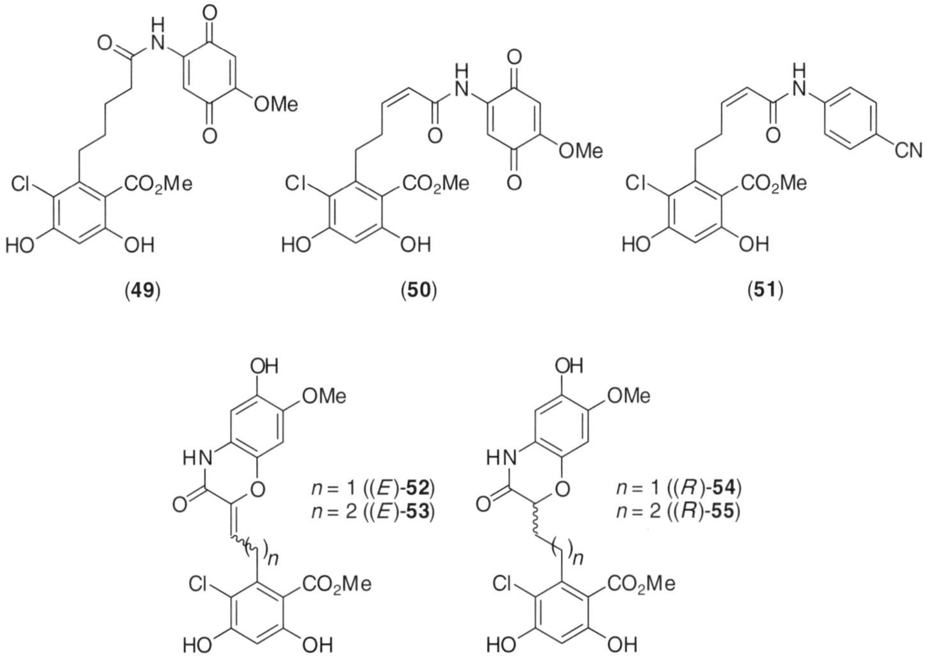

Figure 24. Radamide analogs.

linking their anticancer activity directly to Hsp90 inhibition. The hypothesis that compounds predisposed to a *cis*-amide would exhibit increased affinity for Hsp90 led to the design and evaluation of an additional series of radamide analogs that contain a conformationally constrained *cis*-amide [160]. Analogs containing three- or four-carbon linkers as either an olefin or a saturated chain were prepared and evaluated. For the olefin series, the *trans*-orientation was more active than the corresponding *cis*-analogs (10–27-fold) with the most active analogs exhibiting potent antiproliferative activities (**52**, $IC_{50} = 1.8$–$3.2\,\mu M$; **53**, IC_{50} 1.2–1.5 μM). For the saturated series of analogs, the absolute configuration was also important, as the (*R*)-stereoisomers were 5–40-fold more active than the corresponding (*S*)-stereoisomers (**54**, $IC_{50} = 2.1$–$3.6\,\mu M$; **55**, IC_{50} 2.1–2.7 μM) (Table 5). There was no difference in activity between the three- or the four-carbon linkers suggesting that conformation is the most important feature of these analogs.

Table 5. *In Vitro* **Hsp90 Inhibition of Chimeric Analogs**[a]

Compound	Antiproliferation IC_{50} (MCF-7) (μM)	Her2 Degradation IC_{50} (μM)
Radamide	18.6 ± 0.9	16.3 ± 5.0
Radamide HQ	12.9 ± 0.2	12.1 ± 3.3
Radester	13.9 ± 1.4	10[b]
Radester HQ	7.1 ± 0.3	25[b]
Radanamycin	1.2 ± 0.1	2[b]
42	11.6 ± 1.88	3.14 ± 0.79
43	24.7 ± 2.03	7.69 ± 1.21
44	5.72 ± 0.34	9.45 ± 4.55
45	69.4 ± 6.13	10.7 ± 2.07
46	4.27 ± 0.32	8.24 ± 0.07
47	0.94 ± 0.07	4.54 ± 1.37
48	0.93 ± 0.17	1.25 ± 0.03
49	11.9 ± 0.6	13.0 ± 3.1
50	25.5 ± 7.6	34.9 ± 0.4
51	7.0 ± 1.6	>100
52	3.2 ± 0.1	ND
53	1.5 ± 0.1	5[b]
54	3.6 ± 0.1	ND
55	2.7 ± 0.5	5[b]

[a] Values were gathered from various references and reflect general activities of these compounds.
[b] Values represent approximate IC_{50} values as determined through western blot analysis.
ND: not disclosed.

2.4. Purine-Based Inhibitors

2.4.1. Design of First-Generation Purine-Scaffold Hsp90 Inhibitors

The first synthetic, rationally designed small-molecule inhibitors of Hsp90 were the purine (PU) class of analogs developed by Chiosis and colleagues [161]. Researchers utilized a structure-based approach in which the purine bicyclic ring system of the natural ATP substrate was used as the starting point for deviation. Detailed analyses of GDA, RDC, and ATP bound to the N-terminal ATP-binding site of Hsp90 elucidated specific interactions that were incorporated into the design of the first purine derivative. Based on these modeling studies, the purine analog, PU3 (Fig. 25), was synthesized and evaluated for Hsp90 inhibition. Docking of PU3 in the ATP-binding site of Hsp90 suggested that Lys112 interacts with two of the methoxy substituents and the butyl chain occupies a well-defined hydrophobic environment. At the opposite end, were the key amino acids Asp93 and Ser52 that formed hydrogen bonds with the purine nitrogens. In addition, the crystal structure of PU3 showed that it adopted the predisposed C-shaped conformation necessary for optimal inhibition. This compound demonstrated moderate affinity for the Hsp90 N-terminus ($K_d = 15$–$20\,\mu M$). In addition, PU3 manifested antiproliferative activity against MCF-7 cells at levels comparable to those required for depletion of the Hsp90 client proteins, Her2, the estrogen receptor (ER) and Raf-1. These early studies with PU3 identified a novel class of small-molecule Hsp90 inhibitors for which further SAR were developed.

In order to probe SAR for the purine class of Hsp90 inhibitors, a library of approximately 70 analogs was prepared and evaluated with derivatization focusing on specific moieties: (1) the 9-alkyl chain, (2) substitutions at the 2-position, (3) halogen addition to the trimethoxyphenyl substituent, and (4) the linker between the purine and the aryl rings [162]. Key SAR for the PU class of compounds resulted from these studies. For the 9-alkyl derivatives, secondary linkers at the carbon α to N9 such as aromatic moieties, larger cycloalkyl substituents (as well as heteroaromatic rings), and tertiary amines were unable

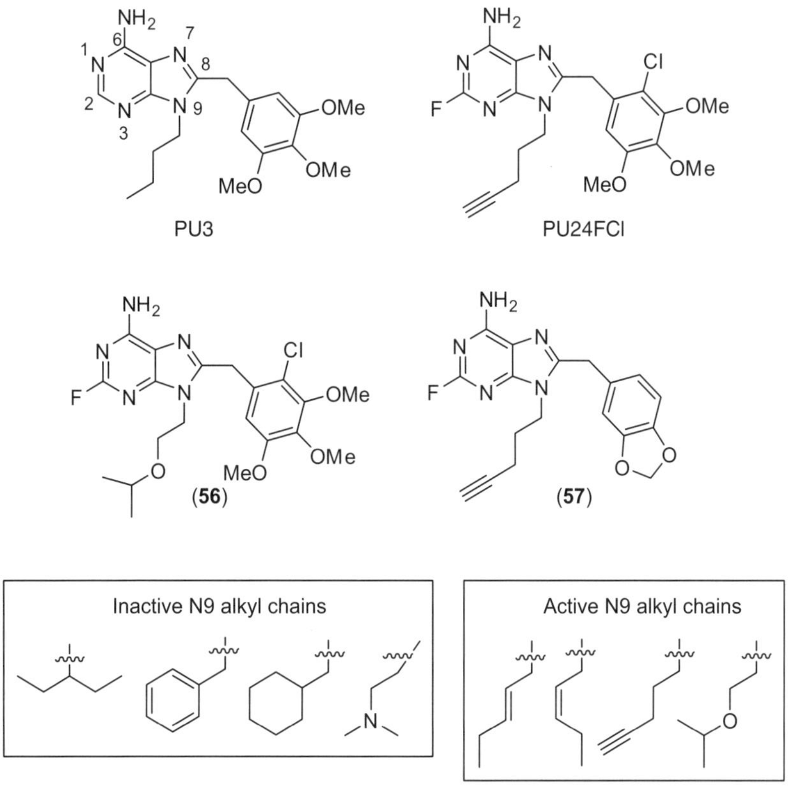

Figure 25. First-generation purine-scaffold Hsp90 inhibitors.

to bind Hsp90. In contrast, smaller alkyl substituents, including straight chains, branched chains, unsaturated chains, and carbocycles maintained binding affinity. The most active N9 substitutions were the pent-4-ynyl and 2-isopropxy-ethyl chains. These compounds bound Hsp90 at low micromolar concentrations ($K_d \sim 1.5\,\mu M$) In addition, both analogs demonstrated improved antiproliferative activity and induced Her2 degradation in human breast cancer cells compared to PU3. Of the numerous substituents incorporated at the C2-position of the purine ring system (cyano, vinyl, iodo, methoxy, amino, and fluoro), only fluorine led to compounds that maintained Hsp90 inhibitory activity *in vitro*. It was shown that addition of fluorine as an electron-withdrawing group at C2, improved the hydrogen-bond donating properties of the C6-amino substituent. The addition of either a single bromine or chlorine to the trimethoxyphenyl ring in the *ortho-* position also produced analogs that exhibited improved inhibitory activity compared to PU3, however, dibromination completely abolished Hsp90 binding. Only conversion of the methylene linker to sulfur was tolerated [163]; O, OCH$_2$, NH, SO, and SO$_2$ linkers either greatly reduced or completely abolished inhibitory activity. After each location of PU3 had been individually optimized, two compounds were produced that incorporated optimal moieties (**PU24FCl** and **56**, Fig. 25). Both compounds bound Hsp90 at submicromolar concentrations (K_d values: **PU24FCl** = 0.45 and **56** = 0.52 μM), comparable to 17-AAG. In addition, they displayed potent inhibitory activity *in vitro*, with ~25-fold improvements in antiproliferation and Her2 degradation activity compared to PU3 [162]. Further characterization of the effects of **PU24FCl** *in vitro* and *in vivo* identified it as a specific Hsp90 inhibitor in transformed cells [164]. **PU24FCl** was evaluated for Hsp90 binding with cellular lysates from both transformed and nontransformed human cells. It demonstrated high-affinity binding for Hsp90 derived from tumor cell lines (including breast, leukemia, and lung cancers) with an average IC_{50} of 0.22 µM. The affinity of **PU24FCl** for Hsp90 isolated from normal tissue was 10–20-fold lower than from transformed cells. **PU24FCl** demonstrated potent antiproliferative effects against a broad range of human cancer cell lines; including breast, prostate, chronic and acute myeloid leukemia, colon, and lung cancers (IC_{50} range of 2.0–7.2 µM). The tumor specific inhibition of Hsp90 was also studied *in vivo* wherein **PU24FCl** was shown to preferentially accumulate in tumor cells in a mouse xenograft model of breast cancer. Oncogenic Hsp90 client proteins were degraded in a time-dependent manner following both intraperitoneal and oral doses in the xenograft models. Finally, scheduled dosing (200 mg/kg every other day) over a period of 30 days resulted in a marked decrease in tumor volume.

Subsequent cocrystallization of PU3 with the *N*-terminal ATP-binding site provided important insight into the orientation that the purine analogs adopt within the binding pocket [165,166]. The purine functionality of PU3 resides in the same region as the adenine of the natural ATP substrate and maintains key binding interactions with Asp93 and three conserved water molecules through the C6-amine. Both GDA, through its primary carbamate, and RDC, through its resorcinols maintain these interactions and bind to this portion of the pocket and are essential for high affinity. While the trimethoxyphenyl ring of PU3 was originally designed to occupy the ribose and phosphate binding regions, the cocrystal structure confirmed this moiety to reside in a lipophilic, π-rich aromatic pocket composed of Phe138, Met98, Val150, Leu103, Leu107, Trp162, and Tyr139 [165]. The aromatic ring rests between Phe138 and Leu107 while forming favorable hydrophobic interactions with Met98. The methyls of the methoxy groups maintain hydrophobic contacts with the aromatic rings of Trp162 and Tyr139 in addition to the aliphatic carbons of Ala111, Leu103, and Val150. Data from the cocrystal structure also predicts that the nature of the methylene linker is important and minor modification could result in inactive compounds. Based on these results, the researchers sought to improve the PU3-binding affinity for Hsp90 by modification of the trimethoxyphenyl ring [165,166]. Interestingly, a single *meta*-methoxy was shown to be more potent than the trimethoxy analogs, while a single *para*-methoxy was inactive. Spatial arrangements of the *meta*- and *para*- methoxy groups suggested that the 2,3-methylenedioxy compound may be active and resulted in **57**, one of the most potent inhibitors, binding Hsp90 with a K_d value of 14 µM.

A series of dimeric purine-scaffold analogs was prepared in an attempt to improve Hsp90 inhibition by occupation of both N-terminal binding sites simultaneously (Fig. 26) [167]. An increase in tether length resulted in increased antiproliferative activity against both MCF-7 and SKBr3 cell lines. The dimer containing a four-carbon tether, **58**, exhibited no

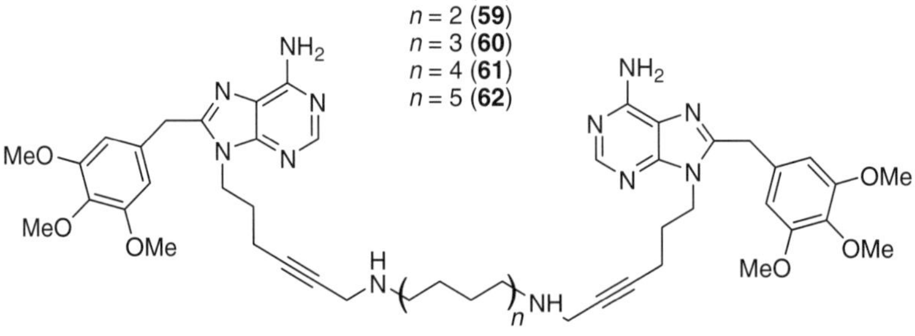

Figure 26. PU3-derived dimers.

effect on cancer cell growth at 100 μM, however, increasing the tether to eight-carbons, **59**, produced active compounds (MCF-7 = 30.6 μM and SKBr3 = 85.7 μM). The most dramatic increase in inhibitory activity was observed when the tether was extended to 12-carbons, with **60** exhibiting 6–17-fold increased activity compared to **59**. A modest increase in antiproliferative activity was seen for **61** and **62** (IC$_{50}$ values ∼ 1 μM), suggesting that upon achievement of the critical tether length, neither enhanced nor decreased inhibitory activity is produced by increasing tether length. In addition, **60** induced a concentration-dependent degradation of Her2.

2.4.2. Second-Generation Purine-Scaffold Hsp90 Inhibitors

Taking into account early SAR revealed for the PU class of Hsp90 inhibitors, several groups sought to identify second-generation analogs that exhibit improved potency and pharmacokinetic properties [168–171]. Similar to early attempts at SAR development, researchers focused their attention on functionalities manifesting the most dramatic alteration of activity, including: (1) modification at each position of the aryl ring, (2) the 9-alkyl chain, (3) the tether between the adenine and the aryl rings, and (4) position 2 of the adenine ring tether.

Attempts to determine a full range of SAR for the aromatic ring had been prohibited by limitation of available synthetic methods to produce the methylene linker. As noted previously, substitution of the methylene linker with sulfur resulted in compounds that retained inhibitory activity. Consequently, a library of sulfanyl, sulfoxyl, and sulfonyl linked PU analogs was prepared and evaluated [168]. Commercially available sulfhydryl aromatic compounds that coupled with the adenine ring provided a thorough assessment of optimal substituents. An iterative process was adopted to analyze a comprehensive set of aryl moities beginning with single substitution at the *ortho*-, *meta*-, or the *para*-position. As previously shown [165,166], monosubstitution at the *meta*-position was favored over *ortho*- or *para*-substitutions with the most active compound containing a methyl ether at the 3-position (**63**, $K_d = 0.97$ μM for Hsp90 binding, Fig. 27). For the disubstituted series

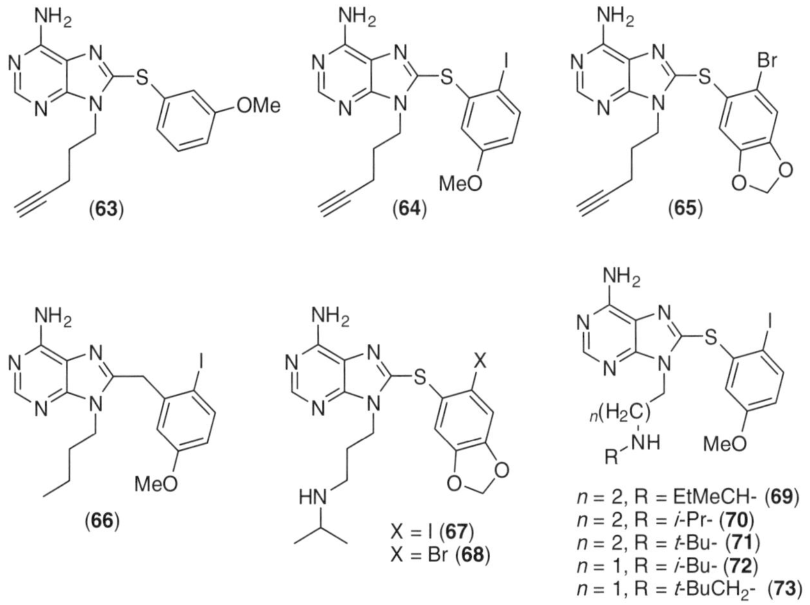

Figure 27. Second-generation purine-scaffold analogs.

of compounds, a preference for 2,5-substituents was observed and this substitution pattern yielded analogs with ~10-fold improvement in activity over the single 3-OMe substituent [168]. The most active disubstituted compound (**64**) contained the 2-iodo-5-methoxy combination as previously identified [163]. A similar series of compounds was also prepared and evaluated simultaneously by independent researchers and showed inhibitory activity to increase as the size of the halogen at the 2-position increased (Cl < Br < I) [170]. Building upon previous SAR for the purine analogs, the trisubstituted compound, **65**, was prepared and evaluated as an extension of the methylenedioxy variant of compound **57** [165,166] combined with the *ortho*-halogen substituent of **PU24FCl**, **64**, and **66** [168]. It bound Hsp90 with high affinity ($K_d = 0.03\,\mu M$), and demonstrated low nanomolar inhibition of SKBr3 cell proliferation ($IC_{50} = 0.2\,\mu M$) and induced Her2 degradation ($IC_{50} = 0.3\,\mu M$). As expected, substitution of an iodo for the *ortho*- bromine led to approximately threefold increase in activity, confirming previous SAR identified for **57** [169].

Following optimization of the aryl ring, modifications to the N9 alkyl chain were undertaken and results showed that activity for the sulfanyl series mimicked SAR previously determined for methylene tethered analogs (pent-4-ynyl > 2-isopropoxy-ethyl > butyl) [168]. Because addition of the 2-iodo substituent significantly decreased water solubility and hampered formulation, N9 alkyl side chains containing ionizable amino groups were incorporated to improve bioavailability [169,170]. The inability of N9 tertiary amines to maintain Hsp90 inhibitory activity coupled with improved water solubility demonstrated by 17-DMAG, led researchers to incorporate secondary amines on the adenine side chain. The sulfanyl linked analogs prepared by Chiosis and colleagues included 3-isopropylamino-propyl substituents at the N9 position; and the Hsp90 inhibitory activity of these compounds was comparable to the aliphatic chain analogs (**67** and **68**) [169]. Compounds **67** and **68** were formulated as both the hydrochloric and the phosphoric acid salts and were soluble up to 40 mM in phosphate buffered saline, pH 7.4. An expanded series of secondary amine analogs demonstrated improved water solubility as well as increased Hsp90 inhibition [170]. The bulkier the terminal alkyl group, the greater its potency (*i*-Pr < *t*-Bu < *i*-Bu < *i*-BuCH$_2$), with the most active compounds, **69–73** demonstrating low nanomolar inhibition of Hsp90 as measured by Her2 degradation (IC_{50} values = 90–210 nM). In addition, conversion of these compounds into the corresponding phosphoric acid salts provided improved water solubility (> 10 mg/mL) comparable to that demonstrated for **67** and **68**.

The oxidation state of sulfur was also probed. Results similar to those previously determined [163] were observed; a sulfoxide linkage produced decreased Hsp90 inhibitory activity, while oxidation to the corresponding sulfone abolished *in vitro* activity [168]. Direct comparison of the methylene and sulfur tether yielded conflicting results dependent upon aryl ring substituents [168–170]. Analogs that incorporated the 2-halogen, 4,5-methylenedioxy substitution pattern showed no significant differences in Hsp90 inhibition [169]. In contrast, analogs with the 2-iodo-5-methoxy aryl ring demonstrated approximately threefold increased potency upon incorporation of the sulfur linkage [163,170]. Finally, it was demonstrated that incorporation of a fluorine at the 2-position of the adenine ring was favorable for the methylene linked analogs, however, increased potency was not observed for the sulfanyl series [169,170].

In vivo antitumor activities of the most active, bioavailable purine-scaffold analogs were evaluated [169,170]. Intraperitoneal (IP) injection of compound **68** at a dose of 25 mg/kg in a mouse xenograft model of breast cancer resulted in pharmacologically active concentrations (IC_{50} value ~ 300 nM) present in the tumor 24 h postadministration [169]. Significant levels of **68** could not be detected in plasma during the same time period, suggesting tumor specific accumulation. The phosphate salt of **68** decreased tumor levels of the Hsp90-dependent client protein, Raf-1, 24 h after IP administration in a dose-dependent fashion. The pharmacokinetic parameters measured for **69–73** after both intravenous (IV) and oral (PO) administration confirmed good bioavailability [170]. In mouse xenograft models of cancer, the phosphoric acid salts of **70** (stomach) and **72** (lung) demonstrated significant degradation of Hsp90-dependent oncogenic signaling proteins following oral administration. In addition, oral dosing of compounds **69**, **70**, and **73** also resulted in decreased tumor volume compared to controls.

Recent disclosure of the 2-amino-6-halopurine series of purine-scaffold Hsp90 inhibitors identified analogs that maintain Hsp90 inhibition after transposition of the aryl moiety to N9 of the adenine ring [171,172]. Rationale for the development of this class of compounds is based on the previously described cocrystal structure of the PU3/Hsp90 complex [166]. Originally, when the aryl moiety was placed at the N9 position, molecules lost the ability to inhibit Hsp90. However, when this modification was carried out while transposing the 6-position amine to the 2-position, compounds regained Hsp90 inhibition, presumably through reestablishment of the six-bond distance between the 6-amine and the aryl functionalities. In addition, incorporation of a chlorine substituent at the 2-position increased the ability of these compounds to inhibit Hsp90 as determined by Her2 degradation [171]. Optimization of the aryl ring resulted in identification of **74**, which contained a heteroaromatic pyridine ring and exhibited potent inhibition ($IC_{50} = 0.02 \mu M$) (Fig. 28). Transposition of the pyridine nitrogen to the 1-position of the aromatic ring and SAR studies on the 4-position substituent provided insights into the functionalities necessary for Hsp90 inhibition (**75**, $IC_{50} = 0.03$ μM). Replacement of the methyl on compound **75** with the corresponding 4-OH abolished activity. In addition, larger alkyl chains, including ethyl and isopropyl groups manifested less active derivatives than **75** (IC_{50} values ~1.5 μM). Halogen substitution at the 4-position of the aromatic ring were well tolerated with an increase in size producing increased inhibitory activity (I > Br > Cl). The most active analog identified was **76**, which contains the 4-iodo substituent and manifested an IC_{50} of 0.009 μM for Her2 degradation, comparable to 17-AAG ($IC_{50} = 0.007$ μM) [172]. Based on their solubility profiles in simulated gastric and intestinal environments, **74** and **75** were chosen for *in vivo* xenograft studies against a human stomach carcinoma. Both compounds reduced tumor growth following oral administration, demonstrating there potential as clinically useful anticancer agents. Promising preclinical activities led compound **75** (CNF2024/BIIB021) to become the first rationally designed Hsp90 inhibitor to enter the clinic, which occurred in 2007. Several Phase I trials have produced promising results leading to the announcement that a Phase II trial evaluating CNF2024 in patients with gastrointestinal stromal tumors (GISTs) would commence in March 2008 [100].

2.4.3. Summary of Key SAR for Purine-Scaffold Hsp90 Inhibitors

A summary of the Hsp90 inhibitory activities for the purine-scaffold derivatives is presented in Table 6. The adenine ring is essential for activity. Halogens at the 2-position of the aryl ring improve activity when the linker is a methylene, however, they are less potent when the sulfanyl linker is present. Potent activity is retained only when the tether consists of a methylene or sulfur, all

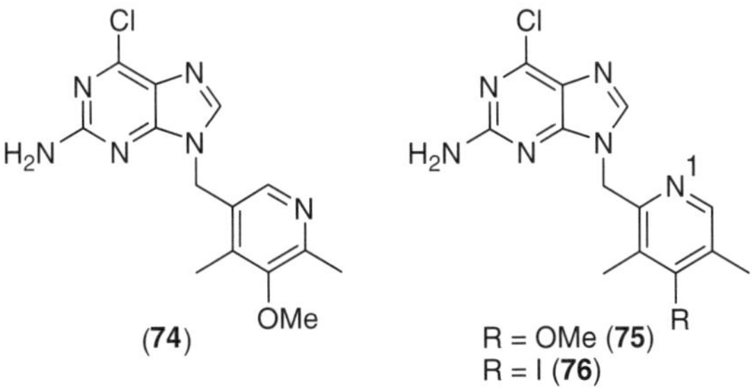

Figure 28. 2-Amino-6-halopurine Hsp90 inhibitors.

Table 6. Biological Activities of Pertinent Purine-Scaffold Inhibitors[a]

Compound	Hsp90 Binding K_d (μM)	Antiproliferation IC_{50} (μM)	Her2 Degradation IC_{50} (μM)	Tumor-Specific Inhibition	In Vivo Activity
PU3					
PU24FCl	0.45	2	2	Yes	Yes
56	0.52	5.4	3	ND	ND
60	ND	4.5	20	ND	ND
63	0.97	36.2	28.3	ND	ND
64	0.05	0.57	0.81	Yes	ND
65	0.03	0.2	0.3	Yes	ND
67	0.01	0.05	0.05	Yes	ND
68	0.045	0.2	0.02	Yes	Yes
73	ND	0.5	0.09	ND	Yes
74	0.018	0.1	0.02	ND	Yes
75	0.02	0.1	0.03	Yes	Yes

[a] Values were gathered from various references and reflect general activities of these compounds.
ND: not disclosed.

other tethers either reduce or abolish activity. Optional N9 alkyl substitutions in the methylene series consist of smaller, shorter alkyl chains. Incorporation of the sulfur linker allows for addition of ionizable secondary amines that improve water solubility. In this case, the bulkier groups resulted in more potent analogs. Finally, for the aryl side chain, 2-halo-4,5-methylenedioxy derivatives were identified as most active.

2.5. Shepherdin

Survivin belongs to the inhibitor of apoptosis (IAP) family of proteins that plays a role in both cell-death machinery and mechanisms of cell-cycle progression and microtubule stability [173]. Survivin expression is undetectable in most normal adult tissues, however, increased levels of the protein have been identified in almost every tumor type studied. Recently, it was shown that through its baculovirus inhibitor of apoptosis repeat, survivin interacts with the ATPase domain of Hsp90 in a manner critical for activity [174]. Disruption of Hsp90-survivin interaction results in destabilization of survivin, initiation of apoptosis, and cell growth inhibition. Focusing on the binding interface between survivin and Hsp90, the short peptide sequence K79–L87 (KHSSGCAFL, shepherdin) was identified as an Hsp90 inhibitor [174,175]. Shepherdin disrupted interactions between the N-terminal domain of Hsp90 and the immobilized ATP, suggesting it interacts directly with the ATP-binding site. Molecular modeling and simulation studies predict shepherdin to adopt a conformation that binds the ATP-binding site with a geometry that correlates well with the GDA-Hsp90 complex [175]. In addition, shepherdin associated in a concentration-dependent manner with recombinant N-terminal Hsp90 and exhibited no binding to the C-terminus.

Further characterization of shepherdin in vitro highlighted potent Hsp90 inhibitory properties [175]. It induced degradation of Hsp90-dependent client proteins, survivin, Akt, CDK-4, and CDK-6 in human prostate cancer cells. Interestingly, shepherdin had no effect on the levels of Hsp90 or Hsp70. Time-dependent experiments demonstrated shepherdin to be fast acting with concentration-dependent cell death arising within 5 h specific to tumor cells. In vivo studies of shepherdin in xenograft models of prostate and breast cancer showed detectable levels of the peptide in tumor tissue following 1 h administration protocols. Significant reduction in tumor volume was seen over the course of treatment and a near complete loss of Akt and survivin levels was observed.

2.6. Small-Molecule Hsp90 Inhibitors Identified Through HTS

The deleterious properties described for GDA and RDC led to identification of new molecular scaffolds that inhibit Hsp90. The development of high-throughput screening (HTS) methods and their implementation, aimed at identifying Hsp90 inhibitors has received considerable attention. As a result, there are now numerous highly reproducible assays capable of screening large libraries of compounds for small molecules that exhibit Hsp90 inhibitory activity. Utilization of these assays has proven useful, as several novel inhibitors have been identified and further characterized for *in vitro* and *in vivo* anticancer activities. The majority of assays developed for Hsp90 and optimized for HTS have focused on inhibitors of the Hsp90 N-terminal ATP-binding site. These include fluorescence intensity [176], fluorescence polarization [177–179] and colorimetric detection of the inherent ATPase activity of Hsp90 [79,180,181]. More recently, an assay based on the Hsp90-dependent refolding of firefly luciferase was optimized and utilized in an HTS format to identify inhibitors of both the N-terminus and the C-terminus of Hsp90 [182]. In addition, structure-based virtual screening has been used to identify new small-molecule Hsp90 inhibitors [183,184].

2.6.1. N-terminal Hsp90 Inhibitors

Pyrazole-Scaffold Hsp90 Inhibitors Researchers at the Institute of Cancer Research performed a screen of ~50,000 small molecules utilizing a malachite green assay to measure inhibition of the inherent N-terminal ATPase activity of yeast Hsp90. This screen identified several biarylpyrazoles as Hsp90 inhibitors with **CCT018159** (Fig. 29) as the most potent compound (79% inhibition at 40 µM, IC$_{50}$ = 8.9 µM) [180,185]. The less potent analogs identified early SAR for this series of compounds and allowed the design and preparation of a more focused library. A *para*-substituted phenol containing the pyrazole linkage was found to be necessary for optimal activity as methylation resulted in reduced inhibitory activity. In addition, researchers observed a sevenfold decrease in activity upon replacement of the ethyl moiety on the resorcinol with hydrogen [185]. A small series of analogs was prepared to probe SAR for this new class of inhibitors. Replacement of the ethyl substituent with a chlorine resulted in increased Hsp90

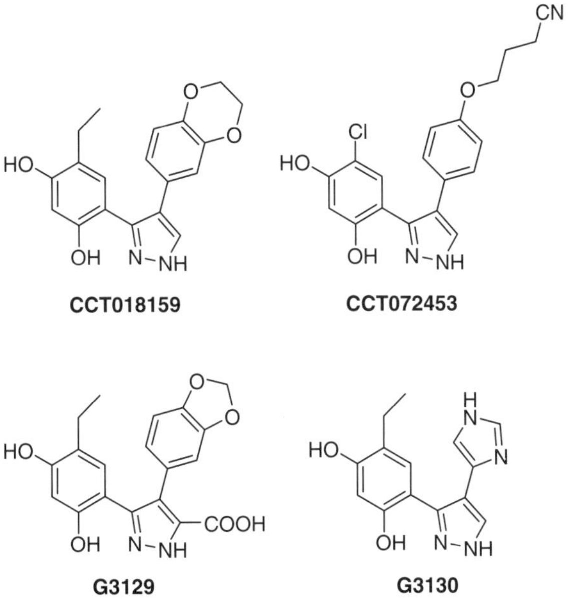

Figure 29. Early biarylpyrazole inhibitors of Hsp90.

binding and was therefore retained in subsequent analogs. Incorporation of large flexible groups in lieu of the dioxane ring were tolerated and in some cases increased potency. In addition to its high affinity for Hsp90 and its potent antiproliferative effects, **CCT018159** induced degradation of Hsp90-dependent client proteins c-Raf and Cdk4 with concomitant upregulation of Hsp70 [185,186].

The cocrystal structure of both **CCT018159** and **CCT072453** bound to the N-terminal ATP-binding site of Hsp90 provided further insight into their mode of inhibition. The resorcinol and the pyrazole amine form H-bonding interactions through conserved water molecules with Leu34, Asp79, Gly83, and Thr71; interactions that are conserved in the cocrystal structures for GDA, RDC, and ADP. In addition, the ethyl and chlorine substituents on the resorcinol ring reside in the same location as the chlorine of RDC. The butyronitrile chain of **CCT072453** extends out of the ATP-binding pocket toward the solvent, supporting SAR that larger, more flexible groups are tolerated at this position [185].

Independent of the research at Institute of Cancer Research and Vernalis, the Genomics Institute of the Novartis Research Foundation (GNF) identified the biarylpyrazole scaffold, in particular **G3129** and **G3130**, Fig. 29 [187,188]. They demonstrated reversible binding kinetics to Hsp90 with K_d values of 0.68 and 0.28 µM, respectively. However, their *in vitro* Hsp90 inhibitory activity as measured by Her2 degradation was lower. While **G3130** exhibited modest activity ($IC_{50} \sim 30$ µM), **G3129** was inactive up to 100 µM. The cocrystal structures containing these compounds were similar to those previously published for the CCT and VER series [188]. The resorcinol resides in the same position as RDC and coordinates H-bond interactions with the protein backbone through conserved water molecules. In addition, the pyrazole NH donates a hydrogen bond directly to Gly97, which is in contrast to the amide of VER-49009. The carboxylate of **G3129** forms a salt bridge with Lys58 and was predicted to provide increased Hsp90 inhibitory activity. However, the loss of water-mediated interactions between the imidazole NH and the Asn51/Asp54 present in **G3130** appears responsible for decreased binding. Further analogs of this scaffold have not been disclosed by GNF. Of note, other groups have also identified the biarylpyrazole and related compounds as Hsp90 inhibitors through various high-throughput screens, however, no additional optimization of these analogs has been reported [181,182].

A distinguishing feature of the cocrystal structure of CCT derivatives bound to Hsp90 was the close proximity of the pyrazole ring to the carbonyl oxygen of Gly97, suggesting that hydrogen-bond donors at this position could form H-bond interactions with this residue [189]. This observation led to preparation of biarylpyrazole analogs that contain amides at the pyrazide s-position, **77–80** (Fig. 30), and demonstrated increased Hsp90 affinity and *in vitro* potency. The most active compound, **79** (now denoted VER-49009), demonstrated submicromolar Hsp90 binding affinity ($K_d = 0.025$ µM), ATPase inhibition ($IC_{50} = 0.14$ µM), and antiproliferative activity ($IC_{50} = 0.26$ µM). The cocrystal structure of VER-49009 bound to Hsp90 confirmed the amide moiety formed a H-bond with Gly97 as predicted, thus explaining the observed increase in potency.

As noted, the cocrystal structures of these analogs complexed with Hsp90 suggested that modification to the second aryl ring would be well tolerated, as this moiety projects toward solvent. It was hypothesized that such modifications could provide improved physicochemical properties while simultaneously allowing access to new regions of the binding pocket. The first analogs to incorporate these modifications were the carboxamides [190]. The phenyl and benzyl carboxamides included substituents (hydroxyl, chlorine, methoxy, fluorine, etc.) at the *meta*- or *para*-position of the aryl ring. Measured for their ability to bind Hsp90, most of the compounds were active in the low micromolar range (K_d values ~ 5–50 µM) and exhibited minimal or no antiproliferative activity. However, two notable exceptions were obtained **81** and **82**, which exhibited submicromolar binding affinities with K_d values of 0.258 and 0.461, respectively. Rationalization for their increased affinities was provided by solution of their cocrystal structures when bound to the ATP-binding site of Hsp90 and suggests that both the acetyl (**81**) and the sulfonamide (**82**) form hydrogen bonds

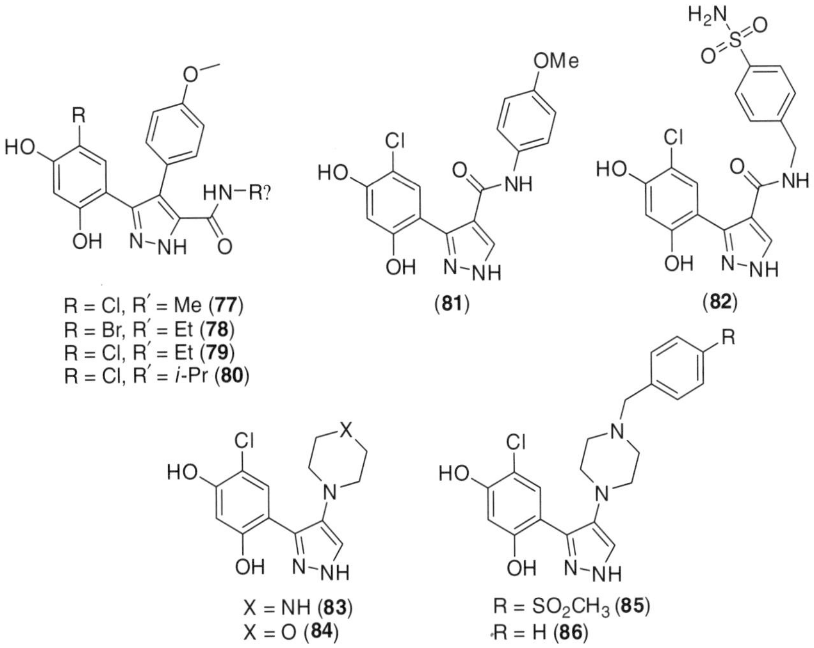

Figure 30. Second-generation pyrazole-scaffold Hsp90 inhibitors.

with Phe138. Of note, the α-phosphate of ADP and the amide oxygen of GDA also form H-bond interactions with Phe138. This increased binding affinity did not translate into improved *in vitro* activity as **81** manifested only modest antiproliferative activity with an IC$_{50}$ of 11.6 μM, while **82** was completely inactive.

A series of piperazinyl, morpholino, and piperidyl derivatives was also prepared and evaluated for Hsp90 inhibitory activity [191]. Piperizine **83** retained Hsp90 inhibitory activity similar to **CCT018159** and exhibited a binding affinity of 2.0 μM. However, a 10-fold loss in activity was observed for morpholine derivative **84**, highlighting the importance of nitrogen at this position. The corresponding piperidine was also less potent than **83** and led to a series of analogs with various substituents on the pyrazine nitrogen. Incorporation of aromatic moieties resulted in the most potent analogs, some of which exhibited submicromolar binding affinities. Benzyl incorporation produced compound **86**, which exhibited IC$_{50}$ values of 2.5 μM for ATPase inhibition and 6.5 μM for antiproliferation of HCT-116 cells. Addition of a *para*-sulfone to the phenyl ring (**85**), resulted in approximately twofold increased inhibition (IC$_{50}$ = 1.3 μM for ATPase and 3.1 μM for antiproliferation). Surprisingly, the addition of an amide to the 5-position of the pyrazole ring, a substitution that previously resulted in the potent analog VER-49009, did not produce increased activity. Compounds **83**, **85**, and **86** induced degradation of Raf-1 *in vitro*, linking Hsp90 inhibition to antiproliferative activity.

Isoxazole-Scaffold Hsp90 Inhibitors Researchers at Vernalis and the Institute of Cancer Research further explored the biaryl-pyrazole class of inhibitors and in 2008 reported their findings [192,193]. Extensive analysis of the cocrystal structure of VER-49009 bound to Hsp90 suggested three approaches toward analog improvement: (1) Addition of a solubilizing functionality to the second aromatic ring, (2) modification of the central pyrazole core, and (3) optimization of the 5-position on the resorcinol moiety [192].

As noted previously, substituents in the *meta*- and *para*-positions of the nonresorcinol ring orient away from the Hsp90 binding pocket and toward the solvent interface. Thus, a series of analogs that incorporated ionizable primary and secondary amines at these positions were pursued (Fig. 31). Analogs that contained aminomethyl groups in the *meta* position demonstrated reduced binding affinity and antiproliferative activity compared to VER-49009, suggesting that improvements in solubility may contradict inhibitory activity. However, analogs with similar substitutions at the *para*-position demonstrated comparable affinity and/or antiproliferative activities to VER-49009. In particular, analog **87**, which contains a morpholino substitution, retained activity in both Hsp90 binding and cellular growth assays ($K_d = 0.037$ µM and $IC_{50} = 0.29$ µM).

The ability of the pyrazole heterocycle to exist in a tautomeric form and the ammonium species to exhibit decreased Hsp90 affinity led to the design of analogs that have the N1 position "fixed" as an H-bond acceptor. This was achieved through incorporation of an isoxazole ring in lieu of the pyrazole. Although the isoxazole is slightly smaller in size, it is expected to be well accommodated in the binding pocket. In addition, the isoxazole cannot be protonated or exist in differing isomeric forms at physiological pH. Preparation and evaluation of VER-49009 isoxazole **88**, confirmed incorporation of this moiety is advantageous for Hsp90 inhibition. Binding affinities for VER-49009 and **88** were similar (K_d values = 25 and 28 nM, respectively), however, improvements in antiproliferative activity (2–10-fold) across a wide-range of cancer cells was seen for **88** [192]. Subsequent incorporation of the ionizable primary and secondary amines at the *meta*- and *para*-positions of the aromatic ring further supported previous SAR trends that *para*-substitutions in general, and morpholino in particular, **89** produced optimal activity. Interestingly, the *meta*-substituted isoxazoles were found to be more potent than the corresponding pyrazoles.

Proximity of the 5-position resorcinol to a lipophilic pocket created by the protein backbone led to incorporation of moieties containing additional sterics to probe SAR for this region of the molecule. Both large (phenyl and phenethyl) and small (ethyl, isopropyl) substituents were well tolerated at this position with only a slight or modest loss in binding affinity and antiproliferative activity. The most potent substitution at this position was

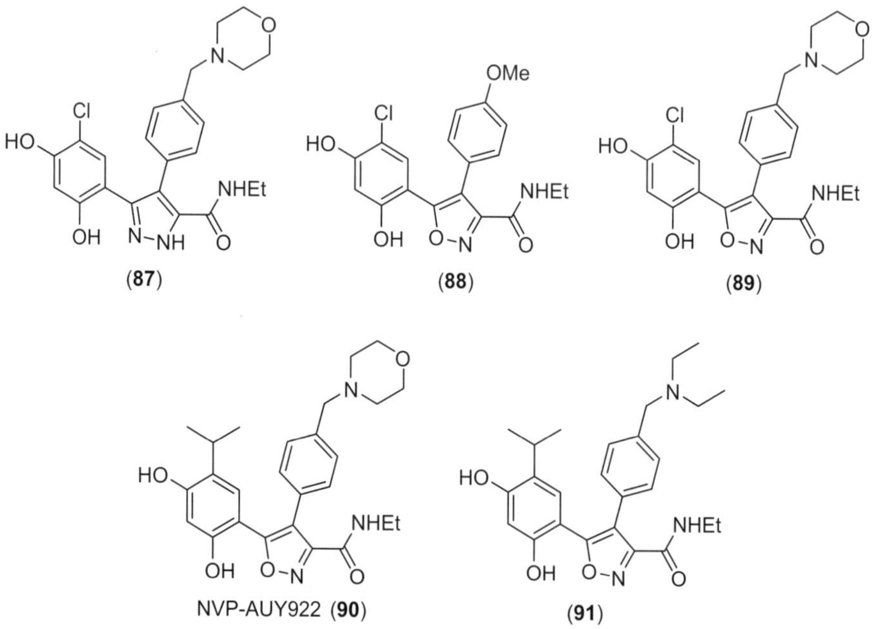

Figure 31. Isoxazole inhibitors of Hsp90.

the isopropyl moiety which, when incorporated along with the previously optimized functionalities resulted in compounds with K_d values for binding Hsp90 of 6–20 nM, (**90** and **91**). The key analogs in each series, **87, 88, 89**, and **90** were evaluated for their ability to induce degradation of Hsp90 client proteins *in vitro*. Each compound induced degradation of Raf-1 and Her2 with concomitant upregulation of Hsp70. Compound **90** (now denoted NVP-AUY922) demonstrated the highest ratio of tumor concentration relative to its IC_{50} value in antiproliferative assays (up to 35-fold) and was therefore chosen as the lead compound for *in vivo* xenograft studies of human colon cancer. Preclinical evaluation of NVP-AUY922 demonstrated that it reduced tumor size in xenograft models of breast, ovarian, prostate, and skin cancer [194,195]. Collectively, these preclinical data supported the Hsp90 inhibitor as acting via several distinct processes, including cytostasis, apoptosis, invasion, angiogenesis, and to inhibit tumor growth and metastasis. Based on this evidence, NVP-AUY922 entered Phase I clinical trials in 2007 [100].

Benzisoxazole Inhibitors of Hsp90 A high-throughput screen followed by database mining efforts by researchers at Wyeth identified the benzisoxazole **92** (Fig. 32) as a potent Hsp90 inhibitor that manifested a K_d value of 0.19 μM [196]. However, **92** exhibited poor water solubility (4 μg/mL) and antiproliferative activity ($IC_{50} \geq 20$ μM in both SKBr3 and HCT-116 cell lines). Cocrystallization of **92** with the Hsp90 N-terminus highlighted key interactions with the protein that were subsequently verified through SAR studies. Both free phenols were involved in H-bond interactions with the binding pocket through conserved water molecules and their methylation resulted in a complete loss of binding affinity ($K_d > 20$ μM). The 5′-bromo group on the resorcinolic ring projected toward a small hydrophobic pocket consisting of Phe138 and Leu107. While chlorine substitution was well tolerated, removal of a hydrophobic group at this position (replacement with hydrogen) led to decreased affinity (IC_{50} 15.2 μM). In addition, the 4- and 5-positions of the benzisoxazole scaffold were oriented toward the solvent interface, suggesting incorporation of solubilizing moieties to improve pharmacokinetic properties. The resulting analog, **93**, demonstrated high affinity for Hsp90, with a K_d value of 0.03 μM. Cocrystallization of **93** to the N-terminal ATP-binding site showed that it bound in the same manner as **92** and that the ethylmorpholino side chain was readily accommodated by rearrangement of the β-sheets. Compound **93** exhibited potent antiproliferative activity against a variety of human cancer cell lines with IC_{50} values ranging from 0.17 to 0.37 μM. It also resulted in specific degradation of the Hsp90 client proteins, Her2 and the androgen receptor, while maintaining no activity against a panel of kinases.

2.7. Hsp90 Inhibitors Identified Through Structure-Based Screening

The improvement of molecular modeling and docking software has led several groups to use virtual screening protocols to identify new Hsp90 inhibitory scaffolds [183,184,197]. Generally speaking, this protocol consists of docking a library of virtual molecules to a predetermined binding site and identifying compounds that fit the best. A smaller, focused library is then evaluated *in vitro* to confirm its ability to bind or inhibit Hsp90.

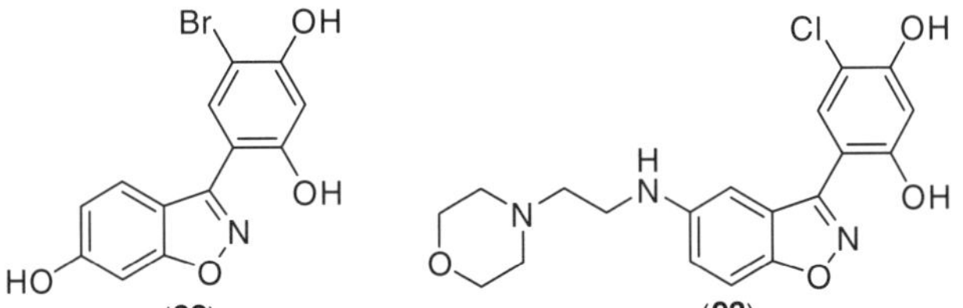

Figure 32. Benzisoxazole inhibitors of Hsp90.

A library of ~700,000 small molecules was screened for their ability to dock to the Hsp90 N-terminal ATP-binding site [183]. Based on the results, 1000 compounds were chosen for subsequent *in vitro* evaluation via Hsp90 binding affinity. This process identified several compounds with the general composition of 1-(2-phenol)-2-naphthol and sulfonamides, as low to submicromolar inhibitors of Hsp90 (Fig. 33). Compounds **94** and **95** demonstrated the greatest affinity for Hsp90 with K_d values of 0.7 and 0.6 µM, respectively. However, their antiproliferative activity was significantly reduced (IC_{50} = 29.0 and 59.8 µM, respectively), demonstrating that these scaffolds have reduced *in vitro* activity. The cocrystal structure of **94** in complex with Hsp90 revealed a binding mode similar to that previously identified for the resorcinol-containing Hsp90 inhibitors. The substituted naphthol binds in the same region as the resorcinol and maintains the key H-bond with Asp93. In addition, the phenol resides in close proximity to Thr184 and a conserved water molecule, suggesting these interactions are also important. The sulfonamide linkage and the dichlorophenyl ring project toward the surface of the binding site and do not form any obvious interactions with the protein. Preparation and evaluation of improved analogs of this scaffold have not yet been disclosed.

An improved virtual screening protocol utilizing a solvation model that included crystallographic water molecules was used to screen ~85,000 compounds selected from a library of 320,000 compounds containing drug-like physicochemical properties [184]. Based on virtual docking results, 285 compounds were tested *in vitro* for their ability to inhibit Hsp90's inherent ATPase activity. Several scaffolds were identified as Hsp90 inhibitors (>70% inhibition at 50 µM). Subsequent studies to determine whether these hits induced degradation of Her2 with concomitant upregulation of Hsp70 resulted in the identification of compounds **96**, **97** and **98**, which contain the 3-phenyl-2-styryl-3*H*-quinazolin-4-one scaffold as low micromolar inhibitors of Hsp90. The K_d and IC_{50} values for binding affinity and antiproliferative activity for all three compounds were between 20 and 35 µM. Molecular modeling simulations demonstrated that H-bonds between the quinazoline N3 and the amide side-chain of Asn51 and the carbonyl oxygen and Asp93 (through a conserved water molecule) are critical for binding.

A different approach was taken by Colombo and colleagues in which the bound conformation of the N-terminal Hsp90 inhibitor, sheperdin, was used as a scaffold to identify three pharmacophores that were then screened

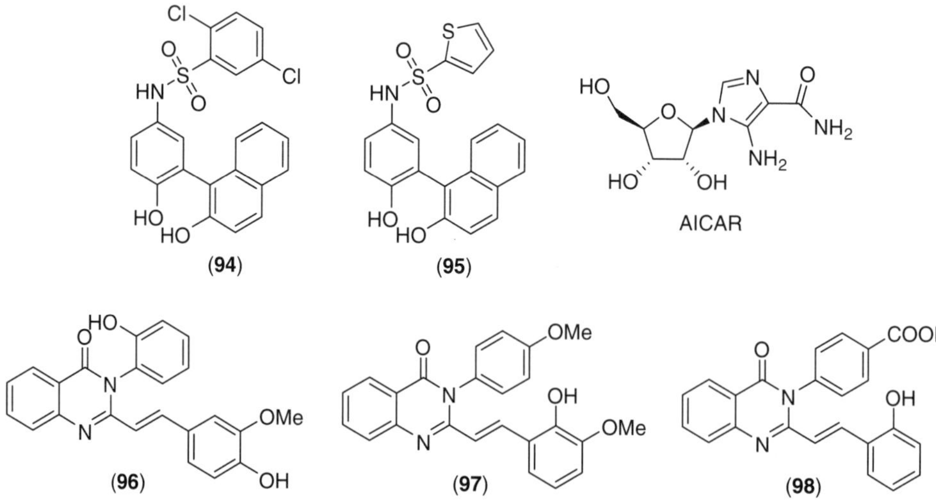

Figure 33. Inhibitors of Hsp90 identified through virtual screening.

against a database of ~160,000 nonpeptidic small molecules for Hsp90 antagonism [197]. Through this process, 5-aminoimidazole-4-carboxamide-1-β-D-ribofuranoside (AICAR) was identified as a novel inhibitor of Hsp90. Further characterization of AICAR as an Hsp90 inhibitor demonstrated that it bound to the N-terminal domain in a specific and saturable manner. Modest antiproliferative activity against three distinct cancer cell lines was observed (IC_{50} range of 59–126 µM). AICAR also induced the degradation of Hsp90 client proteins Her2, Akt, and CDK-6 without alteration of Hsp90 and Hsp70 levels.

2.7.1. Fragment-Based Screening Researchers at Abbott utilized an NMR-based fragment screen of the Hsp90 N-terminal domain to screen a library of 11,520 compounds (average MW = 225 Da) for Hsp90 inhibition [198]. Chemical shifts of Leu, Val, and Ile methyl groups in the presence of compound were used as evidence for Hsp90 binding. Through this screen, two related molecular scaffolds were identified as N-terminal Hsp90 inhibitors, the aminotriazine, **99**, ($K_d = 0.32$ µM) and the aminopyridine, **100** ($K_d = 18$ µM) (Fig. 34). A structure-based approach based upon the cocrystal structures of these compounds bound to Hsp90 was then pursued to optimize leads. The most potent compound identified, **101**, displayed high-affinity binding for Hsp90 with a K_d value of 60 nM. Interestingly, binding of **99** to Hsp90 resulted in a conformational change that opened a larger binding region adjacent to the primary binding site. Because generation of a large library of compounds with the general scaffold of **100** was limited, a second screen using a 3360 compound library (average MW = 150 Da) was undertaken to identify fragments that bound Hsp90 in the presence of saturating amounts of **100**. The most potent hit identified was furanone **102**, which exhibited a binding affinity of 150 µM in the presence of **100**. Binding of **102** was indeed cooperative as the observed K_d in the absence of **100** was >5000 µM. Based upon the cocrystal structures of these two compounds, linkers were incorporated to optimize binding and eventually led to **103** and **104**, $K_d = 1.9$ and 4.0 µM, respectively. The *in vitro* cell-based activity of these compounds has not been reported.

Tetrahydro-4H-Carbazol-4-One Inhibitors of Hsp90 Researchers at Serenex recently disclosed results that were obtained from high-throughput screening of a focused library designed to inhibit proteins that contain purine binding sites [199]. This screen produced a benzamide inhibitor of Hsp90, which was combined with a carbazol-4-one functionality to exhibit a favorable pharmacokinetic profile and yield the potent Hsp90 inhibitor **105** (Fig. 35). While **105** bound with high affinity

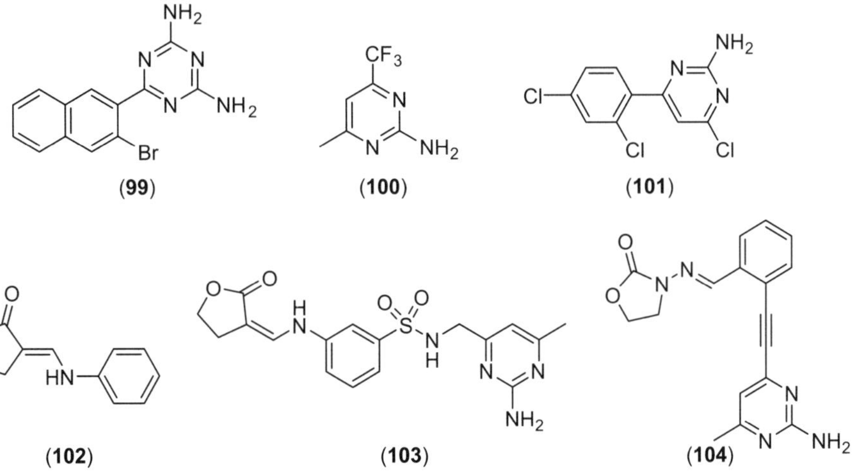

Figure 34. Hsp90 inhibitory fragments identified via NMR screening.

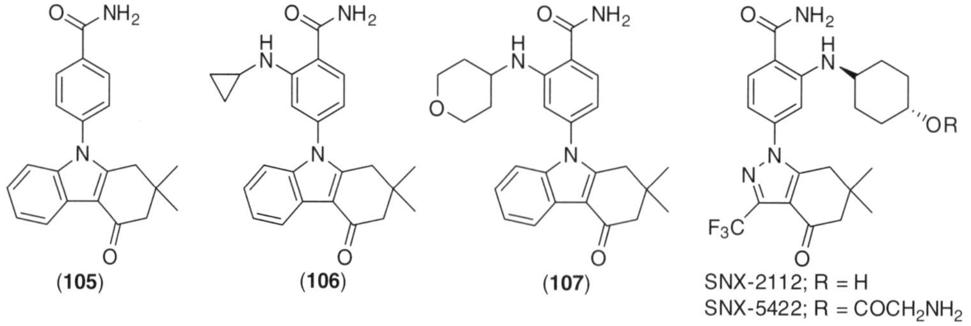

Figure 35. Tetrahydro-4H-carbazol-4-one inhibitors of Hsp90.

to the N-terminus of Hsp90, $K_d = 0.75\,\mu M$, it was inactive in cellular assays (>50 μM, Her2 degradation). Therefore, a series of analogs was prepared that focused upon the incorporation of functionalities *ortho* to the amide (R) with concomitant removal of the gem-dimethyl substituents. Addition of bromo, cyano, or methoxy at this position completely abolished Hsp90 binding, however, an amino functionality demonstrated reduced, but measurable affinity ($K_d = 2.9\,\mu M$). Because the binding model predicted the accommodation of larger substituents, various alkylamines were incorporated. Compound **106** demonstrated enhanced Hsp90 binding affinity and was the first analog to manifest *in vitro* cellular activity, $K_d = 0.35\,\mu M$ and $IC_{50} = 0.61$ μM. Other alkylamines placed at the 2-position to probe SAR included phenyl, pyridinyl, methoxyethyl, and morpholine. Several analogs exhibited submicromolar binding affinity for Hsp90 and low micromolar Her2 degradation. However, only one other compound, **107**, induced degradation of Her2 at submicromolar levels ($IC_{50} = 0.56\,\mu M$). Compound **107** was evaluated for its antiproliferative activity against a variety of human cancer cell lines and demonstrated potent antitumor activity with IC_{50} values for numerous cell lines below 0.4 μM (IC_{50} range 0.26–0.96 μM).

A small-molecule inhibitor of Hsp90 that was not disclosed in this manuscript, but presumably resulted from these studies, is the structurally related compound, SNX-2112 (Fig. 35) [200,201]. SNX-2112 demonstrated potent inhibition of Hsp90 in both cellular assays and xenograft models of Her kinase-dependent cancers [200]. Her2 degradation induced by SNX-2112 resulted in disruption of both Akt and Erk oncogenic signaling pathways both *in vitro* and *in vivo*. Furthermore, a prodrug form of the compound, SNX-5422, possessed increased water solubility and oral bioavailability and after a single oral dose, resulted in preferential tumor accumulation of the active metabolite SNX-2112. SNX-5422 entered clinical trials in May 2007 and has shown promising preliminary results [100]. In March 2008, Pfizer announced its intention to acquire Serenex, Inc., and continue clinical development of this Hsp90 inhibitor.

Dihydroxyphenyl Amide Inhibitors of Hsp90
Researchers at Pfizer have recently reported a class of dihydroxylphenyl amides as inhibitors of the Hsp90 N-terminus [202]. The trihydroxyamide, **108**, bound with high affinity to Hsp90 with a K_d value of 0.2 μM, however, its potency was greatly reduced *in vitro* and produced an IC_{50} for Akt degradation >20 μM, presumably due to poor cell permeability as suggested by the authors (Fig. 36). The cocrystal structure of **108** in complex with the N-terminus was obtained to identify key interactions and regions amenable to further alteration. The amide carbonyl and both the 2′- and the 4′-hydroxyl groups formed H-bond interactions with the protein backbone through conserved water molecules reminiscent of the interactions important for binding GDA and RDC. In contrast, the 6′-hydroxyl formed H-bond interactions with two nonconserved water molecules, suggesting this moiety may not be critical for binding. In addition, the pyrrole of **108** occupied a flexible pocket consisting of conformationally mobile lipophilic amino acids. Consequently, a series of

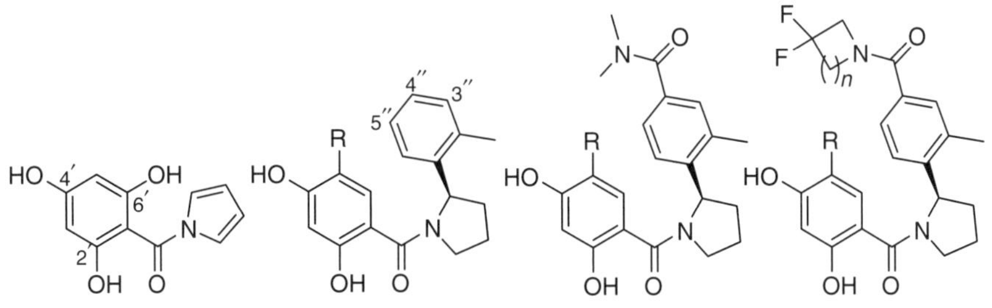

Figure 36. Dihydroxyphenyl amide inhibitors of Hsp90.

amines were incorporated at this location to probe SAR for this region of the molecule. It was determined that secondary amines displayed greater Hsp90 affinity than primary amines and the most active compounds demonstrated Hsp90 inhibition in cell-based assays. In particular, compound **109** bound Hsp90 with a K_d of 0.03 μM and induced Akt degradation with an IC_{50} of 4 μM. The cocrystal structure of **109** in complex with Hsp90 identified a small, readily accessible hydrophobic pocket adjacent to the 5′-position of the resorcinol. Based on the RDC resorcinol and previous SAR for other Hsp90 inhibitors, a chlorine was introduced at this position to produce **110**, which resulted in an approximately three- to four-fold increase in activity ($K_d = 0.01$ μM, $IC_{50} = 1$ μM). A second, slightly larger unfilled region of the binding pocket was also identified adjacent to the phenyl moiety, suggesting simple esters or amides could be incorporated at the 3″-, 4″-, or 5″-positions to probe SAR for this region. Substitutions at the 5″-position were found to be detrimental, however, 3″- or 4″-substitutions were well tolerated, and generally resulted in improvement of both binding and Akt degradation. The most potent compound, **111**, exhibited a K_d of <0.01 μM and IC_{50} of 0.3 μM (Fig. 36). Encouraged by these results, a series of 4″-amide analogs was prepared and evaluated. The majority of these compounds demonstrated high-affinity binding to Hsp90, with K_d values below the assay quantitation limit of 0.01 μM. Derivatives **112** and **113** demonstrated low nanomolar inhibition of Hsp90 as measured by Akt degradation, with IC_{50} values of 0.02 and 0.027 μM, respectively. Further characterization of their Hsp90 inhibitory activity showed that **112** and **113** manifest low nanomolar antiproliferative activity comparable to 17-AAG (IC_{50} values = 5.8–8.9 and 22.6–24.5 nM, respectively).

3. Hsp90 INHIBITORS OF UNIDENTIFIED BINDING SITES

3.1. Derrubone

Through the RRL-based screen for Hsp90 inhibitors recently disclosed [182], derrubone was identified and further characterized as a novel natural product inhibitor of Hsp90 (Fig. 37) [203]. Derrubone was originally isolated and characterized as one of a series of structurally related isoflavonoids derived from the Indian tree, *Derris robusta* [204]. Derrubone demonstrated potent inhibition of Hsp90 with an IC_{50} value of 0.23 ± 0.04 μM. In addition, derrubone manifested antiproliferative activity against MCF-7 and SkBr3 cells with IC_{50} values of 9 ± 0.7 and 12 ± 0.3 μM, respectively. Derrubone induced the degradation of numerous Hsp90-dependent oncogenic signaling proteins including Her2, the estrogen receptor, Akt, and Raf [177]. Studies into the mechanism of Hsp90 inhibition displayed by derrubone suggested it inhibits the Hsp90 chaperone machinery by a mechanism distinct from GDA and novobiocin.

In an attempt to identify more potent derrubone derivatives, a total synthesis of the compound amenable to the generation of diverse analogs was designed and implemen-

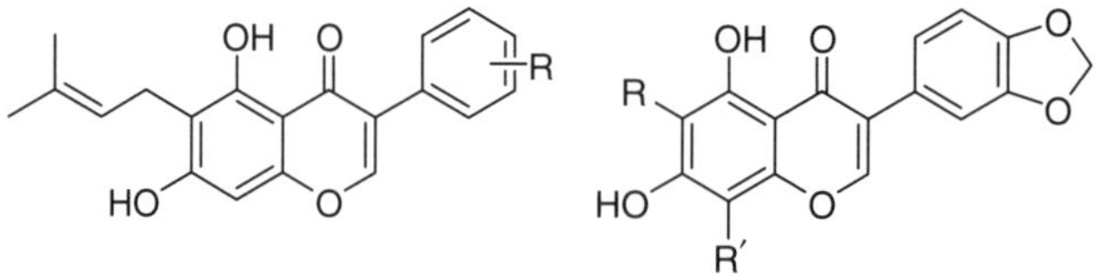

Derrubone; R = 3,4-OCH$_2$O-
R = 4-OMe (**114**)
R = 3-OMe (**115**)
R = 3,4-Cl (**116**)
R = H (**117**)

R = R′ = H (**118**)
R = H, R′ = isoprenyl (**119**)
R = CH$_2$CO$_2$Me, R′ = H (**120**)

Figure 37. Derrubone and analogs.

ted [205]. The derrubone analogs prepared for establishment of SAR focused on two specific areas of the molecule; (1) substitution of the aromatic side chain and (2) modification of the prenyl side chain. The aromatic moiety is essential for activity as its replacement with a chlorine atom resulted in complete loss of Hsp90 inhibitory activity. Removal of the methylenedioxy moiety, **117**, resulted in an analog that retained comparable antiproliferative activity to derrubone against both MCF-7 and HCT-116 cancer cells (IC$_{50}$ = 12.3 and 13.9 μM, respectively). Incorporation of either a 3′,4′-dichloro (**116**, IC$_{50}$ = 5.2–7.3 μM) or 4′-methoxy aromatic ring (**114**, IC$_{50}$ = 5.5–7.3 μM) resulted in approximately twofold increase in activity. In addition, the analog containing only a 3′-methoxy substituent, **115**, was completely inactive, suggesting functionality at the para-position is necessary for Hsp90 inhibition. SAR studies for the isoprenyl side chain demonstrated that removal of this moiety, **118**, (IC$_{50}$ values > 100) or replacement with a polar functionality, **120** (IC$_{50}$ = 55.5.2–62.6 μM), resulted in diminished activity. In contrast, the addition of lipophilic side chains at this position produced compounds with activity comparable to derrubone. Interestingly, transposition of the isoprenyl side chain to the 8-position yielded analog **119**, which demonstrated a modest increase in inhibitory activity and an IC$_{50}$ value of ~10 μM. One analog from each series, **116** and **119**, was chosen for further characterization of its Hsp90 inhibitory properties via its ability to induce Hsp90 client protein degradation. Both compounds reduced Her2 and Raf levels following 24 h incubation.

3.2. Naphthoquinones

The RRL-based screen that identified derrubone as an Hsp90 inhibitor also elucidated the 1,4-naphthoquinone scaffold as an Hsp90 inhibitor [206]. Several derivatives with this core structure were identified from this screen and in particular two compounds, **121** and **122**, demonstrated nanomolar inhibition of Hsp90 (IC$_{50}$ = 0.25 and 0.38 μM, respectively) (Fig. 38). A series of analogs was designed and synthesized in an attempt to simultaneously probe SAR for the phenyl moiety of **121** and the acetamide of **122**. Single phenyl rings with substituents at the ortho-, meta-, and para-positions were well tolerated and induced Her2 degradation at low micromolar concentrations (IC$_{50}$ values ~ 1–10 μM). Amide side chains containing a benzofuran or phenoxyphenol were less active and produced IC$_{50}$ values ranging from 13–100 μM. Based on these results, a series of analogs that contained functionality at the ortho-, meta-, or para-positions of the benzamide were prepared and evaluated. In addition, a biaryl moiety, **123**, was evaluated as a derivative analogous to recently published novobiocin analogs that may bind similarly. The Hsp90 inhibitory activity of these analogs was found to be comparable to the parent benzamide compound. Several analogs from each series were evaluated for Hsp90 inhibition via the RRL assay, and while both the benzamide and the acetamide derivatives were active in cell-based assays, the acetamides were more potent in the luciferase refolding assay, a direct measure of Hsp90 inhibition. In addition, two of the most active compounds, **123** and **124**

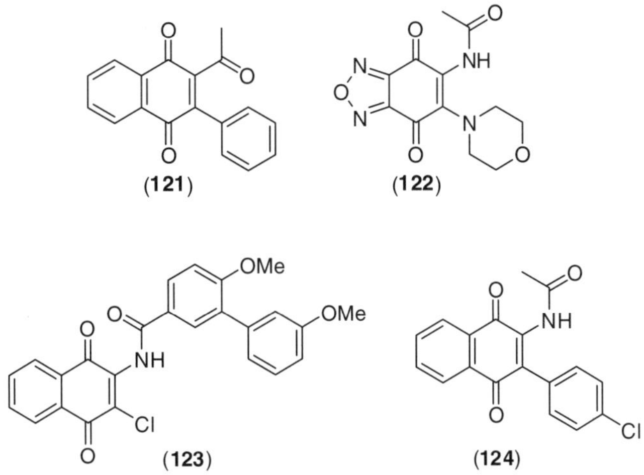

Figure 38. Naphthoquinone inhibitors of Hsp90.

(RRL-assay IC$_{50}$ = 1.6 and 0.2 μM, respectively) induced degradation of Hsp90-dependent client proteins at concentrations similar to their antiproliferative activity.

3.3. Small Molecules That Disrupt Hsp90/HOP Interactions

In order for Hsp90 to correctly fold client proteins into their three-dimensional conformations, requires the assembly of both cochaperones and partner proteins. In particular, the interaction between Hsp90 and the TPR2A domain of the cochaperone Hsp90 organizing protein (HOP) is essential for Hsp90 activity [207]. Previous studies have demonstrated that a designed TPR domain binds to the C-terminus of Hsp90 and can compete for binding, thus preventing formation of the multiprotein complex necessary for Hsp90 activity [208]. A high-throughput screen for small molecules that inhibit the TPR2A-Hsp90 interaction was conducted with a library of ~21,000 compounds [209]. Three compounds with the same core structure, a 7-azapteridine ring system, were identified and characterized as inhibitors. Several other commercially available analogs with the same core scaffold were subsequently tested for Hsp90 inhibitory activity (Fig. 39). Each compound, **125–130**, bound with high affinity to the TPR2A domain of HOP with K_d values ranging between 0.1 and 0.9 μM. No specific interactions between these compounds and Hsp90 were apparent.

Each compound inhibited the growth of BT474 human breast cancer cells at levels consistent with their binding affinities (IC$_{50}$ values = 0.7–4.3 μM). Compound **125** was chosen for subsequent *in vitro* analysis and manifested the ability to induce degradation of Her2 in both BT474 and SKBr3 cells without upregulating Hsp70, suggesting a unique mechanism for Hsp90 inhibition.

3.4. Taxol

Taxol (Fig. 40), a well-known drug for the treatment of cancer, is responsible for stabilization of microtubules and blockage of mitosis [210]. Previous studies have shown that taxol induces transcription factors and kinase activation, mimicking the effects of bacterial lipopolysaccharide (LPS), an attribute unre-

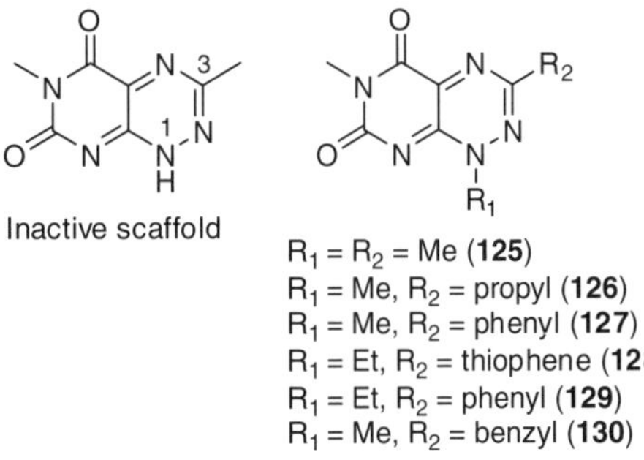

Inactive scaffold

R$_1$ = R$_2$ = Me (**125**)
R$_1$ = Me, R$_2$ = propyl (**126**)
R$_1$ = Me, R$_2$ = phenyl (**127**)
R$_1$ = Et, R$_2$ = thiophene (**128**)
R$_1$ = Et, R$_2$ = phenyl (**129**)
R$_1$ = Me, R$_2$ = benzyl (**130**)

Figure 39. Azapteridine inhibitors of the Hsp90-HOP interaction.

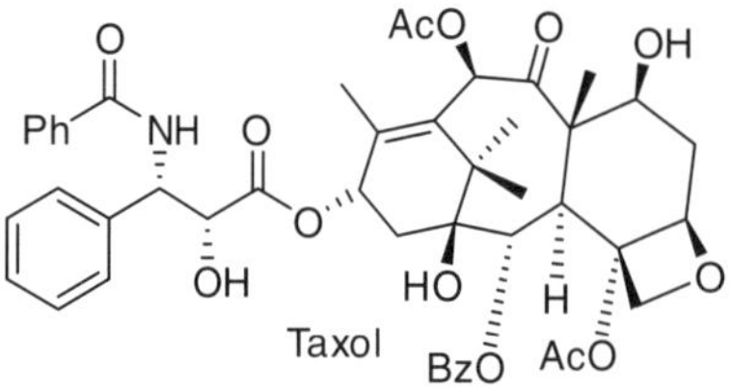

Figure 40. Structure of taxol.

lated to its tubulin binding properties. A significant amount of evidence suggests that the LPS-mimetic activity of taxol is independent of β-tubulin binding. Thus, Rosen and coworkers prepared a biotinylated taxol derivative and performed affinity chromatography experiments with lysates from both mouse brain and macrophage cell lines, which led to the isolation of Hsp70 and Hsp90, which commonly form heteroprotein complexes. In contrast to typical Hsp90 binding drugs, taxol exhibits a stimulatory response, mediating the activation of macrophages and exerting the LPS effects observed [211].

Recently it was reported that the geldanamycin derivative, 17-AAG, behaves synergistically with taxol-induced apoptosis. The mechanism by which these two interact is best explained by the sensitization of tumor cells to taxol-induced apoptosis by 17-AAG through suppression of Akt kinase [212, 213]. The use of Hsp90 inhibitors in combination with proapoptotic therapies represents an exciting new strategy for chemotherapy.

3.5. Gedunin and Celastrol

In recent years, gedunin (Fig. 41), a tetranotriterpenoid isolated from the Indian neem tree *Azadirachta Indica* [214], and structurally related celastrol, a quinone methide triterpene from the *Celastraceae* family of plants, have become compounds of interest due to their antiproliferative and neuroprotective properties [214–217]. Recently, studies by Lamb and coworkers identified these natural products as Hsp90 inhibitors [218–220]. Using a connectivity map– they were able to find high correlation scores between gedunin, celastrol, GDA, 17-AAG, and 17-DMAG, suggesting these natural products exhibit their antiproliferative activity through Hsp90 modulation. A subsequent paper confirmed this hypothesis, however, the exact mechanism of inhibition remained undetermined. Utilizing a fluorescence polarization assay, gedunin, and celastrol failed to displace GDA, indicating the natural products were not binding competitively to the N-terminal ATP-binding site. To explain this phenomenon, Zhang and coworkers provided evidence that celastrol disrupts Hsp90 function by blocking interactions between Hsp90 and its cochaperone, Cdc37, which prevents formation of the Hsp90 heteroprotein complex [215]. Based on structural similarities between gedunin and celastrol, it has been proposed that gedunin utilizes a similar mechanism of action toward Hsp90 inhibition.

In an attempt to elucidate structure–activity relationships between Hsp90 and gedunin, libraries have been synthesized [255]. Although the analogs made thus far have not proven more effective than gedunin in antiproliferation assays, Brandt and coworkers identified key structural features necessary for activity. Steric bulk applied to the C7-position exhibits a pronounced effect on antiproliferative activity, as inhibitory activity is

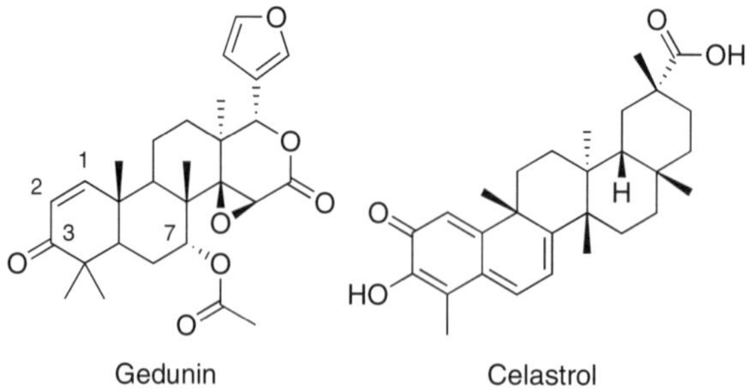

Figure 41. Structures of gedunin and celastrol.

Table 7. Biological Activities of Pertinent Small-Molecule Inhibitors[a]

Compound	Hsp90 Binding K_d (µM)	Antiproliferation IC$_{50}$ (µM)	Client Protein Degradation IC$_{50}$ (µM)
CCT018159		4.1	
VER-49009	0.025	0.26	0.5[b]
81	0.26	11.6	20[b]
85	0.74	3.1	5[b]
87	0.037	0.29	0.4[b]
88	0.028	0.2	0.02[b]
89	0.021	0.083	0.1[b]
NVP-AUY9232	0.021	0.016	0.02[b]
93	0.03	0.2	0.33
94	0.7	29.0	ND
96	ND	25	20[b]
AICAR	ND	75	50[b]
103	1.9	ND	ND
107	0.41	0.2–0.9	0.56
SNX-2112	0.03	0.025	0.05[b]
111	<0.01	0.02	7.6-19.1
106	0.36	0.95	1[b]
Derrubone	ND	10	30[b]
116	ND	6	25[b]
123	ND	2	1.7

[a] Values were gathered from various references and reflect general activities of these compounds.
[b] Values estimated from western blot analysis.
ND: not disclosed.

diminished as size is increased. Although it appears as though the electronic nature is not critical, the presence of a hydrogen-bond acceptor can slightly improve antiproliferative properties. C7-substituents also exhibit an influence on the overall conformation of the molecule, and influence the binding of other substituents. The olefin of the α,β-unsaturated ketone is also essential for activity. Early studies suggested this is a consequence of the electrophilic nature of this moiety, however modifications to and reduction of the ketone itself have proven otherwise. Hydrogen-bond accepting properties at the 3-position, as well as the rigidity of the 1,2-olefin are responsible for retention of activity. Studies are currently underway to further clarify gedunin's structure–activity relationship with Hsp90.

3.6. Summary of Hsp90 Inhibitors Identified Through HTS and Modeling

The optimization and implementation of HTS assays for Hsp90 inhibition has identified numerous molecular scaffolds as Hsp90 inhibitors. SAR for many of these scaffolds has resulted in analogs with potent *in vitro* and *in vivo* anticancer activity. Several of these analogs have entered clinical trials and have shown promising preliminary results to support their use as cancer chemotherapeutics. A summary of key Hsp90 inhibitors identified via HTS is provided in Table 7.

4. INHIBITORS OF THE Hsp90 C-TERMINAL NUCLEOTIDE BINDING SITE

4.1. The GHKL Family of Proteins

The Hsp90 N-terminal ATP-binding site is rather unique compared to that of most ATP-binding proteins. Hsp90 binds ATP/ADP in a bent conformation, which is in stark contrast to the more typical extended conformation utilized to bind nucleotides by most proteins. In fact, only four proteins are known to bind ATP/ADP in this bent conformation and they comprise the GHKL family (DNA *gyrase*,

Figure 42. (a) Cocrystal structure of ADPNP bound to DNA Gyrase. (b) Cocrystal structure of the Hsp90 N-terminus bound to ADP. (See color insert.)

Hsp90, Histidine Kinase, and MutL) [221]. This unique attribute of the GHKL family of proteins allows the development of inhibitors that have the potential to exhibit high selectivity. A side-by-side comparison of the cocrystal structures of DNA gyrase bound to ADPNP [222] and Hsp90 bound to ADP is provided in Fig. 42. As can be seen in both structures, the ribose ring serves as the apex of the ligand, orientating both the phosphodiesters and the base into downward trajectories, resulting in a cup-shaped conformation. In addition to the conformation, key hydrogen-bond interactions with the adenine ring are conserved in these Bergerat folds.

4.2. Elucidation of the Hsp90 C-Terminal Nucleotide Binding Site

Based on similarities between the DNA gyrase and the Hsp90 ATP-binding sites, it was proposed by Marcu and Neckers that inhibitors of other GHKL family members may also serve as Hsp90 inhibitors [223,224]. It was noted that the DNA gyrase ATP-binding site inhibitor, novobiocin (Fig. 43), exhibited toxicity against some cancer cell lines, suggesting that it may also target mammalian proteins in addition to its bacterial target. Upon preparation of a novobiocin-sepaharose column, Neckers and coworkers determined that Hsp90 could be eluted in a concentration-dependent manner, confirming their hypothesis. Furthermore, when a GDA-sepharose column was prepared, novobiocin was able to elute Hsp90 in a similar manner, suggesting the potential for competitive inhibition. However, the reciprocal experiment did not provide similar results. Instead, it was observed that upon loading Hsp90 onto a novobiocin-sepharose column, GDA could not elute Hsp90, suggesting the mechanism of binding varied from that of GDA. Through purified Hsp90 deletion studies, it was determined that, unlike GDA, novobiocin bound to the C-terminus of Hsp90, in a region proximal to the dimerization domain. Furthermore, studies carried out in cellular assays confirmed that the activity of novobiocin was significantly lower than that of GDA. Hsp90 client proteins were degraded at approximately 700 µM, making it more than 1000-fold less active than GDA. Ramifications from this pioneering study were multifold; (1) a new binding site that led to Hsp90 inhibition had been identified, (2) binding at the C-terminus resulted in

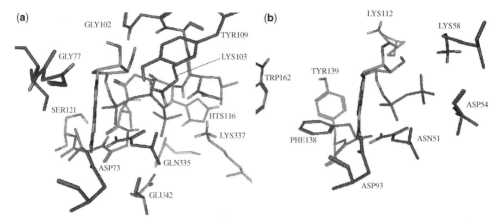

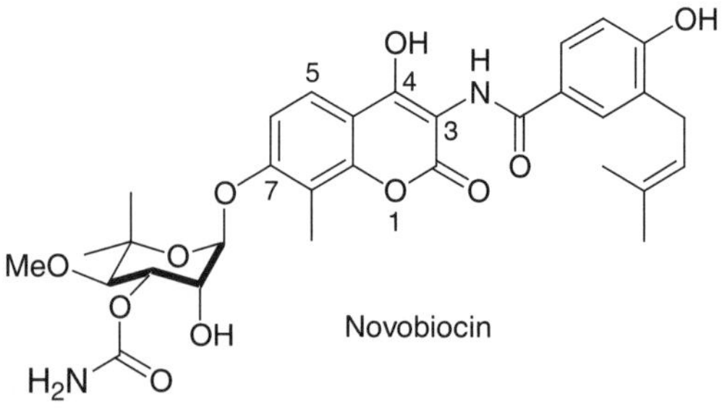

Figure 43. The chemical structure of novobiocin.

ligand displacement from the N-terminus, suggesting crosstalk between the two termini, and (3) a clinically used drug had been identified as the lead scaffold, enabling the development of improved analogs.

Consistent with these studies, Csermely and coworkers experimentally determined that the C-terminal site becomes available following occupancy of the N-terminal nucleotide binding site [225]. Through oxidative nucleotide affinity cleavage studies, they showed that, while the N-terminal binding site is specific for adenine nucleotides, the C-terminus binds both purines and pyrimidines (GTP and UTP preferentially) [226] Subsequent studies by Garnier and coworkers utilizing isothermal titration calorimetry, scanning differential calorimetry and fluorescence spectroscopy further confirmed ATP binding to the purified C-terminal domain, and were the first to suggest this binding site may resemble a Rossman fold [227,228]. In support of Neckers original data, Garnier and coworkers performed additional studies that concluded the C-terminal nucleotide binding site overlaps with the dimerization domain, suggesting a strong relationship between ATP binding, dimerization, and magnesium-dependent oligomerization [227].

Although the Hsp90 C-terminus does not exhibit ATPase activity, it is involved in the conformational rearrangement of Hsp90 upon ATP binding [39,229]. The ATPase activity of Hsp90, which leads to a conformational change of the entire homodimer, is dependent upon the Hsp90 C-terminal region to trap the nucleotide during the ATPase cycle [227]. Yun and coworkers suggested that the conformational switch that occurs upon novobiocin binding causes changes to Hsp90/cochaperone/client interactions and may be responsible for the observed biological activities [39]. Complex interactions between the N-terminus and C-terminus are a critical regulatory component of chaperone function. Garnier and coworkers have alluded to the crosstalk observed as allosteric interactions between the two termini [226,227,230–233]. Garnier has also observed that when nucleotides are bound to the N-terminus, the molecules negatively impact the binding of nucleotides to the C-terminus. Furthermore, affinity of the truncated C-terminus for ATP is higher than that of the entire protein. This result suggests that the N-terminus negatively affects binding to the C-terminus and that interdomain crosstalk occurs, and explains why GDA could not displace novobiocin in Neckers' original studies.

4.3. Novobiocin, the First Hsp90 C-Terminal Inhibitor

The first inhibitor of the Hsp90 C-terminal nucleotide binding pocket, novobiocin, exhibited a poor IC_{50} value as determined by Western blot analyses of Hsp90 clients upon treatment of SKBr3 cells. Novobiocin manifested an IC_{50} value of ~700 µM against these cells, and provided a scaffold that could lead to improved analogs. In an effort to develop more efficacious compounds, the cocrystal structure of DNA gyrase bound to novobiocin was studied as a starting point for further modification. As can be seen in Fig. 44, the carbamate of novobiocin maintains key hydrogen-bond interactions with aspartate 73, which is remi-

N-Terminal domain contains an ATP binding domain GDA/RDC bind at this location.

C-Terminal domain interacts with another monomer to form an active homodimer. Contains the recognition motif for tetratricopeptide repeats found on cochaperones and immunophilins. Location of second ATP (novobiocin) binding site.

Hsp90 Dimer

Figure 44. The Hsp90 homodimer.

niscent of the exocyclic amine of adenine bound to gyrase [234]. In addition to this interaction, it was determined that the coumarin ring lactone provides complementary binding to arginine 136. Assuming the binding sites were similar in nature, it was proposed that new analogs of novobiocin should be generated to maintain these critical interactions.

4.3.1. A Parallel Library Approach Toward the Optimization of Novobiocin
Based upon the DNA gyrase cocrystal structures, a library of novobiocin analogs was prepared to explore structure–activity relationships for Hsp90. As shown in Fig. 45, the coumarins were designed to include a shortened amide side chain and removal of the 4-hydroxy substituent (**A**), removal of both the 4-hydroxy and amide linker (**B**), steric replacements of both the 4-hydroxy and benzamide ring (**C**), and 1,2-positional isomers of the noviosyl linkage (**D** and **E**). Upon synthetic modification, the resulting cyclic carbonates (**A–E1**), 2′-carbamoyl (**A–E2**), 3′-carbamoyl (**A–E3**), and descarbamoyl products (**A–E4**) were produced (Fig. 46) [45].

The most active compound identified was **A4**, which contains an N-acetyl side chain in lieu of the benzamide, lacks the 4-hydroxyl of the coumarin moiety (coumarin **A**), and has an unmodified diol. Structure–activity relationships for these compounds suggested that attachment of noviose at the 7-position of the coumarin ring is critical for biological activity, based on the inactivity of **B**, **D**, and **E**. Incorporation of the amide linker (**A**) resulted in greater inhibitory activity than the unsubstituted derivative, **B**. Furthermore, it was hypothesized that the diol (**4**) mimics the ribose ring in the normal substrate (ATP), and could, explain why replacement with a cyclic carbonate (**1**) or 2′-carbamate (**2**) resulted in a loss of activity. **A4** was shown to induce client protein degradation at approximately 10 µM, but was unique in that it also induced upregulation of Hsp's at the lowest concentration tested, 10 nM. Although it is well known that N-terminal inhibitors induce client protein degradation at the same concentration needed to induce a heat shock response, an Hsp90 inhibitor had not induced a heat shock response at concentrations considerably lower than that needed for degradation. Perhaps, induction of these prosurvival proteins explains why **A4** was found to exhibit no toxicity, but instead was found to exhibit excellent neuroprotective activity [235].

In an effort to determine whether the structure–activity relationships observed for this relatively simple scaffold (**A4**) paralleled those of the natural product, two derivatives of the novobiocin natural product were prepared [68]. These were 4-deshydroxynovobiocin (**DHN1**, Fig. 47), which lacked the vinylogous acid moiety (4-hydroxyl) that is believed to be critical for gyrase inhibition due to its prescribed role in isomerization of the amide bond from *trans* to *cis* upon binding DNA gyrase [236,237]. Also, the 3′-carbamoyl-4-deshydroxynovobiocin analog (**DHN2**) was prepared, based on the knowledge that the 3′-carbamoyl group was shown to exhibit decreased activity in the aforementioned Hsp90 assays. Upon biological evaluation in SKBr3 cells, **DHN1** was shown to induce degradation of both ErbB2 and mutant p53 between 5 and

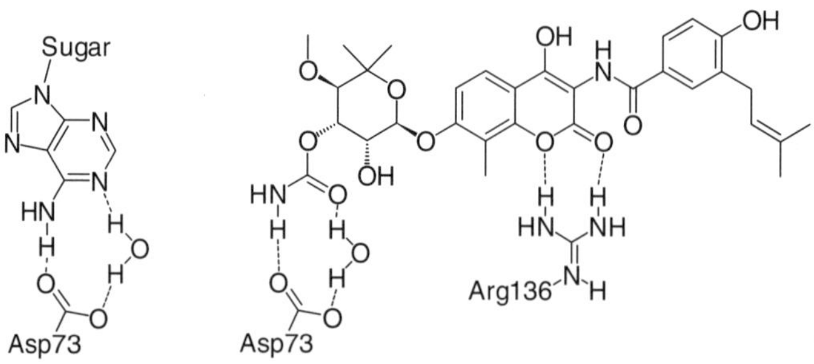

Figure 45. Key interactions observed between DNA gyrase and ADP/novobiocin as determined by analyses of cocrystal structures.

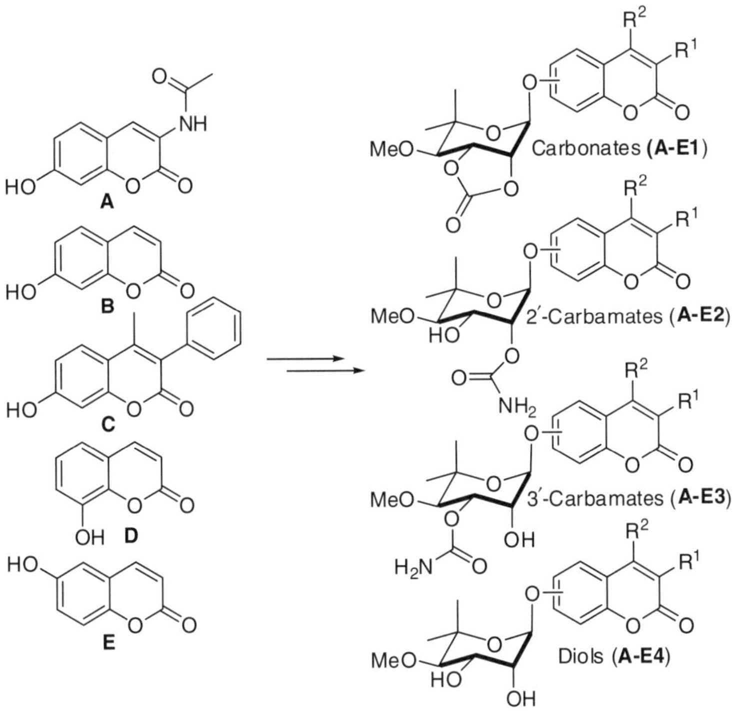

Figure 46. Synthesis of a novobiocin library.

10 µM, whereas **DHN2** induced the degradation of these clients between 0.1 and 1.0 µM. This data indicated that **DHN2** is more effective than **DHN1**, which itself is ~70× more active than novobiocin. When compared to structure–activity relationships obtained from the library approach above, the data from these studies supported the notion that both the 4-hydroxyl substituent on the coumarin ring and the 3′-carbamoyl group on the noviose side chain are detrimental to Hsp90 inhibition. Furthermore, these compounds were shown to exhibit cytotoxicity at the same concentration needed to induce client protein degradation, and manifested activities devoid of DNA gyrase inhibition. However, in con-

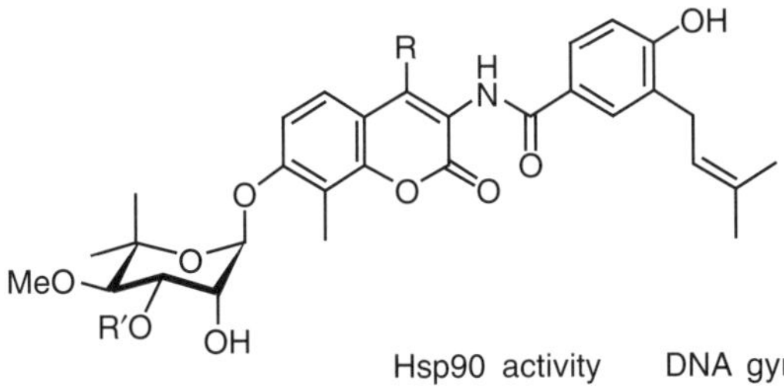

		Hsp90 activity	DNA gyrase activity
R′ = C(O)NH$_2$, R = OH,	Novobiocin	700 µM	0.9 µM
R′ = C(O)NH$_2$, R = H	**DHN1**	~7.5 µM	~250 µM
R′ = H, R = H	**DHN2**	~0.5 µM	~500 µM

Figure 47. Structures of **DHN1** and **DHN2** alongside their inhibitory activities against Hsp90 and DNA gyrase.

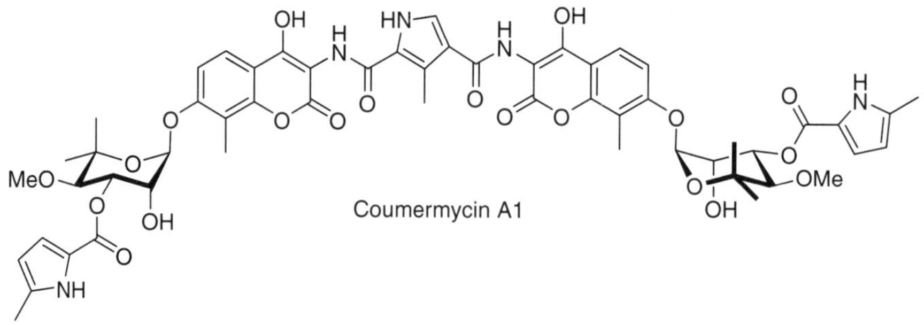

Figure 48. Chemical structure of coumermycin A1.

trast to **A4**, these compounds did not induce a heat shock response, as no induction of Hsp70 or Hsp90 was observed. Finally, these initial SAR studies suggested that the original model used for novobiocin binding to DNA gyrase did not provide a platform on which additional analogs could be pursued rationally.

4.3.2. Novobiocin-Based Dimers Previous studies by the Neckers Laboratory had determined that the dimeric novobiocin compound, coumermycin A1 (IC$_{50}$ ~ 70 µM, Fig. 48), was more effective than the monomeric species (IC$_{50}$ ~ 700 µM) at Hsp90 inhibition [224]. Considering the fact that Hsp90 exists as a homodimer, it is not surprising that such compounds exhibit a ~10-fold increase in inhibitory activity, as both nucleotide binding sites can be simultaneously occupied by the binding of one molecule. As such, it was hypothesized that dimeric variants of **A4** would also exhibit increased activity against the Hsp90 protein folding machinery. In an effort to fully investigate this group of natural product analogs, two approaches were pursued. The first approach involved preparation of **A4** dimers (Fig. 49) that were linked through *meta-* and *para-*phthalic acid, and the second approach utilized the cross-metathesis of olefins to generate a series of compounds that contained various methylene units between the corresponding amides. Unfortunately, the phthalic acid-derived molecules, which closely resemble the bis-pyrrolic ester of coumermycin A1, neither manifested cytotoxicity nor they were able to induce Hsp90-dependent client protein degradation. However, when the metathesis-derived products (**135–138**) were evaluated against MCF-7 and SKBr3 breast cancer cells, a tight SAR was observed, indicating that a tether of six-carbons (**136**) between the amide bonds exhibits optimal activity [238]. Of special note is the observation that even the monomeric species containing an extended alkene group (**131–134**) manifested reasonable antiproliferative activity compared to the nontoxic derivative, **A4**. The rationale denoted for the six-carbon tether was ascribed to the potential distance between the two nucleotide binding sites in the Hsp90 homodimeric species. Like the DHN series, these dimers induced Hsp90-dependent client protein degradation at the same concentration needed to prohibit cell growth, and did not induce Hsp levels. It was therefore hypothesized that substituents on the amide side chain were responsible for mediating the induction of Hsp levels and potentiating cytotoxic effects.

The conversion of a nontoxic molecule (**A4**) into a potent antiproliferative agent (**136**) occurred through modification of the amide side chain, suggesting that further modification of **A4** in a similar manner could produce monomeric species that also exhibit anticancer activity. The identification of such a compound would clearly demonstrate that C-terminal inhibitors function via a mechanism distinct from that of N-terminal inhibitors. Such a finding would provide an opportunity to produce molecules that can be developed specifically for anticancer or neuroprotective activity via modulation of the same biological target.

4.3.3. Optimization of Amide Side Chain In an effort to develop initial structure–activity relationships into more efficacious inhibitors,

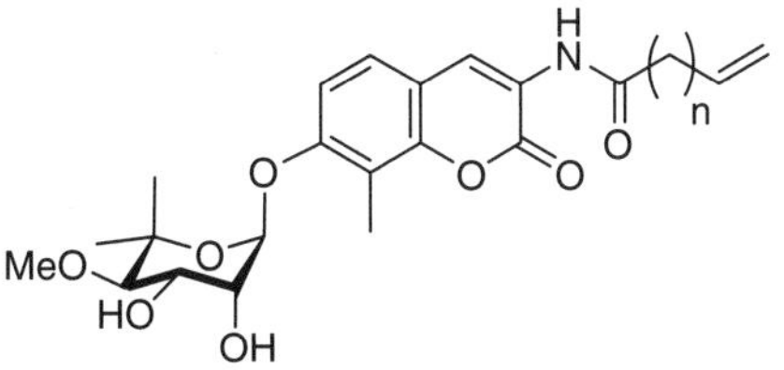

Monomers (IC$_{50}$ μM)

	MCF-7	SKBr3	Her2 ELISA
(**131**) n = 1	26.6 ± 0.7	34.9 ± 11.0mM	>100
(**132**) n = 2	6.7 ± 0.5	5.5 ± 1.5mM	5.0 ± 0.4
(**133**) n = 3	6.2 ± 0.5	8.4 ± 1.8mM	11.9 ± 1.9
(**134**) n = 4	15.6 ± 1.5	28.1 ± 4.6mM	10.3 ± 2.9

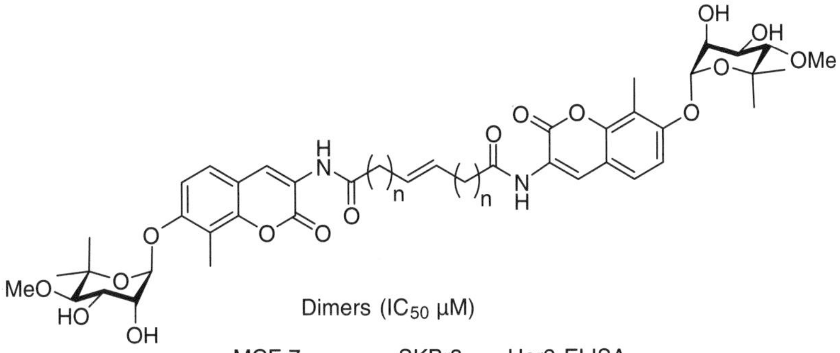

Dimers (IC$_{50}$ μM)

	MCF-7	SKBr3	Her2 ELISA
(**135**) n = 1	53.1 ± 7.1	>100	82.9 ± 4.3
(**136**) n = 2	3.9 ± 0.7	1.5 ± 0.1	5.6 ± 1.3
(**137**) n = 3	13.7 ± 3.1	16.7 ± 7.2	9.6 ± 2.4
(**138**) n = 4	67.4 ± 5.1	>100	10.5 ± 0.3

Figure 49. **A4** dimers and olefinic precursors (monomers).

two additional novobiocin libraries were prepared [239]. The first library explored optimization of the benzamide side chain that contained a *p*-methoxy or *m*-phenyl substituent. The second library focused on the incorporation of heterocycles into the benzamide region in order to investigate hydrogen-bond donor/acceptor interactions and the effects of rigidity, as suggested by the initial findings.

Benzamide **139** (Fig. 50) manifested antiproliferative activity, which is in stark contrast to **A4**, which is nontoxic. Monosubstitution of the ring system led to increased activity due to benzamide modification, as demonstrated by methoxy (**140–142**) and phenyl (**143–145**) derivatives, which produced antiproliferative activity against several cell lines. The data indicated that a *p*-hydrogen-bond acceptor and an *m*-aryl side chain on the benzamide were most effective for activity. Furthermore, replacement of the amide with a sulfonamide led to complete abolishment of activity (**146**), suggesting key hydrogen-bonding interactions are critical for manifesting antiproliferative activity. Concerning spatial requirements of the hydrophobic pocket, a two-carbon spacer between the amide and the phenyl ring (**148**) was found to be more potent than the corresponding methylene linker (**147**), benzyl carbamate (**149**), and benzamide (**139**). Furthermore, the *trans*-cinnamide **150**, which increases rigidity of the ethyl linker, exhibited an approximately threefold increase in inhibitory activity versus the saturated derivative (**147**). The results suggested that the hydrophobic pocket into which the benzamide projects may accommodate larger aro-

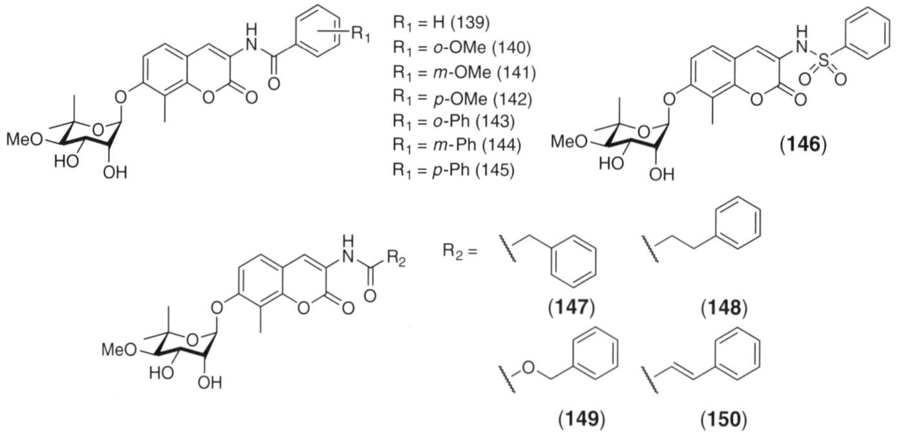

Entry [IC$_{50}$ (μM)]	SKBr3	MCF-7	HCT-116	PL45	LNCaP	PC-3
139	21.5 ± 1.4	20.6 ± 0.4	13.0 ± 2.1	3.4 ± 0.6	72.0 ± 4.0	67.6 ± 9.7
140	>100	5.3 ± 1.3	>100	35.8 ± 3.4	N/T	9.1 ± 0.5
141	>100	5.6 ± 2.5	1.9 ± 0.6	2.8 ± 0.8	21.7 ± 2.0	N/T
142	15.6 ± 4.2	10.3 ± 0.9	15.9 ± 1.9	5.9 ± 2.1	7.3 ± 0.9	17.1 ± 4.3
143	39.1 ± 4.1	18.9 ± 7.0	32.7 ± 1.6	14.4 ± 2.4	17.3 ± 5.2	65.3 ± 5.6
144	13.0 ± 1.4	18.0 ± 3.8	12.8 ± 2.3	1.6 ± 0.2	1.6 ± 0.5	11.6 ± 1.4
145	16.3 ± 1.6	8.1 ± 6.0	3.6 ± 2.0	1.6 ± 0.2	44.9 ± 31.6	19.3 ± 5.1
146	>100	>100	>100	>100	>100	>100
147	21.4 ± 2.2	16.4 ± 0.4	13.2 ± 0.6	8.5 ± 1.7	10.4 ± 0.2	43.3 ± 13.4
148	10.2 ± 2.3	6.9 ± 0.3	5.4 ± 0.6	9.8 ± 0.2	92.8 ± 0.9	22.0 ± 5.8
149	17.8 ± 2.4	12.1 ± 0.1	13.0 ± 0.2	11.4 ± 0.6	N/T	N/T
150	2.6 ± 0.6	4.0 ± 0.3	3.2 ± 0.5	3.0 ± 1.0	4.5 ± 0.7	3.9 ± 0.1

Figure 50. Benzamide derivatives based on **A4**.

matic systems that may exhibit increased affinity. Key structure–function relationships observed for the biaryl benzamide novobiocin derivatives are summarized in Fig. 50.

In addition to these studies, classical SAR studies (Fig. 51) suggested that incorporation of an o-aniline also improved antiproliferative activity. Probing for the optimal benzamide tether length via the *trans*-cinnamide derivative (**150**) and the o-aniline species (**151**), led to identification of the 2-indoleamide (**152**), which exhibited greater than 500-fold increased activity against SKBr3 cells than novobiocin (Fig. 52). 3-Indoleamide was approximately 3–10 times less effective than its 2-indoleamide counterpart. A summary of the observed SAR trends for this class of analogs is provided in Fig. 53.

4.3.4. Optimization of Coumarin Ring System

To further explore SAR, derivatives of **A4** with variations to the coumarin scaffold were designed to probe for interactions typically manifested by the natural substrate, purine ring system [240]. The coumarin-derived motifs were chosen to possess hydrogen-bonding capabilities similar to the nucleotide bases, adenine and guanine, and contain strategically placed hydrogen-bond acceptors/donors and alkyl groups of variable size to probe the size and nature of the binding pocket. Although minor perturbations were made on each analog, the addition of one hydrogen bond is known to produce 1–2 kcal/mol of binding energy and thus increases binding by 10-fold [241]. While it has been demonstrated that the N-terminal site is fairly specific for

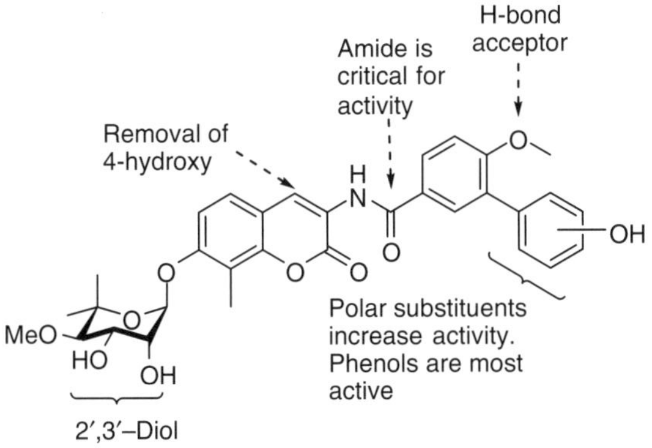

INHIBITORS OF THE Hsp90 C-TERMINAL NUCLEOTIDE BINDING SITE

Figure 51. SAR for the biaryl novobiocin derivatives and Hsp90.

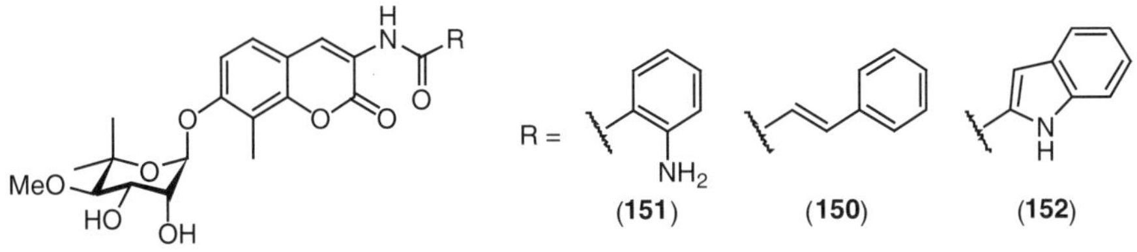

Entry [IC$_{50}$ (μM)]	SKBr3	MCF-7	HCT-116	PL45	LNCaP	PC-3
151	17.7 ± 1.4	17.2 ± 3.3	14.4 ± 0.8	6.2 ± 1.3	12.7 ± 0.8	57.9 ± 10.1
150	2.6 ± 0.6	4.0 ± 0.3	3.2 ± 0.5	3.0 ± 1.0	4.5 ± 0.7	3.9 ± 0.1
152	0.37 ± 0.06	0.57 ± 0.07	0.17 ± 0.01	0.47 ± 0.34	12.2 ± 0.0	22.3 ± 10.1

Figure 52. Identification of an optimal side chain.

adenine nucleotides, the C-terminal site has been shown to be more promiscuous, and bind both purines and pyrimidines. Recalling that adenine specifically binds to the N-terminus, GTP and UTP bind to the C-terminus [226]. Based on these previous studies, mimics of the guanosine nucleus (GTP, Fig. 54) were chosen to take advantage of this differential. Hydrogen-bond acceptors were placed at the 5-, 6-, and 8-positions of the coumarin ring to mimic those at the 6-, 7-, and 3-positions of guanine, respectively. In addition, analogs bearing

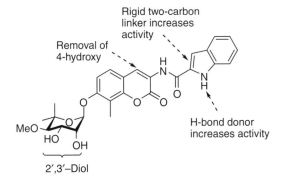

Figure 53. Structure–activity relationships observed for **314** and novobiocin.

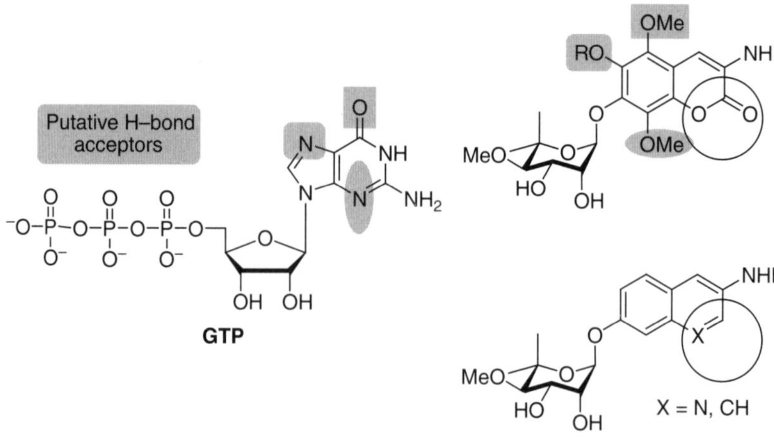

Figure 54. Coumarin analogs that incorporate quanine-type interactors.

modification to the coumarin lactone were constructed to probe the importance of hydrogen-bond donors/acceptors at this position as well as to potentially improve upon solubility.

Alkyl and aryl group were attached at the 5-, 6-, and 8-positions of the coumarin ring to maximize putative hydrophobic interactions. A methoxy group was attached at the 5-position of the coumarin ring, while methoxy, propoxy, and isopropoxy ethers were installed at the 6-position. Methy, ethyl, methoxy, benzyl and phenyl substituents were placed at the 8-position to occupy the pocket normally accommodated by the chlorine of clorobiocin and the methyl substituent of novobiocin. It was found that 6-alkoxy substituents resulted in enhanced antiproliferative activity against cancer cell lines versus compounds containing a hydrogen at this location. In contrast to 6-alkoxy substitution, 5-methoxy functionalities appeared detrimental to antiproliferative activity. 8-Methyl analogs had previously been found to exhibit a ~10-fold improvement in activity compared to 8-hydrogen derivatives. The 8-methoxy analog (**153**) exhibited 2–3-fold improved activity over its 8-methyl counterpart, and fivefold increased activity over the similarly sized 8-ethyl analog (Fig. 55). Replacing the hydrogen-bond accepting group with steric bulk led to compounds that were less active against most cell lines tested.

Noncoumarins **155** and **156** both exhibited low micromolar antiproliferative activity against breast and prostate cancer cells (Fig. 56). These analogs simultaneously exhibited significantly improved activities against AR negative PC-3 prostate cancer cells than the corresponding 8-methylcoumarin and seven- to ninefold reduced activity

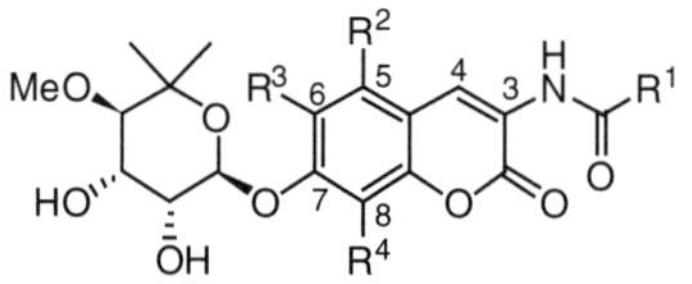

Entry [IC$_{50}$, (μM)]	R^1	R^2	R^3	R^4	MCF-7	SKBr3	PC-3	LnCaP
153	Biaryl	H	H	OMe	9.0 ± 5.4	13.9 ± 1.2	2.3 ± 2.9	1.1 ± 0.1
154	2-Indole	H	OPr	Me	2.1 ± 0.1	2.1 ± 0.8	6.2 ± 1.8	1.8 ± 0.7

Figure 55. Antiproliferation values for modified coumarin analogs.

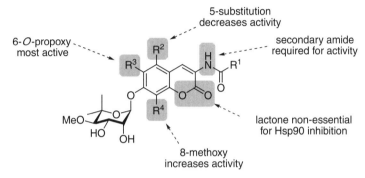

Entry [IC$_{50}$(μM)]	R^1	R^7	X	MCF-7	SkBr3	PC-3	LnCaP
155	Biaryl	H	N	13.1 ± 4.1	16.5 ± 6.2	17.6 ± 4.6	14.2 ± 0.4
156	Biaryl	H	CH	46.4 ± 5.3	38.9 ± 2.4	10.9 ± 0.7	19.6 ± 1.6

Figure 56. Antiproliferation values for noncoumarin analogs.

Figure 57. SAR observed for the coumarin ring system.

against AR positive LnCap prostate cancer cells. Given that both **155** and **156** lack the 8-methyl feature that yields an increased activity of ∼10-fold, it is reasonable to speculate that the quinoline- and naphthalene-derived analogs that include an 8-methyl substituent may exhibit antiproliferative activities between 1 and 5 µM against breast cancer cells and 1–2 µM against prostate cancer cells. These results suggest that, while the lactone moiety may provide hydrogen-bonding interactions with the Hsp90 binding pocket, the lactone itself is not essential for Hsp90 inhibition. As such, incorporation of alternative heterocyclic ring systems in lieu of the naturally occurring coumarin may yield analogs with improved activity. A summary of the observed structure–activity relationships for coumarin modifications is depicted in Fig. 57.

4.3.5. Simplified Coumarin Ring Systems and Tosyl Inclusion The Alami Laboratory has also pursued novobiocin as a lead compound for development as an Hsp90 inhibitor. Initial attempts altered the 7-position of the coumarin ring by preparation of desnoviose analogs [242]. They determined that in compounds containing a *p*-toluene sulfonic ester at the 4-position (**157–160**), the coumarin ring lacking the 7-phenol maintained antiproliferative activity (35–70 µM against MCF-7 cells) (Fig. 58). Furthermore, they observed a tolerance for the benzamide, consistent with findings from the Blagg Laboratory. Most notable was the observation that the 7-hydroxyl and noviose appendages were not critical for inhibitory activity.

Subsequent modification of the novobiocin scaffold resulted in a series of derivatives that reincorporated the 7-phenol and the corresponding methyl ether. As shown in Fig. 59, the most potent compounds obtained were those that contained the *p*-toluenesulfonic ester in either the 4- or the 7-position of the coumarin ring, suggesting a mutually exclusive relationship between these two positions. Furthermore, these studies indicated that incorporation of bicyclic ring systems on the

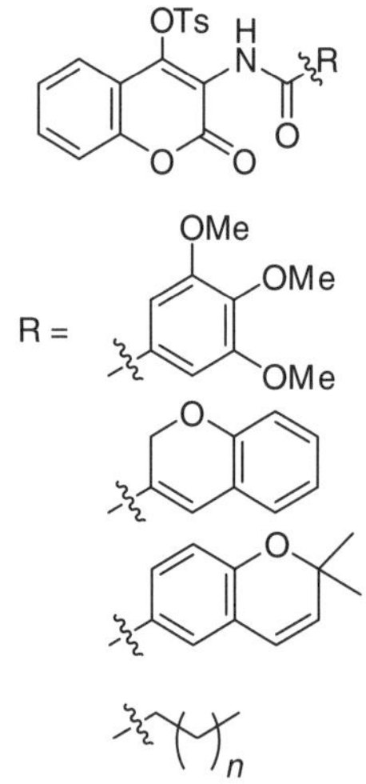

Figure 58. Desnoviose benzamide derivatives of novobiocin.

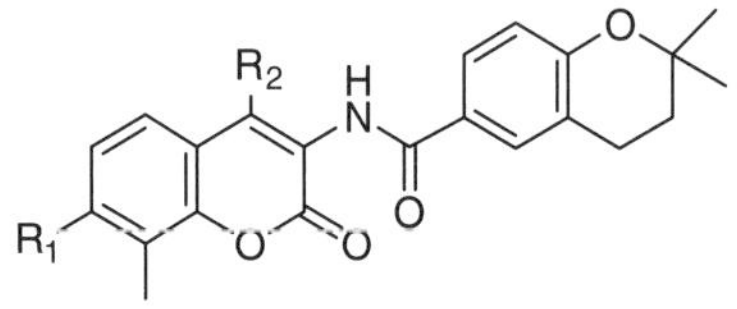

Figure 59. 7-Substituted novobiocin analogs, IC_{50} values reported in micromolar against MCF-7 cells.

R_1 =	OH	OMe	OH	OMe	OH	OTs
R_2 =	OH	OMe	OMe	OH	OTs	OH
IC_{50}	>200	>200	40	>200	50	75

amide side chain also improved inhibitory activity over the native prenylated benzamide side chain [243]. Consistent with earlier studies by the Blagg Laboratory, cytotoxic novobiocin derivatives prepared and evaluated by the Alami Laboratory also demonstrated no heat shock response [244].

4.4. Cisplatin

Cisplatin (Fig. 60) is a platinum-containing chemotherapeutic used clinically to treat various types of cancers, including testicular, ovarian, bladder, and small cell lung cancer [245]. It is well known that cisplatin coordinates to DNA bases, resulting in cross-linked DNA, which prohibits DNA replication [246–248]. In addition to these effects, Sreedhar and coworkers reported that cisplatin binds to the Hsp90 C-terminal domain and interferes with nucleotide binding [55]. Rosenhagen has observed that the hyperactive Hsp90-androgen receptor (AR) complex in prostate cancer is targeted by cisplatin through Hsp90 inhibition. Previous studies have also shown that Hsp90 inhibitors can be used to treat cisplatin-resistance cells that have been transfected with the Hsp90-dependent protein kinase, v-src [23].

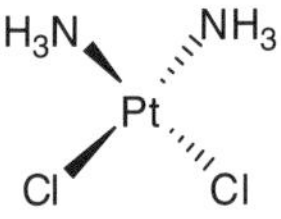

Figure 60. Structure of cisplatin.

Itoh and coworkers reported that cisplatin impairs Hsp90 chaperone activity. By applying bovine brain cytosol to a cisplatin-affinity column, and eluting with cisplatin, they were able to elute Hsp90 in a concentration-dependent manner, indicating cisplatin exhibits high affinity for Hsp90. Subsequent studies utilizing Hsp90 fragments demonstrated that cisplatin binds the C-terminal region [249]. Interestingly, Rosenhagen and coworkers observed degradation of the androgen and glucocorticoid steroid receptors but not other Hsp90 clients, such as raf-1, lck, and c-src upon exposure of neuroblastoma cells with cisplatin, indicating preferential inhibition of the Hsp90/AR complex. The steroid receptor-specific proteolysis observed by Rosenhagen suggests that cisplatin does not complex Hsp90 and other client proteins, but rather it specifically inhibits steroid receptor-Hsp90 complexes [23]. Csermely and coworkers have suggested that the cisplatin binding site is proximal to the C-terminal nucleotide binding site and concluded that cisplatin can be used to efficiently and selectively to block C-terminal nucleotide binding [225].

Acquired resistance to cisplatin can limit its therapeutic potential and many mechanisms of resistance have been encountered [250], including decreased intracellular drug accumulation, enhanced cellular detoxification by glutathione and metallothionein, altered DNA repair and inhibition of apoptosis [250,251]. The inability of certain tumors to

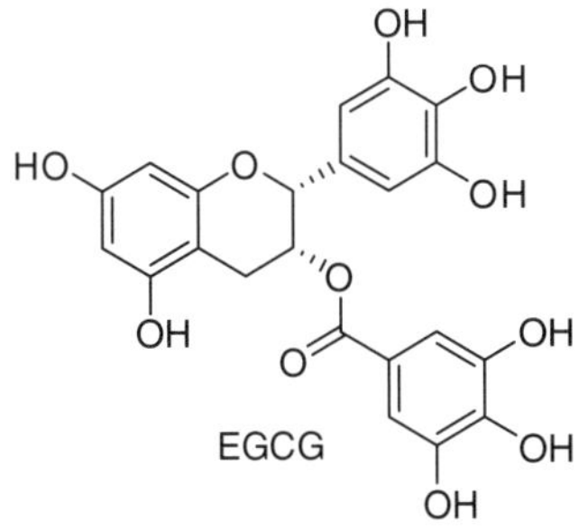

Figure 61. Structure of EGCG.

respond to cisplatin cannot be explained by these mechanisms, suggesting alternative pathways may also be present [252,253]. *In vivo* genomic screening has provided new mechanisms that contribute to the understanding of cisplatin toxicity and resistance [254].

4.5. Epigallocatechin-3-Gallate

EGCG (Fig. 61) is a polyphenolic compound abundant in green tea, along with related structures. EGCG is known to inhibit the activity of many Hsp90 client proteins, including telomerase, multiple kinases, and the aryl hydrocarbon receptor (AhR). EGCG also manifests activities against growth factor signaling, which involves epidermal and vascular endothelial growth factors as well as transcription factors such as AP-1 and NF-κB. Recently Palermo and coworkers demonstrated via affinity chromatography that EGCG manifests its antagonistic activity against AhR through Hsp90 binding [37]. Affinity purification of Hsp90 fragments from immobilized EGCG revealed EGCG manifests its antagonistic activity against AhR through Hsp90 binding [37] at the C-terminus, specifically with amino acids 538–728. Unlike previously identified N-terminal Hsp90 inhibitors, EGCG does not prevent Hsp90 from forming heteroprotein complexes. Studies are currently underway to determine whether EGCG competes with novobiocin or cisplatin binding.

5. APPLICATIONS FOR Hsp90 INHIBITORS

Hsp90's primary objectives include the folding of client proteins and the refolding of aggregated or misfolded proteins. As a result, Hsp90 is an attractive target for various disease states. By inhibiting Hsp90, the normal protein folding machinery is transformed into a catalyst for protein degradation. In contrast, upregulation of Hsp90 and its related homologs can assist in the disaggregation of protein plaques and other aggregated species as illustrated in Fig. 62. Thus, Hsp90 can exert a bifunctional role and be sought for the development of different drug classes.

5.1. Cancer

More than 40 proteins associated with malignant progression are Hsp90-dependent [44]. Therefore, through Hsp90 inhibition, one can simultaneously disrupt all six hallmarks of cancer and offer a unified approach toward cancer [1]. In addition, Hsp90 is overexpressed in malignant cells, and its expression correlates directly with the proliferation of these cells [256–259].

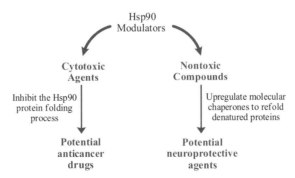

Figure 62. Hsp90 modulation results in two different therapeutic options.

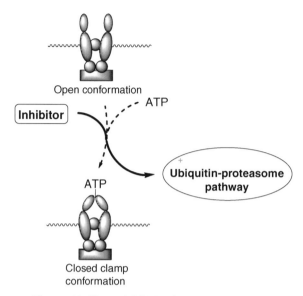

Figure 63. Hsp90 inhibition by anticancer agents.

The mechanism by which Hsp90 inhibitors exert their anticancer activity is by competitively binding to the nucleotide binding sites. After inhibitor binding, the heteroprotein complex is unable to fold or stabilize client proteins and becomes a substrate for degradation via that ubiquitin-proteasome pathway. Figure 63 outlines how disruption of the protein folding process occurs by Hsp90 inhibitors during the critical conformation switch step.

Many of the 40 client proteins associated with oncogenesis are individually sought after anticancer targets for which therapies have been developed [256–260]. While many of these proteins are associated with a specific hallmark of cancer as defined by Weinberg, other examples exist that regulate factors upstream to cancer development. Examples include oncogenic proteins like Mdm2 and SV40 large T-antigen, which are associated with tumor suppressor genes. These Hsp90-dependent proteins play essential roles in regulating p53, a tumor suppressor that is commonly mutated in many cancers. Ral-binding protein 1 is another example of an oncogenic Hsp90-dependent protein. This protein interacts with RalA and RalB, both or which are associated with Ras and many signaling pathways directly related to the malignant phenotype. Hsp90 inhibitors, therefore, offer the potential to treat cancer through disruption of many targets at various stages, further increasing their utility as anticancer agents. The potential to disrupt multiple targets is also what provides Hsp90 inhibition its dual therapeutic potential and promising role for the treatment of neurodegenerative diseases.

5.2. Neurodegenerative Diseases

The accumulation of misfolded proteins that result in plaque formation is responsible for several neurodegenerative diseases, including Alzheimer's, Parkinson's, Huntington's, and prion diseases [261]. Hsp90 is a major molecular chaperone responsible for the rematuration, disaggregation, and resolubilization of these misfolded proteins and their aggregates. As noted previously, Hsp90 inhibitors can induce Hsp levels, resulting in the rematuration of aggregated proteins, which ultimately provides neuroprotection [262]. Figure 64 summarizes the role of Hsp90 as it pertains to neurodegeneration by refolding of protein aggregates.

There are several Hsp90-dependent proteins that play critical roles within the central nervous system that contribute to various disease states. Tau proteins are associated with microtubules and are abundant in neu-

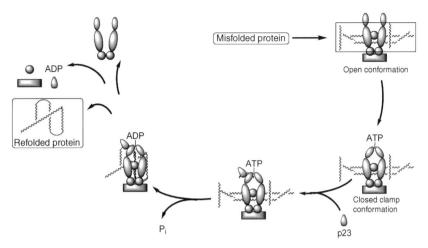

Figure 64. Hsp90 inhibition by neuroprotective agents.

rons within the central nervous system. These proteins promote tubulin assembly into microtubules and different Tau isoforms stabilize these microtubules after phosphorylation. However, hyperphosphorylation of Tau protein results in the self-assembly of filament tangles, which have been shown to be directly involved with the pathogenesis of Alzheimer's disease [263]. The aggregation of Tau protein into neurofibrillary tangles has been proposed to occur with other diseases such as progressive supranuclear palsy, corticobasal degeneration, and Pick's disease [264].

Although GDA and other inhibitors of the Hsp90 N-terminal binding site induce Hsp levels, they also induce client protein degradation at similar concentrations. The net result of such activities is cell arrest and/or death. As discussed earlier, **A4**, a novel C-terminal Hsp90 inhibitor, induces Hsp levels at 1000–10,000-fold lower concentrations than that needed to induce client protein degradation. Since the Hsp family of proteins are pro-survival and are induced at such low concentrations upon treatment with **A4**, no effect on cell growth has been observed. As a result, it was investigated for neuroprotective activity against fresh neurons exposed to lethal levels of Aβ, as a model for Alzheimer's disease. In this study, it was determined that **A4** provided neuroprotection at the same concentration it induced Hsp70 levels. Thus, an Hsp90 modulator that induced prosurvival proteins and exhibited no toxicity was discovered [235]. Utilization of **A4** and improved analogs remain under active investigation for not only Alzheimer's but also the potential treatment of other neurodegenerative diseases.

Soluble protein levels correlate well with high levels of Hsp90. Therefore, high levels of granular Tau oligimers (Tau filaments and intermediates) are observed when Hsp levels are low. It has been suggested that Hsp90 can function to regulate levels of soluble Tau protein, but the chaperone system can become saturated [265]. Chiosis and coworkers studied Tau hyperphosphorylation as the result of the aberrant activation of several kinases, such as cyclin-dependent protein kinase 5 (cdk5) and glycogen synthase kinase-3β. They specifically studied the cdk5/p35 kinase complex, demonstrating that in mice, cdk5 inhibitors reduce Tau hyperphosphorylation and apoptosis in neurons [266,270]. In addition to abnormal phosphorylation of Tau by kinases, the accumulation of aggregated Tau in several tauopathies has been linked to mutations in Tau isoforms found on chromosome 17 [267–269]. Chiosis and coworkers demonstrated that expression of the most common mutation, TauP301L, can be suppressed to inhibit neuronal loss and improve function in mice. Both cdk5/p35 and TauP301L were shown to require Hsp90 assistance for their stability and proper function [270].

Another Hsp90-dependent client associated with neurological disease is alpha-synuclein. This protein is found predominantly at presynaptic terminals in neural tissue, but its primary function remains unknown. Although it is usually a soluble protein, alpha-synuclein can aggregate to form insoluble fibrils in diseases characterized by Lewy bodies, such as Parkinson's disease, dementia with Lewy bodies, and multiple system atrophy. An alpha-synuclein fragment, the non-Abeta component (NAC), is also found in the amyloid plaques associated with Alzheimer's disease [271].

Hsp90 offers a new range of therapies for treating neurodegenerative diseases. Whether through induction of Hsp90 to allow refolding of denatured or aggregated proteins or through directly inhibiting clients related to neurodegeneration, Hsp90 offers an improved target for therapeutic development.

REFERENCES

1. Hanahan D, Weinberg RA. The hallmarks of cancer. Cell 2000;100:57–70.
2. McCarty MF. Targeting multiple signaling pathways as a strategy for managing prostate cancer: multifocal signal modulation therapy. Int Cancer Ther 2004;3:349–380.
3. Bishop S, Burlison J, Blagg BSJ. Hsp90: a novel target for the disruption of multiple signaling cascades. Curr Cancer Drug Targets 2007;7:369–388.
4. Solit DB, Chiosis G. Development and application of Hsp90 inhibitors. Drug Discov Today 2008;13:38–43.
5. Hadden MK, Lubbers D, Blagg BSJ. Geldanamycin, radicicol, and chimeric inhibitors of the Hsp90 N-terminal ATP binding site. Curr Top Med Chem 2006;6:1173–1182.
6. Soo ET, Yip GW, Lwin ZM, Kumar SD, Bay BH. Heat shock proteins as novel therapeutic targets in cancer. In Vivo 2008;22:311–315.
7. Sharp S, Workman P. Inhibitors of the HSP90 molecular chaperone: current status. Adv Cancer Res 2006;95:323–348.
8. Workman P, Burrows F, Neckers L, Rosen N. Drugging the cancer chaperone HSP90: combinatorial therapeutic exploitation of oncogene addiction and tumor stress. Ann NY Acad Sci 2007;1113:202–216.
9. Chiosis G, Huezo H, Rosen N, Mimnaugh E, Whitesell L, Neckers L. 17AAG: low target binding affinity and potent cell activity-finding an explanation. Mol Cancer Ther 2003;2:123–129.
10. Banerji U. Preclinical and clinical activity of the molecular chaperone inhibitor 17-allylamino, 17-demethoxygeldanamycin in malignant melanoma. Proc Am Assoc Cancer Res 2003;44:677.
11. Sausville EA. Clinical development of 17-allylamino, 17-demethoxygeldanamycin. Curr Cancer Drug Targets 2003;3:377–383.
12. Giannini A, Bijlmakers MJ. Regulation of the Src family kinase Lck by Hsp90 and ubiquitination. Mol Cell Biol 2004;24:5667–5676.
13. An WG, Schulte TW, Neckers LM. The heat shock protein 90 antagonist geldanamycin alters chaperone association with p210bcr-abl and v-src proteins before their degradation by the proteasome. Cell Growth Differ 2000;11:355–360.
14. da Rocha Dias S, Friedlos F, Light Y, Springer C, Workman P, Marais R. Activated B-RAF is an Hsp90 client protein that is targeted by the anticancer drug 17-allylamino-17-demethoxygeldanamycin. Cancer Res 2005;65:10686–10691.
15. Grammatikakis N, Lin JH, Grammatikakis A, Tsichlis PN, Cochran BH. p50(cdc37) acting in concert with Hsp90 is required for Raf-1 function. Mol Cell Biol 1999;19:1661–1672.
16. Citri A, Kochupurakkal BS, Yarden Y. The achilles heel of ErbB-2/HER2: regulation by the Hsp90 chaperone machine and potential for pharmacological intervention. Cell Cycle 2004;3:51–60.
17. Zsebik B, Citri A, Isola J, Yarden Y, Szollosi J, Vereb G. Hsp90 inhibitor 17-AAG reduces ErbB2 levels and inhibits proliferation of the trastuzumab resistant breast tumor cell line JIMT-1. Immunol Lett 2006;104:146–155.
18. Burch L, Shimizu H, Smith A, Patterson C, Hupp TR. Expansion of protein interaction maps by phage peptide display using MDM2 as a prototypical conformationally flexible target protein. J Mol Biol 2004;337:129–145.
19. Akalin A, Elmore LW, Forsythe HL, Amaker BA, McCollum ED, Nelson PS, Ware JL, Holt SE. A novel mechanism for chaperone-mediated telomerase regulation during prostate cancer progression. Cancer Res 2001;61:4791–4796.
20. Keppler BR, Grady AT, Jarstfer MB. The biochemical role of the heat shock protein 90 chaperone complex in establishing human tel-

omerase activity. J Biol Chem 2006;281: 19840–19848.
21. Toogun OA, Dezwaan DC, Freeman BC. The hsp90 molecular chaperone modulates multiple telomerase activities. Mol Cell Biol 2008;28:457–467.
22. Fang L, Ricketson D, Getubig L, Darimont B. Unliganded and hormone-bound glucocorticoid receptors interact with distinct hydrophobic sites in the Hsp90 C-terminal domain. Proc Natl Acad Sci USA 2006;103:18487–18492.
23. Rosenhagen MC, Soti C, Schmidt U, Wochnik GM, Hartl FU, Holsboer F, Young JC, Rein T. The heat shock protein 90-targeting drug cisplatin selectively inhibits steroid receptor activation. Mol Endocrinol 2003;17:1991–2001.
24. Cheung J, Smith DF. Molecular chaperone interactions with steroid receptors: an update. Mol Endocrinol 2000;14:939–946.
25. de Carcer G, Avides MC, Lallena MJ, Glover DM, Gonzalez C. Requirement of Hsp90 for centrosomal function reflects its regulation of polo kinase stability. EMBO J 2001;20: 2878–2884.
26. de Carcer G. Heat shock protein 90 regulates the metaphase-anaphase transition in a polo-like kinase-dependent manner. Cancer Res 2004;64:5106–5112.
27. Sato S, Fujita N, Tsuruo T. Modulation of AKT kinase activity by binding to Hsp90. Proc Natl Acad Sci USA 2000;97:10832–10837.
28. Dickey CA, Koren J, Zhang YJ, Xu YF, Jinwal UK, Birnbaum MJ, Monks B, Sun M, Cheng JQ, Patterson C, Bailey RM, Dunmore J, Soresh S, Leon C, Morgan D, Petrucelli L. Akt and CHIP coregulate tau degradation through coordinated interactions. Proc Natl Acad Sci USA 2008;105:3622–3627.
29. Basso AD, Solit DB, Chiosis G, Giri B, Tsichlis P, Rosen N. Akt forms an intracellular complex with heat shock protein 90 (Hsp90) and Cdc37 and is destabilized by inhibitors of Hsp90 function. J Biol Chem 2002;277: 39858–39866.
30. Lewis J, Devin A, Miller A. Disruption of Hsp90 function results in degradation of the death domain kinase, receptor-interacting protein (RIP), and blockage of tumor necrosis factor-induced nuclear factor-kB activation. J Biol Chem 2000;275:10519–10526.
31. Panner A, Murray JC, Berger MS, Pieper RO. Heat shock protein 90alpha recruits FLIPS to the death-inducing signaling complex and contributes to TRAIL resistance in human glioma. Cancer Res 2007;67:9482–9489.
32. Webb CP, Hose CD, Koochekpour S. The geldanamycins are potent inhibitors of the hepatocyte growth factor/scatter factor-met-urokinase plasminogen activator-plasmin proteolytic network. Cancer Res, 2000;60:342–349.
33. Ochel H–J Schulte TW, Nguyen P, Trepel J, Neckers L. The Benzoquinone ansamycin geldanamycin stimulates proteolytic degradation of focal adhesion kinase. Mol Genet Metab 1999;66:24–30.
34. Masson-Gadais B, Houle F, Laferriere J, Huot J. Integrin alphavbeta3, requirement for VEGFR2-mediated activation of SAPK2/p38 and for Hsp90-dependent phosphorylation of focal adhesion kinase in endothelial cells activated by VEGF. Cell Stress Chaperones 2003;8:37–52.
35. Bell DR, Poland A. Binding of aryl hydrocarbon receptor (AhR) to AhR-interacting protein. J Biol Chem 2000;275:36407–36414.
36. Cox MB, Miller CA, 3rd. Cooperation of heat shock protein 90 and p23 in aryl hydrocarbon receptor signaling. Cell Stress Chaperones 2004;9:4–20.
37. Palermo CM, Westlake CA, Gasiewicz TA. Epigallocatechin gallate inhibits aryl hydrocarbon receptor gene transcription through an indirect mechanism involving binding to a 90 kDa heat shock protein. Biochemistry 2005;44: 5041–5052.
38. Donze O, Abbas-Terki T, Picard D. The Hsp90 chaperone complex is both a facilitator and a repressor of the dsRNA kinase PKR. EMBO J 2001;20:3771–3780.
39. Yun BG, Huang W, Leach N, Hartson SD, Matts RL. Novobiocin induces a distinct conformation of Hsp90 and alters Hsp90-cochaperone-client interactions. Biochemistry 2004;43: 8217–8229.
40. Garcia-Cardena G, Fan R, Shah V, Sorrentino R, Cirino G, Papapetropoulos A, Sessa WC. Dynamic activation of endothelial nitric oxide synthase by Hsp90. Nature 1998;392:821–824.
41. Pritchard KA Jr, Ackerman AW, Gross ER, Stepp DW, Shi Y, Fontana JT, Baker JE, Sessa WC. Heat shock protein 90 mediates the balance of nitric oxide and superoxide anion from endothelial nitric-oxide synthase. J Biol Chem 2001;276:17621–17624.
42. Averna M, Stifanese R, De Tullio R, Salamino F, Pontremoli S, Melloni E. In vivo degradation of nitric oxide synthase (NOS) and heat shock protein 90 (HSP90) by calpain is modulated by the formation of a NOS-HSP90 heterocomplex. FEBS J 2008;275:2501–2511.

43. Basto R, Gergely F, Draviam VM, Ohkura H, Liley K, Raff JW. Hsp90 is required to localise cyclin B and Msps/ch-TOG to the mitotic spindle in *Drosophila* and humans. J Cell Sci 2007;120:1278–1287.
44. Adams J, Elliot PJ. New agents in cancer clinical trials. Oncogene 2000;19:6687–6692.
45. Maloney A, Workman P. Hsp90 as a new therapeutic target for cancer therapy: the story unfolds. Expert Opin Biol Ther 2002;2:3–24.
46. Blagosklonny MV. Hsp90-associated oncoproteins: multiple targets of geldanamycin and its analogs. Leukemia 2002;16:455–462.
47. Goetz MP, Toft DO, Ames MM, Erlichman C. The hsp90 chaperone complex as a novel target for cancer therapy. Ann Oncol 2003;14:1169–1176.
48. Richter K, Buchner J. Hsp90: chaperoning signal transduction. J Cell Physiol 2001;188:281–290.
49. Picard D. Heat-shock protein 90, a chaperone for folding and regulation. Cell Mol Life Sci 2002;59:1640–1648.
50. Carrello A, Ingley E, Minchin RF, Tsai A, Ratajczak T. The common tetratricopeptide repeat acceptor site for steroid receptor-associated immunophilins and HOP Is located in the dimerization domain of Hsp90. J Biol Chem 1999;274:2682–2689.
51. Felts SJ, Karnitz LM, Toft DO. Functioning of the Hsp90 machine in chaperoning checkpoint kinase I (Chk1) and the progesterone receptor (PR). Cell Stress Chaperones 2007;12: 353–363.
52. Cliff MJ, Williams MA, Brooke-Smith J, Barford D, Ladbury JE. Molecular recognition via coupled folding and binding in a TPR domain. J Mol Biol 2005;346:717–732.
53. Ramsey AJ, Russell LC, Whitt SR, Chinkers M. Overlapping sites of tetratricopeptide repeat protein binding and chaperone activity in heat shock protein 90. J Biol Chem 2000;275:17857–17862.
54. Pratt WB, Toft DO. Regulation of signaling protein function and trafficking by the Hsp90/Hsp70-based chaperone machinery. Exp Biol Med 2003;228:111–133.
55. Sreedhar AS, Soti C, Csermely P. Inhibition of Hsp90: a new strategy for inhibiting protein kinases. Biochim Biophys Acta 2004;1697: 233–242.
56. Falsone SF, Gesslbauer B, Rek A, Kungl AJ. A proteomic approach towards the Hsp90-dependent ubiquitinylated proteome. Proteomics 2007;7:2375–2383.
57. Xu W, Marcu M, Yuan X, Mimnaugh E, Patterson C, Neckers L. Chaperone-dependent E3 ubiquitin ligase CHIP mediates a degradative pathway for c-ErbB2/Neu. Proc Natl Acad Sci USA 2002;99:12847–12852.
58. Walter S, Buchner J. Molecular chaperone–cellular machines for protein folding. Angew Chem Int Ed 2002;41:1098–1113.
59. Zhang H, Burrows F. Targeting multiple signal transduction pathways through inhibition of Hsp90. J Mol Med 2004;82:488–499.
60. Duvvuri M, Konkar S, Hong KH, Blagg BSJ. A new approach for enhancing differential selectivity of drugs to cancer cells. ACS Chem Bio 2006;1:309–315.
61. Kamal A, Thao L, Sensintaffar J, Zhang L, Boehm MF, Fritz LC, Burrows FJ. A high-affinity conformation of Hsp90 confers tumour selectivity on Hsp90 inhibitors. Nature 2003;425:407–410.
62. Neckers L, Lee Y-S. The rules of attraction. Nature 2003;425:357–359.
63. Ali MMU, Roe SM, Vaughan CK, Meyer P, Panaretou B, Piper PW, Prodromou C, Pearl LH. Nature 2006;440:1013–1017.
64. Panaretou B, Prodromou C, Roe SM, O'Brien R, Ladbury JE, Piper PW, Pearl LH. ATP binding and hydrolysis are essential to the function of the Hsp90 molecular chaperone in vivo. EMBO J 1998;17:4829–4836.
65. Obermann WMJ, Sondermann H, Russo AA, Pavletich NP, Hartl FU. *In vivo* function of Hsp90 is dependent on ATP Binding and ATP hydrolysis. J Cell Biol 1998;143: 901–910.
66. Yu XM, Shen G, Neckers L, Blake H, Holzbeierlein J, Cronk B, Blagg BSJ. Hsp90 inhibitors identified from a library of novobiocin analogues. J Am Chem Soc 2005;127: 12778–12779.
67. Chiosis G, Vilenchik M, Kim J, Solit D. Hsp90: the vulnerable chaperon. Drug Discov Today 2004;9:881–888.
68. Burlison JA, Neckers L, Smith AB, Maxwell A, Blagg BSJ. Novobiocin: redesigning a DNA gyrase inhibitor for selective inhibition of Hsp90. J Am Chem Soc 2006;128: 15529–15536.
69. Toft DO. Recent advances in the study of Hsp90 structure and mechanism of action. Trends Endocrinol Metab 1998;9:238–243.
70. Hawle P, Siepmann M, Harst A, Siderius M, Reusch HP, Obermann WMJ. The middle domain of Hsp90 acts as a discriminator between different types of client proteins. Mol Cell Biol 2006;26:8385–8395.

71. Harst A, Lin H, Obermann WMJ. Aha1 competes with Hop, p50 and p23 for binding to the molecular chaperone Hsp90 and contributes to kinase and hormone receptor activation. Biochem J 2005;387:789–796.

72. Lotz GP, Lin H, Harst A, Obermann WMJ. Aha1 binds to the middle domain of Hsp90, contributes to client protein activation, and stimulates the ATPase activity of the molecular chaperone. J Biol Chem 2003;278:17228–7235.

73. Matsumoto S, Tanaka E, Nemoto TK, Ono T, Takagi T, Imai J, Kimura Y, Yahara I, Kobayakawa T, Ayuse T, Oi K, Mizuno A. Interaction between the N-terminal and middle regions is essential for the *in vivo* function of Hsp90 molecular chaperone. J Biol Chem 2002;277: 34959–34966.

74. Meyer P, Prodromou C, Hu B, Vaughan C, Roe SM, Panaretou B, Piper PW, Pearl LH. Structural and functional analysis of the middle segment of Hsp90: implications for ATP hydrolysis and client protein and cochaperone interactions. Mol Cell 2003;11:647–658.

75. Panaretou B, Siligardi G, Meyer P, Maloney A, Sullivan JK, Singh S, Millson SH, Clarke PA, Naaby-Hansen S, Stein R, Cramer R, Mollapour M, Workman P, Piper PW, Pearl LH, Prodromou C. Activation of the ATPase activity of Hsp90 by the stress-regulated cochaperone Aha1. Mol Cell 2002;10:1307–1318.

76. Chadli A, Bruinsma ES, Stensgard B, Toft D. Analysis of Hsp90 cochaperone interactions reveals a novel mechanism for TPR protein recognition. Biochemistry 2008;47:2850–2857.

77. Roe SM, Prodromou C, O'Brien R, Ladbury JE, Piper PW, Pearl LH. Structural basis for inhibition of the Hsp90 molecular chaperone by the antitumor antibiotics radicicol and geldanamycin. J Med Chem 1999;42: 260–266.

78. Panaretou B, Prodromou C, Roe SM, O'Brien R, Ladbury JE, Piper PW, Pearl LH. ATP binding and hydrolysis are essential to the function of the Hsp90 molecular chaperone *in vivo*. EMBO J 1998;17:4829–4836.

79. Avila C, Kornilayev BA, Blagg BSJ. Development and optimization of a useful assay for determining Hsp90's inherent ATPase activity. Bioorg Med Chem 2006;14:1134–1142.

80. DeBoer C, Meulman PA, Wnuk RJ, Peterson DH. Geldanamycin, a new antibiotic. J Antibiot 1970;23:442–447.

81. Rinehart KL, Sasaki K, Slomp G, Grostic MF, Olson EC, Geldanamycin I. Structure assignment. J Am Chem Soc 1970;92:7591–7593.

82. Uehara Y, Murakami Y, Mizuno S, Kawai S. Inhibition of transforming activity of tyrosine kinase oncogenes by herbimycin A. Virology 1988;164:294–298.

83. Oikawa T, Hirotani K, Shimaura M, Ashino-Fuse H, Iwaguchi T. Powerful antiangiogenic activity of herbimycin A. J Antibiot 1989;42: 1202–1202.

84. Honma Y, Okabe-Kado J, Hozumi M, Uehara Y, Mizuno S. Induction of erythroid differentiation of K562 human leukemic cells by herbimycin A, an inhibitor of tyrosine kinase activity. Cancer Res 1989;49:331–334.

85. Kondo K, Watanabe T, Sasaki H, Uehara Y, Oishi M. Induction of *in vitro* differentiation of mouse embryonal carcinoma (F9) and erythroleukemia cells by herbimycin A, an inhibitor of protein phosphorylation. J Cell Biol 1989;109: 285–293.

86. Uehara Y, Fukazawa H, Murakami Y, Mizuno S. Irreversible inhibition of v-*src* tyrosine kinase activity by herbimycin A and its abrogation by sulfhydryl compounds. Biochem Biophys Res Commun 1989;163:803–809.

87. Uehara Y, Murakami Y, Sugimoto Y, Mizuno S. Mechanism of reversion of Rous sarcoma virus transformation by hebimycin A: reduction of total phosphotyrosine levels due to reduced kinase activity and increased turnover of p60^{v-src1}. Cancer Res 1989;49:780–785.

88. Whitesell L, Shifrin SD, Schwab G, Neckers LM. Benzoquinonoid ansamycins possess selective tumoricidal activity unrelated to src kinase inhibition. Cancer Res 1992;52: 1721–1728.

89. Whitesell L, Mimnaugh EG, De Costa B, Myers CE, Neckers LM. Inhibition of heat shock protein Hsp90-pp60^{v-src} heteroprotein complex formation by benzoquinone ansamycins: essential role for stress proteins in oncogenic transformation. Proc Natl Acad Sci 1994;91:8324–8328.

90. For a complete list of Hsp90 client proteins along with appropriate references, see http://www.picard.ch/downloads/downloads.htm.

91. Stebbins CE, Russo AA, Schneider C, Rosen N, Hartl FU, Pavletich NP. Crystal structure of an Hsp90-geldanamycin complex: targeting of a protein chaperone by an antitumor agent. Cell 1997;89:239–250.

92. Jez JM, Chen JC-H, Rastelli G, Stroud RM, Santi DV. Crystal structure and molecular modeling of 17-DMAG in complex with human Hsp90. Chem Biol 2003;10:361–368.

93. Dikalov S, Landmesser U, Harrison DG. Geldanamycin leads to superoxide formation by

94. Dikalov S, Rumyantseva GV, Piskunov AV, Weiner LM. Role of quinone-iron (III) interaction in NADPH-dependent enzymatic generation of hydroxyl radicals. Biochemistry 1992;31:8947–8953.
95. Schnur RC, Corman ML, Gallaschun RJ, Cooper BA, Dee MF, Doty JL, Muzzi ML, Moyer JD, DiOrio CI, Barbacci EG, Miller PE, O'Brien AT, Morin MJ, Foster BA, Pollack VA, Savage DM, Sloan DE, Pustilnik LR, Moyer MP. Inhibition of the oncogene product p185^{erbB-2} in vitro and in vivo by geldanamycin and dihydrogeldanamycin derivatives. J Med Chem 1995;38:3806–3812.
96. Schnur RC, Corman ML, Gallaschun RJ, Cooper BA, Dee MF, Doty JL, Muzzi ML, DiOrio CI, Barbacci EG, Miller PE, Pollack VA, Savage DM, Sloan DE, Pustilnik LR, Moyer JD, Moyer MP. erbB-2 oncogene inhibition by geldanamycin derivatives: synthesis, mechanism of action, and structure–activity relationships. J Med Chem 1995;38:3813–3820.
97. Schulte TW, Neckers LM. The benzoquinone ansamycin 17-allylamino-17-demethoxygeldanamycin binds to Hsp90 and shares important biologic activities with geldanamycin. Cancer Chemother Pharmacol 1998;42: 273–279.
98. Page J, Heath J, Fulton R, Yalkowsky E, Tabibi E, Tomaszewski J, Smith A, Rodman L. Comparison of geldanamycin (NSC-122750) and 17-allylaminogeldanamycin (NSC-330507D) toxicity in rats. Proc Am Assoc Cancer Res 1997;38:308.
99. For a complete list of current and recently completed clinical trials involving 17-AAG, see www.clinicaltrials.gov.
100. Taldone T, Gozman A, Maharaj R, Chiosis G. Targeting Hsp90: small molecule inhibitors and their clinical development. Curr Opin Pharmacol 2008;8:370–374.
101. Egorin MJ, Lagattuta TF, Hamburger DR, Covey JM, White DR, Musser SM, Eiseman JL. Pharmacokinetics, tissue distribution, and metabolism of 17-(dimethylaminoethylamino)-17-demethoxygeldanamycin (NSC 707545) in CD_2F_1 mice and Fischer 344 rats. Cancer Chemother Pharmacol 2002;49:7–19.
102. Smith V, Suasville EA, Camalier RF, Fiebig H-H, Burger AM. Comparison of 17-dimethylaminoethylamino-17-demethoxy-geldanamycin (17DMAG) and 17-allylamino-17-demethoxygeldanamycin (17AAG) in vitro: effects on Hsp90 and client proteins in melanoma models. Cancer Chemother Pharmacol 2005;56:126–137.
103. Hollingshead M, Alley M, Burger AM, Borgel S, Pacula-Cox C, Fiebig H-H, Sausville EA. In vivo antitumor efficacy of 17-DMAG (17-dimethylaminoethylamino-17-demethoxygeldanamycin hydrochloride), a water soluble geldanamycin derivative. Cancer Chemother Pharmacol 2005;56:115–125.
104. Brazidec J-YL Kamal A, Busch D, Thao L, Zhang L, Timony G, Grecko R, Trent K, Lough R, Salazar T, Khan S, Burrows F, Boehm MF. Synthesis and biological evaluation of a new class of geldanamycin derivatives as potent inhibitors of Hsp90. J Med Chem 2004;47: 3865–3873.
105. Ge J, Normant E, Porter JR, Ali JA, Dembski MS, Gao Y, Georges AT, Grenier L, Pak RH, Patterson J, Sydor JR, Tibbitts TT, Tong JK, Adams J, Palombella VJ. Design, synthesis, and biological evaluation of hydroquinone derivatives of 17-amino-17-demethoxygeldanamycin as potent, water-soluble inhibitors of Hsp90. J Med Chem 2006;49:4606–4615.
106. Sydor JR, Normant E, Pien CS, Porter JR, Ge J, Grenier L, Pak RH, Ali JA, Dembski MS, Hudak J, Patterson J, Penders C, Pink M, Read MA, Sang J, Woodward C, Zhang Y, Grayzel DS, Wright J, Barrett JA, Palombella VJ, Adams J, Tong JK. Development of 17-allylamino-17demethoxygeldanamycin hydroquinone hydrochloride (IPI-504), an anti-cancer agent directed against Hsp90. Proc Natl Acad Sci USA 2006;103:17408–17413.
107. Rastelli G, Tian Z-Q, Wang Z, Myles D, Liu Y. Structure-based design of 7-carbamate analogs of geldanamycin. Bioorg Med Chem Lett 2005;15:5016–5021.
108. McErlean CSP, Proisy N, Davis CJ, Boland NA, Sharp SY, Boxall K, Slawin AMZ, Workman P, Moody CJ. Synthetic ansamycins prepared by a ring-expanding Claisen rearrangement. Synthesis and biological evaluation of ring and conformational analogues of the Hsp90 molecular chaperone inhibitor geldanamycin. Org Biomol Chem 2007;5:531–546.
109. Andrus MB, Meredith EL, Hicken EJ, Simmons BL, Glancey RR, Ma W. Total synthesis of (+)-geldanamycin and (−)-o-guinogeldanamycin: asymmetric glycolate aldol reactions and biological evaluation. J Org Chem 2003;68:8162–8169.
110. Qin H-L, Panek JS. Total synthesis of the Hsp90 inhibitor geldanamycin. Org Lett 2008;10:2477–2479.

111. Hu Z, Liu Y, Tian Z-Q, Ma W, Starks CM, Regentin R, Licari P, Myles DC, Hutchinson CR. Isolation and characterization of novel geldanamycin analogues. J Antibiot 2004;57: 421–428.

112. Patel K, Piagentini M, Rascher A, Tian Z-Q, Buchanan GO, Regentin R, Hu Z, Hutchinson CR, McDaniel R. Engineered biosynthesis of geldanamycin analogs for Hsp90 inhibition. Chem Biol 2004;11:1625–1633.

113. Hong Y-S, Lee D, Kim W, Jeong J-K, Kim C-G, Sohng JK, Lee J-H, Paik S-G, Lee JJ. Inactivation of the carbamoyltransferase gene refines post-polyketide synthase modification steps in the biosynthesis of the antitumor agent geldanamycin. J Am Chem Soc 2004;126: 11142–11143.

114. Lee K, Ryu JS, Jin Y, Kim W, Kaur N, Chung SJ, Jeon Y-J, Park J-T, Bang JS, Lee HS, Kim TY, Lee JJ, Hong YS. Synthesis and anticancer activity of geldanamycin derivatives derived from biosynthetically generated metabolites. Org Biomol Chem 2008;6:340–348.

115. Lee D, Lee K, Cai XF, Dat NT, Boovanahalli SK, Lee M, Shin JC, Kim W, Jeong JK, Lee JS, Lee C-H, Lee J-H, Hong Y-S, Lee JJ. Biosynthesis of the heat-shock protein 90 inhibitor geldanamycin: new insight into the formation of the benzoquinone moiety. ChemBioChem 2006;7:246–248.

116. Kim W, Lee JS, Lee D, Cai XF, Shin JC, Lee K, Lee C-H, Ryu S, Paik S-G, Lee JJ, Hong YS. Mutasynthesis of geldanamycin by the disruption of a gene producing starter unit: generation of structural diversity at the benzoquinone ring. ChemBioChem, 2007;8: 1491–1494.

117. Zheng FF, Kudok SD, Chiosis G, Munster PN, Sepp-Lorenzino L, Danishefsky SJ, Rosen N. Identification of a geldanamycin dimer that induces the selective degradation of HER-family tyrosine kinases. Cancer Res 2000;60: 2090–2094.

118. Zhang H, Yang Y-C, Zhang L, Fan J, Chung D, Choi D, Grecko R, Timony G, Karjian P, Boehm M, Burrows F. Dimeric ansamycins: a new class of antitumor Hsp90 modulators with prolonged inhibitory activity. Int J Cancer 2006;120:918–926.

119. Kuduk SD, Zheng FF, Sepp-Lorenzino L, Rosen N, Danishefsky SJ. Synthesis and evaluation of geldanamycin-estradiol hybrids. Bioorg Med Chem Lett 1999;9:1233–1238.

120. Kuduk SD, Harris CR, Zheng FF, Sepp-Lorenzino L, Ouerfelli Q, Rosen N, Danishefsky SJ. Synthesis and evaluation of geldanamycin-testosterone hydrids. Bioorg Med Chem Lett 2000;10:1303–1306.

121. Chiosis G, Rosen N, Sepp-Lorenzino L. LY294002-geldanamycin heterodimers as selective inhibitors of the PI3K and PI3K-related family. Bioorg Med Chem Lett 2001;11: 909–913.

122. Baselga J. Clinical trials of Herceptin (trastuzumab). Eur J Cancer 2001;37(Suppl 1): S18–S24.

123. Mandler R, Wu C, Sausville EA, Roettinger AJ, Newman DJ, Ho DK, King CR, Yang D, Lippman ME, Landolfi NF, Dadachova E, Brechbiel MW, Waldmann TA. Immunoconjugates of geldanamycin and anti-Her2 monoclonal antibodies: antiproliferative activity on human breast carcinoma cell lines. J Natl Cancer Inst 2000;92:1573–1581.

124. Tanida S, Hasegaqa T, Higashide E. Macbecins I and II, new antitumor antibiotics. I. Producing organism, fermentation and anti-microbial activities. J Antibiot 1980;33: 199–204.

125. Shibata K, satsumabayashi S, Nakagawa A, Omura S. The structure and cytocidal activity of herbimycin C. J Antibiot 1986;39: 1630–1633.

126. Takatsu T, Ohtsuki M, Muramatsu A, Enokita R, Kurakata SI. J Antibiot 2000;53:1310.

127. Tanida S, Muroi M, Hasegawa T. Antibiotic TAN-420, its production and use. 1983; EP 0110710B1.

128. Martin CJ, Gaisser S, Challis IR, Carletti I, Wilkinson B, Gregory M, Prodromou C, Roe SM, Pearl LH, Boyd SM, Zhang M-Q. Molecular characterization of macbecin as an Hsp90 inhibitor. J Med Chem 2008;51:2853–2857.

129. Zhang M-Q, Gaisser S, Nur-E-Alam M, Sheehan LS, Vousden WA, Gaitatzis N, Peck G, Coates NJ, Moss SJ, Radzom M, Foster TA, Sheridan RM, Gregory MA, Roe SM, Prodromou C, Pearl L, Boyd SM, Wilkinson B, Martin CJ. Optimizing natural products by biosynthetic engineering: discovery of nonquinone Hsp90 inhibitors. J Med Chem 2008;51: 5494–5497.

130. Onodera H, Kaneko M, Takahashi Y, Uochi Y, Funahashi J, Nakashima T, Soga S, Suzuki M, Ikeda S, Yamashita Y, Rahayu ES, Kanda Y, Ichimura M. Conformational significance of EH21A1-A4, phenolic derivatives of geldanamycin, for Hsp90 inhibitory activity. Bioorg Med Chem Lett 2008;18:1588–1591.

131. Delmotte P, Delmotte-Plaque J. A new antifungal substance of fungal origin. Nature 1953;171:344.

132. Kwon HJ, Yoshida M, Abe K, Horinouchi S, Beppu T, Radicicol an agent inducing the reversal of transformed phenotypes of src-transformed fibroblasts. Biosci Biotechnol Biochem 1992;56:538–539.
133. Zhao JF, Nakano H, Sharma S. Suppression of RAS and MOS transformation by radicicol. Oncogene 1995;11:161–173.
134. Schulte TW, Akinaga S, Soga S, Sullivan W, Stensgard B, Toft D, Neckers LM. Antibiotic radicicol binds to the N-terminal domain of Hsp90 and shares important biological activities with geldanamycin. Cell Stress Chaperones 1998;3:100–108.
135. Sharma SV, Agatsuma T, Nakano H. Targeting of the protein chaperone, HSP90, by the transformation suppressing agent, radicicol. Oncogene 1998;16:2639–2645.
136. Yamamoto K, Garbaccio RM, Stachel SJ, Solit DB, Chiosis G, Rosen N, Danishefsky SJ. Cyclopropylradicicol: a fully synthetic, fully active point mutant of radicicol: stereochemistry–activity relationships in the radicicol series. Angew Chem Int Ed 2003;42:1280–1284.
137. Agatsuma T, Ogawa H, Akasaka K, Asai A, Yamashita Y, Mizukami T, Akinaga S, Saitoh Y. Halohydrin and oxime derivatives of radicicol: synthesis and antitumor activities. Bioorg Med Chem 2002;10:3445–3454.
138. Ikuina Y, Amishiro N, Miyata M, Narumi H, Ogawa H, Akiyama T, Shiotsu Y, Akinaga S, Murakata C. Synthesis and antitumor activity of novel O-carbamoylmethyloxime derivatives of radicicol. J Med Chem 2003;46:2534–2541.
139. Soga S, Sharma S, Shiotsu Y, Shimizu M, Tahara H, Yamaguchi K, Ikuina Y, Murakata C, Tamaoki T, Kurebayashi J, Schulte TW, Neckers LM, Akinaga S. Stereospecific antitumor activity of radicicol oxime derivatives. Cancer Chemother Pharmacol 2001;48:435–445.
140. Soga S, Neckers LM, Schulte TW, Shiotsu Y, Akasaka K, Narumi H, Agatsuma T, Ikuina Y, Murakata C, Tamaoki T, Akinaga S. KF25706, a novel oxime derivative of radicicol, exhibits in vivo antitumor activity via selective depletion of Hsp90 binding signaling molecules. Cancer Res 1999;59:2931–2938.
141. Lampilas M, Lett R. Convergent stereospecific total synthesis of monocillin and monorden (or radicicol). Tetrahedron Lett 1992;33:773–776.
142. Garbaccio RM, Stachel SJ, Baeschlin DK, Danishefsky SJ. Concise asymmetric syntheses of radicicol and monocillin I. J Am Chem Soc 2001;23:10903–10908.
143. Yamamoto K, Garbaccio RM, Stachel SJ, Solit DB, Chiosis G, Rosen N, Danishefsky SJ. Total synthesis as a resource in the discovery of potentially valuable antitumor agents: cycloproparadicicol. Angew Chem Int Ed 2003;42: 1280–1284.
144. Geng X, Yang Z-Q, Danishefsky SJ. Synthetic development of radicicol and cycloproparadicicol: highly promising anticancer agents targeting Hsp90. Synlett 2004;8:1325–1333.
145. Yang Z-Q, Geng X, Solit D, Pratilas CA, Rosen N, Danishefsky SJ. New efficient synthesis of resorcinylic macrolides via ynolides: establishment of cycloproparadicicol as synthetically feasible preclinical anticancer agent based on Hsp90 as the target. J Am Chem Soc 2004;126: 7881–7889.
146. Moulin E, Zoete V, Barluenga S, Karplus M, Winssinger N. Design, synthesis, and biological evaluation of Hsp90 inhibitors based on conformational analysis of radicicol and its analogues. J Am Chem Soc 2005;127:6999–7004.
147. Proisy N, Sharp SY, Boxall K, Connelly S, Roe SM, Prodromou C, Slawin AM, Pearl LH, Workman P, Moody CJ. Inhibition of Hsp90 with synthetic macrolactones: synthesis and structural and biological evaluation of ring and conformational analogs of radicicol. Chem Biol 2006;13:1203–1215.
148. Isaka M, Suyarnsestakorn C, Tanticharoen M, Kongsaeree P, Thebtaranoth Y. Aigialomycins A-E, new resorcylic macrolides from the marine mangrove fungus *Aigialus parvus*. J Org Chem 2002;67:1561–1566.
149. Geng X, Danishefsky SJ. Total synthesis of aigialomycin D. Org Lett 2004;6:413–416.
150. Barluenga S, Dakas P-Y, Ferandin Y, Meijer L, Winssinger N. Modular asymmetric synthesis of aigialomycin D, a kinase-inhibitory scaffold. Angew Chem Int Ed 2006;45:3951–3954.
151. Hellwig V, Mayer-Bartschmid A, Muller H, Greif G, Kleymann G, Zitzmann W, Tichy H-V, Stadler M, Pochonins A-F. New antiviral and antiparasitic resorcylic acid lactones from Pochonia chlamydosporia var. catenulata. J Nat Prod 2003;66:829–837.
152. Moulin E, Barluenga S, Winssinger N. Concise synthesis of pochinin A, an Hsp90 inhibitor. Org Lett 2005;7:5637–5639.
153. Barluenga S, Wang C, Fontaine J-G, Aouadi K, Beebe K, Tsutsumi S, Neckers L, Winssinger N. Divergent synthesis of a pochonin library targeting Hsp90 and in vivo efficacy of an identified inhibitor. Angew Chem Int Ed 2008;47:4432–4435.

154. Clevenger RC, Blagg BSJ. Design, synthesis, and evaluation of a radicicol and geldanamycin chimera, radamide. Org Lett 2004;6:4459–4462.
155. Shen G, Blagg BSJ. Radester, a novel inhibitor of the Hsp90 protein folding machinery. Org Lett 2005;7:2157–2160.
156. Wang M, Shen G, Blagg BSJ. Radanamycin, a macrocyclic chimera of radicicol and geldanamycin. Bioorg Med Chem Lett 2006;16:2459–2462.
157. Wang M, Blagg BSJ. Synthesis of a versatile metacyclophane macrolactam. Tetrahedron Lett 2008;49:141–144.
158. Shen G, Wang M, Welch TR, Blagg BSJ. Design, synthesis, and structure–activity relationships for chimeric inhibitors of Hsp90. J Org Chem 2006;71:7618–7631.
159. Hadden KM, Blagg BSJ. Synthesis and evaluation of radamide analogues, a chimera of radicicol and geldanamycin. J Org Chem 2009;74:4697–4704.
160. Duerfeldt AS, Brandt GE, Blagg BSJ. Design, synthesis and biological evaluation of conformationally constrained cis-amide Hsp90 inhibitors. Org Lett 2009;11:2353–2356.
161. Chiosis G, Timaul MN, Lucas B, Munster PN, Zheng FF, Sepp-Lorenzino L, Rosen N. A small molecule designed to bind to the adenine nucleotide pocket of Hsp90 causes Her2 degradation and the growth arrest and differentiation of breast cancer cells. Chem Biol 2001;8:289–299.
162. Chiosis G, Lucas B, Shtil A, Huezo H, Rosen N. Development of a purine-scaffold novel class of Hsp90 binders that inhibit the proliferation of cancer cells and induce the degradation of Her2 tyrosine kinase. Bioorg Med Chem 2002;10:3555–3564.
163. Kasibhatla SR, Zhang L, Boehm MF, Fan J, Hong K, Shi J, Biamonte MA. Purine analogues having Hsp90 inhibiting activity. 2003; WO3037860.
164. Vilenchik M, Solit D, Basso A, Huezo H, Lucas B, He H, Rosen N, Spampinato C, Modrich P, Chiosis G. Targeting wide-range oncogenic transformation via PU24FCl, a specific inhibitor of tumor Hsp90. Chem Biol 2004;11:787–797.
165. Wright L, Barril X, Dymock B, Sheridan L, Surgenor A, Beswick M, Drysdale M, Collier A, Massey A, Davies N, Fink A, Fromont C, Aherne W, Boxal K, Sharp S, Workman P, Hubbard RE. Structure–activity relationships in purine-based inhibitor binding to Hsp90 isoforms. Chem Biol 2004;11:775–785.
166. Dymock B, Barril X, Beswick M, Collier A, Davies N, Drysdale M, Fink A, Fromont C, Hubbard RE, Massey A, Surgenor A, Wright L. Adenine derived inhibitors of the molecular chaperone Hsp90: SAR explained through multiple X-ray structures. Bioorg Med Chem Lett 2004;14:325–328.
167. Muranaka K, Sano A, Ichikawa S, Matsuda A. Synthesis of Hsp90 inhibitor dimers as potential antitumor agents. Bioorg Med Chem 2008;16:5862–5870.
168. Llauger L, He H, Kim J, Aguirre J, Rosen N, Peters U, Davies P, Chiosis G. Evaluation of 8-arylsulfanyl, 8-arylsulfoxyl, and 8-arylsulfonly adenine derivatives as inhibitors of the heat shock protein 90. J Med Chem 2005;48:2892–2905.
169. He H, Zatorska D, Kim J, Aguirre J, Llauger L, She Y, Wu N, Immormino RH, Gewirth DT, Chiosis G. Identification of potent water soluble purine-scaffold inhibitors of the heat shock protein 90. J Med Chem 2006;49:381–390.
170. Biamonte MA, Shi J, Hong K, Hurst DC, Zhang L, Fan J, Busch DJ, Karjian PL, Maldonado AA, Sensintaffar JL, Yang Y-C, Kamal A, Lough RE, Lundgren K, Burrows FJ, Timony GA, Boehm MF, Kasibhatla SR. Orally active purine-based inhibitors of the heat shock protein 90. J Med Chem 2006;49:817–828.
171. Kasibhatla SR, Hong KD, Boehm MF, Biamonte MA, Zhang L. 2-Aminopurine analogs having Hsp90-inhibiting activity. 2006; US7138401B2.
172. Kasibhatla SR, Hong K, Biamonte MA, Busch DJ, Karjian PL, Sensintaffar JL, Kamal A, Lough RE, Brekken J, Lundgren K, Grecko R, Timony GA, Ran Y, Mansfield R, Fritz LC, Ulm E, Burrows FJ, Boehm MF. Rationally designed high-affinity 2-amino-6-halopurine heat shock protein 90 inhibitors that exhibit potent antitumor activity. J Med Chem 2007;50:2767–2778.
173. Altieri DC. Validating survivin as a cancer therapeutic target. Nat Rev Cancer 2003;3:46–54.
174. Fortugno P, Beltrami E, Plescia J, Fontana J, Pradhan D, Marchisio PC, Sessa WC, Altieri DC. Regulation of survivin function by Hsp90. Proc Natl Acad Sci USA 2003;100:13791–13796.
175. Plescia J, Salz W, Xia F, Pennati M, Zaffaroni N, Daidone MG, Meli M, dohi T, Fortugno P, Nefedova Y, Gabrilovich DI, Colombo G, Altieri DC. Rational design of shepherdin, a

novel anticancer agent. Cancer Cell 2005;7: 457–468.

176. Schilb A, Riou V, Schoepfer J, Ottl J, Muller K, Chene P, Mayr LM, Filipuzzi I. Development and implementation of a highly miniaturized confocal 2D-FIDA-based high-throughput screening assay to search for active site modulators of the human heat shock protein 90β. J Biomol Screen 2004;9: 569–577.

177. Kim J, Felts S, Llauger L, He H, Huezo H, Rosen N, Chiosis G. Development of a fluorescence polarization assay for the molecular chaperone Hsp90. J Biomol Screen 2004;9: 375–381.

178. Du Y, Moulick K, Rodina A, Aguirre J, Felts S, Dingledine R, Fu H, Chiosis G. High-throughput screening fluorescence polarization assay for tumor-specific Hsp90. J Biomol Screen 2007;12:915–924.

179. Howes R, Barril X, Dymock BW, Grant K, Northfield CJ, Robertson AGS, Surgenor A, Wayne J, Wright L, James K, Matthews T, Cheung K-M, McDonald E, Workman P, Drysdale MJ. A fluorescence polarization assay for inhibitors of Hsp90. Anal Biochem 2006;350: 202–213.

180. Rowlands MG, Newbatt YM, Prodromou C, Pearl LH, Workman P, Aherne W. High-throughput screening for inhibitors of heat-shock protein 90 ATPase activity. Anal Biochem 2004;327:176–183.

181. Avila C, Hadden MK, Ma Z, Kornilayev BA, Ye Q-Z, Blagg BSJ. High-throughput screening for Hsp90 ATPase inhibitors. Bioorg Med Chem Lett 2006;16:3005–3008.

182. Galam L, Hadden MK, Ma Z, Ye Q-Z, Yun B-G, Blagg BSJ, Matts RL. High-throughput assay for the identification of Hsp90 inhibitors based on Hsp90-dependent refolding of firefly luciferase. Bioorg Med Chem 2007;15:1939–1946.

183. Barril X, Brough P, Drysdale M, Hubbard RE, Massey A, Surgenor A, Wright L. Structure-based discovery of a new class of Hsp90 inhibitors. Bioorg Med Chem Lett 2005;15: 5187–5191.

184. Park H, Kim Y-J, Hahn J-S. A novel class of Hsp90 inhibitors isolated by structure-based virtual screening. Bioorg Med Chem Lett 2007;17:6345–6349.

185. Cheung K-MJ Matthews TP, James K, Rowlands MG, Boxall KJ, Sharp SY, Maloney A, Roe SM, Prodromou C, Pearl LH, Aherne GW, McDonald E, Workman P. The identification, synthesis, protein crystal structure and *in vitro* biochemical evaluation of a new 3,4-diarylpyrazole class of Hsp90 inhibitors. Bioorg Med Chem Lett 2005;15:3338–3343.

186. Sharp SY, Boxall K, Rowlands M, Prodromou C, Roe SM, Maloney A, Powers M, Clarke PA, Box G, Sanderson S, Patterson L, Matthews TP, Cheung K-MJ Ball K, Hayes A, Raynaud F, Marais R, Pearl L, Eccles S, Aherne W, McDonald E, Workman P. *In vitro* biological characterization of a novel, synthetic diaryl pyrazole resorcinol class of heat shock protein 90 inhibitors. Cancer Res 2007;67:2206–2216.

187. Zhou V, Han S, Brinker A, Klock H, Caldwell J, Gu X-J. A time-resolved fluorescence resonance energy transfer-based HTS assay and a surface plasmon resonance-based binding assay for heat shock protein 90 inhibitors. Anal Biochem 2004;331:349–357.

188. Kreusch A, Han S, Brinker A, Zhou V, Choi H-S, He Y, Lesley SA, Caldwell J, Gu X-J. Crystal structrues of human Hsp90α-complexed with dihydroxyphenylpyrazoles. Bioorg Med Chem Lett 2005;15:1475–1478.

189. Dymock BW, Barril X, Brough PA, Cansfield JE, Massey A, McDonald E, Hubbard RE, Surgenor A, Roughley SD, Webb P, Workman P, Wright L, Drysdale MJ. Novel, potent small-molecule inhibitors of the molecular chaperone Hsp90 discovered through structure-based design. J Med Chem 2005;48:4212–4215.

190. Brough PA, Barril X, Beswick M, Dymock BW, Drysdale MJ, Wright L, Grant K, Massey A, Surgenor A, Workman P. 3-(5-chloro-2,4-dihydroxyphenyl)-pyrazole-carboxamides as inhibitors of the Hsp90 molecular chaperone. Bioorg Med Chem Lett 2005;15:5197–5201.

191. Barril X, Beswick MC, Collier A, Drysdale MJ, Dymock BW, Fink A, Grant K, Howes R, Jordan A, Massey A, Surgenor A, Wayne J, Workman P, Wright L. 4-Amino derivatives of the Hsp90 inhibitor CCT018159. Bioorg Med Chem Lett 2006;16:2543–2548.

192. Brough PA, Aherne W, Barril X, Borgognoni J, Boxall K, Cansfield JE, Cheung K-MJ Collins I, Davies NGM, Drysdale MJ, Dymock B, Eccles SA, Finch H, Fink A, Hayes A, Howes R, Hubbard RE, James K, Jordan AM, Lockie A, Martins V, Massey A, Matthews TP, McDonald E, Morthfield CJ, Pearl LH, Prodromou C, Ray S, Raynaud FI, Roughley SD, Sharp SY, Surgenor A, Walmsley DL, Webb P, Wood M, Workman P, Wright L. 4,5-Diarylisoxazole Hsp90 chaperone inhibitors: potential therapeutic agents for the treatment of cancer. J Med Chem 2008;51:196–218.

193. Sharp SY, Prodromou C, Boxall K, Powers MV, Holmes JL, Box G, Matthews TP, Cheung K-MJ Kalusa A, James K, Hayes A, Hardcastle A, Dymock B, Brough PA, Barril X, Cansfield JE, Wright L, Surgenor A, Foloppe N, Hubbard RE, Aherne W, Pearl L, Jones K, McDonald E, Raynaud F, Eccles S, Drysdale M, Workman P. Inhibition of the heat shock protein 90 molecular chaperone in vitro and in vivo by novel, synthetic, potent resorcinylic pyrazole.isoxazole amide analogues. Mol Cancer Ther 2007;6:1198–1211.

194. Jensen MR, Schoepfer J, Radimerski T, Massey A, Guy CT, Brueggen J, Quadt C, Buckler A, Cozens R, Drysdale MJ, Garcia-Echeverria C, Chene P. NVP-AUY922: a small molecule Hsp90 inhibitor with potent antitumor activity in preclinical breast cancer models. Breast Cancer Res 10:2 2008; R33.

195. Eccles SA, Massey A, Raynaud FI, Sharp SY, Box G, Valenti M, Patterson L, Brandon AH, Gowan S, Boxall F, Aherne W, Rowlands M, Hayes A, Martins V, Urban F, Boxall K, Prodromou C, Pearl L, James K, Matthews TP, Cheung K-M, Kalusa A, Jones K, McDonald E, Barril X, Brough PA, Cansfield JE, Dymock B, Drysdale MJ, Finch H, Howes R, Hubbard RE, Surgenor S, Webb P, Wood M, Wright L, Workman P. NVP-AUY922: a novel heat shock protein 90 inhibitor active against xenograft tumor growth, angiogenesis, and metastasis. Cancer Res, 2008;68:2850–2860.

196. Gopalsamy A, Shi M, Golas J, Vogan E, Jacob J, Johnson M, Lee F, Nilakantan R, Peterson R, Svenson K, Chopra R, Tam MS, Wen Y, Ellingboe J, Arndt K, Boschelli F. Discovery of benzoisoxazoles as potent inhibitors of chaperone heat shock protein 90. J Med Chem 2008;51:373–375.

197. Meli M, Pennati M, Curto M, Daidone MG, Plescia J, Toba S, Altieri DC, Zaffaroni N, Colombo G. Small-molecule targeting of heat shock protein 90 chaperone function: rational identification of a new anticancer lead. J Med Chem 2006;49:7721–7730.

198. Huth JR, Park C, Petros AM, Kunzer AR, Wendt MD, Wang X, Lynch CL, Mack JC, Swift KM, Judge RA, Chen J, Richardson PL, Jin S, Tahir SK, Matayoshi ED, Dorwin SA, Ladror US, Severin JM, Walter KA, Bartley DM, Fesik SW, Elmore SW, Hajduk PJ. Discovery and design of novel Hsp90 inhibitors using multiple fragment-based design strategies. Chem Biol Drug Des 2007;70:1–12.

199. Barta TE, Veal JM, Rice JW, Partridge JM, Fadden RP, Ma W, Jenks M, Geng L, Hanson GJ, Huang KH, Barabasz AF, Foley BE, Otto J, Hall SE. Discovery of benzamide tetrahydro-4H-carbazol-4-ones as novel small molecule inhibitors of Hsp90. Bioorg Med Chem Lett 2008;18:3517–3521.

200. Chandarlapaty S, Sawai A, Ye Q, Scott A, Silinski M, Huang K, Fadden P, Partdrige J, Hall S, Steed P, Norton L, Rosen N, Solitm DB. SNX2112, a synthetic heat shock protein inhibitor, has potent antitumor activity against HER kinase-dependent cancers. Clin Cancer Res 2008;14:240–248.

201. Hall SE. Discovery and pre-clinical profile of SNX-5422: an orally acitve Hsp90 inhibitor in phase I trials for solid and hematological tumors. Proc Am Assoc Cancer Res 2008; Abstract 2449.

202. Kung P-P, Funk L, Meng J, Collins M, Zhou JZ, Johnson MC, Ekker A, Wang J, Mehta P, Yin M-J, Rodgers C, Davies JF, Bayman E, Smeal T, Meagley KA, Gehring MR. Dihydroxyphenyl amides as inhibitors of the Hsp90 molecular chaperone. Bioorg Med Chem Lett 2008; 18:6273–6278.

203. Hadden MK, Galam L, Gestwicki JE, Matts RL, Blagg BSJ. Derrubone, an inhibitor of the Hsp90 protein folding machinery. J Nat Prod 2007;70:2014–2018.

204. East AJ, Ollis WD, Wheeler RE. Natural occurrence of 3-aryl-4-hydroxycoumarins. Part 1. Phytochemical examination of Derris robusta (Roxb.) Benth. J Chem Soc C Org 1969;3: 365–374.

205. Hastings JM, Hadden MK, Blagg BSJ. Synthesis and evaluation of derrubone and select analogues. J Org Chem 2008;73: 369–373.

206. Hadden KM, Hill SA, Davenport J, Matts RL, Blagg BSJ. Synthesis and evaluation of Hsp90 inhibitors that contain the 1,4-naphthoquinone scaffold. Bioorg Med Chem 2009;17: 634–640.

207. Chen S, Smith DF. HOP as an adaptor in the heat shock protein 70 (Hsp70) and Hsp90 chaperone machinery. J Biol Chem 1998;273: 35194–35200.

208. Cortajarena AL, Yi F, Regan L. Designed TPR modules as novel anticancer agents. ACS Chem Biol 2008;3:161–166.

209. Yi F, Regan L. A novel class of small molecule inhibitors of Hsp90. ACS Chem Biol 2008;3: 645–654.

210. Wani MC, Taylor HL, Wall ME, Coggon P, McPhail AT. Plant antitumor agents. VI. Isolation and structure of taxol, a novel antileu-

kemic and antitumor agent from Taxus brevifolia. J Am Chem Soc 1971;93:2325–2327.
211. Byrd CA, Bornmann W, Erdjument-Bromage H, Tempst P, Pavletich N, Rosen N, Nathan CF, Ding A. Heat shock protein 90 mediates macrophage activation by taxol and bacterial lipopolysaccharide. Proc Natl Acad Sci USA 1999;96:5645–5650.
212. Ding AH, Porteu F, Sanchez E, Nathan CF. Shared actions of endotoxin and taxol on TNF receptors and TNF release. Science 1990;248:370–372.
213. Solit DB, Basso AD, Olshen AB, Scher HI, Rosen N. Inhibition of heat shock protein 90 function down-regulates Akt kinase and sensitizes tumors to taxol. Cancer Res 2003;63: 2139–2144.
214. Khalid SA, Duddeck H, Gonzalez-Sierra M. Isolation and characterization of an antimalarial agent of the neem tree Azadirachta indica. J Nat Prod 1989;52:922–927.
215. Zhang T, Hamza A, Cao X, Wang B, Yu S, Zhan C, Sun D. A novel Hsp90 inhibitor to disrupt Hsp90/Cdc37 complex against pancreatic cancer cells. Mol Cancer Ther 2008;7:162–170.
216. Westerheide SD, Bosman JD, Mbadugha BNA, Kawahara TLA, Matsumoto G, Kim S, Gu W, Devlin JP, Silverman RB, Morimoto RI. Celastrols as inducers of the heat shock response and cytoprotection. J Biol Chem 2004;279:56053–56060.
217. Uddin SJ, Nahar L, Shilpi JA, Shoeb M, Borkowski T, Gibbons S, Middleton M, Byres M, Sarker S. Gedunin, a limonoid from *Xylocarpus granatum*, inhibits the growth of CaCo-2 colon cancer cell line *in vitro*. Phytother Res 2007;21:757–761.
218. Powers MV, Workman P. Inhibitors of the heat shock response: biology and pharmacology. FEBS Lett 2007;581:3758–3769.
219. Hieronymus H, Lamb J, Ross KN, Peng XP, Clement C, Rodina A, Nieto M, Du J, Stegmaier K, Raj SM, Maloney KN, Clardy J, Hahn WC, Chiosis G, Golub TR. Gene expression signature-based chemical genomic prediction identifies a novel class of HSP90 pathway modulators. Cancer Cell 2006;10:321–330.
220. Lamb J, Crawford ED, Peck D, Modell JW, Blat IC, Wrobel MJ, Lerner J, Brunet J, Subramanian A, Ross KN, Reich M, Hieronymus H, Wei G, Armstrong SA, Haggarty SJ, Clemons PA, Wei R, Carr SA, Lander ES, Golub TR. The connectivity map: using gene-expression signatures to connect small molecules, genes, and disease. Science 2006;313:1929–1935.
221. Corbett KD, Berger JM. Structure of the topoisomerase VI-B subunit: implications for type II topoisomerase mechanism and evolution. EMBO J 2003;22:151–163.
222. Holdgate GA, Tunnicliffe A, Ward WHJ, Weston SA, Rosenbrock G, Barth PT, Taylor IWF, Paupit RA, Timms D. The entropic penalty of ordered water accounts for weaker binding of the antibiotic novobiocin to a resistant mutant of DNA gyrase: a thermodynamic and crystallographic study. Biochemistry 1997;36:9663–9673.
223. Marcu MG, Schulte TW, Neckers L. Novobiocin and related coumarins and depletion of heat shock protein 90-dependent signaling proteins. J Natl Cancer Inst 2000;92:242–248.
224. Marcu MG, Chadli A, Bouhouche I, Catelli B, Neckers LM. The heat shock protein 90 antagonist novobiocin interacts with a previously unrecognized ATP-binding domain in the carboxyl terminus of the chaperone. J Biol Chem 2000;275:37181–37186.
225. Soti C, Racz A, Csermely P. A nucleotide-dependent molecular switch controls nucleotide binding at the C-terminal domain of Hsp90. J Biol Chem 2002;277:7066–7075.
226. Soti C, Vermes A, Haystead TAJ, Csermely P. Comparative analysis of the ATP-binding sites of Hsp90 by nucleotide affinity cleavage: a distinct nucleotide specificity of the C-terminal ATP-binding site. Eur J Biochem 2003;270: 2421–2428.
227. Garnier C, Lafitte D, Tsvetkov PO, Barbier P, Leclerc-Devin J, Millot J-M, Briand C, Makarov AA, Catelli MG, Peyrot V. Binding of ATP to Heat shock protein 90. Evidence for an ATP-binding site in the C-terminal domain. J Biol Chem 2002;277:12208–12214.
228. Callebaut I, Catelli MG, Portetelle D, Meng X, Cadepond F, Burny A, Baulieu EE, Mornon JP. Redox mechanism for the chaperone activity of heat shock proteins HSPs 60, 70 and 90 as suggested by hydrophobic cluster analysis: hypothesis. C R Acad Sci III 1994;317: 721–729.
229. Allan RK, Mok D, Ward BK, Ratajczak T. Modulation of chaperone function and cochaperone interaction by novobiocin in the C-terminal domain of Hsp90: evidence that coumarin antibiotics disrupt Hsp90 dimerization. J Biol Chem 2006;281:7161–7171.
230. Scheibel T, Weikl T, Buchner J. Two chaperone sites in Hsp90 differing in substrate specificity and ATP dependence. Proc Natl Acad Sci USA 1998;95:1495–1499.

231. Hartson SD, Thulasiraman V, Huang W, Whitesell L, Matts RL. Molybdate inhibits Hsp90, induces Structural changes in its C-terminal domain, and alters its interactions with substrates. Biochemistry 1999;38:3837–3849.
232. Jibard N, Meng X, Leclerc P, Rajkowski K, Fortin D, Schweizer-Groyer G, Catelli M-G, Baulieu E-E, Cadepond F. Delimitation of two regions in the 90-kDa heat shock protein (Hsp90) able to interact with the glucocorticosteroid receptor (GR). Exp Cell Res 1999;247:461–474.
233. Scheibel T, Siegmund HI, Jaenicke R, Ganz P, Lilie H, Buchner J. The charged region of Hsp90 modulates the function of the N-terminal domain. Proc Natl Acad Sci USA 1999;96:1297–1302.
234. Lewis RJ, Singh OMP, Smith CV, Skarzyknski T, Maxwell A, Wonacott AJ, Wigley DB. The nature of inhibition of DNA gyrase by the coumarins and the cyclothialidines revealed by X-ray crystallography. EMBO J 1996;15:1412–1420.
235. Ansar S, Burlison JA, Hadden MK, Yu X-M, Desino KE, Bean J, Neckers LM, Audus K, Michaelis ML, Blagg BSJ. A non-toxic Hsp90 inhibitor protects neurons from Ab-induced toxicity. Bioorg Med Chem Lett 2007;17:1984–1990.
236. Laurin P, Ferroud D, Schio L, Klich M, Dupuis-Hamelin C, Mauvais P, Lassaigne P, Bonnefoy A, Musicki B. Structure–activity relationship in two series of aminoalkyl substituted coumarin inhibitors of gyrase B. Bioorg Med Chem Lett 1999;9:2875–2880.
237. Laurin P, Ferroud D, Klich M, Dupuis-Hamelin C, Mauvais P, Lassaigne P, Bonnefoy A, Musicki B. Synthesis and in vitro evaluation of novel highly potent coumarin inhibitors of gyrase B. Bioorg Med Chem Lett 1999;9:2079–2084.
238. Burlison JA, Blagg BSJ. Coumermycin A1 analogues that inhibit the Hsp90 protein folding machinery. Org Lett 2006;8:4855–4858.
239. Burlison JA, Avila C, Vielhauer G, Lubbers DJ, Holzbeierlein J, Blagg BSJ. Development of novobiocin analogues that manifest antiproliferative activity against several cancer cell lines. J Org Chem 2008;73:2130–2137.
240. Donnelly AC, Mays JR, Burlison JA, Nelson JT, Vielhauer G, Holzbeierlein J, Blagg BSJ. The design, synthesis and evaulation of coumarin ring derivatives of the novobiocin scaffold that exhibit anti-proliferative actvity. J Org Chem 2008;73:8901–8920.
241. Klaholz BP, Moras D. C–H...O Hydrogen bonds in the nuclear receptor RARγ: a potential tool for drug selectivity. Structure 2002;10:1197–1204.
242. Radanyi C, Le Bras G, Messaoudi S, Bouclier C, Peyrat J-F, Brion J-D, Marsaud V, Renoir J-M, Alami J. Synthesis and biological activity of simplified denoviose-coumarins related to novobiocin as potent inhibitors of heat shock protein 90 (Hsp90). Bioorg Med Chem Lett 2008;18:2495–2498.
243. Le Bras G, Radanyi C, Peyrat J-F, Brion J-D, Alami M, Marsaud V, Stella B, Renoir J-M. New novobiocin analogues as antiproliferative agents in breast cancer cells and potential inhibitors of heat shock protein 90. J Med Chem 2007; 6189–6200.
244. Radanyi C, Le Bras G, Marsaud V, Peyrat J-F, Messaoudi S, Catelli M-G, Brion J-D, Alami M, Renoir J-M. Antiproliferative and apoptotic properties of tosylcyclonovobioic acids as potent heat shock protein 90 inhibitors in human cancer cells. Cancer Lett 2009;274:88–94.
245. Galanski M. Recent developments in the field of anticancer platinum complexes. Recent Pat Anticancer Drug Discov 2006;1:285–295.
246. Goodisman J, Hagrman D, Tacka KA, Souid AK. Analysis of cytotoxicities of platinum compounds. Cancer Chemother Pharmacol 2006;57:257–267.
247. Brabec V, Kasparkova J. Modifications of DNA by platinum complexes. Drug Resist Update 2005;8:131–146.
248. Frankenberg-Schwager M, Kirchermeier D, Greif G, Baer K, Becker M, Frankenberg D. Cisplatin-mediated DNA double-strand breaks in replicating but not in quiescent cells of the yeast Saccharomyces cerevisiae. Toxicology 2005;212:175–184.
249. Itoh H, Ogura M, Komatsuda A, Wakui H, Miura AB, Tashima Y. A novel chaperone-activity-reducing mechanism of the 90-kDa molecular chaperone HSP90. Biochem J 1999;343:697–703.
250. Perez RP. Cellular and molecular determinants of cisplatin resistance. Eur J Cancer 1998;34:1535–1542.
251. Huang RY, Eddy M, Vujcic M, Kowalski D. Genome-wide screen identifies genes whose inactivation confer resistance to cisplatin in Saccharomyces cerevisiae. Cancer Res 2005;65:5890–5897.
252. Schenk PW, Brok M, Boersma AWM, Brandsma JA, Kulk HD, Brok M, Burger H, Stoter G, Brouwer J, Nooter K. Anticancer drug resis-

252. ...tance induced by disruption of the *Saccharomyces cerevisiae NPR2* gene: a novel component involved in cisplatin- and doxorubicin-provoked cell kill. Mol Pharmacol 2003;64:259–268.

253. Niedner H, Christen R, Lin X, Kondo A, Howell SB. Identification of genes that mediate sensitivity to cisplatin. Mol Pharmacol 2001;60:1153–1160.

254. Liao C, Hu B, Arno MJ, Panaretou B. Genomic screening *in vivo* reveals the role played by cacuolar H^+ ATPase and cytosolic acidification in sensitivity to DNA-damaging agents such as cisplatin. Mol Pharmacol 2007;71: 416–425.

255. Brandt GEL, Schmidt MD, Prisinzano TE, Blagg BSJ. Gedunin, a novel Hsp90 inhibitor: semi-synthesis of derivatives and preliminary structure–activity relationships. J Med Chem 2008;51:6495–6502.

256. Csermely P, Schnaider T, Soiti C, Prohaszka Z, Nardi G. The 90-kDa molecular chaperone family: structure, function, and clinical applications. A comprehensive review. Pharmacol Ther 1998;79:129–168.

257. Yufu Y, Nishimura J, Nawata H. High constitutive expression of heat shock protein 90 alpha in human acute leukemia cells. Leuk Res 1992;16:597–605.

258. Franzen B, Linder S, Alaiya AA, Eriksson E, Fujioka K, Bergman AC. Analysis of polypeptide expression in benign and malignant human breast lesions. Electrophoresis 1997;18:582–587.

259. Luparello C, Noel A, Pucci-Minafra I. Intratumoral heterogeneity for hsp90beta mRNA levels in a breast cancer cell line. DNA Cell Biol 1997;16:1231–1236.

260. Chaudhury S, Welch TR, Blagg BSJ. Hsp90 as a target for drug development. ChemMedChem 2006;1:1331–1340.

261. Warrick JM, Chan HY, Chai GBY, Paulson HL, Bonin NM. Suppression of polyglutamine-mediated neurodegeneration in *Drosophila* by the molecular chaperone HSP70. Nat Genet 1999;23:425–428.

262. Kim HR, Kang HS, Kim HD. Geldanamycin induces heat shock protein expression through activation of HSF1 in K562 erythroleukemic cells. IUBMB Life 1999;48:429–433.

263. Alonso AdC, Zaidi T, Novak M, Grundke-Iqbal I, Iqbal K. Hyperphosphorylation induces self-assembly of τ into tangles of paired helical filaments/straight filaments. Proc Natl Acad Sci USA 2001;98:6923–6928.

264. Delacourte A. Tauopathies: recent insights into old diseases. Folia Neuropathol 2005;43:244–257.

265. Sahara N, Maeda S, Yoshiike Y, Mizoroki T, Yamashita S, Murayama M, Park JM, Saito Y, Murayama S, Takashima A. Molecular chaperone-mediated tau protein metabolism counteracts the formation of granular tau oligomers in human brain. J Neurosci Res 2007;85:3098–3108.

266. Lau LF, Schachter JB, Seymour PA, Sanner MA. Tau protein phosphorylation as a therapeutic target in Alzheimer's disease. Curr Top Med Chem 2002;4:395–415.

267. Lee VM, Goedert M, Trojanowski JQ. Neurodegenerative tauopathies. Annu Rev Neurosci 2001;24:1121–1159.

268. Goedert M, Jakes R. Mutations causing neurodegenerative tauopathies. Biochim Biophys Acta 2005;1739:240–250.

269. Zheng YL, Kesavapany S, Gravell M, Hamilton RS, Schubert M, Amin N, Albers W, Grant P, Pant HC. A Cdk5 inhibitory peptide reduces tau hyperphosphorylation and apoptosis in neurons. EMBO J 2005;24:209–220.

270. Luo W, Dou F, Rodina A, Chip S, Kim J, Zhao Q, Moulick K, Aquirre J, Wu N, Greengard P, Chiosis G. Roles of heat-shock protein 90 in maintaining and facilitating the neurodegenerative phenotype in tauopathies. Proc Natl Acad Sci USA 2007;104:9511–9516.

271. Culvenor JG, McLean CA, Cutt S, Campbell BCV, Maher F, Jakala P, Hartmann T, Beyreuther K, Masters CL, Li Q-X. Non-Abeta component of Alzheimer's disease amyloid (NAC) revisited. NAC and alpha-synuclein are not associated with Abeta amyloid. Am J Pathol 1999;155:1173–1181.

GENE THERAPY WITH PLASMID DNA

Surendar Reddy Bathula
Leaf Huang
Division of Molecular Pharmaceutics,
Eshelman School of Pharmacy,
University of North Carolina at Chapel
Hill, Chapel Hill, NC

1. OVERVIEW OF PLASMID-MEDIATED GENE THERAPY

The idea of using genes for therapeutic purpose was initially conceived in the late 1960s and the early 1970s and the concept of delivery of nucleic acids by means of a vector to patients for some therapeutic purpose has termed gene therapy. This methodology was first proposed to manipulate cells at the molecular level in order to cure rare genetic diseases like cystic fibrosis, phenylketonuria, is now being considered for treating acquired disorders, such as cancer and infectious disease such as AIDS.

The ideal method of gene therapy should deliver highly efficient genes to specific target cells without being toxic or immunogenic. The transgene expression should be regulated long-term without causing mutagenic effect. Finally, it should be easy to manufacture and clinically applicable. Gene therapy is largely classified into two modes of delivery, it either utilizes viruses or it does not. Viruses, consisting of small pieces of DNA, RNA, and a few proteins are natural and potent vehicles for gene delivery. They can be genetically altered to remove the pathogenicity, yet still efficiently infect specific cell types and transfer their genomes into the target cells. Examples of viruses used as gene transfer vectors include retrovirus, adenovirus, adeno-associated virus, herpes simplex virus, vaccinia virus, sendai virus, and so on. Though viral vectors are highly efficient in transfecting cells, they suffer from a number of major limitations. For example, adenoviral vectors generate toxic inflammatory responses and can lead to onset of serious host immune responses against viral structural components thereby preventing their readministration in the same host. Additional set-backs include (a) possibility of random integration into the host chromosome leading to subsequent activation of proto-oncogenes; (b) systemic clearance of viral vectors due to complement activation; (c) possibility of generating replication-competent virus through recombination with the host genome; (d) limited insert-size of the virally packaged therapeutic genes, and so on. All these alarming biosafety concerns associated with the use of viral vectors are either absent or presented only as a minor problem, with the nonviral vectors. However, nonviral vectors still need improvement, both in the efficiency and in the duration of the transgene expression. These aspects are the central concerns of the current nonviral vector development. Another important concern for the development of safe gene therapy systems is toxicity. Although it is well known that nonviral gene delivery produces a less severe immune responses than virus-mediated delivery, problems still need to be addressed. Apart from the pure vector toxicity, lipoplex/lipopolyplex toxicity mostly arises from the stimulation of the immune cells by the unmethylated CpG motifs in the plasmid DNA. Various approaches have been taken to reduce this inflammatory toxicity without affecting transfection efficacy, including elimination of CpG motifs in the plasmid DNA, active targeting of the DNA to the endothelium and sequential injection of cationic liposomes followed by naked plasmid DNA.

Most of the nonviral vectors use recombinant plasmid DNA (pDNA) as the carrier for the gene of interest. pDNA is a circular double-stranded DNA molecule of bacterial origin and can be produced on a large scale by fermentation of bacterial cultures. Besides the therapeutic gene(s), pDNA contains other important gene sequences such as promoter/enhancer elements, which are responsible for controlling the transcription and expression levels of the encoded protein once it is introduced into the target cells. Generally, standard pDNA is 4–10 kbp in size (molecular weight $1-3 \times 10^6$) (Fig. 1). pDNA is a strong polyanion because of the phosphate groups in its polynucleotide backbone and mostly it exists in a compact, supercoiled structure. pDNA can elicit immune responses as a result of the

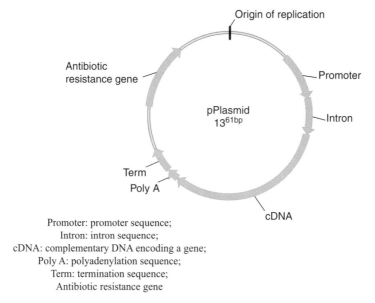

Promoter: promoter sequence;
Intron: intron sequence;
cDNA: complementary DNA encoding a gene;
Poly A: polyadenylation sequence;
Term: termination sequence;
Antibiotic resistance gene

Figure 1. Schematic representation of a typical plasmid DNA. (This figure is available in full color at http://mrw.interscience.wiley.com/emrw/9780471266945/home.)

unmethylated CpG motifs that are abundant in the molecule. While this may be an advantage in gene vaccines, which require small degree of inflammation, it can be a major toxicity in other therapy modalities. These unwanted motifs can be removed by site-directed mutagenesis [1] or the entire bacterial sequence in the plasmid can be excised such as in the case of the mini genes [2]. The major advantage of plasmid expression vector is that the cells upon transfection produce homogeneous, correctly folded, post-translationally modified proteins within itself. This thwarts the attempt to purify and isolate proteins, which often leads to partial degradation or inactivation of the protein.

This chapter is, however, intended to emphasize the role of pDNA as a gene-carrying cask that is delivered to the target cells by nonviral vectors. Plasmid DNA-mediated gene therapy can be broadly classified into the following classes:

(1) Direct delivery of naked pDNA
(2) pDNA delivery by physical methods
(3) pDNA delivery by chemical methods

2. DIRECT DELIVERY WITH NAKED PLASMID DNA

2.1. Gene Transfer with pDNA and Attaining Long-Term Expression

When pDNA is not delivered as a complex with other molecules, it is called "naked" DNA. Naked pDNA provides a promising mechanism for gene delivery as it is less immunogenic than most nonviral vectors currently used. The simplest of all DNA delivery systems to transfect cells is the injection of naked pDNA to the organ or tissue of interest without additional help from either a chemical agent or a physical force. Skeletal muscle was first transfected by intramuscular (IM) injection of naked pDNA [3]. Local injection of pDNA into the muscle, liver, or skin, or airway instillation into the lungs, leads to low-level gene expression [4–8]. However, naked DNA-mediated gene transfer became attractive for airway gene delivery and for the treatment of some acute diseases and DNA-based immunization because of its simplicity and lack of toxicity. These two areas of application are likely to benefit from naked DNA-mediated gene transfer, provided that the delivery efficiency and the duration of transgene expression are improved [9]. The delivery by direct injection of naked DNAs containing one or more anticancer genes was also tested against various tumors with mixed efficacies [10–17].

Sustained therapeutic benefit needs assurance for long-lasting gene expression of the therapeutic gene without eliciting any expression related cytotoxicity. Different strategies were attempted with mixed success. For example, high-expressing human factor IX (hFIX) pDNA upon transfection in the mouse liver yielded therapeutic level of gene expression over 1.5 years, eliciting no expression-related cytotoxicity or therapy related toxicity [18]. However, with liver regeneration, there was a decline in transgene expression. Thus, the decline was associated with the loss of the plasmid that was not integrated into the host genome. Random genomic integration of the transgene is risky because of the possibility of insertional mutagenesis.

For extended transgene expression of a conventional plasmid vector, the plasmid was modified to contain Epstein–Barr virus (EBV) episomal replication elements that consisted of *EBNA1* (EBV nuclear antigen-1) gene, and oriP element. Interestingly, nuclear localization and transcriptional upregulation of the oriP-bearing plasmid was assisted with the help of EBNA-1. This enabled efficacious transgene expression and long-term maintenance of the expression in cells transfected with the EBV/lipoplex. EBV plasmid vector found its important role in various disorders such as congenital, malignant, chronic, and infectious diseases. This plasmid was used in gene therapy against tumor malignancies, such as in reducing tumor size by intratumoral injection of EBV plasmid vector encoding *HSV1-TK* anticancer gene. *EBV* plasmid based interleukin genes, *IL-12* and *IL-18* gene transfer to tumors elicited natural killer (NK) cell activities and cytotoxic T cell (CTL) responses, which lead to tumor growth retardation. Cardiomyopathic animals experienced extended cardiac output upon intracardiomuscular transfer of EBV-based β-adrenergic receptor gene. EBV-based plasmid gene transfer was also used for genetic vaccination against acute viral infections [19,20].

Another school of thought sprouted while developing and fine-tuning the conditions for long-term transgene expression. Retrovirus-mediated gene transfer led to random integration into the host chromosome. This often leads to abnormality in gene expression. However, site-specific genetic integration into inconspicuous genomic sequence might be one of the most-sought-after preconditions for stable, long lasting, and powerful transgene expressions mediated through recombinant plasmids. This became possible with phage integrases. These enzymes mediate unidirectional site-specific recombination between two specific DNA sequences, one containing phage attachment site (attP) and the other bacterial attachment site (attB). These DNA sequences are short, yet specific on a genetic scale. A plasmid containing the bacterial attachment site flanking the gene of interest in both the ends was transfected into cells. The integrases of the serine family showed efficient recombination in mammalian cells. They mediate efficient integration at introduced att sites or native sequences with partial att sites (pseudo-attP). Human α1-antitrypsin and hFIX gene associated with attB cassette, coadministered with an integrase plasmid, upon liver delivery to mice showed elevated long-term expression (therapeutic level of protein obtained for about 8 months). Upon partial hepatectomy, the expression did not decline, indicating that the DNA was integrated to the mouse genome. Similar results were obtained during the functional restoration of human type VII collagen protein in skin cells obtained from recessive dystrophic epidermolysis bullosa (RDEB) patients. Additionally, this technique could be useful in developing large transgenic animals dedicated to the preparation of therapeutic proteins on a large scale [21].

This technique of direct injection has become a useful tool to analyze the gene expression and promoter function in the respective organ. Utilizing viral promoter, such as that from cytomegalovirus (CMV), a high level of initial expression was attained, which, however, declined sharply to near background within 2–3 weeks. Although within 2 days of pDNA injection, maximum decrease in the expression level was observed, but pDNA continued to be scantily detected till the end of 12 weeks (0.2 copies per genome). The major cause of this early decline was due to viral promoter activated immune responses [18].

A wide spectrum of application of naked DNA-mediated gene transfer is limited by the fact that free pDNA is too large and contains too many charges to cross biological barriers. In fact, only cells along the needle track that are damaged by the physical injection are transfected. The barriers include the blood endothelium, the interstitial matrices, the mucus lining and specialized ciliates, tight junctions of epithelial cells, and the plasma membranes of all normal cells. In addition, DNA degradation by intra- and extracellular nucleases further reduces the chance of intact and functional DNA entering the nucleus. It has been shown that after i.v. injection, most of the injected DNA was cleared by the liver from the circulation and by increasing the permeability of the endothelial sheet, it is possible to significantly increase transfection of the hepatocytes [22]. Pharmacokinetic analysis has demonstrated that the hepatic uptake clearance is almost identical to the plasma flow rate in the liver, suggesting highly effective elimination by this organ [22]. In addition, pDNA is taken up preferentially by the liver nonparenchymal cells via a scavenger receptor (SRA)-mediated process, in a manner specific for polyanions [23].

For improving clinical use of naked DNA-based gene transfer, it is important to develop optimized delivery systems, which will enhance the cellular uptake, offer protection from enzymatic degradation and provide cell-specific delivery. The addition of water-immiscible solvents, nonionic polymers or surfactants or the use of hypotonic solution can elevate gene transfer across the cell membrane. Also, several nuclease inhibitors can also enhance naked DNA-mediated gene transfer in cultured cells, muscle, and lung [24–33].

2.2. Therapeutic Uses of Naked Plasmid DNA

Nonviral gene therapy had become a useful tool in generating various therapeutic effects in diverse animal models. A variety of diseases such as cancer, muscular dystrophies, and chronic ischemic limb syndromes were treated by the technique of direct gene transfer [34]. Naked pDNA-based gene transfer had been used in different animal models, for example, in correcting Duchenne muscular dystrophy, to supply sources of therapeutic protein systemically, for genetic vaccination against pathogens and tumor cells [35–41]. However, larger animals showed limited clinical utility owing to inefficient gene expression in these animals.

Naked DNA induces anticancer response in tumor models by introducing anticancer genes. NO synthase II (*NOS II*) gene that catalyzes nitric oxide (NO) production was incorporated in a plasmid under CMV promoter control. This plasmid, upon injecting in medullary thyroid cancer (MTC), elicited anticancer effect in tumorigenic cells. The gene triggered a suicidal effect (apoptosis) on the tumor cells by the production of NO, which specifically activated the macrophages against the fast dividing tumor cells, leaving normal cells untouched. NO was also observed to mediate bystander antitumor effect: The activated *NOS* gene in a tumor cell induced the cytotoxicity in the neighboring tumor cells. The fact that NO-mediated tumorogenic effect does not require transfection of all neoplastic cells promised a capable suicide gene therapy approach to human cancer [15–17].

Fas ligand is another apoptosis inducing molecule that showed similar bystander effect. The membrane protein receptor Fas and its ligand FasL interacting with each other initiate an apoptotic signal in Fas-bearing cells. pDNA could be used to deliver these apoptosis-inducing genes to initiate killing of transfected and nontransfected surrounding cells. On direct injection of FasL encoding pDNA vector into the inflamed thyroid, pathogenic lymphocytes were inhibited to enter into thyroid, leaving the already infiltrated T cells dead [42]. Thus, FasL expressing in thyroicytes might lead to potential remedial therapy for the experimental autoimmune thyroiditis (EAT).

On quantifying the gene-uptake in the muscle, the IM injection showed less than 1% uptake of injected dose and was limited to cells adjacent to the needle track [43]. Using hypertonic sucrose [44], or muscle revitalizers, the efficiency and reproducibility of gene expression could be increased [45,46]. Intraarterial injection of naked pDNA into the femoral arteries of rats showed two orders of magnitude higher transgene-expression level throughout the muscles of hind limb compared to that obtained in direct muscular injection [36]. Myofibers were 10% more transfected through intravascular delivery than with the direct IM injection. This intra-arterial injection technique was extended to the nonhuman primates also [47]. Hypothetically, the pDNA upon intravascular injection extravasates by the intravascular pressure following the convective flow across the endothelium [48], and was soon picked up by the *in situ* muscle cells with the help of membrane receptors. The Possible mechanisms of efficient naked pDNA uptake, especially via intravascular injection are (a) large membrane disruption, (b) small membrane pores, and (c) receptor-mediated endocytosis [49]. Several DNA receptors had been discovered in human leukocytes, peritoneal macrophages, and in wide variety of tissues and tumor cells [50–56]. For the enhancement of IM gene expression, a fresh look is needed toward the poorly understood mechanism of molecular recognition and characterization of the cell surface receptor(s) involved in the binding and internalization of DNA. It is reported that rapid injection of pDNA in a large volume (e.g., 5 μg of DNA/ 20 g mouse in 1.5–2.0 mL of saline solution) through the tail vein left the injected DNA in the inferior vena cava. The DNA flowed back to the tissues linked to this vascular system, primarily the liver. The hydrodynamic pressure forced DNA into the liver cells before it was mixed with the blood. By this process, the liver showed the highest expression of gene; internal organs such as lung, spleen, heart, and kidney were also efficiently transfected [57]. It was also shown that briefly clamping the vena cava following tail vein injection of pDNA in a small volume efficiently transfected both liver and diaphragm [58]. The result is potentially important because diaphragm is barely transfected by hydrodynamics based method. The full-length dystrophin cDNA can be delivered to the diaphragm for the treatment of Duchenne muscular dystrophy. It is well known that patients with Duchenne muscular dystrophy often suffer from fatal respiratory failure due to the dystrophic diaphragm muscle collapse [59].

2.3. Functional Use of Plasmid DNA

Direct pDNA transfer technique could find its use in examining the role of a physiologically related protein-transduction signal pathway toward certain endogenous disease phenotype, such as investigating the functional role of the tissue kallikreinkinin system (KKS) in the central control of blood pressure homeostasis. Kallikrein is a proteinase enzyme that converts kininogen to vasodialative kinin peptides. To understand the role of vasodialative KKS on the pathogenesis of hypertension, the human tissue kallikrein gene *CMV-cHK*, was intracerebroventricularly injected into hypertensive rats. Cortex, cerebellum, brain stem, hippocampus, and hypothalamus identifiably expressed human tissue kallikrein protein. The expression level and its effect could lead to verify the role of the KKS system [60].

Similarly, the method of direct injection of pDNA to rat fibroblast has been used to study the consequences of overexpressing dominant negative forms of the ubiquitous mitogen-activated protein (MAP)-kinase toward inhibition of cell proliferation. Direct pDNA microinjection technique was also used to demonstrate the absolute requirement of the protein Cyclin-A in DNA synthesis, as opposed to the popular belief that the Cyclin-A was uniquely involved in the mitotic activation only. An SV40 expression vector encoding the full-length, antisense Cyclin-A RNA, upon direct microinjection into synchronized mammalian cells, completely suppressed Cyclin-A expression at S-phase, resulting in inhibition of DNA synthesis [21].

2.4. Gene Delivery to Myocardium

Successful tissue-specific transgene expression of injected pDNA in skeletal muscle *in vivo* raised the possibility that myocardial gene delivery could be achieved by this approach. Therapeutic effects have been achieved in animal models using pDNA, and its safe delivery to patients with intractable myocardial ischemia has been demonstrated [61–63]. Myocardial gene delivery via direct injection or through coronary vasculature with naked pDNA usually showed low levels of gene transfection. These studies in clinical trials showed some improvements in symptoms and exercise tolerance, but no increase in angioneogenesis or perfusion has been demonstrated [64]. This may reflect the transgene chosen, but data from animal studies would also suggest that pDNA injection is an inefficient technique for myocardial gene delivery [65]. The low efficiency of plasmid-mediated gene delivery is primarily due to low entry kinetics across the sarcolemma into the cardiomyocyte, cytoplasmic degradation, and low transfer from the cytoplasm into the nucleus, where gene expression is initiated. However, the results so far obtained had invaluable impact in characterizing the role of promoters in cardiac tissue and further helped in examining the effect on transferred foreign gene expression by the influences of naturally occurring mechanical and hormonal stimuli of the myocardium [66].

It was demonstrated that the tissue-specific promoter chimeras injected into the heart could respond accurately to shift in thyroid hormone levels *in vivo*. Injection of pDNA with gene constructs driven by cellular promoters resulted in detectable levels of reporter gene activities [67]. The cellular promoter was derived from the rat α-myosin heavy chain (α-MHC) gene. The expression of this gene *in vivo* is restricted to cardiac muscle and is positively regulated by the thyroid hormone. The regulatory portion of genes expressing specifically in cardiac muscles could be identified using this method [68]. Direct DNA injection had been extended to evaluate and characterize the activation properties of a cardiac-specific promoter/enhancer of the slow/cardiac troponin C (cTnC) gene that express in cardiac striated muscles [68]. Myocardial direct DNA injection was also utilized to analyze the transcriptional regulation of brain creatine kinase (BCK) gene in the developing heart [69]. pDNA constructs containing BCK promoter and *CAT* or luciferase reporter gene was delivered into the left lateral wall and apex of the ventricle on the heart [70]. The study might provide insight into the embryonic gene expressing mechanism during cardiogenesis. Because the *BCK* gene, the major gene for cytoplasmic creatine kinase expressed in the embryonic heart, is downregulated during cardiogenesis, it is reinstated in response to stimuli such as ischemia, hypertrophy, or heart failure in the adult.

The method of direct pDNA injection was used to explore the effect of specific pathophysiological state on cardiac gene expression, such as ischemia, myocardial infarction, reperfusion injury, hypertension, and so on [71]. Ischemia is a disease state that is formed when tissues are starved of blood supply through possibly narrowed or blocked arteries. Sporadic myocardial ischemia is prevalently associated with coronary arterial diseases. Ischemia related disease phenotype needs a therapeutic gene, which could be selectively upregulated by the elevated signals for ischemic activity and consequently downregulated when the activity represses. In this context, Prentice et al. introduced expression plasmids containing muscle-specific α-MHC promoters and hypoxia-responsive enhancer (HRE) elements linked to a reporter gene in cultured cells or into the rabbit myocardium, and measured the regulation of these constructs by hypoxia or experimental ischemia [70]. It was shown that the expression of the reporter gene was induced by both hypoxia *in vitro* and a short interval of ischemia *in vivo*.

There were different reports concerning the stability of plasmid-based transgenes in both skeletal and cardiac muscles. It was shown that the rat cardiac myocytes could express *β-galactosidase* gene under the control of the Rous sarcoma virus promoter, by the injection of pDNA encoding the reporter gene directly into the left ventricular wall. β-galactosidase expressed in cardiac myocytes was detected in rat hearts for at least four weeks after injection of the *β-galactosidase* gene [71]. In postmitotic cardiac and skeletal muscle cells, the transgene expression of the pDNA declines with time, probably due to the episomal localization of the DNA. The reason that the striated muscles showed higher capacity for uptake and expressing of pDNA following direct injection was not clear; the efficient gene transfer might be induced by cellular membrane rupturing and destabilization, followed by inflammation caused by the injection needle [72,73].

2.5. Gene Therapy for Angiogenesis

Angiogenesis is the process of new blood vessel development for the vascularization of various organs, for wound healing, and to allow cancer development and proliferation. Since the relationship between angiogenesis and tumor growth was established by Folkman in 1971, efforts have made to explore the possibility in treating cancer by targeting angiogenesis [74].

Therapeutic angiogenesis involves replenishing angiogenic growth factors by administering recombinant proteins or endothelial growth-factor gene. The recombinant proteins have severe limitation on its usage, as they are expensive and difficult to produce on a large scale. On the other hand, gene therapy provides a systemic and long-term effect with modification in the effective dosage of the therapeutic agent. To evade potential problems of pathological angiogenesis, transient gene expression is usually preferred for this kind of treatment. Tsurumi et al. introduced naked pDNA encoding vascular endothelial growth factor (VEGF) by IM injection into ischemic hind limb muscles in a rabbit model and observed that the vessels and blood capillaries were increased in muscles injected with VEGF compared to the control [75]. An enhanced vascularity-induced perfusion followed by increased blood flow in the ischemic limbs was also observed. In a clinical trial, Simovic et al. introduced naked pDNA encoding human *VEGF* gene by direct IM injection to chronic ischemic limbs of patients, to treat peripheral neuropathy caused by critical limb ischemia. The patients showed decreased neuropathic disability in the treated limbs, indicating that long-term therapy might improve the integrity in tissues of ischemic limb and consequent retrieval of limb [76]. Uesato and coworkers [77] have demonstrated the antitumor and atiangiogenesis effect of mouse antiangiogenesis genes, angiostatin and endostatin, delivered to tumors by low-voltage electroporation in 26 models of mouse colon [78,79].

2.6. Gene Therapy for Autoimmune Diseases

The application of gene therapy in autoimmune disease represents a novel use of this technology. Autoimmune disease is a pathogenic condition in which ones immune system erroneously attacks a person's own cells, tissues, and organs. The most prevalent symptom of this disease is inflammation, which is caused by the abundant presence of a large number of immune cells and molecules in the target site of the body. The goal of gene therapy in the treatment of autoimmune disease is to restore "immune homeostasis" by countering the proinflammatory effects of the CD4(+) T cells in the lesions of autoimmunity. This can be accomplished by adoptive therapy through the delivery of cytokines or cytokine inhibitors. Transduction of autoantigen recognizing CD4(+) T cells, which secrete antiinflammatory products, may become the "magic bullet" to combat the ravages of autoimmune inflammation and tissue destruction. Interferon γ (IFN-γ), interleukin-1 (IL-1, α or β), IL-12, and tumor necrosis factor α (TNF-α) are recurrently addressed inflammatory cytokines in illnesses related to autoimmune/inflammatory diseases. Other than these cytokines, transforming growth factor β (TGF-β) is also a key regulatory cytokine, because TGF-β inhibits T and B cell responses, irregularity of which leads to elevation of autoimmunity invoked disease conditions [80].

In animal models, pDNA constructs with the encoding anti-inflammatory cytokine genes for IL-10, IL-4, and TGF-β1 were injected into either tibialis anterior or rectus femoris muscles in nonobese diabetic (NOD) mouse against autoimmune diabetic disease. Although there was no marked decrease in severity of insulitis, the diabetes was reduced in NOD mice injected with IL-10 as compared to the untreated NOD mice [81,82]. In another experiment, treatment of autoimmunity prone NOD mice with pCMV-TGF-β1 resulted in considerable elevation of TGF-β1 level in the plasma. The increased level of TGF-β1 exerted various immunosuppressive effects such as suppression of delayed-type hypersensitivity (DTH), and prevention of insulitic and diabetic incidence in this kind of mice [83]. TGF-β1, IL-4, and IFN-γ gene coding plasmid vectors were also injected intramuscularly into rodent models for treating experimental allergic encephalomyelitis (EAE), systemic lupus erythematosus (SLE), colitis, and streptococcal cell wall-induced(SCW) arthritis. In *ex vivo* gene transfer nucleofection a direct transfer of DNA into the nucleus and artificial chromosome expression systems was reported for the treatment of rheumatoid arthritis in mouse models with limited efficacy [84]. Poor transfection efficiency is a well-known limitation of naked DNA injection. Transfection is often adequate in muscle, but rarely in other tissues. Further, transfection has often been achieved by complexing the DNA to cationic lipids and/or other polycations [85].

3. PLASMID DNA DELIVERY BY PHYSICAL METHODS

Important physical methods of pDNA transfer into cells include the use of (1) electroporation where DNA is delivered into cells through the transient application of electric field; (2) gene gun in which a high-pressure helium gas stream is used to shoot gold particles coated DNA directly into the cytoplasm; (3) sonoporation in which ultrasound facilitates gene transfer at cellular and tissue levels; (4) hydrodynamic injection; and (5) laser method in which uptake of naked DNA by cells is mediated by introduction of minute holes on the cell membrane by brief pulses of a finely focused laser beam. In general, physical methods provide inefficient transfection and have limited range of applications, although some of these applications are promising and important [9,86–88].

3.1. Improving Plasmid DNA-Mediated Gene Transfer by Electroporation

The first inventive demonstration that pDNA could be introduced into living cells by means of electric pulses was published in 1982 [89]. Electroporation is a process of exposing cells to controlled electric field for the purpose of cellular membrane permeability. The electric pulses, intensely localized to destabilize the membrane, allow molecules to enter cells. Electroporation is a useful method for developing nonviral gene therapies and nucleic acid vaccines (NAVs) that has been extensively tested in many types of tissues *in vivo* (4, 5), among which skin and muscles are the most extensively investigated. The system should work in any tissues into which a pair of electrodes can be inserted. A variety of genetic materials were inserted into the cells *in vitro* by electroporation [90–93].

Electroporation increased the naked pDNA expression two to three orders of magnitude higher than that with pDNA alone. However, the duration of a gene expression following *in vivo* electroporation was dependent on the target tissue, plasmid construct and age of the recipient animal. Despite this fact, a broader variety of cells showed reporter gene expressions. The major advantage of electroporation is *in vivo* tissue-specific gene expression could be accomplished by using specifically designed electrodes. Various disease pathologies could be subjected to electroporation-mediated *in vivo* gene transfer. The successful utilization of this technique was done in targeting hepatic parenchyma, hepatocellular carcinoma (HCC), skin, skeletal muscle, mouse testes, melanoma, human primary myoblast, glomeruli, brain, human primary hematopoietic stem cells, human esophageal tumor, and rat skeletal muscle for correcting anemia of renal failure. Several major drawbacks exist for *in vivo* application of electroporation such as a limited effective range of approximately 1 cm between the electrodes, which makes it difficult to transfect cells in a large area of tissues and surgical procedure is required to place the electrodes deep into the internal organs. High voltage applied to tissues can result in irreversible tissue damage as a result of thermal heating and Ca^{2+} influx. The possibility that the high voltage applied to cells could affect the stability of genomic DNA is an additional safety concern [94]. However, the concern may be resolvable by using a technique called *microelectroporation*. This method contains a novel syringe electrode, which can deliver electric field directly to the cells in which pDNA has been injected. High transfection efficiency was observed with comparatively lower field strength that caused minimal tissue damage [95]. By the microelectroporation technique, DNA molecules were efficiently introduced into the optic vesicle, sensory placodes, surface ectoderm, neuroepithelium of the central nervous system (CNS), and into the somites and limb mesenchyme [9,23,96–115].

3.2. Improving Plasmid DNA Transfer Mediated by Gene Gun

Gene gun is a physical way of administering gene *in vitro* or *in vivo*. This biological and ballistic delivery uses heavy metal particles coated with pDNA, which are dispensed at a high velocity into the target cells. pDNA makes electrostatic complex with gold or tungsten microparticles. The delivery is highly localized to the tissue part [23]. The delivery to different tissue depths and areas is regulated by adjusting the speed and hence impact-pressure of the projectile. The process involves easy and speedy preparation of the delivery vehicle while keeping DNA intact. This technique sometimes allows DNA to gain direct access to nucleus bypassing the endosome/lysosome. These subcellular compartments may induce possible enzymatic degradation of the pDNA. Because of the benefit of accessibility, this technique is especially suitable for gene transfer to skin and for superficial wounds. The gene delivery system using gene gun is useful for the percutaneous administration. This technique was used first *in vitro* in 1987 and then extended to mammalian cells and living tissues in the early 1990s. There are several examples in various animal models, which have shown a high level of transgene expression in the epidermis and dermis of the skin. By the gene gun technique, mouse skin was transfected with IL-6 and hemagglutinin encoded DNA to elicit protective immune responses against equine influenza virus [116]. Several different large animals such as rhesus monkeys, pigs, and horses were also immunized against virus by transfecting genes encoding the viral antigens [117–119]. Detectable transgene activities were also noticed in various nonsuperficial organs, such as lungs, pancreas, kidney, muscle, and cornea. Gene gun was also used to treat against tumor growth. Intradermal tumor upon *IL-12* gene delivery by gene gun gave detectable levels of the gene product at the treatment site. This eventually led to complete tumor regression within seven days [120]. TGF-β1 encoded plasmid bombardment to rat tissue enhanced the tensile strength of the tissue by almost twofold, as compared to the control treatment with non-gene gun-mediated direct injection [121]. Porcine partial-thickness wounds, when transfected with a vector expressing epidermal growth factor (EGF) showed an increased rate of epithelial cell formation. This increased epithelialization shortened the wound healing time by 20% [122,123]. As far as clinical trials are concerned, gene gun applied for the hepatitis [124] and DNA vaccine therapy to treat influenza [125] and melanoma [126] without severe toxicity. Using gene gun, vaccination of NOD mice with pDNA encoding glutamic acid decarboxylase 65 (GAD65) at a late preclinical stage of T1D induced IL-4 secreting CD4(+) T cells and significantly delayed the onset of diabetes [9,23,127–134].

3.3. Improving Plasmid DNA-Mediated Gene Transfer by Ultrasound

This physical method utilizes sound energy to promote gene transfer at cellular and tissue levels. Ultrasound is in general clinical use for both therapeutic and diagnostic purposes. Low-level ultrasound is used for diagnostic imaging, and ultrasonic shock waves are used in the treatment of kidney stones and high-intensity focused ultrasound is used for the thermal destruction of tumors. Ultrasound has emerged as a unique method for pDNA-mediated gene therapy applications. To date, sonoporation has been shown to improve gene expression in muscle, carotid artery, and solid tumors [135–139]. In this method transfection efficiency is greatly influenced by the size and local concentration of plasmid DNA, because the mechanism of gene transfer proceeds through passive diffusion of DNA across the membrane pores created by ultrasound [140,141]. By this method, a 10- to 20-fold enhancement of reporter gene expression over that of naked DNA has been achieved. pDNA transfecting efficiency of this system is determined by several factors, including the frequency, the output strength of the ultrasound applied, the duration of ultrasound treatment [142] and the amount of plasmid DNA used. The efficiency can be enhanced by the use of ultrasound contrast agents or conditions that make membranes more fluidic. The use of contrasting agent is suggested to lower the energy threshold for cavitation, making it feasible in the therapeutic frequencies. The contrast agents are air-filled microbubbles that rapidly expand and shrink under ultrasound radiation, generating local shock waves that transiently permeate the nearby cell membranes. Recently, *in vivo* pDNA transfection mediated by ultrasound has been reported, first using a lithotripter, and then using focused sinusoidal sonoporation [143–145]. Significant enhancement of transfection mediated through sonoporation has been reported in cell culture and *in vivo* when lipoplexes have been used. Since ultrasound can penetrate soft tissue and be applied to a specific area, it could become an ideal method for noninvasive gene transfer into cells of internal organs. Evidence supporting this possibility has been presented: in one study, plasmid DNA was coadministered with a contrasting agent to the blood circulation, followed by ultrasound treatment of a selected tissue [146]. So far, the major problem for ultrasound-facilitated gene delivery is the relatively low gene delivery efficiency [147].

3.4. pDNA Delivery and Hydrodynamic Injection

This method involves a simple injection of DNA solution equivalent to 8–9% of body weight of the animal through tail vein during a short period of time to attain significant amount of gene expression [9]. Hydrodynamic gene delivery is a valuable method for the transfection of cells in highly perfused internal organs (e.g., the liver) and it allows direct transfer of substances into the cytoplasm without endocytosis. The gene delivery efficiency is determined by three main factors given below.

(1) Anatomic structure of the organ
(2) Volume of DNA solution
(3) Speed of injection

For example, optimal conditions for liver transfection in a 20 g mouse are 1.6–1.8 mL of pDNA solution and an injection time of approximately 5 s via the tail vein [148]. Mechanistically, the rapid tail vein injection of a large volume of DNA solution causes a transient overflow of injected solution at the inferior vena cava that exceeds the cardiac output. As a result, the injection induces a flow of DNA solution in retrograde into the liver, a rapid rise of intrahepatic pressure, liver expansion, and reversible disruption of the liver fenestrae. In a single hydrodynamic injection with less than 50 µg of plasmid DNA approximately 30–40% of the hepatocytes are transfected [149]. Along with the plasmid DNA this method could be applicable in the delivery of small interfering RNA, small dye molecules, proteins, and oligonucleotides. This method has been used to express proteins of therapeutic value such as hemophilia factors, alpha-1 antitrypsin, cytokines, hepatic growth factors, and erythropoietin in mouse and rat models. The real challenge for gene transfer by the hydrodynamic method is how to translate this simple and effective procedure to one that is applicable to humans [150,151]. Recent advances in using a computer-controlled injection device and a catheter-guided injection port has greatly improved the safety feature of this promising technique [152].

3.5. pDNA Delivery by Laser Irradiation

Irradiation of electro magnetic radiation produced by the process of stimulated emission helps in gene expression enhancement by creating self reparable short lived transient pores approximately 2 mm in diameter in the cell membrane has been reported [153]. However, the mechanism of gene expression enhancement *in vivo* by laser is not clear. It has been predicted that the thermal effect at the site of laser beam impact causes a difference in osmotic pressure between the cytoplasm and the medium surrounding the cell that changes the permeability of the cell membrane [154]. Recently, it has been reported that gene transfer into muscle could be enhanced using a femtosecond infrared laser [155]. High cost and the critical experimental conditions hindering the development in the field of laser-mediated gene delivery. Further studies will be needed before application to human gene therapy [156].

4. PLASMID DNA DELIVERY BY CHEMICAL METHODS

pDNA-mediated gene therapy largely depends on chemical methods with which the DNA can be delivered into mammalian cells. In this method DNA is formulated with cationic lipids or cationic polymers or their hybrid systems in the form of nanoparticles. Entry into the cells is facilitated by endocytosis, macropinocytosis, or phagocytosis into the endocytic vesicles, or endosomes. The DNA escapes from the endosomes and dissociates from the carrier in the cytoplasm and subsequently migrates into the nucleus, where transgene expression takes place.

4.1. Synthetic Gene Delivery Systems

Most common synthetic gene delivery systems are the lipid-based vectors. The lipid-based delivery system was first utilized on the premise that naked DNA upon injection *in vivo* might be degraded by endogenous DNase. These gene delivery systems encapsulate or complex the pDNA to protect it and modulate its interaction with the biological system. With necessary chemical modification, this vehicle would target the DNA toward specific cellular sites. These lipids are a class of molecules that self-assemble in aqueous or organic media. Depending on its molecular architecture and in the presence or absence of cosolute, they assume structures such as micelle, emulsion, or liposome in aqueous media, and reverse micelle in organic media. However, among all the systems, liposomal systems found most biological usage, probably owing to its close resemblance with the cellular membrane. They constitute a bilayered structure, which can encapsulate water-soluble molecules in its aqueous hydrophilic core or water insoluble molecules in its hydrophobic bilayer. Liposome-based drug delivery systems have shown promise in clinical use and several products have already been approved by the FDA. Fraley et al. had shown for the first time that encapsulating DNA in liposomal aqueous core could also deliver DNA to cellular targets [21]. They used a composite liposomal system containing anionic and nonionic lipids. The DNA encapsulation efficiency for this system was low. Toward enhancing gene delivery and targeting ability of the liposome to specific tissue site, anionic liposomes encapsulating DNA were coated with target-specific antibody to make pH sensitive immunoliposomes. These liposomes upon intraperitoneal (IP) injection, showed specific transfection to tumor cells in an ascites tumor model in nude mice [157]. PEGylated immunoliposomes conjugated with antibody to rat transferrin receptor showed brain targeting via the transcytosis of transferrin receptor [25].

Among the synthetic gene delivery systems cationic lipids or cationic polymers (or combinations thereof) are widely studied. They show a strong electrostatic interaction with the anionic DNA. Due to this strong favorable interaction, the DNA is condensed to form nanoparticles and becomes a part of the carrier system.

4.2. Cationic Lipids and Lipoplexes

Cationic lipids are amphiphiles that self-assemble into molecular aggregates such as vesicles. Since the first description of successful *in vitro* transfection using a cationic lipid by Felgner et al. [158] in 1987, substantial progress in cationic lipid-mediated gene delivery has been made. Lipoplexes are widely used for transferring genes to various *in vivo* targets with comparative ease and least toxic effect to the host. Use of lipoplexes in gene therapy clinical trials comprises almost one-fourth of the total gene therapy clinical trials to date. pDNA containing various therapeutic genes were used to treat a variety of disease pathologies. A vast amount of literature is available on the *in vitro*, preclinical, and clinical evaluations [159]. Lipoplexes enjoy the property of easy administration in the body without eliciting any extra discomfort to the patient or the animal. Intravenous, intradermal, intraperitoneal, and direct injections are the common ways through which lipoplexes are injected into the body. In the targeting front, several ligand-associated cationic liposomes were developed to target genes to specific cell targets such as cancer. Recently, cationic liposomal stealth gene delivery systems were developed utilizing a sigma and estrogen receptor ligands associated polyethylene glycol-based lipid [160,161]. These systems could target and deliver genes specifically to respective receptor overexpressing breast adenocarcinoma cells but not the normal cells. Additionally, the stealth lipoplexes were serum stable. In case of sigma receptor targeting the PEG lipid was chemically modified with a neuroleptic drug, haloperidol (used for over three decades to treat psychotic patients), to target cancer cells that over express the sigma receptor [162,163]. No conclusive links were ascertained between cancer pathogenicity and overexpression of the sigma receptor in certain cancers. But, this versatile sigma receptor targeting stealth cationic liposomal system could be potentially used to target therapeutic anticancer genes to breast cancer or other cancer models that were implicated with overexpressing the sigma receptor. However, in case of development of the estrogen receptor targeting stealth liposomal systems, 17β-estradiol a well-established endogenous estrogen receptor ligand was covalently conjugated to the digital end of PEG lipid to target estrogen receptor positive breast cancer cells [164–166].

4.2.1. Structures of First Generation Cationic Lipids

In the process of finding out the best cationic transfection lipid, several groups throughout the world, over the last 25 years, explored different chemistries to evolve different chemical structures, which made quite a large library of molecules. It is very interesting to know that several architectural differences in the molecular infrastructure of cationic lipids emerged, which lead to the formation of cationic lipids of varied transfection ability. The cationic lipids used in nonviral gene delivery consist mainly of three domains: hydrophobic domain [167], head group domain [168], and backbone, or linker domain [169].

4.2.2. Hydrophobic Domain

The common hydrophobic domains include steroid hydrophobic domain, aliphatic chain and fluorinated hydrocarbon domain. DC-Chol as well as newer lipids such as BGTC, cholestane, and lithocholic acid (Fig. 2) are examples of cholesterol lipids. Both lipids and colipids containing fluorinated lipids [170] have been synthesized. Examples of aliphatic chains are bis-guanidium, diacetylene single tailed lipid.

4.2.3. Head Group

The cationic lipids used for gene delivery are also classified according to the structure of head groups in quaternary ammonium lipids, lipopolyamine, cationic lipids bearing both quaternary ammonium and polyamine moieties, aminidium and guanidium salt lipids and heterocyclic cationic lipids.

Quaternary Ammonium Head Group Quaternary ammonium head group [171] has been widely used in designing many of the contemporary cationic lipids. Important parameters that determine transfection efficiency of lipids with quaternary ammonium head groups are the steric hindrance at nitrogen atom, electronic effects of the substituents and presence of hydrophilic head groups, more specifically hydroxyls [172].

Primary, Secondary, and Tertiary Ammonium Head Groups There are fewer lipids containing a primary, secondary or tertiary amino head group. An example is DC-Chol, with a tertiary amine head group linked to cholesterol (Fig. 3). The helper lipid DOPE contains a primary amine in its head group [173].

Polyamine Head Group Lipopolyamines are one of the most successful classes of cationic lipids. DOSPA and DOGS are the best examples. Cationic lipids containing bis-guanidine, guanidine [174] have also been synthesized.

Heterocyclic Head Group Lipids containing aliphatic and aromatic heterocyclic groups in the head group have been designed. As aliphatic heterocyclic functionalities, morpholine and piperazine were conjugated to cholesterol via a spacer [175]. A series of pyridinium amphiphile called SAINT has also been synthesized [176].

4.2.4. Backbone Domain

Backbone separates the head group and acts as scaffold on which the cationic lipid is built.

Glycerol Backbone Most of the commercially available lipids share the glycerol backbone. These consist of the DOTMA analogs where the cationic head group is linked to the hydrophobic tail through a glycerol-backbone linker. DOTMA, DOTAP, DOSPA all share the glycerol backbone [177]. A representative list

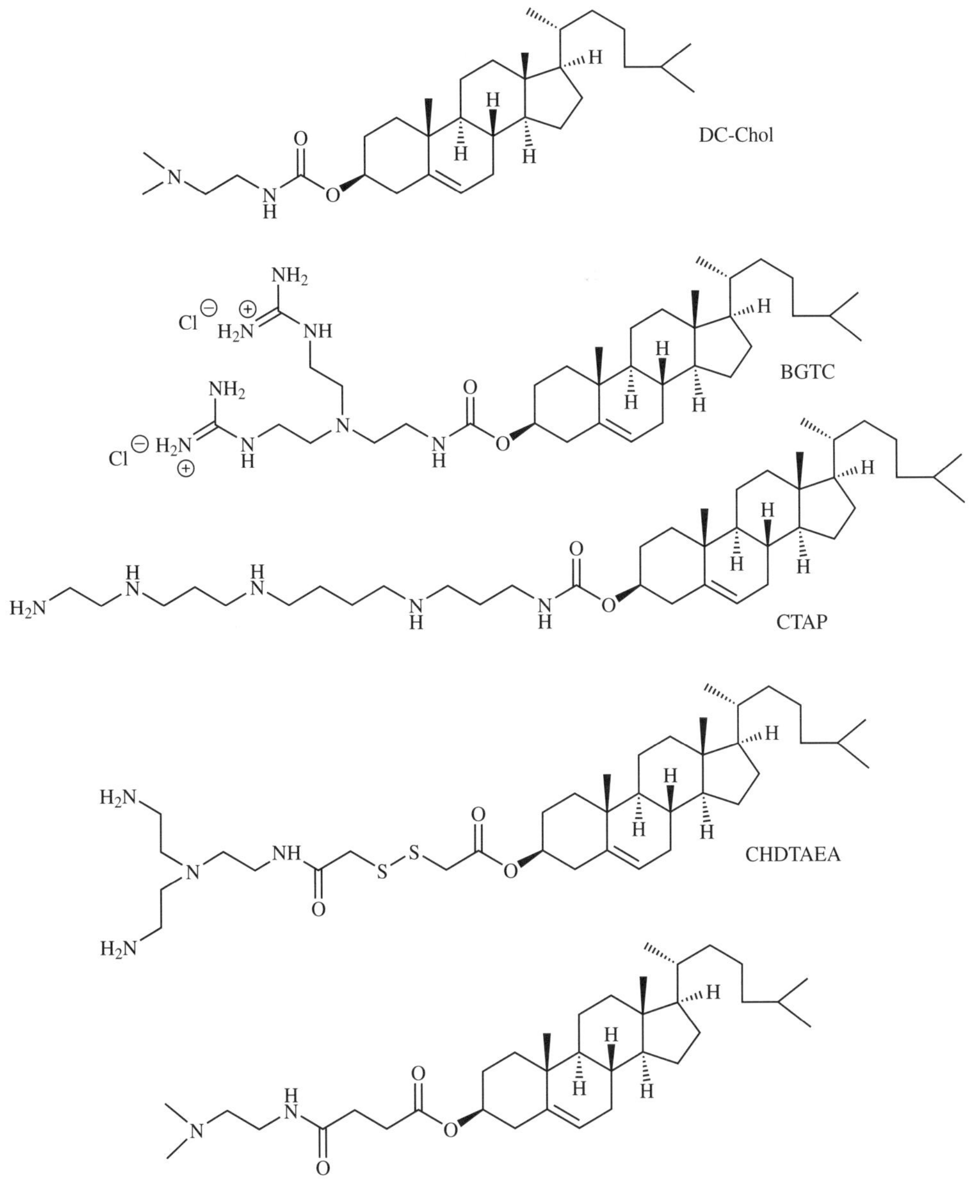

Figure 2. Representative structures of cholesterol-based efficient cationic transfection lipids.

of such glycerol-backbone based cationic lipids is shown in Fig. 4.

Nonglycerol-Based Cationic Lipids Design, synthesis, and enhanced transfection efficiencies of other simple cationic transfection lipids lacking the glycerol-based backbone have been reported by several groups [172,183]. These include a wide variety of lipids having different linkers and head groups linked to the hydrophobic tail. Some important examples are DOGS, di-C14-amidine, DOTIM, SAINT, and so on. [176]. A representative list of such nonglycerol-backbone based cationic lipids is shown in Fig. 3.

Other Backbones Other backbones include pentaerithrol[178],trislinkage[179],glycosidic,

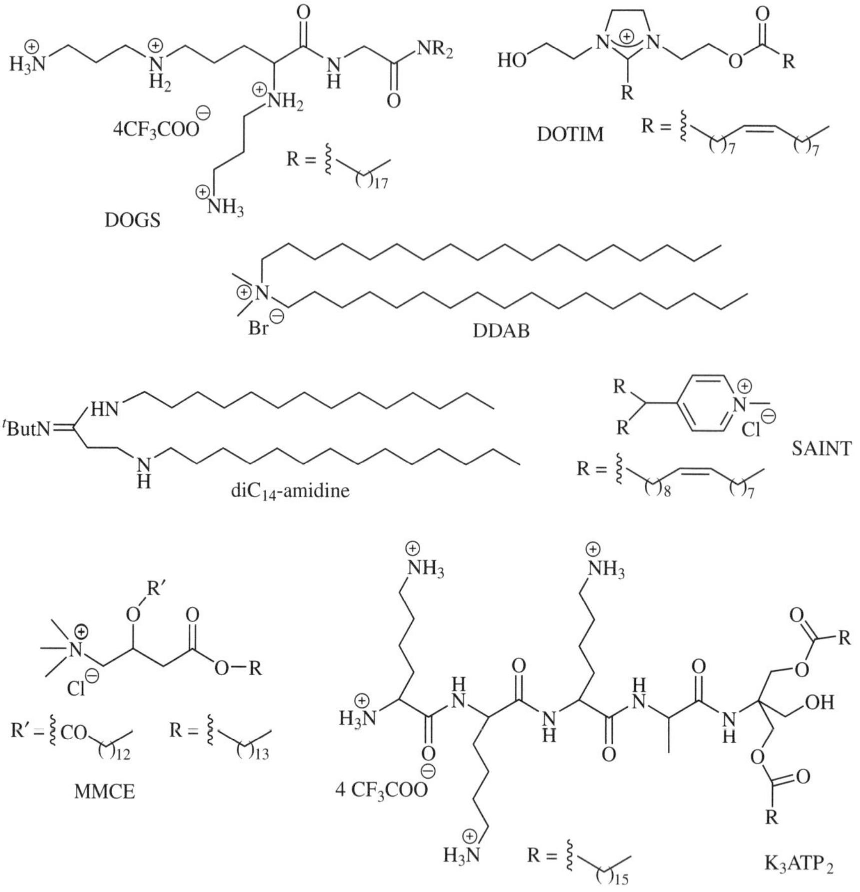

Figure 3. Representative structures of nonglycerol based efficient cationic transfection lipids containing simple monocationic, polycationic, aromatic, and amidine head groups.

phosphate or phosphonate [174,178], and aromatic. Examples of aromatic backbones include 3,5-dihydroxybenzyl alcohol [180] and 3,4-dihydroxybenzoic backbone [181].

There were many other lipids that were synthesized and tested for gene transfection. Only a few of these lipids had shown consistent and relatively high level of transfection in various cell lines and in animal models *in vivo*. Most of these first generation formulations were not long circulating in the blood, owing to the fact that the DNA/cationic lipid complex aggregates or disintegrates in the presence of negatively charged blood proteins. Additionally, they showed no specific targeting of cells. Finally, in spite of such a large library of molecules, no consensus thumb-rule could be predicted or ascertained regarding the magic structure of lipid, which might show consistently high transfection in different kinds of cells. This is because there were so many parameters involved in the whole process of transfection. Through understanding of these transfection pathways helps to develop the potentially ideal nonviral gene delivery system. Here, we will briefly discuss the mechanism of transfection.

4.2.5. Mechanism for Transfection To develop an efficient gene delivery system, it seems necessary to understand the extra- and intracellular processes involved in the overall transfection mechanism. For this purpose, cationic liposomes and pDNA are used widely to understand the cellular mechanism involved in the transfection. Currently believed

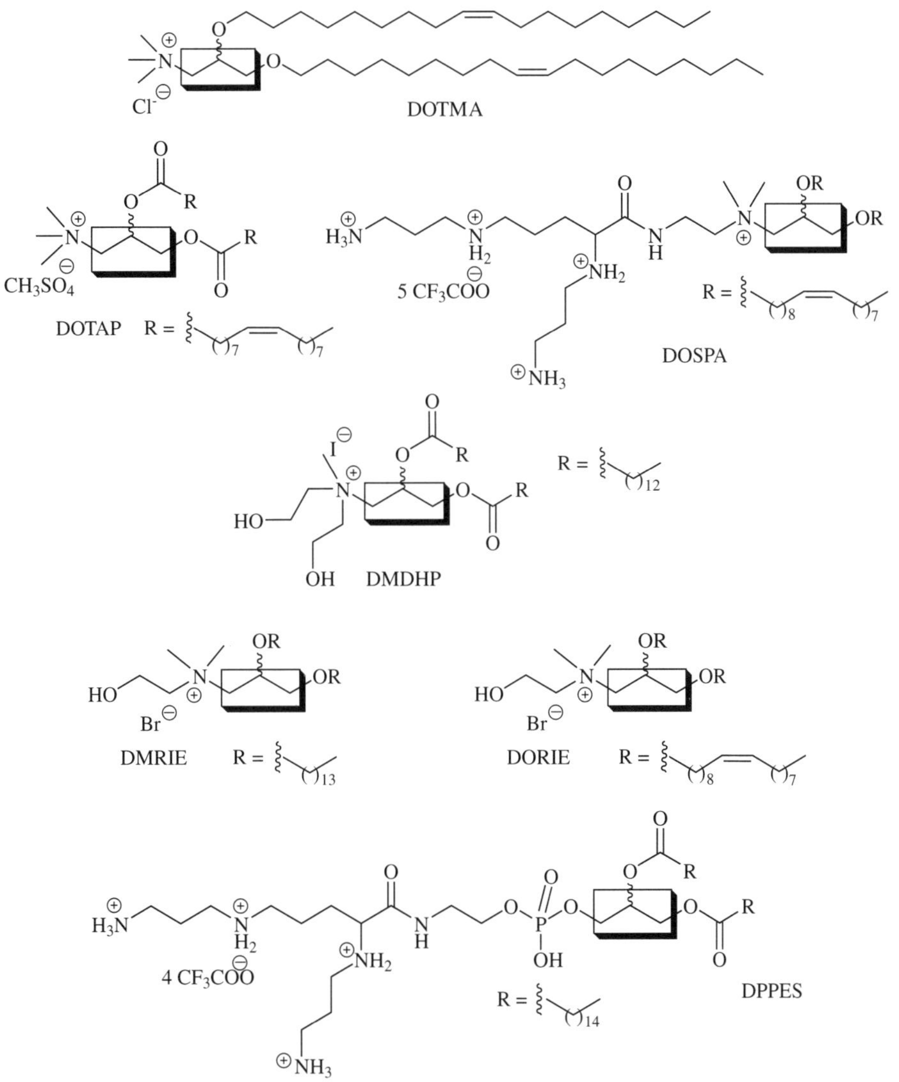

Figure 4. Representative structures of glycerol-based efficient cationic transfection lipids. The glycerol backbones of all the cationic amphiphiles are shown within the rectangular boxes.

lipofection (lipoplex-mediated intracellular transfection) pathways [182] involve (a) formation of lipid-DNA complex (lipoplex); (b) initial binding of the lipoplex to the cell surface; (c) endocytotic internalization of the lipoplex; (d) trafficking in the endosome/lysosome compartment and escape of DNA/lipoplex from the endosome/lysosome compartment to the cytosol; (e) transport of the endosomally released DNA to the nucleus followed by its transgene expression (Fig. 5).

Formation of Lipoplex The driving force for the formation of lipid-DNA complexes is the spontaneous electrostatic interaction of the positively charged cationic liposomes and the negatively charged DNA. Usually these complexes have a heterogeneous size distribution. The structures of lipid-DNA complexes have been characterized using techniques such as freeze-fracture electron microscopy, cryo-transmission electron microscopy, small-angle X-ray scattering, and so on. These

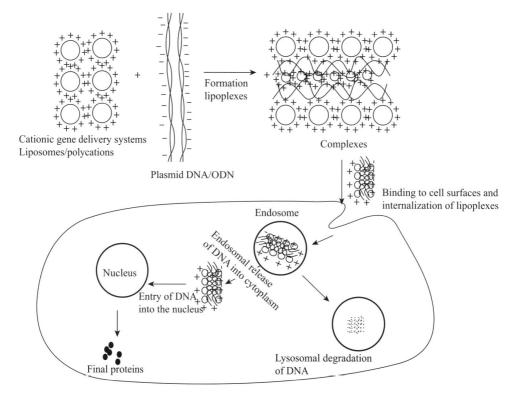

Figure 5. Schematic illustration of processes involved in gene delivery and expression.

nvestigations have revealed diverse structures including spaghetti and meatball structures, entrapped DNA into aggregated multilamellar structures, multilamellar structure with DNA intercalating within the lipid bilayers, and so on for different lipid formulations [184].

Internalization of Lipoplex The cell membrane is negatively charged owing to its high content of glycoproteins and glycolipids containing negatively charged sialic acid residues that can display, depending on the function of the cell, various types of receptors and antigens. In the absence of a targeting ligand, the driving force for the binding of the lipid-DNA complex to the cell membrane is mainly electrostatic. Electron microscopic studies [182] have shown that internalization of the lipid-DNA complexes occur mainly through endocytosis. The lipid-DNA complex is engulfed into lower pH compartment called "early endosomes" in the perimembranous region. Late endosomes are formed as the pH continues to drop to 5–6.0 and it also moves deeper into the cytoplasm.

Endosomal Release of DNA into Cytoplasm
Under normal cellular conditions, endosomal contents are carried into lysosomes, where they are degraded by the various degradative enzymes. For transgene expression to take place, DNA has to escape from the endosomal compartments to the cytosol before endosome-lysosome fusion takes place. Fusogenic lipids such as DOPE are often used as a helper lipid in facilitating endosomal disruption that, in turn, leads to efficient release of DNA to the cytoplasm [185]. pH-sensitive cationic lipids have also been used to facilitate endosome disruption leading to efficient release of endosomally trapped lipid-DNA complex and subsequent enhanced transfection efficiency [186].

Entry of DNA into the Nucleus Entry of DNA into the cell nucleus after its endosomal release into the cytosol is a prerequisite for gene expression. The precise mechanism of nuclear entry of the endosomally released DNA still remains elusive but is believed to be an inefficient process. The presence of a single nuclear localization signal (NLS) peptide linked to one end of a DNA has been reported to increase the *in vitro* transfection efficacies by 1000-fold compared to DNA lacking the NLS sequence [187]. Some studies have reported the correlation between mitotic activity and transfection by cationic lipid-DNA complex. Though the exact mechanism of enhancement in gene transfer has not been elucidated, it is believed that the breakdown of the nuclear membrane during mitosis could facilitate entry of DNA into the nucleus [188]. It is also not clear if pDNA enters into the nucleus as a lipoplex or as a naked DNA.

4.2.6. *In Vivo* Administration of Lipoplex

In vivo gene transfer by nonviral vectors is subjected to anatomical constraints depending on the route of administration. Administration methods including direct injection into an organ, intravenous injection, airway administration of lipoplex, and intraperitonial injection are widely used techniques in transfection of cells *in vivo* using lipoplex. Intratumoral injection of lipoplex containing either E6 or E7 antisense plasmid resulted in significant growth inhibition of C3 tumors grown in a syngeneic mouse model [189]. A direct injection of recombinant pDNA containing murine class I major histocompatibility (*MHC*) gene into localized arterial segment of various major organs showed that the direct gene transfer by lipoplex did not lead to treatment related toxicity, autoimmunity, or gonadal localization of the transgene in mice [190]. The toxicity of gene delivery by lipoplex was also analyzed in pigs and rabbits *in vivo*. There were no clinically significant immunopathology in major organs such as, brain, heart, lung, liver, kidney, spleen, and skeletal muscles. To stimulate local tumor immunity in patients with stage IV melanoma, Nabel et al. [191] injected pDNA encoding MHC class I protein complexed with DC-Chol directly into the cutaneous tumor nodules. Treated lesions exhibited the presence of T cell, followed by an enhanced reactivity of tumor infiltrating lymphocytes. As a result, local inhibition of tumor followed by complete diminution of tumor was observed in some of the treated patients. Mohr et al. [192] had shown that direct lipoplex injection to intrahepatic hepatocellular carcinoma produced by human HCC cells appeared far superior to systemic administration for gene therapy for localized intrahepatic tumors, because the direct administration to tumors left the surrounding normal hepatic cells untouched. In another typical example, *in vivo* direct intratumor injection of plasmid containing the coding sequence for the human *IL-2* gene complexed with cationic lipid formulation resulted in retention of intact pDNA in the tumor tissue and IL-2 secretion by cell cultures derived from the injected tumors. Formulation of this lipid with the cationic lipid inhibited DNA degradation and enhanced *in vivo* transfection efficiency over pDNA alone [193].

Intravenous injection is a widely used mode of gene delivery in animals. On i.v. lipoplex injection *in vivo*, the residence of DNA primarily in heart and lungs even after 9 days with minimal toxicity was observed. Intravenous injected lipoplex expressed transgene in almost all organs including lung, kidney, heart, spleen, liver, brain, and so on, and the expression stayed for nine weeks with apparently no treatment related toxicity [194]. Toxicity and antitumor response was evaluated on mice and pigs with high doses of lipoplex containing *MHC* gene incorporated pDNA.

Recently, more work has been done to increase the overall transfection efficiency with much higher targeting capability and reproducibility of the liposomal delivery system. Intravenously administered lipoplex avidly reacts with blood components. So, it is necessary to keep the complex intact in the blood till it reaches the organ of interest. It was shown that in the presence of erythrocytes, cationic lipid/cholesterol formulation did not induce fusion between erythrocytes, whereas the cationic lipid/DOPE formulation possessing high fluidity in its structure induced fusion between the erythrocytes after a short incubation period [195]. This offered an explanation as to why cholesterol makes a superior formulation with cationic lipids for *in vivo* purpose. A repeated systemic i.v. injection of cytokine gene (*IFNβ1*) by lipoplex gave a systemic expression of human interferon-β in mice, thereby increasing the possibility of cytokines used for therapeutic purposes in a systemic manner [196]. An enhanced and highly selective liver targeting by i.v. injection of cationic lipoplex containing β-sitosterol beta-D-glucoside (*Sit-G*) gene was also observed [197,198]. Intravenous injections showed gene expression in all major organs including heart, lung, liver, spleen, and kidney, with lung being most efficiently transfected. For efficient targeting and gene expression in the lung, i.v. injection was favored over intratracheal instillation. Uyechi et al. [199] had shown by injecting fluorescently tagged lipoplex through the vein that the entire lung lobe was homogeneously fluorescent, whereas intratracheal administration resulted in regional distribution of lipoplex, concentrated around bronchioles and distal airways [200–214].

Airway administration of lipoplex was used for the treatment of pulmonary diseases including cystic fibrosis (CF). Cationic lipoplex showed no adverse effect toward airway epithelial integrity; hence, the cationic lipid-based delivery system proved to be appropriate for use in human trials for CF [205,206]. A series of preclinical trials were done in CF patients with intranasal instillation to evaluate the risk factors associated with the treatment. Since there was no apparent toxicity associated with the lipoplex as was seen from these trials, progress has been made in delivering the complex to the entire lung by aerosol in CF patients. By nebulization, lipoplex was delivered into the airways of mutant mice to obtain human cystic fibrosis transmembrane conductance regulator (CFTR) cDNA expression in the respiratory tract. A study conducted on CF patients revealed that pCMV-CFTR/cationic liposome complex on administration to the nasal epithelium gave no evidence of excess nasal inflammation, or any adverse events related to active treatment. Gene transfer and expression assayed by PCR revealed the presence of transgene DNA in seven of the eight treated patients up to 28 days after treatment. Intranasal instillation technique was also used in the mouse model, to incorporate cationic lipoplexes. The defective CFTR gene gives rise to multiple defects in the airway epithelia in CF patients, one of them being altered Cl^- and Na^+ permeability. Zhang et al. [207,208] had shown that the goblet cells were more efficiently targeted with lipoplex than any other cells in the entire spectrum of lung airway epithelia. This was ascertained by the fact that an efficiently reduced mucous sulfation to levels seen in non-CF airways was observed with lipoplex in spite of low levels of *CFTR* gene expression in lung epithelial cells in human bronchial xenograft model of mice as compared to non-CF airways of control mice [209,210]. The apparent complexity in CFTR function presented challenges in the design of different lipoplex formulations that were capable of generating the endogenous patterns of *CFTR* gene expression in specific lung epithelial cells [211–214].

IP injection of lipoplex was done to transfect cells in the peritoneum region. Nude mice bearing disseminated human ovarian tumors derived from the p185-overexpressing SKOV-3 ovarian cancer cells were injected with *E1A* gene/lipoplex intraperitoneally [215,216]. These tumors resemble stage III of human ovarian cancer. The expression of E1A protein decreased the expression of p185 oncoprotein, and hence increased the survival rate of mice. Seventy percent of the treated group survived for 1.5 years from the last injection, but the untreated group barely survived more than 16 weeks. The treatment of complex containing 1/13 of the original lipid dose also worked as efficiently as the normal dose. There was no apparent toxicity or major organ pathologic change. There was no trace of E1A DNA in the liver, lung, heart, spleen, brain, uterus, and ovaries of the treated mice after 1.5 years [206].

4.3. Cationic Polymers in pDNA Delivery

Another class of synthetic vectors widely used as a gene delivery vector is cationic polymer [217]. The polymer reversibly condenses DNA by electrostatic interactions into compact, ordered particles known as polyplex. Many different types of polymers have been used as gene delivery systems of which PEI (polyethyleneimine) is the most promising [196]. Wu et al. for the first time utilized polylysine-asialo-orosomucoid conjugate to condense pDNA and targeted the pDNA to the liver. pDNA condensed with protamine/polylysine conjugate of iron transport protein, transferrin, was efficiently delivered to eukaryotic cells [207]. Conjugation of specific cell-binding molecule to the polymer imparts targeting ability to specific cell lines [218]. Gene transfer mechanism of the regular polyplex made out of polymers with different size and architectural properties, into eukaryotic cells is similar to that of lipoplex and it involves multiple steps. These distinct steps are the condensation of DNA; cellular uptake by endocytosis; release from the endosome; nuclear transport; and vector unpacking and transcription. A thorough understanding of these steps is important for improving polycation-mediated gene delivery. Cationic polymers used for pDNA delivery are of two types, natural polymers including chitosan [219,220] and atelocollagen [221,222] or synthetic polymers including poly(L-lysine) (PLL) [223,224], PEI [25,225], and dendrimer [226–230]. Chemical structures of representative polycations are depicted in Fig. 6. Jean Paul Behr's group first introduced PEI as an efficient and economic synthetic polymeric gene transfer agent [231]. It is a widely used organic polymer in manufacturing industry. PEI is a highly positively charged dense polymer with high number of protanatable nitrogens. It consists of all three different primary; secondary; and tertiary amines and the number of nitrogens are almost one-third of the total atoms in the polymer [232]. Transfection activity and toxicity of both the linear and the branched PEIs are molecular weight dependent with small molecular weight PEI always less toxic and more active [231]. Superior gene transfer activity *in vitro* or *in vivo* with the PEI and its derivatives can be explained by the proton sponge hypothesis [233]. The original hypothesis was that PEI buffering, leading to osmotic rupture and subsequent escape, occurred in lysosomes. The high buffering capacity over a broad pH range aids delivery of plasmid DNA to a variety of cell types *in vitro* and *in vivo* without the addition of any membrane-disruption agents [234,235]. Godbey et al. challenged the proton sponge hypothesis based on their findings of a lack of lysosomal involvement in PEI-mediated gene transfer [236]. However, a version of the proton sponge hypothesis, whereby PEI buffering leading to osmotic rupture occurs in endosomes prior to fusion with lysosomes [237], is consistent with the findings of Godbey et al. Further, Sonawane and coworkers have recently shown that the concentration of endosomal chloride ions increases upon vesicle acidification and have observed the swelling/lysis of endosomes containing PEI polyplexes [238,239]. While these studies support the proton sponge hypothesis and an endosomal site of action, Bieber et al. have provided evidence for lysosomal accumulation of PEI polyplexes and have observed lysosomal membrane disruption attributable to either osmotic rupture or physical binding of PEI aggregates to the lysosomal membrane [240,241]. The mechanism behind the transfection activity of the cationic polymers and lipids, not very efficient in triggering endosome rupture, is still not clear. Using PEI as vector considerable amount of gene expression has showed in different *in vivo* models including rat kidney, mouse brain, mouse tumors and rabbit lungs but systemic administration always leads the predominant transgene expression in the lungs [242], just like lipoplex. The major stumbling block of PEI polymer is its *in vivo* toxicity. PEGylation reduces the toxicity of PEI polyplex to some extent and it also added some degree of specificity upon intravenous administration [218]. Although PEI vectors mediate effective gene transfer, the activity is not tissue specific, hence limiting the suitability of this vector for clinical applications. Targeting has, however, been introduced by the addition of ligands, such as transferrin, RGD peptides, anti-CD3, and epidermal growth factor, either with or without the added PEG shield [243].

Polyamidoamine cascade polymers, or Starburst dendrimers are another class of commercially available pH sensitive polycations with high transfection potential [244] and they have been shown to transfect corneal endothelium with tumor necrosis factor receptor immunoglobulin (TNFR-Ig) encoding

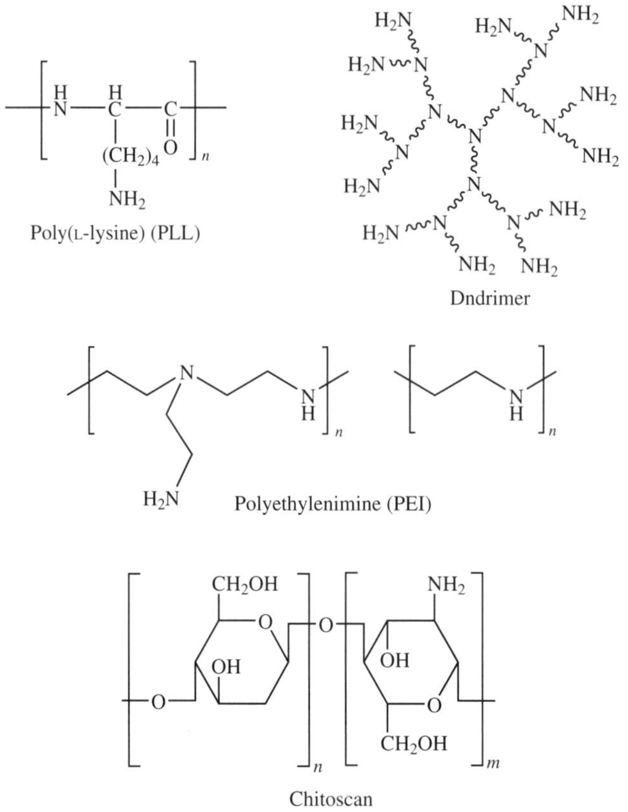

Figure 6. Structures of representative common polycations.

plasmid to block TNF action and reduce corneal allograft rejection [245]. Dendrimers have also been used in a murine cardiac transplantation model to deliver viral IL-10, resulting in prolongation of graft survival [246,247]. Most important drawback associated with the use of the efficient high molecular weight polycation transfection agents such as PLL, PEI, and PDMAEMA in plasmid DNA-mediated gene therapy is their *in vivo* toxicity and cellular accumulation after repeated administration. Lower molecular weight agents of these polymers, although less toxic, are inefficient in condensing DNA and hence active in transfection. Consequently, the use of biodegradable polymers for gene delivery has gained increasingly interest during the past 5–10 years. The potential advantages of biodegradable carriers are their reduced toxicity because their degraded products are mostly nontoxic and do not accumulate in the cells. Moreover, the degradation of the polymer can be used as a tool to release the plasmid DNA into the cytosol. PHP [poly(4-hydroxy-L-proline ester)] and PAGA [poly(γ-(4-aminobutyl)-L-glycolic acid)] are water-soluble biodegradable cationic polyamine polyesters with a threefold higher transfection activity *in vitro* than PLL and displayed no cytotoxicity in the concentration range tested [248–250]. In animal studies, tail vein injection of PAGA/PCAGGS mouse IL-10 [251] polyplex in 3-week-old NOD mice showed high serum IL-10 levels up to 5 days after injection, and remained detectable for more than 9 weeks [252,253]. Poly(phosphazenes) (PPZs), poly(phosphoesters) (PPEs), and poly(phosphoramidates) (PPAs) are phosphor-containing biodegradable polymers. Major plus points of these polymers are that they are fully degradable; they can easily be custom modified for targeting strategies and their high *in vivo* transfection efficiency [254]. Drawbacks are the relatively slow degradation of these polymers and rapid blood clearance after injection. The transfection efficiencies these polyplexes were about two or three orders of magnitude higher than PLL-mediated transfection, but in general not as high as the PEI-based systems [255,256]. PEI-based polymers with degradable or reducible linkers are very promising. Upon modification with cross linking agents such as dithiobis(succinimidylpropionate) (DSP), dimethyl-3,3′-dithiobispropionimidate (DTBP), bifunctional PEG, degradable oligo (L-lactic acid-co-succinic acid), and 1,3-butanediacrylate or 1,6-hexanediacrylate, low molecular weight (LMW) PEI-mediated gene expression was two- to threefold less efficiently than PEI 25 kDa, but exhibited less toxicity to cultured cells [257–259]. LMW PEI and PEGdiacrylate based polymers (MW 13 kDa) degrades in a controlled manner, with a half-life of 4.5–5 days at pH 7.4 and 37°C, showed low cytotoxicity and enhanced gene transfer efficiency in HepG2 and MG63 cells as compared to PEI 25 kDa [260]. The highest reporter gene expression was observed for pDNA/PEI-1.2 K (MW 1200) complex having an *in vitro* transfection activity 15–25-fold higher than PEI-25 K based systems. The polyplex was also capable of transfecting cells *in vivo* after airway administration as an aerosol [261]. These cationic polymers are also excellent potential candidates for the intracellular delivery of siRNA therapeutics [262].

4.4. Peptide-Based Gene Delivery Systems

Peptides are the third group of compounds, which have been explored as a synthetic gene delivery system [263]. These carriers take advantage of the existence of one or more functional domains, for example, DNA binding, condensation, receptor-specific uptake, endosomal escape, and also nuclear targeting. Peptides can be incorporated into multicomponent gene delivery complexes for specific purposes, such as for DNA condensation, cell-specific targeting, endosomolysis, or nuclear transport [264].

4.5. Targeted Delivery Systems

Tissue specific *in vivo* DNA delivery is still a major challenge for gene therapy. Biodistribution of lipoplex and polyplex mainly depends upon the physicochemical properties, and the size and the charge of the complex. After intravenous administration of a cationic DNA complex, transgene expression primarily occurs in the lung, secondarily in the liver and the tumor neovasculature. In the case of the liver, the sinusoidal fenestrations may play an important role in limiting the gene medicine accessing the parenchyma. Increased hydrodynamic pressure has been used to widen the fenestrations and thus improve delivery [265]. For the lung, it is not clear what factors are involved in overcoming the endothelial barriers, but an active transcellular transport has been observed for PEI polyplex in the lung [266]. Some gaps or small pores on the wall of the tumor neovasculature may allow the accumulation of gene medicine in this organ. Due to the relatively small size of the lipoplex, they can pass through the neovasculature to access the tumor cells. This phenomenon is known as "enhanced permeability and retention" (EPR) effect [267].

To date, tissue-specific targeting of lipoplex and polyplex has been accomplished by two distinct techniques. The first method involves transfection of selected tissues, such as nasal epithelium, arterial endothelium, lung, or tumors by locally administering the complexes within the defined region. This method has proven to be a viable option for the clinical treatment of several diseases, including cystic fibrosis and cancer. A second method used to enhance the specificity of gene therapy is by coupling cell-binding ligands, such as folic acid (FA), transferring, carbohydrates, estrogens, receptor-specific drugs, receptor-specific molecules or antibodies to liposomes, or polymeric systems for the purpose of combining the intrinsic activities of lipids with the receptor-mediated uptake properties of the applied ligand [268–274]. These surface modifications of lipoplex or polyplex prevent their intrinsic, nonspecific interactions and provide extrinsic targeting ability. The approach is called "active targeting."

The incorporation of a glycolipid into the liposomal composition determines its tropism to hepatocytes and to liver endothelial cells [275]. Campbell et al. [276] found that cationic liposomes, stabilized with the addition of a 5 mol% of PEG, accumulated more in the vasculature of tumor than the neutral liposomes. PEG was used to increase circulation lifetime of the positively charged liposomes. Inclusion of PEGylated lipids in the vesicles has the added advantage of reducing aggregate formation, thus increasing both yield and injectability of the complex. Unmodified lipoplex has a relatively short circulation half-life of less than 5 min. Furthermore, when the percentage of cationic lipid was increased from 10 to 50 mol%, the accumulation in tumor endothelial cells increased twofold [277].

The use of the transferrin (Tf) ligand to target a lipoplex delivery system resulted in a significant increase in the transfection efficiency. Delivery of wild-type (wt) p53 to a radiation-resistant squamous cell carcinoma of the head and neck (SCCHN) cell line via this ligand-targeted lipoplex was also able to revert the radiation resistant phenotype of these cells *in vitro*. The Tf-targeted lipoplex showed high gene transfer efficiency and efficacy with human head and neck cancer *in vitro* and *in vivo* [278,279].

Xu et al. [280] described a novel cationic immunolipoplex system that showed high *in vivo* gene transfer efficiency and antitumor efficacy when used for systemic p53 gene therapy of cancer. The novel cationic immunolipoplex incorporating a biosynthetically lipid-tagged, antitransferrin receptor single-chain antibody (TfRscFv) targeted tumor cells both *in vitro* and *in vivo*.

Folate receptor is overexpressed in majority of tumors. Folate receptor is internalized through caveolae in a process termed potolysis [281], which is different from endocytosis. Gottschalt et al. [271] achieved targeted DNA delivery by conjugating folate to poly-L-lysine. Folate-targeted pH-sensitive liposomes complexed to polylysine-condensed DNA have been formulated as LPDII particles for the receptor-specific cellular delivery of plasmid and antisense oligonucleotides [282]. Several other groups also showed targeted delivery of DNA to the folate receptor expressing cancer cells [283,284].

Recently our laboratory has developed ligand (anisamide) targeted, PEGylated LPD formulation that showed significant increase in cellular uptake in sigma receptor overexpressing cells both *in vitro* and *in vivo* via specific receptor-mediated pathway. This targeted formulation also demonstrated its strong gene-silencing effect mediated by RNAi [282].

4.6. Second-Generation Lipidic Delivery System (Lipopolyplex)

4.6.1. Polycation-Condensed DNA Entrapped in Cationic Liposome: LPD-I Formulation
The idea of using cationic liposomes together with cationic polymer was generated by Huang lab in 1996 and hypothesis behind this idea is the physical mixing of cationic polymers with the lipoplex at appropriate ratios will enhance both the cellular uptake and the nuclear transportation of pDNA by forming a compact structure. The overall complex so formed was termed liposome-polycation-DNA (LPD). There are two versions of LPD. LPD-I refers to nanoparticles containing a cationic surface membrane [285] and LPD-II refers to those containing an anionic surface membrane [282]. Liposome modified PLL polyplex showed higher transfection activity (2–28-folds) than the corresponding lipoplex in several cell lines [286]. One of the further studied systems incorporates DOTAP:cholesterol cationic liposome and protamine to encapsulate DNA for the formation of LPD-I. Protamine sulfate is a commercially available, FDA approved, naturally occurring peptide with high positive charge. It is very basic in nature due to the presence of 21 arginine residues, which contains an NLS. Both protamine sulfate and PLL were compared for their transfection efficiencies [279,287]. With the same amount of DNA and cationic lipid, PLL reached its efficiency plateau at an amount that was half the amount of protamine sulfate, but the overall efficiency of protamine sulfate was two- to sevenfold higher than PLL in different cell lines. Upon i.v. injection, LPD-I made with

DOTAP had shown very high gene expression in heart, lung, liver, spleen, and kidney, with the highest expression found in the lung [286]. Additionally, studies have shown that modified LPD vectors may be useful in cancer vaccination [187]. Systemic administration of the LPD rapidly initiates nonspecific immunostimulation associated with tumoristatic effects and produces several T helper type 1 (Th1) cytokines, most notably TNF-α, IL-12, and IFN-γ. LPD-I lipopolyplex particles upon i.v. injection were also distributed to the spleen along with other major organs, where they were endocytosed and released to the cytoplasm. Administrations of these particles containing unmethylated CpG motif in pDNA elicit immunostimulation that subsequently showed tumoristatic effect. The effect of CpG will be discussed in detail below. Furthermore, the spleen tropism was utilized to deliver antigenic peptides entrapped in LPD particles to the cytoplasm of splenic antigen presenting cells. It is known that *E6* and *E7* are two oncogenes responsible for the maintenance of the malignant state of HPV-positive tumors. Therefore, an arginine containing small peptide epitope (E749–57) derived from the tumor antigen, HPV-E7 protein, was entrapped in LPD-I. The particle upon i.v. injection was delivered to the splenic APCs and thereby induced E7-specific immune response against syngeneic HPV-induced tum- To fully exploit the therapeutic potential of emerging RNA interference technology, the development of safe and efficient reagents to deliver siRNA is a necessary requirement. RNA interference, discovered in *Caenorhabditis elegans*, inhibits gene expression through the activation of RNA-induced silencing complex (RISC) by interaction with siRNAs [289–291]. Nonviral vectors used for gene delivery, are promising methods of choice for siRNA transfection *in vitro* and have also been used for *in vivo* delivery of siRNA.

We have modified our LPD-I formulation to deliver siRNA and found that it is efficient in delivering siRNA to the tumor [292]. Modified LPD-I formulation comprises the cationic liposomes, protamine and nucleic acids (mixture of DNA and siRNA) at a fixed ratio. The PEG shield and the sigma receptor targeting ligand (anisamide) were then incorporated into the naked nanoparticles via the postinsertion method. The resultant formulation size was around 100 nm in diameter with 90% encapsulation efficiency for siRNA and the zeta potential close to neutral. PEGylation completely shielded the charge; stabilized the particles in serum and abolished the reticuloendothelial uptake in the isolated liver perfusion experiment [293]. Upon i.v. injection into nude mice bearing human lung tumor (NCI-H460 xenograft model), the LPD-encaptulated EGFR siRNA delivered 70–80% injected siRNA/g of tissue into the tumor, while the normal organs only showed a moderate uptake (10–20% injected dose per gram) [294]. After the conjugation of a targeting ligand, anisamide, at the distal end of the PEG, the intracellular delivery of siRNA into the sigma receptor expressing H460 tumor was significantly enhanced. This led to efficient EGFR silencing, significant apoptosis induction and 40% tumor growth inhibition at the dose of 1.2 mg siRNA/kg for three consecutive injections [295]. Complete growth inhibition lasted for 1 week when combined with cisplatin injection. The serum level of liver enzymes and body weight monitoring during the treatment indicated a low level of toxicity of the formulation. LPD-I targeted with anisamide also delivered combined sequences of siRNA against MDM2, c-myc, and VEGF into a lung metastasis model of B16F10, sigma receptor-expressing murine melanoma cells and caused simultaneous silencing of each of the oncogenes in the metastatic nodules [296]. It led to 70–80% tumor load reduction and 30% prolongation in animal lifespan. The results promise the potential use of this formulation clinically [295,297].

4.6.2. Polycation-Condensed DNA in Anionic Liposomes: LPD-II Formulation
Some disadvantages always accompany the cationic lipid-mediated gene transfection. Cytotoxic effect in cells/tissues is the primary concern that requires early attention. Liposomes composed of cationic lipids exhibit relatively large size, while complexation with DNA, provide suboptimal DNA condensation. Consequently, cationic lipoplexes are relatively large in size, show limited transfection efficiency and lack tissue specificity. Anionic liposomes used

in gene therapy show poor encapsulation efficiency due to the large size and excessive negative charge of the uncondensed pDNA. The utilization of anionic liposome was revisited with a different approach, which also utilized the concept of condensing DNA by cationic polymers. A delivery vector was developed wherein polylysine-condensed pDNA was entrapped into folate-targeted anionic liposomes via charge interaction. It had structural similarity as LPD-I and was named LPD-II. It differed from LPD-I in that anionic lipids instead of cationic lipids were used. This novel vector was more efficient and less cytotoxic compared to conventional cationic liposomal vectors, including LPD-I. Folate-targeted LPD-II particles were generated by mixing anionic liposomes composed of DOPE/CHEMS/folate-PEG-DOPE and the cationic DNA-polylysine (1:0.75, w/w) complex [287]. Structural analysis of LPD-II by negative-stain EM showed that the DNA-polylysine (which appears as rod shaped individually) and lipid complex appeared as a highly electron dense, spherical core with a low-density coating. The mean diameter of these particles was 74 ± 14 nm, that is, smaller than the empty liposomes. KB cells expressing folate receptors were transfected with LPD-II particles containing luciferase reporter gene and high transfection efficiency was observed. The activity could be inhibited by the presence of excess free folate. Control LPD-II particles generated with nontargeted liposomes was only active at low lipid/DNA ratios, suggesting that the transfection by LPD-II particles was only receptor dependent when the overall charge was negative. Compared with DC-Chol/DOPE/DNA liposome complexes, LPD-II showed 20–30-fold more transfection activity. On replacing DOPE with DOPC in the original formulation, the transfection was severely reduced. This indicated that the fusogenic activity of DOPE was essential for the transfection activity of LPD-II particles. Their low physiological stability, inefficient cellular uptake, and the lack of tissue specificity currently limit the therapeutic applications of antisense oligonucleotides (ODN). The use of various vectors renders phosphodiester (PO) ODN resistant to enzymatic digestion. KB cells, which overexpress folate binding protein, were also transfected with LPD-II containing targeting ligand folate to deliver ODN against epidermal growth-factor receptor (EGFR) [286]. This resulted in downregulation of EGFR and growth inhibition of KB cells. The modified backbone-ODN, namely, phosphorothioate (PS) and monomethylphosphonate (MP) are more stable to enzymatic degradation compared to PO ODN, but they suffer from increased toxicity and decreased specificity. In a study, PO ODN against EGFR had shown growth inhibitory effect to KB cells compared to that of PS/PO ODN when delivered with LPD-II, indicating that LPD-II could also protect PO ODN from attack by enzymes inside the cells [287,298].

5. PLASMID DNA DELIVERY BY EMULSIONS

Emulsion, a homogeneous mixture of two immiscible liquids, are one of the most widely studied systems for the delivery of lipid-soluble drugs. The most direct way of developing a drug emulsion is to base the formulation on nutritional emulsion products. Lipid dispersion formulations, such as liposomes and oil-in-water (o/w) emulsions, are attractive carriers for lipophilic drugs and/or pDNA. Emulsions are considered superior due to their suitability for industrial scale production, stability on storage, biocompatibility, and the incorporation efficacy for lipophilic drugs. However, *in vivo* stability of emulsions are still questioned since emulsions are readily "attacked" by plasma proteins, high-density lipoproteins, and other biological components, such that the pharmacokinetics of the carried genes is not favorable for the diseased cells [299,300].

The oil-in-water emulsion is made of oil dispersed in an aqueous phase with a suitable emulsifier such as phospholipids, nonionic or ionic surfactants. Castor oil or soybean oil is predominantly used as the core oil phase. Nonionic surfactants such as Tween, Span, Brij, and pluronic copolymers are often used as coemulsifiers. The ionic coemulsifiers are phospholipids or cationic lipids. A number of structure/activity studies had been done with different emulsion formulations, which were subsequently used for gene delivery *in vitro* and *in vivo*. The nonionic surfactants such as Tween, Span, Brij, and pluronic copolymers were found to be excellent coemulsifiers when used along with castor oil, DC-Chol, and DOPE. The *in vitro* transfection study on BL-6 cells showed that the Tween surfactant containing formulations had more serum resistivity and exhibited higher transfection in serum-containing media than in the absence of serum [301]. One of the Brij containing formulations, that is, one with 2-oxyethylene chains, showed the highest transfection efficiency in the presence of serum. The toxicity of each formulation was minimal. In a DOTAP, soybean oil, and pDNA emulsion complex, it was observed that in spite of the change of zeta potential with varying amounts of DNA, the structure and the size of the emulsion complex remained mostly unchanged [302]. The stability of this emulsion complex was good and it protected pDNA from DNase-I digestion. In serum-containing media, the emulsion showed much higher transfection efficiency compared to lipofectamine/DNA transfection complex. On inclusion of polyethyleneglycol-PE in the emulsion complex, a high-level transfection was observed even in the media containing 90% of serum [303]. This result suggested that *in vivo* transfection could be done with this emulsion complex. Another soybean oil-DOTAP emulsion was used to transfer genes to the epithelial cells of the mouse nasal cavity via intranasal instillation [302,304]. The emulsion showed enhanced stability against heparin exchange and exhibited higher level of transfection than a commercially available transfection reagent in the nasal cavity mucosa.

Gene expression level in the liver after intraportal injection of pDNA incorporated into reconstituted chylomicron remnants (RCRs) was almost 100 times greater than that of the naked pDNA [304–307]. Chylomicrons are triglyceride-rich lipoproteins that are slowly modified during circulation in the blood. In addition, gene expression in the liver after intraportal injection of pDNA incorporated into RCR was reduced to about one-third by preinjection of RCR. These results suggest that pDNA incorporated into RCR is taken up via the chylomicron remnant receptors on the liver. Incorporation of olive oil emulcified lipoplexes of TC-Chol, a quaternary ammonium analog of DC-Chol, into the internal oil space of RCR gave the transgene luciferase expression in the liver 100-fold higher than with naked DNA injection. Addition of PEG to the surface of RCR imparts the long circulation property and improved the formulation efficiency [308].

6. TOXICITY OF NONVIRAL VECTORS

Nonviral vectors, although less toxic than viral vectors, may still elicit a strong, nonspecific immune response. A major limitation of the use of gene therapy vectors is the innate immune responses, such as the production of inflammatory cytokines, triggered by systemic administration of such vectors. It is essential to overcome such toxicity, because it not only shortens the period of gene expression but also leads to serious side effects. It is known from the 1980s that the bacterial DNA stimulates the formation of cytotoxic IFN-α, β, and IL-12 when the DNA is taken up by macrophages [309]. It in turn leads to NK cell activation and production of proinflammatory cytokine IFN-γ [310–312]. This is accompanied by the proliferation of B cell and hence reduction of apoptosis and release of IL-6 and IL-12. The response is due to the existence of some immunostimulatory sequences in prokaryotic DNA. The most potent motif contains unmethylated CpG dinucleotide flanked by two 5'-purines and two 3'-pyrimidines [313]. pDNA, which is derived from bacterial DNA, readily induces these immune responses [314]. The unmethylated CpG motif-containing sequence occurs four times more frequently in prokaryote DNA than in eukaryotic DNA. Moreover, the CpG motifs are usually 75% more methylated in mammalian DNA than in prokaryotic DNA. On methylation of the cytosine bases in the plasmid, the immunostimulatory effect is decreased considerably. Immature dendritic cells tend to produce proinflammatory IFN-α, β, IL-6, IL-12, and TNF-α on the exposure to CpG containing oligonucleotide or bacterial plasmid [315–317]. These immunestimulatory effects leading to inflammatory cytokine productions had negative impact on the systemic gene delivery of cationic lipoplex. Since the circulating lipoplex is mainly taken up by splenic and other macrophages, they recognize the CpG motifs and elicit an inflammatory response. A high level of proinflammatory cytokine also led to inactivation of several promoters, resulting in a decrease in transgene expression [318–321]. The death of animals by high dose lipoplex injection for obtaining high transgene expression could be attributed both to the high concentration of lipid and to pDNA-mediated toxicity. In the case of local lipoplex administration in animals, minimal toxicity was observed. However, on i.v. injection or intratracheal instillation, high levels of IFN-γ and TNF-α were observed [322,323]. Pretreating cytokine-neutralizing antibodies during i.v. injection of lipoplex, CpG triggered inflammation and immune responses were minimized and prolonged gene expression was obtained. Repeated dosing without any antibody treatment led to silenced transgene expression for 1 or 2 weeks [322]. It is not clear though how cytokine production decreases the transgene expression, but various ongoing efforts to minimize CpG related immune responses and toxicity, and to enhance transgene expression is worth mentioning. Hofman et al. [324] had used PCR-amplified fragments containing encoded therapeutic gene and regulatory elements for preparing LPD [325]. On delivering, a similar level of gene expression comparable to pDNA lipoplex was obtained. However, a much lower level of cytokine response was observed, which sustained the gene expression for a longer period than pDNA. PCR fragment contains fewer CpG motifs than the full-length pDNA, which led to reduced CpG triggered adverse effects. Yew et al. [326] had shown that on mutating CpG or its flanking motifs in the plasmid, a decreased level of cytokine and increased transgene expression could be obtained. A limited interaction of plasmid with immune cells could also lead to decreased cytokine response. It could be achieved by sequentially injecting cationic liposomes and free pDNA. Song et al. [327] used this process for the purpose of efficiently transfecting the lung by prolonging the residency time and interaction of DNA with pulmonary endothelium. Tan et al. [328] used the same concept to show that sequential injection led to the formation of a lower level of cytokines as compared to lipoplex injection. It is evident from the above efforts that understanding the detailed mechanism of CpG induced immune response is required for increasing the efficacy of pDNA-mediated gene delivery. Further efforts to reduce toxicity have involved the incorporation of molecules like dexamethasone known to suppress the production of the cytokine, NF-κB [329]. These "safeplexes" were able to maintain low levels of TNF-α as compared to lipoplex alone, while achieving comparable levels of gene expression [329]. It has also been demonstrated that lipopolysaccharide-induced TNF-α production is suppressed with the preinjection of NF-κB decoy, whose double-stranded oligonucleotides contain an NF-κB binding sequence, suggesting that NF-κB decoy might be another suppressor of lipoplex-induced cytokine production. Thus, significant advances have been made toward decreasing the toxicity of these nonviral vectors [328,330].

7. CONCLUSION

Numerous nonviral gene delivery systems have already been developed from the first attempt of utilizing pDNA for gene delivery to cells. A number of gene therapy clinical trials were also performed. Although no single vector is superior to other vectors, each *in vivo* gene transfer application will find its way for optimal performance. Successful gene therapy depends on the development of efficient delivery systems. Continuous effort to improve currently available systems and to develop new methods of gene delivery is needed and could lead to safer and more efficient nonviral gene delivery. It is essential to understand the cellular barriers for delivery and then design possible means to overcome the barriers. The future remains bright for nonviral gene therapy.

ACKNOWLEDGMENT

The work performed in authors' laboratory was supported by NIH grants CA129421 and CA129835.

GLOSSARY

Gene Therapy — A therapeutic technique to eradicate genetic malfunctions by introducing new gene or repairing defective gene in the host cell. The therapy, initially aimed toward treating inheritable disorders, is also used in treating acquired disorders such as AIDS, cancer, and so on.

Lipoplex — A charged complex made of DNA and cationic lipid. The complex is formed by electrostatic combination of negatively charged DNA with positively charged cationic lipids. Preformed cationic liposomes made of cationic lipids and/or colipids interact and condense DNA by cooperative interaction of positively charged head group of aggregated surface of lipids with negatively charged DNA.

Lipopolyplex — An entity formed by mixing the preformed liposomes made of cationic, anionic, or neutral lipids, and/or colipids with cationic polymer induced precondensed DNA. Electron microscopic studies revealed a lipid-based envelope, carrying the polyplex-containing core.

Nonviral Gene Delivery — A special gene delivery technique, which does not use viral mechanism to deliver genes to cells. It involves various physical techniques, chemical methods, and biomimetic synthetic agents to introduce genes to cells.

Polyplex — Another charged complex as lipoplex, but here the cationic entities are polymers. Cationic polymers associated with net positive charges also condense negatively charged DNA by electrostatic combination to form poly-plex.

ABBREVIATIONS

CHEMS — cholesteryl hemisuccinate

CTAP — N^{15}cholesteryloxycarbonyl-3,7,12-triazapentadecane-1,15-diamine

CTAP — N^{15}cholesteryloxycarbonyl-3,7,12-triazapentadecane-1,diamine

DC-Chol — 3β-[N-(N',N'-dimethylaminoethane)-carbamoyl] cholesterol

DDAB — dimethyldioctadecylammonium bromide

14Dea2 — O,O'-ditetradecanolyl-N-(trimethylammonioacetyl) diethanolamine chloride

DHDEAB — N,N-di-n-hexadecyl-N,N-dihydroxyethylammonium bromide

DMDHP — N,N-(2-hydroxyethyl)-N-methyl-2,3-bis(myristoyloxy)-1-propanaminium iodide

DMRIE — N-(2-hydroxyethyl)-N,N-dimethyl-2,3-bis(tetradecyloxy)-1-propanaminium chloride

DOPE — 1,2-dioeoyl-sn-glycero-3-phosphatidy lethanolamine

DORIE — N-[1-(2,3-dioleyloxy)propyl]-N-hydroxyethyl N,Ndimethylammonium chloride

DOSPA	2,3-dioleyloxy-N-[2(spermi-necar-boxamido)ethyl]-N,N-dimethyl-1-propanaminium trifluoroacetate
DOTAP	N-[1-(2,3-dioleyloxy)propyl]-N,N,N-trimethylammonium chloride
DOTIM	1-[2-[9-(Z)-octadecenoyloxy]ethyl]]-2-[8](Z)-heptadecenyl]-3-[hydroxyethyl]imidazolinium chloride
DOTMA	N-[1-(2,3-dioleyloxy)propyl]-N,N,N-trimethylammonium chloride
GAP-DLRIE	N-(3-aminopropyl)-N,N-dimethyl-2,3-bis(dodecyloxy)-1-propaniminium bromide
LPLL	lipopoly(L-lysine)
MP	monomethylphosphonate
PAMAM	polyamidoamine
PEG	polyethyleneglycol
PEI	polyethyleneimines
PLL	poly-L-lysine
PO	phosphodiester
PS	phosphorothioate
TC-Chol	3β-[N-(N',N',N'-trimethylaminoethane)carbamoyl] cholesterol

REFERENCES

1. Yew NS, Zhao H, Wu IH, Song A, Tousignant JD, Przybylska M, Cheng SH. Reduced inflammatory response to plasmid DNA vectors by elimination and inhibition of immunostimulatory CpG motifs. Mol Ther 2000;1(3):255–262.
2. Bigger BW, Tolmachov O, Collombet JM, Fragkos M, Palaszewski I, Coutelle C. An araC-controlled bacterial cre expression system to produce DNA minicircle vectors for nuclear and mitochondrial gene therapy. J Biol Chem 2001;276(25):23018–23027.
3. Wolff JA, Malone RW, Williams P, Chong W, Acsadi G, Jani A, Felgner PL. Direct gene transfer into mouse muscle *in vivo*. Science 1990;247(4949 Pt 1):1465–1468.
4. Hickman MA, Malone RW, Lehmann-Bruinsma K, Sih TR, Knoell D, Szoka FC, Walzem R, Carlson DM, Powell JS. Gene expression following direct injection of DNA into liver. Hum Gene Ther 1994;5(12):1477–1483.
5. Zhang G, Vargo D, Budker V, Armstrong N, Knechtle S, Wolff JA. Expression of naked plasmid DNA injected into the afferent and efferent vessels of rodent and dog livers. Hum Gene Ther 1997;8(15):1763–1772.
6. Budker V, Zhang G, Knechtle S, Wolff JA. Naked DNA delivered intraportally expresses efficiently in hepatocytes. Gene Ther 1996;3(7):593–598.
7. Choate KA, Khavari PA. Direct cutaneous gene delivery in a human genetic skin disease. Hum Gene Ther 1997;8(14):1659–1665.
8. Meyer KB, Thompson MM, Levy MY, Barron LG, Szoka FC Jr. Intratracheal gene delivery to the mouse airway: characterization of plasmid DNA expression and pharmacokinetics. Gene Ther 1995;2(7):450–460.
9. Xiang G, Keun-Sik K, Dexi L. Nonviral gene delivery: what we know and what is next. AAPS J 2007;9(1):92–104.
10. Herweijer H, Zhang G, Subbotin VM, Budker V, Williams P, Wolff JA. Time course of gene expression after plasmid DNA gene transfer to the liver. J Gene Med 2001;3(3):280–291.
11. Sikes ML, O'Malley BW Jr, Finegold MJ, Ledley FD. *In vivo* gene transfer into rabbit thyroid follicular cells by direct DNA injection. Hum Gene Ther 1994;5(7):837–844.
12. Ardehali A, Fyfe A, Laks H, Drinkwater DC Jr, Qiao JH, Lusis AJ. Direct gene transfer into donor hearts at the time of harvest. J Thorac Cardiovasc Surg 1995;109(4):716–719; discussion 719–720.
13. Schwartz B, Benoist C, Abdallah B, Rangara R, Hassan A, Scherman D, Demeneix BA. Gene transfer by naked DNA into adult mouse brain. Gene Ther 1996;3(5):405–411.
14. Yoo JJ, Soker S, Lin LF, Mehegan K, Guthrie PD, Atala A. Direct *in vivo* gene transfer to urological organs. J Urol 1999;162(3 Pt 2):1115–1118.
15. Soler MN, Bobe P, Benihoud K, Lemaire G, Roos BA, Lausson S. Gene therapy of rat medullary thyroid cancer by naked nitric oxide synthase II DNA injection. J Gene Med 2000;2(5):344–352.
16. Xie K, Huang S, Dong Z, Juang SH, Wang Y, Fidler IJ. Destruction of bystander cells by tumor cells transfected with inducible nitric

oxide (NO) synthase gene. J Natl Cancer Inst 1997;89(6):421–427.

17. Xie K, Huang S, Dong Z, Juang SH, Gutman M, Xie QW, Nathan C, Fidler IJ. Transfection with the inducible nitric oxide synthase gene suppresses tumorigenicity and abrogates metastasis by K-1735 murine melanoma cells. J00 Exp Med 1995;181:1333–1343.

18. Miao CH, Thompson AR, Loeb K, Ye X. Long-term and therapeutic-level hepatic gene expression of human factor IX after naked plasmid transfer in vivo. Mol Ther 2001;3(6):947–957.

19. Mazda O. Improvement of nonviral gene therapy by Epstein–Barr virus (EBV)-based plasmid vectors. Curr Gene Ther 2002;2(3):379–392.

20. Wolff JA, Naked DNA gene transfer in mammalian cell. In: Friedmann T, editor. Development of Human Gene Therapy. New York: Cold Spring Harbor Laboratory; 1999. p 279.

21. Banerjee R, Huang L. Vector system: plasmid DNA. In: Robert A Mayers, editor. Encyclopedia of Molecular Cell Biology and Molecular Medicine. 2nd ed. Wiley VCH publishers; 2003. p 287–337. Fraley R., Subramani S, Berg P, Papahadjopoulos D. Introduction of liposome-encapsulated SV40 DNA into cells. J Biol Chem 1980;255:10431–10435.

22. Liu F, Shollenberger LM, Conwell CC, Yuan X, Huang L. Mechanism of naked DNA clearance after intravenous injection. J Gene Med 2007; 9(7):613–619.

23. Kawakami S, Higuchi Y, Hashida M. Nonviral approaches for targeted delivery of plasmid DNA and oligonucleotide. J Pharm Sci 2008; 97(2):726–745.

24. Li S, Huang L. Nonviral gene therapy: promises and challenges. Gene Ther 2000;7(1):31–34.

25. Christine CC, Huang L. Recent progress in non-viral gene delivery. In: Taira K, Kataoka K, Niidome T, editors. Non-viral Gene Therapy Gene Design and Delivery, Springer Link; 2005. p 3–8.

26. Schughart K, Rasmussen UB. Solvoplex synthetic vector for intrapulmonary gene delivery Preparation and use. Methods Mol Med 2002;69:83–94.

27. Schughart K, Bischoff R, Rasmussen UB, Hadji DA, Perraud F, Accart N, Boussif O, Akinc A, Thomas M, Klibanov AM, Langer R. Exploring polyethylenimine-mediated DNA transfection and the proton sponge hypothesis. J Gene Med 2005;7(5):657–663.

28. Desigaux L, Gourden C, Bello-Roufai M, Richard P, Oudrhiri N, Lehn P, Escande D, Pollard H, Pitard B. Nonionic amphiphilic block copolymers promote gene transfer to the lung. Hum Gene Ther 2005;16(7):821–829.

29. Freeman DJ, Niven RW. The influence of sodium glycocholate and other additives on the in vivo transfection of plasmid DNA in the lungs. Pharm Res 1996;13(2):202–209.

30. Lemoine JL, Farley R, Huang L. Mechanism of efficient transfection of the nasal airway epithelium by hypotonic shock. Gene Ther 2005;12(16):1275–1282.

31. Ross GF, Bruno MD, Uyeda M, Suzuki K, Nagao K, Whitsett JA, Korfhagen TR. Enhanced reporter gene expression in cells transfected in the presence of DMI-2, an acid nuclease inhibitor. Gene Ther 1998;5 (9):1244–1250.

32. Walther W, Stein U, Siegel R, Fichtner I, Schlag PM. Use of the nuclease inhibitor aurintricarboxylic acid (ATA) for improved nonviral intratumoral in vivo gene transfer by jet-injection. J Gene Med 2005;7(4):477–485.

33. Glasspool-Malone J, Malone RW. Marked enhancement of direct respiratory tissue transfection by aurintricarboxylic acid. Hum Gene Ther 1999;10(10):1703–1713.

34. Marshall DJ, Leiden JM. Recent advances in skeletal-muscle-based gene therapy. Curr Opin Genet Dev 1998;8(3):360–365.

35. Wolff JA, Ludtke JJ, Acsadi G, Williams P, Jani A. Long-term persistence of plasmid DNA and foreign gene expression in mouse muscle. Hum Mol Genet 1992;1(6):363–369.

36. Acsadi G, Dickson G, Love DR, Jani A, Walsh FS, Gurusinghe A, Wolff JA, Davies KE. Human dystrophin expression in mdx mice after intramuscular injection of DNA constructs. Nature 1991;352(6338):815–818.

37. Fazio VM, Fazio S, Rinaldi M, Catani MV, Zotti S, Ciafre SA, Seripa D, Ricci G, Farace MG. Accumulation of human apolipoprotein-E in rat plasma after in vivo intramuscular injection of naked DNA. Biochem Biophys Res Commun 1994;200(1):298–305.

38. Anwer K, Shi M, French MF, Muller SR, Chen W, Liu Q, Proctor BL, Wang J, Mumper RJ, Singhal A, Rolland AP, Alila HW. Systemic effect of human growth hormone after intramuscular injection of a single dose of a muscle-specific gene medicine. Hum Gene Ther 1998;9 (5):659–670.

39. Davis NL, Brown KW, Johnston RE. A viral vaccine vector that expresses foreign genes in lymph nodes and protects against mucosal challenge. J Virol 1996;70(6):3781–3787.

40. Conry RM, LoBuglio AF, Loechel F, Moore SE, Sumerel LA, Barlow DL, Curiel DT. A carcinoembryonic antigen polynucleotide vaccine has in vivo antitumor activity. Gene Ther 1995;2(1):59–65.

41. Spooner RA, Deonarain MP, Epenetos AA. DNA vaccination for cancer treatment. Gene Ther 1995;2(3):173–180.

42. Batteux F, Tourneur L, Trebeden H, Charreire J, Chiocchia G. Gene therapy of experimental autoimmune thyroiditis by in vivo administration of plasmid DNA coding for Fas ligand. J Immunol 1999;162(1):603–608.

43. Wolff JA, Williams P, Acsadi G, Jiao S, Jani A, Chong W. Conditions affecting direct gene transfer into rodent muscle in vivo. Biotechniques 1991;11(4):474–485.

44. Davis HL, Whalen RG, Demeneix BA. Direct gene transfer into skeletal muscle in vivo: factors affecting efficiency of transfer and stability of expression. Hum Gene Ther 1993;4(2):151–159.

45. Wells DJ. Improved gene transfer by direct plasmid injection associated with regeneration in mouse skeletal muscle. FEBS Lett 1993;332(1-2):179–182.

46. Danko I, Fritz JD, Jiao S, Hogan K, Latendresse JS, Wolff JA. Pharmacological enhancement of in vivo foreign gene expression in muscle. Gene Ther 1994;1(2):114–121.

47. Zhang G, Budker V, Williams P, Subbotin V, Wolff JA. Efficient expression of naked DNA delivered intra-arterially to limb muscles of nonhuman primates. Hum Gene Ther 2001;12(4):427–438.

48. Wolff J. Naked DNA gene transfer in mammalian cell. Friedmann T, editor. Development of Human Gene Therapy. Cold Spring Harbor Monograph Series 36 Cold Spring Harbor, NY: Cold Spring Harbor Laboratory; 1999. p 279.

49. Budker V, Budker T, Zhang G, Subbotin V, Loomis A, Wolff JA. Hypothesis: naked plasmid DNA is taken up by cells in vivo by a receptor-mediated process. J Gene Med 2000;2(2):76–88.

50. Bennett RM, Gabor GT, Merritt MM. DNA binding to human leukocytes. Evidence for a receptor-mediated association, internalization, and degradation of DNA. J Clin Invest 1985;76(6):2182–2190.

51. Takagi T, Hashiguchi M, Mahato RI, Tokuda H, Takakura Y, Hashida M. Involvement of specific mechanism in plasmid DNA uptake by mouse peritoneal macrophages. Biochem Biophys Res Commun 1998; 245(3):729–733.

52. Hefeneider SH, Cornell KA, Brown LE, Bakke AC, McCoy SL, Bennett RM, Nucleosomes DNA bind to specific cell-surface molecules on murine cells and induce cytokine production. Clin Immunol Immunopathol 1992;63(3):245–251.

53. Prabhakar BS, Allaway GP, Srinivasappa J, Notkins AL. Cell surface expression of the 70-kD component of Ku, a DNA-binding nuclear autoantigen. J Clin Invest 1990;86(4):1301–1305.

54. Kabakov AE, Saenko VA, Poverenny AM. LDL-mediated interaction of DNA and DNA-anti-DNA immune complexes with cell surface. Clin Exp Immunol 1991;83(3):359–363.

55. Stein CA, Tonkinson JL, Zhang LM, Yakubov L, Gervasoni J, Taub R, Rotenberg SA. Dynamics of the internalization of phosphodiester oligodeoxynucleotides in HL60 cells. Biochemistry 1993;32(18):4855–4861.

56. Siess DC, Vedder CT, Merkens LS, Tanaka T, Freed AC, McCoy SL, Heinrich MC, Deffebach ME, Bennett RM, Hefeneider SH. A human gene coding for a membrane-associated nucleic acid-binding protein. J Biol Chem 2000;275(43):33655–33662.

57. Liu F, Song Y, Liu D. Hydrodynamics-based transfection in animals by systemic administration of plasmid DNA. Gene Ther 1999;6(7):1258–1266.

58. Zhang G, Budker V, Wolff JA. High levels of foreign gene expression in hepatocytes after tail vein injections of naked plasmid DNA. Hum Gene Ther 1999;10(10):1735–1737.

59. Liu F, Nishikawa M, Clemens PR, Huang L. Transfer of full-length Dmd to the diaphragm muscle of Dmd(mdx/mdx) mice through systemic administration of plasmid DNA. Mol Ther 2001;4(1):45–51.

60. Wang C, Chao C, Madeddu P, Chao L, Chao J. Central delivery of human tissue kallikrein gene reduces blood pressure in hypertensive rats. Biochem Biophys Res Commun 1998;244(2):449–454.

61. Losordo DW, Vale PR, Symes JF, Dunnington CH, Esakof DD, Maysky M, Ashare AB, Lathi K, Isner JM. Gene therapy for myocardial angiogenesis: initial clinical results with direct myocardial injection of phVEGF165 as sole therapy for myocardial ischemia. Circulation 1998;98(25):2800–2804.

62. Losordo DW, Vale PR, Hendel RC, Milliken CE, Fortuin FD, Cummings N, Schatz RA, Asahara T, Isner JM, Kuntz RE. Phase 1/2 placebo-controlled, double-blind, dose-escalating trial of myocardial vascular endothelial growth factor 2 gene transfer by catheter delivery in patients

with chronic myocardial ischemia. Circulation 2002;105(17): 2012–2018.
63. Kastrup J, Jorgensen E, Ruck A, Tagil K, Glogar D, Ruzyllo W, Botker HE, Dudek D, Drvota V, Hesse B, Thuesen L, Blomberg P, Gyongyosi M, Sylven C. Direct intramyocardial plasmid vascular endothelial growth factor-A165 gene therapy in patients with stable severe angina pectoris A randomized double-blind placebo-controlled study: the Euroinject One trial. J Am Coll Cardiol 2005;45(7):982–988.
64. Wright MJ, Wightman LM, Lilley C, de Alwis M, Hart SL, Miller A, Coffin RS, Thrasher A, Latchman DS, Marber MS. *In vivo* myocardial gene transfer: optimization, evaluation and direct comparison of gene transfer vectors. Basic Res Cardiol 2001;96(3):227–236.
65. Rutanen J, Rissanen TT, Markkanen JE, Gruchala M, Silvennoinen P, Kivela A, Hedman A, Hedman M, Heikura T, Orden MR, Stacker SA, Achen MG, Hartikainen J, Yla-Herttuala S. Adenoviral catheter-mediated intramyocardial gene transfer using the mature form of vascular endothelial growth factor-D induces transmural angiogenesis in porcine heart. Circulation 2004;109(8):1029–1035.
66. Lyon AR, Sato M, Hajjar RJ, Samulski RJ, Harding SE. Gene therapy: targeting the myocardium. Heart 2008;94(1):89–99.
67. Webster K, Prentice H, Discher DJ, Hicks MC, Bisphoric NH. In: Wheelan WJ, editor. Molecular Biology in the Conquest of Disease, Oxford, UK: Oxford University Press; 1998. p 37.
68. Parmacek MS, Ip HS, Jung F, Shen T, Martin JF, Vora AJ, Olson EN, Leiden JM. A novel myogenic regulatory circuit controls slow/cardiac troponin C gene transcription in skeletal muscle. Mol Cell Biol 1994;14 (3):1870–1885.
69. Ritchie ME. Characterization of human B creatine kinase gene regulation in the heart *in vitro* and *in vivo*. J Biol Chem 1996;271(41): 25485–25491.
70. Prentice H, Bishopric NH, Hicks MN, Discher DJ, Wu X, Wylie AA, Webster KA. Regulated expression of a foreign gene targeted to the ischaemic myocardium. Cardiovasc Res 1997;35(3):567–574.
71. Lin H, Parmacek MS, Morle G, Bolling S, Leiden JM. Expression of recombinant genes in myocardium *in vivo* after direct injection of DNA. Circulation 1990;82(6):2217–2221.
72. McDonald P, Hicks MN, Cobbe SM, Prentice H. Gene transfer in models of myocardial ischemia. Ann NY Acad Sci 1995;752:455–459.
73. McNeil PL, Khakee R. Disruptions of muscle fiber plasma membranes. Role in exercise-induced damage. Am J Pathol 1992;140(5):1097–1109.
74. Liu CC, Shen Z, Kung HF, Lin MC. Cancer gene therapy targeting angiogenesis: an updated review. World J Gastroenterol 2006;12 (43):6941–6948.
75. Tsurumi Y, Takeshita S, Chen D, Kearney M, Rossow ST, Passeri J, Horowitz JR, Symes JF, Isner JM. Direct intramuscular gene transfer of naked DNA encoding vascular endothelial growth factor augments collateral development and tissue perfusion. Circulation 1996;94(12):3281–3290.
76. Simovic D, Jeffrey MI, Allan HR, Ann Pieczek, David HW. Improvement in Chronic Ischemic Neuropathy after Intramuscular phVEGF$_{165}$ Gene Transfer in Patients with Critical Limb Ischemia. Arch Neurol 2001;58:761–768. Prud'homme GJ. Gene therapy of autoimmune diseases with vectors encoding regulatory cytokines or inflammatory cytokine inhibitors. J Gene Med 2000;2(4):222–232.
77. Uesato M, Gunji Y, Tomonaga T, Miyazaki S, Shiratori T, Matsubara H, Kouzu T, Shimada H, Nomura F, Ochiai T. Synergistic antitumor effect of antiangiogenic factor genes on colon 26 produced by low-voltage electroporation. Cancer Gene Ther 2004;11(9):625–632.
78. Folkman J, Merler E, Abernathy C, Williams G. Isolation of a tumor factor responsible for angiogenesis. J Exp Med 1971;133(2):275–288.
79. Horiki M, Yamato E, Noso S, Ikegami H, Ogihara T, Miyazaki J. High-level expression of interleukin-4 following electroporation-mediated gene transfer accelerates Type 1 diabetes in NOD mice. J Autoimmun 2003; 20(2):111–117.
80. Chang Y, Prud'homme GJ. Intramuscular administration of expression plasmids encoding interferon-gamma receptor/IgG1 or IL-4/IgG1 chimeric proteins protects from autoimmunity. J Gene Med 1999;1(6):415–423.
81. Piccirillo CA, Chang Y, Prud'homme GJ. TGF-beta1 somatic gene therapy prevents autoimmune disease in nonobese diabetic mice. J Immunol 1998;161(8):3950–3956.
82. Piccirillo CA, Prud'homme GJ. Prevention of experimental allergic encephalomyelitis by intramuscular gene transfer with cytokine-encoding plasmid vectors. Hum Gene Ther 1999;10(12):1915–1922.
83. Raz E, Dudler J, Lotz M, Baird SM, Berry CC, Eisenberg RA, Carson DA. Modulation of disease activity in murine systemic lupus

erythematosus by cytokine gene delivery. Lupus 1995;4(4):286–292.
84. Giladi E, Raz E, Karmeli F, Okon E, Rachmilewitz D. Transforming growth factor-beta gene therapy ameliorates experimental colitis in rats. Eur J Gastroenterol Hepatol 1995;7(4):341–347.
85. Song XY, Gu M, Jin WW, Klinman DM, Wahl SM. Plasmid DNA encoding transforming growth factor-beta1 suppresses chronic disease in a streptococcal cell wall-induced arthritis model. J Clin Invest 1998;101(12): 2615–2621.
86. Klein TM, Fromm M, Weissinger A, Tomes D, Schaaf S, Sletten M, Sanford JC. Transfer of foreign genes into intact maize cells with high-velocity microprojectiles. Proc Natl Acad Sci USA 1988;85(12):4305–4309.
87. Capecchi MR. High efficiency transformation by direct microinjection of DNA into cultured mammalian cells. Cell 1980;22(2 Pt 2):479–488.
88. Heller LC, Ugen K, Heller R. Electroporation for targeted gene transfer. Expert Opin Drug Deliv 2005;2(2):255–268. Yang NS, Burkholder J, Roberts B, Martinell B, McCabe D. In vivo and in vitro gene transfer to mammalian somatic cells by particle bombardment. Proc Natl Acad Sci USA 1990;87(24):9568–9572. Kurata S, Tsukakoshi M, Kasuya T, Ikawa Y. The laser method for efficient introduction of foreign DNA into cultured cells. Exp Cell Res 1986;162(2):372–378.
89. Zheng QA, Chang DC. High-efficiency gene transfection by in situ electroporation of cultured cells. Biochim Biophys Acta 1991;1088(1):104–110.
90. Coster HG. A quantitative analysis of the voltage–current relationships of fixed charge membranes and the associated property of "punch-through". Biophys J 1965;5(5): 669–686.
91. Sukharev SI, Klenchin VA, Serov SM, Chernomordik LV, Chizmadzhev Yu A. Electroporation and electrophoretic DNA transfer into cells. The effect of DNA interaction with electropores. Biophys J 1992;63(5):1320–1327.
92. Rols MP, Bachaud JM, Giraud P, Chevreau C, Roche H, Teissie J. Electrochemotherapy of cutaneous metastases in malignant melanoma. Melanoma Res 2000;10(5):468–474.
93. Rols MP, Delteil C, Golzio M, Dumond P, Cros S, Teissie J. In vivo electrically mediated protein and gene transfer in murine melanoma. Nat Biotechnol 1998;16(2):168–171.
94. Jaroszeski MJ, Gilbert R, Nicolau C, Heller R. In vivo gene delivery by electroporation. Adv Drug Deliv Rev 1999;35(1):131–137.
95. Yasuda K, Momose T, Takahashi Y. Applications of microelectroporation for studies of chick embryogenesis. Dev Growth Differ 2001;42(3):203–206.
96. Somiari S, Glasspool-Malone J, Drabick JJ, Gilbert RA, Heller R, Jaroszeski MJ, Malone RW. Theory and in vivo application of electroporative gene delivery. Mol Ther 2000;2(3):178–187.
97. Mir LM, Bureau MF, Rangara R, Schwartz B, Scherman D. Long-term, high level in vivo gene expression after electric pulse-mediated gene transfer into skeletal muscle. CR Acad Sci III 1998;321(11):893–899.
98. Aihara H, Miyazaki J. Gene transfer into muscle by electroporation in vivo. Nat Biotechnol 1998;16(9):867–870.
99. Heller R, Jaroszeski M, Atkin A, Moradpour D, Gilbert R, Wands J, Nicolau C. In vivo gene electroinjection and expression in rat liver. FEBS Lett 1996;389(3):225–228.
100. Suzuki T, Shin BC, Fujikura K, Matsuzaki T, Takata K. Direct gene transfer into rat liver cells by in vivo electroporation. FEBS Lett 1998;425(3):436–440.
101. Heller L, Jaroszeski MJ, Coppola D, Pottinger C, Gilbert R, Heller R. Electrically mediated plasmid DNA delivery to hepatocellular carcinomas in vivo. Gene Ther 2000;7(10):826–829.
102. Regnier V, De Morre N, Jadoul A, Preat V. Mechanisms of a phosphorothioate oligonucleotide delivery by skin electroporation. Int J Pharm 1999;184(2):147–156.
103. Dujardin N, Van Der Smissen P, Preat V. Topical gene transfer into rat skin using electroporation. Pharm Res 2001;18(1):61–66.
104. Rizzuto G, Cappelletti M, Maione D, Savino R, Lazzaro D, Costa P, Mathiesen I, Cortese R, Ciliberto G, Laufer R. La Monica N. Fattori E Efficient and regulated erythropoietin production by naked DNA injection and muscle electroporation. Proc Natl Acad Sci USA 1999;96(11):6417–6422.
105. Bettan M, Emmanuel F, Darteil R, Caillaud JM, Soubrier F, Delaere P, Branelec D, Mahfoudi A, Duverger N, Scherman D. High-level protein secretion into blood circulation after electric pulse-mediated gene transfer into skeletal muscle. Mol Ther 2000;2(3):204–210.
106. Muramatsu T, Shibata O, Ryoki S, Ohmori Y, Okumura J. Foreign gene expression in the mouse testis by localized in vivo gene transfer. Biochem Biophys Res Commun 1997;233(1):45–49.

107. Espinos E, Liu JH, Bader CR, Bernheim L. Efficient non-viral DNA-mediated gene transfer to human primary myoblasts using electroporation. Neuromuscul Disord 2001;11(4):341–349.

108. Tsujie M, Isaka Y, Nakamura H, Imai E, Hori M. Electroporation-mediated gene transfer that targets glomeruli. J Am Soc Nephrol 2001;12(5):949–954.

109. Tabata H, Nakajima K. Efficient in utero gene transfer system to the developing mouse brain using electroporation: visualization of neuronal migration in the developing cortex. Neuroscience 2001;103(4):865–872.

110. Li LH, McCarthy P, Hui SW. High-efficiency electrotransfection of human primary hematopoietic stem cells. FASEB J. 2001Mar; 15(3):586-588.

111. Matsubara H, Maeda T, Gunji Y, Koide Y, Asano T, Ochiai T, Sakiyama S, Tagawa M. Combinatory anti-tumor effects of electroporation-mediated chemotherapy and wild-type p53 gene transfer to human esophageal cancer cells. Int J Oncol 2001Apr; 18(4):825–829.

112. Rizzuto G, Cappelletti M, Mennuni C, Wiznerowicz M, DeMartis A, Maione D, Ciliberto G, La Monica N, Fattori E. Gene electrotransfer results in a high-level transduction of rat skeletal muscle and corrects anemia of renal failure. Hum Gene Ther 2000;11(13):1891–1900.

113. Momose T, Tonegawa A, Takeuchi J, Ogawa H, Umesono K, Yasuda K. Efficient targeting of gene expression in chick embryos by microelectroporation. Dev Growth Differ 1999;41(3):335–344.

114. Nakamura H, Watanabe Y, Funahashi J. Misexpression of genes in brain vesicles by in ovo electroporation. Dev Growth Differ 2000;42(3):199–201.

115. Takeuchi JK, Koshiba-Takeuchi K, Matsumoto K, Vogel-Hopker A, Naitoh-Matsuo M, Ogura K, Takahashi N, Yasuda K, Ogura T, Tbx5 Tbx4 genes determine the wing/leg identity of limb buds. Nature 1999;398(6730):810–814.

116. Larsen DL, Dybdahl-Sissoko N, McGregor MW, Drape R, Neumann V, Swain WF, Lunn DP, Olsen CW. Coadministration of DNA encoding interleukin-6 and hemagglutinin confers protection from influenza virus challenge in mice. J Virol 1998;72(2):1704–1708.

117. Fuller DH, Murphey-Corb M, Clements J, Barnett S, Haynes JR. Induction of immunodeficiency virus-specific immune responses in rhesus monkeys following gene gun-mediated DNA vaccination. J Med Primatol 1996;25(3):236–241.

118. Macklin MD, McCabe D, McGregor MW, Neumann V, Meyer T, Callan R, Hinshaw VS, Swain WF. Immunization of pigs with a particle-mediated DNA vaccine to influenza A virus protects against challenge with homologous virus. J Virol 1998;72(2):1491–1496.

119. Lunn DP, Soboll G, Schram BR, Quass J, McGregor MW, Drape RJ, Macklin MD, McCabe DE, Swain WF, Olsen CW. Antibody responses to DNA vaccination of horses using the influenza virus hemagglutinin gene. Vaccine 1999;17(18):2245–2258.

120. Rakhmilevich AL, Turner J, Ford MJ, McCabe D, Sun WH, Sondel PM, Grota K, Yang NS. Gene gun-mediated skin transfection with interleukin 12 gene results in regression of established primary and metastatic murine tumors. Proc Natl Acad Sci USA 1996;93(13):6291–6296.

121. Petrie NC, Yao F, Eriksson E. Gene therapy in wound healing. Surg Clin North Am 2003;83(3):597–1616, vii.

122. Benn SI, Whitsitt JS, Broadley KN, Nanney LB, Perkins D, He L, Patel M, Morgan JR, Swain WF, Davidson JM. Particle-mediated gene transfer with transforming growth factor-beta1 cDNAs enhances wound repair in rat skin. J Clin Invest 1996;98(12):2894–2902.

123. Andree C, Swain WF, Page CP, Macklin MD, Slama J, Hatzis D, Eriksson E. In vivo transfer and expression of a human epidermal growth factor gene accelerates wound repair. Proc Natl Acad Sci USA 1994;91(25):12188–12192.

124. Roberts LK, Barr LJ, Fuller DH, McMahon CW, Leese PT, Jones S. Clinical safety and efficacy of a powdered Hepatitis B nucleic acid vaccine delivered to the epidermis by a commercial prototype device. Vaccine 2005;23(40):4867–4878.

125. Drape RJ, Macklin MD, Barr LJ, Jones S, Haynes JR, Dean HJ. Epidermal DNA vaccine for influenza is immunogenic in humans. Vaccine 2006;24(21):4475–4481.

126. Cassaday RD, Sondel PM, King DM, Macklin MD, Gan J, Warner TF, Zuleger CL, Bridges AJ, Schalch HG, Kim KM, Hank JA, Mahvi DM, Albertini MR. A phase I study of immunization using particle-mediated epidermal delivery of genes for gp100 and GM-CSF into uninvolved skin of melanoma patients. Clin Cancer Res 2007;13(2 Pt 1):540–549.

127. Klein RM, Wolf ED, Wu R, Sanford JC. High-velocity microprojectiles for delivering nucleic

acids into living cells, 1987. Biotechnology 1992;24:384–386.
128. Eisenbraun MD, Fuller DH, Haynes JR. Examination of parameters affecting the elicitation of humoral immune responses by particle bombardment-mediated genetic immunization. DNA Cell Biol 1993;12(9):791–797.
129. Lin MT, Pulkkinen L, Uitto J, Yoon K. The gene gun: current applications in cutaneous gene therapy. Int J Dermatol 2000;39(3):161–170.
130. Wang S, Joshi S, Lu S. Delivery of DNA to skin by particle bombardment. Methods Mol Biol 2004;245:185–196.
131. Alvarez D, Harder G, Fattouh R, Sun J, Goncharova S, Stampfli MR, Coyle AJ, Bramson JL, Jordana M. Cutaneous antigen priming via gene gun leads to skin-selective Th2 immune-inflammatory responses. J Immunol 2005;174(3):1664–1674.
132. Muangmoonchai R, Wong SC, Smirlis D, Phillips IR, Shephardl EA. Transfection of liver in vivo by biolistic particle delivery: its use in the investigation of cytochrome P450 gene regulation. Mol Biotechnol 2002;20(2):145–151.
133. Kuriyama S, Mitoro A, Tsujinoue H, Nakatani T, Yoshiji H, Tsujimoto T, Yamazaki M, Fukui H. Particle-mediated gene transfer into murine livers using a newly developed gene gun. Gene Ther 2000;7(13):1132–1136.
134. Sato H, Hattori S, Kawamoto S, Kudoh I, Hayashi A, Yamamoto I, Yoshinari M, Minami M, Kanno H. In vivo gene gun-mediated DNA delivery into rodent brain tissue. Biochem Biophys Res Commun 2000;270(1):163–170.
135. Taniyama Y, Tachibana K, Hiraoka K, Aoki M, Yamamoto S, Matsumoto K, Nakamura T, Ogihara T, Kaneda Y, Morishita R. Development of safe and efficient novel nonviral gene transfer using ultrasound: enhancement of transfection efficiency of naked plasmid DNA in skeletal muscle. Gene Ther 2002;9(6):372–380.
136. Lu QL, Liang HD, Partridge T, Blomley MJ. Microbubble ultrasound improves the efficiency of gene transduction in skeletal muscle in vivo with reduced tissue damage. Gene Ther 2003;10(5):396–405.
137. Chen S, Shohet RV, Bekeredjian R, Frenkel P, Grayburn PA. Optimization of ultrasound parameters for cardiac gene delivery of adenoviral or plasmid deoxyribonucleic acid by ultrasound-targeted microbubble destruction. J Am Coll Cardiol 2003;42(2):301–308.
138. Taniyama Y, Tachibana K, Hiraoka K, Namba T, Yamasaki K, Hashiya N, Aoki M, Ogihara T, Yasufumi K, Morishita R. Local delivery of plasmid DNA into rat carotid artery using ultrasound. Circulation 2002;105(10): 1233–1239.
139. Miller DL, Song J. Tumor growth reduction and DNA transfer by cavitation-enhanced high-intensity focused ultrasound in vivo. Ultrasound Med Biol 2003;29(6):887–893.
140. Kim HJ, Greenleaf JF, Kinnick RR, Bronk JT, Bolander ME. Ultrasound-mediated transfection of mammalian cells. Hum Gene Ther 1996;7(11):1339–1346.
141. Lin AJ, Slack NL, Ahmad A, Koltover I, George CX, Samuel CE, Safinya CR. Structure and structure–function studies of lipid/plasmid DNA complexes. J Drug Target 2000;8 (1):13–27.
142. Huber PE, Jenne J, Debus J, Wannenmacher MF, Pfisterer P. A comparison of shock wave and sinusoidal-focused ultrasound-induced localized transfection of HeLa cells. Ultrasound Med Biol 1999;25(9):1451–1457.
143. Nozaki T, Ogawa R, Feril LB Jr, Kagiya G, Fuse H, Kondo T. Enhancement of ultrasound-mediated gene transfection by membrane modification. J Gene Med 2003;5(12): 1046–1055.
144. Ogawa R, Kagiya G, Feril LB Jr, Nakaya N, Nozaki T, Fuse H, Kondo T. Ultrasound mediated intravesical transfection enhanced by treatment with lidocaine or heat. J Urol 2004;172(4 Pt 1):1469–1473.
145. Koch S, Pohl P, Cobet U, Rainov NG. Ultrasound enhancement of liposome-mediated cell transfection is caused by cavitation effects. Ultrasound Med Biol 2000;26(5):897–903.
146. Anwer K, Kao G, Proctor B, Anscombe I, Florack V, Earls R, Wilson E, McCreery T, Unger E, Rolland A, Sullivan SM. Ultrasound enhancement of cationic lipid-mediated gene transfer to primary tumors following systemic administration. Gene Ther 2000;7(21):1833–1839.
147. Unger EC, Hersh E, Vannan M, Matsunaga TO, McCreery T. Local drug and gene delivery through microbubbles. Prog Cardiovasc Dis 2001;44(1):45–54.
148. Magin-Lachmann C, Kotzamanis G, D'Aiuto L, Cooke H, Huxley C, Wagner E. In vitro and in vivo delivery of intact BAC DNA: comparison of different methods. J Gene Med 2004;6 (2):195–209.
149. Miao CH, Thompson AR, Loeb K, Ye X. Long-term and therapeutic-level hepatic gene expression of human factor IX after naked plasmid transfer in vivo. Mol Ther 2001;3 (6):947–957.

150. Miao CH, Ye X, Thompson AR. High-level factor VIII gene expression *in vivo* achieved by nonviral liver-specific gene therapy vectors. Hum Gene Ther 2003;14(14):1297–1305.

151. Maruyama H, Higuchi N, Kameda S, Miyazaki J, Gejyo F. Rat liver-targeted naked plasmid DNA transfer by tail vein injection. Mol Biotechnol 2004;26(2):165–172.

152. Al-Dosari MS, Knapp JE, Liu D. Hydrodynamic delivery. Adv Genet 2005;54:65–82. Kamimura K, Suda T, Xu W, Zhang G, Liu D. Image-guided, lobe-specific hydrodynamic gene delivery to swine liver. Mol Ther 2009;17 (3):491–499.

153. Kurata S, Tsukakoshi M, Kasuya T, Ikawa Y. The laser method for efficient introduction of foreign DNA into cultured cells. Exp Cell Res 1986;162(2):372–378.

154. Shirahata Y, Ohkohchi N, Itagak H, Satomi S. New technique for gene transfection using laser irradiation. J Investig Med 2001;49 (2):184–190.

155. Palumbo G, Caruso M, Crescenzi E, Tecce MF, Roberti G, Colasanti A. Targeted gene transfer in eucaryotic cells by dye-assisted laser optoporation. J Photochem Photobiol B 1996;36 (1):41–46.

156. Zeira E, Manevitch A, Khatchatouriants A, Pappo O, Hyam E, Darash-Yahana M, Tavor E, Honigman A, Lewis A, Galun E. Femtosecond infrared laser-an efficient and safe *in vivo* gene delivery system for prolonged expression. Mol Ther 2003;8(2):342–350.

157. Turner C, Weir N, Catterall C, Baker TS, Jones MN. The transfection of Jurkat T-leukemic cells by use of pH-sensitive immunoliposomes. J Liposome Res 2002;12(1-2):45–50.

158. Felgner PL, Gadek TR, Holm M, Roman R, Chan HW, Wenz M, Northrop JP, Ringold GM, Danielsen M. Lipofection: a highly efficient, lipid-mediated DNA-transfection procedure. Proc Natl Acad Sci USA 1987;84(21):7413–7417.

159. Karmali PP, Chaudhuri A. Cationic liposomes as non-viral carriers of gene medicines: resolved issues, open questions, and future promises. Med Res Rev 2007;27(5):696–722.

160. Reddy BS, Banerjee R. 17Beta-estradiol-associated stealth-liposomal delivery of anticancer gene to breast cancer cells. Angew Chem Int Ed Engl 2005;44(41):6723–6727.

161. Mukherjee A, Prasad TK, Rao NM, Banerjee R. Haloperidol-associated stealth liposomes: a potent carrier for delivering genes to human breast cancer cells. J Biol Chem 2005;280 (16):15619–15627.

162. Mounkes LC, Zhong W, Cipres-Palacin G, Heath TD, Debs RJ. Proteoglycans mediate cationic liposome-DNA complex-based gene delivery *in vitro* and *in vivo*. J Biol Chem 1998;273(40):26164–26170.

163. Majeti BK, Karmali PP, Reddy BS, Chaudhuri A. *In vitro* gene transfer efficacies of *N,N*-dialkylpyrrolidinium chlorides: a structure–activity investigation J Med Chem 2005;48 (11):3784–3795.

164. Fraley R, Subramani S, Berg P, Papahadjopoulos D. Introduction of liposome-encapsulated SV40 DNA into cells. J Biol Chem 1980;255(21):10431–10435.

165. Wang CY, Huang L. pH-sensitive immunoliposomes mediate target-cell-specific delivery and controlled expression of a foreign gene in mouse. Proc Natl Acad Sci USA 1987;84 (22):7851–7855.

166. Shi N, Pardridge WM. Noninvasive gene targeting to the brain. Proc Natl Acad Sci USA 2000;97(13):7567–7572.

167. Tatum EL. Molecular biology, nucleic acids, and the future of medicine. Perspect Biol Med 1966;10(1):19–32.

168. Sinsheimer RL. The prospect for designed genetic change. Am Sci 1969;57(1):134–142.

169. Davis BD. Prospects for genetic intervention in man. Science 1970;170(964):1279–1283.

170. Gaucheron J, Santaella C, Vierling P. Highly fluorinated lipospermines for gene transfer: synthesis and evaluation of their *in vitro* transfection efficiency. Bioconjug Chem 2001;12 (1):114–128.

171. Miller AD. Cationic liposome systems in gene therapy. IDrugs 1998;1(5):574–583. Kumar VV, Singh RS, Chaudhuri A. Cationic transfection lipids in gene therapy: successes, setbacks, challenges and promises. Curr Med Chem 2003;10(14):1297–1306.

172. Singh RS, Mukherjee K, Banerjee R, Chaudhuri A, Hait SK, Moulik SP, Ramadas Y, Vijayalakshmi A, Rao NM. Anchor dependency for non-glycerol based cationic lipofectins: mixed bag of regular and anomalous transfection profiles. Chemistry 2002;8 (4):900–909. Banerjee R, Das PK, Srilakshmi GV, Chaudhuri A, Rao NM. Novel series of non-glycerol-based cationic transfection lipids for use in liposomal gene delivery. J Med Chem 1999;42(21):4292–4299.

173. MacDonald RC, Ashley GW, Shida MM, Rakhmanova VA, Tarahovsky YS, Pantazatos DP, Kennedy MT, Pozharski EV, Baker KA, Jones RD, Rosenzweig HS, Choi KL, Qiu R, McIntosh

TJ. Physical and biological properties of cationic triesters of phosphatidylcholine. Biophys J 1999;77(5):2612–2629.Mukherjee K, Sen J, Chaudhuri A. Common co-lipids, in synergy, impart high gene transfer properties to transfection-incompetent cationic lipids. FEBS Lett 2005;579(5):1291–1300.

174. Behr JP, Demeneix B, Loeffler JP, Perez- Mutul J. Efficient gene transfer into mammalian primary endocrine cells with lipopolyamine-coated DNA. Proc Natl Acad Sci USA 1989;86 (18):6982–6986.Byk G, Dubertret C, Escriou V, Frederic M, Jaslin G, Rangara R, Pitard B, Crouzet J, Wils P, Schwartz B, Scherman D. Synthesis, activity, and structure–activity relationship studies of novel cationic lipids for DNA transfer. J Med Chem 1998;41(2):229–235. Floch V, Bolch GL, Gable-Guillaume C, Bris NL, Yaouanc JJ, Abbayes HD, Ferec C, Clement JC. Phosphonolipids as non-viral vectors for gene therapy. Eur J Med Chem 1998;33(12):923–934. Wang J, Guo X, Xu Y, Barron L, Szoka FC Jr. Synthesis and characterization of long chain alkyl acyl carnitine esters, Potentially biodegradable cationic lipids for use in gene delivery. J Med Chem 1998;41(13):2207–2215.

175. Gao H, Hui KM. Synthesis of a novel series of cationic lipids that can act as efficient gene delivery vehicles through systematic heterocyclic substitution of cholesterol derivatives. Gene Ther 2001;8(11):855–863.

176. Solodin I, Brown CS, Bruno MS, Chow CY, Jang EH, Debs RJ, Heath TD. A novel series of amphiphilic imidazolinium compounds for *in vitro* and *in vivo* gene delivery. Biochemistry 1995;34(41):13537–13544.Zhang G, Gurtu V, Smith TH, Nelson P, Kain SR. A cationic lipid for rapid and efficient delivery of plasmid DNA into mammalian cells. Biochem Biophys Res Commun 1997;236(1):126–129.

177. Felgner JH, Kumar R, Sridhar CN, Wheeler CJ, Tsai YJ, Border R, Ramsey P, Martin M, Felgner PL. Enhanced gene delivery and mechanism studies with a novel series of cationic lipid formulations. J Biol Chem 1994;269(4):2550–2561.

178. Aberle AM, Tablin F, Zhu J, Walker NJ, Gruenert DC, Nantz MH. A novel tetraester construct that reduces cationic lipid-associated cytotoxicity. Implications for the onset of cytotoxicity Biochemistry 1998;37(18):6533–6540.

179. Cameron FH, Moghaddam MJ, Bender VJ, Whittaker RG, Mott M, Lockett TJ. A transfection compound series based on a versatile Tris linkage. Biochim Biophys Acta 1999;1417(1):37–50.

180. Ren T, Zhang G, Song YK, Liu D. Synthesis and characterization of aromatic ring-based cationic lipids for gene delivery *in vitro* and *in vivo*. J Drug Target 1999;7(4):285–292.

181. Schulze U, Schmidt HW, Safinya CR. Synthesis of novel cationic poly(ethylene glycol) containing lipids. Bioconjug Chem 1999;10(3):548–552.

182. Zabner J, Fasbender AJ, Moninger T, Poellinger KA, Welsh MJ. Cellular and molecular barriers to gene transfer by a cationic lipid. J Biol Chem 1995;270(32):18997–19007.Xu Y, Szoka FC Jr. Mechanism of DNA release from cationic liposome/DNA complexes used in cell transfection. Biochemistry 1996;35(18): 5616–5623.

183. Wolff JA, Malone RW, Williams P, Chong W, Acsadi G, Jani A, Felgner PL. Direct gene transfer into mouse muscle *in vivo*. Science 1990;247(4949 Pt 1):1465–1468.Liu F, Shollenberger LM, Conwell CC, Yuan X, Huang L. Mechanism of naked DNA clearance after intravenous injection. J Gene Med 2007;9(7):613–619.

184. Sternberg B, Sorgi FL, Huang L. New structures in complex formation between DNA and cationic liposomes visualized by freeze-fracture electron microscopy. FEBS Lett 1994;356(2-3):361–366.Gustafsson J, Arvidson G, Karlsson G, Almgren M. Complexes between cationic liposomes and DNA visualized by cryo-TEM. Biochim Biophys Acta 1995;1235(2):305–312.Radler JO, Koltover I, Salditt T, Safinya CR. Structure of DNA-cationic liposome complexes: DNA intercalation in multilamellar membranes in distinct interhelical packing regimes. Science 1997;275(5301):810–814.

185. Farhood H, Serbina N, Huang L. The role of dioleoyl phosphatidylethanolamine in cationic liposome mediated gene transfer. Biochim Biophys Acta 1995;1235(2):289–295.

186. Budker V, Gurevich V, Hagstrom JE, Bortzov F, Wolff JA. pH-sensitive, cationic liposomes: a new synthetic virus-like vector. Nat Biotechnol 1996;14(6):760–764.

187. Zanta MA, Belguise-Valladier P, Behr JP. Gene delivery: a single nuclear localization signal peptide is sufficient to carry DNA to the cell nucleus. Proc Natl Acad Sci USA 1999;96(1):91–96.Ludtke JJ, Zhang G, Sebestyen MG, Wolff JA. A nuclear localization signal can enhance both the nuclear transport and expression of 1 kb DNA. J Cell Sci 1999;112(Pt 12):2033–2041.

188. Tseng WC, Haselton FR, Giorgio TD. Mitosis enhances transgene expression of plasmid delivered by cationic liposomes. Biochim Biophys Acta, 1999;1445(1):53–64. Mortimer I, Tam P, MacLachlan I, Graham RW, Saravolac EG, Joshi PB. Cationic lipid-mediated transfection of cells in culture requires mitotic activity. Gene Ther 1999;6(3):403–411.

189. He Y, Huang L. Growth inhibition of human papillomavirus 16 DNA-positive mouse tumor by antisense RNA transcribed from U6 promoter. Cancer Res 1997;57(18):3993–3999.

190. Nabel EG, Gordon D, Yang ZY, Xu L, San H, Plautz GE, Wu BY, Gao X, Huang L, Nabel GJ. Gene transfer *in vivo* with DNA-liposome complexes: lack of autoimmunity and gonadal localization. Hum Gene Ther 1992;3(6):649–656.

191. Nabel GJ, Gordon D, Bishop DK, Nickoloff BJ, Yang ZY, Aruga A, Cameron MJ, Nabel EG, Chang AE. Immune response in human melanoma after transfer of an allogeneic class I major histocompatibility complex gene with DNA-liposome complexes. Proc Natl Acad Sci USA 1996;93(26):15388–15393.

192. Mohr L, Yoon SK, Eastman SJ, Chu Q, Scheule RK, Scaglioni PP, Geissler M, Heintges T, Blum HE, Wands JR. Cationic liposome-mediated gene delivery to the liver and to hepatocellular carcinomas in mice. Hum Gene Ther 2001;12(7):799–809.

193. Parker SE, Khatibi S, Margalith M, Anderson D, Yankauckas M, Gromkowski SH, Latimer T, Lew D, Marquet M, Manthorpe M, Hobart P, Hersh E, Stopeck AT, Norman J. Plasmid DNA gene therapy: studies with the human interleukin-2 gene in tumor cells *in vitro* and in the murine B16 melanoma model *in vivo*. Cancer Gene Ther 1996;3(3):175–185.

194. Stewart MJ, Plautz GE. Del Buono L, Yang ZY, Xu L, Gao X, Huang L, Nabel EG. Nabel GJ. Gene transfer *in vivo* with DNA-liposome complexes: safety and acute toxicity in mice. Hum Gene Ther 1992;3(3):267–275.

195. Templeton NS, Lasic DD, Frederik PM, Strey HH, Roberts DD, Pavlakis GN. Improved DNA: liposome complexes for increased systemic delivery and gene expression. Nat Biotechnol 1997;15(7):647–652.

196. Zhu N, Liggitt D, Liu Y, Debs R. Systemic gene expression after intravenous DNA delivery into adult mice. Science 1993;261(5118):209–211.

197. Hwang SH, Hayashi K, Takayama K, Maitani Y. Liver-targeted gene transfer into a human hepatoblastoma cell line and *in vivo* by sterylglucoside-containing cationic liposomes. Gene Ther 2001;8(16):1276–1280.

198. Hong K, Zheng W, Baker A, Papahadjopoulos D. Stabilization of cationic liposome-plasmid DNA complexes by polyamines and poly(ethylene glycol)-phospholipid conjugates for efficient *in vivo* gene delivery. FEBS Lett 1997;400(2):233–237.

199. Uyechi LS, Gagne L, Thurston G. Szoka FC Jr. Mechanism of lipoplex gene delivery in mouse lung: binding and internalization of fluorescent lipid and DNA components. Gene Ther 2001;8(11):828–836.

200. San H, Yang ZY, Pompili VJ, Jaffe ML, Plautz GE, Xu L, Felgner JH, Wheeler CJ, Felgner PL, Gao X,et al. Safety and short-term toxicity of a novel cationic lipid formulation for human gene therapy. Hum Gene Ther 1993;4(6):781–788.

201. Li S, Huang L. *In vivo* gene transfer via intravenous administration of cationic lipid-protamine-DNA (LPD) complexes. Gene Ther 1997;4(9):891–900.

202. Liu F, Qi H, Huang L, Liu D. Factors controlling the efficiency of cationic lipid-mediated transfection *in vivo* via intravenous administration. Gene Ther 1997;4(6):517–523.

203. Sakurai F, Nishioka T, Saito H, Baba T, Okuda A, Matsumoto O, Taga T, Yamashita F, Takakura Y, Hashida M. Interaction between DNA-cationic liposome complexes and erythrocytes is an important factor in systemic gene transfer via the intravenous route in mice: the role of the neutral helper lipid. Gene Ther 2001;8(9):677–686.

204. Meyer O, Schughart K, Pavirani A, Kolbe HV. Multiple systemic expression of human interferon-beta in mice can be achieved upon repeated administration of optimized pcTG90-lipoplex. Gene Ther 2000;7(18):1606–1611.

205. Leventis R, Silvius JR. Interactions of mammalian cells with lipid dispersions containing novel metabolizable cationic amphiphiles. Biochim Biophys Acta 1990;1023(1):124–132.

206. Middleton PG, Caplen NJ, Gao X, Huang L, Gaya H, Geddes DM, Alton EW. Nasal application of the cationic liposome DC-Chol:DOPE does not alter ion transport, lung function or bacterial growth. Eur Respir J 1994;7(3):442–445.

207. Zhang Y, Jiang Q, Dudus L, Yankaskas JR, Engelhardt JF. Vector-specific complementation profiles of two independent primary defects in cystic fibrosis airways. Hum Gen Ther 1998;9(5):635–648. Wu CH, Walton CM,

Wu GY. Targeted gene transfer to liver using protein–DNA complexes. Methods Mol Med 2002;69:15–23.
208. Caplen NJ, Alton EW, Middleton PG, Dorin JR, Stevenson BJ, Gao X, Durham SR, Jeffery PK, Hodson ME, Coutelle C, et al. Liposome-mediated CFTR gene transfer to the nasal epithelium of patients with cystic fibrosis. Nat Med 1995;1(1):39–46.
209. Porteous DJ, Dorin JR, McLachlan G, Davidson-Smith H, Davidson H, Stevenson BJ, Carothers AD, Wallace WA, Moralee S, Hoenes C, Kallmeyer G, Michaelis U, Naujoks K, Ho LP, Samways JM, Imrie M, Greening AP, Innes JA. Evidence for safety and efficacy of DOTAP cationic liposome mediated CFTR gene transfer to the nasal epithelium of patients with cystic fibrosis. Gene Ther 1997;4(3):210–218.
210. Gill DR, Southern KW, Mofford KA, Seddon T, Huang L, Sorgi F, Thomson A, MacVinish LJ, Ratcliff R, Bilton D, Lane DJ, Littlewood JM, Webb AK, Middleton PG, Colledge WH, Cuthbert AW, Evans MJ, Higgins CF, Hyde SC. A placebo-controlled study of liposome-mediated gene transfer to the nasal epithelium of patients with cystic fibrosis. Gene Ther 1997;4(3):199–209.
211. Chadwick SL, Kingston HD, Stern M, Cook RM, O'Connor BJ, Lukasson M, Balfour RP, Rosenberg M, Cheng SH, Smith AE, Meeker DP, Geddes DM, Alton EW. Safety of a single aerosol administration of escalating doses of the cationic lipid GL-67/DOPE/DMPE-PEG5000 formulation to the lungs of normal volunteers. Gene Ther 1997;4(9):937–942.
212. Zabner J, Cheng SH, Meeker D, Launspach J, Balfour R, Perricone MA, Morris JE, Marshall J, Fasbender A, Smith AE, Welsh MJ. Comparison of DNA-lipid complexes and DNA alone for gene transfer to cystic fibrosis airway epithelia *in vivo*. J Clin Invest 1997;100(6):1529–1537.
213. Alton EW, Middleton PG, Caplen NJ, Smith SN, Steel DM, Munkonge FM, Jeffery PK, Geddes DM, Hart SL, Williamson R, et al. Non-invasive liposome-mediated gene delivery can correct the ion transport defect in cystic fibrosis mutant mice. Nat Genet 1993;5(2):135–142.
214. Yoshimura K, Rosenfeld MA, Nakamura H, Scherer EM, Pavirani A, Lecocq JP, Crystal RG. Expression of the human cystic fibrosis transmembrane conductance regulator gene in the mouse lung after *in vivo* intratracheal plasmid-mediated gene transfer. Nucleic Acids Res 1992;20(12):3233–3240.
215. Zhang Y, Jiang Q, Dudus L, Yankaskas JR, Engelhardt JF. Vector-specific complementation profiles of two independent primary defects in cystic fibrosis airways. Hum Gene Ther 1998;9(5):635–648.
216. Yu D, Matin A, Xia W, Sorgi F, Huang L, Hung MC. Liposome-mediated *in vivo* E1A gene transfer suppressed dissemination of ovarian cancer cells that overexpress HER-2/neu. Oncogene 1995;11(7):1383–1388.
217. Wagner E, Zenke M, Cotten M, Beug H, Birnstiel ML. Transferrin-polycation conjugates as carriers for DNA uptake into cells. Proc Natl Acad Sci USA 1990;87(9):3410–3414. Goula D, Benoist C, Mantero S, Merlo G, Levi G, Demeneix BA. Polyethylenimine-based intravenous delivery of transgenes to mouse lung. Gene Ther 1998;5(9):1291–1295. Toncheva V, Wolfert MA, Dash PR, Oupicky D, Ulbrich K, Seymour LW, Schacht EH. Novel vectors for gene delivery formed by self-assembly of DNA with poly(L-lysine) grafted with hydrophilic polymers. Biochim Biophys Acta 1998;1380(3):354–368. Aral C, Akbuga J. Preparation and *in vitro* transfection efficiency of chitosan microspheres containing plasmid DNA:poly(L-lysine) complexes. J Pharm Pharm Sci 2003;6(3):321–326. Leong KW, Mao HQ, Truong-Le VL, Roy K, Walsh SM, August JT. DNA-polycation nanospheres as non-viral gene delivery vehicles. J Control Release 1998;53(1-3):183–193.
218. Erbacher P, Remy JS, Behr JP. Gene transfer with synthetic virus-like particles via the integrin-mediated endocytosis pathway. Gene Ther 1999;6(1):138–145.
219. Jayakumar R, Chennazhi KP, Muzzarelli RAA, Tamura H, Nair SV, Selvamurugan N. Chitosan conjugated DNA nanoparticles in gene therapy. Carbohydr Polym 2010;79(1):1–8.
220. Mansouri S, Lavigne P, Corsi K, Benderdour M, Beaumont E, Fernandes JC. Chitosan-DNA nanoparticles as non-viral vectors in gene therapy: strategies to improve transfection efficacy. Eur J Pharm Biopharm 2004;57(1):1–8.
221. Sano A, Maeda M, Nagahara S, Ochiya T, Honma K, Itoh H, Miyata T, Fujioka K. Atelocollagen for protein and gene delivery. Adv Drug Deliv Rev 2003;55(12):1651–1677.
222. Ochiya T, Nagahara S, Sano A, Itoh H, Terada M. Biomaterials for gene delivery:

atelocollagen-mediated controlled release of molecular medicines. Curr Gene Ther 2001; 1(1):31–52.
223. Hashida M, Takemura S, Nishikawa M, Takakura Y. Targeted delivery of plasmid DNA complexed with galactosylated poly(L-lysine). J Control Release 1998;53(1-3):301–310.
224. Oupicky D, Howard KA, Konak C, Dash PR, Ulbrich K, Seymour LW. Steric stabilization of poly-L-Lysine/DNA complexes by the covalent attachment of semitelechelic poly[N-(2-hydroxypropyl)methacrylamide]. Bioconjug Chem 2000;11(4):492–501.
225. Boussif O, Lezoualc'h F, Zanta MA, Mergny MD, Scherman D, Demeneix B, Behr JP. A versatile vector for gene and oligonucleotide transfer into cells in culture and in vivo: polyethylenimine. Proc Natl Acad Sci USA 1995;92 (16):7297–7301.
226. Kunath K, von Harpe A, Petersen H, Fischer D, Voigt K, Kissel T, Bickel U. The structure of PEG-modified poly(ethylene imines) influences biodistribution and pharmacokinetics of their complexes with NF-kappaB decoy in mice. Pharm Res 2002; 19(6):810–817.
227. Kukowska-Latallo JF, Bielinska AU, Johnson J, Spindler R, Tomalia DA, Baker JR Jr. Efficient transfer of genetic material into mammalian cells using Starburst polyamidoamine dendrimers. Proc Natl Acad Sci USA 1996;93 (10):4897–4902.
228. Bielinska AU, Yen A, Wu HL, Zahos KM, Sun R, Weiner ND, Baker JR Jr, Roessler BJ. Application of membrane-based dendrimer/DNA complexes for solid phase transfection in vitro and in vivo. Biomaterials 2000;21 (9):877–887.
229. Ratner BD, Bryant SJ. Biomaterials: where we have been and where we are going. Annu Rev Biomed Eng 2004;6:41–75.
230. Borchard G. Chitosans for gene delivery. Adv Drug Deliv Rev 2001;52(2):145–150.
231. Rettig GR, Rice KG. Non-viral gene delivery: from the needle to the nucleus. Expert Opin Biol Ther 2007;7(6):799–808.
232. von Harpe A, Petersen H, Li Y, Kissel T. Characterization of commercially available and synthesized polyethylenimines for gene delivery. J Control Release 2000;69(2): 309–322.
233. Kichler A, Leborgne C, Coeytaux E, Danos O. Polyethylenimine-mediated gene delivery: a mechanistic study. J Gene Med 2001;3 (2):135–144.
234. Sonawane ND, Szoka FC Jr, Verkman AS. Chloride accumulation and swelling in endosomes enhances DNA transfer by polyamine-DNA polyplexes. J Biol Chem 2003;278 (45):44826–44831.
235. Morimoto K, Nishikawa M, Kawakami S, Nakano T, Hattori Y, Fumoto S, Yamashita F, Hashida M. Molecular weight-dependent gene transfection activity of unmodified and galactosylated polyethyleneimine on hepatoma cells and mouse liver. Mol Ther 2003;7(2):254–261.
236. Godbey WT, Barry MA, Saggau P, Wu KK, Mikos AG. Poly(ethylenimine)-mediated transfection: a new paradigm for gene delivery. J Biomed Mater Res 2000;51(3):321–328.
237. Behr JP. The proton sponge: a trick to enter cells the viruses did not exploit. CHIMIA Int J Chem 1997;51(2):34–36.
238. Midoux P, Pichon C, Yaouanc JJ, Jaffres PA. Chemical vectors for gene delivery: a current review on polymers, peptides and lipids containing histidine or imidazole as nucleic acids carriers. Br J Pharmacol 2009;157(2):166–178.
239. Sonawane ND, Thiagarajah JR, Verkman AS. Chloride concentration in endosomes measured using a ratioable fluorescent Cl- indicator: evidence for chloride accumulation during acidification. J Biol Chem 2002;277(7):5506–5513.
240. Bieber T, Meissner W, Kostin S, Niemann A, Elsasser HP. Intracellular route and transcriptional competence of polyethylenimine-DNA complexes. J Control Release 2002;82 (2-3):441–454.
241. Akinc A, Thomas M, Klibanov AM, Langer R. Exploring polyethylenimine-mediated DNA transfection and the proton sponge hypothesis. J Gene Med 2005;7(5):657–663.
242. Kircheis R, Kichler A, Wallner G, Kursa M, Ogris M, Felzmann T, Buchberger M, Wagner E. Coupling of cell-binding ligands to polyethylenimine for targeted gene delivery. Gene Ther 1997;4(5):409–418.
243. Blessing T, Kursa M, Holzhauser R, Kircheis R, Wagner E. Different strategies for formation of pegylated EGF-conjugated PEI/DNA complexes for targeted gene delivery. Bioconjug Chem 2001;12(4):529–537.Thomas M, Lu JJ, Ge Q, Zhang C, Chen J, Klibanov AM. Full deacylation of polyethylenimine dramatically boosts its gene delivery efficiency and specificity to mouse lung. Proc Natl Acad Sci USA 2005;102(16):5679–5684.Kloeckner J, Bruzzano S, Ogris M, Wagner E. Gene carriers based on hexanediol diacrylate linked oligoethylenimine: effect of chemical structure of polymer on

244. Tang MX, Redemann CT, Szoka FC Jr. In vitro gene delivery by degraded polyamidoamine dendrimers. Bioconjug Chem 1996;7(6):703–714.

245. Hudde T, Rayner SA, Comer RM, Weber M, Isaacs JD, Waldmann H, Larkin DF, George AJ. Activated polyamidoamine dendrimers, a non-viral vector for gene transfer to the corneal endothelium. Gene Ther 1999;6(5):939–943.

246. Qin L, Pahud DR, Ding Y, Bielinska AU, Kukowska-Latallo JF, Baker JR Jr, Bromberg JS. Efficient transfer of genes into murine cardiac grafts by Starburst polyamidoamine dendrimers. Hum Gene Ther 1998;9(4):553–560.

247. Louise C. Nonviral vectors. Methods Mol Biol 2006;333:201–226.

248. Lim Y, Choi YH, Park JS, JS. A self-destroying polycationic polymer: biodegradable poly(4-hydroxy-L-proline ester). J Am Chem Soc 1999;121(24):5633–5639.

249. Lim Y, Kim CH, Kim K, Kim SW, Park JS. Development of a safe gene delivery system using biodegradable polymer, poly[α-(4-aminobutyl)-L-glycolic acid]. J Am Chem Soc 2000;122(27):6524–6525.

250. Lim YB, Han SO, Kong HU, Lee Y, Park JS, Jeong B, Kim SW. Biodegradable polyester, poly[alpha-(4-aminobutyl)-L-glycolic acid], as a non-toxic gene carrier. Pharm Res 2000;17(7):811–816.

251. Nitta Y, Tashiro F, Tokui M, Shimada A, Takei I, Tabayashi K, Miyazaki J. Systemic delivery of interleukin 10 by intramuscular injection of expression plasmid DNA prevents autoimmune diabetes in nonobese diabetic mice. Hum Gene Ther 1998;9(12):1701–1707.

252. Koh JJ, Ko KS, Lee M, Han S, Park JS, Kim SW. Degradable polymeric carrier for the delivery of IL-10 plasmid DNA to prevent autoimmune insulitis of NOD mice. Gene Ther 2000;7(24):2099–2104.

253. Lee M, Koh JJ, Han SO, Ko KS, Ki SW. Prevention of autoimmune insulitis by delivery of interleukin-4 plasmid using a soluble and biodegradable polymeric carrier. Pharm Res 2002;19(3):246–249.

254. Wang J, Zhang PC, Lu HF, Ma N, Wang S, Mao HQ, Leong KW. New polyphosphoramidate with a spermidine side chain as a gene carrier. J Control Release 2002;83(1):157–168.

255. Wang J, Mao HQ, Leong KW. A novel biodegradable gene carrier based on polyphosphoester. J Am Chem Soc 2001;123(38):9480–9481.

256. Wang J, Huang SW, Zhang PC, Mao HQ, Leong KW. Effect of side-chain structures on gene transfer efficiency of biodegradable cationic polyphosphoesters. Int J Pharm 2003;265(1-2):75–84.

257. Gosselin MA, Guo W, Lee RJ. Efficient gene transfer using reversibly cross-linked low molecular weight polyethylenimine. Bioconjug Chem 2001;12(6):989–994.

258. Petersen H, Merdan T, Kunath K, Fischer D, Kissel T. Poly(ethylenimine-co-L-lactamide-co-succinamide): a biodegradable polyethylenimine derivative with an advantageous pH-dependent hydrolytic degradation for gene delivery. Bioconjug Chem 2002;13(4):812–821.

259. Forrest ML, Koerber JT, Pack DW. A degradable polyethylenimine derivative with low toxicity for highly efficient gene delivery. Bioconjug Chem 2003;14(5):934–940.

260. Park MR, Han KO, Han IK, Cho MH, Nah JW, Choi YJ, Cho CS. Degradable polyethylenimine-alt-poly(ethylene glycol) copolymers as novel gene carriers. J Control Release 2005;105(3):367–380.

261. Arote R, Kim TH, Kim YK, Hwang SK, Jiang HL, Song HH, Nah JW, Cho MH, Cho CS. A biodegradable poly(ester amine) based on polycaprolactone and polyethylenimine as a gene carrier. Biomaterials 2007;28(4):735–744.

262. Luten J, van Nostrum CF, De Smedt SC, Hennink WE. Biodegradable polymers as non-viral carriers for plasmid DNA delivery. J Control Release 2008;126(2):97–110.

263. Simeoni F, Morris MC, Heitz F, Divita G. Insight into the mechanism of the peptide-based gene delivery system MPG: implications for delivery of siRNA into mammalian cells. Nucleic Acids Res 2003;31(11):2717–2724. Sen J, Chaudhuri A. Gene transfer efficacies of novel cationic amphiphiles with alanine, beta-alanine, and serine head groups: a structure–activity investigation. Bioconjug Chem 2005;16(4):903–912. Choi HS, Kim HH, Yang JM, Shin S. An insight into the gene delivery mechanism of the arginine peptide system: role of the peptide/DNA complex size. Biochim Biophys Acta 2006;1760(11):1604–1612.

264. Mahato RI. Non-viral peptide-based approaches to gene delivery. J Drug Target 1999;7(4):249–268.

265. Herweijer H, Wolff JA. Gene therapy progress and prospects: hydrodynamic gene delivery. Gene Ther 2007;14(2):99–107.

266. Goula D, Becker N, Lemkine GF, Normandie P, Rodrigues J, Mantero S, Levi G, Demeneix

BA. Rapid crossing of the pulmonary endothelial barrier by polyethylenimine/DNA complexes. Gene Ther 2000;7(6):499–504.
267. Schatzlein AG. Targeting of synthetic gene delivery systems. J Biomed Biotechnol 2003;2003(2):149–158.
268. Caplen NJ, Alton EW, Middleton PG, Dorin JR, Stevenson BJ, Gao X, Durham SR, Jeffery PK, Hodson ME, Coutelle C, et al. Liposome-mediated CFTR gene transfer to the nasal epithelium of patients with cystic fibrosis. Nat Med 1995;1(1):39–46.
269. Stephan DJ, Yang ZY, San H, Simari RD, Wheeler CJ, Felgner PL, Gordon D, Nabel GJ, Nabel EG. A new cationic liposome DNA complex enhances the efficiency of arterial gene transfer *in vivo*. Hum Gene Ther 1996;7(15):1803–1812.
270. Wheeler CJ, Felgner PL, Tsai YJ, Marshall J, Sukhu L, Doh SG, Hartikka J, Nietupski J, Manthorpe M, Nichols M, Plewe M, Liang X, Norman J, Smith A, Cheng SH. A novel cationic lipid greatly enhances plasmid DNA delivery and expression in mouse lung. Proc Natl Acad Sci USA 1996;93(21):11454–11459.
271. Gottschalk S, Cristiano RJ, Smith LC, Woo SL. Folate receptor mediated DNA delivery into tumor cells: potosomal disruption results in enhanced gene expression. Gene Ther 1994;1(3):185–191. Flotte TR, Laube BL. Gene therapy in cystic fibrosis. Chest 2001;120(Suppl 3):124S–131S.
272. Figlin RA, Parker SE, Horton HM. Technology evaluation: interleukin-2 gene therapy for the treatment of renal cell carcinoma. Curr Opin Mol Ther 1999;1(2):271–278.
273. Stopeck AT, Hersh EM, Akporiaye ET, Harris DT, Grogan T, Unger E, Warneke J, Schluter SF, Stahl S. Phase I study of direct gene transfer of an allogeneic histocompatibility antigen, HLA-B7, in patients with metastatic melanoma. J Clin Oncol 1997;15(1):341–349.
274. Rubin J, Galanis E, Pitot HC, Richardson RL, Burch PA, Charboneau JW, Reading CC, Lewis BD, Stahl S, Akporiaye ET, Harris DT. Phase I study of immunotherapy of hepatic metastases of colorectal carcinoma by direct gene transfer of an allogeneic histocompatibility antigen. HLA-B7 Gene Ther 1997;4(5):419–425.
275. Soriano P, Dijkstra J, Legrand A, Spanjer H, Londos-Gagliardi D, Roerdink F, Scherphof G, Nicolau C. Targeted and nontargeted liposomes for *in vivo* transfer to rat liver cells of a plasmid containing the preproinsulin I gene. Proc Natl Acad Sci USA 1983;80(23):7128–7131.
276. Campbell RB, Fukumura D, Brown EB, Mazzola LM, Izumi Y, Jain RK, Torchilin VP, Munn LL. Cationic charge determines the distribution of liposomes between the vascular and extravascular compartments of tumors. Cancer Res 2002;62(23):6831–6836.
277. Meyer O, Kirpotin D, Hong K, Sternberg B, Park JW, Woodle MC, Papahadjopoulos D. Cationic liposomes coated with polyethylene glycol as carriers for oligonucleotides. J Biol Chem 1998;273(25):15621–15627.
278. Xu L, Pirollo KF, Chang EH. Transferrin-liposome-mediated p53 sensitization of squamous cell carcinoma of the head and neck to radiation *in vitro*. Hum Gene Ther 1997;8(4):467–475.
279. Xu L, Pirollo KF, Tang WH, Rait A, Chang EH. Transferrin-liposome-mediated systemic p53 gene therapy in combination with radiation results in regression of human head and neck cancer xenografts. Hum Gene Ther 1999;10(18):2941–2952.
280. Xu L, Tang WH, Huang CC, Alexander W, Xiang LM, Pirollo KF, Rait A, Chang EH. Systemic p53 gene therapy of cancer with immunolipoplexes targeted by anti-transferrin receptor scFv. Mol Med 2001;7(10): 723–734.
281. Dachs GU, Dougherty GJ, Stratford IJ, Chaplin DJ. Targeting gene therapy to cancer: a review. Oncol Res 1997;9(6-7):313–325.
282. Gao X, Huang L. Potentiation of cationic liposome-mediated gene delivery by polycations. Biochemistry 1996;35(3):1027–1036.
283. Sudimack J, Lee RJ. Targeted drug delivery via the folate receptor. Adv Drug Deliv Rev 200041(2):147–162. Zhao XB, Lee RJ. Tumor-selective targeted delivery of genes and antisense oligodeoxyribonucleotides via the folate receptor. Adv Drug Deliv Rev 200456(8):1193–1204. Gosselin M, Lee RJ Biotech Ann Rev 2002;8:103–131.
284. Lee RJ, Low PS. Delivery of liposomes into cultured KB cells via folate receptor-mediated endocytosis. J Biol Chem 1994;269(5): 3198–3204.
285. Gao X, Huang L. Potentiation of cationic liposome-mediated gene delivery by polycations. Biochemistry 1996;35(3):1027–1036.
286. Sorgi FL, Bhattacharya S, Huang L. Protamine sulfate enhances lipid-mediated gene transfer. Gene Ther 1997;4(9):961–968.
287. Lee RJ, Huang L. Folate-targeted, anionic liposome-entrapped polylysine-condensed DNA

287. for tumor cell-specific gene transfer. J Biol Chem 1996;271(14):8481–8487.
288. Cui Z, Huang L. Liposome-polycation- DNA (LPD) particle as a carrier and adjuvant for protein-based vaccines: therapeutic effect against cervical cancer. Cancer Immunol Immunother 2005;54(12): 1180–1190.
289. Siomi H, Siomi MC. On the road to reading the RNA-interference code. Nature 2009;457 (7228):396–404.
290. Hannon GJ. RNA interference. Nature 2002;418(6894):244–251.
291. Meister G, Tuschl T. Mechanisms of gene silencing by double-stranded RNA. Nature 2004;431(7006):343–349.
292. Fire A, Xu S, Montgomery MK, Kostas SA, Driver SE, Mello CC. Potent and specific genetic interference by double-stranded RNA in Caenorhabditis elegans. Nature 1998;391 (6669):806–811.
293. Chono S, Li SD, Conwell CC, Huang L. An efficient and low immunostimulatory nanoparticle formulation for systemic siRNA delivery to the tumor. J Control Release 2008;131 (1):64–69.
294. Li SD, Huang L. Pharmacokinetics and biodistribution of nanoparticles. Mol Pharm 2008;5 (4):496–504.
295. Chen Y, Sen J, Bathula SR, Yang Q, Fittipaldi R, Huang L. Novel cationic lipid that delivers siRNA and enhances therapeutic effect in lung cancer cells. Mol Pharm 2009;6(3): 696–705.
296. Li SD, Huang L. Surface-modified LPD nanoparticles for tumor targeting. Ann NY Acad Sci 2006;1082:1–8.
297. Li SD, Huang L. Targeted delivery of antisense oligodeoxynucleotide and small interference RNA into lung cancer cells. Mol Pharm 2006;3(5):579–588.
298. Benita S, Levy MY. Submicron emulsions as colloidal drug carriers for intravenous administration: comprehensive physicochemical characterization. J Pharm Sci 1993;82(11): 1069–1079.
299. Singh M, Ravin LJ. Parenteral emulsions as drug carrier systems. J Parenter Sci Technol 1986;40(1):34–41.
300. Yi SW, Yune TY, Kim TW, Chung H, Choi YW, Kwon IC, Lee EB, Jeong SY. A cationic lipid emulsion/DNA complex as a physically stable and serum-resistant gene delivery system. Pharm Res 2000;17(3):314–320.
301. Kim TW, Chung H, Kwon IC, Sung HC, Jeong SY. In vivo gene transfer to the mouse nasal cavity mucosa using a stable cationic lipid emulsion. Mol Cells 2000;10(2):142–147.
302. Huettinger M, Retzek H, Eder M, Goldenberg H. Characteristics of chylomicron remnant uptake into rat liver. Clin Biochem 1988;21(2):87–92.
303. Hussain MM, Maxfield FR, Mas-Oliva J, Tabas I, Ji ZS, Innerarity TL, Mahley RW. Clearance of chylomicron remnants by the low density lipoprotein receptor-related protein/alpha 2-macroglobulin receptor. J Biol Chem 1991;266(21):13936–13940.
304. Rensen PC, van Dijk MC, Havenaar EC, Bijsterbosch MK, Kruijt JK, van Berkel TJ. Selective liver targeting of antivirals by recombinant chylomicrons: a new therapeutic approach to hepatitis B. Nat Med 1995;1(3):221–225.
305. Hara T, Tan Y, Huang L. In vivo gene delivery to the liver using reconstituted chylomicron remnants as a novel nonviral vector. Proc Natl Acad Sci USA 1997;94(26):14547–14552.
306. Vrancken Peeters MJ, Perkins AL, Kay MA. Method for multiple portal vein infusions in mice: quantitation of adenovirus-mediated hepatic gene transfer. Biotechniques 1996;20 (2):278–285.
307. Chesnoy S, Durand D, Doucet J, Stolz DB, Huang L. Improved DNA/emulsion complex stabilized by poly(ethylene glycol) conjugated phospholipid. Pharm Res 2001;18(10): 1480–1484.
308. Yamamoto S, Kuramoto E, Shimada S, Tokunaga T. In vitro augmentation of natural killer cell activity and production of interferon-alpha/beta and -gamma with deoxyribonucleic acid fraction from Mycobacterium bovis BCG. Jpn J Cancer Res 1988;79 (7):866–873.
309. Sakurai H, Kawabata K, Sakurai F, Nakagawa S, Mizuguchi H. Innate immune response induced by gene delivery vectors. Int J Pharm 2008;354(1-2):9–15.
310. Yi AK, Chace JH, Cowdery JS, Krieg AM. IFN-gamma promotes IL-6 and IgM secretion in response to CpG motifs in bacterial DNA and oligodeoxynucleotides. J Immunol 1996;156 (2):558–564.
311. Cowdery JS, Chace JH, Yi AK, Krieg AM. Bacterial DNA induces NK cells to produce IFN-gamma in vivo and increases the toxicity of lipopolysaccharides. J Immunol 1996;156 (12):4570–4575.
312. Whitmore M, Li S, Huang L. LPD lipopolyplex initiates a potent cytokine response and inhibits tumor growth. Gene Ther 1999;6 (11):1867–1875.

313. Krieg AM, Yi AK, Matson S, Waldschmidt TJ, Bishop GA, Teasdale R, Koretzky GA, Klinman DM. CpG motifs in bacterial DNA trigger direct B-cell activation. Nature 1995;374 (6522): 546–549.

314. Klinman DM, Yamshchikov G, Ishigatsubo Y. Contribution of CpG motifs to the immunogenicity of DNA vaccines. J Immunol 1997; 158(8):3635–3639.

315. Sparwasser T, Koch ES, Vabulas RM, Heeg K, Lipford GB, Ellwart JW, Wagner H, Bacterial DNA and immunostimulatory CpG oligonucleotides trigger maturation and activation of murine dendritic cells. Eur J Immunol 1998;28 (6):2045–2054.

316. Jakob T, Walker PS, Krieg AM, Udey MC, Vogel JC. Activation of cutaneous dendritic cells by CpG-containing oligodeoxynucleotides: a role for dendritic cells in the augmentation of Th1 responses by immunostimulatory DNA. J Immunol 1998;161(6):3042–3049.

317. Tan Y, Li S, Pitt BR, Huang L. The inhibitory role of CpG immunostimulatory motifs in cationic lipid vector-mediated transgene expression *in vivo*. Hum Gene Ther 1999;10(13):2153–2161.

318. Qin L, Ding Y, Pahud DR, Chang E, Imperiale MJ, Bromberg JS. Promoter attenuation in gene therapy: interferon-gamma and tumor necrosis factor-alpha inhibit transgene expression. Hum Gene Ther 1997;8(17):2019–2029.

319. Li S, Wu SP, Whitmore M, Loeffert EJ, Wang L, Watkins SC, Pitt BR, Huang L. Effect of immune response on gene transfer to the lung via systemic administration of cationic lipidic vectors. Am J Physiol 1999;276(5 Pt 1):L796–804.

320. Huang L, Li S. Liposomal gene delivery: a complex package. Nat Biotechnol 1997;15 (7):620–621.

321. Yew NS, Wang KX, Przybylska M, Bagley RG, Stedman M, Marshall J, Scheule RK, Cheng SH. Contribution of plasmid DNA to inflammation in the lung after administration of cationic lipid:pDNA complexes. Hum Gene Ther 1999;10(2):223–234.

322. Li S, Tseng WC, Stolz DB, Wu SP, Watkins SC, Huang L. Dynamic changes in the characteristics of cationic lipidic vectors after exposure to mouse serum: implications for intravenous lipofection. Gene Ther 1999;6(4):585–594.

323. Tousignant JD, Gates AL, Ingram LA, Johnson CL, Nietupski JB, Cheng SH, Eastman SJ, Scheule RK. Comprehensive analysis of the acute toxicities induced by systemic administration of cationic lipid:plasmid DNA complexes in mice. Hum Gene Ther 2000;11 (18):2493–2513.

324. Hofman CR, Dileo JP, Li Z, Li S, Huang L. Efficient *in vivo* gene transfer by PCR amplified fragment with reduced inflammatory activity. Gene Ther 2001;8(1):71–74.

325. Bird AP. CpG-rich islands and the function of DNA methylation. Nature 1986;321(6067): 209–213.

326. Yew NS, Zhao H, Wu IH, Song A, Tousignant JD, Przybylska M, Cheng SH. Reduced inflammatory response to plasmid DNA vectors by elimination and inhibition of immunostimulatory CpG motifs. Mol Ther 2000;1(3):255–262.

327. Song YK, Liu F, Liu D. Enhanced gene expression in mouse lung by prolonging the retention time of intravenously injected plasmid DNA. Gene Ther 1998;5(11):1531–1537.

328. Tan Y, Liu F, Li Z, Li S, Huang L. Sequential Injection of Cationic Liposome and Plasmid DNA Effectively Transfects the Lung with Minimal Inflammatory Toxicity. Mol Ther 2001;3:673–682. Higuchi Y, Kawakami S, Oka M, Yamashita F, Hashida M. Suppression of TNFalpha production in LPS induced liver failure in mice after intravenous injection of cationic liposomes/NFkappaB decoy complex. Pharmazie 2006;61(2):144–147.

329. Liu F, Shollenberger LM, Huang L. Non-immunostimulatory nonviral vectors. FASEB J 2004;18(14):1779–1781.

330. Higuchi Y, Kawakami S, Nishikawa M, Yamashita F, Hashida M. Intracellular distribution of NFkappaB decoy and its inhibitory effect on TNFalpha production by LPS stimulated RAW 264.7 cells. J Control Release 2005;107 (2):373–382.

INDEX

ABT-737 compound, brain tumor therapy, 258
Actinomycins, DNA-targeted compounds, dactinomycin, 3–8
Acute myelogenous leukemia (AML), Flt3 inhibitor trerapy, 301
2-Acylaminothiazole derivative, cyclin-dependent kinase inhibitors, 313–314
Adenosine triphosphate (ATP), kinase inhibitors and, 295–296
ADP-ribose (ADR), poly(ADP-ribosyl)ation biochemistry, 151–157
Akt signaling pathway:
 brain tumor therapy, 249–250
 survival signaling inhibitors, 323–324
Alemtuzumab, resistance to, 369
Alkaloids:
 dual topoisomerase I/II inhibitors, 103–105
 vinca alkaloids, 35–40
  chemical structure, 35–37
  vinblastine, 37–39
  vincristine, 39
  vinorelbine, 39–40
Alkylating agents, DNA-targeted chemotherapeutic compounds, 83–95
 cyclopropylindoles, 91–93
 mustards, 83–89
 nitrosoureas, 93–94
 platinum complexes, 89–91
 triazenes, 94–95

Alvocidib cyclin-dependent kinase inhibitor, medicinal chemistry and classification, 310–315
Amides:
 geldanamycin Hsp90 inhibitor derivatives, 389–391
 novobiocin Hsp90 inhibitor, side chain optimization, 434–436
 poly(ADP-ribose)polymerase inhibitors, 163–167
1-Amino-4-benzylphthalazines, brain tumor therapy, hedgehog signaling pathway, 260–261
5-Amino-isoqauinolinone (AIQ), poly(ADP-ribose) polymerase inhibitors, 164–167
Aminopterin, cancer therapy, 106–109
Amrubicin, topoisomerase II inhibitors, 100–102
Amsacrine, topoisomerase II inhibitors, 100–102
Analog design:
 dual topoisomerase I/II inhibitors, bis analogs, 105–106
 geldanamycin Hsp90 inhibitors, 392–393
 radicicol Hsp90 inhibitor, 399–405
Anaplastic astrocytomas, histologic classification, 230–231

Anastrozole, breast cancer resistance, 369–370
Angiogenesis. *See also* Antiangiogenic inhibitors
 brain tumors, 242–243
Anilides, histone deacetylase inhibitors, 66–70
 chidamide, 68, 70
 entinostat (MS-275), 67
 mocetinostat (MGCD-0103), 67–70
 tacedinaline (CI-994), 66–67
Animal studies, brain cancer modeling, 272–273
Anthracyclines, anti-cancer activity, 17–22
 daunorubicin, 19–20
 doxorubicin, 20–22
 epirubicine, 22
 valrubicin, 22–23
Anthrapyrazoles, topoisomerase II inhibitors, 102
Antiangiogenic inhibitors:
 resistance to, 368
 vascular endothelial growth factor, 307–310
Antiangiogenic therapies, central nervous system tumors, 272
Antibody-directed enzyme prodrug therapy (ADEPT), DNA-targeted therapeutics, 120–123
Antifolates, DNA-targeted chemotherapeutic compounds, 106–109

Antigen-presenting cells (APCs), central nervous system tumors, vaccine development, 269–270
Antimetabolites, DNA-targeted chemotherapeutic compounds, 106–113
  antifolates, 106–109
  purine analogs, 112–113
  pyrimidine analogs, 109–112
Apoptosis, protein targeting, multiple drug resistance, 370–371
ARRY-520 kinesin spindle protein inhibitor, 209–210
Aryl compounds, geldanamycin Hsp90 inhibitor derivatives, 389–391
Astrocytic tumors (astrocytomas), histologic classification, 230–234
AT7519 cyclin-dependent kinase inhibitor, 314–315
ATP binding cassette (ABC) transporters, resistance targeting and mediation, 371–376
  breast cancer resistance protein, 372–373
  membrane structure, 374–376
  multidrug resistance protein 1, 372
  P-glycoprotein, 371–372
  structure-activity and quantitative structure-activity relationship studies, 374
ATP-competitive kinesin spindle protein inhibitor, 211–214
  GlaxoSmithKline/Cytokinetics compounds, 211–213
  Merck compounds, 213–214
Atrasentan, brain tumor therapy, 268
Aurora kinase inhibitors:
  drug resistance, 366
  medicinal chemistry and clinical trials, 316–321
Axitinib inhibitor, antiangiogenic properties, 309–310

AZD4877 kinesin spindle protein inhibitor, ispenesib and, 203–204

Banoxantrone, hypoxia-activated bioreductive prodrugs, 117–120
Base-excision repair (BER) pathway:
  poly(ADP-ribose)polymerase-1 (PARP-1) single-strand break repair, 160–162
  poly(ADP)-ribosylation biochemistry, 151–157
BBR 3464 platinum complex, cross-linked interstrand DNA targeting, 91
Bcr-Abl inhibitors:
  chronic myelogenous leukemia, 298–301
  drug-resistant mutation, 354–356
  small-molecule imatinib mesylate, resistance to, 364–366
Becatecarin, topoisomerase II inhibitors, 102
Belinostat, structure and development, 58–59
Belotecan, biological activity and side effects, 97–98
Bendamustine, cancer therapy, 83–84
Benzamides, poly(ADP-ribose) polymerase inhibitors, 163–167
Benzimidazole, poly(ADP-ribose) polymerase inhibitors, 164–167
Benzisoxazole Hsp90 inhibitors, 431
Benzoquinone ansamycins, geldanamycin Hsp90 inhibitor, 397–399
Benzoxazole derivatives, poly (ADP-ribose) polymerase inhibitors, 164–167
Bevacizumab:
  glioblastoma therapies, 232–233
  resistance to, 368–369
  VEGFR2 inhibitors, antiangiogenic properties, 307–310

BIBW-2992 erbB family inhibitor, 302–305
Binding pockets:
  Hsp90 inhibitors, N-terminal ATP-binding pocket, 386–425
    chimeric geldanamycin and radicicol, 405–410
    geldanamycin, 386–399
    high-throughput small-molecule inhibitors, 417–421
    purine-based inhibitors, 410–416
    radicicol analogs, 399–405
    shepherdin, 416
    structure-based design, 421–425
  kinase inhibitors, multiconformational catalytic domain, 346–349
  kinesin spindle protein inhibitors:
    EMD-534085, 210–211
    monastrol, 193–197
Binding sites, Hsp90 inhibitors, 383–386, 425–429
  C-terminal nucleotide binding site:
    cisplatin, 440–441
    elucidation, 430–431
    epigallocatechin-3-gallate, 441
    GHLK protein family, 429–430
    novobiocin inhibitor, 431–440
  derrubone, 425–426
  gedunin and celastrol, 428–429
  high-throughput screening and modeling, 429
  naphthoquinones, 426–427
  small-molecule Hsp90/HOP interactions, 427
  taxol, 427–428
Bioluminescence imaging, brain tumor modeling, 273–274
Biosynthetic pathways:
  anthracyclines, 22–24
  antifolates, 106–109
  DNA-targeted compounds:
    bleomycin, 11
    dactinomycin, 6–8
    plicamycin, 17
  docetaxel/taxotere, 34–36

Biricodar kinase inhibitor, drug resistance, 372
Bis analogs, dual topoisomerase I/II inhibitors, 105–106
Blenoxane. *See* Bleomycin
Bleomycins:
 biosynthesis, 11
 chemical structure, 8
 contraindications and side effects, 8
 current and future research issues, 12
 molecular mechanisms, 12
 pharmacokinetics, 8–11
 therapeutic applications, 8
Blood-brain barrier (BBB), central nervous system tumors, 270–271
BMS-387032 cyclin-dependent kinase inhibitor, medicinal chemistry and classification, 311–312, 314
Bone morphogenetic proteins (BMPs), brain tumor therapy, 267
Boronic acid derivatives, proteosome inhibitors, bortezomib clinical trials, 177–180
Bortezomib, discovery and development, clinical trials, 177–180
Bosutinib:
 second-generation development, drug resistance and, 364–365
 tumor metastases inhibition, 325–327
B-Raf inhibitors, structure and properties, 305–306
Brain cancer. *See also* specific tumors, e.g. Glioma
 angiogenesis, 242–243
 astrocytic tumors, 230
 ependymoma, 233
 epidemiology and prevalence, 223–224
 imaging studies, 273–274
 invasiveness, 241–242
 meningiomas, 237
 metastatic, 240
 modeling approaches, 272–273

neuroepithelial tumors, 229–230
oligodendroglial tumors, 233–234
primary brain tumors, 227–234
therapies, 244–270
 blood-brain barrier, 270–271
 bone morphogenetic proteins, 267
 chemokine receptors, 264–265
 cytotoxic agents, 244
 epidermal growth factor receptors, 246–247
 estrogen receptors, 265
 growth factors, receptors, and signaling pathways, 245
 GSK3b inhibitors, 263–264
 heat shock protein inhibitors, 267–268
 hedgehog signaling pathway, 259–263
 HGF/MET sytems, 247–248
 hypoxia inducible factor pathway, 253–255
 integrins, 255–256
 kinase inhibitors, 248–249
 MDM2/MDMX pathway, 257–258
 mTOR signaling, 250–252
 nuclear hormone receptors, 265
 obstacles to, 270
 p53 signaling pathway, 256–257
 peroxisome proliferator-activated receptors, 265–267
 phosphatase and tensin homolog (PTEN), 258–259
 phosphodiesterase 4, 265
 phosphoinositide 3-kinase inhibitors, 249–250
 PI3K/AKT signaling pathway, 249
 platelet-derived growth factor and receptor, 249
 protein kinase C, 245–246
 RAS/RAF/MEK/ERK signaling, 252
 retinoid X receptor, 267
 Rho/ROCK and MAPK signaling pathways, 252–253

 small-molecule compounds, 268
 targeted delivery, 271–272
 transforming growth factor-b, 267
 vaccines for, 268–270
 vascular endothelial growth factor inhibitors, 248–249
 Wnt signaling pathway, 263
tumor classification, 228–229
 astrocytic tumors, 230
 ependymoma, 233
 neuroepithelial tumors, 229–230
 oligodendroglial tumors, 233–234
 WHO grading system (I-IV), 230–233
tumor initiation, 224–227
WHO grading system (I-IV), 230–233
BRCA1/BRCA2 genes, poly(ADP-ribose)polymerase (PARP) inhibitors, cancer therapy, 168–170
Breast cancer:
 endocrine resistance in, 369–370
 kinesin spindle protein inhibitors, ispenesib, 198–204
Breast cancer resistance protein (MXR), drug resistance mechanisms, 372–373
Brivanib, antiangiogenic properties, 308–310
BSI-201 chemotherapeutic agent, poly(ADP-ribose) polymerase (PARP) inhibitor coadjuvants, 169–170

Calcium channel blockers, ATP binding cassette (ABC) transporters, 371–372
Camptothecins:
 enzyme inhibition, 23–29
 irinotecan, 26
 metabolism, 25–26
 topotecan, 26–29
 topoisomerase inhibitors, 96–106
 Topo I inhibitors, 96–98

Cancer therapy:
  brain tumors:
    angiogenesis, 242–243
    astrocytic tumors, 230
    ependymoma, 233
    imaging studies, 273–274
    invasiveness, 241–242
    metastatic, 240
    modeling approaches, 272–273
    neuroepithelial tumors, 229–230
    oligodendroglial tumors, 233–234
    WHO grading system (I-IV), 230–233
  central nervous system cancers, 243–270
    blood-brain barrier and compound design, 270–271
    cranial/spinal nerve tumors, 237–239
    embryonal tumors, 234–235
    epidemiology and prevalence, 223–224
    germ cell tumors, 235
    histological classification, 228–229
    meningioma, 237
    microenvironment, 227
    obstacles to, 270
    pituitary tumors, 235–237
    primary lymphoma, 239–240
    recurrent tumors, 240–241
    signaling pathways, 227–228
    targeted delivery, 271–272
  DNA-targeted therapeutics:
    alkylating agents, 83–95
      cyclopropylindoles, 91–93
      mustards, 83–89
      nitrosoureas, 93–94
      platinum complexes, 89–91
      triazenes, 94–95
    antimetabolites, 106–113
      antifolates, 106–109
      purine analogs, 112–113
      pyrimidine analogs, 109–112
    cytotoxic agents, 3–17
      bleomycin, 8–12
      dactinomycin, 3–8
      enzyme inhibitors, 17–31
        anthracyclines, 17–23
        camptothecins, 23–29
        isopodophyllotoxins, 29–31
    future research issues, 40–41
    mitomycin (mutamycin), 12–16
    plicamycin, 16–17
    research background, 1–3
    tubulin polymerization/depolymerization inhibition, 31–40
      dimeric vinca alkaloids, 35–40
      taxus diterpenes, 32–35
    research background, 83
    topoisomerase inhibitors, 95–106
      dual topo I/II inhibitors, 103–106
      Topo II inhibitors, 98–102
      Topo I inhibitors, 96–98
    tumor-activated prodrugs, 113–124
      ADEPT prodrugs, 120–123
      hypoxia-activated bioreductive prodrugs, 114–120
  drug resistance mechanisms:
    endocrine resistance, breast cancer, 369–370
    multiple drug resistance, 370–376
      ABC transporter mediation, 371–376
      apoptosis protein targeting, 370–371
      stem cells, 376–377
    new strategies for, 377–378
    research background, 361–363
    small-molecule tyrosine kinase inhibitor resistance, 364–369
    stem cells and multiple drug resistance, 376–377
    targeted drug development, 363–370
    tyrosine kinase inhibitor resistance, 364
  Hsp90 inhibitors, 441–442
  kinesin spindle protein inhibitors:
    ATP-competitive inhibitors, 211–214
    dihydropyrazoles, 204–210
    future research issues, 214–215
    "induced fit" pocket inhibitors, 210–211
    ispinesib, 198–204
    mitosis targeting, 191–192
    monastrol, 193–197
    quinazolinone-based inhibitors, 197–202
  peroxisome proliferator-activated receptors, 160–162
    chemotherapy/radiotherapy coadjuvants, 168–170
    future research issues, 170–171
    inhibitors, 168
    monotherapy, 170
    poly(ADP-ribose) glycohydrolase and inhibitors, 170
    poly(ADP-ribose) glycohydrolase, 170
    ubiquitin-proteasome pathway, 176–177
Capecitabine:
  ispenesib combined therapy, 200–204
  pyrimidine antimetabolite analogs, 110–112
Carbamates, antibody-directed enzyme prodrug therapy, 121–123
Carboplatin, ispenesib combined therapy, 200–204
Cardiotoxicity biomarkers, topoisomerase II inhibitors, 101–102
Carfilzomib proteasome inhibitor, clinical trials, 180
Catalytic cleft, multiconformational kinase catalytic domain, 346–349
Cediranib:
  antiangiogenic properties, 308–310
  brain tumor angiogenesis, 242–243
  glioblastoma therapies, 232–233
Celastrol, Hsp90 inhibitor binding, 428–429
Cell cycle kinase inhibitors, 310–321
  aurora kinase inhibitors, 316–321
  checkpoint kinase inhibitors, 315–316

cyclin-dependent kinase inhibitors, 310–315
polo-like kinase inhibitors, 316
Cell line development, brain cancer modeling, 272–273
Cellular adhesion molecules, brain cancer therapy, hypoxia-inducible factor pathway, 254–255
Central nervous system (CNS), cancers
brain tumors:
angiogenesis, 242–243
astrocytic tumors, 230
ependymoma, 233
imaging studies, 273–274
invasiveness, 241–242
metastatic, 240
modeling approaches, 272–273
neuroepithelial tumors, 229–230
oligodendroglial tumors, 233–234
WHO grading system (I-IV), 230–233
cranial/spinal nerve tumors, 237–239
embryonal tumors, 234–235
epidemiology and prevalence, 223–224
germ cell tumors, 235
histological classification, 228–229
meningioma, 237
microenvironment, 227
pituitary tumors, 235–237
primary lymphoma, 239–240
recurrent tumors, 240–241
signaling pathways, 227–228
therapies, 243–270
blood-brain barrier and compound design, 270–271
obstacles to, 270
targeted delivery, 271–272
CEP-9722 chemotherapeutic agent, poly(ADP-ribose)polymerase (PARP) inhibitor coadjuvants, 169–170
Cerubidine. See Daunorubicin
Cetuximab:
growth factor signaling inhibitor, 302–305

resistance to, 369
Checkpoint kinase inhibitors, medicinal chemistry and clinical trials, 315–316
Chemokine receptors, brain tumor therapy, 264–265
Chemotherapy:
brain cancer, ependymoma, 233
poly(ADP-ribose)polymerase (PARP) inhibitor coadjuvants, 169–170
Chidamide, histone deacetylase inhibitors, 68, 70
Chimeric structures, Hsp90 inhibitors, geldanamycin and radicicol chimeras, 405–410
Chlorambucil, cancer therapy, 83–84
Chronic lymphocytic leukemia (CLL), cyclin-dependent kinase inhibitor therapies, 311–315
Chronic myelogenous leukemia (CML):
Bcr-Abl inhibitor therapy, 298–301
small-molecule imatinib mesylate, resistance to, 364–366
CI-994. See Tacedinaline
Cisplatin:
DNA-targeted chemotherapeutic compounds, 89–91
Hsp90 inhibitor, C-terminal binding sight, 440–441
CK929866SB-743921 kinesin spindle protein inhibitor, ispenesib and, 202–203
Cladribine, antimetabolite analogs, 112–113
Clinical trials:
aurora kinase inhibitors, 316–321
checkpoint kinase inhibitors, 315–316
kinesin spindle protein inhibitors, ispenesib, 198–204
polo-like kinase inhibitors, 316–317

proteosome inhibitors, 177–180
Cloretazine, biological activity and side effects, 94
c-Met kinase inhibitors:
brain tumor therapy, 247–248
structure-based design, hydrogen bonding, 353–354
tumor metastases, 326, 328–330
Cocrystallization, Hsp90 inhibitors:
chimeric radicicol and geldanamycin structures, 405–410
pyrazole-scaffold Hsp90 inhibitors, 418–419
radicicol Hsp90 inhibitor, 400–405
Combrestatin phosphate, antibody-directed enzyme prodrug therapy, DNA-targeted therapeutics, 120–123
Conjugated derivatives, geldanamycin Hsp90 inhibitor, 393–397
Convection-enhanced delivery, central nervous system tumor therapies, 271–272
Cosmegen. See Dactinomycin
Coumarin ring system, novobiocin Hsp90 inhibitor optimization, 436–440
Cranial nerve tumors, classification and therapy, 237–239
Cross-linked DNA-targeted compounds, platinum complexes, 91
C-terminal nucleotide binding site, Hsp90 inhibitors:
cisplatin, 440–441
elucidation, 430–431
epigallocatechin-3-gallate, 441
GHLK protein family, 429–430
novobiocin inhibitor, 431–440
CXCR4 receptor antagonists, brain tumor therapy, 264–265
Cyclic amides, poly(ADP-ribose) polymerase inhibitors, 164–167

Cyclin-dependent kinase inhibitors:
  medicinal chemistry and classification, 310–315
  multidrug resistance and, 377–378
Cyclopamine, brain tumor therapy, 260
Cyclophosphamide, cancer therapy, DNA-targeted mechanisms, 83–84, 86
Cyclopropoaradicicol, analog design, 402–403
Cyclopropylindoles, DNA-targeted chemotherapeutics, 91–93
CYP3A4 enzyme, drug resistance inhibition, 372
Cytokinetics ATP-competitive kinesin spindle protein inhibitors, 211–213
Cytosine arabinose, antimetabolite analogs, 109–112
Cytosine deaminase, gene-directed enzyme prodrug therapy, 123–124
Cytotoxic agents:
  brain tumors:
    stem cell initiation, 225–227
    temozolomide, 244
  DNA targeting compounds, 3–17
    bleomycin, 8–12
    dactinomycin, 3–8
    mitomycin (mutamycin), 12–16
    plicamycin, 16–17
  enzyme inhibitors, 17–31
    anthracyclines, 17–23
    camptothecins, 23–29
    isopodophyllotoxins, 29–31
  future research issues, 40–41
  research background, 1–3
  tubulin polymerization/depolymerization inhibition, 31–40
    dimeric vinca alkaloids, 35–40
    taxus diterpenes, 32–35

Dacarbazine, DNA-targeted compounds, 94–95

Dactinomycin:
  biosynthesis, 6–8
  chemical structure, 3
  contraindications and side effects, 4
  medicinal chemistry, 4–5
  molecular mechanisms, 5–7
  pharmacokinetics, 4–7
  therapeutic applications, 3–4
Dasatinib:
  chronic myelogenous leukemia therapy, Bcr-Abl inhibitors, 299–301
  second-generation development, drug resistance and, 364–365
  structure-based design:
    multiconformational catalytic domain, 346–349
    protein targeting, 352–354
Daunorubicin (Daunomycin), chemical structure and therapeutic applications, 19–21
Dendritic cells, central nervous system tumors, vaccine development, 269–270
Derived acridine-4-carboxamide analog (DACA), dual topoisomerase I/II inhibitors, 104–105
Derrubone Hsp90 inhibitor, 425–426
Desnoviase benzamide derivative, novobiocin Hsp90 inhibitor optimization, 439–440
DFG motif, multiconformational kinase catalytic domain, 346–349
Diflomotecan, biological activity and side effects, 98
Dihydrogeldanamycin derivatives, geldanamycin Hsp90 inhibitor, 388–391
Dihydropyrazoles, kinesin spindle protein inhibitors, 204–210
  ARRY-520, 209–210
  LY2523355 compound, 210
  MK-0731, 204–209

Dihydropyrroles, kinesin spindle protein inhibitors, 204–209
Dihydroxyphenyl amide Hsp inhibitors, fragment-based design, 424–425
Dimeric structures:
  geldanamycin Hsp90 inhibitor, 393–397
  novobiocin Hsp90 inhibitor, 434
  vinca alkaloids, 35–40
DNA binding domain (DBD), poly (ADP-ribose) polymerase-1 (PARP-1), 159–160
DNA methylation, pyrimidine antimetabolite analogs, 111–112
DNA-targeted therapeutics:
  cytotoxic agents, 3–17
    bleomycin, 8–12
    dactinomycin, 3–8
    mitomycin (mutamycin), 12–16
    plicamycin, 16–17
  synthetic chemotherapeutic compounds:
    alkylating agents, 83–95
    cyclopropylindoles, 91–93
    mustards, 83–89
    nitrosoureas, 93–94
    platinum complexes, 89–91
    triazenes, 94–95
  antimetabolites, 106–113
    antifolates, 106–109
    purine analogs, 112–113
    pyrimidine analogs, 109–112
  research background, 83
  topoisomerase inhibitors, 95–106
    dual topo I/II inhibitors, 103–106
    Topo II inhibitors, 98–102
    Topo I inhibitors, 96–98
  tumor-activated prodrugs, 113–124
    ADEPT prodrugs, 120–123
    hypoxia-activated bioreductive prodrugs, 114–120
Docetaxel:
  ispenesib combined therapy, 200–204
  structure and therapeutic applications, 34–35

Donor-acceptor motifs, kinase inhibitors, 295–296
Doramapimod kinase inhibitor, structure-based design, protein targeting, 351–354
Double-stranded breaks, poly (ADP-ribose) polymerase-1 (PARP-1), 161–162
Doxorubicin:
 antibody-directed enzyme prodrug therapy, peptidase enzymes, 120–121
 structure and therapeutic effects, 20–22
 topoisomerase II inhibitors, 98–102
Drug resistance:
 cancer therapy:
  endocrine resistance, breast cancer, 369–370
  mechanisms, 361–363
  multiple drug resistance, 370–376
   ABC transporter mediation, 371–376
   apoptosis protein targeting, 370–371
   stem cells, 376–377
   new strategies for, 377–378
  small-molecule tyrosine kinase inhibitor resistance, 364–369
  targeted drug development, 363–370
  tyrosine kinase inhibitor resistance, 364
  second-line kinase inhibitors, conformational targeting, 354–356

E3 ubiquitin ligases, proteasome inhibitors, 177
Edatrexate, antifolate DNA-targeted chemotherapeutics, 106–109
Elinafide, dual topoisomerase I/II inhibitors, bis analogs, 105–106
EMD-534085 kinesin spindle protein inhibitor, "induced fit" binding pockets, 210–211
Endocrine resistance, breast cancer, 369–370
Entinostat, histone deacetylase inhibitors, 67
Enzastaurin, brain tumor therapy, 245–246
Enzyme inhibitors:
 cytotoxic agents, 17–31
  anthracyclines, 17–23
  camptothecins, 23–29
  isopodophyllotoxins, 29–31
 histone deacetylases:
  anilides, 66–70
  crystalline structure, 51–53
  fatty acids, 73
  future research issues, 73–75
  hydroxamic acid-based compounds, 53–66
  natural products, 70–72
  research background, 49–50
 topoisomerase inhibitors, 95–106
  dual topo I/II inhibitors, 103–106
  Topo II inhibitors, 98–102
  Topo I inhibitors, 96–98
Ependymoma, histologic classification, 233
Epidermal growth factor receptor (EGFR):
 brain tumor therapy, 246–247
 erlotinib, 251
 vaccine development, 269–270
 growth factor signaling inhibitors, 302–305
 kinase inhibitors, structure-based design:
  protein targeting, 350–354
  resistance, 366–368
Epigallocatechin-3-gallate (EGCG), Hsp90 inhibitor, C-terminal binding sight, 441
Epirubicin, structure and therapeutic effects, 22–23
erbB-family kinases, growth factor signaling inhibitors, 302–305
Erlotinib:
 brain tumor therapy, 247–248
 EGFR-driven gliomas, 251
 growth factor signaling inhibitors, 302–305
 structure-based design, protein targeting, 350–354
Estrogens, receptors:
 brain tumor therapy, 265
 breast cancer resistance, 369–370
Etoposide:
 antibody-directed enzyme prodrug therapy, 120–123
 structure and therapeutic applications, 30–31
Everolimus, brain cancer therapy, 251–252
Exatecan, biological activity and side effects, 98
Exemestane, breast cancer resistance, 369–370

Fatty acids, histone deacetylase inhibition, 73
FK-228. *See* Romidepsin
Flt3 inhibitors, acute myelogenous leukemia, 301–302
Fludarabine, antimetabolite analogs, 112–113
5-Fluorouracil, antimetabolite analogs, 109–112
Focal adhesion kinase (FAK) inhibitors, tumor metastases, 326, 328
Fragment-based approach, Hsp90 inhibitors, 423–425
 dihydroxyphenyl amide inhibitors, 424–425
 tetrahydro-4H-carbazol-4-one inhibitors, 423–424
Fulvestrant, breast cancer resistance, 369–370

GDC-0449, brain tumor therapy, hedgehog signaling pathway, 262
GDC-0941 PI3K inhibitor, survival signaling pathways, 322–323
Gedunin, Hsp90 inhibitor binding, 428–429
Geftinib:
 brain tumor therapy, 247
 growth factor signaling inhibitors, 302–305

Geldanamycin Hsp90 inhibitor, 386–399
  analog design, 392–395
  benzoquinone ansamycins, 397–399
  chemical structure, 386–388
  chimeric structures, 405–410
  dihydrogeldanamycin derivatives, 388–391
  semisynthetic derivatives, 391–392
Gene-directed enzyme prodrug therapy, DNA-targeted compounds, 123–124
Gene therapy, plasmid DNA:
  angiogenesis, 463
  autoimmune diseases, 463–464
  chemical delivery, 467–480
  electroporation gene transfer, 464–465
  emulsion delivery, 480–481
  functional applications, 461
  gene gun mechanisms, 465
  gene transfer and long-term expression, 458–460
  hydrodynamic injection, 466
  laser irradiation, 466
  myocardial delivery, 461–463
  naked direct delivery, 458–464
  nonviral vector toxicity, 481–482
  physical delivery, 464–466
  research background, 457–458
  therapeutic applications, 460–461
  ultrasound gene transfer, 465–466
Germ cell tumors, classification and therapy, 235
Germinomas, central nervous system, 235
GHKL protein family, Hsp90 inhibitors, C-terminal nucleotide binding site, 429–430
Gimatecan, biological activity and side effects, 98
GlaxoSmithKline ATP-competitive kinesin spindle protein inhibitors, 211–213
Gleevec, cranial and spinal nerve tumors, 238–239
Glioblastoma:
  histologic classification, 231–233

hypoxia inducible factor (HIF) pathway, 253–255
mTOR therapies, 251–252
PI3K/AKT signaling pathway therapies, 249–250
stem cell initiation, 225–227
Glioblastoma multiforme (GBM):
  histologic classification, 231–233
  hypoxia-inducible factor pathway, 255
  temozolomide therapy, 244
  Wnt signaling pathway, 263
Glioma:
  classification, 228–230
  hypoxia inducible factor (HIF) pathway, 253–255
  invasiveness, 241–242
  stem cell initiation, 225–227
Gli transcription factor, brain tumor therapy, hedgehog signaling pathway, 261–263
Glucuronidases, antibody-directed enzyme prodrug therapy, 121–122
Glycoproteins. See also P-glycoprotein
a-Glicosidic bond, poly(ADP-ribosyl)ation biochemistry, 153–157
G-protein coupled receptors (GPCR), brain tumor therapy, Wnt signaling pathway, 263
Growth factors:
  brain tumor therapy, 245
  RAS/RAF/MEK/ERK signaling, 252
  breast cancer resistance, 369–370
  signaling inhibitors, 301–307
  B-Raf inhibitors, 305–306
  erbB-family kinases, 302–305
  MEK inhibitors, 306–307
GSK3b inhibitors, brain tumor therapy, 263–264
GSK690693 protein kinase B inhibitor, 323–324

Hairpin polyamide concept, minor groove targeting compounds, 88

Heat shock proteins (Hsps):
  brain tumor therapy, inhibitors, 267–268
  drug resistance, 366
  Hsp90 inhibitors:
    applications, 441–444
    cancer therapy, 441–442
    neurodegenerative diseases, 442–444
  cancer markers, 384
  C-terminal nucleotide binding site
    cisplatin, 440–441
    elucidation, 430–431
    epigallocatechin-3-gallate, 441
    GHLK protein family, 429–430
    novobiocin inhibitor, 431–440
  N-terminal ATP-binding pocket, 386–425
    benzisoxazole inhibitors, 421
    chimeric geldanamycin and radicicol, 405–410
    fragment-based screening, 423–425
    geldanamycin, 386–399
    high-throughput small-molecule inhibitors, 417–421
    isoxazole-scaffold inhibitors, 419–421
    purine-based inhibitors, 410–416
    pyrazole-scaffold inhibitors, 417–419
    radicicol analogs, 399–405
    shepherdin, 416
    structure-based design, 421–425
  protein folding process, 384–385
  protein structure and function, 385–386
  research background, 383
  tumor cell selectivity, 385
  unidentified binding site inhibitors, 425–429
    derrubone, 425–426
    gedunin and celastrol, 428–429
    high-throughput screening and modeling, 429
    naphthoquinones, 426–427

small-molecule Hsp90/
    HOP interactions, 427
  taxol, 427–428
Hedgehog signaling pathway:
  brain tumor therapy, 259–263
  central nervous system tumors,
    228
Hepatocyte growth factor/
    mesenchymal-
    epithelial transition
    factor (HGF/MET),
    brain tumor therapy,
    247–248
Hexhydropyranoquinoline
    (HHPQ) scaffold,
    EMD-534085 kinesin
    spindle protein
    inhibitor, "induced fit"
    binding pockets,
    210–211
High-throughput screening
    (HTS), Hsp90
    inhibitors:
  modeling and identification,
    429
  small-molecule compounds, 417
Histone deacetylase (HDAC)
    inhibitors:
  anilides, 66–70
    chidamide, 68, 70
    entinostat (MS-275), 67
    mocetinostat (MGCD-0103),
      67–70
    tacedinaline (CI-994), 66–67
  brain tumor therapy, hedgehog
    signaling pathway,
    262–263
  classification, 50–51
  crystalline structure, 51–53
  drug resistance, 366
  fatty acids, 73
  future research issues, 73–75
  hydroxamic acid compounds,
    53–66
    current clinical trials and
      drug development,
      53–54
    ITF-2357, 60–61
    JNJ-26481585 second-
      generation HDACi,
      64–66
    NVP-LAQ-824 and LBH-589,
      55, 57–58
    PCI-024781, 60–62
    PXD-101, 58–59
    R306465 first-generation
      HDACi, 62–64

SB939, 59–60
suberoyl hydroxamic acid
  analogs, 53, 55–56
natural products, 70–72
  largazole, 72
  romidepsin (FK-228), 70–72
research background, 49–50
Homologous recombination, poly
    (ADP-ribose)
    polymerase (PARP)
    inhibitors:
  cancer therapy, 168
  PARP-1 structures, 161–162
Homology modeling, ATP-binding
    cassette transporters,
    375–376
Hsp90 organizing protein (HOP),
    Hsp90 inhibitors
    interruption, 427
Hydrogen bonds, kinase
    inhibitors, structure-
    based design, protein
    targeting, 353–354
Hydroquinone derivatives,
    geldanamycin Hsp-90
    inhibitor, 390–391
Hydroxamic acid compounds,
    histone deacetylase
    inhibitors, 53–66
  current clinical trials and drug
    development, 53–54
  ITF-2357, 60–61
  JNJ-26481585 second-
    generation HDACi,
    64–66
  NVP-LAQ-824 and LBH-589,
    55, 57–58
  PCI-024781, 60–62
  PXD-101, 58–59
  R306465 first-generation
    HDACi, 62–64
  SB939, 59–60
  suberoyl hydroxamic acid
    analogs, 53, 55–56
Hypoxia-activated bioreductive
    prodrugs, DNA-
    targeted therapeutics,
    tumor-activation,
    114–120
Hypoxia inducible factor (HIF)
    pathway, brain tumor
    therapy, 253–255

Imatinib:
  chronice myelogenous
    leukemia therapy,
    small-molecule

imatinib mesylate,
    resistance to, 364–366
  chronic myelogenous leukemia
    therapy, 298–301
  structure-based design:
    drug-resistant mutation,
      354–356
    multiconformational
      catalytic domain,
      346–349
    protein targeting, 351–354
Imidazoacridinones,
    topoisomerase II
    inhibitors, 102
Immunoproteasome-specific
    inhibitors:
  basic properties, 175
  mechanism of action, 184–185
"Induced fit" binding pockets,
    kinesin spindle
    protein inhibitors:
  EMD-534085, 210–211
  monastrol, 193–197
In silico methods, chimeric
    radicicol and
    geldanamycin Hsp90
    inhibitors, 406–410
Insulin-like growth factor-1
    receptor kinase
    inhibitors, survival
    signaling pathways,
    321–322
Integrin-linked kinase (ILK),
    brain cancer therapy,
    250
Integrins, brain tumor
    invasiveness, 241–242
Intoplicine, dual topoisomerase I/
    II inhibitors, 103–105
IPI-504 Hsp90 inhibitor,
    geldanamycin
    derivatives, 390–391
IPI-926 compound, brain tumor
    therapy, hedgehog
    signaling pathway,
    260
Irinotecan (CPT-11):
  meningioma therapy, 237
  structure and applications, 26
Isopodophyllotoxins, enzyme
    inhibition, 29–31
Isoquinolinones, poly(ADP-
    ribose)polymerase
    inhibitors, 164–167
Isoxazole-scaffold inhibitors,
    Hsp90 inhibitors,
    419–421

Ispenesib, kinesin spindle protein inhibitors, 198–204
CK929866SB-743921 and, 202–203
ITF-2357 histone deacetylase inhibitor, structure and development, 60–61

JNJ-26481585 second-generation histone deacetylase inhibitor, 64–66
JSM6427, brain cancer therapy, hypoxia-inducible factor pathway, 254–255

Karenitecin, biological activity and side effects, 98
Kinase catalytic domain, small-molecule inhibitor design, 346–349
Kinase inhibitors. *See also* Protease inhibitors; Tyrosine kinase inhibitors
  Bcr-Abl inhibitors, chronic myelogenous leukemia therapy, 298–301
  brain tumor therapy, 245
    vascular endothelial growth factor, 248–249
  cell cycle inhibitors, 310–321
    aurora kinase inhibitors, 316–321
    checkpoint kinase inhibitors, 315–316
    cyclin-dependent kinase inhibitors, 310–315
    polo-like kinase inhibitors, 316
  classification and properties, 295–298
  clinical candidates *vs.* marketed drugs, 297–298
  donor-acceptor motifs, 295–296
  FDA-approved compounds, 296–297
  Flt3 inhibitors, acute myelogenous leukemia, 301
  future research issues, 330–331
  growth factor signaling inhibitors, 301–307, 302–307

B-Raf inhibitors, 305–306
erbB-family kinases, 302–305
MEK inhibitors, 306–307
structure-based design:
  future research issues, 356–357
  multiconformational catalytic domain binding cleft, 346–348
  binding pockets, 348–349
  protein targeting, 350–354
  second-line inhibitors, drug-resistant mutation, 354–356
  research background, 345–346
survival signaling inhibitors, 321–325
  insulin-like growth factor-1 receptor inhibitors, 321
  mTOR inhibitors, 324–325
  phosphoinoside-3 kinase inhibitors, 321–323
  protein kinase B inhibitors, 323–324
tumor invasion/metastasis inhibitors, 325–330
  c-Met kinase inhibitors, 326–330
  focal adhesion kinase inhibitors, 326
  Src kinase inhibitors, 325–326
VEGFR2 inhibitors, antiangiogenic agents, 307–310
Kinases, gene-directed enzyme prodrug therapy, 123–124
Kinesin spindle protein (KSP): basic properties, 191
inhibitors:
  ATP-competitive inhibitors, 211–214
  dihydropyrazoles, 204–210
  future research issues, 214–215
  "induced fit" pocket inhibitors, 210–211
  ispinesib, 198–204
  monastrol, 193–197
  quinazolinone-based inhibitors, 197–202
mitosis-targeted therapy, 191–192

structure and mechanism, 192–193
Kit tyrosine kinase, structure-based design:
  drug-resistant mutation, 354–356, 367–368
  protein targeting, 351–354

Lactacystin, proteasome inhibitors, 183–184
b-Lactamases, antibody-directed enzyme prodrug therapy, 121–123
Lactams, poly(ADP-ribose) polymerase (PARP) inhibitor coadjuvants, 169–170
Lapatinib:
  cranial and spinal nerve tumors, 238–239
  growth factor signaling inhibitors, 302–305
  structure-based design, multiconformational catalytic domain, 346–349
Largezole, histone deacetylase inhibition, 72
LBH-589 histone deacetylase inhibitor, structure and development, 55, 57–58
LDE-225 small molecule, brain tumor therapy, hedgehog signaling pathway, 262
Lestaurtinib, acute myelogenous leukemia therapy, 301–302
Letrozole, breast cancer resistance, 369–370
Linifanib, brain cancer therapy, hypoxia inducible factor (HIF) pathway, 254–255
Linker compounds, histone deacetylase inhibitiors, siberoyl hydroxamic acid structures, 53–56
Lipophilicity, antifolate DNA-targeted chemotherapeutics, 109
LY2523355 kinesin spindle protein inhibitor, 210

Lymphomas, primary central nervous system tumors, 239–240

Macebecin Hsp90 inhibitors, 397–399
Macrocyclic compounds, proteasome inhibitors, 183–184
Macrolides, radicicol Hsp90 inhibitor, 402–405
Magnetic resonance imaging, brain tumor modeling, 273–274
Malignant peripheral nerve sheath tumors (MPNSTs), classification and therapy, 238–239
Matrix metalloproteinase (MMP) inhibitors, brain cancer therapy, hypoxia-inducible factor pathway, 254–255
MDM2/MDM4, brain cancer therapy, 256–258
MDMX protein, brain cancer therapy, 257–258
Medicinal chemistry:
camptothecins, 28–29
DNA-targeted compounds:
bleomycin, 9–10
dactinomycin, 4
mitomycin, 13–14, 15–16
plicamycin, 17
proteasome inhibitors, 181–184
vinblastine, 39
Medulloblastoma:
classification and therapy, 234–235
hedgehog signaling pathway, 259–260
MEK inhibitors, structure and properties, 306–307
Melphalan, cancer therapy, 83–84
Meningiomas, classification and therapy, 237
6-Mercaptopurine, antimetabolite analogs, 112–113
Merck ATP-competitive kinesin spindle protein inhibitor, 213–214
Metastases:
brain tumors, 240
animal models, 273
kinase inhibitors, 325–330

c-Met kinase inhibitors, 326, 328–330
focal adhesion kinase inhibitors, 326, 328
Src kinase inhibitors, 325–328
Methchloramine, cancer therapy, 83–85
Methotrexate, folic acid uptake and metabolism, antifolate cancer therapy, 106–109
Methylguanine methyltransferase (MGMT), glioblastoma therapy, 231–232
temozolomide, 244
MGCD-0103. See Mocetinostat
MI-63 compound, brain tumor therapy, 257–258
Microenvironmental factors, central nervous system tumors, 227
Microtubules, kinesin spindle protein structure:
crystallographic analysis, 192–193
mitosis-targeted therapies, 191–192
Midostaurin, acute myelogenous leukemia therapy, 301–302
Minor groove targeting, DNA-targeted chemotherapeutic compounds:
cyclopropylindoles, 91–93
mustards, 87–88
Mithracin/Mithramycin. See Plicamycin
Mitogen-activated protein kinase (MAPK) inhibitors:
brain tumor therapy, signaling pathways, 252–253
drug resistance, 366
growth factor signaling pathways, 301–307
Mitomycin:
chemical structure, 12–13
contraindications and side effects, 13
hypoxia-activated bioreductive prodrugs, 116–120
medicinal chemistry, 13–16
molecular mechanisms, 14–15
pharmacokinetics, 13
therapeutic applications, 13

Mitosis, therapeutic targeting, 191–192
Mitoxantrone, topoisomerase II inhibitors, 99–102
Mitozolomide, DNA-targeted compounds, 94–95
MK-0731 kinesin spindle protein inhibitor, 204–209
Mocetinostat, histone deacetylase inhibitors, 67–70
Molecular biology:
camptothecins, 27–28
DNA-targeted chemotherapeutic compounds, mustards, 87
DNA-targeted compounds:
bleomycin, 11
dactinomycin, 5–7
mitomycin, 14–15
plicamycin, 17
peroxisome proliferator-activated receptors, 157–159
vinblastine, 39
Monastrol kinesin spindle protein inhibitor, "induced fit" binding pockets, 193–197
Monoclonal antibodies (mAb), targeted chemotherapy, resistance and, 363–370
tyrosine kinase targeting, 368–369
Motesanib, antiangiogenic properties, 309–310
MS-275. See Entinostat
MsbA lipid transporter, homology modeling, 374–376
mTOR signaling pathway:
brain cancer therapy, 250–252
cranial and spinal nerve tumors, 238–239
inhibitors, 324–325
Multidrug resistance protein 1 (MRP1; ABCC1), drug resistance mechanisms, 372
Multiple drug resistance:
apoptosis-based protein targeting, 370–371
cancer therapies, 370–376
ATP binding cassette (ABC) transporters, 371–376
stem cells, 376–377

Mustards:
DNA-targeted chemotherapeutic compounds, 83–89
antibody-directed enzyme prodrug therapy, peptidase enzymes, 120–121
biology and side effects, 87
mechanism and structure-activity relationship, 85–87
minor-groove targeting, 87–88
hypoxia-activated bioreductive prodrugs, 117–120
Mutamycin. *See* Mitomycin

Naphthalamide, poly(ADP-ribose)polymerase inhibitors, 167
Naphthoquinones, Hsp 90 inhibitors, 426–427
Natural products:
cytotoxic agents:
DNA targeting compounds, 3–17
bleomycin, 8–12
dactinomycin, 3–8
mitomycin (mutamycin), 12–16
plicamycin, 16–17
enzyme inhibitors, 17–31
anthracyclines, 17–23
camptothecins, 23–29
isopodophyllotoxins, 29–31
future research issues, 40–41
research background, 1–3
tubulin polymerization/depolymerization inhibition, 31–40
dimeric vinca alkaloids, 35–40
taxus diterpenes, 32–35
histone deacetylase inhibitors, 70–72
largazole, 72
romidepsin (FK-228), 70–72
proteasome inhibitors:
clinical trials, 179–180
medicinal chemistry, 183–184
Navelbine. *See* Vinrelbine
Neratinib erbB family inhibitor, 302–305
Neural stem cells, brain tumor initiation, 225–226
Neurodegenerative diseases, Hsp90 inhibitor therapy, 442–444
Neuroectodermal tumors, classification and therapy, 234–235
Neuroepithelial tumors, classification, 230–231, 235
Neurofibroma, classification and therapy, 237–239
Neurofibromatosis, central nervous system tumors, 238–239
Neuroprotectants, Hsp90 inhibitor therapy, 442–444
NFkB, proteasome inhibitors, 177
Nicotinamide:
poly(ADP-ribose)polymerase-1 (PARP-1) domain organization, 160
poly(ADP-ribosyl)ation biochemistry, 153–157
Nicotinamide adenine dinucleotide (NAD):
binding sites, 162–167
poly(ADP-ribosyl)ation biochemistry, 151–157
Nilotinib:
chronic myelogenous leukemia therapy, Bcr-Abl inhibitors, 298–301
second-generation development, drug resistance and, 364–365
structure-based design, drug-resistant mutation, 355–356
Nitacrine-*N*-oxide, hypoxia-activated bioreductive prodrugs, 115–120
Nitidine, dual topoisomerase I/II inhibitors, 103–105
Nitroaromatics, hypoxia-activated bioreductive prodrugs, 115–120
Nitrosoureas, DNA-targeted compounds, 93–94
Nolatrexed, lipophilic antifolate DNA-targeted chemotherapeutics, 109
Non-Hodgkin's lymphoma, primary central nervous system tumors, 239–240
Nonhomologous end-joining (NHEJ) pathway, poly(ADP-ribose) polymerase-1 (PARP-1), 161–162
Novobiocin:
chemical structure, 430–431
Hsp90 C-terminal inhibition, 431–440
parallel library approach, 432–433
NSC 333003 compound, brain tumor therapy, 258
N-terminal ATP-binding pocket, Hsp90 inhibitors, 386–425
chimeric geldanamycin and radicicol, 405–410
geldanamycin, 386–399
high-throughput small-molecule inhibitors, 417–421
isoxazole-scaffold inhibitors, 419–421
purine-based inhibitors, 410–416
pyrazole-scaffold inhibitors, 417–419
radicicol analogs, 399–405
shepherdin, 416
structure-based design, 421–425
N-terminal nucleophilic (NTN) hydrolases, proteasome inhibitors, 175
Nuclear hormone receptors, brain tumor therapy, 265–267
Nuclear localization signal (NLS), poly(ADP-ribose) polymerase-1 (PARP-1) domain organization, 159–160
Nucleotide binding domains (NBDs), ATP binding cassette (ABC) transporters, resistance targeting, 371
Nutilin-3 compound, brain cancer therapy, 257–258
NVP-BEZ235 inhibitor:
brain cancer therapy, 250–251

survival signaling pathways, 322–323
NVP-LAQ-824 histone deacetylase inhibitior, structure and development, 55, 57–58

Olaparib, poly(ADP-ribose) polymerase (PARP) inhibitor coadjuvants, 169–170
Oligodendrogliomas:
classification and therapy, 233–234
stem cell initiation, 225–227
Oncovin. See Vincristine
Ortho-biphenyl carboxamides, brain tumor therapy, hedgehog signaling pathway, 260–261
Oxacarbenium ions, poly(ADP-ribosyl) ation biochemistry, 153–157
Oxidative enzymes, gene-directed enzyme prodrug therapy, 124
Oxime derivatives, radicicol Hsp90 inhibitor, 400–405

p53 signaling pathway, brain cancer therapy, 256–257
Paclitaxel, structure and therapeutic applications, 32–33
Panobinostat histone deacetylase inhibitor, structure and development, 58
Pazopanib inhibitor, antiangiogenic properties, 309–310
PCI-024781 histone deacetylase inhibitor, structure and development, 60–62
PD 166326 kinase inhibitor, structure-based design, protein targeting, 352–354
PD 0332991 cyclin-dependent kinase inhibitor, medicinal chemistry and classification, 313–314

PDE4 inhibitors, brain tumor therapy, 265
Pemetrexed, antifolate DNA-targeted chemotherapeutics, 107–109
Pentostatin, antimetabolite analogs, 112–113
Peptidases, antibody-directed enzyme prodrug therapy, 120–121
Peptide aldehydes, proteasome inhibitors, medicinal chemistry, 181–184
Peptide epoxyketone proteasome inhibitor:
clinical trials, 180
medicinal chemistry, 183–184
Peptide vinylsulfone proteasome inhibitors, medicinal chemistry, 181–184
Peroxisome proliferator-activated receptors (PPARs), brain tumor therapy, 265–267
P-glycoprotein:
ATP binding cassette (ABC) transporters, resistance mechanisms, 371–374
homology modeling, 374–376
PHA-739358 aurora kinase inhibitor, 320–321
Pharmacokinetics:
DNA-targeted compounds:
bleomycin, 8–9
dactinomycin, 4–5
mitomycin, 13
plicamycin, 16–17
docetaxel/taxotere, 34–35
enzyme inhibitors:
daunorubicin, 19–21
doxorubicin, 22
paclitaxel/taxol, 33
topotecan, 26–27
vinblastine, 39
Pharmacophores:
poly(ADP-ribose)polymerase inhibitors, 162–167
proteasome inhibitors:
bortezomib clinical trials, 178–180
medicinal chemistry, 181–184

Phenantridinone (PND), poly (ADP-ribose) polymerase inhibitors, 166–167
Philadelphia chromosome, chronic myelogenous leukemia therapy, small-molecule tyrosine kinase inhibitors, resistance mechanism, 264–266
Phosphatase and tensin homolog (PTEN), brain tumor therapy, 258–259
signaling pathways, 245, 251
Wnt signaling pathway, 263
Phosphatases, antibody-directed enzyme prodrug therapy, DNA-targeted therapeutics, 120–123
Phosphoinositide 3-kinase inhibitors:
brain cancer therapy, 249–250
resistance mechanisms, 367–368
survival signaling pathways, 321–323
Phytochemicals, proteasome inhibitors, 184
Pilocytic astrocytomas, histologic classification, 230
Piritrexim, lipophilic antifolate DNA-targeted chemotherapeutics, 109
Pituitary tumors, classification and therapy, 235–237
Pixantrone, topoisomerase II inhibitors, 101–102
Platelet-derived growth factor (PDGF) receptor, brain tumor therapy, 249
Platinum complexes, DNA-targeted chemotherapeutic compounds, 89–91
Plicamycin, structure and properties, 16–17
Polo-like kinase 1 inhibitors:
medicinal chemistry and clinical trials, 316–317
multidrug resistance and, 377–378

Poly(ADP-ribose)glycohydrolase (PARG):
  cancer therapy, 170
  poly(ADP) ribosylation biochemistry, 151–157
  research background, 151
Poly(ADP-ribose)polymerase (PARP) inhibition:
  cancer therapy, 160–162, 168
  chemotherapy/radiotherapy coadjuvants, 168–170
  future research issues, 170–171
  inhibitors, 168
  monotherapy, 170
  poly(ADP-ribose) glycohydrolase and inhibitors, 170
  chemotherapy/radiotherapy coadjuvants, 168–170
  molecular biology, 157–159
  monotherapy, 170
  PARP-1:
    biochemistry, 151–157
    cancer therapy, 160–162
    domain organization and catalytic site structure, 159–160
    domain organization and structure, 159–160
    molecular biology, 157–159
    research background, 151
    selectivity, 167–168
  PARP-2, molecular biology, 158–159
  PARP-3, molecular biology, 158–159
  poly(ADP-ribosylation) biochemistry, 151–157
  research background, 151
  selective inhibitors, 167–168
  structure and classification, 162–167
Polymers:
  central nervous system tumors, targeted drug delivery, 271–272
  poly(ADP-ribosyl)ation biochemistry, branching reaction, 155–157
  tubulin polymerization/ depolymerization inhibition, 31–40
  dimeric vinca alkaloids, 35–40
  taxus diterpenes, 32–35

Primary central nervous system lymphoma (PCNSL), classification and therapy, 239–240
Prodrugs, DNA-targeted therapeutics, tumor-activation, 113–124
  ADEPT prodrugs, 120–123
  hypoxia-activated bioreductive prodrugs, 114–120
Prolactin secreting adenomas, classification and therapy, 236–237
Proteasome, basic structure, 175
Proteasome inhibitors:
  clinical trials, 177–180
  future research issues, 184–185
  immunoproteasome-specific inhibitors, 184
  medicinal chemistry, 181–184
Protein folding, Hsp90 inhibitors, 384–385
Protein kinase B (PKB):
  brain tumor therapy, 249–250
  survival signaling inhibitors, 323–324
Protein kinase C (PKC):
  brain tumor therapy, 245–246
  chronic myelogenous leukemia therapy, Bcr-Abl inhibitors, 298–301
Protein(s). See also specific proteins, e.g., Heat-shock proteins
  kinase inhibitor targeting, structure-based design, 350–354
Purine-based Hsp90 inhibitors, N-terminal ATP-binding pocket, 410–416
Purine nucleoside phosphorylase (PNP), antimetabolite analogs, 113
Purines, antimetabolite analogs, 112–113
PX-866 PI3K inhibitor, survival signaling pathways, 323
PXD-101 histone deacetylase inhibitor, structure and development, 58–59
Pyrazole-scaffold Hsp90 inhibitors, 417–419

Pyrazoloacridine, dual topoisomerase I/II inhibitors, 103–105
Pyrazolols, poly(ADP-ribose) polymerase inhibitors, 166–167
Pyrimidines, antimetabolite analogs, 109–112

Quantitative structure-activity relationships (QSARs):
  ATP binding casette transporter inhibitors, 374
  camptothecins, 29
Quaternary salts, dual topoisomerase I/II inhibitors, 103–105
Quinazolines, kinesin spindle protein inhibitors, 197–200
Quinones, hypoxia-activated bioreductive prodrugs, 115–120

R547 cyclin-dependent kinase inhibitor, 313–315
R306465 first-generation histone deacetylase inhibitor, 62–64
Radamide analogs, chimeric radicicol and geldanamycin Hsp90 inhibitors, 408–410
Radanamycin analogs, chimeric radicicol and geldanamycin Hsp90 inhibitors, 408–410
Radicicol Hsp90 inhibitor:
  analog design, 399–405
  brain tumor therapy, 268
  chimeric structures, 405–410
Radiotherapy, poly(ADP-ribose) polymerase (PARP) inhibitor coadjuvants, 169–170
Raf kinases, signaling inhibitors, 305–306
Raltitrexed, antifolate DNA-targeted chemotherapeutics, 107–109
Rapamycins:
  cranial and spinal nerve tumors, 238–239
  mTOR inhibitors, 324–325

RAS/RAF/MEK/ERK signaling
pathway:
brain cancer therapies, 252
growth factor signaling
inhibitors, 301–307
Ras signaling pathway, brain
cancer therapy, 252
Receptor tyrosine kinases
(RTKs):
brain tumor therapy, 245
PDGF receptor, 249
RAS/RAF/MEK/ERK
signaling, 252
EGFR inhibitor, resistance
mechanisms,
367–368
Recurrent cancer, central nervous
system tumors,
240–241
Reductase enzymes, gene-
directed enzyme
prodrug therapy, 124
Resorcinolic macrolides, radicicol
Hsp90 inhibitor and,
402–405
Retinoic X receptor (RXR), brain
tumor therapy, 267
Rho/ROCK signaling pathway,
brain tumor therapy,
252–253
Ring systems:
epidermal growth factor
receptor signaling
inhibitors, 304–305
hypoxia-activated bioreductive
prodrugs, 119–120
novobiocin Hsp90 inhibitor
optimization, 436–440
Rituximab, resistance
mechanisms, 369
Rolipram, brain tumor therapy,
265
Romidepsin, histone deacetylase
inhibitors, 70–72
Rubidomycin. See Daunorubicin
Rubitecan, biological activity and
side effects, 97–98

Sabarubicin, topoisomerase II
inhibitors, 100–102
Salinosporamide A proteasome
inhibitor, clinical
trials, 179–180
SANT-2 small molecule, brain
tumor therapy,
hedgehog signaling
pathway, 260–261

Saracatinib, tumor metastases
inhibition, 325–327
SB939 histone deacetylase
inhibitor, structure
and development,
59–60
Scaffold structures:
Hsp90 inhibitors:
novobiocin C-terminal
inhibition, parallel
library approach,
432–433
purine-based Hsp90
inhibitors, 410–416
pyrazole-scaffold inhibitors,
417–419
poly(ADP-ribose)polymerase
inhibitors, 162–167
Schwannoma, classification and
therapy, 237–239
Schwannomatosis, classification
and therapy,
238–239
Secondary metabolites, cytotoxic
agents, research
background, 1–3
Second-generation compounds,
kinase inhibitors:
drug resistance:
conformational targetig,
354–356
P-glycoprotein inhibition,
372
small-molecule tyrosine
kinase inhibitors,
364–366
purine-based Hsp90 inhibitors,
413–416
pyrazole-scaffold Hsp90
inhibitors, 419
Seleciclib cyclin-dependent
kinase inhibitor,
medicinal chemistry
and classification,
310–315
Selective estrogen receptor
modulators (SERMs),
breast cancer
resistance, 369–370
Selectivity:
Hsp90 inhibitors, tumor cell
selectivity, 385
poly(ADP-ribose)polymerase
(PARP) inhibitors,
167–168
Shepherdin Hsp90 inhibitor,
416

Side chain modification,
novobiocin Hsp90
inhibitor, 434–436
Side effects:
DNA-targeted compounds:
antibody-directed enzyme
prodrug therapy, 123
antifolates, 108–109
bleomycin, 8
cyclopropylindoles, 92–93
dactinomycin, 4
mitomycin, 13
mustards, 87
nitrosoureas, 94
platinum complexes, 91
plicamycin, 16
pyrimidine antimetabolite
analogs, 110–112
triazines, 95
docetaxel/taxotere, 34–35
dual topoisomerase I/II
inhibitors, 104–105
enzyme inhibitors:
daunorubicin, 19–20
doxorubicin, 21–22
gene-directed enzyme prodrug
therapy, 124
hypoxia-activated bioreductive
prodrugs, 118–120
paclitaxel/taxol, 33
topoisomerase inhibitors, Topo
I inhibitors, 97–98
topotecan, 26–27
vinblastine, 38–39
vinorelbine, 40
Side population, cancer stem cells,
multidrug resistance
and, 376–377
Signaling pathways:
central nervous system tumors,
227–228
growth factors:
brain tumor therapy, 245
kinase inhibitors, 301–307
hedgehog signaling pathway,
brain tumor therapy,
259–263
hypoxia-induced pathway,
253–255
kinase inhibitors:
growth factors, 301–307
survival signaling inhibitors,
321–325
insulin-like growth factor-1
receptor inhibitors,
321
mTOR inhibitors, 324–325

516 INDEX

Signaling pathways: (*Continued*)
  phosphoinoside-3 kinase inhibitors, 321–323
  protein kinase B inhibitors, 323–324
 monoclonal antibodies, resistance mechanisms, 369
 mTOR signaling pathway:
  brain cancer therapy, 250–252
  cranial and spinal nerve tumors, 238–239
  p53 protein, brain cancer therapy, 256–257
  PI3K/AKT pathway, brain tumor therapy, 249–250
  RAS/RAF/MEK/ERK signaling, brain tumor therapy, 252
  Rho/ROCK and MAPK signaling, brain tumor therapy, 252–253
Single-strand break (SSB) repair, poly(ADP-ribose) polymerase-1 (PARP-1), 160–162
SJ749 antagonist, brain cancer therapy, hypoxia-inducible factor pathway, 254–255
Small-molecule compounds:
 brain tumor therapy:
  atrasentan, 268
  hedgehog signaling pathway, 260–263
  HGF/MET compounds, 247–248
  hypoxia-inducible factor pathway, 254–255
  MDM2/MDMX, 257–258
  PI3K/AKT pathway, 249–250
 cancer therapy:
  targeted chemotherapy, resistance and, 363–370
  tumor metastases, c-Met kinase inhibitors, 328–330
  tyrosine kinase inhibitors, resistance mechanisms, 364–366
 Hsp90 inhibitors, 427–429
  high-throughput screening identification, 417
 kinesin spindle protein inhibitors, crystallographic analysis, 192–193
 protein kinase drugs, structure-based design, 345–346
Somatostatin, pituitary adenoma therapy, analog compounds, 236–237
Sorafenib signaling inhibitor:
 antiangiogenic properties, 307–310
 resistance to, 368
 structure and properties, 305–306
Sphingosine 1-phosphate (S1P) and receptors, brain cancer therapy, hypoxia-inducible factor pathway, 254–255
Spinal nerve tumors, classification and therapy, 237–239
Src kinase inhibitors, tumor metastases, 325–328
Stem cells:
 brain tumor initiation, 224–227
 cancer therapy, multiple drug resistance, 376–377
Streptozotocin, biological activity and side effects, 94
Structure-activity relationships (SARs):
 antibody-directed enzyme prodrug therapy, DNA-targeted therapeutics, 120–121
 antifolate DNA-targeted chemotherapeutics, 108
 ATP binding casette transporter inhibitors, 374
 DNA-targeted chemotherapeutic compounds:
  cyclopropylindoles, 92–93
  mustards, 85–87
  nitrosoureas, 94
  platinum complexes, 90–91
 gene-directed enzyme prodrug therapy, 123–124
 Hsp90 inhibitors:
  geldanamycin Hsp90 inhibitor, 389–391
  novobiocin Hsp90 inhibitor optimization, 436–440
  purine-based Hsp90 inhibitors, 415–416
 hypoxia-activated bioreductive prodrugs, 116–120
 purine antimetabolite analogs, 112–113
 pyrimidine antimetabolite analogs, 109–112
 radicicol Hsp90 inhibitor, 405
 topoisomerase II inhibitors, 99–102
Structure-based design (SBD):
 Hsp90 inhibitors, 421–423
 kinase inhibitors:
  future research issues, 356–357
  multiconformational catalytic domain binding cleft, 346–348
  binding pockets, 348–349
  protein targeting, 350–354
  second-line inhibitors, drug-resistant mutation, 354–356
  research background, 345–346
Suberoyl hydroxamic acid (SAHA), histone deacetylase inhibition, 53, 55–56
Suicide hypothesis, poly(ADP-ribosyl)ation biochemistry, 155–157
Sunitinib:
 antiangiogenic properties, 307–310
 brain tumor angiogenesis, 242–243
 resistance to, 368
 structure-based design, protein targeting, 352–354
Survival signaling inhibitors, 321–325
 insulin-like growth factor-1 receptor inhibitors, 321
 mTOR inhibitors, 324–325
 phosphoinoside-3 kinase inhibitors, 321–323
 protein kinase B inhibitors, 323–324
Synthetic drug development:
 Geldanamycin Hsp90 inhibitors, 391–392
 proteasome inhibitors, 182–183

Tacedinaline, histone deacetylase inhibitors, 66–67
Tacrolimus, drug resistance, biricodar kinase inhibitor, 372
Tafluposide, dual topoisomerase I/II inhibitors, 103–105
Tallimustine, minor groove targeting compound, 88
Tamoxifen, breast cancer resistance, 369–370
Tandutinib, acute myelogenous leukemia therapy, 301–302
Tankyrases, molecular biology, 158–159
Targeted drug development:
  cancer chemotherapy, resistance and, 363–370
  ATP binding cassette (ABC) transporters, 371–376
  central nervous system tumors, 271–272
  DNA-targeted compounds, cytotoxic agents, 3–17
    bleomycin, 8–12
    dactinomycin, 3–8
    mitomycin (mutamycin), 12–16
    plicamycin, 16–17
  kinase inhibitors, structure-based design, 350–354
  mitosis targeting, 191–192
  multiple drug resistance, apoptosis-based protein targeting, 370–371
TAS 103 dual topoisomerase I/II inhibitors, 103–105
Taxol:
  Hsp90 inhibitor binding, 427–428
  structure and therapeutic applications, 32–33, 36
Taxotere, structure and therapeutic applications, 34–35
Taxus diterpenes:
  docetaxel/taxotere, 34–35
  paclitaxel/taxol, 32–33
Temozolomide:
  brain tumors, 244
  DNA-targeted compounds, 94–95
  glioblastoma therapy, 231–233
  poly(ADP-ribose)polymerase (PARP) inhibitor coadjuvants, 169–170
Temsirolimus:
  brain cancer therapy, 250–251
  mTOR inhibitors, 324–325
Teniposide, structure and therapeutic applications, 31
Tetracyclic compounds:
  poly(ADP-ribose)polymerase inhibitors, 167
  topoisomerase II inhibitors, 102
Tetrahydro-4H-carbazol-4-one Hsp90 inhibitors, fragment-based design, 423–424
Thiazolidinediones, survival signaling pathways, 323
6-Thioguanine, antimetabolite analogs, 112–113
Third generation kinase inhibitors, multidrug resistance, 372
Thymidylate synthase (TS):
  antifolate DNA-targeted chemotherapeutics, 106–109
  pyrimidine antimetabolite analogs, 109–110
Tirapazamine, hypoxia-activated bioreductive prodrugs, 115–120
Tissue associated antigens (TIAs), central nervous system tumors, vaccine development, 269–270
Toceranib, brain tumor therapy, 249
Topoisomerase inhibitors, synthetic chemotherapeutic compounds, 95–106
  dual topo I/II inhibitors, 103–106
  Topo II inhibitors, 98–102
  Topo I inhibitors, 96–98
Topotecan:
  brain cancer therapy, 254–255
  poly(ADP-ribose)polymerase (PARP) inhibitor coadjuvants, 169–170
  structure and applications, 26–29
Tosyl compounds, novobiocin Hsp90 inhibitor optimization, 439–440
Transforming growth factor-b, brain tumor therapy, 267
Translocation pathway, P-glycoprotein, 376
Transmembrane domains (TBDs), ATP binding cassette (ABC) transporters, 371
Trastuzumab:
  geldanamycin Hsp90 inhibitor, 396–397
  growth factor signaling inhibitors, 302–305
  resistance to, 369
Triazines, DNA-targeted compounds, 94–95
Trichostatin A, histone deacetylase inhibitor binding, 51–53
Trimetrexate, lipophilic antifolate DNA-targeted chemotherapeutics, 109
Tubacin, discovery of, 74–75
Tubulin polymerization/depolymerization inhibition, cytotoxic agents, 31–40
  dimeric vinca alkaloids, 35–40
  taxus diterpenes, 32–35
Tumor-activated prodrugs, DNA-targeted therapeutics, 113–124
  ADEPT prodrugs, 120–123
  hypoxia-activated bioreductive prodrugs, 114–120
Tumorigenesis, poly(ADP)-ribosylation, 161–162
Tumor initiation, brain tumor stem cells, 224–227
Tyrosine kinase inhibitors:
  angiogenic properties, 307–310
  cancer therapy, resistance to, 364–369
  glioblastoma therapy, 232–233
  growth factor signaling pathways, 301–307

Ubiquitin-conjugating enzymes, proteasome inhibitors, 175
Ubiquitin-proteasome pathway (UPP):
  basic structure, 175
  targeted therapy, cancer and other disease, 176–177
Urea derivatives, geldanamycin Hsp90 inhibitor, 389–391

Vaccine development, central nervous system tumors, 268–270
Valproic acid, histone deacetylase inhibition, 73
Valrubicin, structure and therapeutic effects, 22–23
Valspodar, drug resistance inhibition, 372
Vandetanib:
  antiangiogenic properties, 308–310
  brain tumor angiogenesis, 242–243
Vascular endothelial growth factor (VEGF):
  antiangiogenic inhibitors, 307–310
  resistance to, 368
  glioblastoma therapies, 232–233
Vascular endothelial growth factor receptor (VEGFR), brain tumor therapy:
  inhibitors, 248
  kinase inhibitors, 248–249
Vatalanib:
  antiangiogenic properties, 309–310
  brain tumor angiogenesis, 242–243
Vault poly(ADP-ribose) polymerase (V-PARP), molecular biology, 158–159
Verapamil, ATP binding cassette (ABC) transporters, 371–372
Vinblastine, structure and therapeutic applications, 37–39
Vinca alkaloids:
  chemical structure, 35–37
  vinblastine, 37–39
  vincristine, 39
  vinorelbine, 39–40
Vincasar PFS. *See* Vincristine
Vincristine, structure and therapeutic applications, 38–39
Vindisine, analog design, 40
Vinflunine, analog design, 40
Vinorelbine, structure and therapeutic applications, 39–40

Vorinostat, histone deacetylase inhibition, 53, 55–56

WHO histological classification system, central nervous system tumors, 230–234
Wnt signaling pathway:
  brain tumor therapy, 263
  central nervous system tumors, 228

X-ray crystallography, kinesin spindle protein structure, 192–193
  dihydropyrazole inhibitors, 204–210
  "induced fit" binding pockets, 193–197

Ynolides, cycloproparadicicol analogs, 402–403

Zinc binding group (ZBG), histone deacetylase inhibitors, 53–54
  future drug development, 74–75
Zinc finger proteins, poly(ADP-ribose)polymerase-1 (PARP-1) domain organization, 159–160